Integrated Pest Management

Concepts, Tactics, Strategies and Case Studies

Integrated Pest Management (IPM) is an effective and environmentally sensitive approach to pest management. It uses natural predators, pest-resistant plants and other methods to preserve a healthy environment in an effort to decrease reliance on harmful pesticides. Featuring 40 chapters written by leading experts, this textbook covers a broad and comprehensive range of topics in Integrated Pest Management, focused primarily on theory and concepts. It is complemented by two award-winning websites, which are regularly updated and emphasize specific IPM tactics, their application, and IPM case studies:

Radcliffe's IPM World Textbook –
http://ipmworld. umn.edu
VegEdge – www.vegedge.umn.edu

The two products are fully cross-referenced and form a unique and highly valuable resource. Written with an international audience in mind, this text is suitable for advanced undergraduate and graduate courses on Integrated Pest Management, Insect or Arthropod Pest Management. It is also a valuable resource for researchers, extension specialists and IPM practitioners worldwide.

Edward B. Radcliffe is a Professor in the Department of Entomology, University of Minnesota, St. Paul, where he has taught IPM since 1966.

William D. Hutchison is Professor and Extension Entomologist in the Department of Entomology, University of Minnesota, St. Paul.

Rafael E. Cancelado is an independent crop consultant working with vegetable growers in Lara Region, Venezuela.

Integrated Pest Management

Concepts, Tactics, Strategies and Case Studies

Edited by

Edward B. Radcliffe
University of Minnesota, St. Paul

William D. Hutchison
University of Minnesota, St. Paul

Rafael E. Cancelado
Venezuela

CAMBRIDGE UNIVERSITY PRESS
Cambridge, New York, Melbourne, Madrid, Cape Town, Singapore, São Paulo, Delhi

Cambridge University Press
The Edinburgh Building, Cambridge CB2 8RU, UK

Published in the United States of America by Cambridge University Press, New York

www.cambridge.org
Information on this title: www.cambridge.org/9780521875950

First published 2009

Printed in the United Kingdom at the University Press, Cambridge

A catalog record for this publication is available from the British Library

Library of Congress Cataloging in Publication data
Integrated pest management : concepts, tactics, strategies & case studies /
edited by Edward B. Radcliffe, William D. Hutchison, Rafael E. Cancelado.
p. cm.
Includes index.
ISBN 978-0-521-87595-0 (hardback)
1. Pests – Integrated control. I. Radcliffe, Edward B., 1936– II. Hutchison,
William D. III. Cancelado, Rafael E. IV. Title.
SB950.I4572 2008
632′.9 – dc22 2008042307

ISBN 978-0-521-87595-0 hardback
ISBN 978-0-521-69931-0 paperback

Dedication

This Cambridge University Press IPM textbook is dedicated to the memory of Robert (Bob) J. O'Neil (1955–2008), who died on February 6 following an almost year-long battle against bladder cancer. Bob worked on the chapter he co-authored with John Obrycki for this book throughout his illness. During his professional career at Purdue University, Bob established an internationally recognized research, teaching and outreach program in biological control. His research and outreach accomplishments serve as a model for how undertaking fundamental research on ecologically important questions can lead to solutions of environmental problems. One of Bob's greatest joys and lasting contributions will be his work with students from the Pan American School of Agriculture (Zamorano) in Honduras. His efforts resulted in opportunities for hundreds of Latin American students to work study and pursue advanced degrees at Purdue and several universities in the United States and Europe.

Bob was the creative spark who acted as the catalyst to bring together individuals from several universities in the Midwestern USA to work together on biological control projects. His idea to create collaborative summer workshops for graduate students through the Midwest Biological Control Institute is now in its 18th year. His recognition of the need to enhance the level of knowledge about biological control among extension agents resulted in several multi-state extension workshops on biological control. Bob knew that biological control could contribute to the management of pests in the Midwest and formed cooperative research teams to find new natural enemies to address these pest problems. Working together with Bob, this group of individuals was able to significantly advance implementation of biological control in the Midwestern USA. Bob's vision, can-do attitude, humor and optimistic view of the world (he was a lifelong Boston Red Sox fan who occasionally rooted for the Cubs) will be greatly missed by his colleagues and friends.

Memoriam

Professor Chris Curtis, one of the world's leading medical entomologists, died unexpectedly in May 2007. Chris was completely dedicated to the cause of useful science in the service of practical public health. In his early career, he worked on genetic methods of insect control, but most of his professional life was devoted to development of low technology methods for mosquito control.

In the last two decades, Chris played a major role in developing and promoting use of insecticide-treated mosquito nets (ITN) for the prevention of malaria transmission by *Anopheles* mosquitoes, as this technique gradually moved to center-stage for practical malaria control. His team demonstrated that, when ITNs are used by most people in a village, there is a "mass effect" on the local mosquito population that reduces its ability to transmit malaria and gives extra protection to all, including those without nets. Chris was a tireless and highly influential campaigner for the cause of "free nets," against the notion that nets should be sold, or targeted only at subgroups most vulnerable to malaria. In doing so, he contributed to a significant strengthening of political will in developed countries, and thus to vastly increased donor funding for malaria control. To date, about 50 million treated nets have been given away, preventing tens of thousands of deaths due to malaria among African children.

As a person, Chris was exceptionally gentle, honest and kind. As a teacher he was positively luminous: he gave lasting inspiration to countless students, and there is a worldwide community of students and colleagues (including the writer) who are linked by what Chris taught them about how to do science, how to value it, and how to enjoy it.

JO LINES
London School of Hygiene and Tropical Medicine

Contents

Contributors

1 **The IPM paradigm: concepts, strategies and tactics**

Michael E. Gray
University of Illinois, Crop Sciences, Urbana-Champaign, Illinois, USA

Susan T. Ratcliffe
University of Illinois, Crop Sciences, Urbana-Champaign, Illinois, USA

Marlin E. Rice
Iowa State University, Department of Entomology, Ames, Iowa, USA

2 **Economic impacts of IPM**

Scott M. Swinton
Michigan State University, Department of Agricultural Economics, East Lansing, Michigan, USA

George W. Norton
Virginia Polytechnic Institute and State University, Agricultural and Applied Economics, Blacksburg, Virginia, USA

3 **Economic decision rules for IPM**

Leon G. Higley
University of Nebraska, Department of Entomology, Lincoln, Nebraska, USA

Robert K. D. Peterson
Montana State University, Department of Land Resources and Environmental Sciences, Bozeman, Montana, USA

4 **Decision making and economic risk in IPM**

Paul D. Mitchell
University of Wisconsin, Department of Agricultural and Applied Economics, Madison, Wisconsin, USA

William D. Hutchison
University of Minnesota, Department of Entomology, St. Paul, Minnesota, USA

5 **IPM as applied ecology: the biological precepts**

David J. Horn
The Ohio State University, Department of Entomology, Columbus, Ohio, USA

6 **Population dynamics and species interactions**

William E. Snyder
Washington State University, Department of Entomology, Pullman, Washington, USA

Anthony R. Ives
University of Wisconsin, Department of Zoology, Madison, Wisconsin, USA

7 **Sampling for detection, estimation and IPM decision making**

Roger D. Moon
University of Minnesota, Department of Entomology, St. Paul, Minnesota, USA

L. T. (Ted) Wilson
Texas A&M, University, Agricultural Research and Extension Center, Beaumont, Texas, USA

8 **Application of aerobiology to IPM**

Scott A. Isard
The Pennsylvania State University, Department of Plant Pathology, University Park, Pennsylvania, USA

David A. Mortensen
The Pennsylvania State University, Department of Agronomy, University Park, Pennsylvania, USA

Shelby J. Fleischer
The Pennsylvania State University, Department of Entomology, University Park, Pennsylvania, USA

Erick D. De Wolf
The Pennsylvania State University, Department of Plant Pathology, University

Park, Pennsylvania, USA, currently with Kansas State University, Department of Plant Pathology, Manhattan, Kansas, USA

9 Introduction and augmentation of biological control agents

Robert J. O'Neil (deceased)
Purdue University, Department of Entomology, West Lafayette, Indiana, USA

John J. Obrycki
University of Kentucky, Department of Entomology, Lexington, Kentucky, USA

10 Crop diversification strategies for pest regulation in IPM systems

Miguel A. Altieri
University of California, Department of Environmental Science, Policy and Management, Division of Insect Biology, Berkeley, California, USA

Clara I. Nicholls
University of California, Department of Environmental Science, Policy and Management, Division of Insect Biology, Berkeley, California, USA

Luigi Ponti
University of California, Department of Environmental Science, Policy and Management, Division of Insect Biology, Berkeley, California, USA

11 Manipulation of arthropod pathogens for IPM

Stephen P. Wraight
United States Department of Agriculture, Agricultural Research Service, Plant Protection Research Unit, USA

Ann E. Hajek
Cornell University, Department of Entomology, Ithaca, NY, USA

12 Integrating conservation biological control into IPM systems

Mary M. Gardiner
Michigan State University, Department of Entomology, East Lansing, Michigan, USA

Anna K. Fiedler
Michigan State University, Department of Entomology, East Lansing, Michigan, USA

Alejandro C. Costamagna
Michigan State University, Department of Entomology, East Lansing, Michigan, currently with University of Minnesota, Department of Entomology, St. Paul, Minnesota, USA

Douglas A. Landis
Michigan State University, Department of Entomology, East Lansing, Michigan, USA

13 Barriers to adoption of biological control agents and biological pesticides

Pamela G. Marrone
Marrone Organic Innovations, Davis, California, USA

14 Integrating pesticides with biotic and biological control for arthropod pest management

Richard A. Weinzierl
University of Illinois, Crop Sciences, Urbana–Champaign, Illinois, USA

15 Pesticide resistance management

Casey W. Hoy
The Ohio State University, Ohio Agricultural Research and Development Center, Department of Entomology, Wooster, Ohio, USA

16 Assessing environmental risks of pesticides

Paul C. Jepson
Oregon State University, Environmental and Molecular Toxicology Department, Corvallis, Oregon, USA

17 Assessing pesticide risks to humans: putting science into practice

Brian Hughes
Dow Chemical Company, Midland, Michigan, USA

Larry G. Olsen
Michigan State University, Department of Entomology, East Lansing, Michigan, USA

Fred Whitford
Purdue University, Department of Entomology, West Lafayette, Indiana, USA

18 Advances in breeding for host plant resistance

C. Michael Smith
Kansas State University, Department of Entomology, Manhattan, Kansas, USA

19 Resistance management to transgenic insecticidal plants

Anthony M. Shelton
Cornell University, Department of Entomology, Geneva, New York, USA

Jian-Zhou Zhao
Cornell University, Department of Entomology, Geneva, New York, USA, currently with Pioneer Hi-Bred International, Inc., a DuPont Company, Johnson, Iowa, USA

20 Role of biotechnology in sustainable agriculture

Jarrad R. Prasifka
Energy Biosystems Institute, Institute for Genome Biology, University of Illinois, Urbana, Illinois, USA

Richard L. Hellmich
United States Department of Agriculture, Agricultural Research Service, Corn Insects and Crop Genetics Research Unit, Ames, Iowa, USA

Michael J. Weiss
Golden Harvest Seeds, Decorah, Iowa, USA

21 Use of pheromones in IPM

Thomas C. Baker
The Pennsylvania State University, Department of Entomology, University Park, Pennsylvania, USA

22 Insect endocrinology and hormone-based pest control products in IPM

Daniel Doucet
Canadian Forest Service, Great Lakes Forestry Centre, Sault Ste. Marie, Ontario, Canada

Michel Cusson
Canadian Forest Service, Laurentian Forestry Centre, Quebec City, Canada

Arthur Retnakaran
Emeritus, Canadian Forest Service, Great Lakes Forestry Centre, Sault Ste. Marie, Ontario, Canada

23 Eradication : strategies and tactics

Michelle L. Walters
United States Department of Agriculture, Animal Plant Health and Inspections Service, Center for Plant Health Science and Technology, Phoenix, Arizona, USA

Ron Sequeira
United States Department of Agriculture, Animal Plant Health and Inspection Service, Center for Plant Health Science and Technology, North Carolina, USA

Robert Staten
Emeritus, United States Department of Agriculture, Animal Plant Health and Inspection Service Center for Plant Health Science and Technology, Phoenix, Arizona, USA

Osama El-Lissy
United States Department of Agriculture, Animal Plant Health and Inspections Service, Invasive Species and Pest Management, Riverdale, Maryland, USA

Nathan Moses-Gonzales
United States Department of Agriculture, Animal Plant Health and Inspections Service

Center for Plant Health Science and Technology, Phoenix, Arizona, USA

24 **Insect management with physical methods in pre- and post-harvest situations**

Charles Vincent
Agriculture and Agri-Food Canada, Horticultural Research and Development Centre, Saint-Jean-sur-Richelieu, Quebec, Canada

Phyllis G. Weintraub
Agricultural Research Organization, Gilat Research Center, Israel, D. N. Negev, Israel

Guy J. Hallman
United States Department of Agriculture, Agricultural Research Service, Subtropical Agricultural Research Center, Weslaco, Texas, USA

Francis Fleurat-Lessard
INRA, Laboratory for Post-Harvest Biology and Technology, Villenave-d'Ornon, France

25 **Cotton arthropod IPM**

Steven E. Naranjo
United States Department of Agriculture, Agricultural Research Service, Arid-Land Agricultural Research Center, Maricopa, Arizona, USA

Randall G. Luttrell
University of Arkansas, Department of Entomology, Fayetteville, Arkansas, USA

26 **Citrus IPM**

Richard F. Lee
United States Department of Agriculture, Agricultural Research Service, Germplasm Resources Information Network, National Clonal Germplasm Repository for Citrus and Dates, Riverside, California, USA

27 **IPM in greenhouse vegetables and ornamentals**

Joop C. van Lenteren
Wageningen University, Laboratory of Entomology, Wageningen, The Netherlands

28 **Vector and virus IPM for seed potato production**

Jeffrey A. Davis
University of Minnesota, Department of Entomology, St. Paul, Minnesota, USA, currently with Louisiana State University, Department of Entomology, Baton Rouge, Louisiana, USA

Edward B. Radcliffe
University of Minnesota, Department of Entomology, St. Paul, Minnesota, USA

David W. Ragsdale
University of Minnesota, Department of Entomology, St. Paul, Minnesota, USA

Willem Schrage
Minnesota Department of Agriculture, Potato Program, East Grand Forks, Minnesota, USA, currently with North Dakota State Seed Department, Fargo, North Dakota, USA

29 **IPM in structural habitats**

Stephen A. Kells
University of Minnesota, Department of Entomology, St. Paul, Minnesota, USA

30 **Fire ant IPM**

David H. Oi
United States Department of Agriculture, Agricultural Research Service, Center for Medical, Agricultural, and Veterinary Entomology, Gainesville, Florida, USA

Bastiaan (Bart) M. Drees
Texas A&M University, Department of Entomology, College Station, Texas, USA

31 **Integrated vector management for malaria**

Chris F. Curtis (deceased)
London School of Hygiene and Tropical Medicine, London, UK

32 **Gypsy moth IPM**

Michael L. McManus
Emeritus, United States Forest Service, Northeast Research Station, Hamden, Connecticut, USA

Andrew M. Liebhold
United States Forest Service, Northeastern Area, Morgantown, West Virginia, USA

33 **IPM for invasive species**

Robert C. Venette
United States Forest Service, Northern Research Station, Biological and Environmental Influences on Forest Health and Productivity, St. Paul, Minnesota, USA

Robert L. Koch
University of Minnesota, Department of Entomology, St. Paul, Minnesota, USA

34 **IPM information technology**

John K. VanDyk
Iowa State University, Department of Entomology, Ames, Iowa, USA

35 **Private-sector roles in advancing IPM adoption**

Thomas A. Green
IPM Institute of North America, Inc., Madison, Wisconsin, USA

36 **IPM: ideals and realities in developing countries**

Stephen Morse
University of Reading, International Development Centre, Applied Development Studies, Department of Geography, Reading, UK

37 **The USA National IPM Road Map**

Harold D. Coble
United States Department of Agriculture, Office of Pest Management Policy, North Carolina State University, Raleigh, North Carolina, USA

Eldon E. Ortman
Emeritus, Purdue University, Department of Entomology, West Lafayette, Indiana, USA

38 **The role of assessment and evaluation in IPM implementation**

Carol L. Pilcher
Iowa State University, Department of Entomology, Ames, Iowa, USA

Edwin G. Rajotte
The Pennsylvania State University, Department of Entomology, University Park, Pennsylvania, USA

39 **From IPM to organic and sustainable agriculture**

John Aselage
Amy's Kitchen, Santa Rosa, California, USA

Donn T. Johnson
University of Arkansas, Department of Entomology, Fayetteville, Arkansas, USA

40 **Future of IPM: a worldwide perspective**

E. A. (Short) Heinrichs
Emeritus, Department of Entomology, University of Nebraska, Lincoln, Nebraska, USA

Karim M. Maredia
Michigan State University, Institute of International Agriculture and Department of Entomology, East Lansing, Michigan, USA

Subbarayalu Mohankumar
Tamil Nadu Agricultural University, Department of Plant Molecular Biology and Biotechnology, Coimbatore, India

Preface

Integrated Pest Management (IPM) has been taught in the Department of Entomology at the University of Minnesota since 1966. Over the years, we've used many different textbooks for this course, supplementing these with primary references and more recently with web resources. We've never lacked for quality information resources to use in teaching our course, especially so in recent years, but we've never felt satisfied that any one textbook provided the breath of coverage of all the IPM related topics we think need to be included in a university-level course. We recognized that our expectations might be unrealistic since such broad coverage could make for a book of such size and cost that it wouldn't be appropriate to adopt as a required textbook.

We attempted to overcome these challenges by developing our own online textbook, *Radcliffe's IPM World Textbook*, http://ipmworld.umn.edu. Our concept for creating this website was that we'd solicit content from a cadre of internationally recognized experts with the goal eventually of a comprehensive online IPM resource having "chapters" covering all aspects of IPM. Our primary objectives in creating this website were to provide (1) a venue for easily maintaining and updating "state of the art" information from the world's leading experts on all aspects of IPM, and (2) a resource economically deliverable anywhere in the world that could be freely downloaded for use by students, teachers and IPM practitioners. Since 1996, we've used this resource, supplemented with an electronic library of primary references and links to other IPM websites, as the textbook for our teaching of IPM. This website has achieved considerable success and recognition, but coverage of topics is still uneven, and we believe there is still need for a comprehensive, printed IPM textbook.

In late 2005, Cambridge University Press Commissioning Editor Jacqueline Garget suggested that we consider submitting a proposal to the Press for the development of a printed IPM textbook. The concept we developed, and that we believed would make for a book unique among its peers, was that the new printed textbook and our existing online textbook should be complementary and cross-referenced. Our idea was that the printed textbook would focus on theory, i.e., concepts and guiding principles, and that it would provide information of general application that would not become quickly dated, whereas information and specific examples that are more time-sensitive or situation-specific would be posted online. Again, we proposed creating a multi-authored textbook with the contributed chapters following the outline of a typical IPM course. To achieve that, we invited contributors to this book to write their chapters in the style of a classroom lecture. We asked that authors emphasize those key concepts they would want to communicate were they invited to present a guest lecture on their chapter topic to an undergraduate/graduate-level IPM class. To keep this book to a reasonable size, the chapters in this work are shorter and generally contain fewer specifics and/or examples than is typical of chapters in more traditionally organized IPM textbooks. The complementary online textbook allows us to make available supplemental material including colored illustrations, searchable lists, detailed case studies and much more, all of which being online can be conveniently updated as appropriate.

The terminology "Integrated Control" entered the lexicon of economic entomology almost 50 years ago. The concepts of Integrated Control, soon renamed Integrated Pest Management, were quickly embraced by the scientific community and officially accepted as the operative pest management paradigm by most governments and international organizations. Nevertheless, pesticide use continues to grow and to be the tactic of primary reliance for most pest management practitioners. It is appropriate to ask why this is so, and why IPM has not been more fully adopted. The authors of this book have addressed many of the constraints that have slowed IPM adoption, but they also present convincing arguments that IPM remains the most robust, ecologically sound and socially desirable approach to addressing pest control challenges.

In summary, we hope that readers from many perspectives will find this book interesting and of practical value. Specifically, we trust that the book, along with the complementary *IPM World Textbook* website, will continue to be of value to students and faculty as an IPM resource for advanced undergraduate- and graduate-level courses in IPM, and for courses that examine alternative IPM systems. We also believe that the text will be useful to IPM practitioners, extension and outreach specialists and industry colleagues worldwide who have responsibilities for implementing sustainable IPM programs and policies. There is also much here that should be of interest to an audience of those concerned with a broad range of issues relating to agriculture production and/or environmental protection.

Acknowledgements

We thank Dr. Mark Ascerno, our Department Head at the University of Minnesota, for encouraging us to undertake this project. We thank Betty Radcliffe, surprisingly still married to the senior editor, for her many hours of expert copy-editing. We offer particular thanks to the 90 authors whose contributions and considerable expertise made this work a reality. That these individuals not only contributed their time, talent, and insight, but also shared our enthusiasm for this project is greatly appreciated. Lastly, we thank Jacqueline Garget for proposing this project, and for her guidance and patience as we brought it to fruition.

Chapter 1

The IPM paradigm: concepts, strategies and tactics

Michael E. Gray, Susan T. Ratcliffe and Marlin E. Rice

Pests compete with humans for food, fiber and shelter and may be found within a broad assemblage of organisms that includes insects, plant pathogens and weeds. Some insect pests serve as vectors of diseases caused by bacteria, filarial nematodes, protozoans and viruses. Densities of many pests are regulated by density-independent factors, particularly under fluctuating environmental extremes (e.g. temperature, precipitation). Biotic components within a pest's life system also may serve as important population regulation factors, such as interactions with predators and parasitoids. Some ecologists have theorized that competition (interspecific and/or intraspecific) for resources ultimately limits the densities and distributions of organisms, including those that are anthropocentrically categorized as pests.

1.1 Historical perspectives

Humans have been in direct competition with a myriad of pests from our ancestral beginnings. Competition with pests for food intensified when humans began to cultivate plants and domesticate animals at the beginnings of agriculture, 10 000 to 16 000 years ago (Perkins, 2002; Thacker, 2002; Bird, 2003). As humans became more competent in producing crops used for food and fiber, human densities began to increase and were organized in larger groupings such as villages. This increased concentration of humans in close proximity to their livestock is believed to have facilitated the mutation and spread of diseases across species in some instances. The earliest attempts at agricultural pest control were likely very direct and included handpicking and crushing insects, pulling or cutting weeds and discarding rotting food sources. Some pest control activities were inadvertent and included rotation or movement of crops (primarily planting crops in more fertile areas) and selection of plants for seed that had the greatest yields for sowing the following growing season.

The reasoned use of pesticides is centuries old (2500 BC) dating back to when sulfur was directed at the control of mites and insects (Bird, 2003; Kogan & Prokopy, 2003). The ancient Egyptians also are credited with the use of compounds extracted from plants to aid in the control of insects and approximately 2000 years ago, Pliny listed arsenic and olive oil as pesticides. (Thacker, 2002). In AD 307, biological control was utilized in Chinese citrus orchards (Bird, 2003) and in AD 1100 soap was being used as an insecticide in China (Kogan & Prokopy, 2003). Perkins (2002) asserted that pest control began to transform significantly about four centuries ago:

Integrated Pest Management, ed. Edward B. Radcliffe, William D. Hutchison and Rafael E. Cancelado. Published by Cambridge University Press.

About 400 years ago in Western Europe, a set of transformations completely changed economic life and, with it, pest control. New machines and new ways of making metals enabled industrialization. The new industrial processes were themselves linked to a new philosophy of nature, in which humans learned to manipulate natural processes more powerfully, particularly energy resources.

Thacker (2002) provides a list of insecticidal plants and their active compounds discovered by Europeans following the sixteenth century: sabadilla (*Sabadilla officinarum*) (*c*. 1500s); nicotine (*Nicotiana tabacum*) (late 1500s); quassin (*Quassia amara*) (late 1700s); heliopsin (*Heliopsis longipes*) (early 1800s); ryanodine (*Ryania speciosa*) (1940s); naphthoquinones (*Calceolaria andina*) (1990s); and derris (*Derris chinensis*) (mid-1990s). Many of these insecticidal plants were already being used for pest control purposes by native cultures prior to European exploration of the New World (Thacker, 2002).

In the late 1800s, inorganic compounds were discovered that offered impressive insecticidal and fungicidal properties. In 1865, the Colorado potato beetle (*Leptinotarsa decemlineata*) was controlled by Paris green (cupric acetoarsenite), the first synthetic insecticide (Metcalf, 1994). Prior to the introduction of potatoes by settlers (1850s) into the western plains of the USA, this beetle fed primarily on the buffalo burr (*Solanum rostratum*). This insect soon found potatoes to be an excellent host. Lead arsenate replaced Paris green and was used extensively for Colorado potato beetle control until DDT became more readily available. Plant pathologists also determined (1880s) that synthetic compounds such as Bordeaux mixture (copper sulfate and hydrated lime) reduced the severity of downy mildew in grape vineyards (Perkins, 2002). In subsequent years, other metabolic inhibitory fungicidal compounds were utilized, such as those containing mercury. Weed control was largely dependent upon plowing and hoeing until the introduction (early 1940s) of 2,4-dichlorophenoxy acetic acid (Perkins, 2002). In addition to these early chemical approaches to pest control, farmers relied upon their rudimentary knowledge (Webster, 1913) of pest life cycles and the use of cultural tactics to limit crop losses.

In 1939, the insecticidal properties of DDT (dichlorodiphenyltrichloroethane) were discovered by Paul Herman Müeller, a scientist with the Geigy Chemical Company. Most entomologists view this development as the beginning of the modern insecticide era. The pest control benefits of this new insecticide were regarded initially as miraculous. Some referred to DDT as the "wonder" insecticide (Metcalf, 1994). During World War II, DDT was used extensively to prevent epidemics of several insect-vectored diseases such as yellow fever, typhus, elephantiasis and malaria. The use of DDT for insect control in the production of crops, protection of livestock, in forestry and in urban and public health arenas soared in the late 1940s and 1950s. In 1946, DDT-resistant strains of the house fly (*Musca domestica*) were reported in Sweden and Denmark (Metcalf, 1994). Despite this "chink" in the armor of DDT, the promise of chemicals to deliver economical and effective pest control (including that of plant diseases and weeds) heralded in an atmosphere characterized by an over-reliance on pesticides throughout the 1950s and 1960s. This over-reliance on insecticides soon led to many significant ecological backlashes such as insecticide resistance, concentration of chlorinated hydrocarbon insecticides in the food chain, significant declines in densities of natural enemy (predators and parasitoids) populations, secondary outbreaks of pests, resurgence of primary pests and unwanted insecticide residues on fruits and vegetables. Critics of this over-reliance on pesticides argued that basic biological research on pest ecology and alternative management strategies were being ignored. Entomologists engaged in biological control efforts in California, cotton production in North and South America and production of fruit in orchard systems (Canada, Europe and the USA) were among the first to recognize many of the acute ecological problems associated with indiscriminate pesticide use (Kogan, 1998).

1.2 Early conceptual efforts in IPM development

In 1959, University of California entomologists at Berkeley, Vernon Stern, Ray Smith, Robert van den Bosch and Kenneth Hagen, published a seminal paper entitled "The Integration of Chemical and

Biological Control of the Spotted Alfalfa Aphid." In this paper, they offered the following statement concerning the integrated control concept:

> Whatever the reasons for our increased pest problems, it is becoming more and more evident that an integrated approach, utilizing both biological and chemical control, must be developed in many of our pest problems if we are to rectify the mistakes of the past and avoid similar ones in the future.

Many terms and concepts, now well known by entomologists, plant pathologists, weed scientists and IPM practitioners, were defined by these authors such as economic threshold, economic injury level and general equilibrium position. The following definitions are provided from Stern *et al.* (1959):

economic injury level	The lowest population density that will cause economic damage.
economic threshold	The density at which control measures should be determined to prevent an increasing pest population from reaching the economic injury level
general equilibrium position	The average density of a population over a period of time (usually lengthy) in the absence of permanent environmental change

Integrated control was defined as *applied pest control which combines and integrates biological and chemical control* and employed the use of economic thresholds to determine when chemical control should be utilized to prevent pests from reaching the economic injury level. The integrated control concept has evolved into the IPM concept that includes insects, plant pathogens, weeds and vertebrate pests. Since the initial tenets of the integrated control concept were developed in response to insect pests, not all of the early basics fit well with regard to the practical management of weeds, plant pathogens and vertebrate pests. Knake & Downs (1979) indicated that IPM should be an interdisciplinary approach rather than simply combining various control options within one discipline: "Weeds harbor insects and diseases, diseases may kill insects and weeds, and insects can be used to control other insects and weeds." Ford (1979) described three threshold types for plant pathology IPM programs: (1) a threshold addressing detection, (2) a threshold for prevention due to zero injury tolerance and (3) the more standardized economic injury threshold. Integrated vertebrate pest control applies ecology and only supports destruction of individual vertebrates as a last option to address animal damage (Timm, 1979). The impact of pest management implementation requires careful examination of potential benefits, costs and risks. While increased producer productivity is often considered a benefit, if it is obtained at a high environmental cost, the true economic impact may be obscured (Carlson & Castle, 1972). Higley & Wintersteen (1992) suggested that the traditional use of economic thresholds and injury levels are insufficient in estimating the hidden environmental externalities associated with insecticide use.

Some debate persisted among academics throughout the 1960s and into the 1980s regarding the perceived differences between "pest management" and "integrated control" (Kogan, 1998). Smith & Reynolds (1966) presented the concept of integrated pest control as a multifaceted, flexible, evolving system that blends and harmonizes control practices in an organized way. They believed the system must integrate all control procedures and production practices into an ecologically based system approach aimed at producing high quality products in a profitable manner. While this debate ensued, Rachel Carson published *Silent Spring* in 1962. This book galvanized sentiment among the general public against the abuses of pesticide applications. She was criticized by some for her use of emotionally charged passages such as (from Chapter 3, "Elixirs of Death"):

> For the first time in the history of the world, every human being is now subjected to contact with dangerous chemicals, from the moment of conception until death. In the less than two decades of their use, the synthetic pesticides have been so thoroughly distributed throughout the animate and inanimate world that they occur virtually everywhere.

She is given deserved credit for inspiring a generation of environmentalists and forcing the scientific community and governmental agencies to more closely scrutinize pesticide use and registration requirements. Eight years after *Silent Spring* was published, the US Congress mandated that the administration and enforcement of the Federal Insecticide, Fungicide, and Rodenticide Act (FIFRA) be transferred from the US Department of Agriculture (USDA) to a newly created federal entity, the US Environmental Protection Agency (EPA). The passage of FIFRA amendments over the past 30 years has resulted in policies aimed at reducing environmental and human health and safety risks that are linked with pesticide use (Gray, 2002). Kogan (1998) indicated the following with respect to the popularization of IPM:

> Not until 1972, however, were "integrated pest management" and its acronym IPM incorporated into the English literature and accepted by the scientific community. A February 1972 message from President Nixon to the US Congress, transmitting a program for environmental protection, included a paragraph on IPM.

Kogan (1998) further added that broad agreement had by then been reached on several key points regarding IPM:

> (1) **integration** meant the harmonious use of multiple methods to control single pests as well as the impacts of multiple pests; (2) **pests** were any organism detrimental to humans, including invertebrate and vertebrate animals, pathogens, and weeds; (3) IPM was a multidisciplinary endeavor; and (4) **management** referred to a set of decision rules based on ecological principles and economic and social considerations.

Some continue to debate the definition of IPM; however, the key components of this concept can be found in these four elements. More recently, in response to a national review of the federally supported US IPM Program (US General Accounting Office, 2001), and considerable stakeholder input, the USDA developed the "IPM Road Map" (see Chapter 37) with the ultimate objective of increasing IPM implementation by practitioners such as "land managers, growers, structural pest managers, and public and wildlife health officials." The IPM Road Map (2003) offers a definition of IPM that includes the historical elements of IPM reviewed by Kogan (1998), and in many ways extends the concept to focus on reducing the risks of economic and environmental losses. Within the IPM Road Map (May, 2004 version) IPM is defined as:

> . . . a long-standing, science-based, decision-making process that identifies and reduces risks from pests and pest management related strategies. It coordinates the use of pest biology, environmental information, and available technology to prevent unacceptable levels of pest damage by the most economical means, while posing the least possible risk to people, property, resources, and the environment. IPM provides an effective strategy for managing pests in all arenas from developed residential and public areas to wild lands. IPM serves as an umbrella to provide an effective, all encompassing, low-risk approach to protect resources and people from pests.

1.3 Kinds of pests

The selection of a strategy and components of an IPM program are largely influenced by the status of a pest in relationship to its host. Four pest types are commonly recognized by IPM practitioners: (1) *subeconomic*, (2) *occasional*, (3) *perennial* and (4) *severe* (Pedigo & Rice, 2006).

(1) The general equilibrium position of a *subeconomic pest* is always below the economic injury level, even during its highest population peaks. An insect in this category may cause direct losses but if the host (crop) values are modest, and the pest densities are always low, then it is not appropriate to initiate control practices whose costs exceed the value of host damage.
(2) The general equilibrium position of an *occasional pest* is nearly always below the economic injury level but occasionally population peaks exceed this level. The occasional pest is a very common type of pest. It may be present on or near a host nearly every year, but only sporadically does it cause economic damage.
(3) The general equilibrium position of a *perennial pest* is below the economic injury level but peak populations occur with such frequency that economic damage usually occurs yearly.

(4) A *severe pest* has a general equilibrium position that is always above the economic injury level so that when they occur in or on a host, economic damage is always the end result. As might be expected, perennial and severe pests cause the most serious damage and difficult challenges in an IPM program.

1.4 Pest management strategies and tactics

A pest management strategy is the total approach to eliminate or reduce a real or perceived pest problem. The development of a particular strategy will be greatly influenced by the biology and ecology of the pest and its interaction with a host or environment. The goal should be to reduce pest status when addressing problems using pest management. Because both the pest and host determine pest status, modification of either or both of these may be emphasized in a management program. Therefore, four types of strategies (Pedigo & Rice, 2006) could be developed based on pest characteristics and economics of management: (1) do nothing, (2) reduce pest numbers, (3) reduce host susceptibility to pest injury and (4) combine reduced pest populations with reduced host susceptibility. Once a pest management strategy has been developed, the methods of implementing the strategy can be chosen. These methods are called pest management tactics, and several tactics may be used to implement a management strategy.

1.4.1 Do-nothing strategy

All pest injury does not cause an economic loss to a host. Many hosts, especially plants and occasionally animals, are able to tolerate small amounts of injury without suffering economic damage. It is not uncommon for trivial insect injury to be mistaken for economically significant injury. This is most likely to occur when the pest population density is not considered in relationship to an economic threshold. If the pest density is below the economic threshold, then the do-nothing strategy is the correct approach; otherwise money would be expended on control that would not result in a net benefit. The do-nothing strategy is frequently used when insects cause indirect injury to a host, or when a successful pest management program reduces the pest population and only surveillance of the remaining population is necessary. No tactics are used in the do-nothing strategy, but this does not imply that no effort is necessary or that pest suppression is not occurring. Sampling of the pest population is required to determine that the do-nothing strategy is the appropriate response, and environmental influences may reduce the population, resulting in pest suppression.

1.4.2 Reduce pest numbers

Reducing pest densities to alleviate or prevent problems is probably the most frequently used strategy in pest management. This strategy is often employed in a therapeutic manner when populations reach the economic threshold or in a preventive manner based on historical problems (Pedigo & Rice, 2006). Two objectives may be desirable in attempting to reduce pest densities. If the pest's long-term average density, or general equilibrium position, is low compared with the economic threshold, then the best approach would be to diminish the population peaks of the pest. This action would not appreciably change the pest's general equilibrium position, but it should prevent damage from occurring during pest outbreaks. If, however, the pest population's general equilibrium position is near or above the economic threshold, then the general equilibrium position must be lowered so that the highest peak populations never reach the economic threshold. This may be done by either reducing the carrying capacity of the environment, or by reducing the inherited reproductive and/or survival potential of the population (Pedigo & Rice, 2006). There are many tactics that can be used to reduce pest numbers including resistant hosts, insecticides, pheromones, mechanical trapping, natural enemies, insect growth regulators, release of sterilized insects and modification of the environment.

1.4.3 Reduce host susceptibility to pest injury

One of the most environmentally compatible and effective strategies is to reduce host susceptibility to pest injury. This strategy does not modify the pest population; instead the host or host's relationship and interaction with the pest is changed to make it less susceptible to a potentially damaging pest population. A common form of this strategy is where plant cultivars or animal breeds are developed with a type of resistance, known as tolerance, which provides greater impunity to a pest than a similar plant or animal without the tolerance. The tolerance expressed by a plant or animal does not reduce the attacking pest population, but the injury caused by the pests has less of a detrimental affect on the host (i.e. yield loss in plants or weight loss in animals) than it does on a similar host without the tolerance. The other component to this strategy, ecological modification of factors that influence the distribution or abundance of a pest, also can reduce host susceptibility. Examples of this strategy would be reducing livestock exposure to a pest insect by moving them from an outdoor environment to an indoor facility or adjusting a crop planting date to create an asynchrony between a pest and a susceptible plant stage.

1.4.4 Combine reduced pest populations with reduced host susceptibility

A strategy that combines the objectives of the previous strategies is a logical step in the development of a pest management program. A multifaceted approach is more likely to produce greater consistency than a single strategy using a single tactic. Experience has shown that a single strategy is more likely to fail when either, slowly or quickly, a single tactic approach falters. With the multifaceted approach, if one tactic fails, then other tactics operate to help modulate losses. The use of multiple strategies and tactics is the basic principle in developing an IPM program.

1.5 Funding IPM research and implementation

Since the early 1970s, the USDA, the EPA and the National Science Foundation (NSF) have been the primary governmental agencies in the USA that have provided competitive and formula-based funding for research and extension IPM programs. The majority of these IPM research and extension programs are conducted by investigators located at land-grant universities (Morrill Land-Grant Acts, 1862, 1890). Two of the most visible and comprehensive IPM pilot efforts included the Huffaker (1972–1979, $US 13 million in funding, EPA, NSF, USDA) and Adkisson (1979–1984, $US 15 million in funding, EPA, USDA) projects (Allen & Rajotte, 1990). The Huffaker Project concentrated on the development of IPM tactics for insect pests in cotton, soybeans, alfalfa, citrus fruits, and pome and stone fruits. The Adkisson Project expanded its range of targeted pests to include diseases, insects and weeds in alfalfa, apples, cotton and soybeans. In 1978, a USDA report from the Extension Committee on Organization and Policy recommended that $US 58 million be spent on extension IPM programs. This goal was never achieved and federal funding for extension IPM programs began to falter reaching approximately $US 7.0 million in the early 1980s (Allen & Rajotte, 1990). By 2006, federal formula funds [Smith-Lever 3(d)] allocated across the USA for extension IPM programs had risen to a modest $US 9.86 million, or roughly $US 200 000 per state. Reasons are diverse for the weakening political support and funding for new and large-scale IPM initiatives in the USA (Gray, 1995). These reasons include the perception that implementation of IPM would lead to greater overall reductions in pesticide use than has occurred in some cropping systems, political support for "older" programs often wanes over time in lieu of new initiatives, continued difficulty in quantifying successes and impact of IPM implementation, struggles of IPM leadership to clearly articulate the goals of IPM implementation, and increasing popularity of organic production practices.

Funding of IPM research and implementation programs in developing countries is increasingly important as food production and environmental concerns intensify in many densely populated areas around the globe. Some key organizations and programs that fund and promote these IPM efforts include: Food and Agriculture Organization of the United Nations (FAO), United Nations Environment Program (UNEP) and the United Nations Development Program (UNDP). In 1995, the Global IPM Facility was established and is housed in FAO Headquarters in Rome, Italy. Co-sponsors of the Facility include FAO, UNEP,

UNDP and the World Bank (Stemerding & Nacro, 2003). It was hoped that the Facility would ultimately result in more lending operations that would support IPM implementation. Thus far, the impact of the Global IPM Facility has been assessed as "mixed" (Schillhorn van Veen, 2003). Other key organizations that fund and promote IPM globally include the Integrated Pest Management Collaborative Research Support Program (IPM CRSP) This program was started in 1993 with the financial assistance of the US Agency for International Development (USAID). Current sites include: Albania, Bangladesh, Ecuador, Guatemala, Jamaica Mali, Philippines and Uganda. Several USA institutions (Virginia Tech, Ohio State University, Purdue University) provide personnel who collaborate with scientists at the host institutions Successful IPM programs that have been developed through this effort include: rice and vegetable cropping systems in the Philippines, maize and bean cropping systems in Africa, horticultural export crops in Latin America and sweet potato production in the Caribbean (Gebrekidan, 2003). Significant international contributions in host plant resistance to a variety of pests in crops have been achieved through support of the Consultative Group on International Agricultural Research (CGIAR) centers. These centers support the implementation of systemwide programs on IPM in several international "target zones" such as Africa, Asia and Latin America (James *et al.*, 2003).

1.6 Measuring IPM implementation

Assessing the level of IPM implementation has historically presented a challenge to policy makers, governmental agencies and scientists (Wearing, 1988). In an era of increasing pressure to ensure accountability, continued governmental support of IPM programs (research and extension) is contingent upon documenting increasing levels of IPM adoption and proving impact (economic, environmental and human health and safety benefits). Not all scientists, policy makers or practitioners of IPM agree that the primary goal of IPM is to reduce pesticide use (Gray, 1995; Ratcliffe & Gray, 2004). The US Council on Environmental Quality (1972) described IPM as "an approach that employs a combination of techniques to control the wide variety of potential pests that may threaten crops." It suggests numerous economic pests can be managed by "maximum reliance" on natural pest controls with the incorporation of key elements including cultural methods, pest-specific diseases, resistant crop varieties, sterile insects and attractants together with the use of biological control and reduced risk, species-specific chemical controls as part of an IPM program. Risk management and the fear of crop loss is often overemphasized, but coupled with the lack of implementation incentives many producers choose to only adopt limited aspects of IPM rather than a whole system approach (US Council on Environmental Quality, 1972).

In September 1993 (US Congress, 1993) the Clinton Administration set a goal for 75 percent implementation of IPM practices, by 2000, on managed agricultural areas in the USA. A National Agricultural Statistics Service (2001) report indicated that by 2000, IPM adoption levels for many crops had met or exceeded this goal. However, in 2001, the United States General Accounting Office (GAO) published a document that was critical of the coordination and management of federal IPM efforts (across more than a dozen federal agencies). In addition, some criticism within the GAO report was directed at the lack of measurement and evaluation tools (environmental and economic) for assessing the level of IPM implementation. Since 2000, four regional IPM Centers within the USA have sought to improve the coordination of IPM implementation efforts utilizing a National Road Map for IPM (first articulated at the 4th National IPM Symposium, Indianapolis, IN, April 2003; see Chapter 37) as a blueprint (Ratcliffe & Gray, 2004). Bajwa & Kogan (2003) provide a very good assessment of IPM adoption in Africa, Americas (other than USA), Asia, Australia, Europe and the USA for many crops. The percentage of farmers who have adopted IPM practices is very high in many cases, such as: pear production in Belgium (98 percent), cotton production in Australia (90 percent), pome fruits in British Columbia (75 percent), and sugarcane production in

Colombia (100 percent). Despite these advances in IPM implementation, pesticide usage has increased in many developing countries throughout the 1990s and remains the exclusive tactic to control pests. Bajwa & Kogan (2003) remind us that "IPM is a tangible reality in some privileged regions of the world, but still remains a distant dream for many others."

1.7 Examples of successful implementation of IPM

1.7.1 Ecological management of environment: push–pull polycropping in Africa

Push–pull strategies use a combination of behavior-modifying stimuli to manipulate the distribution and abundance of pest or beneficial insects in pest management with the goal of pest reduction on the protected host or resource (Cook *et al.*, 2007). Pests are repelled or deterred away from the resource (push) by using stimuli that mask host apparency or are deterrent or repellant. Pests are simultaneously attracted (pull), using highly apparent and attractive stimuli, such as trap crops, where they are concentrated, facilitating their elimination (Cook *et al.*, 2007).

The most successful push–pull strategy was developed for subsistence farmers in east Africa. Maize (*Zea mays*) and sorghum (*Sorghum bicolor*) two principal foods in east Africa, are attacked by lepidopteran stem borers, e.g. *Busseola fuscus*, *Chilo partellus*, *Eldana saccharina* and *Sesamia calamistis*, that cause 10–50 percent yield losses (Cook *et al.*, 2007). Farmers combine the use of intercrops and trap crops, using plants that are appropriate for the farmers and exploit natural enemies. Stem borers are repelled from the maize and sorghum by non-hosts such as greenleaf desmodium (*Desmodium intortum*), silverleaf desmodium (*Desmodium uncinatum*) and molasses grass (*Melinis minutiflora*), which are interplanted with the maize or sorghum (the push). Around the field edges are planted trap crops, mostly Napier grass (*Pennisetum purpureum*) and Sudan grass (*Sorghum vulgare sudanense*), which attract and concentrate the pests (the pull). These grasses have a dual purpose as they are also used as forage for livestock. Molasses grass, as an intercrop, reduces stem borer populations by producing stem borer repellent volatiles; it also increases parasitism by a parasitoid wasp. *Desmodium* also produces similar repellent volatiles; but also produces sesquiterpenes that suppress the parasitic African witchweed (*Striga hermonthica*), a major yield constraint of cropland in east Africa. The desmodium compounds stimulate germination of witchweed seeds and subsequent mortality of the seedlings. The push–pull strategy has contributed to increased grain yields and livestock production in east Africa, resulting in significant impact on food security (Cook *et al.*, 2007).

1.7.2 Biological control: prickly pear cactus and cactus moths in Australia

Prickly pears, or prickly pear cactus (*Opuntia* spp.), are native to the Americas but have become serious invasive weeds in suitable habitats around the world. Around 1840, cuttings of prickly pears were brought to Queensland, Australia for use as a hedge around fields and homesteads, as a botanical curiosity, and for production of cochineal – a dark reddish dye produced by scale insects that feed on the plant. Livestock and native birds quickly spread prickly pear seeds across overgrazed grasslands, where competition was reduced during droughts, whereas during heavy rainfall, broken pieces of prickly pears were carried into the interior on westward-flowing rivers (DeFelice, 2004). The climate and soil of eastern Australia was ideal for prickly pear and the weed quickly spread. Attempts were made by farmers and ranchers in the 1880s to control the weed, but were without success. In 1893, it was declared a noxious weed in Queensland.

By 1913, prickly pear was estimated to cover 1.4 million ha with dense infestations and another 4.9 million ha with scattered infestations. By 1926, the prickly pear had infested 24 million ha in Queensland and New South Wales and was spreading at the rate of 1 million ha annually (DeFelice, 2004). Attempts at controlling the prickly pear using mechanical, chemical and cultural methods completely failed to stop the spread of the weed, mostly because control was poorly supported and many government policies only conspired to worsen the problem (DeFelice, 2004).

The infestation was so dense the 12 million ha were rendered useless, resulting in worthless grazing land and the abandonment of many farms and homesteads.

In 1927, hope appeared in the form of an imported parasitic insect from South America – the cactus moth (*Cactoblastis cactorum*). This insect was evaluated and confirmed to only feed on prickly pear. Over 220 million eggs were reared and distributed and three years later 200 000 ha of prickly pear were destroyed. The insect rapidly spread and by the end of 1931, millions of hectares of prickly pears were a mass of rotting vegetation (DeFelice, 2004). Land that had been useless for decades was cleared and restored to rangelands and agricultural production. The prickly pear experience in Australia was one of the most frightening cases in history of ecological destruction by an invasive plant and also one of the most successful biological control campaigns ever mounted against a pest.

1.7.3 Sterile insect technique: screwworm eradication in North and Central America

The classic achievement of success with the sterile insect technique was the eradication of the screwworm (*Cochliomyia hominivorax*) from the USA, Mexico and Central America. The screwworm is an obligate parasite of livestock and has occasionally attacked humans. The adult fly lays up to 450 eggs in open wounds where the larvae feed on tissues and enlarge the wound (Krafsur *et al.*, 1987). Feeding by the larvae attracts other flies to oviposit in the wound, thereby aggravating the damage to the animal. Heavily parasitized livestock may be killed within 10 days. Historical livestock losses to this pest were astronomical. Prior to the sterile release program, losses were estimated at $US 70–100 million annually across the southern USA from Florida to California. A severe pest outbreak occurred in this region in 1935, with 1.2 million cases of infestation and 180 000 livestock deaths.

The sterile insect technique involves the intentional release of large numbers of sterilized insects to compete with wild insects for mates (Krafsur *et al.*, 1987). The sterile insect technique with screwworms involves the mass rearing of larvae on a specialized liquefied diet of bovine blood and powdered egg. The pupae are collected from the rearing containers and at five days of age are irradiated with cesium. Female flies irradiated with this process fail to undergo vitellogenesis and therefore do not deposit eggs. Male flies likewise are sterilized and when they mate with a wild-type female, no viable eggs are produced.

The concept of the sterile insect technique was put to the test in a pilot program on Sanibel Island, Florida and produced positive results. A larger test was initiated in 1954 on Curaçao, an 444-km^2 island off the coast of Venezuela, where 400 sterilized males per 2.6 km^2 were released for three months. The effort resulted in the complete eradication of the screwworm from the island and demonstrated the potential of the technique. The technique was then applied to livestock in Florida and southern Georgia and Alabama in 1958. More than 2000 million sterilized flies were released from airplanes during an 18-month period, resulting in complete eradication from the region. The program was then moved to southwestern USA in the early 1960s where sterile flies were released along the international border with Mexico. This resulted in a fly-free zone nearly 3200 km long and 500 to 800 km deep which prevented the flies from moving north into the USA. Fly infestation reports dropped from more than 50 000 in 1962 to 150 by 1970. Unfortunately, infestations did not remain low so a cooperative agreement between the USA and Mexican governments worked together to push the screwworm further south in Mexico. By 1986, Mexico was declared free of screwworm. The fly-free zone was continually moved south, eradicating the pest from numerous Central American countries. A fly-free barrier is currently maintained in Panama to prevent reinfestations from South America. In 1992, Raymond Bushland and Edward Knipling received the World Food Prize for their collaborative achievements in developing the sterile insect technique for eradicating or suppressing the threat posed by pests to crops and livestock.

1.7.4 Transgenic plants: control of European corn borer in North America

The European corn borer (*Ostrinia nubilalis*) has been considered by some (Ostlie *et al.*, 1997) to be the most damaging pest of maize in North

America with damage and control costs exceeding $US 1000 million during the early to mid-1990s. Insecticides were occasionally used by growers to prevent stalk tunneling, kernel damage and fallen ears in maize but often they were reluctant to embrace chemical control (Rice & Ostlie, 1997). Reasons for reluctance included the fact that larval damage was hidden, large infestations are unpredictable, fields had to be scouted multiple times requiring time and skill, insecticides were expensive and raised environmental and health concerns and benefits of insecticide control were uncertain. These concerns paved the way for a novel way of managing this pest through the use of transgenic plants.

In 1996, Mycogen Seeds and Novartis Seeds introduced the first commercial *Bt* maize hybrids. The *Bt* hybrids were genetically transformed to express a gene from the soil bacterium, *Bacillus thuringiensis*, which produces a protein that is toxic to European corn borer larvae. Most larvae die after taking only a few bites of maize leaf tissue. Consequently, *Bt* maize provides extremely high levels of larval mortality resulting in exceptional yield protection even during heavy infestations of European corn borer (Ostlie *et al.*, 1997). In 2005, approximately 35 percent of the maize hectares were planted to a corn borer resistant transgenic hybrid with the result being that during the past ten years, the European corn borer has had a steady decline in the severity of populations, thereby leading some to conclude that the insect has become a secondary pest (Gray, 2006). An additional effect was that the percent of farmers who decreased their insecticide use doubled during the first three years of planting a transgenic maize hybrid resulting in less broad-spectrum insecticide applied to the fields (Pilcher *et al.*, 2002). Maize growers perceive that less exposure to insecticides and less insecticide in the environment are the two primary benefits of planting transgenic maize hybrids (Wilson *et al.*, 2005). The success of commercial transgenic *Bt* maize has lead to the development of triple-stacked hybrids that may express a protein for corn borers, a different protein specific for corn rootworms (*Diabrotica* spp.) and resistance to herbicides.

1.7.5 Insect growth regulators: termite control in North America

Termites are destructive pests of wooden structures and the latest industry estimates place the annual cost of damage and treatment at $US 5000 million worldwide (National Pest Management Association, 2005). Termite control generally consists of five types of treatment programs: liquid termiticides, bait systems, wood preservatives, mechanical barriers and biological termiticides (Hu, 2005). Each type of program has its advantages and disadvantages, but the bait system is the most novel as it uses an insect growth regulator to control the termite colony.

The bait system is a relatively new tool for termite control. Instead of applying a chemical barrier designed to exclude termites from a wooden structure, termites are offered food in the form of baits (Hu *et al.*, 2001). Treatment baits have two components: a termite food source, such as a block of wood in the soil, and a slow-acting termiticide, often an insect growth regulator. The insect growth regulator (diflubenzuron, hexaflumuron or noviflumuron) is a slow-acting, non-repellent toxicant that prevents the formation of chitin in the insect cuticle. Termites feeding on the bait are not killed immediately, but through colony recruitment when worker termites find the bait the insect growth regulator is passed to other colony members, ultimately leading to decline or perhaps elimination of the colony. The advantage of baiting is that the system is non-intrusive, consumer friendly, safer than most of the soil-applied insecticides, specifically targets termites and dramatically reduces the amount of chemical needed to protect a structure. However, a disadvantage is that the process may take weeks or months to knock down termite populations.

1.8 IPM within a transgenic era

In 1996, transgenic crops were commercialized on a limited basis in the USA for the first time. In ten years, the use of transgenic crops has seemingly transformed the IPM paradigm, particularly in the major field crops arenas. The primary transgenic tools include the planting of herbicide-tolerant varieties of cotton, maize and soybeans (primarily to the herbicides glyphosate and glufosinate-ammonium) as well as *Bt* cotton and maize that express proteins derived from various strains of the bacterium *Bacillus thuringiensis*. Fernandez-Cornejo *et al.* (2006) reported that 87 percent and 60 percent of soybean and cotton hectares, respectively, in the USA were planted to herbicide-tolerant varieties in 2005. The use of *Bt* cotton and maize also was impressive, estimated at 52 percent and 35 percent, respectively, on USA hectares in 2005. Although the USA is estimated to account for 55 percent of the global area devoted to transgenic crops, 21 other nations in 2005 were planting transgenic crops as well. Nations that are heavily engaged in the production of transgenic crops (James, 2005) include the USA (49.8 million ha), Argentina (17.1 million ha), Brazil (9.4 million ha), Canada (5.8 million ha) and China (3.3 million ha).

Increasingly the dialogue among scientists engaged in IPM research and extension programs in field crops has turned towards resistance management. For many of these scientists, their focus has shifted to that of evaluating models and recommending to producers the best deployment of transgenic crops across the agricultural landscape to delay the onset of resistance to these new tools (Gould, 1998). The philosophical debate will continue to rage for many years regarding the "fit" of transgenic crops within the IPM framework. For supporters, transgenic crops fit within the host plant resistance pillar of IPM. For others, transgenic crops, such as *Bt* maize and cotton, are little different than using a broadcast insecticide application on a prophylactic basis. While global adoption rates of transgenic crops are expected to increase, the philosophical debate within the academic community will continue on what truly constitutes IPM.

References

Allen, W. A. & Rajotte, E. G. (1990). The changing role of extension entomology in the IPM era. *Annual Review of Entomology*, **35**, 379–397.

Bajwa, W. I. & Kogan, M. (2003). Integrated pest management adoption by the global community. In *Integrated Pest Management in the Global Arena* eds. K. M. Maredia, D. Dakouo & D. Mota-Sanchez, pp. 97–107. Wallingford, UK: CABI Publishing.

Bird, G. W. (2003). Role of integrated pest management in sustainable development. In *Integrated Pest Management in the Global Arena*, eds. K. M. Maredia, D. Dakouo & D. Mota-Sanchez, pp. 73–85. Wallingford, UK: CABI Publishing.

Carlson, G. A. & Castle, E. N. (1972). Economics of pest control. II. Systems approach to pest control. In *Pest Control Strategies for the Future*, pp. 79–99. Washington, DC: National Academy of Sciences.

Carson, R. L. (1962). *Silent Spring*. Boston, MA: Houghton Mifflin Company.

Cook, S. M., Khan, Z. R. & Pickett, J. A. (2007). The use of push–pull strategies in integrated pest management. *Annual Review of Entomology*, **52**, 375–400.

DeFelice, M. S. (2004). Prickly pear cactus, *Opuntia* spp.: a spine-tingling tale. *Weed Technology*, **18**, 869–877.

Fernandez-Cornejo, J., Caswell, M., Mitchell, L., Golan, E. & Kuchler, F. (2006). *The First Decade of Genetically Engineered Crops in the United States*, USDA Economic Information Bulletin No. 11. Washington, DC: US Government Printing Office.

Ford, R. E. (1979). The role of plant pathology in integrated pest management. In *Integrated Pest Management North Central Region Workshop Proceedings*, St. Louis, MO, December 11–13, 1979, ed. S. Elwynn Taylor, Section I, pp. 23–33.

Gebrekidan, B. (2003). Integrated pest management collaborative research support program (USAID–IPM CRSP): highlights of its global experience. In *Integrated Pest Management in the Global Arena*, eds. K. M. Maredia, D. Dakouo & D. Mota-Sanchez, pp. 407–418. Wallingford, UK: CABI Publishing.

Gould, F. (1998). Sustainability of transgenic insecticidal cultivars: integrating pest genetics and ecology. *Annual Review of Entomology*, **43**, 701–726.

Gray, M. E. (1995). Status of CES–IPM programs: results of a national IPM coordinators survey. *American Entomologist*, **41**, 136–138.

Gray, M. E. (2002). Federal Insecticide, Fungicide, and Rodenticide Act. In *Encyclopedia of Pest Management*, ed. D Pimentel, pp. 261–262, New York: Marcel Dekker.

Gray, M. (2006). European corn borer: a secondary pest for now? *Pest Management and Crop Development Bulletin*, **9**, 261–262. Available at www.ipm.uiuc.edu/bulletin/article.php?id=530.

Higley, L. G. & Wintersteen, W. K. (1992). A novel approach to environmental risk assessment of pesticides as a basis for incorporating environmental costs into economic injury levels. *American Entomologist*, **38**(2), 34–39.

Hu, X. P. (2005). *Subterranean Termite Control Products for Alabamians*. Alabama Cooperative Extension System, Auburn University. Available at www.aces.edu/pubs/docs/A/ANR-1252/.

Hu, X. P., Appel, A. G., Oi, F. M. & Shelton, T. G. (2001). *IPM Tactics for Subterranean Termite Control*, ANR-1022. Alabama Cooperative Extension System, Auburn University. Available at www.aces.edu/pubs/docs/A/ANR-1022/.

James, B., Neuenschwander, P., Markham, R. *et al.* (2003). Bridging the gap with the CGIAR systemwide program on integrated pest management. In *Integrated Pest Management in the Global Arena*, eds. K. M. Maredia, D. Dakouo & D. Mota-Sanchez, pp. 419–434. Wallingford, UK: CABI Publishing.

James, C. (2005). Executive summary. In *Global Status of Commercialized Biotech/GM Crops: 2005*, ISAAA Briefs No. 34. Ithaca, NY: International Service for the Acquisition of Agri-biotech Applications.

Knake, E. L. & Downs, J. P. (1979). The weed science phase of pest management. In *Integrated Pest Management North Central Region Workshop Proceedings*, St. Louis, MO, December 11–13, 1979, ed. S. Elwynn Taylor, Section I, pp. 33–37.

Kogan, M. (1998). Integrated pest management: historical perspectives and contemporary developments. *Annual Review of Entomology*, **43**, 243–270.

Kogan, M. & Prokopy, R. (2003). Agricultural entomology. In *Encylopedia of Insects*, eds. V. H. Resh & R. T. Cardé, pp. 4–9. San Diego, CA: Academic Press.

Krafsur, E. S., Whitten, C. J. & Novy, J. E. (1987). Screwworm eradication in North and Central America. *Parasitology Today*, **3**, 131–137.

Metcalf, R. L. (1994). Insecticides in pest management. In *Introduction to Insect Pest Management*, 3rd edn, eds. R. L. Metcalf & W. H. Luckmann, pp. 245–314. New York: John Wiley.

National Agricultural Statistics Service (2001). *Pest Management Practices: 2000 Summary, Sp Cr 1 (01)*. Washington, DC: US Department of Agriculture.

National Pest Management Association (2005). *The Big Bite of Termites: $5 Billion a Year in Damages*. Available at www.pestworld.org/Database/article.asp?ArticleID=22.

National Roadmap for Integrated Pest Management (2003). In *Proceedings, Integrated Pest Management for Our Environment – for Our Future, 4th National Integrated Pest Management Symposium*, Indianapolis, IN, April 8–10, 2003, pp. 9–11. Urbana, IL: University of Illinois.

Ostlie, K. R., Hutchison, W. D., Hellmich, R. L. *et al.* (1997). Bt-Corn and European Corn Borer: Long-Term Success Through Resistance Management. North Central Regional Extension Publication NCR 602. St. Paul, MN: University of Minnesota.

Pedigo, L. P. & Rice, M. E. (2006). *Entomology and Pest Management*, 5th edn. Upper Saddle River, NJ: Pearson Prentice Hall.

Perkins, J. H. (2002). History. In *Encyclopedia of Pest Management*, ed. D. Pimentel, pp. 368–372. New York: Marcel Dekker.

Pilcher, C. D., Rice, M. E., Higgins, R. A. *et al.* (2002). Biotechnology and the European corn borer: measuring historical farmer perceptions and the adoption of transgenic Bt corn as a pest management strategy. *Journal of Economic Entomology*, **95**, 878–892.

Ratcliffe, S. T. & Gray, M. E. (2004). Will the USDA IPM centers and the national IPM roadmap increase IPM accountability? – responses to the 2001 General Accounting Office report. *American Entomologist*, **50**(1), 6–9.

Rice, M. E. & Ostlie, K. R. (1997). European corn borer management in field corn: a survey of perceptions and practices in Iowa and Minnesota. *Journal of Production Agriculture*, **10**, 628–634.

Schillhorn van Veen, T. W. (2003). The World Bank and pest management. In *Integrated Pest Management in the Global Arena*, eds. K. M. Maredia, D. Dakouo & D. Mota-Sanchez, pp. 435–440. Wallingford, UK: CABI Publishing.

Smith, R. F. & Reynolds, H. T. (1966). Principles, definitions and scope of integrated pest control. In *Proceedings of the FAO Symposium on Integrated Pest Control*, Rome, Italy, October 11–15, 1965, **1**, 11–18. Rome, Italy: Food and Agriculture Organization of the United Nations.

Stemerding, P. & Nacro, S. (2003). FAO integrated pest management programs: experiences of participatory IPM in West Africa. In *Integrated Pest Management in the Global Arena*, eds. K. M. Maredia, D. Dakouo & D. Mota-Sanchez, pp. 397–406. Wallingford, UK: CABI Publishing.

Stern, V. M., Smith, R. F., van den Bosh, R. & Hagen, K. S. (1959). the integration of chemical and biological control of the spotted alfalfa aphid (the integrated control concept). *Hilgardia*, **29**(2), 81–101.

Thacker, J. R. M. (2002). *An Introduction to Arthropod Pest Control*. Cambridge, UK: Cambridge University Press.

Timm, R. M. (1979). Vertebrate zoology's role in integrated pest management. In *Integrated Pest Management North Central Region Workshop Proceedings*, St. Louis, MO, December 11–13, 1979, ed. S. Elwynn Taylor, Section I, pp. 7–15.

US Congress (1993). Testimony of Carol M. Browner, Administrator EPA; Richard Rominger, Deputy Secretary of Agriculture; and David Kessler, Commissioner of FDA. Hearings before the Committee on Labor and Human Resources, US Senate, and Subcommittee on Health and the Environment, Committee on Energy and Commerce, US House of Representatives, 22 September, 1993.

US Council on Environmental Quality (1972). *Integrated Pest Management*. No. 4111–0010, pp. 9–15. Washington, DC: US Government Printing Office.

US General Accounting Office (2001). *Agricultural pesticides: management improvements needed to further promote integrated pest management. Report GAO-01–815*. Washington, DC: US Government Printing Office.

Wearing, C. H. (1988). Evaluating the IPM implementation process. *Annual Review of Entomology*, **33**, 17–38.

Webster, F. M. (1913). Bringing applied entomology to the farmer. In *Department of Agriculture Yearbook*, pp. 75–92. Washington, DC: US Government Printing Office.

Wilson, T. A., Rice, M. E., Tollefson, J. J. & Pilcher, C. D. (2005). Transgenic corn for control of the European corn borer and corn rootworms: a survey of Midwestern farmer practices and perceptions. *Journal of Economic Entomology*, **98**, 237–247.

Chapter 2

Economic impacts of IPM

Scott M. Swinton and George W. Norton

Economic impact analyses of IPM programs measure the economic effects on producers and consumers that can be attributed to IPM programs and practices. Good impact assessments are tailored to the objectives of the programs they are evaluating. Because IPM program objectives and approaches can vary widely, there is no one-size-fits-all method for IPM impact assessment. Some IPM activities are narrowly focused, such as a new methodology for measuring pest density. Others are broad, such as a national training program in pest recognition. Small programs may have narrow impacts, while large programs may have repercussions great enough to change prices at the regional or even national level.

Despite the diversity of approaches and objectives, virtually all IPM programs aim to influence economic and health or environmental outcomes. Economic outcomes may be measured at the level of the individual management unit (e.g. a farm) or at the aggregate level of all producers and consumers in a given market. Environmental and health outcomes may be measured using indicative, average approaches or using site-specific data about environmental vulnerability. Decision makers often wish to explore the trade-offs between economic and environmental/health outcomes and methods exist for that purpose. This chapter offers an overview of economic impact analysis methods, including ways to incorporate environmental and health effects; in part this chapter also draws upon previous summaries of impact assessment methods by Norton & Mullen, 1994; Norton *et al.*, 2001; Cuyno *et al.*, 2005; and Norton *et al.*, 2005. Recognizing that economic impact analysis is often performed under tight budget constraints, the methods proposed range from basic and somewhat unreliable but inexpensive, at one extreme, to nuanced and reliable, but costly, at the other extreme.

2.1 Measuring IPM adoption

Impact assessments must measure and attribute impacts before the costs and benefits of those impacts can be calculated. The measurement of IPM impacts typically begins with estimating the extent of IPM adoption. Unlike simple technologies whose adoption might be measured by a yes/no question (e.g. planting seed of an improved crop variety), IPM often involves a suite of different practices that are not necessarily adopted jointly. Recognizing different degrees of IPM practice adoption is central to properly measuring IPM adoption. The way that the degree of adoption is incorporated will depend upon whether adoption is estimated by expert opinion or by survey methods.

Integrated Pest Management, ed. Edward B. Radcliffe, William D. Hutchison and Rafael E. Cancelado. Published by Cambridge University Press.

Short of relying wholly on secondary data, the simplest, least costly approach to measuring IPM adoption is to rely upon expert opinion. For this purpose, IPM practices can be bundled together into ideal typical groupings to represent a range from no adoption to advanced IPM adoption. An expert-opinion-based estimate of IPM adoption might categorize adoption levels into: (1) no IPM, (2) weekly pest scouting, (3) weekly pest scouting with use of a threshold-based decision rule for treatment and (4) weekly pest scouting with a threshold-based decision rule and pesticide rates reduced by some specified level, say 50 percent.

The chief limitation to relying upon expert opinion to estimate IPM adoption levels is that even the best experts often have a biased perspective. They are familiar with IPM adoption levels among those with whom they have contact. But they do not have contact with a random sample of the population; instead, they tend to meet individuals interested in IPM. One comparison of IPM adoption estimates by extension agent experts versus scientific survey results found that while the experts estimated that 98% of Michigan tart cherry producers had adopted some form of IPM by 1999, a randomly sampled farm survey found that only 64% had done so.[1]

For survey purposes, IPM adoption must be described in terms that are simple and explicitly additive. Whereas the ideal types of adopters used for expert opinion purposes may involve a detailed mix of practices at each level, real producers will constitute a continuum of adoption levels that is likely to include surprising combinations or absences of practices. Hence, surveys must base adoption measures on binary choices that are exhaustive. To contrast with the ideal type adoption categories above, a survey would need to rely on simpler categories such as: (1) no IPM, (2) scouted for pests at least one time, (3) scouted for pests and used a threshold decision rule at least once and (4) did (3) and sprayed with a median rate less than or equal to 50% of the label pesticide rate. Note each of these adoption levels is based on the presence or absence of a general condition, with those conditions being additive (scouting, threshold rule, reduced rate), which permits them to be summed up to 100% of all producers.

In addition to estimating the level of IPM adoption, impact assessments must sometimes be able to attribute adoption to a particular program. For an extension program, it will be important to show how much adoption can be attributed to that program, versus the amount that would have occurred anyway in its absence. Evaluators refer to this benchmark case as the "counterfactual," that is, what would have occurred without the program. When impact assessments are planned from the outset, the attribution task can be facilitated in two ways. First, a benchmark survey before the program begins can establish the degree of adoption, as well as other baseline conditions. Second, a "no treatment" control area may be planned as a basis for comparison with the program intervention area. Such an arrangement explicitly builds in a counterfactual case that is subject to other evolutionary effects outside of the IPM program.

When a no-treatment control area is not planned from the outset, the next best means to decide whether adoption is attributable to the program is to use a statistical regression model. The first step is to survey adopters and non-adopters. Using their responses, a statistical regression model is developed where the dependent variable is the decision whether or not to adopt (or some variant, such as the number of IPM practices adopted). The explanatory variables include all those that might explain adoption, including variables unrelated to the IPM program of interest. Economists typically include input and output prices, important producer and business traits, and local environmental characteristics, as well as a measure of IPM program exposure. Such a model can help to predict the degree of influence by the IPM program, when taking into account other factors that might also have influenced IPM adoption (McNamara *et al.*, 1991; Fernandez-Cornejo, 1998).

Those who participate in IPM programs may be different in some respects than those who do not

[1] Unpublished data from a Michigan Agricultural Statistics Service survey conducted by S. M. Swinton in 1999.

Additions to Net Revenue	Reductions in Net Revenue
Increased Returns:	*Decreased Returns:*
1. ________ $_____	*4.* ________ $_____
2. ________ $_____	*5.* ________ $_____
3. ________ $_____	*6.* ________ $_____
Total $_____ (A)	*Total $_____ (B)*
Decreased Costs:	*Increased Costs:*
7. ________ $_____	*10.* ________ $_____
8. ________ $_____	*11.* ________ $_____
9. ________ $_____	*12.* ________ $_____
Total $_____ (C)	*Total $_____ (D)*
A+C = $_____ (E)	*B+D = $_____ (F)*
Change in Net Returns = E - F = $_______	

Fig. 2.1 Partial budget form.

participate, and as a result the estimated effects of IPM exposure in the regression described above may be biased. To reduce the possibility of bias, a second regression can be run in which IPM program participation or exposure is regressed on a set of producer characteristics or other variables. The results of this regression can be used to generate a predicted IPM program exposure variable which is used in the adoption regression in place of the simple IPM exposure variable to remove the bias (Feder *et al.*, 2004).

2.2 Individual economic impacts to the IPM user

2.2.1 Analyzing typical adoption cases using budgets

Economic impacts are realized by individual IPM users (e.g. producers, households, schools) when they adopt IPM practices and programs. The level and variability of costs and/or returns may change with IPM. Enterprise budgets or partial budgets are commonly used to assess relative costs and returns of non-IPM versus IPM practices. An enterprise budget lists all estimated expenses and receipts associated with a particular enterprise, while a partial budget only includes changes in yield, prices and costs to assess whether benefits (due to increased revenues and reduced costs) exceed burdens (from reduced revenues or increased costs). Budgets for various pest management alternatives can be compared using data from replicated experiments or from producer surveys. When using experimental data, enterprise budgets are commonly simplified to gross margin analysis, which includes only those costs that vary across treatments (Centro Internacional de Mejoramiento de Maíz y Trigo, 1988). Budget analysis assumes typical conditions, no carry-over effects from one period to the next, predictable prices, costs and yield effects of pests, and that the decision maker's only objective is to maximize profits (Swinton & Day, 2003).

Partial budgets are the most common and practical type of budget used for assessing IPM impacts. Budgets can be constructed for each adoption level (group of practices). A typical partial budget form is presented in Fig. 2.1.

By developing a budget for each level of adoption, changes in net revenue can be associated with levels of IPM adoption. Data are required on inputs, outputs, and their prices. These data can be obtained from on-farm trials, user surveys or from taking existing budgets and adjusting the cost categories. When based on replicated experiments, budget results such as gross margins can be subjected to analysis of variance to test for differences in mean profitability by treatment (Swinton *et al.*, 2002).

Results of budgeting analysis can be used by scientists and extension workers to judge the profitability of practices they are developing or recommending to farmers or of practices already adopted. A second major use of budget information is as an input into a market or societal level assessment of the economic benefits and costs of an IPM program as discussed below.

2.2.2 Analyzing optimal IPM decisions

Economic evaluations of IPM programs at the user level may be concerned not only with the effects of IPM adoption, but also with the optimal level of pest control or the optimal IPM mix at a whole-farm scale. When profit maximization is the goal, optimal use of an IPM practice occurs when the marginal increase in net returns from applying another unit of the practice equals the marginal cost of its application. Entomologists and some weed scientists have applied this concept when identifying economic thresholds for pest densities (Pedigo *et al.*, 1986; Cousens, 1987). An economic threshold is the pest population that produces incremental damage equal to the cost of preventing that damage. If pest density is below the threshold, no treatment is warranted. If it is above, treatment should occur to knock back pest density to below the economic threshold. IPM programs often involve scouting to inform producers about pest densities in relation to the threshold. Economic thresholds may be influenced by many factors including pesticide costs, output prices, the development of pesticide resistance, and the relationship between pest and predator levels and crop losses. If risk or environmental costs are considered, thresholds will be influenced by these factors as well (e.g. Szmedra *et al.*, 1990).

Optimal use of pest management practices can also be examined using mathematical programming techniques. For example, linear programming can be used to maximize an objective, such as net returns from a set of cropping activities that include IPM, subject to constraints on factors such as land, labor, capital, water and pesticide runoff or residuals. Martin *et al.* (1991) provides an example, with an analysis of alternative tillage systems, crop rotations and herbicide use on east-central USA Corn Belt farms. An application of non-linear programming to a pest management problem related to pesticide resistance is found in Gutierrez *et al.* (1979). Dynamic programming allows for examination of optimal pest control strategies when time is included in the models, and variables such as plant product, pest population density and the stock of pest susceptibility to pesticides are functions of time. Zacharias & Grube (1986) provide an example of applying such a model to examine optimal control of corn rootworms (*Diabrotica barberi* and *D. longicornis longicornis*) and soybean cyst nematode (*Heterodera glycines*) in Illinois.

Producers who consider adopting IPM strategies are often interested in the degree of risk as well as profitability. Risk may arise from biological, technical, or economic factors. The attractiveness of alternative pest management practices in the presence of risk can be assessed with a method called stochastic dominance (SD). With SD one can compare pairs of alternative pest management strategies for various sets of producers. These sets of producers are defined by their degrees of risk aversion. Examples of using SD in economic evaluation and comparison of IPM strategies with other strategies are found in Musser *et al.* (1981), Moffitt *et al.* (1983), and Greene *et al.* (1985); see also Chapter 4.

2.3 Aggregate or market-level economic impacts of IPM

When many growers all adopt changed pest management methods, there may be non-linear impacts, such as changes in crop supplies that are large enough to affect prices. Likewise, the timing of impacts may affect their total value, as effects nearer in time are generally valued more than ones in the more distant future. Properly measuring the aggregate economic impacts of IPM across space and time requires some care to account for the possibility of price effects and discounted future values (Norton *et al.*, 2005).

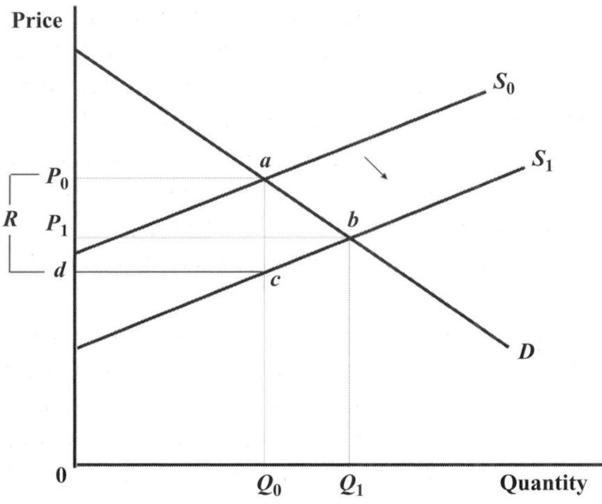

Fig. 2.2 IPM benefits measured as changes in economic surplus when supply shifts.

2.3.1 Capturing price effects when IPM impacts are large

When widespread adoption of IPM occurs, changes in crop prices, cropping patterns, producer profits and societal welfare can occur. These changes arise because costs differ and because supplies may increase, affecting prices for producers and consumers. These changes are illustrated in Fig. 2.2. In this figure, S_0 represents the supply curve before adoption of an IPM strategy, and D represents the demand curve. The initial price and quantity are P_0 and Q_0. Suppose IPM leads to savings of R in the average and marginal cost of production, reflected in a shift down in the supply curve to S_1. This shift leads to an increase in production and consumption to Q_1 (by $\Delta Q = Q_1 - Q_0$) and the market price falls to P_1 (by $\Delta P = P_0 - P_1$).

Consumers are better off because they can consume more of the commodity at a lower price. Consumers benefit from the lower price by an amount equal to their cost saving on the original quantity ($Q_0 \times \Delta P$) plus their net benefits from the gain in quantity consumed. Their total benefit is represented by the area P_0abP_1.

Although they may receive a lower price per unit, producers may be better off too, because their cost has fallen by R per unit, an amount greater than the fall in price. Producers gain the increase in profits on the original quantity ($Q_0 \times (R - \Delta P)$) plus the profits earned on the additional output, for a total producer gain of P_1bcd. Total benefits are the sum of producer and consumer benefits.

The distribution of benefits between producers and consumers depends on the size of the fall in price (ΔP) relative to the fall in costs (R), the price elasticities of supply and demand, and on the nature of the supply shift. For example, if a commodity is internationally traded, production has less of an effect on price, so more benefits may to accrue to producers. Or, if the supply curve shifts in more of a pivotal fashion as opposed to a parallel fashion as shown in Fig. 2.3, the benefits to producers would be reduced. Examples of IPM evaluation using this type of model, often called the "economic surplus" model, are found in Napit *et al.* (1988). Formulas for calculating consumer and producer gains for a variety of market situations are found in Alston *et al.* (1995).[2]

2.3.2 Aggregating economic impacts over time

Research and outreach programs in IPM often last many years, and their impacts may endure even longer. Benefit–cost analysis provides a framework for aggregating or projecting economic surplus over time by calculating net present values, internal rates of return or benefit–cost ratios. The benefits are the change in total economic surplus calculated for each year, and the costs are the expenditures on the IPM program. The primary purpose of benefit–cost analysis is to account for the fact that the sooner benefits and costs occur, the more they are worth. The net present value (NPV) of discounted benefits and costs can be calculated as follows:

$$\text{NPV} = \sum_{t=1}^{T} \frac{R_t - C_t}{(1+i)^t} \qquad (2.1)$$

[2] For example the formula to measure the total economic benefits to producers and consumers in Fig. 2.2, which assumes no trade, is $KP_0Q_0(1 + 0.5Zn)$, where: K = the proportionate cost change, P_0 = initial price, Q_0 = initial quantity, $Z = Ke/(e + n)$, e = the supply elasticity, and n = the demand elasticity. Other formulas would be appropriate for other market situations.

where:

R_t = the return in year t
T = (change in economic surplus)
C_t = the cost in year t (the IPM program costs)
i = the discount rate.

A tool closely related to NPV is the internal rate of return (IRR), which measures the rate of return that would render benefits just equal to costs (the implied discount rate that would make NPV = 0).

2.4 Valuing health and environmental impacts

Traditionally, reporting on changes in use of pesticide active ingredients associated with changes in pest management practices or just listing the number of pesticide applications have been common methods for assessing health and environmental (HE) impacts of IPM. However, specific pesticides and the means, timing and location of their application differentially affect health and the environment. Several methods have been developed to provide more refined evaluation of the effects of IPM on reducing HE risks. Some methods estimate monetary values associated with hazard reductions (Cuyno *et al.*, 2005), while others present HE measures as trade-offs with profitability measures.

Measuring HE benefits of IPM is difficult for various reasons. First, it is a challenge to assess the physical and biological effects of pesticide use that occur under different levels of IPM. Second, pesticides can have many distinct acute and long-term effects on sub-components of the health and the environment such as mammals, birds, aquatic life and beneficial organisms. Third, because the economic value associated with HE effects is generally not priced in the market, it is difficult to know how heavily to weight the various HE effects compared to one another and compared to profitability measures.

2.4.1 Hazard indexes and scoring models

Location-specific models have been used that require detailed field information such as soil type, irrigation system, slope and weather and produce information on the fate of chemicals applied. For example, the Chemical Environmental Index (CINDEX), based on another model GLEAMS, was developed by Teague *et al.* (1995) to describe the effects of pesticides on ground and surface water. CINDEX values are defined for individual pesticide use strategies. Calculations are based on the 96-hour fish LC_{50}, lifetime Health Advisory Level (HAL) value, the US Environmental Protection Agency (EPA) Carcinogenic Risk Category and the runoff and percolation potential for each pesticide used in the strategy under consideration. Location specific models can be used to provide information on trade-offs between HE effects and income effects for pest management practices.

Non-location-specific models have been more commonly used for impact assessment because they require less effort and data. These models require information on the pesticides applied and the method of application and produce indicators of risks by HE category as well as weighted total risk for the pesticide applications. Examples are the Environmental Impact Quotient (EIQ) developed by Kovach *et al.* (1992), the Pesticide Index (PI) of Penrose *et al.* (1994) and a multi-attribute toxicity index developed by Benbrook *et al.* (2002).

Each indexing method is a type of scoring model that involves subjective weighting of risks across environmental categories. These methods perform two tasks (Norton *et al.*, 2001): the first is to identify the risks of pesticides to the individual categories of health and the environment, such as groundwater, birds, beneficial insects and humans, and the second is to aggregate and weight those impacts across categories. The first task is complicated by the desire to identify mutually exclusive categories, especially ones with available data. The categories in most models contain a mixture of non-target organisms (e.g. humans, birds, aquatic organisms, beneficial insects, wildlife) and modes of exposure (e.g. groundwater, surface water). The second task is challenging because of the inherent subjectivity of the weights.

A widely used non-location-specific index of HE impacts of pesticide use is the EIQ developed by Kovach *et al.* (1992). The EIQ uses a discrete ranking scale in each of ten categories to identify a single rating for each pesticide active ingredient (a.i.). The categories include acute toxicity[3] to non-target species (birds, fish and bees), acute dermal toxicity, long-term health effects, residue half-life (soil and plant surface), toxicity to beneficial organisms and groundwater/runoff potential. The EIQ groups the ten categories into three broad areas of pesticide action: farm worker risk, consumer exposure potential and ecological risk. The EIQ is then calculated as the average impact of a pesticide (AI) over these three broad areas and is reported as a single number.

The EIQ is defined for specific pesticide active ingredients. In order to assess the actual damage from pesticide use on a specific field, the EIQ can be converted into a "field use rating." If only one pesticide is applied, this rating is obtained by multiplying the pesticide's EIQ by its percent a.i. and by the rate at which the pesticide was applied.

Benbrook *et al.* (2002) developed an indexing method to monitor progress in reducing the use of high-risk pesticides. For pesticides used in Wisconsin potato production, multi-attribute toxicity factors were calculated that reflect each pesticide's acute and chronic toxicity to mammals, birds, fish and small aquatic organisms and compatibility with bio-intensive IPM. These factors were multiplied by the pounds of active ingredients of the pesticides applied to estimate pesticide-specific toxicity units. These units can be tracked over time or related to use of IPM.

2.4.2 Monetary valuation of health and environmental effects

One mechanism that can be used to reduce the subjectivity on the weights used in the EIQ or in studies that apply multi-attribute toxicity factors is to elicit information on individuals' willingness to pay for risk reduction for the various HE components. There is usually no market price for reduced HE risk that can be used to provide these willingness-to-pay weights, but alternative monetary valuation techniques can be used such as contingent valuation (CV), experimental auctions and hedonic pricing (Champ *et al.*, 2003; Freeman, 2003). These methods can be costly and time consuming to perform well, so they are best executed by experts in non-market valuation.

Perhaps the most direct market-based way to estimate the human health impact of environmental damage or illness is to measure the associated costs. Studies have been completed on the cost of medical treatment and productivity losses due to pesticides in the Philippines and Ecuador (Antle & Pingali, 1994; Pingali *et al.*, 1994; Crissman *et al.*, 1998). Data were collected on pesticide use, demographics and ailments that might have been related to pesticide use by pesticide applicators. Medical doctors made the health assessments, and the costs of ailments were regressed on pesticide use and other variables so the cost of pesticide-related illness could be estimated. Models were also used to estimate the relationship between labor productivity and health problems related to pesticides.

Another cost-based approach includes assessing the cost of repairing environmental damage, for example the cost of treating pesticide-affected groundwater to make it potable. Another is to measure the cost of avoiding exposure to the environmental risk. For example, Abdalla *et al.* (1992) calculated the extra cost of purchasing drinking water in order to avoid drinking contaminated groundwater. Such measures tend to underestimate total environmental impacts because they omit consumer surplus (satisfaction above price paid) and they only focus on avoiding human exposure.

Contingent valuation uses survey methods to collect data on people's stated willingness to pay (WTP) to receive a benefit or their willingness to accept compensation for a loss. In the context of pest management, respondents might be asked how much they would be willing to pay to reduce the risk of pesticides to various categories of HE assets. The WTP data could later be linked to pesticide use data to arrive at a value for a change in pesticide use. Higley & Wintersteen (1992), Mullen *et al.* (1997) Swinton *et al.* (1999) and Cuyno *et al.* (2001) provide examples of using CV for such an assessment.

Contingent valuation is subject to potential biases due to the way the survey is designed or

[3] See Chapter 17 for details on toxicity indicators.

administered, or to the hypothetical nature of the questions, so it should be used with care. An experimental auction technique can be used to minimize hypothetical bias. In an experimental auction, people use real money to bid on health or environmental improvements, and then pay their bids at the end of the experiment (Brookshire & Coursey, 1987).

Hedonic pricing attempts to infer willingness to pay for environmental amenities from the prices of other goods based on the characteristics of those goods. This approach was used by Beach & Carlson (1993), who took a data set of herbicide attributes and estimated the value of herbicide safety from the effect of safety attributes on the herbicide prices.

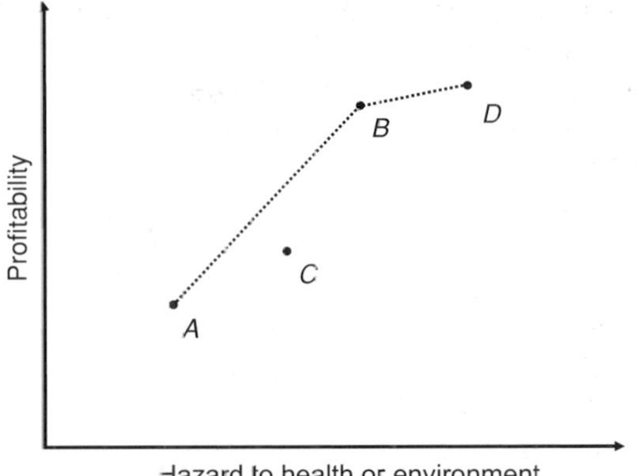

Fig. 2.3 Illustrative trade-off frontier of profitability versus a hazard to health or environment.

2.4.3 Trade-off approaches to compare economic and non-economic impacts

Trade-off approaches avoid the complications of monetary valuation of pesticide risk, but they sacrifice the potential to aggregate market and non-market benefits in a single monetary measure. Trade-off analysis generally begins by plotting a given practice in two dimensions. Figure 2.3 illustrates a profitability measure plotted against a measure of health or environmental hazard. Each point represents a specific IPM practice or treatment, plotted according to its profitability and HE hazard level. Because profitability is desirable but HE hazards are not, the desired direction of movement is toward the upper left (northwest) of the figure. Points *A*, *B* and *D* represent increasing levels of profitability accompanied by increasing environmental hazard. Moving from point *A* to point *B*, there is a trade-off between a gain in profitability and an increase in HE hazard. Compared to points *A* and *B*, point *C* increases HE hazard without the gain in profitability that could be had from a mixture of practices *A* and *B*. Hence, point *C* is "inefficient" in the sense that more profitability per unit of environmental hazard could be had by a mixture of *A* and *B*, which lie on the trade-off frontier.

Trade-off analysis can be used in at least three ways. First, it can be used to identify practices that are not efficient, because alternative practices could achieve both objectives more effectively. Hoag & Hornsby (1992) plotted cost versus groundwater hazard index for various pesticides, finding that most were inefficient at attaining the twin objectives of lower cost and reduced groundwater hazard.[4] Second, moves along the frontier can be used to measure the implied cost to a profit-maximizing farmer of adopting a more costly IPM practice with reduced HE hazards. For example, moving from point *D* to *B* or from *B* to *A* in Fig. 2.3 would entail sacrificing some profitability for the sake of reducing HE hazards. Third, trade-off analysis can be used to inform policy debates about who should bear the costs and enjoy the benefits of changing IPM practices (Antle *et al.*, 2003).

2.5 Choosing an impact assessment method within a budget

A balanced economic assessment of IPM impacts can divided into five basic parts: (1) definition of the IPM measure, (2) measurement of IPM adoption, (3) estimation of individual market-based economic effects, (4) aggregation to market effects and (5) estimation of health and environmental (non-market) impacts. These parts can be evaluated with different amounts of detail,

[4] The analytical tool of Hoag & Hornsby (1992) has been generalized by Nofziger *et al.* (2002).

Table 2.1 Information needs for basic and extended IPM assessment protocols

Key assessment steps	Basic protocol *(extended protocol in italic)*
Define IPM measure	Identify IPM practices with direct, measurable economic and Health and Environmental (HE) impacts.
	Group practices into levels of IPM adoption or assign points in an adoption scale.
Measure adoption	Secondary trend data (e.g. national statistics).
	Ask experts to estimate maximum adoption, years to maximum adoption, and whether practices will become obsolete.
	Fit logistic curve to predict adoption over time.
	Ask farmers to keep detailed records on input use and yields over defined period of time.
	Conduct a grower adoption survey based on random sample to be extrapolated to population. Include questions on pest pressure, reasons for adoption, cost and yield, price received, and farm and farmer characteristics.
	Analyze determinants of adoption.
Estimate individual economic impacts	Construct budgets (partial or enterprise).
	Build multi-year net present value (NPV) analysis.
	Regress profitability and crop yield on IPM practices and other explanatory variables to measure IPM effect.
	Risk analysis of IPM practices using stochastic dominance.
	Optimal pest control practices or whole-farm IPM effects using math programming.
Estimate aggregate impacts	Multiply per-ha change in costs and returns by total ha under IPM (assumes IPM does not affect prices). Subtract annual IPM public costs.
	Conduct economic surplus analysis (allows price effects from aggregate IPM adoption), based on supply shift.
	Calculate NPV and internal rate of return based on annual net benefits from economic surplus and total IPM research and outreach costs.
Estimate health and environmental impacts	Identify changes in pesticide use as a result of adopting IPM. Do by pesticide a.i., rate, times sprayed, when applied, method of application, form applied.
	Regress pesticide use and aggregate hazard on IPM practices and other explanatory variables to measure IPM effect.
	Assess pesticide risk to water quality using spatial model (e.g. GLEAMS, CINDEX, etc.)
	Assess human health risks based on published, survey recall or medical visit data.
	Assess other non-target pesticide effects (e.g. to arthropods, birds, etc.)
	Weight HE effects in a monetary or non-monetary index or conduct trade-off analysis between profitability and HE criteria.

depending on the available time, money and expertise. Table 2.1 combines the methods described in the previous sections into elements for basic and extended protocols for IPM impact evaluation. The basic level covers data and methods for conducting the simplest elements of impact evaluation, while the extended protocol items offer ways to address multiple farmer or consumer objectives, risk and non-market valuation.

2.6 Conclusions

Economic impacts of IPM programs can be measured at the level of the individual management unit or at the market level. Economic and environmental or health outcomes can be measured with differing levels of detail depending on time and budget constraints. This chapter briefly summarizes the major approaches to IPM impact assessment, guiding the reader to the key references for details.

References

Abdalla, C. W., Roach, B. A. & Epp D. J. (1992). Valuing environmental quality changes using averting expenditures: an application to groundwater contamination. *Land Economics*, **68**(2), 163–169.

Alston, J. M., Norton, G. W. & Pardey P. G. (1995). *Science under Scarcity: Principles and Practice for Agricultural Research Evaluation and Priority Setting*. Ithaca, NY: Cornell University Press.

Antle, J. M. & Pingali, P. L. (1994). Pesticides, productivity and farmer health: a Philippine case study. *American Journal of Agricultural Economics*, **75**, 418–430.

Antle, J., Stoorvogel, J., Bowen, W., Crissman, C. & Yanggen, D. (2003). The tradeoff analysis approach: lessons from Ecuador and Peru. *Quarterly Journal of International Agriculture*, **42**, 189–206.

Beach, D. & Carlson, G. A. (1993). Hedonic analysis of herbicides: do user safety and water quality matter? *American Journal of Agricultural Economics*, **75**, 612–623.

Benbrook, C. M., Sexson, D. L., Wyman, J. A. *et al.* (2002). Developing a pesticide risk assessment tool to monitor progress in reducing reliance on high-risk pesticides. *American Journal of Potato Research*, **79**, 183–199.

Brookshire, D. S. & Coursey, D. L. (1987). Measuring the value of a public good: an empirical comparison of elicitation procedures. *American Economic Review*, **77** 554–566.

Champ, P. A., Boyle, K. J. & Brown, T. C. (eds.) (2003). *A Primer on Nonmarket Valuation*. Dordrecht, Netherlands: Kluwer.

Centro Internacional de Mejoramiento de Maíz y Trigo (1988). *From Agronomic Data to Farmer Recommendations: An Economics Training Manual*, completely revised edn. Mexico City, Mexico: Centro Internacional de Mejoramiento de Maíz y Trigo (CIMMYT).

Cousens, R. (1987). Theory and reality of weed control thresholds. *Plant Protection Quarterly*, **2**, 13–20.

Crissman, C. C., Antle, J. M. & Capalbo, S. M. (1998). *Economic, Environmental and Health Tradeoffs in Agriculture: Pesticides and the Sustainability of Andean Potato Production*. Dordrecht, Netherlands: Kluwer.

Cuyno, L. C. M., Norton, G. W. & Rola, A. (2001). Economic analysis of environmental benefits of integrated pest management: a Philippines case study. *Agricultural Economics*, **25**, 227–234.

Cuyno, L., Norton, G. W., Crissman, C. C. & Rola, A. (2005). Evaluating the health and environmental impacts of IPM. In *Globalizing Integrated Pest Management*, eds. G. W. Norton, E. A. Heinrichs, G. C. Luther & M. E. Irwin, pp. 245–262. Ames, IA: Blackwell.

Feder, G., Murgai, R. & Quizon, J. B. (2004). Sending farmers back to school: the impact of farmer field schools in Indonesia. *Review of Agricultural Economics*, **26**, 45–62.

Fernandez-Cornejo, J. (1998). Environmental and economic consequences of technology adoption: IPM in viticulture. *Agricultural Economics*, **18**, 145–155.

Freeman, A. M., III (2003). *The Measurement of Environmental and Resource Values: Theory and Methods*, 2nd edn. Washington, DC: Resources for the Future.

Greene, C. R., Kramer, R. A., Norton, G. W., Rajotte, E. G. & McPherson, R. M. (1985). An economic analysis of soybean integrated pest management. *American Journal of Agricultural Economics*, **67**, 567–572.

Gutierrez, A. P., Regev, U. & Shalit, H. (1979). An economic optimization model of pesticide resistance: alfalfa and Egyptian alfalfa weevil. *Environmental Entomology*, **8**, 101–107.

Higley, L. G & Wintersteen, W. K. (1992). A novel approach to environmental risk assessment of pesticides as a basis for incorporating environmental costs into economic injury levels. *American Entomologist*, **38**, 34–39.

Hoag, D. L. & Hornsby, A. G. (1992). Coupling groundwater contamination with economic returns when applying farm pesticides. *Journal of Environmental Quality*, **21**, 579–586.

Kovach, J., Petzoldt, C., Degni, J. & Tette, J. (1992). *A Method to Measure the Environmental Impact of Pesticides*, New York Food and Life Sciences Bulletin No. 139. Geneva, NY: Cornell University. New York State Agricultural Experiment Station. Available at http://ecommons.library.cornell.edu/handle/1813/5203/.

Martin, M. A., Schreiber, M. M., Riepe, J. R. & Bahr, J. R. (1991). The economics of alternative tillage systems, crop rotations, and herbicide use on three representative east central Corn Belt farms. *Weed Science*, **39**, 299–397.

McNamara, K. T., Wetzstein, M. E. & Douce, G. K. (1991). Factors affecting peanut producer adoption of integrated pest management. *Review of Agricultural Economics*, **13**, 129–139.

Moffitt, L. J., Tanagosh, L. K. & Baritelle, J. L. (1983). Incorporating risk in comparisons of alternative pest management methods. *Environmental Entomology*, **12**, 1003–1111.

Mullen, J. D., Norton, G. W. & Reaves, D. W. (1997). Economic analysis of environmental benefits of integrated pest management. *Journal of Agricultural and Applied Economics*, **29**, 243–253.

Musser, W. N., Tew, B. V. & Epperson, J. E. (1981). An economic examination of an integrated pest management production system with a contrast between E-V and stochastic dominance analysis. *Southern Journal of Agricultural Economics*, **13**, 119–124.

Napit, K. B., Norton, G. W., Kazmierczak, R. F. Jr. & Rajotte, E. G. (1988). Economic impacts of extension integrated pest management programs in several states. *Journal of Economic Entomology*, **81**, 251–256.

Norton, G. & Mullen, J. (1994). *Economic Evaluation of Integrated Pest Management Programs: A Literature Review*, Virginia Cooperative Extension Publication No. 448-120. Blacksburg, VA: Virginia Tech.

Norton, G. W., Swinton, S. M, Riha, S. *et al.* (2001). *Impact Assessment of Integrated Pest Management Programs*. Blacksburg, VA: Department of Agricultural and Applied Economics, Virginia Tech.

Norton, G. W., Moore, K., Quishpe, D. *et al.* (2005). Evaluating socio-economic impacts of IPM. In *Globalizing Integrated Pest Management*, eds. G. W. Norton, E. A. Heinrichs, G. C. Luther & M. E. Irwin, pp. 225–244. Ames, IA: Blackwell.

Pedigo, L. P., Hutchins, S. H. & Higley, L. G. (1986). Economic injury levels in theory and practice. *Annual Review of Entomology*, **31**, 341–368.

Penrose, L. J., Thwaite, W. G. & Bower, C. C. (1994). Rating index as a basis for decision making on pesticide use reduction for accreditation of fruit produced under integrated pest management. *Crop Protection*, **13**, 146–152.

Pingali, P. L., Marquez, C. B. & Palis, F. G. (1994). Pesticides and Philippine rice farmer health: a medical and economic analysis. *American Journal of Agricultural Economics*, **76**, 587–592.

Swinton, S. M. & Day, E. (2003). Economics in the design, assessment, adoption, and policy analysis of IPM. In *Integrated Pest Management: Current and Future Strategies*, R-140, ed. K. R. Barker, pp. 196–206. Ames, IA: Council for Agricultural Science and Technology.

Swinton, S. M., Owens, N. N. & van Ravenswaay, E. O. (1999). Health risk information to reduce water pollution. In *Flexible Incentives for the Adoption of Environmental Technologies in Agriculture*, eds. F. Casey, A. Schmitz, S. Swinton & D. Zilberman, pp. 263–271. Boston, MA: Kluwer.

Swinton, S. M., Renner, K. A. & Kells, J. J. (2002). On-farm comparison of three postemergence weed management decision aids in Michigan. *Weed Technology*, **16**, 691–698.

Szmedra, P. I., Wetzstein, M. E. & McClendon, R. (1990). Economic threshold under risk: a case study of soybean production. *Journal of Economic Entomology*, **83**, 641–646.

Teague, M. L., Mapp, H. P. & Bernardo, D. J. (1995). Risk indices for economic and water quality tradeoffs: an application to Great Plains agriculture. *Journal of Production Agriculture*, **8**, 405–415.

Zacharias, T. P. & Grube, A. H. (1986). Integrated pest management strategies for approximately optimal control of corn rootworm and soybean cyst nematode. *American Journal of Agricultural Economics*, **68**, 704–715.

Chapter 3

Economic decision rules for IPM

Leon G. Higley and Robert K. D. Peterson

The year 2009 marks the 50th anniversary of the elaboration of the economic injury level (EIL) and the economic threshold (ET) concepts by Stern *et al.* (1959). The EIL and ET are widely recognized as the most important concepts in IPM. Because IPM is posited on the premise that certain levels of pests and pest injury are tolerable, the EIL and ET represent a crucial underpinning for any theory of IPM. Given the centrality of economic decision rule concepts to IPM, it follows that every IPM program should be based on these concepts. But this clearly is not the case. Why? In this chapter, we will discuss the historical development of economic decision levels, current approaches, and limitations of the EIL and ET. Finally, we will argue that EILs should be incorporated much more into IPM programs and that EILs are central to the continued development of environmental and economic sustainability concepts so important to IPM.

3.1 Economic decision levels

The concept of tolerating pest injury was not introduced by Stern *et al.* (1959); it was discussed at least as early as 1934 (Pierce, 1934) but does not seem to have been developed further until 1959 (Kogan, 1998). In response to failures in pest control in California because of insecticide resistance by pests and insecticide mortality of natural enemies, Stern *et al.* (1959) first proposed the fundamental concepts of the EIL, the ET, economic damage and pest status. The EIL is often defined as the lowest population density of pests that will cause economic damage. Economic damage is defined as the amount of injury that will justify the cost of control. The ET is defined as the density of pests at which control measures should be taken to prevent the pest population from reaching the EIL.

In our remaining discussion we look at the EIL and ET in detail. However, these are not the only approaches for pest management decision making – many other sorts of decision models are possible. However, the factors used in establishing the EIL and ET are the same as those that must be considered in any management model. Moreover, the EIL and ET are often incorporated into more complex management models. Thus, whether as stand-alone criteria, or as components of more sophisticated and complex models, the EIL and ET have central roles in pest management decision making.

3.1.1 The EIL

The EIL may be the simplest concept in all of applied ecology, yet it and the related concept of the ET continue to be misunderstood even by many IPM researchers and practitioners. The EIL is a straightforward cost–benefit equation in which

Integrated Pest Management, ed. Edward B. Radcliffe, William D. Hutchison and Rafael E. Cancelado. Published by Cambridge University Press.

the costs (losses associated with managing the pest) are balanced with the benefits (losses prevented by managing the pest). Despite its simplicity, it was not until 1972 that a formula for calculating the EIL was introduced (Stone & Pedigo, 1972); see Pedigo *et al.* (1986) and Higley & Pedigo (1996) for detailed discussions of the historical evolution of the EIL equation.

The EIL actually represents a level of injury, not a density of pests. However, numbers of pests per unit area are often used as an index for injury because injury can be very difficult to sample and measure. Using pest numbers as an index, EILs may be expressed as "larvae/plant," "beetles/sweep," "grass weeds/m^2" or "moths/trap" (Peterson & Higley, 2002). The most frequently used equation to determine the EIL is:

$$\mathrm{EIL} = C \div VIDK \tag{3.1}$$

where C = management costs per production unit (e.g. \$/ha), V = market value per production unit (\$/kg), I = injury per pest equivalent, D = damage per unit injury (kg reduction/ha/injury unit) and K = proportional reduction in injury with management (Pedigo *et al.*, 1986).

Despite the simplicity of the EIL concept, there is considerable complexity in the biologic components of the equation (Peterson & Higley, 2002). This is because injury per pest (I) and yield loss per unit injury (D) can be very difficult and costly to determine (Higley & Peterson, 1996). It is important that these components are known so that pests can be managed effectively within an IPM program, but because of the difficulty of determining I and D separately, both components often are determined together as simply loss per pest (Higley & Peterson, 1996; Peterson & Higley, 2002).

When insects injure a plant through direct tissue loss, I may represent insect consumption rates. In contrast, with sucking insects direct measures of loss of photosynthate or nutrients may be impossible or impractical. Also, with many sucking insects it is necessary to characterize injury as a combination of quantity and time. Thus, an EIL may be defined for aphid-days (the injury–time combination), rather than an aphid density (although the ET may still be defined as a density, requiring an additional conversion from the EIL).

Variation in estimates of the damage relationship (how injury influences yield) follow from factors altering yield, most especially nutrient and water availability. The damage relationship, D, is the most variable component of the EIL (Higley, 2001; Peterson & Hunt, 2003). Because injury–yield loss relationships can change most strikingly between wet and dry conditions, where sufficient data are available separate EILs for normal and drought conditions may be determined (Hammond & Pedigo, 1982; Ostlie & Pedigo, 1985; Haile, 1999).

Determining the damage relationship is the most difficult and limiting aspect of EIL development, not only because of environmental variability in yield responses. Experimentally, injury–yield loss determinations require quantification of injury, treatments consisting of different levels of injury and replication over a minimum of two years. Additionally, to determine a curvilinear relationship, at least three (and preferably four or more) levels of injury are necessary (Hammond, 1996; Higley & Peterson, 1996; Higley, 2001).

Despite these difficulties, our theoretical understanding of injury–yield loss relationships is sound and evolving. The general nature of injury and yield, the damage curve, was originally defined by Tammes (1961) and has been demonstrated empirically in many systems (see Higley & Peterson, 1996 for a review). Subsequently, Pedigo and co-workers (Pedigo *et al.*, 1986; Higley *et al.*, 1993; Higley & Pedigo, 1996) assigned physiological interpretations to various portions of the damage curve and related these to different types of injury (Fig. 3.1). One application of the physiological approach is through the development of injury–yield loss relationships and associated EILs driven by an understanding of how defoliation by insects reduces light interception in soybean (Higley, 1992; Hammond *et al.*, 2000; Malone *et al.*, 2002).

The EIL value determines the injury level, most often in the form of pest density, at which the pest management cost equals the cost from yield loss if no management occurs. For example, if the EIL is 5 larvae/plant, then economic damage is occurring at 5 larvae/plant. Producers and other decision makers should not wait until the EIL has been reached because at that level economic loss

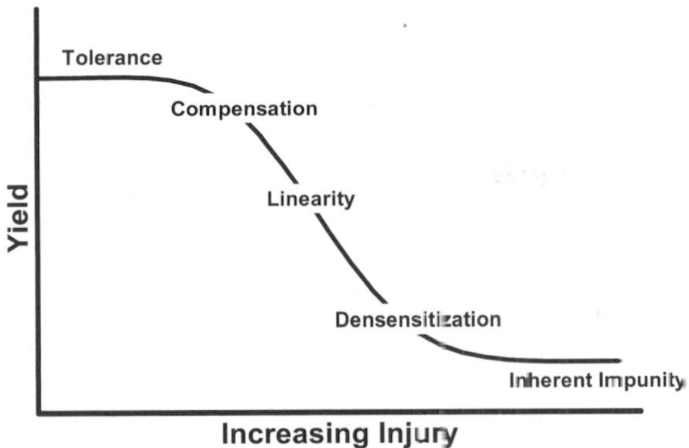

Fig. 3.1 The damage curve, the general relationship between yield and increasing injury, and specific regions of the curve. A given injury–yield relationship may include all or part of this general function.

is already occurring. This is a common misunderstanding about the EIL. Therefore, the decision to initiate management activities, such as pesticide application, must be made before the EIL is reached so that economic damage can be prevented. Indeed, prevention of economic damage is, for all intents and purposes, the sole rationale underlying the EIL concept.

3.1.2 The ET

In many IPM programs, the decision to initiate management action is based on the ET (Pedigo *et al.*, 1986; Pedigo, 1996). Thus, the ET is the most widely used decision tool in IPM and is sometimes called the "action threshold." Although defined in terms of the pest density at which management action should be taken, the ET is actually an index for *when* to implement pest management activities. For example, if an EIL is 10 larvae/plant, then an ET may be 8 larvae/plant. Action would be taken when 8 larvae/plant are sampled, not because that density represents an economic loss, but rather because it provides a window of time to take action before the pest density or injury increases to produce an economic loss (Peterson & Higley, 2002).

Although it is often difficult to determine EILs, it is unarguably more difficult to calculate ETs. This is because of the risks associated with predicting when a population will exceed the EIL and with the variability in time delays for management action (Pedigo *et al.*, 1986; Peterson, 1996).

Pedigo *et al.* (1989) divided ETs into two categories: subjective and objective. Subjective ETs are not based on calculated EILs, but rather on human experience. Because of this, they are practically always static values (e.g. 2 larvae/sweep-net sample regardless of changes in market value of the commodity and control cost). These subjective thresholds have been termed "nominal thresholds" by Poston *et al.* (1983). Despite their limitations and questionable accuracy in many situations, subjective ETs remain the most commonly used thresholds.

Objective ETs are based on calculated EILs and therefore are inherently dynamic. This type of threshold has been further divided into three categories: fixed, descriptive and dichotomous (Pedigo *et al.*, 1989). Fixed ETs are set at some percentage of the EIL and change proportionally with it. An example of a fixed ET is 80% of the EIL value. Descriptive ETs are based on estimates of pest population growth and dynamics and rely on accurate sampling to determine if the population will be likely to exceed the EIL (Wilson, 1985; Ostlie & Pedigo, 1987). Dichotomous ETs are based on statistical procedures to classify a pest population as "economic" or "non-economic." The time-sequential sampling technique is the best-known example of a dichotomous ET (Pedigo & van Schaik, 1984).

In defense of subjective ETs, some pest situations do not readily fit the EIL model, because the pest-to-damage relationship is obscure or complex, experimental determination of economic damage is difficult or impossible, or pest tolerances are extremely low. In these situations, various thresholds similar to an ET but not explicitly related to an EIL have been proposed. Often these are termed action thresholds (AT), although we find this term problematic because it has been used both as a synonym for ET and as a different form of indicator for management action. In some situations entirely different decision rules based on non-economic criteria have been developed. For example, in instances where pest injury causes cosmetic damage, aesthetic thresholds have been developed (Sadof & Raupp 1996), some of which are economically based and are analogous to conventional ETs. Another example of an alternative decision rule is the sensory threshold

Galvin *et al.* (2007) developed based on changes in wine taste associated with the presence of multicolored Asian lady beetle, *Harmonia axyridis*, in grapes.

Because an ET represents a time to take action, it is best suited for use with a regular sampling program. In some formal sampling/decision making procedures, like sequential sampling, the ET is essential for establishing decision points. With less sophisticated procedures, the ET still provides a benchmark against which pest densities are assessed. Because the EIL and ET concepts depend upon the principle of preventing injury, early assessments are important. Because natural mortality can impact final densities, various means for incorporating pest survivorship information into ETs have been developed (e.g. Ostlie & Pedigo, 1987; Brown, 1997; Barrigossi *et al.*, 2003). These ETs provide a greatly improved assessment of the potential injuriousness of a population, and recognize the potential importance of age-specific mortality.

3.2 Development of EILs and ETs

Development of calculated EILs has increased dramatically since 1972, the year the first calculated EIL was presented by Stone & Pedigo. Today, there are hundreds of published articles on EILs and ETs. Peterson (1996) reviewed the development of economic decision levels from 1959 through 1993, primarily by examining the scientific literature.

Economic decision levels have been determined for more than 40 commodities, including most of the world's major food and fiber crops. Greater than 80% of those decision levels have been for insect pests, with about 10%, 6% and 4% on mites, weeds and plant pathogens, respectively. Within the arthropod pests, nearly 50% of EILs have been determined for lepidopteran species, 17% for homopterans and 14% for coleopterans (Peterson, 1996). Based on injury type, 50% of EILs are for defoliators, 29% for assimilate sappers, 11% for mesophyll feeders (selective leaf and fruit feeders) and 10% for turgor reducers (root and stem feeders). More EILs have been determined for cabbage looper (*Trichoplusia ni*) than any other species. EILs for two-spotted spider mite (*Tetranychus urticae*) have been determined for apple, common bean, cotton, grape and strawberry. Weed species for which EILs have been determined include *Avena fatua, Datura stramonium, Helianthus annuus, Abutilon theophrasti, Amaranthus tuberculatus* and *Xanthium strumarium* (Mortensen & Coble, 1996; Peterson, 1996).

Why are there so many more decision levels for arthropods compared to weeds and plant pathogens? The EIL and ET were first defined and used by entomologists primarily because of the emerging need for insecticide resistance management and conservation of natural enemies. Also, arthropods were, and still are, largely amenable to curative management techniques, making economic decision levels for them relatively easy to incorporate into IPM programs. Plant diseases and, to some extent, weeds traditionally have been managed preventatively, largely precluding the need for economic decision levels. The rapid and widespread adoption of herbicide-resistant transgenic crop varieties has dramatically moved field-crop weed management to an ever greater reliance on preventative herbicide use, further reducing the need for curative decision tools, like EILs.

3.3 Current approaches

Because of its applicability to many situations, advances in the EIL concept have occurred primarily through extensions of the model advanced by Pedigo *et al.* (1986). Aesthetic injury levels (AILs) have been determined based on attributes not readily definable in economic terms, such as form, color, texture and beauty. Examples of resources in which aesthetic injury levels could be used include lawns, ornamental plants, homes and public buildings. To ascertain value of these attributes, researchers have obtained input from owners or the general public using techniques such as the contingent valuation method (Sadof & Raupp, 1996). Using the contingent valuation approach, the damage per unit of pest injury has been determined in economic terms. Most results have revealed that public tolerance of pest injury is low, resulting in low AILs with acceptable levels of injury at or less than 10% (Raupp *et al.*, 1988).

But, the salient point is that thresholds can be developed based on aesthetic considerations.

In the early 1990s, environmentally based EILs were developed. Each variable in the EIL equation reflects management activities that could be manipulated to potentially enhance environmental sustainability. In particular, researchers have suggested incorporating environmental costs into the management cost variable, *C* (Higley & Wintersteen, 1992). The resulting EILs have been termed environmental economic injury levels (EEILs). Although the notion of reflecting environmental costs in farmer-level decision making received much interest, particularly among economists (e.g. Lohr *et al.*, 1999; Brethour & Weersink, 2001; Florax *et al.*, 2005), EEILs have not been implemented. Work by agricultural economists and others (e.g. Kovach *et al.*, 1992) suggest that the value of the EEIL is in providing relative risk information to users and providing an economic context for comparing management options, rather than as a use/no-use criterion (like conventional EILs).

Important conceptual advances in the ET occurred in the 1980s. An important theoretical and practical advance has been the conversion of insect population estimates into insect-injury equivalents. An insect-injury equivalent is the total injury potential of an individual pest if it were to survive through all injurious life stages (Ostlie & Pedigo, 1987). Injury equivalents are determined from estimates of pest population structure, pest density and injury potential. Incorporation of insect larval survivorship has led to a further refinement of the injury equivalency concept.

Probably the most challenging goal for IPM is the establishment of multiple-species EILs. These EILs represent a potentially significant advance in IPM because they can provide decision makers with the ability to manage a complex of pests instead of managing single pest species (Peterson & Higley, 2002). The primary advances in this area have involved integrating pest injury from different species by determining if the multiple-species injury has similar effects on the host. If different species produce injuries resulting in similar physiological responses by the host, then the pest species can be grouped into injury guilds (see Welter, 1989; Higley *et al.*, 1993; Peterson & Higley, 1993; Peterson & Higley, 2001 for more information on insect injury guilds). The injury guilds can then be used to characterize damage functions. One approach to determine multiple-species EILs has been to combine the injury guild concept with the injury equivalency concept (Hutchins *et al.*, 1988). To develop multiple-species EILs using this concept, pest species must: (1) produce a similar type of injury, (2) produce injury within the same physiological time-frame of the host, (3) produce injury of a similar intensity and (4) affect the same plant part (Hutchins & Funderburk, 1991).

A recent development in EIL theory is the probabilistic EIL (PEIL). The EIL typically is calculated by taking mean values for the parameters *C, V, I* and *D*. The *K* value often is set to one to indicate 100% efficacy for the control tactic. Despite the recognition that the EIL is determined by dynamic biological and economic parameters (Pedigo *et al.*, 1986; Peterson, 1996), which can be highly variable and uncertain, there has been little effort to quantify uncertainty and to use estimates of uncertainty in the determination of EILs (Peterson, 1996). Therefore, Peterson & Hunt (2003) defined the PEIL as an EIL that reflects its probability of occurrence. The probability of occurrence is determined by incorporating the variability and uncertainty associated with the input variables used to calculate the EIL. Peterson & Hunt (2003) used Monte Carlo simulation, a random sampling technique in which each input variable in the model was sampled repeatedly from a range of possible values based on each variable's probability distribution. Then, the variability for each input was propagated into the output of the model so that the model output reflected the probability of values that could occur.

How, then, can PEILs be used? The practical value of the PEIL is that multiple EIL values ranked as percentiles as a result of the Monte Carlo distributional analysis allow the decision maker to choose her or his tolerable level of risk within an IPM program (Peterson & Hunt, 2003). For example, if the decision maker is risk averse (i.e. she does not want to risk economic damage even if it means spraying in the absence of economic damage) and needs to decide which threshold to use for bean leaf beetle (*Cerotoma trifurcata*) in seedling-stage soybean, she may choose a PEIL of 7.7 adults/plant. This PEIL represents the 25th percentile of values as determined from the model output. The use of 7.7 adults/plant as the PEIL ensures that an EIL <7.7 will occur only 25% of the time; therefore, the decision maker will use a sufficiently conservative EIL 75% of the time.

Because conventional EILs and ETs are tied to single-event decision making, there has long been a need to develop new applications where multiple pest generations occur in a single season. One approach to this issue has been the development of various types of management models which largely focus on pest population dynamics. Saphores (2000) proposed a different approach to this question by developing a stochastic model for determining ETs through the application of optimal stopping models. This approach accommodates uncertainty and allows for decisions at multiple times, albeit at the cost of considerable complexity in the model.

3.4 Limitations to EILs and ETs

Economic decision levels traditionally have been best suited for IPM approaches that involve the ability to take curative action, such as pest population suppression via pesticide application (Pedigo *et al.*, 1986; Higley & Pedigo, 1996; Peterson & Higley, 2002). Therefore, it is not surprising that the use of thresholds has been closely tied to the use of therapeutic pesticides.

Peterson & Higley (2002) stated that "the use of EILs and ETs is much more limited when pests almost always cause economic damage, when reliable sampling of injury or pests is difficult, or when curative action is difficult or not available." Most pests that almost always cause economic damage, often called perennial pests, directly infest the marketable product. The damage they produce, in addition to the relatively high market value for the products, results in EILs that are so low as to be practically unusable. Sampling is difficult for many pests, such as plant pathogens (Backman & Jacobi, 1996). Difficulties in quantifying plant pathogens, in addition to few therapeutic management options, have hindered EIL development.

There are also considerable difficulties in assigning economic values to the market value of some host plants such as ornamentals or landscape trees. The economic value of a reduction in a host's aesthetics is subjective and difficult to quantify, but this has been done (e.g. aesthetic injury levels). Other limitations include the relatively high cost of conducting the research necessary to determine EILs, and a lack of knowledge about the interaction between biotic and abiotic stresses on the host, especially for plants (Peterson & Higley, 2002).

Development of EILs and ETs for veterinary pests has proven difficult because quantifying injury is difficult. Further, ETs for livestock pests, such as house or stable flies, may be considerably above levels that would result in unacceptable annoyance to humans (i.e. nuisance thresholds). Economic decision levels for pests that affect human health, such as disease-carrying mosquitoes, are virtually impossible to use because they would require a market value for human life. Therefore, ETs would be based on the acceptance of a level of morbidity or mortality before management is taken. This is clearly unacceptable from ethical perspectives (Peterson, 1996). However, it is possible to use non-economic threshold approaches for disease-vectoring insects. For example, the Centers for Disease Control in the USA recommends a series of thresholds for mosquito management in areas that could experience West Nile virus (Centers for Disease Control, 2003). Further, many mosquito control districts in the USA use "management thresholds" for mosquito larvae and adults based on regular sampling intervals.

3.5 Conclusions

As the development and rapid adoption of genetically engineered plant varieties illustrate, pest management has been and remains largely driven by technology. When that primary technology is chemical pesticides, decision making must emphasize use/no-use and timing questions. When the primary technology is plant resistance, the primary decision is cultivar choice. The change in weed management to a primarily preventive tactic (herbicide-resistant varieties) is one reason EILs and ETs are less used in weed management. Similarly, as therapeutic options decline for the management of plant diseases, the immediate benefits of EILs and ETs are similarly reduced. Changes in tactics certainly imply that the primary focus and use of conventional EILs and ETs will be for insect management.

Changes in tactics not withstanding, the EIL and ET remain at the heart of insect management programs. Researchers continue to add to our understanding of insect–yield loss relationships, and continue to develop new directions in the calculation and use of EILs and ETs. Undoubtedly new tactics, new pests and new environmental conditions will require the development of new decision tools for pest management. But one clear lesson from the past 50 years of research and practice is that the EIL and ET remain vital practical concepts now and for future pest management.

References

Backman, P. A. & Jacobi, J. C. (1996). Thresholds for plant-disease management. In *Economic Thresholds for Integrated Pest Management*, eds. L. G. Higley & L. P. Pedigo, pp. 114–127. Lincoln, NE: University of Nebraska Press.

Barrigossi, J. A. F., Hein, G. L. & Higley, L. G. (2003). Economic injury levels and sequential sampling plans for Mexican bean beetle (Coleoptera: Coccinellidae) on dry beans. *Journal of Economic Entomology*, **96**, 1160–1167.

Brethour, C. & Weersink, A. A. (2001). An economic evaluation of the environmental benefits from pesticide reduction. *Agricultural Economics*, **25**, 219–226.

Brown, G. C. (1997). Simple models of natural enemy action and economic thresholds. *American Entomologist*, **43**, 117–124.

Centers for Disease Control (2003). *Epidemic/Epizootic West Nile Virus in the United States: Guidelines for Surveillance, Prevention, and Control*. Fort Collins, CO: US Department of Health and Human Services, CDC, National Center for Infectious Diseases, Division of Vector-Borne Infectious Diseases.

Florax, R. J. G. M., Travisi, C. M. & Nijkamp, P. (2005). A meta-analysis of the willingness to pay for reductions in pesticide risk exposure. *European Review of Agricultural Economics*, **32**, 441–467.

Galvin, T. L., Burkness, E. C., Vickers, Z. *et al.* (2007). Sensory-based action threshold for the multicolored Asian lady beetle-related taint in wine grapes. *American Journal of Enology and Viticulture*, **58**, 518–522.

Haile, F. J. (1999). Physiology of plant tolerance to arthropod injury. Unpublished Ph.D. dissertation, University of Nebraska, Lincoln, NE.

Hammond, R. B. (1996). Limitations to economic injury levels and thresholds. In *Biotic Stress and Yield Loss*, eds. R. K. D. Peterson & L. G. Higley, pp. 58–73. Boca Raton, FL: CRC Press.

Hammond, R. B. & Pedigo, L. P. (1982). Determination of yield-loss relationships for two soybean defoliators by using simulated insect-defoliation techniques. *Journal of Economic Entomology*, **75**, 102–107.

Hammond, R. B., Higley, L. G., Pedigo, L. P. *et al.* (2000). Simulated insect defoliation on soybean: influence of row width. *Journal of Economic Entomology*, **93**, 1429–1436.

Higley, L. G. (1992). New understandings of soybean defoliation and their implications for pest management. In *Pest Management of Soybean*, eds. L. G. Copping, M. B. Green & R. T. Rees, pp. 56–65. Amsterdam, Netherlands: Elsevier.

Higley, L. G. (2001). Yield loss and pest management. In *Biotic Stress and Yield Loss*, eds. R. K. D. Peterson & L. G. Higley, pp. 13–22. Boca Raton, FL: CRC Press.

Higley, L. G. & Pedigo, L. P. (1996). The EIL concept. In *Economic Thresholds for Integrated Pest Management*, eds. L. G. Higley & L. P. Pedigo, pp. 9–21. Lincoln, NE: University of Nebraska Press.

Higley, L. G. & Peterson, R. K. D. (1996). The biological basis of the economic injury level. In *Economic Thresholds for Integrated Pest Management*, eds. L. G. Higley & L. P. Pedigo, pp. 22–40. Lincoln, NE: University of Nebraska Press.

Higley, L. G. & Wintersteen, W. K. (1992). A novel approach to environmental risk assessment of pesticides as a basis for incorporating environmental costs into economic injury levels. *American Entomologist*, **38**, 34–39.

Higley, L. G., Browde, J. A. & Higley, P. M. (1993). Moving towards new understandings of biotic stress and stress interactions. In *International Crop Science I*, eds. D. R. Buxton *et al.*, pp. 749–754. Madison, WI: Crop Science Society of America.

Hutchins, S. H. & Funderburk, J. E. (1991). Injury guilds: a practical approach for managing pest losses in soybean. *Agricultural Zoology Review*, **4**, 1–21.

Hutchins, S. H., Higley, L. G. & Pedigo, L. G. (1988). Injury equivalency as a basis for developing multiple-species economic injury levels. *Journal of Economic Entomology*, **81**, 1–8.

Kogan, M. (1998). Integrated pest management: historical perspectives and contemporary developments. *Annual Review of Entomology*, **43**, 243–270.

Kovach, J., Petzold, C., Degni, J. & Tette, J. (1992). *A Method to Measure the Environmental Impact of Pesticides.* New York Food and Life Sciences Bulletin No. 139. Geneva, NY: Cornell University, New York Agricultural Experiment Station. Available at http://ecommons.library.cornell.edn/handle/1813/5203.

Lohr, L., Park, T. & Higley, L. (1999). Farmer risk assessment for voluntary insecticide reduction. *Ecological Economics*, **30**, 121–130.

Malone, S., Herbert, D. A. Jr. & Holshouser, D. L. (2002). Relationship between leaf area index and yield in double-crop and full-season soybean systems. *Journal of Economic Entomology*, **95**, 945–951.

Mortensen, D. A. & Coble, H. D. (1996). Economic thresholds for weed management. In *Economic Thresholds for Integrated Pest Management*, eds. L. G. Higley & L. P. Pedigo, pp. 89–113. Lincoln, NE: University of Nebraska Press.

Ostlie, K. R. & Pedigo, L. P. (1985). Soybean response to simulated green cloverworm (Lepidoptera: Noctuidae) defoliation: progress toward determining comprehensive economic injury levels. *Journal of Economic Entomology*, **78**, 437–444.

Ostlie, K. R. & Pedigo, L. P. (1987). Incorporating pest survivorship into economic thresholds for pest management. *Journal of Economic Entomology*, **80**, 297–303.

Pedigo, L. P. (1996). General models of economic thresholds. In *Economic Thresholds for Integrated Pest Management*, eds. L. G. Higley & L. P. Pedigo, pp. 41–57. Lincoln, NE: University of Nebraska Press.

Pedigo, L. P. & van Schaik, J. W. (1984). Time-sequential sampling: a new use of the sequential probability ratio test for pest management decisions. *Bulletin of the Entomological Society of America*, **30**, 32–36.

Pedigo, L. P., Hutchins, S. H. & Higley, L. G. (1986). Economic injury levels in theory and practice. *Annual Review of Entomology*, **31**, 341–368.

Pedigo, L. P., Higley, L. G. & Davis, P. M. (1989). Concepts and advances in economic thresholds for soybean entomology. In *Proceedings of the 4th World Soybean Research Conference*, vol. 3, ed. A. J. Pascale, pp. 1487–1493. Buenos Aires, Argentina: Asociación Argentina de la Soja.

Peterson, R. K. D. (1996). The status of economic decision level development. In *Economic Thresholds for Integrated Pest Management*, eds. L. G. Higley & L. P. Pedigo, pp. 151–178. Lincoln, NE: University of Nebraska Press.

Peterson, R. K. D. & Higley, L. G. (1993). Arthropod injury and plant gas exchange: current understandings and approaches for synthesis. *Trends in Agricultural Sciences*, **1**, 93–100.

Peterson, R. K. D. & Higley, L. G. (2001). Illuminating the black box: the relationship between injury and yield. In *Biotic Stress and Yield Loss*, eds. R. K. D. Peterson & L. G. Higley, pp. 1–12. Boca Raton, FL: CRC Press.

Peterson, R. K. D. & Higley, L. G. (2002). Economic decision levels. In *Encyclopedia of Pest Management*, ed. D. Pimentel, pp. 228–230. New York: Marcel Dekker.

Peterson, R. K. D. & Hunt, T. E. (2003). The probabilistic economic injury level: incorporating uncertainty into pest management decision-making. *Journal of Economic Entomology*, **98**, 536–542.

Pierce, W. D. (1934). At what point does insect attack become damage? *Entomological News*, **45**, 1–4.

Poston, F. L., Pedigo, L. P. & Welch, S. M. (1983). Economic injury levels: reality and practicality. *Bulletin of the Entomological Society of America*, **29**, 49–53.

Raupp, M. J., Davidson, J. A., Koehler, C. S., Sadof, C. S. & Reichelderfer, K. (1988). Decision-making considerations for aesthetic damage caused by pests. *Bulletin of the Entomological Society of America*, **34**, 27–32.

Sadof, C. S. & Raupp, M. J. (1996). Aesthetic thresholds and their development. In *Economic Thresholds for Integrated Pest Management*, eds. L. G. Higley & L. P. Pedigo, pp. 203–226. Lincoln, NE: University of Nebraska Press.

Saphores, J.-D. M. (2000). The economic threshold with a stochastic pest population: a real options approach. *American Journal of Agricultural Economics*, **82**, 541–555.

Stern, V. M., Smith, R. F., van den Bosch, R. & Hagen, K. S. (1959). The integrated control concept. *Hilgardia*, **29**, 81–101.

Stone, J. D. & Pedigo, L. P. (1972). Development and economic-injury level of the green cloverworm on soybean in Iowa. *Journal of Economic Entomology*, **65**, 197–201.

Tammes, P. M. L. (1961). Studies of yield losses. II. Injury as a limiting factor of yield. *Tijdschrift over Plantenziekten*, **67**, 257–263.

Welter, S. C. (1989). Arthropod impact on plant gas exchange. In *Insect–Plant Interactions*, vol. 1, ed. E. A. Bernays, pp. 135–151. Boca Raton, FL: CRC Press.

Wilson, L. T. 1985. Developing economic thresholds in cotton. In *Integrated Pest Management in Major Agricultural Systems*, eds. R. E. Frisbie & P. L. Adkisson, pp. 308–344. College Station, TX: Texas Agricultural Experiment Station.

Chapter 4

Decision making and economic risk in IPM

Paul D. Mitchell and William D. Hutchison

Understanding and communicating the value of an IPM program or specific IPM tactics in agriculture, forestry and other venues has historically been difficult for a number of reasons (Grieshop *et al.*, 1988; Wearing, 1988; Kogan, 1998; Swinton & Day, 2003). Some of the more common barriers to adoption include a lack of practical sampling/monitoring tools (Wearing, 1988), challenges to fully integrating biologically based tactics (Kogan, 1998; Ehler & Bottrell, 2000), the need for multiple pest–damage relationships for multiple insect pests per crop (Pedigo *et al.*, 1986), changing economic conditions (with or without government subsidy programs) and the fact that multiple human audiences with diverse backgrounds and motivations are on the receiving end of new IPM programs (e.g. Bechinski, 1994; Cuperus & Berberet, 1994; Bacic *et al.*, 2006; Hammond *et al.*, 2006), including known variability in the adoption of new technologies (e.g. Mumford & Norton, 1984; Grieshop *et al.*, 1988; Rogers, 1995). Equally important barriers, however, could be the perceived complexity of IPM compared to current conventional pest approaches (e.g. Bechinski, 1994; Cuperus & Berberet, 1994; Grieshop *et al.*, 1988), or a lack of up-front consultation with targeted audiences prior to the R&D investment for developing IPM programs (Norton *et al.*, 2005; Bacic *et al.*, 2006). Although many of the concepts discussed in this chapter are relevant to IPM audiences in forestry or residential-urban pest management, our focus will primarily be targeted to decision makers in agricultural systems, and primarily arthropod management in crops.

Germane to understanding the target audience of farmers and crop consultants, it is generally accepted that farmers deal with a variety of risks on a regular basis, including at least five broad categories: financial, marketing, legal, human and production risk (Olson, 2004). In addition, risks associated with pest management reflect just one of many sources of uncertainty under the umbrella of production risk (Olson, 2004). Within the context of what is generally accepted as one of the most uncertain and risky business enterprises, it is not too surprising that farmers may initially perceive new IPM programs or tactics (not tested on their farms) with some skepticism, or perceive that the new system may be more "risky" than current pest management methods. Indeed, risk, or risk perception, is commonly reported as a major reason why farmers do not adopt profit-enhancing Best Management Practices (BMPs), including IPM (Nowak, 1992; Feather & Cooper, 1995; Hrubovcak *et al.*, 1999). Clearly, production risk is important for farmers, as agricultural crops have highly uncertain

Integrated Pest Management, ed. Edward B. Radcliffe, William D. Hutchison and Rafael E. Cancelado. Published by Cambridge University Press.

outcomes each year, subject to annual fluctuations in weather, market prices and pest attack (Barry, 1984; Boehlje & Eidman, 1984; Mumford & Norton, 1984; Hardaker *et al.*, 2004). It is within this context of uncertainty that we revisit some of the long-held views of decision making in IPM from both economic and pest discipline perspectives.

In this chapter, we propose that an additional barrier to IPM adoption is the limited use of risk management and strategic planning tools that can provide IPM developers and clients with objective information on the value and risk of IPM programs, relative to alternative pest management strategies. For example, those of us in the pest and economic disciplines may not always provide the necessary information that potential IPM decision makers need to better understand the benefits (value) and variability (risk) of IPM outcomes, including economic and environmental goals, nor do we compare these metrics to alternative or conventional pest management systems effectively (Mumford & Norton, 1984; Hutchins, 1997; Swinton & Day, 2003; Hutchison *et al.*, 2006a; see also Chapter 2). As such, IPM programs, or the necessary elements (e.g. use of a new sampling plan), are often delivered to clientele without including adequate performance data, or clear value and risk data, so the decision maker can fully understand how a new system may impact net returns short or long term. Consequently, the potential IPM client may perceive the proposed system as being too risky, too time consuming (as the long-term value of sampling is not illustrated) and thus not compatible with other crop/farm components or time demands (Foster *et al.*, 1986; Grieshop *et al.*, 1988; Bechinski, 1994; Cuperus & Berberet, 1994).

Mumford & Norton (1984) reviewed four major decision making tools that are applicable to IPM. Here, we revisit the advantages and disadvantages of these tools and discuss other alternatives from applied economics. We then review those tools that have historically been used in a deterministic way, those that account for uncertainty, and the implications of risk attitudes among IPM clientele. Finally, we provide examples with one case study using a high-value horticultural crop.

4.1 Defining IPM within a management context

Although at least 64 definitions of IPM or "integrated control" had been published by 1997 (Kogan, 1998), contributing to a unique history of the concept, we highlight two well-known definitions, the most recent endorsed by multiple stakeholders and the US Department of Agriculture (USDA) (Chapters 1 and 37). First, following a historical review of the development of the IPM concept over a 35-year period, Kogan (1998) provided:

> IPM is a decision support system for the selection and use of pest control tactics, singly or harmoniously coordinated into a management strategy, based on cost/benefit analyses that take into account the interests of and impacts on producers, society, and the environment.

More recently the USDA, expanding on both the potential value and risk of IPM, and following nationwide input in the USA, concluded in the IPM Road Map, that:

> IPM is a long-standing, science-based, decision-making process that identifies and reduces risks from pests and pest management related strategies.

In addition,

> [IPM] coordinates the use of pest biology, environmental information, and available technology to prevent unacceptable levels of pest damage by the most economical means, while posing the least possible risk to people, property, resources, and the environment. IPM provides an effective strategy for managing pests in all arenas from developed residential and public areas to wildlands. IPM serves as an umbrella to provide an effective, all encompassing, low-risk approach to protect resources and people from pests . . .
>
> (see also Chapter 37)

Both definitions continue to uphold the essence of integrating multiple ecological or biologically based tactics and consistently emphasize three key goals: economic sustainability, environmental integrity and societal acceptance; the latter definition expands the societal component to specifically address health risks. In addition, the IPM Road Map effectively builds upon essential IPM concepts, while providing a useful paradigm for developing and implementing IPM programs in the future. This definition articulates the reality that in addition to achieving economic, environmental and societal goals, a key objective is to also reduce the risk of economic loss, environmental damage and negative health impacts to farm workers and society. Moreover, this definition recognizes that IPM is implemented in systems where uncertainty abounds, whether it is pest density, yield, commodity prices, inclement weather, or farmer and consumer expectations (Barry, 1984; Boehlje & Eidman, 1984).

What is somewhat unique to these definitions, however, is the importance of the "M" in IPM, i.e. *Management* as a decision making process, with risks involved; the clear implication is that decisions are often made under uncertainty. Although these are not new concepts, we contend that for many practical reasons, the management aspect of IPM has received too little attention. Many of us in the pest disciplines have necessarily focused on addressing immediate concerns about *Pest* biology, ecology, pest resistance crises, or how to improve the *Integration* of multiple tactics. One exception to this trend has been the use of simulation models as a strategic tool to assess alternative scenarios to delay pest resistance, where important variables are unknown and management approaches are not set (e.g. Onstad *et al.*, 2003; Hurley *et al.*, 2004; Crowder *et al.*, 2005; Mitchell & Onstad, 2005). Complex optimization methods using dynamic programming and simulation modeling have also been used to assess IPM strategies (e.g. Shoemaker, 1982; Mumford & Norton, 1984; Onstad & Rabbinge, 1985; Stone *et al.*, 1986), with or without risk aspects included. Although simulation approaches are valuable for strategic decision making, these methods require expertise that may not be available to all IPM practioners and are beyond the scope of this chapter (see also Gutierrez & Baumgärtner, 2007).

With regard to the development and application of IPM programs, the management concept has primarily been limited to decision making within a given growing season and making one or more "treat/no-treat" decisions using the economic injury level (EIL) and associated economic thresholds (ET) (Pedigo *et al.*, 1986; see Chapter 3). Although the EIL is foundational to IPM, there are several other approaches to decision making that end-users could consider when evaluating IPM programs. These other approaches can have considerable influence on the perceived benefits and risks of IPM, as well as the subsequent use of IPM, and perhaps the future sustainability of IPM. We agree that the EIL has clearly contributed to the advancement of IPM over the past 40 years. However, from a management perspective, this is just one aspect of decision making that users must consider when deciding to adopt a new IPM program. For example, within a decision science context (Norton, 1982; Mumford & Norton, 1984; Plant & Stone, 1991), the use of an EIL is generally viewed as a *tactical* or *operational* decision rule for immediate use within a field season in response to pest population increase (*ex post*). Use of the EIL is typically most successful where the pest(s) can be effectively monitored, the EIL can be evaluated, and treatment implemented when the cost of control equals or exceeds the anticipated damage or crop yield loss prevented by the treatment. Thus in most cases, the EIL decision rule is most effective as a responsive, therapeutic management action within the growing season for the current year's crop (Pedigo *et al.*, 1986). However, these are not the only types of management actions agricultural decision makers must consider.

An increasingly important aspect of decision making in agriculture, including pest management, includes the strategic long-term analysis of whether to use different types of IPM programs, or whether to use a transgenic crop or seed treatment, where the decisions must be made well in advance of an upcoming field season (*ex ante*). Such strategic decisions can be made at the farm or policy level (Mumford & Norton, 1984; Swinton & Day, 2003; Hardaker *et al.*, 2004; see Chapter 2), but in each case these decisions require performance data for the systems being considered, and an awareness of the decision making tools available (Mumford & Norton, 1984; Fox *et al.*,

1991; see Chapter 2). Moreover, as most IPM decisions are made in a world of uncertainty, the decision maker can make more informed decisions by including estimates of the risks involved (based on subjective or objective probabilities) for each pest management strategy. The IPM decision making principles discussed here should be relevant for applications in many contexts (forestry, urban, livestock, horticultural, agronomic, etc.). However, this chapter focuses on agricultural crop production systems.

4.2 Economic injury level (EIL) and economic threshold (ET)

As previously noted, the primary focus of IPM decision making has been built upon the economic injury level (EIL) concept (Pedigo *et al.*, 1986; Pedigo & Rice, 2006; see Chapter 3). In brief, the EIL is the pest population density at which the *projected* value of the yield loss equals the cost of control (Onstad, 1987). Following Pedigo *et al.* (1986), the EIL is defined as:

$$\mathrm{EIL} = \frac{C}{VIDK}, \tag{4.1}$$

where, on a per unit area basis, C is the cost of the pest control treatment and V is the projected crop value; I is crop injury per pest and D is yield loss per unit of crop injury, which together determine the crop loss or damage caused by the pest density. Finally K is the efficacy of the pest control treatment – the proportion of a population killed or otherwise neutralized to no longer cause crop damage or yield loss. However, for practical field use, the economic threshold (ET) is generally recommended – the pest density at which control action should be taken to prevent the pest population from reaching or exceeding the EIL (Pedigo & Rice, 2006). The ET is usually arbitrarily set to some reasonable level below the EIL to allow sufficient time for making the treatment decision and scheduling control activity (e.g. Pedigo & Rice, 2006; Ragsdale *et al.*, 2007; see Chapter 3).

One indication of the importance of the EIL in IPM decision making is that over 200 EILs have been published in refereed journals (Peterson & Hunt, 2003). Moreover, the EIL and associated ET have been successfully implemented for many crop systems, and have undoubtedly led to reductions in pesticide use and/or more economical pest management (see Chapter 3). Historically, however, the EIL and associated ET have generally been developed and implemented without considering uncertainty (Feder, 1979; Mumford & Norton, 1984; Plant, 1986; Peterson & Hunt, 2003; Mitchell, 2008).

4.3 Risk and IPM

Regarding uncertainty and the EIL, at least two relevant issues emerge. First, the EIL-based decision to treat/not treat assumes that, on average, if one does not treat, the pest population will continue to increase and cause significant (economic) injury and yield loss. This assumption arises from the deterministic EIL, where the key phrase is "on average." However, for any given pest, one can anticipate situations where an intense rainfall event or an insect pathogenic outbreak causes the pest population to "crash" a few days after an ET decision is made. Such outcomes cannot be predicted and result in the possible use of an insecticide when one is not needed. The reverse situation can also occur – generally dependable pathogens or parasitoids decline to unusually low levels, or the pest population has unusually favorable growing conditions so that the pest population "explodes" a few days after a population is observed to be well below the ET. Again, such outcomes cannot be predicted and in this case result in not using pest control when the actual realized crop losses justify it. The point is that the pest population observed at any given time is not a perfect predictor of the future pest population, which creates uncertainty regarding the value of EIL-based decisions to treat or not to treat.

Second, in most applications, some or all of the five parameters determining the EIL reflect the mean response only. In some cases, multiple values for parameters are used as a sort of sensitivity analysis in tabular format to capture changes, e.g. multiple crop values or control costs to adjust the EIL for changes in market or crop conditions (Hammond, 1996; Ragsdale *et al.*, 2007). Although this approach allows for more dynamic thresholds, it does not fully account for the parameter variability or correlations among parameters (Peterson & Hunt, 2003; Mitchell, 2008).

Incorporating the uncertainty in the observed pest population as a decision making signal, or the uncertainty in the parameters determining the EIL, raises an important question: what should an IPM decision maker do when faced with this uncertainty in outcomes? This problem of decision making under uncertainty is broader than just applications to IPM, with large and active areas of research and applications in many disciplines that are beyond the scope of this chapter (see Gollier, 2001; Chavas, 2004; Hardaker *et al.*, 2004). Hence, the methods, criteria and tools discussed here have been applied in a variety of contexts, but we present and interpret them in the context of IPM.

Finally, before continuing, we distinguish between the terms "risk" and "uncertainty." In economics and other fields, uncertainty is converted into risk for analysis (see Box 4.1). Specifically, uncertainty describes situations with incomplete knowledge concerning the possible outcomes and their probabilities, while risk describes situations where all possible events are known and have known probabilities. For example, uncertainty describes the effect of a new invasive crop pest or pathogen, such as the case for the soybean aphid in 2001 (Venette & Ragsdale, 2004) or soybean rust in 2004 (Isard *et al.*, 2005) in the USA – no one really knew all the possible outcomes and how likely each was. Risk describes the effects of these pests/pathogens today – managers now have a better idea of what level of damage is possible and the likelihood of these outcomes after more experience, research and data are collected by scientists (Box 4.1). Although the distinction between risk and uncertainty is not universally adopted and many people and much of the literature (both scientific and popular) use the terms interchangeably (see also Boehlje & Eidman, 1984), we find the clarification in terminology useful.

Box 4.1 | Uncertainty, risk and decision making definitions

Uncertainty The possibilities of outcomes that are difficult to quantify for predictive purposes (e.g. yield loss from a new invasive pest or the effect of an ag-bioterrorism attack on crop prices).

Risk Uncertainty that can be estimated by measuring key outcomes over time and summarizing the data as a probability distribution (e.g net returns from crop production, timing or intensity of weather events, commodity prices).

Probability distribution Set of all possible outcomes and their likelihood (e.g. the number of years in the past decade that a grower experienced light, moderate or heavy European corn borer (*Ostrinia nubilalis*) pressure). The probability distribution defines the complete range of all possible outcomes and the probability that each will occur; these outcomes and probabilities may be objective or subjective.

Objective probability Use of historical and/or experimental data to summarize outcomes and create an outcome probability distribution.

Subjective probability Use of personal experience to create a probability distribution of outcomes and their probabilities.

Risk perceptions An individual's beliefs concerning the probability distribution of possible outcomes (e.g. net returns from crop production, nitrate leaching). These perceptions may vary depending on an individual's source of information or data.

Risk preferences/attitudes Given a known set of outcomes and their probabilities (whether subjective or objective), how one responds to this risk. Different

classifications and measures of an individual's risk preferences/attitudes exist, but people are typically characterized as risk takers, risk averse or risk neutral. Given the choice between a certain outcome and a risky outcome with an average of the same value, a *risk taker* would choose the risky outcome to have the chance to earn more than the average; the *risk averse* person would choose the certain outcome to avoid the risk of earning less than average; and the *risk neutral* person would be indifferent between the certain and risky outcome.

Decision making and management The process of selecting the best choice of resources (e.g. people, goods, services) to achieve goals within pertinent constraints, given an individual's risk perceptions and risk preferences. In the context of IPM, pertinent constraints often include time, government regulation and financial limits. Goals and risk perceptions may include financial, health, environmental and social outcomes.

Source: Adapted from Boehlje & Eidman (1984), Hardaker *et al.* (2004) and Hutchison *et al.* (2006a).

In the remainder of the chapter we first describe and discuss various methods for representing risky situations and measuring risk in the context of IPM. Next, we describe and discuss various decision criteria available for managers to use to choose among a variety of risky options and how these decision criteria are combined with methods for representing risky situations to create decision making tools that can be used for IPM.

4.4 Representing risky situations

Risky situations such as those faced in IPM have the following three elements – actions by the decision maker, events (also called states of nature) and outcomes that depend on the chosen actions and the events that occur. The decision maker chooses from among the available options, then random events occur that imply different outcomes depending on the events that occurred and the action chosen. This risky situation can be represented as a payoff matrix or a decision tree, or by a statistical function (probability density function or cumulative distribution function) describing possible outcomes with parameters depending on the chosen action (Boehlje & Eidman, 1984; Mumford & Norton, 1984; Hardaker *et al.*, 2004). These representations are mathematically equivalent, so to illustrate, we use them to describe the same situation faced by an IPM decision maker.

Table 4.1 presents a payoff matrix for a simple hypothetical EIL-based IPM decision with risk. The decision maker observes the current pest population and chooses one of two actions – to treat or not to treat, with high costs and low damage if a treatment is used. Next, an event occurs – rain or no rain, with the pest population and damage greatly decreased by rain. Four outcomes are possible, with cost either high or low and pest damage either high or low, depending on the treatment decision and the occurrence of rain. For this example, these outcomes are assigned hypothetical net returns (in parentheses). Alternatively, an environmental outcome such as the amount of pesticide runoff could be assigned. Probabilities are also assigned to the two events (also in parentheses). These probabilities can be objective (based on collected data) or subjective (based on personal

Table 4.1 Hypothetical payoff matrix for EIL-based IPM

	Event (state of nature)	
Action	Rain (0.10)	No rain (0.90)
Treat	high cost, low damage ($US 50/ha)	high cost, low damage ($US 50/ha)
Do not treat	low cost, low damage ($US 100/ha)	low cost, high damage ($US 10/ha)

Table 4.2 Payoff matrix representation of net return outcomes ($US/ha) for the Minnesota 1998–2001 cabbage IPM case study[a]

Event (state of nature)		Action to choose	
Pest intensity (*T. ni*) (number of sprays)	Probability[b]	Biologically based IPM	Conventional grower system
Low (1–2 sprays)	0.70	1795	285
Moderate (3–4 sprays)	0.20	775	−366
High (>5 sprays)	0.10	1381	871
Mean		1550	213
Simple average		1317	263

[a] Hutchison *et al.* (2006a, b); Burkness & Hutchison (2008).
[b] Subjective probabilities based on ten-year experience of IPM specialist (E. Burkness) with cabbage.

experience; Box 4.1). With numerical outcomes and probabilities, the situation is risky in the sense that all outcomes and probabilities are known, including how the decision maker's actions affect the outcomes.

This simplistic payoff matrix (or an equivalent decision tree or statistical function) could be constructed to represent one of the many existing EIL-based IPM programs, with the actions, events and outcomes expanded to better match real situations (Mumford & Norton, 1984). However, we want to broaden the conceptual scope to illustrate how these same representations can be used to describe and analyze risk and IPM from a more strategic perspective, as opposed to the more common tactical/operational approach of EIL-based IPM. Hence we examine management alternatives (biologically based IPM versus a conventional grower system) for cabbage looper (*Trichoplusia ni*) in cabbage (*Brassica oleracea*), based on a Minnesota cabbage IPM case study of Hutchison *et al.* (2006a, b) and Burkness & Hutchison (2008).

Table 4.2 is a payoff matrix representation for the case study. The action is the grower's pest management system (IPM or conventional) and events are the pest intensity measured by the number of sprays needed with IPM. Outcomes (net returns $/ha) are based on observed yields and prices, while event probabilities are based on the ten-year experience of a Minnesota IPM specialist (E. Burkness) working with cabbage (Hutchison *et al.*, 2006b, Burkness & Hutchison, 2008). For the same case study, Fig. 4.1 is the equivalent decision tree representation. Figure 4.2 is the probability density function and cumulative distribution function for the random net return outcomes. Commonly, these functions are smooth curves such as the normal (bell) curve, but here, with only three outcomes, Fig. 4.2 is the result. In later discussion, we develop a representation of the case study with more outcomes, which is cumbersome to represent as a payoff matrix or decision tree, but gives smoother functions.

The payoff matrix and decision tree representations show that most of the time, pest pressure is low, which implies low net returns for the conventional (scheduled) system because of unneeded insecticide applications, but high returns with IPM because of the cost savings from not applying unneeded sprays. Moderate pest pressure in about one in five years implies higher returns with IPM than with the conventional system, primarily because consistent scouting reliably detects pest intensity and uses timely control, whereas the conventional grower misses these events often enough either to incur significant losses or to apply too late or when not needed. Finally, high pest pressure in about one in ten years implies higher returns for both cases, but IPM outperforms the conventional system because it more

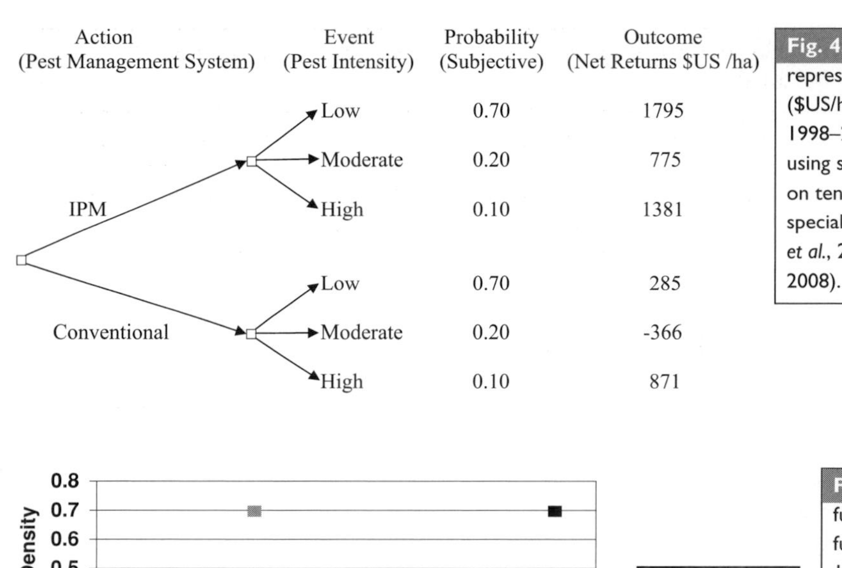

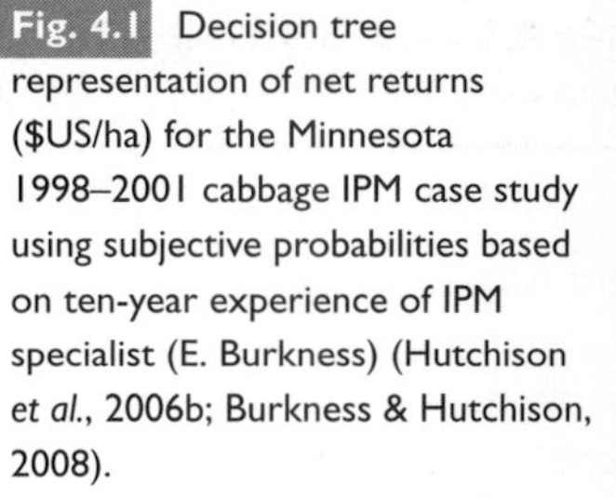

Fig. 4.1 Decision tree representation of net returns ($US/ha) for the Minnesota 1998–2001 cabbage IPM case study using subjective probabilities based on ten-year experience of IPM specialist (E. Burkness) (Hutchison *et al.*, 2006b; Burkness & Hutchison, 2008).

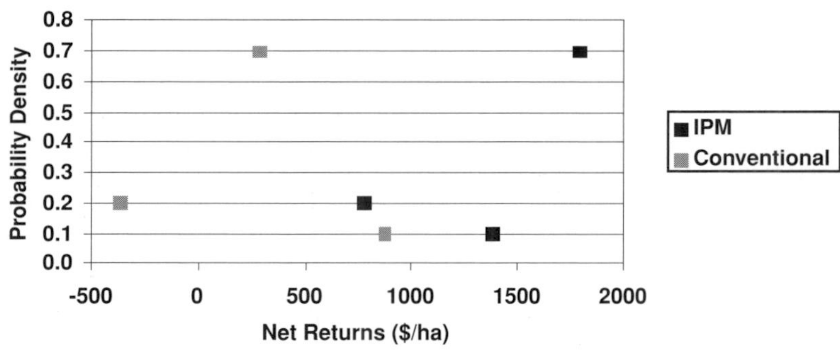

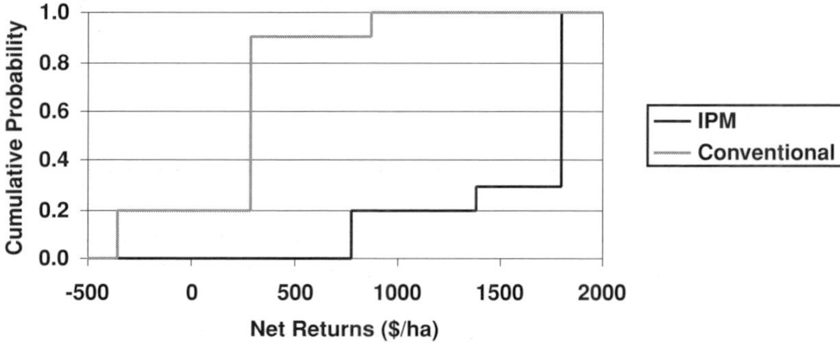

Fig. 4.2 Probability density function and cumulative distribution function for the Minnesota 1998–2001 cabbage IPM case study using subjective probabilities based on ten-year experience of IPM specialist (E. Burkness) working in cabbage pest management (Hutchison *et al.*, 2006a, b; Burkness & Hutchison, 2008).

closely tracks the pest pressure and actually uses more applications, since they are justified. Figure 4.2 shows the probability density function for the conventional system is shifted left because low outcomes are more likely, while shifted right for IPM because high outcomes are more likely. These shifts imply that the cumulative distribution function for the conventional system rises more quickly to 1.0 and is everywhere to the left of the cumulative distribution function for IPM.

For decision making, however, more specific measures of the value and/or riskiness of these two pest management systems seem warranted, as these qualitative descriptions are somewhat unsatisfactory. How does one say a certain case is more or less risky than another?

Several measures exist that allow quantitative comparison of risky cases. First are measures of central tendency or location that indicate where the risky outcome tends to result. Most common among these is the statistical mean or expected

Table 4.3 Risk measures for net returns ($US/ha) for the Minnesota 1998–2001 cabbage IPM case study[a] with both subjective and objective probabilities[b]

Measure	Subjective probabilities		Objective probabilities	
	Biologically based IPM	Conventional system	Biologically based IPM	Conventional system
Probabilities unknown				
Minimum	775	−366	−1 411	−3 998
Maximum	1 795	871	1 949	2 050
Simple average	1 317	263	1 007	−128
Simple variance	263 172	382 894	1 049 102	3 656 829
Simple standard deviation	513	619	1024	1912
Probabilities known				
Mean	1 550	213	1 177	186
Median	1 795	285	1 560	771
Mode	1 795	285	1 896	1 107 & 1272
Variance	164 998	113 973	940 522	2 919 525
Standard deviation	406	338	970	1 709
Coefficient of variation	0.26	1.58	0.82	9.20
Return–risk ratio[c]	3.81	0.63	1.21	0.11
Break-even probability	1.0	0.8	0.88	0.71
Probability of $1000/ha	0.8	0.0	0.75	0.50
Certainty equivalents				
Mean–Variance ($\beta = 0.0005$)	1467	156	706	−1274
Constant absolute risk aversion (CARA) ($r = 0.0001$)	1541	208	1127	28
Constant relative risk aversion (CRRA) ($\rho = 1.2$)	1461	–	–	–

[a] Hutchison *et al.* (2006a, b); Burkness & Hutchison (2008).
[b] Subjective probabilities based on ten-year experience of IPM specialist (E. Burkness) working in cabbage pest management. Objective probabilities based on four-year data set of Hutchison *et al.* (2006a, b); Burkness & Hutchison (2008).
[c] Return–risk ratio is similar in concept to the Sharpe ratio (Sharpe, 1994), but in this case the ratio is based on the probabilistic mean and standard deviation of net returns; values >1 are preferred (see text).

value. For outcomes z_i each with probability p_i, the definition of the mean is $\mu_z = \sum_i p_i z_i$ (Boehlje & Eidman, 1984; Freund, 1992). The mean is sometimes called the probability weighted average of the outcomes or the probabilistic mean to distinguish it from the simple average of the outcomes: $\bar{z} = \frac{1}{n}\sum_i z_i$. The first two columns of Table 4.3 report these values for both systems in the case study for the subjective probabilities of the three outcomes. The means imply that on average, IPM generates greater net returns than the conventional system ($US 1550/ha versus $US 213/ha). These means do not equal the simple average of the three outcomes, since the probabilities of each outcome are not equal.

Two other common measures of central tendency are the median and the mode (Freund, 1992). With symmetric distributions, the median and mode will be similar in magnitude to the mean; a large difference between these and the mean indicates a skewed distribution. The median is the middle outcome – the outcome with half the outcomes above and half below. When outcome probabilities are unequal (the case here), the median is the value on the horizontal axis where the cumulative distribution function equals 0.50 (also called the 50th percentile). As shown in Table 4.3, the median with IPM is $US 1795/ha and $US 285/ha for the conventional system. The mode is the most likely outcome, which here is the same as the median (Table 4.3).

For risky outcomes, measures of dispersion are used to provide some indication of the variability of outcomes. Most common are the variance and standard deviation, though the coefficient of variation and return–risk ratio are useful as well. For outcomes z_i each with probability p_i and mean μ_z, the variance is $\sigma_z^2 = \sum_i p_i(z_i - \mu_z)^2$ and the standard deviation is the positive square root of the variance (Boehlje & Eidman, 1984; Freund, 1992). The standard deviation has the same units as the mean and can be approximately interpreted as the probability weighted average deviation of outcomes from the mean. The variance is sometimes called the probabilistic variance to distinguish it from the simple variance of outcomes: $\frac{1}{n}\sum_i (z_i - \bar{z})^2$, with the comparable distinction made for the probabilistic standard deviation.

Table 4.3 reports these measures, showing that the standard deviation of net returns is greater with IPM than with the conventional system ($US 406/ha versus $US 338/ha), implying that net returns with IPM are "riskier" since they are more variable or dispersed. However, some would argue against this interpretation since mean net returns are larger with IPM as well, so the greater standard deviation is relatively less important. Hence, the coefficient of variation (CV) and the risk–return ratio are measures of risk that normalize for differences in the mean. They are unit-less measures and thus useful for comparing risk-adjusted returns for alternative IPM systems.

The CV is the standard deviation divided by the mean (σ_z/μ_z) and the risk–return ratio adapted to the IPM context is the mean divided by the standard deviation (μ_z/σ_z). The CVs in Table 4.3 indicate that IPM is less risky since the standard deviation is much smaller relative to mean returns (0.26 versus 1.58). The risk–return ratio is used in finance to measure the expected return per unit of risk, where expected returns are net of the risk-free return (e.g. the bond market) and risk is measured by the volatility (the standard deviation of the change in returns) (Sharpe, 1994). In the context of IPM, no equivalent of the risk-free return exists, so the mean return is used, and the standard deviation of returns replaces the volatility, as the change in net returns is usually not tracked continuously. In Table 4.3 the risk–return ratio also indicates that IPM is less risky than the conventional system, since IPM has greater return for the associated risk (3.81 versus 0.63).

A problem with these measures is that they treat risk symmetrically – variability from higher than expected returns is treated the same as variability from lower than expected returns. In many contexts, including IPM, more specific risk measures are useful to differentiate between higher than expected returns (upside risk) and lower than expected returns (downside risk). Statistical measures of skewness describe asymmetric probability distributions (Freund, 1992), but are little used in IPM. However, the Value at Risk (VaR) is a more practical measure of risk borrowed from finance to address these issues (Manfredo & Leuthold, 1999). Value at Risk focuses on downside risk (the occurrence of lower than expected returns) by reporting the returns associated with a chosen critical probability; specifically, with critical probability α_c and cumulative distribution function $F(z)$ for outcomes z, $\text{VaR}(\alpha_c) = F^{-1}(\alpha_c)$. Figure 4.3 (plot A) illustrates the derivation of the VaR. For example, if the VaR(5%) is $US 50/ha, then net returns will be $US 50/ha or less with 5% probability. Here, the larger the VaR for a given probability the less risky the alternative, which is opposite from the original finance context for portfolio values. Also, note that the median is the VaR for a critical probability of 0.50.

An issue with using VaR is what probability to use. In finance, 5% and 1% are common (Manfredo & Leuthold, 1999), but in agriculture and IPM, the appropriate probability is not clear. For example, how useful is it to know that net returns will be $US 50/ha or less with 5% probability? A closely related measure that seems more

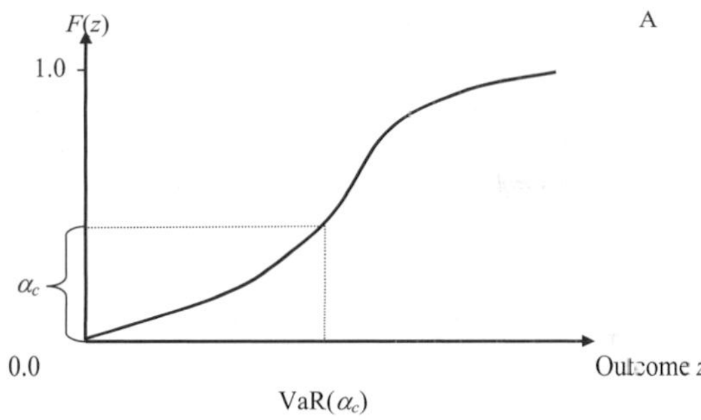

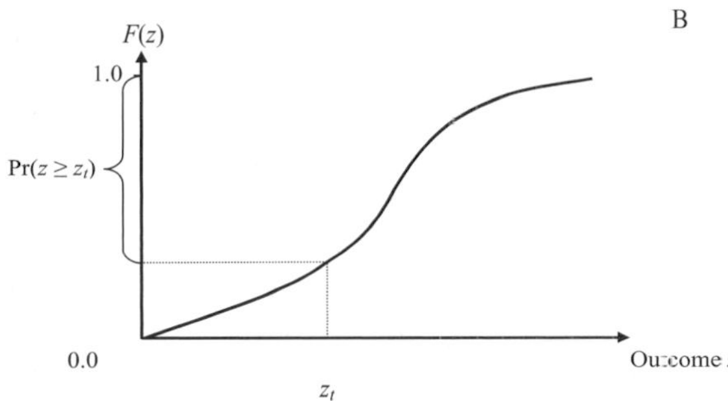

Fig. 4.3 Derivation of Value at Risk (VaR) for a given critical probability α_c (plot A) and of the probability of equaling or exceeding the target outcome $z_t = \Pr(z \geq z_t)$ (plot B), both in terms of the cumulative distribution function $F(z)$ for outcomes z.

appropriate for crop production is to reverse the VaR process and report the probability for a critical outcome (rather than the VaR's outcome for a critical probability). A simple example is the break-even probability – the probability that net returns will be zero or better (e.g. Mitchell, 2005). However, using zero for the critical outcome is arbitrary; decision makers may be interested in the probability of achieving other target outcomes or profit goals. This probability, the probability of equaling or exceeding the target outcome z_t, is defined in terms of the cumulative distribution function as $\Pr(z \geq z_t) = 1 - F(z_t)$. Figure 4.3 (plot B) illustrates the derivation of this probability. For example, if the probability of equaling or exceeding the target outcome of $US 100/ha is 0.10, then in one out of ten years, the grower will earn at least $US 100/ha. For a given target outcome, the larger the probability of equaling or exceeding the target outcome, the lower the risk – a 0.1 probability of equaling or exceeding $100/ha is less risky than a 0.05 probability.

Deriving the VaR and probability of achieving target outcomes for the current specification of the cabbage IPM case study is less useful because with only three outcomes, the cumulative distribution functions have long vertical and horizontal sections implying constant VaR for wide ranges of critical probabilities and constant probabilities for wide ranges of target outcomes (Fig. 4.2). Nevertheless, Table 4.3 reports the break-even probability and the probability of returns reaching at least $US 1000/ha for the case study.

To smooth these curves, we expand the number of outcomes using experimental data (Hutchison *et al.*, 2006b; Burkness & Hutchison, 2008) to switch from subjective probabilities to objective probabilities. Subjective probabilities are useful when few formal data exist for the situation other than personal experience. In such cases, having only a few events as in the case study is reasonable, but implies functions as plotted in Fig. 4.2. However, in research contexts, more formal data can be collected and analyzed, giving smoother functions with objective probabilities. For this case study, data from 24 observations (six fields per year × four years) of net returns for each system give smoother functions to use to illustrate these measures of risk (Hutchison *et al.*, 2006b; Burkness & Hutchison, 2008). Figure 4.4 shows the traditional histograms of net returns (plotting the probability as points as in Fig. 4.2 is difficult to interpret) and the cumulative distribution functions. The last two columns in Table 4.3 report all previously discussed measures of risk, including the break-even probability and probability of returns reaching at least $US 1000/ha.

Figure 4.4 shows a negative skew for the returns of both density functions, which drives the large difference between the means and the medians/modes for both systems (Table 4.3). Interestingly, the conventional system is now "riskier" than IPM based on both absolute measures of risk (variance, standard deviation) and relative measures (CV, return–risk ratio), as well as the two probability-based measures (break-even probability, probability of achieving at least $US 1000/ha).

This use of more outcomes and smoother functions can continue until the density and distribution functions are continuous plots similar to

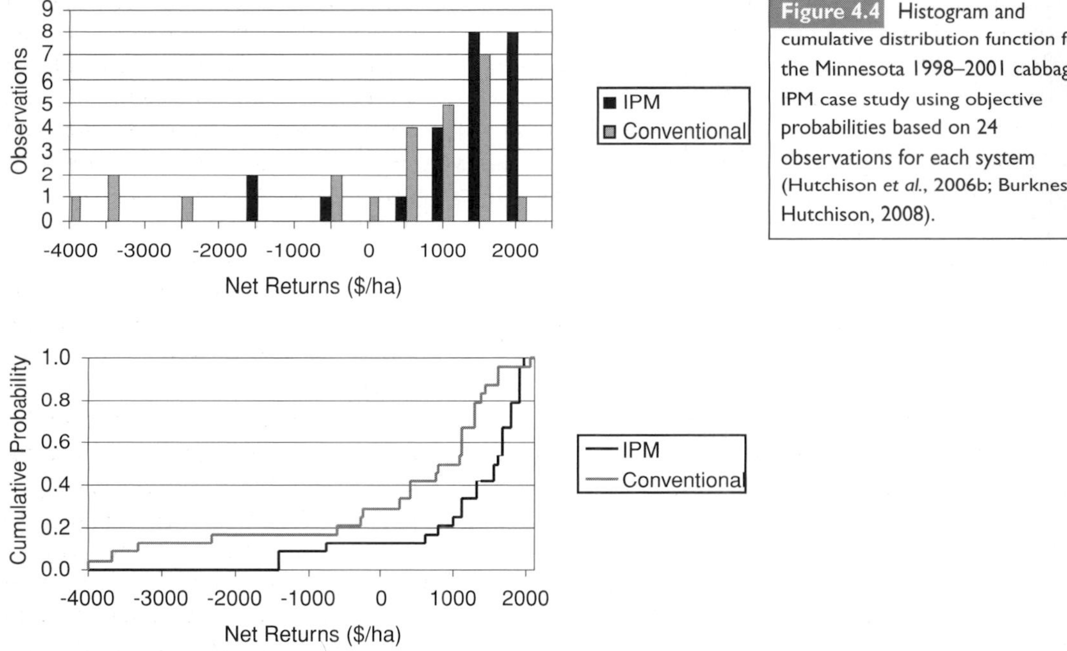

Figure 4.4 Histogram and cumulative distribution function for the Minnesota 1998–2001 cabbage IPM case study using objective probabilities based on 24 observations for each system (Hutchison *et al.*, 2006b; Burkness & Hutchison, 2008).

Fig. 4.3. The only major change is that calculating the risk measures in Table 4.3 requires use of different formulas than given previously (integration rather than summation). For common distributions, texts give formulas for the mean, standard deviation, etc. (e.g. Evans *et al.*, 2000) and software packages give numerical solutions for values of the cumulative distribution and/or its inverse when closed form expressions do not exist, such as for the normal distribution.

4.5 Decision making criteria and tools

These representations of risky pest management situations and the associated measures of the risk under different actions still leave unanswered the question of which action to choose. Combining a representation of the risky situation with a decision criterion gives a decision making tool, a method to identify which actions are optimal given the specific criteria. Clearly, the optimal action depends on the criterion. Here we describe criteria commonly used in agriculture and pest management. The criteria can first be separated into those that require probabilities and those that do not (Boehlje & Eidman, 1984).

Maximin (also called minimax), a criterion adapted from game theory not using probabilities, identifies the worst outcome for each action and chooses the action with the best outcome among these worst-case scenarios. The names arise because the criterion maximizes the minimum outcome (or minimizes the maximum loss), a fairly conservative approach (Boehlje & Eidman, 1984). Maximax, a related criterion little used in practical applications, but useful as an overly optimistic benchmark, identifies the best outcome for each action and chooses the action with the best outcome among these best-case scenarios. Finally, another criterion ignores all probability assessments and chooses the action with the greatest simple average for its outcomes. In some sense, this criterion is comparable to an uninformed prior in Bayesian statistics – initially all outcomes are equally possible (Boehlje & Eidman, 1984).

Table 4.3 shows that IPM is optimal with both the maximin and the maximax criteria, whether using the three-state or 24-state example, the minimum (and maximum) outcome for IPM exceeds the minimum (and maximum) outcome for the conventional system. The simple average of returns is also greater for IPM. We are unaware of published examples using these criteria for IPM.

The most common criterion using probabilities optimizes the expected (mean) outcome, for example choosing the action with the greatest mean net return (as explained below, this decision maker is called "risk neutral"). For our case study, Table 4.3 shows that IPM is the optimal action under this criterion – mean returns with IPM exceed mean returns with the conventional system using subjective or objective probabilities. Applications of this criterion in pest management, with or without assumed probabilities, are often implied in IPM applications, but the probabilities are not typically defined (e.g. Mumford & Norton, 1984; Burkness & Hutchison, 2008).

Various safety-first criteria exist, for example minimizing the probability that returns fall below a critical level. This criteria places no value on mean returns, but other safety-first criteria do, such as maximizing mean returns, subject to ensuring a specific probability that a certain minimum return is achieved (e.g. choosing the action with greatest mean returns subject to maintaining a 5% probability that returns are at least $US100/ha). Safety-first criteria are useful for IPM with limited income farmers such as in developing countries where a certain income level is needed for survival (e.g. Norton *et al.*, 2005).

Maximizing mean returns (or expected profit) ignores variability – actions with the same mean returns are equal under the criterion, even if their variability differs. However, most people are willing to trade off between the mean and the variability of returns, preferring actions with lower mean returns because they are less variable. For example, many people voluntarily buy insurance because the indemnity reduces the variability of returns, though the premium reduces their mean return. How people trade between certain and variable outcomes and between the mean and variability of returns is an active area of empirical and theoretical research in many fields beyond the scope of this chapter (e.g. Camerer, 2003; Eeckhoudt *et al.*, 2005; Ernst & Paulus, 2005; Fecteau *et al.*, 2007; Mitchell & Onstad, 2008). Here we describe and illustrate some of the more common methods used in agricultural contexts.

4.5.1 Risk preferences defined

A fundamental assumption of these methods is that individuals implicitly impose a cost on mean returns to adjust for their variability, similar to a discount factor imposing a cost on future returns. "Risk preferences" is the term used in economics and related disciplines to describe how individuals trade between risk and returns (see Box 4.1). In many contexts, people are risk averse – willing to give up some returns to obtain reduced variability, for example by buying insurance. However, in other contexts, people will give up mean returns to obtain increased variability, for example by buying a lottery ticket. In the first case, the person is called "risk averse," and "risk taking" in the second case (Chavas, 2004; see Box 4.1).[1] Depending on the context, the same person can exhibit both behaviors (e.g. buy insurance and lottery tickets).

Two concepts are often used to understand this trading between risk and returns. For a person facing a risky outcome, the certainty equivalent is the non-random payment that makes the person indifferent between taking the risky outcome and this certain payment. The risk premium is the difference between the expected outcome and this certainty equivalent. For example, if a risky outcome has a mean return of $US 500 and a person's certainty equivalent is $US 450, then this person's risk premium is $US 50. If the certainty equivalent is less than the mean return (or the risk premium is positive), the person is risk averse.

Usually this trading between risk and returns is described using a preference (or utility) function to convert risky outcomes into a measure describing the benefit after accounting for risk; a common example is mean–variance preferences.

[1] Unfortunately, this terminology is not fully standardized. Many use "risk loving" for "risk taking," while Hardaker *et al.* 2004 use "risk preference" for "risk loving" and "risk attitudes" when most use "risk preferences" (Chavas, 2004; Eeckhoudt *et al.*, 2005).

For a random outcome z, a person's outcome is $u(z) = \mu_z - \beta\sigma_z^2$, where μ_z and σ_z^2 are the respective mean and variance of z, β is a parameter describing how the individual adjusts the mean for risk as measured by the variance, and $u(\cdot)$ is called the utility function. For this case, $u(z)$ is the certainty equivalent and $\beta\sigma_z^2$ is the risk premium. If $\beta > 0$, the person is risk averse, if $\beta < 0$, the person is risk loving. If $\beta = 0$, then $u(z) = \mu_z$) and the person is "risk neutral" – outcome variability does not matter, only the mean return.

Within risk averse preferences, further subtypes of risk preferences are defined. Two of the most common are constant absolute risk aversion (CARA) and constant relative risk aversion (CRRA). CARA implies that the risk premium is independent of the initial wealth level. However, the risk premium with CARA depends on the units of measure (e.g. € versus $, or $/acre versus $/ha) (Chavas, 2004). Hence, as a convenient normalization, the relative risk premium expresses the risk premium as a percentage of expected final wealth. CRRA implies that this relative risk premium is constant with respect to wealth, though the absolute risk premium is not. However, the general functional form implying CRRA is undefined for negative outcomes, limiting its use for some applications without adjustments. Mathematically, the CARA utility function is $u(z) = -\exp(-rz)$, where $r > 0$ is the coefficient of absolute risk aversion, while the CRRA utility function is $u(z) = z^{1-\rho}$ when $\rho > 1$, $u(z) = -z^{1-\rho}$ when $\rho < 1$ and $\ln(z)$ when $\rho = 1$, where ρ is the coefficient of relative risk version (Chavas, 2004). Finally, note that concavity of the utility function $u(z)$ over the range of outcomes z implies risk aversion (i.e. a positive risk premium, or a certainty equivalent less than the expected outcome μ_z).

These descriptions are brief and far from comprehensive, plus all presented cases follow the von Neumann–Morganstern expected utility model, which has well-known limits for accurately describing risky decisions (Chavas, 2004). Other types of utility functions based on alternative decision criteria or theories exist, including rank-dependent expected utility, ambiguity aversion, and loss aversion/prospect theory. See Chavas (2004), Eeckhoudt *et al.* (2005) and Hardaker *et al.* (2004) for more comprehensive and detailed descriptions and examples of these.

Table 4.3 summarizes certainty equivalents for mean–variance, CARA and CRRA preferences (when possible) for both systems with both sets of probabilities, with specific parameter values used for utility functions reported in parentheses. These parameters were chosen purely for illustration; different parameters would generate different results. Babcock *et al.* (1993) and McCarl & Bessler (1989) provide guidance on parameter choices. IPM is optimal for both sets of probabilities with these parameters. This is not surprising, given the notably larger mean and lower variability with IPM. For these data, it would be difficult to identify a reasonable model of risk averse preferences that made the conventional pest management approach optimal. Applications of these expected utility criteria in pest management include Hurley *et al.* (2004), Mitchell *et al.* (2004) and Mitchell (2008).

Stochastic dominance is an alternative criterion that is theoretically attractive as it imposes little structure on preferences (i.e. no utility function is required) other than positive slope and possibly concavity. Stochastic dominance ranks the various actions a decision maker has by comparing the cumulative distribution functions of outcomes that result with each action. Without explaining the theoretical foundations (see Boehlje & Eidman, 1984; Chavas, 2004; Hardaker *et al.*, 2004), first-order stochastic dominance (FOSD)[2] implies that, if the cumulative distribution with action A is everywhere less than or equal to the cumulative distribution with action B, then action A is preferred to action B by all decision makers with positively sloped utility functions (i.e. people who prefer more income to less). Alternatively, A first order stochastically dominates B if the entire cumulative distribution function for A is to the right of the function for B. Mathematically, FOSD requires $F_A(z) \leq F_B(z)$ for all outcomes z, where $F_i(\cdot)$ is the cumulative distribution for $i \in \{A, B\}$. Figure 4.2 shows that IPM first order

[2] Some use the terminology first degree stochastic dominance (Hardaker *et al.*, 2004).

stochastically dominates the conventional system for the case study with subjective probabilities. The implication is that almost everyone would prefer IPM to the conventional system in this case.

Second-order stochastic dominance (SOSD) is defined in terms of the area between the cumulative distribution for actions A and B. Specifically, if $\int_{z_{\min}}^{z^*} F_B(z) - F_A(z)dz \geq 0$ for all $z_{\min} \leq z^* \leq z_{\max}$, where $z_{\min}$ and $z_{\max}$ are the minimum and maximum possible values for z, then action A second order stochastically dominates action B so that all decision makers with concave utility functions (i.e. who are risk averse) prefer action A to action B. SOSD requires that the area between the cumulative distributions (above A and below B) be positive, beginning with the minimum outcome $z_{\min}$ and up to any outcome z^*, for all possible z^*. Figure 4.4 shows that IPM second order stochastically dominates the conventional system for the case study with objective probabilities because the cumulative distributions cross only once, $F_A(\cdot)$ is initially above $F_B(\cdot)$, and the area when $F_B(\cdot)$ is above $F_A(\cdot)$ is much larger than the area when $F_A(\cdot)$ is above $F_B(\cdot)$. The implication is that all risk averse individuals would prefer IPM to the conventional system in this case. Finally, note that stochastic dominance is only a partial ordering. Cumulative distributions can occur that cannot be ordered by FOSD or SOSD (Chavas, 2004; Hardaker *et al.*, 2004). Also, Hardaker *et al.* (2004) discuss additional types of stochastic dominance.

Besides the brief examples provided here by the case study, previous applications of the expected value (mean) and variance (or standard deviation) in IPM include: plant pathogens (Carlson, 1970; Norton, 1976), arthropods (Burkness *et al.*, 2002) and alternative weed management systems (Hoverstad *et al.*, 2004). Stochastic dominance has been used to assess alternative IPM tactics or programs for arthropods (Moffitt *et al.*, 1983; Burkness *et al.*, 2002), weed management (Hoverstad *et al.*, 2004) and multiple pest IPM (Musser *et al.*, 1981; see review by Fox *et al.*, 1991).

4.6 Conclusions

In summary, we have shown that although the EIL is foundational to successful IPM programs, it represents just one of several "decision tools" or approaches to decision making. Other tools compare the economic performance, or risk adjusted economic returns for IPM versus other pest management systems (e.g. organic or conventional). The EIL is most often used as a therapeutic (*ex post*) decision criterion for managing pest populations during a given growing season. We have summarized alternative decision making approaches that are applicable to IPM that can be used with the appropriate information to assess the value and risk of IPM in advance of a growing season (*ex ante*). As such, these methods facilitate objective comparison of IPM performance with alternative production systems. Most of the methods reviewed also incorporate some measure of economic risk, such as the variance or standard deviation of net returns, to facilitate decision making under uncertainty. More work is needed to examine the economic impact of multiple arthropods, weeds and/or pathogens on the value and risk of IPM for a given crop, and for two or more crops at the farm or enterprise level (see Fox *et al.*, 1991; Chapter 2). As seed treatments and biotechnology become increasingly popular (e.g. maize growers in the USA), we anticipate that these methods may become more common in future *ex ante* assessments of IPM. In addition, similar approaches could be taken to better understand the value and risk of environmental impacts of IPM versus alternative systems. The increasing use of environmental indices to evaluate pest management programs has been beneficial (e.g. Kovach *et al.*, 1992; Benbrook *et al.*, 2002). However, more studies like those of Mullen *et al.* (1997) and Edson *et al.* (2003), and expanded methods for environmental analysis (e.g. Shiferaw *et al.*, 2004; Jepson, 2007), are needed so growers and policy makers can fully assess both the value and risk of economic and environmental outcomes for IPM and competing pest management systems.

References

Babcock, B. A., Choi, E. K. & Feinerman, E. (1993). Risk and probability premiums for CARA utility functions. *Journal of Agricultural and Resource Economics*, **18**, 17–24.

Bacic, I. L. Z., Bregt, A. K. & Rositer, D. G. (2006). A participatory approach for integrating risk assessment into rural decision-making: a case study in Santa Catarina, Brazil. *Agricultural Systems*, **87**, 229–244.

Barry, P. (ed.) (1984). *Risk Management in Agriculture*. Ames, IA: Iowa State University Press.

Bechinski, E. J. (1994). Designing and delivering in-field scouting programs. In *Handbook of Sampling Methods for Arthropods in Agriculture*, eds. L. P. Pedigo & B. D. Buntin, pp. 683–706. Boca Raton, FL: CRC Press.

Benbrook, C. M., Sexson, D. L., Wyman, J. A. *et al.* (2002). Developing a pesticide risk assessment tool to monitor progress in reducing reliance on high-risk pesticides. *American Journal of Potato Research*, **79**, 183–200.

Boehlje, M. D. & Eidman, V. R. (1984). *Farm Management*. New York: John Wiley.

Burkness, E. C. & Hutchison, W. D. (2008). Implementing reduced-risk IPM in fresh-market cabbage: Improved net returns via scouting and timing of effective management. *Journal of Economic Entomology*, **101**, 461–471.

Burkness, E. C., Hutchison, W. D., Weinzierl, R. A. *et al.* (2002). Efficacy and risk efficiency of sweet corn hybrids expressing a *Bacillus thuringiensis* toxin for lepidopteran pest management in the Midwestern U.S. *Crop Protection*, **21**, 157–169.

Camerer, C. (2003). *Behavioral Game Theory: Experiments in Strategic Interaction*. Princeton, NJ: Princeton University Press.

Carlson, G. A., (1970). The decision theoretic approach to crop disease prediction and control. *American Journal of Agricultural Economics*, **52**, 216–223.

Chavas, J. P. (2004). *Risk Analysis in Theory and Practice*. New York: Elsevier.

Crowder, D. W., Onstad, D. W., Gray, M. E. *et al.* (2005). Economic analysis of dynamic management strategies utilizing transgenic corn for control of western corn rootworm (Coleoptera: Chrysomelidae). *Journal of Economic Entomology*, **98**, 961–975.

Cuperus, G. W. & Berberet, R. C. (1994). Training specialists in sampling procedures. In *Handbook of Sampling Methods for Arthropods in Agriculture*, eds. L. P. Pedigo & B. D. Buntin, pp. 669–681. Boca Raton, FL: CRC Press.

Edson, C., Swinton, S., Nugent, J. *et al.* (2003). *Cherry Orchard Floor Management: Opportunities to Improve Profit and Stewardship*, MSU Ext. Bull. E-2890. East Lansing, MI: Michigan State University.

Eeckhoudt, L., Gollier, C. & Schlesinger, H. (2005). *Economic and Financial Decisions under Risk*. Princeton, NJ: Princeton University Press.

Ehler, L. E. & Bottrell, D. G., (2000). The illusion of integrated pest management. *Issues in Science and Technology*, **16**, 61–64.

Ernst, M. & Paulus, M. P. (2005). Neurobiology of decision making: a selective review from neurocognitive and clinical perspective. *Biological Psychiatry*, **58**, 597–604.

Evans, M., Hastings, N. & Peacock, B. (2000). *Statistical Distributions*, 3rd edn. New York: John Wiley.

Feather, P. & Cooper, J. (1995). *Voluntary Incentives for Reducing Agricultural Nonpoint Source Water Pollution*, Agricultural Information Bulletin No. 716. Washington, DC: US Department of Agriculture, Economic Research Service.

Fecteau, S., Pascual-Leone, A., Zald, D. H. *et al.* (2007). Activation of prefrontal cortex by transcranial direct current stimulation reduces appetite for risk during ambiguous decision making. *Journal of Neuroscience*, **27**, 6212–6218.

Feder, G. (1979). Pesticides, information, and pest management under uncertainty. *American Journal of Agricultural Economics*, **62**, 97–103.

Foster, R. E., Tollefson, J. J., Nyrop, J. P. & Hein, G. L. (1986). Value of adult corn rootworm (Coleoptera: Chrysomelidae) population estimates in pest management decision making. *Journal of Economic Entomology*, **79**, 303–310.

Fox, G., Weersink, A., Sarwar, G., Duff, S. & Deen, B. (1991). Comparative economics of alternative agricultural production systems: a review. *Northeastern Journal of Agricultural and Resource Economics*, **20**, 124–142.

Freund, J. E. (1992). *Mathematical Statistics*, 5th edn. Englewood Cliffs, NJ: Prentice Hall.

Gollier, C. (2001). *The Economics of Risk and Time*. Cambridge, MA: MIT Press.

Grieshop, J. I., Zalom, F. G. & Miyao, G. (1988). Adoption and diffusion of Integrated Pest Management Innovations in Agriculture. *Bulletin of the Entomological Society of America*, **34**, 77–78.

Gutierrez, A. P. & Baumgärtner, J. (2007). Modeling the dynamics of tritrophic population interactions, In *Perspectives in Ecological Theory and Integrated Pest Management*, eds. M. Kogan & P. Jepson, pp. 301–360. Cambridge, UK: Cambridge University Press.

Hammond, C. M., Luschel, E. C., Boerboom, C. M. & Nowak, P. J. (2006). Adoption of integrated pest management tactics by Wisconsin farmers. *Weed Technology*, **20**, 756–767.

Hammond, R. B. (1996). Limitations to EILs and thresholds, In *Economic Thresholds for Integrated Pest Management*, eds. L. G. Higley & L. P. Pedigo, pp. 58–73. Lincoln, NE: University of Nebraska Press.

Hardaker, J. B., Huirne, B. R. M., Anderson, J. R. & Lien, G. (2004). *Coping with Risk in Agriculture*, 2nd edn. Wallingford, UK: CABI Publishing.

Hoverstad, T. R., Gunsolus, J. L., Johnson, G. A. & King, R. P. (2004). Risk-efficiency criteria for evaluating economics of herbicide-based weed management systems in corn. *Weed Technology*, **18**, 687–697.

Hrubovcak, J., Vasavada, U. & Aldy, J. E. (1999). *Green Technologies for a More Sustainable Agriculture*, Agricultural Information Bulletin No. 752. Washington, DC: US Department of Agriculture, Economic Research Service.

Hurley, T. M., Mitchell, P. D. & Rice, M. E. (2004). Risk and the value of Bt corn. *American Journal of Agricultural Economics*, **86**, 345–358.

Hutchins, S. H. (1997). IPM: opportunities and challenges for the private sector. In *Radcliffe's IPM World Textbook*, eds. E. B. Radcliffe & W. D. Hutchison. St. Paul, MN: University of Minnesota. Available at http://ipmworld.umn.edu.

Hutchison, W. D., Burkness, E. C., Carrillo, M. A., Galvan, T. L., Mitchell, P. D. & Hurley, T. M. (2006a). *IPM: A Risk Management Framework to Improve Decision-making*, Public. No. 08229. St. Paul, MN: University of Minnesota Extension Service.

Hutchison, W. D., Burkness, E. C., Carrillo, M. A., Hurley, T. M. & Pahl, G. (2006b). *Fresh-Market Cabbage: Increasing Economic Returns while Reducing Risk*, Public. No. 08230. St. Paul, MN: University of Minnesota Extension Service.

Isard, S. A., Gage, S. H., Comtois, P. & Russo, J. M. (2005). Principles of the atmospheric pathway for invasive species applied to soybean rust. *BioScience*, **55**, 851–861.

Jepson, P. C. (2007). Ecotoxicology: the ecology of interactions between pesticides and non-target organisms. In *Perspectives in Ecological Theory and Integrated Pest Management*, eds. M. Kogan & P. Jepson, pp. 522–551. Cambridge, UK: Cambridge University Press.

Kogan, M. (1998). Integrated pest management: historical perspectives and contemporary developments. *Annual Review of Entomology*, **43**, 243–270.

Kovach, J., Petzoldt, C., Degni, J. & Tette, J. (1992). *A Method to Measure the Environmental Impact of Pesticides*, New York Food and Life Science Bulletin No. 139. Geneva, NY: Cornell University, New York Agricultural Experiment Station. Available at http://ecommons.library.cornell.edu/handle/1813/5203.

Manfredo, M. R. & Leuthold, R. M. (1999). Value-at-risk analysis: a review and the potential for agricultural applications. *Review of Agricultural Economics*, **21**, 99–111.

McCarl, B. & Bessler, D. (1989). Estimating an upper bound on the Pratt risk aversion coefficient when the utility function is unknown. *Australian Journal of Agricultural Economics*, **33**, 56–63.

Mitchell, P. D. (2005). *The Expected Net Benefit and Break-Even Probability for Bt Corn in Wisconsin*, UWEX Information Bulletin with companion spreadsheet, October. Madison, WI: University of Wisconsin Extension. Available at www.aae.wisc.edu/mitchell/extension.htm.

Mitchell, P. D. (2008). *Risk, Farmer Returns and Integrated Pest Management (IPM)*, Agricultural and Applied Economics Staff Paper No. 526. Madison, WI: University of Wisconsin–Madison.

Mitchell, P. D. & Onstad, D. W. (2005). Effect of extended diapause on the evolution of resistance to transgenic *Bacillus thuringiensis* corn by northern corn rootworm (*Coleoptera: Chrysomelidae*). *Journal of Economic Entomology*, **98**, 2220–2234.

Mitchell, P. D. & Onstad, D. W. (2008). Valuing insect resistance in an uncertain future. In *Insect Resistance Management: Biology, Economics, and Prediction*, ed. D. W. Onstad, pp. 17–38. San Diego, CA: Academic Press.

Mitchell, P. D., Gray, M. E. & Steffey, K. L. (2004). A composed error model for estimating pest-damage functions and the impact of the western corn rootworm soybean variant in Illinois. *American Journal of Agricultural Economics*, **86**, 332–344.

Moffitt, L. J., Tanagosh, L. K. & Baritelle, J. L. (1983). Incorporating risk in comparisons of alternative pest management methods. *Environmental Entomology*, **12**, 1003–1111.

Mullen, J. D., Norton, G. W. & Reaves, D. W. (1997). Economic analysis of environmental benefits of integrated pest management. *Journal of Agriculture and Applied Economics*, **29**, 243–253.

Mumford, J. D. & Norton, G. A. (1984). Economics of decision making in pest management. *Annual Review of Entomology*, **29**, 157–174.

Musser, W. N., Tew, B. V. & Epperson, J. E. (1981). An economic examination of an integrated pest management production system with a contrast between E-V and stochastic dominance analysis. *Southern Journal of Agricultural Economics*, **13**, 119–124.

Norton, G. A. (1976). Analysis of decision making in crop protection. *Agroecosystems*, **3**, 27–44.

Norton, G. A. (1982). A decision analysis approach to integrated pest control. *Crop Protection*, **1**, 147–164.

Norton, G. A., Heinrichs, E. A., Luther, G. C. & Irwin, M. E. (eds.) (2005). *Globalizing Integrated Pest Management: A Participatory Research Approach*, Ames, IA: Blackwell.

Nowak, P. (1992). Why farmers adopt production technology. *Journal of Soil and Water Conservation*, **47**, 14–16.

Olson, K. D., (2004). *Farm Management: Principles and Strategies*. Ames, IA: Iowa State University Press.

Onstad, D. W. (1987). Calculation of economic injury levels and economic thresholds for pest management. *Journal of Economic Entomology*, **80**, 297–303.

Onstad, D. W. & Rabbinge, R. (1985). Dynamic programming and the computation of economic injury levels for crop disease control. *Agricultural Systems*, **18**, 207–226.

Onstad, D. W., Crowder, D. W., Mitchell, P. D. *et al.* (2003). Economics versus alleles: balancing IPM and IRM for rotation-resistant western corn rootworm (*Coleoptera: Chrysomelidae*). *Journal of Economic Entomology*, **96**, 1872–1885.

Pedigo, L. P. & Rice, M. E. (2006). *Entomology and Pest Management*, 5th edn. Upper Saddle River, NJ: Prentice Hall.

Pedigo, L. P., Hutchins, S. H. & Higley, L. G. (1986). Economic injury levels in theory and practice. *Annual Review of Entomology*, **31**, 341–368.

Peterson, R. K. D. & Hunt, T. E. (2003). The probabilistic economic injury level: incorporating uncertainty into pest management decision-making. *Journal of Economic Entomology*, **96**, 536–542.

Plant, R. E. (1986). Uncertainty and the economic threshold. *Journal of Economic Entomology*, **79**, 1–6.

Plant, R. E. & Stone, N. D. (1991). *Knowledge-Based Systems in Agriculture*. New York: McGraw-Hill.

Ragsdale, D. W., McCornack, B. P., Venette, R. C. *et al.* (2007). Economic threshold for soybean aphid. *Journal of Economic Entomology*, **100**, 1258–1267.

Rogers, E. M. (1995). *Diffusion of Innovations*, 4th edition. New York: The Free Press.

Sharpe, W. F. (1994). The Sharpe ratio. *Journal of Portfolio Management*, **21**, 49–58.

Shiferaw, B., Freeman, H. A. & Swinton, S. M. (eds.) (2004). *Natural Resource Management in Agriculture: Methods for Assessing Economic and Environmental Impacts*. Wallingford: UK: CABI Publishing.

Shoemaker, C. (1982). Optimal integrated control of univoltine pest populations with age structure. *Operations Research*, **30**, 40–61.

Stone, N. D., Gutierrez, A. P., Getz, W. M. & Norgaard, R. (1986). III. Strategies for pink bollworm control in southwestern desert cotton: an economic simulation study. *Hilgardia*, **54**, 42–56.

Swinton, S. M. & Day, E. (2003). Economics in the design, assessment, adoption, and policy analysis of IPM. In *Integrated Pest Management: Current and Future Strategies*, R-140, ed. K. R. Barker. Ames, IA: Council for Agricultural Science and Technology.

Venette, R. C. & Ragsdale, D. W. (2004). Assessing the invasion by soybean aphid (Homoptera: Aphididae): where will it end? *Annals of the Entomological Society of America*, **97**, 219–226.

Wearing, C. H. (1988). Evaluating the IPM implementation process. *Annual Review of Entomology*, **33**, 17–38.

Chapter 5

IPM as applied ecology: the biological precepts

David J. Horn

Any insect (or other) pest exists within an ecosystem, consisting of the surrounding biological and physical environment with which it interacts. The interactions between a pest population and its ecosystem are highly complex, and in many cases several pests with different biologies need to be simultaneously managed on a single crop. Ecological issues are exacerbated as the scale of management increases. On a typical farm in midwestern USA we might find fields producing maize (corn), soybeans, hay and perhaps small grains or canola, plus several species of vegetables in a family garden, several kinds of livestock and poultry, stored feed and seed, landscaping plantings, weeds, wildlife and the farmer and his/her household, any and all of which might harbor populations of one or more pests. The farm ecosystem occurs in a matrix of surrounding systems each with its own communities including pests. Ecological processes within surrounding habitats influence events within adjacent areas. In our efforts to maintain high yields and maximize profits, we often oversimplify and override ecosystem processes and unknowingly disrupt whatever naturally occurring pest population regulation there may be. Kogan (1995) and others have noted that even successful IPM programs may pay little heed to the complexity and unpredictability of ecological processes. Our pest management efforts therefore are often disruptive of ecosystem functions. In order to develop more ecologically based IPM systems we need greater understanding of ecological processes. The present chapter introduces some of these fundamental ecological processes as they impact pest populations.

While the past few decades have witnessed general acceptance of considering ecology in developing pest management systems, there remains little agreement as to what ecological paradigms are most applicable to pest management systems (Kogan, 1995; Carson *et al.*, 2004). Ecology draws on ideas and data from across biology. As the information base expands, ecological ideas are revised. Ecologists rarely agree completely on such issues as the importance of equilibrium in population regulation, or whether there is a relationship between species diversity and community stability. Those who design and implement pest management systems may be frustrated by the apparent flux in ecological theory. The challenge of applying ecological theory to pest management is intensified by several differences between "natural" ecosystems (such as abandoned fields or forests) and more closely managed, "artificial" ecosystems (such as agricultural fields, orchards or highly manicured landscapes). Ecological ideas generated from studies of natural ecosystems may not be applicable to many managed ecosystems. Even small and isolated ecosystems are enormously complex and variable. Field

Integrated Pest Management, ed. Edward B. Radcliffe, William D. Hutchison and Rafael E. Cancelado. Published by Cambridge University Press.

experiments are subject to widely varying outputs and results may be difficult to interpret. These and other issues supply the context of an ecological approach to IPM.

5.1 Life systems

Because a pest population interacts with the surrounding ecosystem it cannot be considered apart from that ecosystem. Clark *et al.* (1967) developed the "life system" concept to reinforce this reality, and the life system paradigm remains useful 40 years later. A life system consists of the pest population plus its "effective environment," the sum of the surrounding ecological factors that impact the population positively or negatively. The effective environment includes food, competitors, predators, pathogens and the physical surroundings – in short, anything that reduces or increases survivorship, fecundity and movement of the pest (or any other organism). The scale of the life system is arbitrary and the impact on the effective environment varies over space and time; the life system will be different according to whether one views the surrounding ecosystem as a single plant, a single field or orchard, a regional landscape or an ecoregion. An applied ecological approach to IPM needs to consider an appropriate scale. Insects such as the alfalfa snout beetle (*Otiorhynchus ligustici*) that do not fly may be limited to a local area for many decades whereas to understand the ecology of long-distance migrants such as the beet armyworm (*Spodoptera exigua*) or migratory locust (*Schistocerca gregaria*) may require consideration of an effective environment over hundreds of square kilometers. In most cases the effective environment needs to be considered at least to the level of the local landscape (Duelli, 1997; Collins & Qualset, 1999; Landis *et al.*, 2000). An implication of the life system concept is that agricultural or forestry activities such as tilling, thinning or harvesting either disrupt or enhance ecosystem functions resulting in a more or less favorable environment for a pest population. This in turn leads to either an increase or a decrease of its population. Such management activities may have significant unintended influence on the most carefully designed ecological IPM systems.

5.2 Pest population dynamics

One approach to improve our understanding of the complex interaction between a pest population and its effective environment is to use relatively simple population models. Even highly simplified population models can provide a variety of outputs illustrating general ecological principles. As an introductory example, we can denote population density with a single value N and (initially) ignore the obvious fact that members of a population vary regarding a wide variety of biological traits (see section "Population structure and life tables" below). Here I use the single term N for convenience to denote population density, although it actually represents a range of individuals assumed identical only for study and preliminary analysis. To increase realism, Ehrlich *et al.*, 1975 suggested that all populations have the following general characteristics. (1) Populations and their effective environments are always changing in time and space, and a model of a population at a single location and time interval likely does not adequately represent events even in the same population at another place and time. The concept of the "metapopulation" (discussed below) addresses this issue. (2) Practical realities may dictate that we need to manage localized populations (e.g. within a single infested field) but management plans should be developed after an understanding of the pest's population dynamics on a landscape level or (ideally) over its entire geographic distribution. (3) Variation in demographic characteristics (birth rate, emigration, etc.) within a local population may exceed variation in the same parameters among adjacent or distant populations of the same species. (4) Immigration does not always increase gene flow and changes in gene frequency do not necessarily follow immigration. For instance, corn earworm (*Helicoverpa zea*) moving from south to north in the USA may or may not carry genes for insecticide resistance resulting from intensive insecticide use on crops in which they originated.

The Lotka–Volterra model of logistic population growth (Lotka, 1920) remains a useful mathematical model to illustrate the role of equilibrium in population dynamics. This classical model

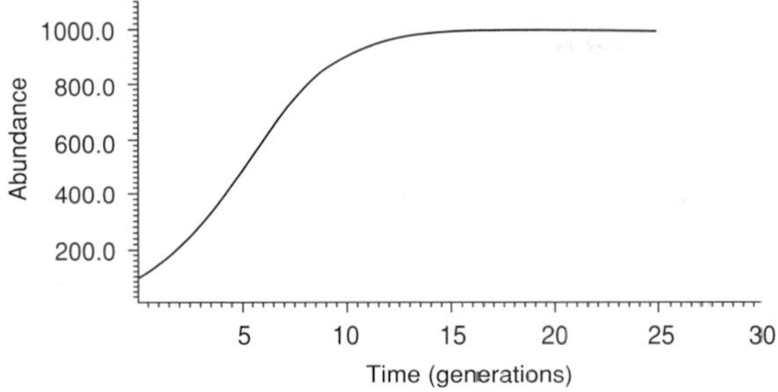

Fig. 5.1 Logistic population growth for a population with $r = 1.5$, $K = 1000$ and initial population $= 100$. Simulation using software from Akçakaya *et al.* (1999).

presupposes that many populations tend to be regulated about an equilibrium set by their effective environment (the causes for equilibrium may not be clear in many cases, and the model may be misleading in a practical sense, as many populations only *appear* to be in equilibrium). In the Lotka–Volterra model, K represents the environmental carrying capacity and acts to regulate population growth according to the following relationship:

$$N_{t+1} - N_t = N_t(b - d)(K - N_t)/K \qquad (5.1)$$

where N = population density
t = time interval
b = birth rate
d = death rate
K = carrying capacity.

This equation gives the familiar and intuitively satisfying sigmoid curve of population growth (Fig. 5.1). (The above is a difference equation; in discussion of ecological models often seen in textbooks, the model may be in differential form, integrated to:

$$N = K/(1 - e^{-rt}) \qquad (5.2)$$

where $r = (b - d)$ and e = base of natural logarithms.)

The difference and differential forms of the model describe essentially the same phenomenon, but the difference equation form is capable of a large array of outputs due to impact of time delays (Horn, 1988). Figure 5.2 illustrates this. From an IPM perspective, if b is large relative to d (characterizing a population with a high intrinsic rate of increase), there is a tendency for great oscillations about K, resulting in apparent instability even though there is equilibrium in the mathematical sense. A population with high r may approach K with such speed that at the next iteration it exceeds K and the population then declines, precipitously if $N > K$

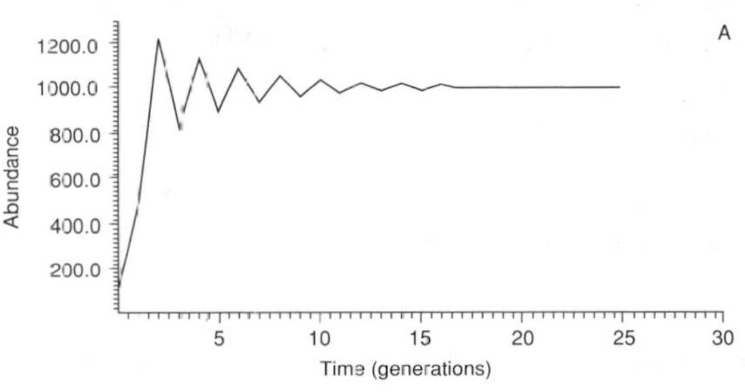

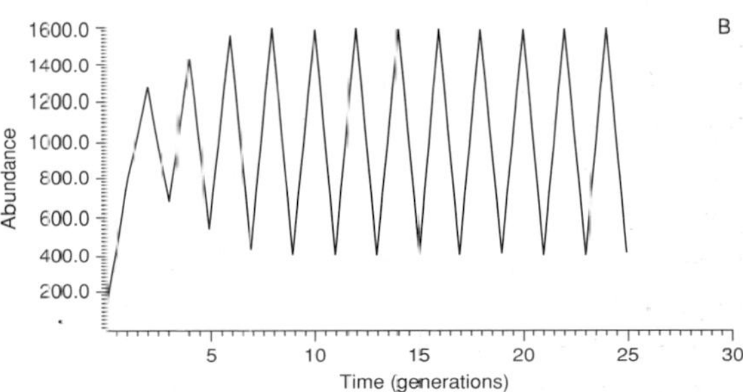

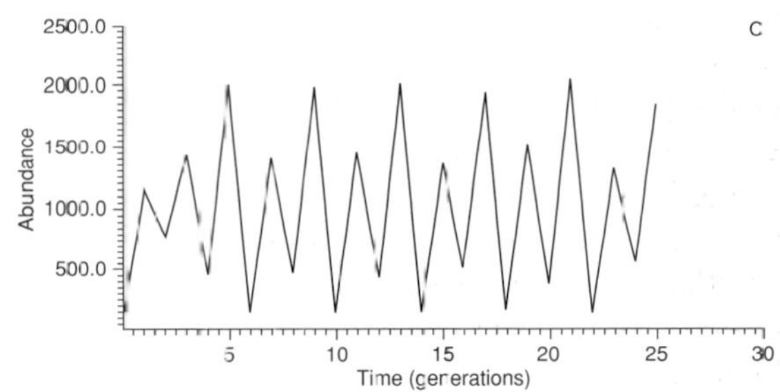

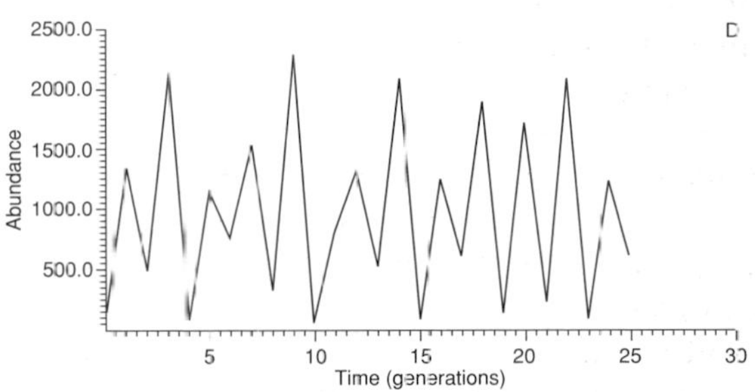

Fig. 5.2 Results of simulations for populations with $K = 1000$ and initial population $= 100$ (as in Fig. 5.1) and progressively increasing growth rate. (A) $r = 6$; (B) $r = 10$; (C) $r = 15$; (D) $r = 18$. Simulation using software from Akçakaya *et al.* (1999).

(Fig. 5.2B). As one increases *b* while holding *d* and *K* constant in computer simulations of this model, one obtains an array of results: low-amplitude cyclic oscillations about *K*, high-amplitude stable limit cycles (Fig. 5.2C) and/or cycles whose periodicity cannot be distinguished from random (Fig. 5.2D). From a modeling standpoint, these results are simple functions of the ratio of *b* to *d* and/or the relationship of the initial *N* to *K*. Such outputs do reflect actual results of studies on local populations of arthropods with high fecundity and short generation time, such as aphids, spider mites and houseflies. The model thus predicts that local insect populations with high fecundity and short generation time may fluctuate wildly and unpredictably, never appearing to be in equilibrium locally but rather exhibiting spectacular instability. Such populations also reach economically damaging levels much more rapidly than do those with lower *r*. The array of adaptations (high fecundity, short generation time, low competitive ability and high dispersal) has been termed "*r*-selection" (MacArthur & Wilson, 1967), and results from selection in environments favoring maximum growth, such as temporarily available habitats that occur early in ecological succession. On a local level such *r*-selected species may quickly over exploit their resources and local extinction follows. The two-spotted spider mite *(Tetranychus urticae)* is an example; females lay a large number of eggs and the resulting nymphs can reach adulthood in six days. Spider mites can overwhelm and destroy an untended house plant in a few weeks.

In contrast to *r*-selection, "*K*-selection" is typical of species that occupy habitats with longer spatial and temporal stability. These species exhibit lower fecundity, longer generation times, low dispersal tendency and high competitive ability. Although not an agricultural pest, the tsetse flies *(Glossina* spp.) are examples of extreme *K*-selection; females give birth to one mature larva at a time after a long incubation period. Characteristics of both *r*- and *K*-selection may occur within a species and may show seasonal variation. Saltmarsh planthoppers (*Prokelisia* spp.) may be short-winged during the growing season. Dispersal is limited and high fecundity results from feeding on vigorous hosts. As their food supply dwindles later in the season, long-winged morphs develop forms with lower fecundity and highly dispersive, relatively *K*-selected (Denno, 1994).

Typically, agricultural pests exhibit adaptations consistent with *r*-selection. Annual agricultural crops are disrupted due to tilling and harvesting and ecological succession is frequently reset to its starting point annually. This seems to select in favor of phytophagous insects that can locate and exploit their food quickly and efficiently. Adaptations of colonizing species of plants and insects are consistent with *r*-selection, i.e. efficient dispersal and a tendency to increase numbers quickly after locating favorable habitat. Many crop plants (or their ancestors) are typical of the pioneering stages of plant succession, as are their associated insect pests. Conventional agriculture, especially when undergoing frequent tillage, invites early-successional pest species that are more likely to exhibit outbreaks simply due to their *r*-selected adaptations. Populations of such pests may not display equilibrium over the time interval that the crop is in production. There may not be enough time during the growing season for the population to increase to its carrying capacity. The model describes this situation with high *r*, i.e. birth rate greatly exceeds death rate (until harvest, when the insects all emigrate or die). Equilibrium around *K* might be more likely in longer-lasting systems such as forests and orchards. Also, pest population fluctuations in these more complex ecosystems are partly buffered by the complexity of interactions within food webs, and outbreak of any particular pest species is less likely (see below under "Population stability and species diversity").

The logistic model describes density-*dependent* population regulation, which by definition is the major way to *regulate* a population about equilibrium. The effect of a density-dependent regulating factor is a function of the numbers (*N*) within a population. At low density the impact is less while at high density the impact is extreme. Density-dependent factors include predation, parasitism and competition. Density-*independent* factors, by contrast, may *control* a population but by definition do not regulate. Weather, pesticides, tillage and harvesting are all density-independent factors. In the life systems of most pest populations, both density-dependent and density-independent factors have impacts and the relative importance of each differs, sometimes leading to an impression that one or the other is the dominant or even the exclusive influence in determining population density (Horn, 1968). Density-*dependent* models assume that equilibrium exists and that one among many factors is the one that *regulates* (Hunter, 1991). Others (e.g. Strong, 1986; Stiling, 1988) argue that density dependence may not be very important in determining numbers of most animal populations. Chesson (1981) suggested that density-dependent regulation might occur most often at extremes of abundance and that in most populations at medium ranges of density the influence of density-dependent regulating factors was undetectable. Many more recent studies of natural populations seem to support this view (Cappuccino & Price, 1995; Frank van Veen *et al.*, 2006).

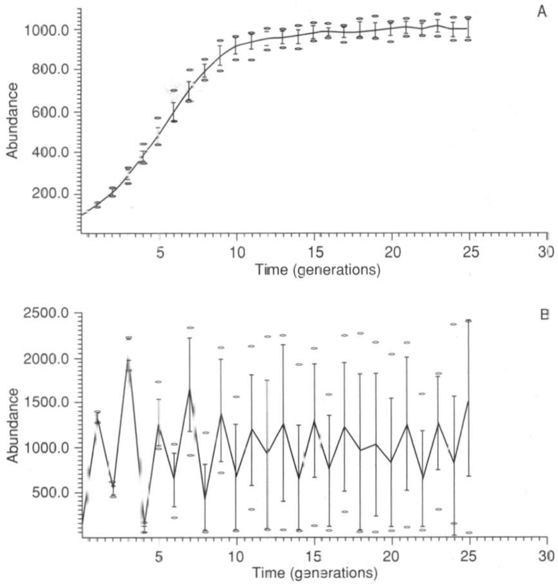

Fig. 5.3 Results of stochastic simulation of logistic growth model, ten iterations. Solid line is mean value; vertical lines represent standard deviation. (A) $r = 1.5$ (as in Fig. 5.1); (B) $r = 18$ (as in Fig. 5.2D). Simulation using software from Akçakaya *et al.*, (1999).

5.3 Stochastic models and metapopulations

The more realistic a model is in accurately describing real events, the more complicated it usually is. *Stochastic* models assign probability functions to birth rate, death rate, carrying capacity and the other components of the life system, so that each term in the logistic model above becomes a probability function that represents fluctuations about a mean (Fig. 5.3). Such probabilistic models are less mathematically tractable and results may be less intuitively understandable than are deterministic models, but such models do supply greater realism in describing actual population events and can be more useful in insect pest management. Use of computer simulations has removed a major obstacle to applying stochastic models in pest management, although experimental validation of such models is tedious (Pearl *et al.*, 1989).

To address the shortcomings of local equilibrium models in describing population events over an area of interest beyond a single crop to the regional and landscape levels, we may consider the demography of populations of the same species throughout a region by including dispersal as a factor involved in population regulation. The interacting system of local populations over a region is a *metapopulation* (Gilpin & Hanski, 1991). The metapopulation occupies both favorable and unfavorable places. Where the environment is favorable (a "source" area), the population is usually increasing ($b > d$) and the excess disperses to other regions including "sink" areas, where $b < d$. The local population in a sink would disappear were it not supplemented by immigration.

The relationship among sources, sinks and dispersal maintains the population in a more or less steady state. Movement among sources and sinks can create an impression that the metapopulation is in equilibrium throughout a wide area even though there is no local equilibrium evident in any one area (Murdoch, 1994). Usually our local area occupies our greatest interest when we apply IPM practically. The ideal goal of IPM is to reduce pest numbers more or less permanently below the economic injury level (EIL). This implies that equilibrium exists. The classic EIL model (Stern *et al.*, 1959) depends on models of population dynamics that presuppose equilibrium. These models are

Table 5.1 Hypothetical life table for an insect population

Stage	Survivorship	Mortality	Mortality rate
Eggs	1000	0	0.00
Larvae	1000	700	0.70
Pupae	300	120	0.40
Reproducing adult females	120	24	0.20
Old adults	24	24	1.00
Average number of eggs produced per female = 50			

understandable, tractable and intuitively satisfying, but in real populations there may not be a general equilibrium position for density of many and perhaps most pest species.

5.4 Population structure and life tables

As noted above, individuals in a population vary so the use of a single term N to denote a population is an oversimplification that can be misleading. One tool to illustrate the importance of such issues as age structure and reproductive ability is through the use of life tables. A life table is a summary of the vital statistics of a population. Life tables can be highly detailed and very useful analytical tools but to construct one involves a huge amount of sampling. Table 5.1 presents an oversimplified, hypothetical life table for illustration. The life table contains a census of the numbers in each age class (here, instars), along with survivorship, mortality and mortality rate within each age class. Having a life table allows us to conduct simulations to predict the impact of various mortality factors. This is illustrated in Fig. 5.4. Clearly mortality among adults has a greater impact on population persistence in these examples. Obviously a population initially consisting entirely of reproducing females increases at a much higher rate than does a population dominated by non-reproductive individuals. To most quickly bring a population under control,

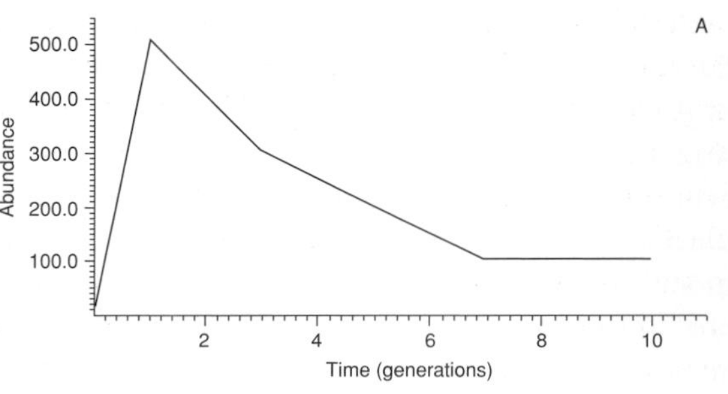

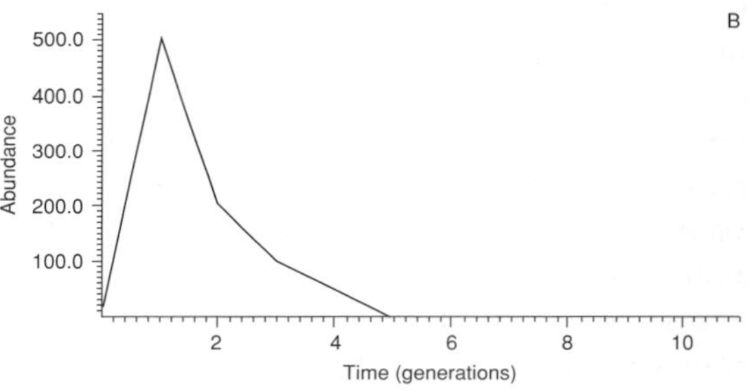

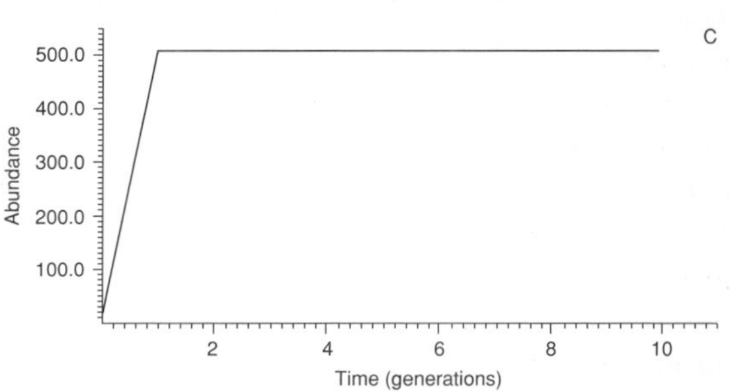

Fig. 5.4 Simulations based on hypothetical life tables, initial population of 10 reproducing females. (A) Mortality rates as in Table 5.1; population numbers level at 100; (B) as in Table 5.1 but larval mortality halved and adult mortality doubled; population extirpated after 5 generations; (C) as in Table 5.1 but larval mortality doubled and adult mortality eliminated; population numbers level at 500. Simulation using software from Akçakaya *et al.* (1999).

pre-reproductive females should be targeted to increase mortality.

5.5 Population stability and species diversity

As noted, the effective environment includes competitors, predators, parasites, pathogens – all ecosystem components that have an impact on a particular species. Food webs for even very simple habitats exhibit a large array of species. The invertebrate fauna associated with cabbage in Minnesota includes at least 11 leaf feeders, 10 sap feeders, 4 root feeders, 21 feeders on decaying plant matter and 79 saccharophiles (feeding on sugar produced by the plant or by sap-feeding Hemiptera) for a total of 125 species of primary consumers (herbivores in the most inclusive sense) (Weires & Chiang, 1973). Additionally Weires and Chiang documented 85 species of arthropod predators and parasitoids. This local diversity is not unusual; I have recorded well over 1200 species of arthropods in my modest urban house and yard. At the crop ecosystem or landscape level the diversity is bewildering and again we may resort to simplified models to improve understanding.

Species diversity is a function of the numbers and proportion of each species present in an ecosystem. It is assumed that species diversity in turn reflects the actual number of interconnections in a food web. *Biodiversity* has become a popular term in discussions about complexity in agricultural and other human-dominated ecosystems (Stinner *et al.*, 1997). Altieri & Nicholls (1999) defined biodiversity as "all species of plants, animals and microorganisms existing and interacting within an ecosystem." In agroecosystems this includes phytophagous species (pests and non-pests), predators, parasitoids, parasites, pollinators and decomposers. Debate continues among theoretical and applied ecologists as to the nature of any relationship between species diversity and stability of individual populations within an ecosystem. It is often assumed that pest outbreaks are generally suppressed in more complex (and therefore more diverse) ecosystems. This so-called "diversity–stability hypothesis" holds that communities with higher species diversity (greater biodiversity) are more stable because outbreaks of any particular species are ameliorated by the checks and balances and alternative food web pathways that exist within a large and integrated ecosystem. Andow (1991) assembled an exhaustive list of studies of agricultural systems addressing the diversity–stability hypothesis and he found that herbivores were less abundant in diverse plantings in 52% of cases. Most of these studies intermixed other plant species with the primary host of a specialist herbivore leading to reduced populations of the specialist herbivore (Risch *et al.*, 1983; Altieri, 1994). Root (1973) formulated this as the *resource concentration hypothesis*:

> herbivores are more likely to find and remain on hosts that are growing in dense or nearly pure stands; the most specialized species frequently attain higher densities in simple environments. As a result, biomass tends to become concentrated in a few species, causing a decrease in the diversity of herbivores in pure stands.

Observed increases in herbivore populations in crop monocultures seem to result from higher rates of colonization and reproduction along with reduced emigration, predation and parasitism. Other studies (e.g. Tilman *et al.*, 1996) have shown experimentally that soil nutrients are more completely cycled and productivity increases in more diverse ecosystems. Structural diversity (Andow & Prokrym, 1990) is important too; cropping systems with taller plants (such as maize among cucurbits and beans) provide more physical space to arthropods and this seems to increase diversity at all trophic levels (Altieri, 1994).

A positive relationship between diversity and stability is intuitively satisfying yet difficult to prove experimentally. Altieri & Nicholls (1999) believed that as biodiversity increased within an agroecosystem, more internal links develop within food webs and these links result in greater population stability and fewer pest outbreaks. This presupposes that most of the interconnecting trophic web is comprised of density-dependent links, which may or may not be the case. The detailed trophic structure of agricultural systems has only recently begun to be analyzed at sufficient detail to adequately address this question (e.g. Janssen & Sabelis, 2004; Jackson *et al.*, 2007). Southwood & Way (1970) suggested that stability of insect populations in agroecosystems depended on the "precision" of density-dependent responses within the food web, and the precision in turn depends on four major ecosystem parameters: (1) surrounding and within-crop vegetation, (2) permanence of the crop in space and time, (3) intensity of management including frequency of disruptions like pesticide applications and tillage and (4) degree of isolation of the agroecosystem from surrounding unmanaged vegetation.

The overall stability of any ecosystem is certainly a function of the sum of interactions among plant, pests, natural enemies and pathogens, but the specific details are often elusive.

Vandermeer (1995) proposed that biodiversity in agroecosystems has two components, planned and associated. *Planned* biodiversity consists of cultivated crops, livestock and other organisms (such as biological control agents) that are purposely introduced into an agroecosystem for direct economic benefit. Planned biodiversity is normally intensively managed, primarily to produce high yields. *Associated* biodiversity includes all the other plants, herbivores, carnivores and microbes that either pre-exist in or immigrate into the agroecosystem from the surrounding landscapes. Whether associated biodiversity persists within an agroecosystem depends on whether the ecological requirements of each organism are met over the time that the agroecosystem exists. Vandermeer (1995) proposed that a high amount of associated biodiversity is essential to maintaining stability of arthropod populations that otherwise might negatively impact planned biodiversity. Consideration of the relationship between planned and associated biodiversity is helpful in developing pest management practices that enhance overall biodiversity. This can lead to greater sustainability due to increased impact of biological control agents, reduced soil loss and enhanced on-site nutrient cycling.

5.6 Open and closed ecosystems

Open (or subsidized) ecosystems depend on importation of nutrients and energy from outside, whereas *closed* ecosystems are relatively self-contained with respect to nutrients and energy. From open ecosystems there is periodic removal of a large proportion of nutrients. A field of maize under conventional tillage is an example, with heavy importation of fertilizer at planting and subsequent energy inputs associated with tilling, pesticide application and so forth (over 20 million tonnes of chemical fertilizer are used annually in the USA, much of it on maize). Yield and crop residue comprise most of the nutrients in a maize field and these are removed at harvest. In addition, the species comprising the food web in maize are largely artificial and recent; interspecific associations are not long-standing and little coevolution has occurred. Maize is native to Mesoamerica whereas many of its major insect pests, e.g. European corn borer (*Ostrinia nubilalis*) are from other regions of the world. It takes a long time for native natural enemies to expand their host or prey range to include exotic organisms.

In a closed ecosystem, such as a deciduous forest, much nutrient cycling occurs on-site with far less importation from elsewhere. The nutrients and energy in the canopy and understory fall to the ground as leaves, insect frass or dead insects, or are converted into arthropods which are eaten by larger predators. Much of the flora and fauna does not leave the ecosystem. Migratory species do leave, of course, but often they return. Overall, there is rather little movement of nutrients into and out of the system. Many closed ecosystems (like the deciduous forest) have existed as assemblages of species for millennia, and the resulting food webs contain many coevolved trophic links and close ecological associations among mostly native species. Such ecosystems can be somewhat resistant to invasion by exotic species. Of course, there are exceptions, as the gypsy moth (*Lymantria dispar*) and emerald ash borer (*Agrilus planipennis*) have demonstrated in forests of the USA.

Most agroecosystems are open ecosystems and artificial assemblages, and because our main objective is to extract a usable product at a profit, it is often economical to maintain these systems in their present state despite a potential increase in pest activity (Lowrance *et al.*, 1984). Even in an artificial landscaped ecosystem that we do not harvest, we fertilize (providing nutrient input) and remove fallen leaves and lawn waste, exporting productivity in order to maintain a pleasing appearance. The species assemblage in planned landscapes often includes a preponderance of exotic species; for instance, from my home in Ohio I can view Chinese ginkgo, Colorado blue spruce, English walnut, Norway maple and Siberian elm trees.

"Open" and "closed" are arbitrary designations and the two types of ecosystems are points on a continuum. There are distinct differences between the two in the level of impact of pests and of management procedures (Altieri, 1987 1994). Especially in commercial agriculture, open and simplified agroecosystems are often devoted to a single crop resulting in higher pest populations and reduced species diversity. Generally the greater the modification and perturbation in the direction of ecosystem simplification and energy subsidy, the more abundant are insect pests. These reductions in biodiversity may extend beyond the local level, impacting the normal functioning of surrounding ecosystems with potential negative consequences for successful IPM (Flint & Roberts 1988).

5.7 Scale and ecologically based pest management

As mentioned earlier, the scale of the area involved is an important consideration in assessing a pest problem and developing an ecological approach to its management. The perception of pest problems, estimation of economic injury levels and approach to pest management can vary greatly depending on scale. One can view an ecosystem at the level of an individual plant, a small research plot, a large crop field or an entire farm with its regional "agropastoral" (Altieri 1994) landscape, the local watershed and so forth. The wider the area under consideration, the more complex the interactions are likely to be (Wilkinson & Landis, 2005; Marshall *et al.*, 2006). At the metapopulation level the regional dynamics of a life system may be very different from the local dynamics, and as noted earlier a pest population may appear to be in equilibrium throughout a regional landscape, even when there is no discernible equilibrium at the local level of interest to a pest manager. Results from small plot research may not be applicable to a larger scale (Kemp *et al.*, 1990). Local movement of pests may be less important at smaller scales (e.g. the individual plant) but very influential in population dynamics at a regional landscape level especially if this includes surrounding unmanaged habitat. For instance, Mexican bean beetle (*Epilachna varivestis* overwinters in hedgerows and along field edges, so that soybean fields closest to these overwintering sites are likely to become infested earlier and subsequent Mexican bean beetle populations will be higher. Soybeans located near pole and bush beans (more favorable hosts for the Mexican bean beetle) are more likely to develop economically important infestations earlier (Stinner *et al.*, 1983). Natural enemies often move from unmanaged edge habitat and nearby abandoned fields and forests into adjacent farm fields and the nature of this movement may be very important to local suppression of pests (Wilkinson & Landis, 2005).

An increase in intensive agriculture often leads to reduction in the size and diversity of surrounding unmanaged communities with their rich store of associated biodiversity including natural enemies. Elimination of hedgerows at field edges in the USA is an example. Many studies have shown that there are increased numbers and activity of predators and parasitoids near field borders when there is sufficient natural habitat to provide cover and alternate prey and hosts, as well as nectar and pollen for adult parasitoids to eat. Unmanaged border areas significantly enhance biological control (van Emden, 1965; Marino & Landis, 1996).

5.8 Conclusions

All ecosystems are complex, even apparently "simplified" monocultures. While simple intuitive equilibrial models may help us to understand ecological processes, the actual impact of alternate crops, weeds, competitors, natural enemies and other associated organisms on the life systems of pest species may be inconsistent and highly variable. The challenge is to develop ecological approaches to pest management in the face of this complexity, and to proceed with an appreciation of ecological processes despite gaps in our understanding.

References

Akçakaya, H. R., Burgman, M. A. & Ginzburg, L. R. (1999) *Applied Population Ecology Principles and Computer Exercises*. Sunderland, MA: Sinauer Associates.

Altieri, M. A. (1987). *Agroecology: The Scientific Basis of Alternative Agriculture*. Boulder, CO: Westview Press.

Altieri, M. A. (1994). *Biodiversity and Pest Management in Agroecosystems*. New York: Food Products Press.

Altieri, M. A. & Nicholls, C. I. (1999). Biodiversity, ecosystem function, and insect pest management in agricultural systems. In *Biodiversity in Agroecosystems*, eds. W. W. Collins & C. O. Qualset, pp. 69–84. Boca Raton, FL: CRC Press.

Andow, D. A. (1991). Vegetational diversity and arthropod population response. *Annual Review of Entomology*, **36**, 561–586.

Andow, D. A. & Prokrym, D. R. (1990). Plant structural complexity and host finding by a parasitoid. *Oecologia*, **62**, 162–165.

Cappuccino, N. & Price, P. W. (eds.) (1995). *Population Dynamics: New Approaches and Synthesis*. New York: Academic Press.

Carson, W. P., Cronin, J. P. & Long, Z. T. (2004) A general rule for predicting when insects will have strong top down effects on plant communities: on the relation between insect outbreaks and host concentration. In *Insects and Ecosystem Function*, eds. W. W. Weisser & E. J. Siemann, pp. 193–211. Berlin, Germany: Springer-Verlag.

Chesson, P. L. (1981). Models for spatially distributed populations: the effect of within-patch variability. *Theoretical Population Biology*, **19**, 288–325.

Clark, L. R., Hughes, R. D., Geier, P. W. & Morris, R. F. (1967). *The Ecology of Insect Populations in Theory and Practice*. London: Methuen.

Collins, W. W. & Qualset, C. O. (eds.) (1999). *Biodiversity in Agroecosystems*. Boca Raton, FL: CRC Press.

Denno, R. F. (1994). Life history variation in planthoppers. In *Planthoppers, Their Ecology and Management*, eds. R. F. Denno & T. J. Perfect, pp. 163–215. New York: Chapman & Hall.

Duelli, P. (1997). Biodiversity evaluation in agricultural landscapes: an approach at two different scales. *Agriculture Ecosystems and Environment*, **62**, 81–91.

Ehrlich, P. R., White, R. R., Singer, M. C., McKechnie, S. W. & Gilbert, L. I. (1975). Checkerspot butterflies: a historical perspective. *Science*, **188**, 221–228.

Flint, M. L. & Roberts, P. A. (1988). Using crop diversity to manage pest problems: some California examples. *American Journal of Alternative Agriculture*, **3**, 164–167.

Frank van Veen, F. J., Morris, R. J. & Godfray, H. C. J. (2006). Apparent competition, quantitative food web shifts and the structure of phytophagous insect communities. *Annual Review of Entomology*, **51**, 187–208.

Gilpin, M. & Hanski, I. (1991). *Metapopulation Dynamics: Empirical and Theoretical Investigations*. San Diego, CA: Academic Press.

Horn, H. S. (1968). Regulation of animal numbers: a model counter-example. *Ecology*, **49**, 776–778.

Horn, D. J. (1988). *Ecological Approach to Pest Management*. New York: Guilford Press.

Hunter, A. F. (1991). Traits that distinguish outbreaking and non-outbreaking Macrolepidoptera feeding on northern hardwood trees. *Oikos*, **60**, 275–282.

Jackson, L. E., Pascual, U. & Hodgkin, T. (2007), Utilizing and conserving agrobiodiversity in agricultural landscapes. *Agriculture Ecosystems and Environment*, **121**, 196–210.

Janssen, A. & Sabelis, M. W. (2004). Food web interactions and ecosystem processes. In *Insects and Ecosystem Function*, eds. W. W. Weisser & E. J. Siemann, pp. 193–211. Berlin, Germany: Springer-Verlag.

Kemp, W. P., Harvey, S. J. & O'Neill, K. M. (1990). Patterns of vegetation and grasshopper community composition. *Oecologia*, **83**, 299–308.

Kogan, M. (1995). IPM: historical perspectives and contemporary development. *Annual Review of Entomology*, **43**, 243–270.

Landis, D. A., Wratten, D. & Gurr, G. M. (2000). Habitat management to conserve natural enemies of arthropod pests in agriculture. *Annual Review of Entomology*, **45**, 175–201.

Lotka, A. J. (1920). Analytical notes on certain rhythmic relations in organic systems. *Proceedings of the National Academy of Sciences of the USA*, **7**, 410–415.

Lowrance, R., Stinner, B. R. & House, G. J. (eds.) (1984). *Agricultural Ecosystems: Unifying Concepts*. New York: John Wiley.

MacArthur, R. H. & Wilson, E. O. (1967). *The Theory of Island Biogeography*. Princeton, NJ: Princeton University Press.

Marino, P. C. & Landis, D. A. (1996). Effects of landscape structure on parasitoid diversity in agroecosystems. *Ecological Applications*, **6**, 276–284.

Marshall, E. J. P., West, T. M. & Kleijn, D. (2006). Impacts of an agri-environment field margin prescription on the flora and fauna of arable farmland in different landscapes. *Agriculture Ecosystems and Environment*, **113**, 36–44.

Murdoch, W. W. (1994). Population regulation in theory and practice. *Ecology*, **75**, 271–287.

Pearl, D. K., Bartoszynski, R. & Horn, D. J. (1989). A stochastic model for simulation of interactions between phytophagous spider mites and their phytoseiid predators. *Experimental and Applied Acarology*, **7**, 143–151.

Risch, S. J., Andow, D. & Altieri, M. A. (1983). Agroecosystem diversity and pest control: data, tentative conclusions and new research directions. *Environmental Entomology*, **12**, 625–629.

Root, R. B. (1973). Organization of a plant–arthropod association in simple and diverse habitats: the fauna of collards (*Brassica oleracea*). *Ecological Monographs*, **43**, 95–124.

Southwood, T. R. E. & Way, M. J. (1970). Ecological background to pest management. In *Concepts of Pest Management*, eds. R. L. Rabb & F. E. Guthrie, pp. 6–29. Raleigh, NC: North Carolina State University Press.

Stern, V. M., Smith, R. F., van den Bosch R. & Hagen, K. S. (1959). The integration of chemical and biological control of the spotted alfalfa aphid: the integrated control concept. *Hilgardia*, **29**, 81–101.

Stiling, P. (1988). Density-dependent processes and key factors in insect populations. *Journal of Animal Ecology*, **57**, 581–594.

Stinner, R. E., Bayfield, C. S., Stomach, J. L. & Dose, L. (1983). Dispersal and movement of insect pests. *Annual Review of Entomology*, **28**, 319–335.

Stinner, D. H., Stinner, B. R. & Martsolf, E. (1997). Biodiversity as an organizing principle in agroecosystem management: case studies of holistic resource management practitioners in the USA. *Agriculture Ecosystems and Environment*, **62**, 199–213.

Strong, D. R. (1986) Density-vague population change. *Trends in Ecology and Evolution*, **1**, 39–42.

Tilman, D., Walden, D. & Knops, J. (1996). Productivity and sustainability influenced by biodiversity in grassland ecosystems. *Nature*, **379**, 718–720.

Vandermeer J. (1995). The ecological basis of alternative agriculture. *Annual Review of Ecology and Systematics*, **26**, 201–224.

van Emden, H. F. (1965). The role of uncultivated land in the biology of crop pests and beneficial insects. *Scientific Horticulture*, **17**, 121–126.

Weires, R. W. & Chiang, H. C. (1973). *Integrated Control Prospects of Major Cabbage Insect Pests in Minnesota: Based on the Faunistic, Host Varietal and Trophic Relationships*, Technical Bulletin No. 291. St. Paul, MN: University of Minnesota, Agricultural Experiment Station.

Wilkinson, T. K. & Landis, D. A. (2005). The role of plant resources and habitat diversification in biological control. In *Plant Provided Food and Plant–Carnivore Mutualism*, eds. F. L. Wackers, P. C. J. Van Rijn & J. Bruin, pp. 305–325. Cambridge, UK: Cambridge University Press.

Chapter 6

Population dynamics and species interactions

William E. Snyder and Anthony R. Ives

Agricultural monocultures are often thought to be more prone to herbivore outbreaks than natural systems, and early agroecologists posited that the lack of biodiversity in agricultural systems contributes to their instability (Pimentel, 1961; van Emden & Williams, 1974). In contrast, some detailed reviews have concluded that perhaps one or two particularly effective natural enemies are all that is needed for effective pest control (Hawkins *et al.*, 1999). Such issues come to the fore when a decision must be made in classical biological control about whether to introduce one or several natural enemy species in an effort to control exotic pests (Myers *et al.*, 1989; Denoth *et al.*, 2002), and when designing schemes to conserve indigenous natural enemies by modifying cultural practices (Landis *et al.*, 2000; Tscharntke *et al.*, 2005). Here, we first review the major classes of natural enemies – specialists and generalists – and the traits of each that are likely to contribute to (or detract from) their effectiveness as biological control agents. We then discuss interactions within diverse communities of natural enemies that are likely to affect biological control.

6.1 Specialist natural enemies: the best biological control agents?

Biological control practitioners have long debated the question: what are the traits of an effective biological control agent? General consensus seems to focus around a few traits that a successful agent will possess (see Chapter 9). For example, Debach & Rosen (1991) describe three key attributes of an effective natural enemy that we condense to:

(1) a high degree of prey specificity;
(2) a reproductive rate as high as that of the target prey (the pest); and
(3) a tolerance of environmental conditions similar to that of the pest.

The reason for the third of these attributes is obvious, as a biological control agent must be able to survive in the same environment where the target pest occurs. However, the logic underlying the first two attributes is more complex and subtle.

Integrated Pest Management, ed. Edward B. Radcliffe, William D. Hutchison and Rafael E. Cancelado. Published by Cambridge University Press.

The first attribute describes a biological control agent that is a *specialist* attacking only one (or perhaps just a few) prey species. The second attribute describes a natural enemy that has a reproductive rate as rapid as its prey, so that the natural enemy is able to reproduce at a pace similar to that of the pest. These traits together can allow a natural enemy to exert *density-dependent population regulation*, in which mortality from the specialist increases as pest populations grow through time, thereby squelching pest outbreaks (Hassell, 1978; Turchin, 2003). Below, we give a more detailed discussion of the disadvantages and advantages of specialist natural enemies. We focus on parasitoids, which in the literature on the biological control of insect pests are almost synonymous with specialist natural enemies. We will describe the disadvantages and advantages of specialist biological control agents using two well-studied systems to carry the discussion: pea aphid (*Acyrthosiphon pisum*) in alfalfa and red scale (*Aonidiella aurantii*) in citrus crops.

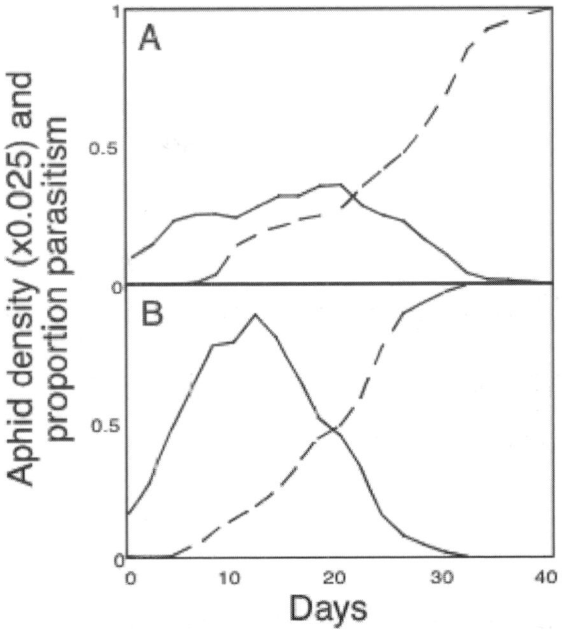

Fig. 6.1 Pea aphid density (solid lines) and parasitism by *A. ervi* (dashed lines) in greenhouse cages (A) without and (B) with bridges connecting 48 alfalfa plants. Pea aphid density is the average number of aphids per plant times 0.025 to fit within the same panels as proportion parasitism. Proportion parasitism is measured by the ratio $m/(m + a)$ where m is the number of new mummies observed between samples taken 2 days apart, and a is the number of pea aphids. Data from Ives (1995).

6.1.1 Boom-and-bust predator–prey cycles

Although many specialist natural enemies show density-dependent responses to increasing pest populations, this generally occurs with a time lag roughly equal to the generation time of the specialist parasitoids. This time lag occurs because it takes a generation for increasing pest abundance to give rise to new adult parasitoids. The time lag can cause boom-and-bust cycles in pest densities (Nicholson & Bailey, 1935; Huffaker, 1958; Hassell, 1978); as pest populations grow, they can reach high densities before the parasitoid population becomes large enough to suppress the pest, but once suppression starts, it continues until the pest population is reduced to very low abundances.

The tendency of specialist parasitoids to cause boom-and-bust cycles with their prey is illustrated by a simple experiment with pea aphid feeding on alfalfa (Ives, 1995). Pea aphids are attacked by the parasitoid *Aphidius ervi*, a solitary braconid wasp that was introduced into North America as a biological control agent of pea aphid (Mackauer & Kambhampati, 1986). In greenhouse cages, we introduced pea aphids and then *A. ervi* adults into each of two cages containing 48 alfalfa plants (Fig. 6.1). The cages differed, with one cage having wire "bridges" between plants on which non-winged aphids could cross, while the other cage had barriers between plants that limited between-plant aphid movement to winged aphid adults. The bridges allowed aphids to disperse more evenly among plants, which increased the rate of aphid population growth. In both cages, pea aphid populations continued to increase for about 15 days following parasitoid introduction, which is roughly one parasitoid generation. After the aphid population peaked it declined rapidly, with the aphid population crashing as parasitism reached 100%. Although bridges among plants increased the population growth rate of the aphids, this did not prolong the time until their population crashed; in fact, it sped the rate of collapse.

Aphidius ervi was clearly a very effective natural enemy, able to kill all pea aphids in the cages very quickly regardless of the initial growth rate of the aphid population. This control was density dependent, because percent parasitism increased as pea aphid density increased through time. The success of *A. ervi* can be attributed to the first two attributes for an effective biological control agent listed above: high specificity and rapid population growth. However, although *A. ervi* did exterminate pea aphids from the cages, this did not occur before aphids reached high densities, increasing for 12–20 days following the introduction of parasitoids. Furthermore, the extinction of pea aphids from cages naturally caused the extinction of parasitoids. In the field, this would then make the crops susceptible to subsequent pest invasions. Thus, high levels of parasitism came at the expense of stability, with the specialist parasitoid generating boom-and-bust population cycles (Hassell & May, 1973).

6.1.2 Stable pest control by specialists

Theoretical models have identified several mechanisms that allow highly specialized parasitoids with high reproductive rates to nonetheless provide stable control of host (pest) populations, without boom-and-bust cycling (Briggs & Hoopes, 2004). A well-studied mechanism is parasitoid spatial aggregation (Hassell & May, 1974; Hassell *et al.*, 1991; Ives, 1992b). If pests are distributed in a patchy fashion, for example in colonies on individual plants, then parasitoids must search among patches for pests. If parasitoid searching causes some hosts to be more or less susceptible to parasitism, then host–parasitoid dynamics can be stabilized. This occurs when, for example, parasitoids aggregate in patches with high host density (Hassell & May, 1974). It also occurs when parasitoids search independently of host density but nonetheless aggregate their searching in a subset of patches (Hassell *et al.*, 1991; Ives, 1992b). The reason this mechanism stops host–parasitoid cycles is that, by focusing parasitoid attacks on a subset of hosts, it increases competition among parasitoids as percent parasitism increases. This stops parasitoid populations from reaching very high densities that then lead to host population crashes (Hassell *et al.*, 1991). There is a cost to this stabilizing effect of parasitoid aggregation, however; due to parasitoid competition, the average host population size is relatively high when there is parasitoid aggregation, creating a trade-off between the effectiveness of host suppression and the ability of the parasitoid to maintain stable host population densities (Ives, 1992b). Despite a lot of interest in the ecological literature, we know of no clear empirical demonstrations of aggregation stabilizing host–parasitoid dynamics, and there are considerable logistical difficulties to testing this idea (Gross & Ives, 1999; Olson *et al.*, 2000).

Another mechanism enabling parasitoids to achieve low and stable pest population control without cycling is selective parasitism on only some of the developmental stages of hosts, leaving a long period of invulnerability during development (Murdoch *et al.*, 1992, 2003). Murdoch and colleagues (2005) have demonstrated the ability of the parasitoid *Aphytis melinus* to control California red scale experimentally in the field (Fig. 6.2A). They caged in entirety 14 citrus trees, and in four cages they experimentally augmented the abundance of red scale to 80 times the abundance in control cages. Within two red scale generations, experimentally augmented populations were brought back to control cage levels as parasitism rates reached 95% (compared to 66% in controls). Using a detailed, system-specific model that includes all developmental stages of red scale, Murdoch and colleagues (2005) demonstrated two key components for the stable suppression of red scale: the very rapid reproduction rate of *A. melinus* whose generation time is shorter than that of red scale, and the long invulnerable adult stage of female red scale. These two features reduce the time lag of the response of parasitoids to increasing host density, and reduce the potential for red scale population crash as parasitoid populations peak. The most remarkable feature of this experiment is that stable pest control is achieved at small spatial scales – single trees.

A final mechanism that can stabilize the inherent tendency for host–parasitoid dynamics to cycle is immigration of hosts and/or parasitoids when they are at low densities and emigration when they are at high densities (Reeve, 1988; Godfray & Pacala, 1992; Ives, 1992a; Murdoch & Briggs, 1997). When this is the case, peak densities of either host or parasitoid may be damped

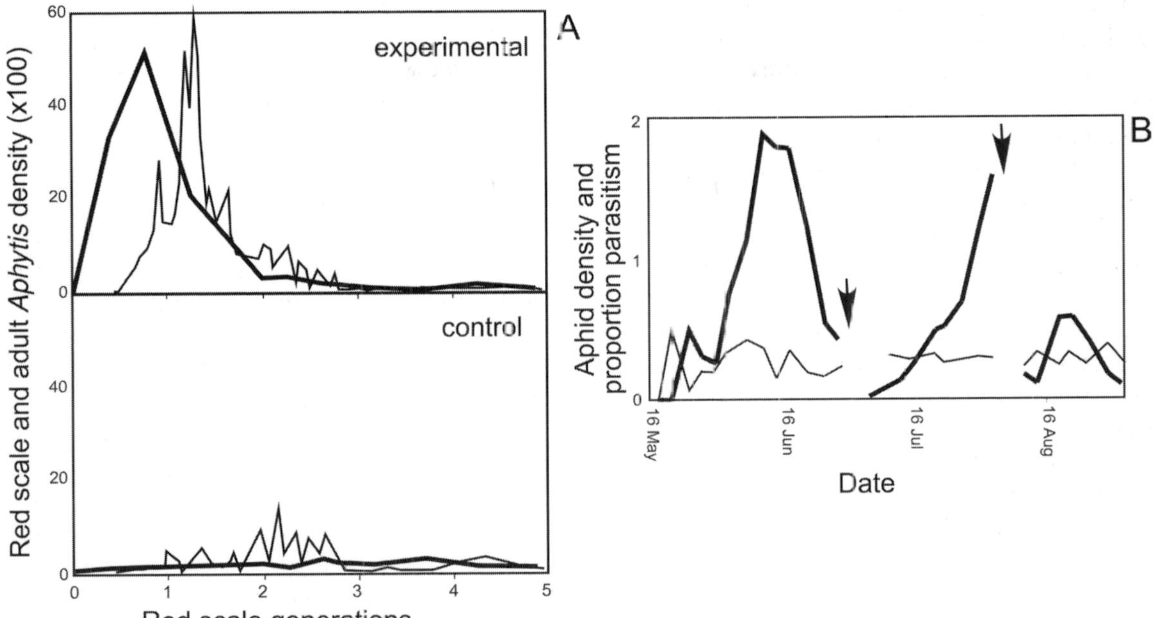

Fig. 6.2 (A) Densities of red scale (solid lines) and adult *Aphytis melinus* (dashed lines) averaged among four experimental trees (upper panel) on which red scale densities were experimentally increased by 80-fold, and ten control trees (lower panel). Data from Murdoch *et al.* (2005). (B) Pea aphid densities and percent parasitism in an alfalfa field over the course of a growing season. Pea aphid densities are given as numbers per stem, and percent parasitism was determined by dissection. Arrows mark alfalfa harvesting. Data from Gross *et al.* (2005).

and low population densities may be rescued. This appears to occur in the alfalfa–pea aphid–*A. ervi* system in the field (Gross *et al.*, 2005), in contrast to in greenhouse cages. Alfalfa fields are harvested two to four times per growing season in Wisconsin, causing 2–3 orders of magnitude reductions in pea aphid densities (Fig. 6.2B). Following harvest, pea aphid populations rebound quickly from both the few survivors and immigrant winged aphids (Rauwald & Ives, 2001). After a few pea aphid generations (a few weeks), aphid populations often plateau or even decline. The pea aphid dynamics can look remarkably similar to those observed in the greenhouse cages (Fig. 6.1), with increases and declines occurring over a 40-day time period (Fig. 6.2B, first harvesting cycle). Nonetheless, detailed statistical analyses of these and similar data show that the plateauing and sometimes decline of pea aphid densities may occur without increased parasitism (Gross *et al.*, 2005), as percent parasitism remains roughly constant over the course of a harvesting cycle (Fig. 6.2B). Lack of temporal changes in parasitism with changes in aphid density is likely due to the high movement rates of *A. ervi*, as females move readily among fields and do not appear to show a strong response to spatial variation in pea aphid densities among fields. When parasitism does not increase with increasing pea aphid density, *A. ervi* do not drive population cycles like those observed in greenhouse cages, nor do they provide density-dependent control of pea aphid populations within fields. Density-dependent control of pea aphid populations may involve a variety of generalist predators that immigrate into alfalfa fields as pea aphid densities increase (Cardinale *et al.*, 2003; Snyder & Ives, 2003). Even when *A. ervi* do not provide density-dependent control, however, the impact of parasitism may be high; in the example illustrated in Fig. 6.2B, on average, 20% of non-parasitized aphids were parasitized every day, and in the absence of parasitism, pea aphid population densities would likely increase to roughly ten times the observed densities over the course of a harvesting cycle (Gross *et al.*, 2005).

These examples show that, while effective biological control agents can be specialists with high reproduction rates, stable suppression of pests likely requires attributes in addition to the three proposed by Debach & Rosen (1991). In the red scale–*Aphytis melinus* system, stability occurred on a very small spatial scale (single trees) due to an invulnerable adult stage of red scale. In contrast, the pea aphid–*Aphidius ervi* system is very "open," with aphids and parasitoids moving readily among fields. While this prohibits *A. ervi* from exterminating pea aphid populations within fields, parasitism is nonetheless a major source of aphid mortality that, in combination with predation, maintains low densities of pea aphids that rarely exceed one aphid per alfalfa stem.

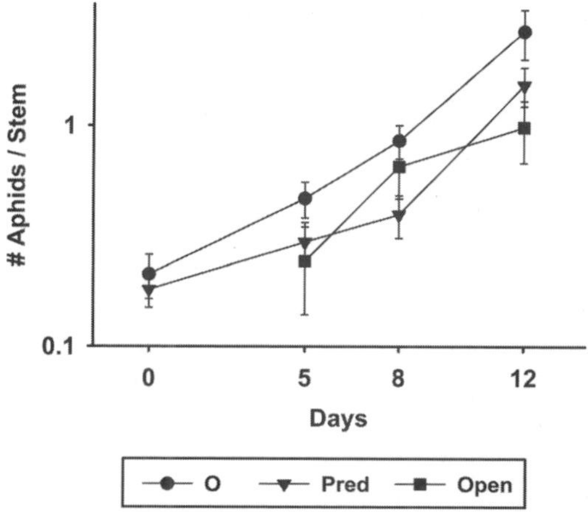

Fig. 6.3 Population dynamics of pea aphids on alfalfa, redrawn from data presented in Snyder & Ives (2003). Predators suppressed aphid densities (Pred) compared to controls lacking predators (O). However, aphid densities in both treatments continued to grow throughout the experiment. Uncaged reference plots (Open) followed aphid dynamics in the alfalfa field surrounding the cages.

6.2 Generalist predators: the good, the bad and the ugly

At the other end of the diet-breadth spectrum from specialists lie the *generalists*. Generalists feed on several or many prey species. A given pest species is often attacked by a small group of specialists, but a diverse community of generalists. The numerous natural enemies of pea aphids in Wisconsin provide an example (Snyder & Ives, 2003). The most common generalists include predatory damsel and minute-pirate bugs (*Nabis* spp. and *Orius* spp.) and ground and ladybird beetles (in the families Carabidae and Coccinellidae). The alfalfa system raises two points about generalist predators. First, while for simplicity we present a dichotomy between specialists and generalists, in reality many species blur these distinctions. Damsel bugs, for example, are broad generalists that feed upon any soft-bodied prey that they encounter and can subdue (Lattin, 1989). In contrast, the ladybird beetles common to this community, *Coccinella septempunctata* and *Harmonia axyridis*, feed primarily on aphids, albeit aphids of many different species, while also opportunistically feeding upon the eggs of other insects, pollen or other easily captured or vulnerable prey (Evans, 1991; Hodek & Honek, 1996; LaMana & Miller, 1996; Yasuda *et al.*, 2004). Second, in the alfalfa system all of the generalists are predators in that they feed on multiple prey items during their lifetime. Some parasitoid species can also be quite diverse in their feeding habits, and in some cases their impacts on and interactions with other species may be quite like those of generalist predators (e.g. Montoya *et al.*, 2003). However, predators that are generalists have been particularly well studied and thus we focus this section of our chapter on generalist predators.

Generalist predators feed on multiple prey species and often have relatively long generation times compared to their prey. Thus, these predators do not fit two of the key attributes Debach & Rosen (1991) propose for effective biocontrol agents. Generalized feeders that reproduce slowly will be unable to mount a density-dependent population response to increasing pest densities. We found this to be the case when examining the impact of the guild of generalist predators of pea aphids in alfalfa, in the absence of the specialist *A. ervi*. We established cages housing alfalfa plants infested with pea aphids, with or without the diverse community of generalist predators also present, and then followed pea aphid dynamics through time. Predators reduced pea aphid densities initially compared to no-predator controls, a reduction maintained through time (Fig. 6.3). However, the subsequent aphid population growth rate was little changed by the addition of predators, and aphid densities in both treatments continued to grow throughout the 12-day course of the experiment (Fig. 6.3). Thus, while predators caused considerable aphid mortality, they did not control aphid densities in a density-dependent fashion.

In addition to having low reproduction rates and hence slow population responses to pest outbreaks, generalists sometimes attack one another rather than pests. Ecologists call predation of one natural enemy by another *intraguild predation* (Polis *et al.*, 1989). When strong, intraguild predation can negate any benefits of generalist predators for biological control (Rosenheim *et al.*, 1995). Rosenheim *et al.* (1993) present a classic example of intraguild predation's disruptive effects. They examined control of cotton aphid (*Aphis gossypii*) on cotton by a community of generalist predators (Fig. 6.4). Key among these natural enemies were lacewing (*Chrysoperla carnea*) larvae, voracious predators of aphids. When alone, lacewings consumed far more aphids per capita than did any other predator species in the community. Unfortunately, the other predators were quite effective at capturing and eating lacewings, so that pairings with other predators generally led to intraguild predation of lacewings and weaker aphid control (Rosenheim *et al.*, 1993). In the field lacewings co-occur with many other natural enemies, so aphid biological control likely will always fall below the ideal level of suppression that lacewings alone might otherwise achieve.

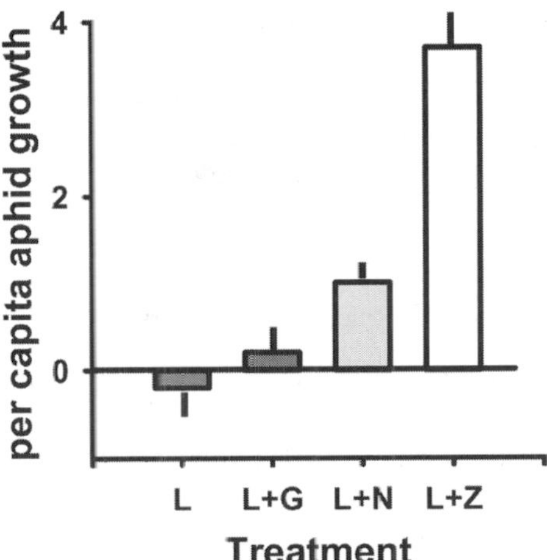

Fig. 6.4 Population growth rates for cotton aphids on cotton plants, redrawn from Rosenheim *et al.* (1993). Aphid densities declined when lacewing larvae (L) were alone. However, aphid populations grew when this single most effective aphid predator species was paired with *Geocoris* (L+G), *Nabis* (L+N) or *Zelus* (L+Z) bugs, other common predators in the cotton system and intraguild predators of lacewing larvae.

6.2.1 Effectiveness of biological control by generalists

Despite the two disadvantages described above – low population growth and intraguild predation – generalists may have several attributes that allow them to be effective biological control agents. Although the often low population growth rate of generalists relative to their prey prohibits generalists from mounting a strong density-dependent response to increasing pest abundance through reproduction, they may nonetheless show a behavioral response, immigrating into fields with high pest densities and remaining longer to consume their prey (Frazer & Gill, 1981; Evans & Youssef, 1992). This behavioral response of generalist predators apparently underlies the control of pea aphids in open field conditions (Fig. 6.2B). In contrast to the case in field cages, open fields allow the immigration of generalist predators (e.g. Östman & Ives, 2003) that in combination suppress aphid populations in a density-dependent manner.

Because generalists feed on several different prey species, their densities in a crop will not be entirely tied to the density of any particular pest (Harmon & Andow, 2004). For this reason generalists may already be present in the field, feeding on non-pest prey, when pests first invade (Settle *et al.*, 1996). Thus, generalists can serve as the "first line of defense" against pest invasion, while the relatively long life cycles of many generalist predators allow them to span peak densities of any single prey item (Symondson *et al.*, 2002). *Alternative prey* are those species present in the crop other than the target pest that generalist predators will also feed upon. Alternative prey that serve to provide an additional prey resource, or that provide key nutrients otherwise limiting predator growth, can bolster predator populations and thus improve biological control (Holt, 1977).

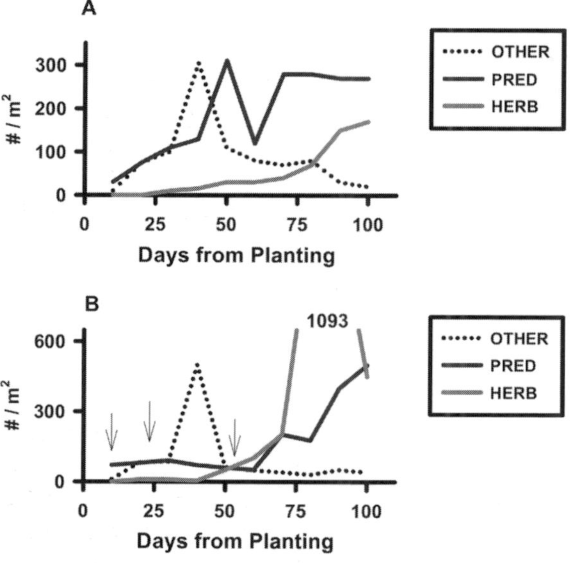

Fig. 6.5 (A) Rice plots that did not receive pesticide applications saw an early build-up of both alternative prey and generalist predators, such that when brown planthoppers arrived predator densities were sufficient to maintain low planthopper densities. (B) In contrast, when plots received three early-season pesticide applications (indicated with arrows), build-up of alternative prey was delayed, and because predators failed to reach high densities early in the season their densities were too low to control brown planthopper later in the season. Figure depicts densities of alternative prey (OTHER), generalist predators (PRED) and brown planthopper (HERB). Note broken line in panel B; planthopper densities peaked at 1093 per m^2 in this treatment. Data from Settle *et al.* (1996).

Settle *et al.* (1996) present a study demonstrating how the presence of alternative prey can build generalist predator densities to the benefit of pest suppression. These authors were working in rice crops in Indonesia, with brown planthopper (*Nilaparvata lugens*) as the target pest. A diverse group of generalist predators inhabits rice paddies, consisting primarily of spiders and predatory bugs. In the absence of early season insecticide application, rice crops were colonized by a diverse community of detritivores feeding on plant debris from the previous year's crop. These detritivores served as alternative prey for the generalist predator community, allowing generalist densities to grow dramatically (Fig. 6.5A). Detritivore densities fell later in the season, just as planthoppers began to invade the crop. The generalists then switched from attacking detritivores to attacking planthoppers, suppressing the pest below damaging levels (Fig. 6.5A). However, when insecticides were applied and detritivores were killed, generalist predator densities never built up, and planthoppers found a largely predator-free crop to colonize and reproduce in (Settle *et al.*, 1996) (Fig. 6.5B).

When predators are omnivores that include both animals and plants in their diets, non-animal foods also can have a beneficial effect on biological control. Eubanks & Denno (2000) examined biological control of pea aphids on lima beans by an omnivorous big-eyed bug (*Geocoris punctipes*). Big-eyed bugs feed primarily on animal prey, but will also feed on some high-quality plant parts such as bean pods. In the laboratory big-eyed bugs will reduce their feeding on aphids when in the presence of bean pods. Surprisingly though, open field plots of bean plants possessing pods exhibited aphid densities lower than those seen in plots lacking pods. Control improved in the presence of bean pods because they attracted and/or retained higher densities of big-eyed bugs, which then also fed upon aphids. Apparently, any distraction that plant feeding provided to individual predators was counteracted by the higher overall predator densities in these plots (Eubanks & Denno, 2000).

With generalists nothing is simple, and the presence of alternative prey can sometimes disrupt, rather than improve, biological control. Prasad & Snyder (2006) present an example of such disruption due to the presence of alternative prey. Here, cabbage root maggot (*Delia radicum*) were the focal pest, and a community of surface-foraging ground and rove beetles (in the families Carabidae and Staphylinidae) were the biological control agents. Densities of ground beetles increased twofold following the installation of grassy strips into cropping fields; the beetles use the grassy strips to overwinter and escape pesticide applications and tilling. However, root maggot control was not improved following this dramatic doubling in predator density. Subsequent field and laboratory experiments revealed that aphids commonly occurred in *Brassica* fields and served as alternative prey. The ground predators found aphids to be tastier prey than root maggots, and when given

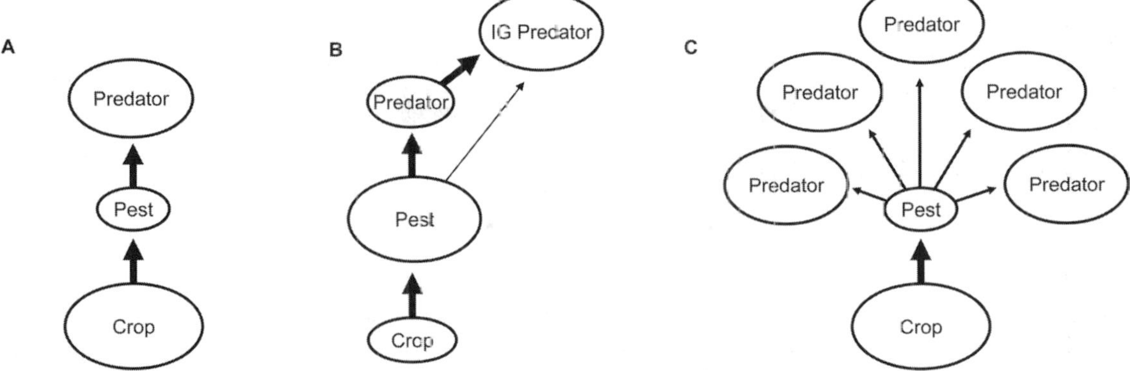

Fig. 6.6 The green world hypothesis envisions distinct trophic levels (A), with predators suppressing pests and in turn protecting the crop from herbivory through trophic cascades. In contrast, when intraguild predation is common (B), predators interfere with one another, disrupting the trophic cascade and helping free the pest from control. Agroecologists have proposed a third scenario (C), wherein multiple, complementary natural enemies work together to suppress pests. Figure depicts the crop (Crop), pest (Pest), primary predator(s) (Predator) and intraguild predator (IG Predator). Arrows depict energy flow and thus point from resource to consumer, and are scaled to reflect the strength of that interaction.

this choice beetles feasted on aphids and ignored root maggots (Prasad & Snyder, 2006).

6.3 Natural enemy biodiversity and biological control

As growers reduce their use of broad-spectrum insecticides, the abundance and diversity of natural enemies increases (Bengtsson *et al.*, 2005; Hole *et al.*, 2005). Presumably a greater abundance of natural enemies is helpful for biological control, but the importance of species diversity is less clear. Here we first discuss different ways that ecologists have envisioned the relationship between natural enemy diversity and the effectiveness of biological control. Next, we discuss the positive interactions among natural enemies that impact how pest control is influenced by natural enemy biodiversity.

Ecologists have advanced three hypotheses concerning the relationship between the number of predator species present and the success of biological control. Hairston *et al.* (1960) argued that predators can be grouped into a cohesive third trophic level, acting in concert to regulate herbivores. As predators kill herbivores they indirectly protect plants through so-called *trophic cascades* (Fig. 6.6A). This theory is sometimes referred to as the *green world hypothesis*, and it implies that the top-down impact of predators is consistently strong regardless of how many species are present (Hairston & Hairston, 1993, 1997). However, communities often include many species of generalist predators. Generalist predators usually feed on members of several trophic levels, and these catholic feeding habits might blur the distinction of discrete trophic levels, thus diffusing trophic cascades (Polis, 1991; Strong, 1992; Polis & Strong, 1996) (Fig. 6.6B). This viewpoint is known as the *trophic level omnivory hypothesis*, and the prediction here is that while trophic cascades may occur in some simple systems (e.g. small islands or agricultural monocultures), top-down effects will weaken as species diversity increases. Biological control theory provides a third view of the relationship between predator biodiversity and predator impact, predicting that diversifying agroecosystems, for example by intercropping, should lead to a diversified prey base and thus a more abundant and diverse predator community (Pimentel, 1961; van Emden & Williams, 1974) (Fig. 6.6C). With growing diversity predators are expected to provide more consistent prey suppression, what Root (1973) called the *enemies hypothesis*.

Thus, frustratingly, theory offers three conflicting viewpoints predicting unchanging, decreasing, or increasing strength of predator impact with increasing biodiversity. Worse still, recent experimental studies wherein predator diversity has been manipulated have uncovered stronger, weaker and unchanged pest suppression with greater predator biodiversity (Cardinale *et al.*, 2006; Straub *et al.*, 2008). How pest control varies with diversity in any particular system may reflect the balance between negative and positive interactions among constituent species (Ives *et al.*, 2005). Negative relationships between predator biodiversity and herbivore suppression are generally attributed to intraguild predation (e.g. Finke & Denno, 2004). While less is known about the positive relationships among predators that underlie improved herbivore suppression with greater predator diversity (Ives *et al.*, 2005), positive predator–predator interactions have begun to receive some attention, and we next review the beneficial effects of greater natural enemy biodiversity.

6.3.1 Natural enemies that complement one another

It is thought that adding species to a community can improve biological control when those predators differ in where, how or when they attack the pest, such that each predator fills a unique role (Wilby & Thomas, 2002). When this occurs, multiple enemy species can *complement* one another, with a sufficiently diverse predator community leaving herbivores with no safe refuge in time or space (Casula *et al.*, 2006). Further, predators that use different microhabitats are less likely to encounter and interfere with one another than are predators that use the same microhabitat (Schmitz, 2005).

Wilby *et al.* (2005) provide a nice example of how complementary relationships among predators may develop, working with two pests of rice, rice leaf-folder (*Marasmia patnalis*, a moth) and brown planthopper. Control was compared among single versus three-species assemblages of natural enemies, chosen from among a group of predators including a wolf spider (*Pardosa pseudoannulata*), a ladybird beetle (*Micraspis crocea*), a predatory cricket (*Metioche vittaticollis*) and a plant bug (*Cyrtorhinus lividipennis*). Predator densities were scaled across all treatments such that predation pressure was equalized, meaning that any differences between single and multi-species treatments would result from an emergent benefit of species diversity *per se*. Biological control improved with greater predator diversity for the moth, but not for the planthopper. The moth had a complex life cycle with stages that differed dramatically in morphology and behavior, whereas the leafhopper had a simple life cycle with successive stages closely resembling one another. The authors conclude that predator diversity was important for moth control because different predators attacked the different moth stages; i.e. the predators complemented one another. For the planthopper, in contrast, all predators impacted the different pest stages similarly, such that natural enemies had identical, redundant effects and thus did not complement one another (Wilby *et al.*, 2005).

Snyder & Ives (2003) examined complementary interactions of a different sort among natural enemies. We compared biological control by the specialist parasitoid *A. ervi* to that exerted by a diverse community of generalist predators within field cages (Fig. 6.7). The parasitoid *A. ervi* on its own was capable of mounting a strong numerical response to aphid increase, but this occurred at a time lag such that aphids reached a relatively high peak density before eventually declining. The generalist predators exerted significant suppression of aphids early on, but their impact did not increase as aphid densities grew and high peak aphid densities again were reached. However, when both the parasitoid and predators were present together, resulting biological control combined the beneficial attributes of both enemy classes, with both early suppression of aphids by predators and later suppression by the parasitoid, thereby leading to lower aphid densities throughout the experiment (Snyder & Ives, 2003). Thus, the density-independent impact of predators complemented the density-dependent response of the parasitoid.

6.3.2 Predators that facilitate one another's prey capture

A second mechanism thought to contribute to improved pest suppression with greater natural enemy diversity is predator–predator

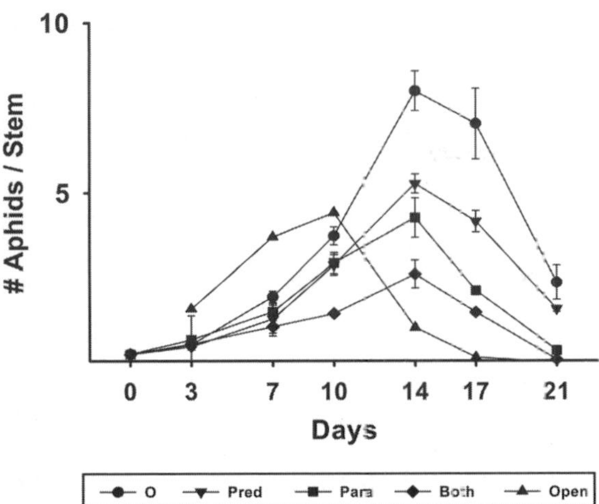

Fig. 6.7 Population dynamics of pea aphid on alfalfa, redrawn from data presented in Snyder & Ives (2003). Peak aphid densities were highest in controls that included no natural enemies (O). When alone the parasitoid *A. ervi* (Para) had little impact on aphid population growth early in the experiment, but drove down aphid densities after day 14. The diverse community of generalist predators alone (Pred) immediately reduced aphid densities, but thereafter aphid densities generally ran parallel to those in no-enemy controls. Aphid suppression was most effective with both parasitoid and predators present (Both), because the early suppression exerted by generalists was combined with later control by parasitoids. In this way parasitoid and predators complemented one another. Uncaged reference plots (Open) followed aphid dynamics in the alfalfa field surrounding the cages.

facilitation. Facilitation occurs when predators indirectly enhance one another's success in prey capture. This occurs, for example, when prey leave a habitat to escape one predator, only to instead fall victim to a second predator species in the would-be refuge (Sih *et al.*, 1998). In these cases prey find themselves "between a rock and hard place," with one predator species chasing prey into the waiting jaws of a second predator. In one well-known study from the alfalfa–pea aphid system (Losey & Denno, 1998), aphids dropped to the ground to escape ladybird beetles (*Coccinella septempunctata*) foraging in the foliage. However, once on the ground the aphids were then vulnerable to attack by ground beetles (e.g. *Harpalus pennsylvanicus*) foraging on the soil surface; in the absence of ladybird beetles to chase aphids to the soil the ground beetles rarely encountered aphids. In this way the combined impact of ladybirds and ground beetles on aphids, when both species were present simultaneously, was greater than could be predicted based on the effectiveness of either species in isolation (Losey & Denno, 1998).

6.4 Summary

Specialists attack a narrow range of pest species and often reproduce at a rate greater than that of their prey. These traits lead to specialists' ability to mount a density-dependent population response to rising prey densities, thought to be a trait important for an effective biological control agent. However, these same attributes also increase the chances that specialists drive boom-and-bust cycles, rather than providing stable control of prey densities. Generalists generally lack the ability to mount a density-dependent population response but are able to switch to attacking pests as they become abundant, either by immigrating into fields with pest outbreaks or switching from alternative prey within the same fields. Thus, both specialists and generalists have advantages and limitations as biological control agents. Different theories about the suppression of herbivores by their natural enemies predict stronger, weaker or unchanged pest suppression with greater natural-enemy diversity. How predator biodiversity influences pest control likely reflects a balance between positive and negative predator–predator interactions. Intraguild predation can disrupt biological control, and when intraguild predation is common pest suppression is expected to grow weaker when more natural enemy species are present. In contrast, when predators complement one another by attacking the pest in unique ways, or facilitate one another's prey capture, then pest suppression is expected to improve when more natural enemy species are present.

Acknowledgments

W. E. Snyder was supported during preparation of this chapter by grant #2004-01215 from the National Research Initiative of the US Department of Agriculture (USDA), Cooperative Research, Education and Extension Service (CSREES). Much of the work on pea aphids was supported by grants from the USDA and National Science Foundation (NSF) to A. R. Ives and by a USDA–National Research Initiative (NRI) Postdoctoral Fellowship to W. E. Snyder.

References

Bengtsson, J., Ahnström, J. & Weibull, A. (2005). The effects of organic agriculture on biodiversity and abundance: a meta-analysis. *Journal of Applied Ecology*, **42**, 261–269.

Briggs, C. J. & Hoopes, M. F. (2004). Stabilizing effects in spatial parasitoid–host and predator–prey models: a review. *Theoretical Population Biology*, **65**, 299–315.

Cardinale, B. J., Harvey, C. T., Gross, K. & Ives, A. R. (2003). Biodiversity and biocontrol: emergent impacts of a multi-enemy assemblage on pest suppression and crop yield in an agroecosystem. *Ecology Letters*, **6**, 857–865.

Cardinale, B. J., Srivastava, D., Duffy, J. E. *et al.* (2006). Effects of biodiversity on the functioning of trophic groups and ecosystems. *Nature*, **443**, 989–992.

Casula, P., Wilby, A. & Thomas, M. B. (2006). Understanding biodiversity effects on prey in multi-enemy systems. *Ecology Letters*, **9**, 995–1004.

Debach, P. & Rosen, D. (1991). *Biological Control by Natural Enemies*. New York: Cambridge University Press.

Denoth, M., Frid, L. & Myers, J. H. (2002). Multiple agents in biological control: improving the odds? *Biological Control*, **24**, 20–30.

Eubanks, M. D. & Denno, R. F. (2000). Host plants mediate ominivore–herbivore interactions and influence prey suppression. *Ecology*, **81**, 936–947.

Evans, E. W. (1991). Intra versus interspecific interactions of ladybeetles (Coleoptera: Coccinellidae) attacking aphids. *Oecologia*, **87**, 401–408.

Evans, E. W. & Youssef, N. N. (1992). Numerical responses of aphid predators to varying prey density among Utah alfalfa fields. *Journal of the Kansas Entomological Society*, **65**, 30–38.

Finke, D. L. & Denno, R. F. (2004). Predator diversity dampens trophic cascades. *Nature*, **429**, 407–410.

Frazer, B. D. & Gill, B. (1981). Hunger, movement and predation of *Coccinella californica* on pea aphids in the laboratory and in the field. *Canadian Entomologist*, **113**, 1025–1033.

Godfray, H. C. J. & Pacala, S. W. (1992). Aggregation and the population dynamics of parasitoids and predators. *American Naturalist*, **140**, 30–40.

Gross, K. & Ives, A. R. (1999). Inferring host–parasitoid stability from patterns of parasitism among patches. *American Naturalist*, **154**, 489–496.

Gross, K., Ives, A. R. & Nordheim, E. V. (2005). Estimating time-varying vital rates from observation time series: a case study in aphid biological control. *Ecology*, **86**, 740–752.

Hairston, N. G. Jr. & Hairston, N. G. Sr. (1993). Cause–effect relationships in energy flow, trophic structure, and interspecies interactions. *American Naturalist*, **142**, 379–411.

Hairston, N. G. Jr. & Hairston, N. G. Sr. (1997). Does food-web complexity eliminate trophic-level dynamics? *American Naturalist*, **149**, 1001–1007.

Hairston, N. G., Smith, F. E. & Slobodkin, L. B. (1960). Community structure, population control and competition. *American Naturalist*, **94**, 421–425.

Harmon, J. P. & Andow, D. A. (2004). Indirect effects between shared prey: predictions for biological control. *BioControl*, **49**, 605–626.

Hassell, M. P. (1978). *The Dynamics of Arthropod Predator–Prey Systems*. Princeton, NJ: Princeton University Press.

Hassell, M. P. & May, R. M. (1973). Stability in insect host–parasite models. *Journal of Animal Ecology*, **42**, 693–736.

Hassell, M. P. & May, R. M. (1974). Aggregation in predators and insect parasites and its effect on stability. *Journal of Animal Ecology*, **43**, 567–594.

Hassell, M. P., May, R. M., Pacala, S. W. & Chesson, P. L. (1991). The persistence of host–parasitoid associations in patchy environments. I. A general criterion. *American Naturalist*, **138**, 568–583.

Hawkins, B.A., Mills, N. J., Jervis, M. A. & Price, P. W. (1999). Is the biological control of insects a natural phenomenon? *Oikos*, **86**, 493–506.

Hodek, I. & Honek, A. (1996). *Ecology of Coccinellidae*. Boston, MA: Kluwer.

Hole, D. G., Perkins, A. J., Wilson, J. D. *et al.* (2005). Does organic farming benefit biodiversity? *Biological Conservation*, **122**, 113–139.

Holt, R. D. (1977). Predation, apparent competition, and the structure of prey communities. *Theoretical Population Biology*, **12**, 197–229.

Huffaker, C. B. (1958). Experimental studies on predation: dispersion factors and predator–prey oscillations. *Hilgardia*, **27**, 343–383.

Ives, A. R. (1992a). Continuous-time models of host–parasitoid interactions. *American Naturalist*, **140**, 1–29.

Ives, A. R. (1992b). Density-dependent and density-independent parasitoid aggregation in model host–parasitoid systems. *American Naturalist*, **140**, 912–937.

Ives, A. R. (1995). Spatial heterogeneity and host–parasitoid population dynamics: do we need to study behavior? *Oikos*, **74**, 366–376.

Ives, A. R., Cardinale, B. J. & Snyder, W. E. (2005). A synthesis of subdisciplines: predator–prey interactions, and biodiversity and ecosystem functioning. *Ecology Letters*, **8**, 102–116.

LaMana, M. L. & Miller, J. C. (1996). Field observations on *Harmonia axyridis* Pallas (Coleoptera: Coccinellidae) in Oregon. *Biological Control*, **6**, 232–237.

Landis, D. A., Wratten, S. D. & Gurr, G. M. (2000). Habitat management to conserve natural enemies of arthropod pests in agriculture. *Annual Review of Entomology*, **45**, 175–201.

Lattin, J. D. (1989). Bionomics of the Nabidae. *Annual Review of Entomology*, **34**, 383–400.

Losey, J. E. & Denno, R. F. (1998). Positive predator–predator interactions: enhanced predation rates and synergistic suppression of aphid populations. *Ecology*, **79**, 2143–2152.

Mackauer, M. & Kambhampati, S. (1986). Structural changes in the parasite guild attacking the pea aphid in North America. In *Ecology of Aphidophaga*, ed. I. Hodek, pp. 347–356. Prague, Czechoslovakia: Academia.

Montoya J. M., Rodríguez, M. A. & Hawkins, B. A. (2003). Food web complexity and higher-level ecosystem services. *Ecology Letters*, **6**, 587–593.

Murdoch, W. W. & Briggs, C. J. (1997). Theory for biological control: recent developments. *Ecology*, **77**, 2001–2013.

Murdoch, W. W., Nisbet, R. M., Luck, R. F., Godfray, H. C. J. & Gurney, W. S. C. (1992). Size-selective sex-allocation and host feeding in a parasitoid host model. *Journal of Animal Ecology*, **61**, 533–541.

Murdoch, W. W., Briggs, C. J. & Nisbet, R. M. (2003). *Consumer-Resource Dynamics*. Princeton, NJ: Princeton University Press.

Murdoch, W., Briggs, C. J. & Swarbrick, S. (2005). Host suppression and stability in a parasitoid–host system: experimental demonstration. *Science*, **309**, 610–613.

Myers, J. H., Higgins, C. & Kovacs, E. (1989). How many insect species are necessary for the biological control of insects? *Environmental Entomology*, **18**, 541–547.

Nicholson, A. J. & Bailey, V. A. (1935). The balance of animal populations. *Proceedings of the Zoological Society of London*, **43**, 551–598.

Olson, A. C., Ives, A. R. & Gross, K. (2000). Spatially aggregated parasitism on pea aphids, *Acyrthosiphon pisum*, caused by random foraging behavior of the parasitoid *Aphidius ervi*. *Oikos*, **91**, 66–76.

Östman, Ö. & Ives, A. R. (2003). Scale-dependent indirect interactions between two prey species through a shared predator. *Oikos*, **102**, 505–514.

Pimentel, D. (1961). Species diversity and insect population outbreaks. *Annals of the Entomological Society of America*, **54**, 76–86.

Polis, G. A. (1991). Complex trophic interactions in deserts: an empirical critique of food-web theory. *American Naturalist*, **138**, 123–155.

Polis, G. A. & Strong, D. R. (1996). Food web complexity and community dynamics. *American Naturalist*, **147**, 813–846.

Polis, G. A., Myers, C. A. & Holt, R. D. (1989). The ecology and evolution of intraguild predation: potential competitors that eat each other. *Annual Review of Ecology and Systematics*, **20**, 297–330.

Prasad, R. P. & Snyder, W. E. (2006). Polyphagy complicates conservation biological control that targets generalist predators. *Journal of Applied Ecology*, **43**, 343–352.

Rauwald, K. S. & Ives, A. R. (2001). Biological control in disturbed agricultural systems and the rapid re-establishment of parasitoids. *Ecological Applications*, **11**, 1224–1234.

Reeve, J. D. (1988). Environmental variability, migration and persistence in host–parasitoid systems. *American Naturalist*, **132**, 810–836.

Root, R. B. (1973). Organization of a plant–arthropod association in simple and diverse habitats: the fauna of collards. *Ecological Monographs*, **43**, 95–124.

Rosenheim, J. A., Wilhoit, L. R. & Armer, C. A. (1993). Influence of intraguild predation among generalist insect predators on the suppression of an herbivore population. *Oecologia*, **96**, 439–449.

Rosenheim, J. A., Kaya, H. K., Ehler, L. E., Marois, J. J. & Jaffee, B. A. (1995). Intraguild predation among biological-control agents: theory and practice. *Biological Control*, **5**, 303–335.

Schmitz, O. J. (2005). Behaviors of predators and prey and links with population-level processes. In *Ecology of Predator–Prey Interactions*, eds. P. Barbosa & I. Castellanos, pp. 256–278. New York: Oxford University Press.

Settle, W. H., Ariawan, H., Astuti, E. T. *et al.* (1996). Managing tropical rice pests through conservation of generalist natural enemies and alternative prey. *Ecology*, **77**, 1975–1988.

Sih, A., Englund, G. & Wooster, D. (1998). Emergent impacts of multiple predators on prey. *Trends in Ecology and Evolution*, **13**, 350–355.

Snyder, W. E. & Ives, A. R. (2003). Interactions between specialist and generalist natural enemies: parasitoids, predators, and pea aphid biocontrol. *Ecology*, **84**, 91–107.

Straub, C. S., Finke, D. L. & Snyder, W. E. (2008) Are the conservation of natural enemy biodiversity and biological control compatible goals? *Biological Control*, **45**, 225–237.

Strong, D. R. (1992). Are trophic cascades all wet? Differentiation and donor-control in speciose systems. *Ecology*, **73**, 747–754.

Symondson, W. O. C., Sunderland, K. D. & Greenstone, M. H. (2002). Can generalist predators be effective biocontrol agents? *Annual Review of Entomology*, **47**, 561–594.

Tscharntke, T., Klein, A., M. Kruess, A., Steffan-Dewenter, I. & Thies, C. (2005). Landscape perspectives on agricultural intensification and biodiversity – ecosystem service management. *Ecology Letters*, **8**, 857–874.

Turchin, P. (2003). *Complex Population Dynamics: A Theoretical/Empirical Synthesis*. Princeton, NJ: Princeton University Press.

van Emden, H. F. & Williams, G. F. (1974). Insect stability and diversity in agro-ecosystems. *Annual Review of Entomology*, **19**, 455–475.

Wilby, A. & Thomas, M. B. (2002). Natural enemy diversity and pest control: patterns of pest emergence with agricultural intensification. *Ecology Letters*, **45**, 353–360.

Wilby, A., Villareal, S. C., Lan, L. P., Heong, K. L. & Thomas, M. B. (2005). Functional benefits of predator species diversity depend on prey identity. *Ecological Entomology*, **30**, 497–501.

Yasuda, H., Evans, E. W., Kajita, Y., Urakawa, K. & Takizawa, T. (2004). Asymmetric larval interactions between introduced and indigenous ladybirds in North America. *Oecologia*, **141**, 722–731.

Chapter 7

Sampling for detection, estimation and IPM decision making

Roger D. Moon and L. T. (Ted) Wilson

IPM requires information about populations of pest and beneficial organisms in managed and constructed habitats. A recurring question is whether potentially injurious pests are abundant enough to warrant intervention, or whether beneficial organisms or other factors are likely to maintain control. Because most managed habitats are too large to be examined completely, practitioners must sample them and draw an inference about the whole.

Sampling plans in IPM can be grouped into three categories, depending on the sampler's goal. First, detection sampling is used in surveillance and regulatory applications, where the critical density of a target organism is effectively zero. Detection plans are designed to control the chance that the organism is erroneously missed. Second, estimation sampling is used where the goal is to quantify abundance, usually with desired levels of precision and reliability. Estimation sampling is used mainly in research, but it can also be used to evaluate IPM implementation and effectiveness. Finally, decision sampling is used where a choice to intervene with one or more management tactics depends on whether abundance has or will soon exceed a threshold density. Rather than estimate density, the goal is more simply to classify the habitat as needing or not needing intervention.

In all three situations, the basic process is the same. A sampler selects a set of sample units from the habitat using a defined procedure, assesses each for presence or abundance of the target organism, and then draws a conclusion based on the results. Fundamental sources on sampling theory and survey design are Cochran (1977) and Lohr (1999). Kogan & Herzog (1980), Pedigo & Buntin (1994) and Southwood & Henderson (2000) review sampling techniques used in ecology and agriculture. Binns *et al.* (2000) review theory and techniques for design and analysis of sampling plans for decision making in crop IPM. Sampling has elements of art and science, but the underlying foundation is mathematical. The goals of this chapter are to provide an overview of sampling theory and design techniques that are the basis of practical IPM sampling.

7.1 Basics of sampling

Sampling of all three kinds begins with a clear understanding of the attribute to be measured. In IPM, pest abundance is commonly indexed by the proportion of units that are infested, or as mean density (number of individuals) per sample unit. Of course, other variables may be of interest,

Integrated Pest Management, ed. Edward B. Radcliffe, William D. Hutchison and Rafael E. Cancelado. Published by Cambridge University Press.

too, depending on management context. Next, the sample universe must be defined. It is usually a specific habitat with limits in space and time.

The sample unit is the basic entity that is actually observed. A set of units constitutes a sample. To be used operationally, sample units should be mutually exclusive, convenient to locate and easy to process. A general recommendation is to choose units that are physically small, because more can be processed affordably, and the resulting estimates tend to be more precise than if larger units are used.

Sample units can be direct, physical subdivisions of a habitat, such as soil cores or quadrats of known area. These provide measures of abundance per unit of habitat area. If a universe of area U is sampled with a unit of area a, then there will be $U/a = N$ possible units in the universe. In contrast, units may have dimensions that are more ambiguous, as is the case with leaves, whole plants, or aerial collection devices. One problem with these kinds of sample units is that N may be difficult to calculate because unit size is unknown. This problem will be unimportant as long as unit size and number can be assumed to remain constant. However, in cases where size, number and occupancy by target organisms might change through time, then results will need to be interpreted with caution. Estimates of abundance from one kind of sample unit can be converted to another kind if prior research has established a basis for calibration.

To develop notation to be used in this chapter, the sampler inspects $i = 1 \ldots n$ of the N units, and records a series of x_i observations. The ratio $f = n/N$ is known as the sampling fraction, and in most cases will be trivially small. Sampling is almost always done without replacement, meaning individual sample units are observed only once. As long as $N > \sim 500$, reliability of a sample is determined by the number of units examined (n), and not by the fraction examined (f).

7.2 Detection sampling

This form of sampling is used to test the hypothesis that a pest is absent in a habitat. Examples include trapping to map geographic limits of invading forest pests, sampling to certify absence of disease in seed lots and inspection of cargo containers at ports of entry to exclude exotic fruit or vegetable insects. Sample universes and sample units in these situations can range from a large number ($N > 100\,000$) of possible trap locations in a region, down to a relatively small number ($N \sim 100$) of cartons in a shipping container.

Detection sampling is done sequentially, where individual units are examined one at a time, beginning with units most likely to be infested. When the first infested one is found, sampling stops and the habitat is declared to be infested. In contrast, as long as all units are clean, sampling could continue until all N units are examined, which will be impractical if N is large. Absence can never be proven, but sampling can be designed to control the chance of detecting an infestation when proportion infested is greater than a specified cut-off level, p_c (Venette *et al.*, 2002).

7.2.1 Necessary sample size

A design problem is to determine how many pest-free units one would need to conclude with desired reliability that the true proportion infested is not above the cut-off, p_c. Statistical theory indicates that if a sample of n units is taken without replacement from a universe of size N, and each unit is tested with a procedure that is 100% sensitive and specific, then the probability of observing no infested units can be calculated as the “zero-term” of a hypergeometric distribution (Kuno, 1991). The number of non-infested units, n_0, needed to conclude that the true proportion is not above the cut-off level is

$$n_0 = N(1 - \alpha^{1/Np_c}). \tag{7.1}$$

The term α is the probability of missing an infested unit when one is actually present, and is customarily set at 0.05, but can be liberal, 0.2 or greater, or conservative, 0.01 or lower. That probability translates into a level of reliability, which is the $100(1 - \alpha)\%$ of sampling occasions that can be expected to detect an infestation. Values for p_c and α are chosen when a detection plan is designed, and will be based on an economic balance between cost of sampling and potential losses of failing to detect a pest when it is actually present.

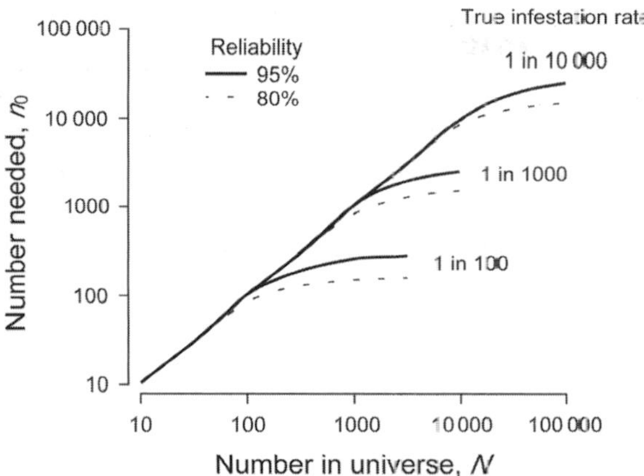

Fig. 7.1 Number of non-infested sample units (n_0) needed to declare with two levels of reliability that true infestation rate is below three chosen cut-off levels (p_c), over a range of universe sizes, via Eq. (7.1). Note n_0 and N in $\log_{10}$ scale.

Solutions of Eq. (7.1) with arbitrary cut-off and reliability levels over a range of Ns are illustrated in Fig. 7.1. In small universes (relatively low N), virtually all units will need to be examined, regardless of cut-off level, to be 80% or more certain that abundance is not above the cut-off level. In larger universes, sample size depends mainly on chosen cut-off level and reliability. If p_c is set at a liberal 1% (1 in 100 units infested), then over the range of N, sample sizes needed to achieve 80% or 95% reliability reach upper asymptotes of 160 or 300, respectively. If p_c is a more conservative 0.01% (1 in 10 000), then maxima of ~16 000 or 30 000 will be required.

7.2.2 Chance of detection

A complementary problem is to evaluate retrospectively the chance that an infestation would have been missed, given that a fixed number of non-infested units had been observed. Rearrangement of Eq. (7.1) leads to

$$\alpha^* = \left(1 - \frac{n_0}{N}\right)^{Np_c}, \tag{7.2}$$

where α^* is the calculated chance that an infestation would have been missed with a sample size of n_0. To illustrate, if a universe contained N = 1000 units, the cut-off level was set at 1% infested, and $n_0 = 10$ units were found to be non-infested, then the chance of missing the infestation would be 0.904. Stated another way, the declaration of absence would have a reliability of 100(1 − 0.904)%, just under 10%. Increasing n_0 to 100 would raise reliability to 65%. Retrospective sampling to determine the probability of a failure of detection is used by the State of Minnesota, Department of Transportation (MN-DOT) to monitor abundance of the roadside weed leafy spurge (*Euphorbia esula*) (Moon, 2007b).

7.3 Estimation sampling

This type of sampling is designed to measure the level of a trait of interest. Equations for calculating sample statistics for common IPM variables are summarized in Table 7.1. Relevant statistics are the sample mean ($\bar{x}$, an estimate of true mean, μ), the sample variance (s^2, an estimate of σ^2) and the standard error of the mean (SE). The term $(1 - f)$ in the formulas for SE adjusts downward for sampling without replacement from a finite population. The estimators in Table 7.1 are appropriate for simple random sampling, where every sample unit has an equal chance of being observed. This sampling design and others will be discussed later.

7.3.1 Precision and reliability of estimates

Samplers can control both the precision and reliability of estimated means. Precision can be represented in absolute terms by h = Student's $t \times$ SE, which is in the same measurement units as the mean. Alternatively, precision can be expressed in relative terms, as a unitless fraction (or percentage) of the mean, d = Student's $t \times \text{SE}/\bar{x}$. Either way, precision is calculated as a confidence interval for an estimated mean, CI $= \bar{x} \pm h$ or $\bar{x} \pm d\bar{x}$: the narrower the confidence interval, the more precise the estimate.

Reliability of a confidence interval is governed by Student's t, which is chosen in accord with permissible type I error rate (α). In estimation sampling, a type I error occurs when a confidence interval fails to contain the true mean by chance alone. Convention is for samplers to set α at 0.05 or 0.10, allowing true means to fall outside an interval in 1/20 or 1/10 sampling events, respectively. For $n > 30$, values of Student's t will be ~1.0, 1.7 or 2.0 for α = 0.33, 0.10 and 0.05, respectively.

Table 7.1 Equations for estimating sample statistics for variables commonly measured in IPM

Variable of interest	Mean	Variance	Standard error[a]
Continuous, amount per unit, (normal distribution)	$\bar{x} = \frac{1}{n}\sum_{i=1}^{n} x_i$	$s^2 = \frac{\sum_{i=1}^{n}(x_i - \bar{x})^2}{n-1}$	$SE = \sqrt{\frac{s^2}{n}(1-f)}$
Proportion infested (binomial)	$\hat{p} = \frac{1}{n}\sum_{i=1}^{n} xi$	$s^2 = \hat{p}(1-\hat{p})$	$SE \approx \sqrt{\frac{s^2}{n-1}(1-f)} + \frac{1}{2n}$, if $n\hat{p} > 15$, else[b]
Density, count per unit, random (Poisson)	$\hat{\lambda} = \frac{1}{n}\sum_{i=1}^{n} x_i$	$s^2 = \frac{\sum_{i=1}^{n}(x_i - \hat{\lambda})^2}{n-1}$	$SE = \sqrt{\frac{s^2}{n}(1-f)}$ if $n \geq 30$, else[b] $SE = \sqrt{\frac{\hat{\lambda}}{n}(1-f)}$ if $n\hat{\lambda} \geq 30$, else[b]
Density, count per unit, aggregated (negative binomial)	$\bar{x} = \frac{1}{n}\sum_{i=1}^{n} x_i$	$s^2 = \bar{x}(\bar{x} + \hat{k})$	$SE = \sqrt{\frac{s^2}{n}(1-f)}$
Density, count per unit (Taylor's power relation)	$\bar{x} = \frac{1}{n}\sum_{i=1}^{n} x_i$	$s^2 = A\,\bar{x}^b$	$SE = \sqrt{\frac{s^2}{n}(1-f)}$

[a] Term $(1-f)$ adjusts for finite population, where $f = n/N$ is fraction of available units actually examined.
[b] Exact values obtained from statistical tables, or iteratively with computer software.

The Central Limit theorem states that regardless of the underlying sampling distribution, as long as the variance is finite, sample means will tend toward being normal as sample size (n) increases. This theorem is the basis of faith that actual error rates will equal the nominal α, as long as sample size is adequate.

The estimators in Table 7.1 are statistically unbiased in the sense that their values will on average equal the true but unknown values in the sample universe (Cochran, 1977). This does not mean, however, that a single estimate can not be badly biased - low or high - due to chance alone. Of greater concern, estimates can be biased consistently if there is systematic error in sample collection or measurement technique, or unequal availability of organisms due to adverse weather or differences in location due to behavior or life history.

7.3.2 Sampling distributions

Four sampling distributions are commonly used to design sampling plans and to calculate associated statistics.

Normal distribution

Measured variables can be continuous (real numbers) as with length, area, volume or mass. In these cases, the response variable can be modeled as having come from the familiar bell-shaped normal distribution, and estimation of a universe's mean (μ) or total ($N \times \mu$) is straightforward (Table 7.1). In cases where the underlying distribution is skewed to the right by an excess of high values, the distribution can be modeled by a lognormal distribution (not illustrated).

Binomial distribution

In many IPM situations, sample units can be scored simply for presence or absence of a target organism. In this case, each x_i will have a value of 1 or 0, respectively. Providing the universe is large (N effectively infinite), and the probability (p) of being positive remains constant throughout the universe, then the response can be modeled as a binomial variable with parameters p and n (Table 7.1). A property of the binomial distribution is that h for $\hat{p}$ will be greatest when $\hat{p}$ is 0.5, but will get smaller as $\hat{p}$ approaches 0 or 1.0. Universes with clustering or a spatially variable pattern in p

can be modeled with a beta-binomial distribution (see Madden & Hughes, 1999; Binns *et al.*, 2000).

Random distribution

Most frequently in IPM sampling, target organisms are counted per sample unit, and this is often referred to as enumerative sampling. Counts can be modeled as a random (Poisson) process, as might occur when fungal spores land on sticky slides or immigrating aphids land in pan traps. A random distribution arises if the organisms act independently of each other, and all sample units are equally likely to be occupied. A property of the Poisson distribution is that its mean is equal to its variance. Consequently, SE and h will be proportional to the mean, $\hat{\lambda}$ (Table 7.1).

Aggregation

In many cases, samples where organisms are counted have more low and high counts than expected from a Poisson distribution. This is a result of spatial or temporal clustering of organisms among sample units. Previous studies have shown that the large majority of species have an aggregated spatial pattern (e.g. Taylor *et al.*, 1978; Wilson & Room, 1983). Aggregation is usually a result of unequal habitat quality and behavioral aspects of a species, but also may be affected by choice of sample unit. For examples, armyworms and stink bugs oviposit their eggs in masses. When sampling individual leaves on host plants, each leaf will contain either zero or several dozen to a few hundred eggs. At the level of a leaf, these species will be highly aggregated. After eggs hatch, mortality and dispersal will reduce aggregation.

Aggregation can be modeled statistically in different ways, but a common approach is to use the negative binomial distribution (NBD), which has a mean and a second parameter k that is inversely related to extent of aggregation. Methods for estimating k and fitting alternative distributions can be found in Pedigo & Buntin (1994) and Binns *et al.* (2000). Generally, sample sizes must be well in excess of $n = 50$ to obtain reliable estimates of k, and to discriminate among alternative distributions. Experience with a variety of universes and target organisms indicates that values of k can change substantially as densities change (see Southwood & Henderson, 2000). Furthermore, values of k and other measures of aggregation are sensitive to size of the sample unit. Aggregation tends to be greatest where sample unit size coincides with the spatial scale of an organism's clusters. Hence, samplers can overcome aggregation (and thereby reduce h) by changing the physical size of the sample unit.

Taylor's variance–mean relation

An alternative to adopting a specific statistical model for a sampling distribution is to make use of the empirical relation between observed sample variances and means among multiple samples that span a range in density. Taylor (1961) observed that sample variances and means were related as a power function (Table 7.1). By applying logs to both sides of the power function,

$$\log_{10}\left(s_i^2\right) = a + b\log_{10}(\bar{x}_i), \tag{7.3}$$

variance–mean relations among many data sets can be analyzed with simple linear regression (or more sophisticated methods: see Perry, 1981). An example is in Fig. 7.2A. The intercept, $a = \log_{10}(A)$, depends on sample unit size. Slope (b) is interpreted as an index of aggregation. If a is not significantly different from zero, then a test of the hypothesis that slope $= 1.0$ formally compares the observed variance–mean pattern with random (Poisson) expectations. Many studies indicate that values for a and b for individual species are reasonably stable across different universes, providing the same sample unit has been used.

Relation between binomial distribution and density

The binomial distribution is useful for describing the relation between proportion (p) of sample units infested by a species and the mean density (μ) of the species. As density increases, so does the proportion infested, and the rate at which p increases depends on extent of aggregation (Fig. 7.2B). For species that are uniformly distributed among sample units, $\sigma^2/\mu < 1.0$, p increases more rapidly with increasing density. In contrast, for species with an aggregated spatial pattern, $\sigma^2/\mu > 1.0$, p increases more slowly.

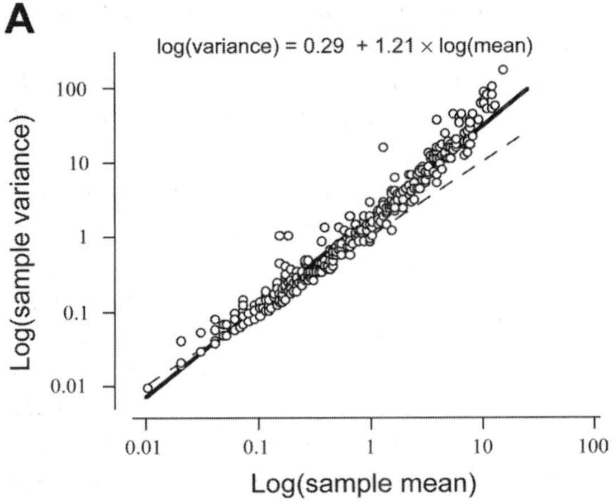

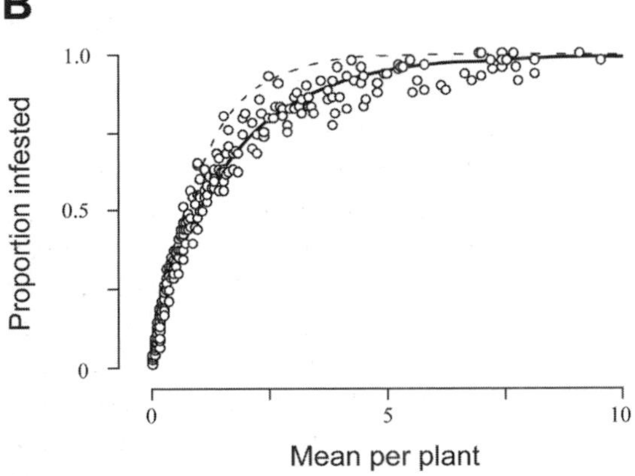

Fig. 7.2 Evidence for moderate aggregation by vegetable leafhopper (*Austroasca viridigrisea*) in cotton fields sampled over three growing seasons in Australia (see Wilson & Room, 1983). Each point represents a sample of vegetable leafhopper, counted visually on individual plants, $n = 96$ plants per date, for total of 474 samples and 45 504 plants. (A) Taylor's variance–mean relation, log-transformed sample variances and means; dashed line random expectations, solid line least squares fit to the data (Eq. 7.3). (B) Same data expressed as proportion of plants in each sample that were infested; lines as above.

The relation between proportion infested (p) and mean density ($\bar{x}$) among multiple samples can be expressed generally as

$$p = 1 - \exp(-\bar{x} \times \log_e[s^2/\bar{x}]/[s^2/\bar{x} - 1]), \quad (7.4a)$$

where "exp" means exponentiation of natural base e. The same relation can be expressed with variance estimated from Taylor's mean–variance relation (Wilson & Room, 1983),

$$p = 1 - \exp(-\bar{x} \times \log_e[a\bar{x}^{b-1}]/[a\bar{x}^{b-1} - 1]). \quad (7.4b)$$

7.3.3 Sample size

Besides setting α, samplers can control precision mainly through choice of sample size (n), which governs numerical values of both Student's t (through degrees of freedom, df) and SE (see Table 7.1). Increasing sample size from small to moderate can substantially improve precision, but the relationship is one of diminishing returns. To illustrate, if s^2 for a normal variable is 100, and α is set to 0.05, doubling n from 5 to 10 decreases $h = t \times \text{SE}$ from 12.4 to 7.2, and doubling again to $n = 20$ decreases it to 4.7. Providing that cost of sampling is not limiting, sample variance will ultimately limit achievable precision. Further refinement will require that sample variance be reduced, perhaps by changing the definition of the sample unit, by stratifying to direct more effort to sources of variability, or both.

The simplest approach to setting sample size is to calculate required sample size (RSS), the sample size that would be needed to achieve desired precision. This procedure is useful when a sampling plan is being developed. Using chosen values for α, and either $h = t \times \text{SE}$ or $d = t \times \text{SE}/\bar{x}$,

$$\text{RSS} = \frac{s^2}{h^2}, \text{ or } = \frac{s^2}{d^2\bar{x}^2}. \quad (7.5a, b)$$

From these general formulae, Karandinos (1976) and Ruesink (1980) derived RSS formulas for binomial, random and aggregated sampling distributions, by substituting corresponding definitions of SE (Table 7.1) into definitions of h or d. Four points are notable.

First, RSS will be proportional to sample variance and inversely proportional to desired precision. The more variable the universe and the greater the precision desired, the larger the sample size needed to achieve a chosen level of precision.

Second, RSS can change substantially, depending on how precision is defined (as h or d) (Fig. 7.3, A vs. B). Small samples may suffice when means are low and precision is a positive number (h). However, at the same density, RSS can be prohibitively large if relative precision (d) is desired.

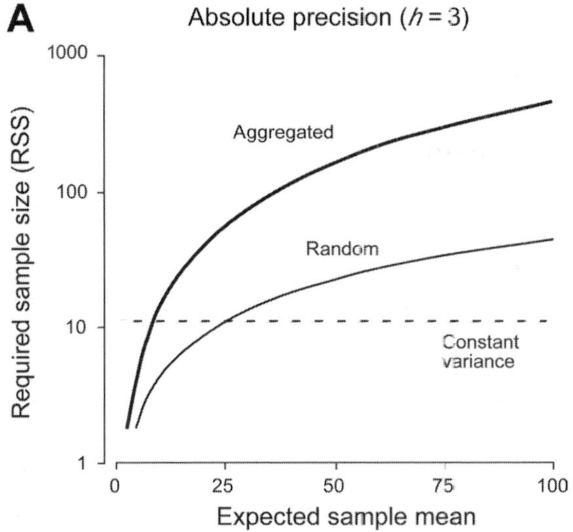

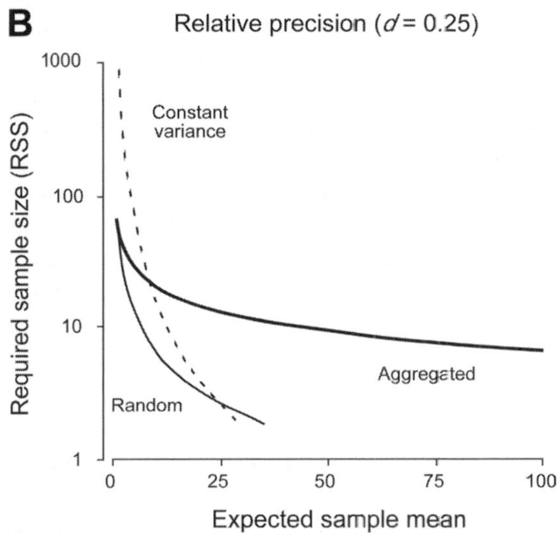

Fig. 7.3 Number of sample units required (RSS) to estimate a mean density with desired precision. (A) Desired precision defined as positive number (h). (B) Precision as proportion of the mean (d). Curves calculated with Eqs. (7.5a, b), with $\alpha = 0.05$. Constant variance, $s^2 = 25$. Random (Poisson), variance = mean. Aggregated, using Taylor's variance–mean relation, $A = 1$, $b = 1.5$.

Third, where sample variance increases with density, as will occur when counted organisms are random or aggregated, RSS to achieve absolute precision (h) will increase as density increases, but it will decrease if relative precision (d) is desired (Fig. 7.3). Furthermore, for a given mean density, a consequence of aggregation is to increase RSS over what would be required if the organisms are distributed randomly among sample units.

Finally, whichever way precision is specified, RSS will be proportional to sample variance, and a preliminary estimate of sample variance is required. Adequacy of calculated sample size will depend largely on accuracy of the estimated variance. In habitats where the variance is not changing rapidly, and sample units can be processed quickly, then Stein's two-stage approach can be used. One takes a small preliminary sample to estimate variance, calculates RSS, and then quickly returns in a second stage to obtain the remaining sample units. Where the same habitat is being sampled over a series of dates, then results from one date can serve to project a value for variance on the next date.

7.3.4 Sequential sampling for desired precision

An extension of Stein's approach is possible in situations where sample units can be processed while sampling is in progress. The basic idea is to use the information in the growing sample to terminate sampling just when desired precision has been achieved, and to avoid excessive effort and expense. This approach is feasible where sample units can be processed quickly, and results analyzed statistically to determine when desired precision has been achieved. In cases where units must be evaluated in an off-site laboratory, then a compromise approach is to calculate RSS as above, to collect a few more units than are expected to be needed, but to process the units sequentially and thereby limit excess processing expense.

Before hand-held calculators became available, much research was done to develop sequential sampling plans for use in IPM sampling. Known as stop rules, guidelines for terminating sampling were coded either in tables or graphs that conveyed a critical total of counts in relation to increasing n. A scout would begin with a total of 0, and then add additional counts as subsequent units were examined. Sampling would stop when the accumulating total first exceeded the critical stop value for the corresponding n. Equations for stop lines are based on chosen levels of precision, confidence, and sampling distribution, and have been reviewed by Hutchison (1994). We will return below to discuss sequential procedures in decision sampling, where they are more widely adopted.

7.3.5 Choice of sample units

Size of sample unit can affect sampling cost and achievable precision. To illustrate, a sample unit might be defined as a single leaf, a cluster of three leaves, or a whole plant. Smaller units tend to be cheaper, because once located, they require less labor and materials to process. Also, depending on extent and spatial scale of aggregation, units of different sizes may yield different variances. To compare units of different size (u), Cochran (1977) defined relative net precision as

$$\mathrm{RNP_u} = \frac{M_\mathrm{u}}{C_\mathrm{u}} \times \frac{M_\mathrm{u}}{s_\mathrm{u}^2}, \quad (7.6)$$

where M_u is sample unit size (mass), C_u is cost of processing a unit, and $s^2{}_u$ is sample variance. The size with greatest RNP will be most efficient in operational use. Costs and variances can be estimated from a sampling experiment designed to compare units of different sizes.

For some pests, qualitatively different sample units might be appropriate. For example, mobile predators can be sampled with a sweep net or by visual examination of plant parts. Which one should be used? The better technique can be chosen based on their relative costs to achieve the same sample precision (Wilson, 1994), presuming results with the different methods can be expressed in the same unit or measurement. Cost to achieve desired precision with a given method (i) can be estimated as

$$C_i = n_i(\theta_i + \phi_i), \quad (7.7a)$$

where C_i is cost in time (or money), n_i is number of units required to achieve desired precision and confidence, and θ_i and ϕ_i are times required to locate and process an individual unit, respectively. Sample size can be estimated as RSS (Eq. 7.5a or b) for estimation, or maximum average sample number (ASN) for decision sampling (see below). The technique with the lowest C_i would be the one of choice.

To compensate for different levels of acceptability by commercial scouts, Espino *et al.* (2007) proposed that each technique's cost could be weighted by its relative acceptability,

$$C_i = \frac{n_i(\theta_i + \phi_i)}{\psi_i}, \quad (7.7b)$$

where ψ_i is the estimated proportion of users willing to use the ith technique. Espino *et al.* (2007) showed that while a sweep net was more cost-reliable than visual inspection for sampling stink bugs in rice, low acceptance ($\Psi = 0.4$, or 40%) by rice scouts rendered visual inspection superior.

7.3.6 Sample selection procedures

The method used to select sample units has important implications for convenience, cost, reliability and precision. Cochran (1977) distinguished between probability and non-probability selection methods. A probability method is one where the chance that an individual unit is included in a sample is known, whereas probability of inclusion is not known with non-probability methods. Probability methods are more desirable because resulting estimates of means, variances and SEs (Table 7.1) can be assumed to be unbiased in most cases.

In IPM sampling, it is convenient for scouts to collect a haphazard sample of units, or a systematic one by walking a habitat in an "V," "W" or "X" pattern in an effort to get a "representative" sample. While these may be practical routines, they allow many opportunities for bias. Of particular concern are "edge effects," where the target organism may be concentrated in field margins or centers. Also, unrecognized spatial patterns in density may coincide with the walked path. These non-probability methods can seriously under- or overestimate mean field abundance (see Alexander *et al.*, 2005). Wherever practical, enhanced reliability of probability methods justifies their use.

Simple or unrestricted random sampling

This design (e.g. Fig. 7.4A) is the simplest and most widely used of the probability designs. Each sample unit has an equal chance ($1/N$) of being observed. The change over haphazard selection is that a randomization method is used to choose units. A variety of mechanical or digital methods can be used to select units. Points in two-dimensional habitats can be located by randomly

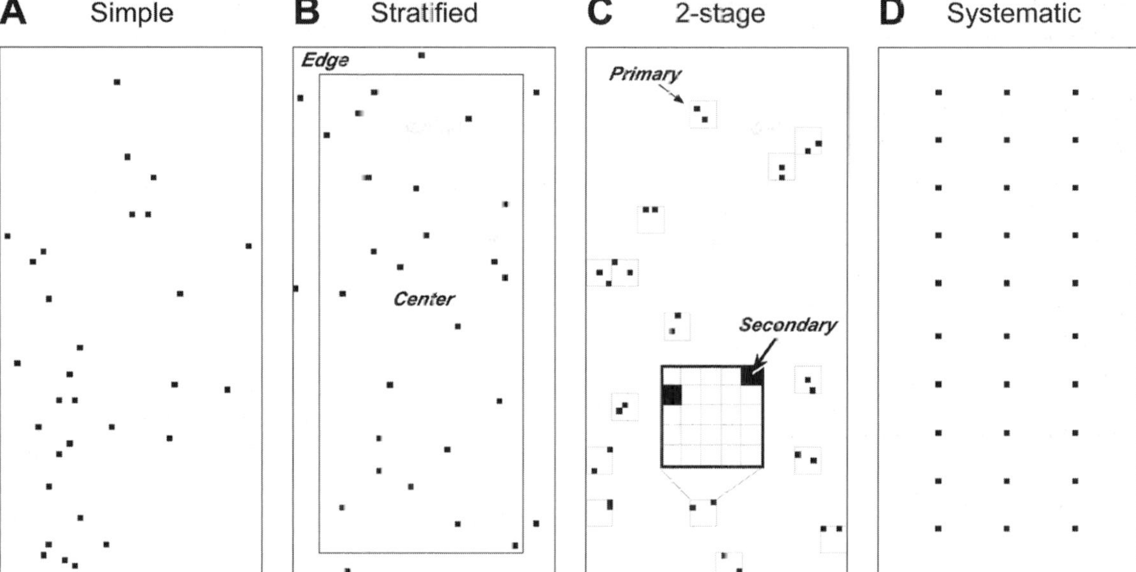

Fig. 7.4 Illustrations of four sample selection procedures applied to a hypothetical 50 × 100 m field, subdivided into sample units of 1-m² quadrats; $N = 5000$, $n = 30$. (A) Simple (unrestricted) random sample chosen with uniform random numbers for X–Y coordinates. Note parts of field undersampled. (B) Stratified sample with universe divided into edge ($N_e = 1080$, $n_e = 7$) and center ($N_c = 3920$, $n_c = 23$) to control for edge effect. (C) Two-stage sample of $n = 15$ primary units and $m = 2$ secondary units to reduce travel costs. (D) Systematic sample of units chosen in grid pattern, every 9th vertically, 12th or 13th horizontally.

choosing X–Y coordinates, mapping them (e.g. Fig. 7.4A), and then navigating through the spatially ordered list. If estimation or decision sampling are being done sequentially, then units can be mapped and processed in overlapping subgroups, each of which will require a separate pass through the habitat.

Stratified random sampling

This design (e.g. Fig. 7.4B) is a powerful alternative to simple random sampling, because it assures coverage, makes use of the sampler's knowledge, and often gives greater control over precision and cost. The universe is divided into an arbitrary number of parts (strata, $h = 1 \ldots L$) that are mutually exclusive subgroups of units of known or estimated number, N_h. Strata can be horizontal, vertical or both; they can be of different sizes and irregular in shape, and units in a stratum do not need to be contiguous. The goal is to form groups of units with means as different as possible, and whose internal variances are as small as possible. Unless the sampler is interested in comparing the strata themselves, division of a universe into more than six strata generally produces marginal improvements in overall precision and efficiency (Cochran, 1977).

Once the universe is divided, each stratum is sampled using simple random sampling. Means and variances from each are calculated (as in Table 7.1) and then combined to obtain a weighted mean for the whole universe:

$$\bar{x}_{st} = \sum_{h=1}^{L} w_h \bar{x}_h = w_1 \bar{x}_1 + w_2 \bar{x}_2 + \cdots w_L \bar{x}_L, \quad (7.8a)$$

with

$$\mathrm{SE}_{st} = \sqrt{\sum_{h=1}^{L} \left\{ w_h^2 \frac{s_h^2}{n_h}(1 - f_h) \right\}}. \quad (7.8b)$$

The weights, $w_h = N_h/N$, are the proportions of total N that are in the different strata. Effective number of degrees of freedom (df) for SE_{st} will depend on sampling fractions and stratum variances, and will be between the smallest $(n_h - 1)$ and $(n - 1)$ (Cochran, 1977).

To use a stratified design, a sampler allocates a total n among the strata. The simplest arrangement is known as proportional allocation, where the number of units per stratum is proportional to stratum size. Proportional allocation is a good choice if preliminary estimates of stratum variances are not available, and cost of processing a unit is approximately the same everywhere.

Many field monitoring protocols instruct scouts to divide fields into quarters, and to inspect multiple points in each one. This is actually a stratified design, intended to assure field coverage, with proportional allocation of equal sample sizes (n_h) in four equally sized strata. However, this prescribed design may not be the most efficient one if units in the quarters have different variances, and if time to travel to the different quarters varies widely.

A refinement over proportional allocation is known as optimal allocation, which adjusts for differences in stratum variances and sampling costs, and yields the greatest precision per unit sampling cost. Presuming preliminary estimates of variances and costs are available, sample size for individual strata are calculated from

$$\hat{n}_h = n \times \frac{w_h s_h \sqrt{c_h}}{\sum_{h=1}^{L}(w_h s_h \sqrt{c_h})}, \tag{7.9}$$

where s_h is square root of variance (standard deviation) in stratum h, and c_h is cost of obtaining and processing a unit in stratum h. Inspection of Eq. (7.8) indicates that the optimal allocation calls for more units in a stratum if it is larger than the others, if it is more variable internally, or if its units are cheaper to obtain. In circumstances where a stratum's variance is exceptionally great, the optimal allocation may call for every unit in that stratum to be examined, i.e. set $n_h = N_h$. For example, stratified sampling and optimal allocation is used by the State of Minnesota, Department of Transportation (MN–DOT) to estimate abundance of Canada thistle (*Cirsium arvense*) (Moon, 2007a).

Cluster sampling and multi-stage subsampling

These designs were originally developed to survey humans in populations too large to be enumerated and sampled at random (see Lohr, 1999). Rather, blocks of households could be selected randomly from maps, and surveyors could travel to chosen blocks to interview all residents in each block. This approach is now known as cluster sampling, where a simple random sample of blocks (= primary sample unit, 1°) is chosen, and then each resident (secondary sample unit, 2°) within the chosen blocks is surveyed.

A variant of cluster sampling is known as multi-stage subsampling, where chosen clusters are randomly subsampled rather than examined in their entirety. Nesting with subsampling at each level is easily extended to three or more levels, as might occur with trees (1°s) in a plantation, branches (2°s) within trees, twigs (3°s) in branches and ultimate sample units of leaves (4°s) on twigs. For brevity, we will illustrate a two-stage design with primary units of equal size, as might occur if a field were divided into blocks of equal size (e.g. Fig. 7.4C). Readers should consult Cochran (1977) and Lohr (1999) for guidance if primary or lower units are unequal in size, or if three or more stages are required.

Estimators for a two-stage nested design are complicated by the fact that sampling and variation occur at both 1° and 2° levels. To develop more notation, let $x_{i,j}$ represent an observation from $i = 1 \ldots n \leq N$ 1° units, and $j = 1 \ldots m \leq M$ possible 2°s per 1°. The overall sample mean is

$$\bar{x} \frac{1}{nm} \sum_{i=1}^{n} \sum_{j=1}^{m} x_{i,j}, \tag{7.10a}$$

$$\text{with SE} = \sqrt{(1 - f_1)\frac{s_1^2}{n} + \frac{n}{N}(1 - f_2)\frac{s_2^2}{nm}}. \tag{7.10b}$$

Subscripts 1 or 2 for sampling fractions and sample variances refer to 1° or 2° units, respectively. The two variance components arise from variation among 1° units and among 2° units within 1°s; both components are estimated with a nested analysis of variance (ANOVA).

Benefits of cluster sampling and subsampling in IPM are that the sampler can control sampling costs by reducing travel time between the n 1° units, collecting m 2°s in each one, for a total of nm observations in all. A shortcoming, though, is that precision with two-stage sampling is often less than if the same nm units were chosen independently with simple random sampling. This occurs because units in the same cluster tend to be positively correlated, and consequently the extra effort to process more than a small subsample adds little information. Depending on the spatial scale of variability in the habitat, precision in multi-stage designs is most frequently determined by variability among 1° units.

Guidance on the optimum number, m_{opt} of 2° units per 1° unit that will yield the greatest precision per total cost can be obtained from

$$m_{\text{opt}} = \frac{s_2}{\sqrt{s_1^2 - \frac{s_2^2}{M}}}\sqrt{\frac{c_1}{c_2}} \approx \sqrt{\frac{s_2^2}{s_1^2} \times \frac{c_1}{c_2}}. \quad (7.11)$$

Here, c_1 is cost to locate the average 1°unit, and c_2 is cost to process the average 2° unit. Total cost would be $C = nc_1 + nmc_2$. The optimal number of 2° units per 1° depends on the ratios of the two variance components and the two cost components; the more variable the 2°s and the cheaper they are to process, the larger m_{opt} will be. Modifications of Eqs. (7.10 and 7.11) have been developed to incorporate different variance–mean relations among 1° and 2° levels, and variable c_2, which can also depend on mean density (Hutchison, 1994; Binns *et al.*, 2000). In rare cases, calculated m_{opt} can exceed *M*, and will occur if variance among 2° units exceeds *M* times the variance among 1°s. In this case, set $m_{\text{opt}} = M$; variability among 2° units is so great, full cluster sampling will be more efficient than subsampling.

Systematic designs

These designs are convenient alternatives to simple random sampling, because they greatly simplify the task of physically locating sample units. They are also useful when the objective is to map pest density (Fleischer *et al.*, 1999). If the *N* units can be arranged in a linear order, and the sampler desires a sample of size *n*, then the interval *t* between chosen units will be the next integer after *N*/*n*. To choose a specific random start sample, one chooses a random integer *R* within $1 \ldots t$, and then actually observes units numbered $R, R + t, R + 2t, \ldots, R + (n-1)t$ (Lohr, 1999). An equivalent procedure in two-dimensional universes leads to grid patterns (e.g. Fig. 7.4D).

If pest density is effectively independent of order, then a systematic sample will behave much like a simple random sample, and the sample mean can be calculated as in Table 7.1. However, in situations where there is an underlying gradient or trend in density, then a systematic sample can yield a biased estimate of the mean. Worse, if there is a cyclical or periodic pattern that coincides with the interval *t*, then potential for bias is great. A further limitation of systematic samples is that, because not every sample of size *n* has an equal chance of being observed, sample variance can be biased, and resulting confidence intervals can be unreliable.

Combinations of the four basic designs

Simple random, stratified, multi-stage and systematic designs can be used in combination. For example, a field could be stratified to achieve coverage, and then the strata could be sampled using a two-stage nested design, with systematic selection of 2° units within randomly chosen 1° units. As long as numbers of units in each subdivision of a universe are known, then results can be weighted appropriately to obtain unbiased estimates. Interested readers should consult Cochran (1977) or Lohr (1999) for guidance.

7.4 Decision sampling

This third form of sampling is fundamentally different from detection and estimation. Rather than detect a pest or estimate its abundance, the goal is to decide if abundance at a given time is safely below a critical threshold, or if it is above and justifies remedial action. For simplicity, we will refer to this decision as "act" versus "no-act," and action may involve a variety of remedial responses. Thresholds can be based on expert opinion or deeper knowledge (see Chapter 3). The utility of decision sampling is greatest in managed habitats where pests are intermittently below or above threshold, where losses can be great and where remedial actions are costly or potentially disruptive to the habitat or to crop market value. Assuming the threshold is correct, decision sampling will lead to a correct decision with a predetermined error rate (e.g. 5%), and will do so with a minimum sample size (Binns *et al.*, 2000). Decision sampling has been shown to reduce sampling costs by 40% to

60%, compared with fixed sample size (RSS) methods with equal type I error rate.

Decision sampling was first developed by Abraham Wald to determine if the quality of munitions manufactured during World War II met acceptable standards. At the time, testers thought they could accept or reject a product well before the requisite number of tests had been completed, especially when failures were very rare or frequent. In response, Wald developed sequential probability ratio tests (SPRTs) that formalized the sequential decision process. His procedures were later adapted to make spray or no-spray decisions in forest pest management in the 1950s, and in row crops in the 1960s (see Binns, 1994). More recently, sequential decision methods have been extended to make three-choice decisions, to allow more flexibility in characterization of biological sampling distributions, and to use computers rather than analytical methods to evaluate performance and practicality of proposed plans. One approach is with Monte Carlo methods that use random number generators for specified probability distributions to simulate performance of proposed sequential plans (see Binns *et al.*, 2000). A second approach is to use resampling methods to evaluate and adjust sampling plans, based on repeated resampling of data sets obtained from extensive field samples (Naranjo & Hutchison, 1997).

7.4.1 General process

Decision sampling requires a sample unit that is biologically appropriate and convenient, a defined threshold, and rules for terminating sampling and making a decision with specified error rates. We will designate a threshold as a critical proportion of units infested, p_c, or as number of individuals per sample unit, μ_c, depending on whether organisms in units are scored for presence/absence or are fully counted.

In practice, the sampler begins by examining a minimum number of units, n_{min}, scores each, calculates the total, $T_n = \sum x_i$, and then updates T_n as additional units are examined. Sampling stops and a decision is made when T_n first exceeds the range between lower or upper "stop limits," as listed in a table or graph. When abundance is well below or above the threshold, a decision can be made early, because T_n quickly exceeds the corresponding limit. In contrast, when true abundance is near the threshold, a decision may not be possible without a very large sample. To prevent sample size from becoming prohibitively large, most sequential plans set an upper limit on sample size, n_{max}. If reached before a decision is made, then the recommendation is to take no action, but return soon to resample.

7.4.2 Stop limits

Wald framed a decision problem as two competing one-tailed tests, Ho: $\mu < \mu_1$ and Ha: $\mu \geq \mu_2$, where $(\mu_1 + \mu_2)/2 = \mu_c$. If Ho is true, no action will be taken, whereas if Ha is true, then action would be warranted. Unfortunately, chance sampling makes it possible that two kinds of errors can occur. An estimated mean ($= T_n/n$) can be above μ_2 when the true mean is below μ_1; the consequence would be to waste action when not warranted. The converse error is that an estimate can appear to be below μ_1 when truly above μ_2. In this case, the decision would be to wrongly do nothing, and incur a loss that could have been prevented. These errors are known as a type I and type II errors, respectively, and convention is to designate their probabilities as α and β.

Wald derived stop line equations for binomial, normal, Poisson and negative binomial distributions, and these are tabulated in Binns *et al.* (2000). For all four distributions, the stop lines are increasing functions of n. When graphed in relation to n (Fig. 7.5A), they appear as parallel lines straddling $n \times p_c$ or $n \times \mu_c$ and are separated vertically by an amount that depends on chosen levels for α and β.

A conceptual limitation of Wald's approach is that his μ_1 and μ_2 are arbitrary lower and upper critical levels, whereas critical limits in IPM are singular. As an alternative to Wald's SPRTs, Iwao (1975) proposed that upper and lower limits could be derived from confidence intervals around the threshold. For p_c, Iwao's upper and lower decision lines are

$$U_n \approx n\left(p_c + z_{\alpha+2}\sqrt{s^2/n}\right) \tag{7.12a}$$

and

$$L_n \approx n(p_c - z_{\alpha/2}\sqrt{s^2/n}), \tag{7.12b}$$

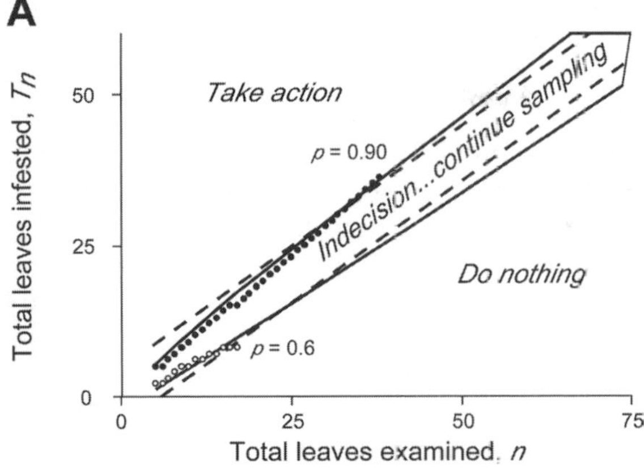

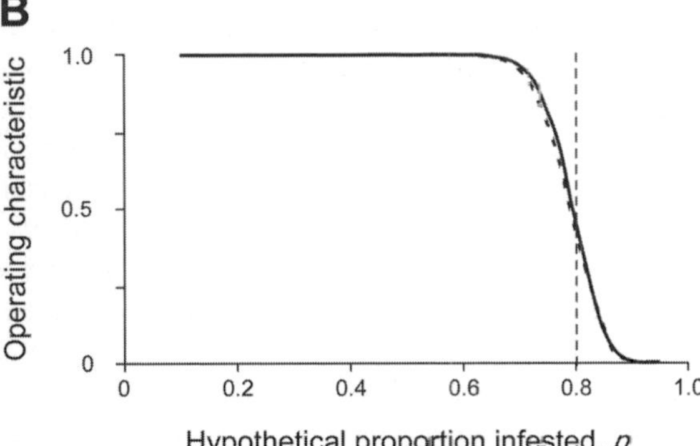

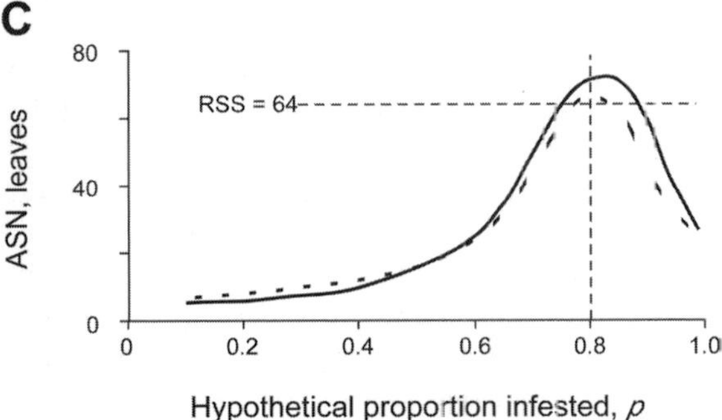

Fig. 7.5 Properties of two sequential decision plans for spider mites on cotton, with $p_c = 0.80$. Dotted lines: Wald's binomial SPRT, $p_0 = 0.75$, $p_1 = 0.85$, $\alpha = \beta = 0.05$. Solid lines: Iwao's procedure $p_c = 0.80$, $\alpha = 0.05$. Limits of n_{min} set at 5 leaves and n_{max} at 75. (A) Calculated stop limits, with results of two simulations with hypothetical p(infested) $= 0.9$ or 0.6. (B) Operating characteristic (OC) curves. (C) Average sample number (ASN) curves. OC and ASN curves estimated with simulations of 1000 trials at hypothetical infestation levels between 0.1 to 1.0.

which also are centered around the diagonal line, $n \times p_c$, but diverge slightly (Fig. 7.5A). The standard normal two-tailed deviate for chosen α does not reliably control type I error rate, but it does govern distance between the two boundaries. Iwao's method does not explicitly control type II error rate. Estimated variance, s^2 is for p at p_c. Other sampling distributions can be accommodated by substituting appropriate estimators for s^2 (see Table 7.1).

7.4.3 Performance of decision sampling plans

A workable plan must lead to a decision over a range of true pest densities surrounding the threshold. This aspect of performance is displayed by an operating characteristic (OC) curve (Fig. 7.5B), which shows the probability of taking no action over a range in abundance straddling the threshold; the steeper the curve at the threshold, the better the plan. A workable plan must also be affordable over the anticipated range in pest density. This aspect is revealed in a plot of average sample number (ASN) versus hypothetical density (Fig. 7.5C), which is the expected number of sample units that will be required to reach a decision at a given density. To maintain chosen error rates (= reliability), expected sample size must increase as abundance nears the threshold. Wald provided equations for calculating OC and ASN curves as functions of density for his four SPRTs, whereas curves for Iwao's procedure must be derived through computer simulation (see Binns *et al.*, 2000).

Important elements of a specific decision sampling plan are its error rates. How do you set an acceptable level for α, the probability of erroneously taking action, and β, the probability of erroneously not taking action? Wilson (1982) offered guidance. For α, the appropriate error rate occurs when the combined cost of sampling, remedial action, and ecological disruption are at a minimum. For β, the appropriate error rate occurs when the combined cost of sampling, action, disruption and loss due to pest injury are at a minimum.

To illustrate, three species of spider mites are commonly found in cotton in the USA, and their damage potential depends on when populations become established in a given field. Spider mites can complete as many as 16 generations per season, depending on season length. A practical sample unit for spider mites is an individual cotton leaf, one per randomly chosen plant. Leaves are scored as clean or infested, because mites are far too numerous to be counted individually. The economic injury level for spider mites varies with species and crop phenology, but a workable threshold is $p_c = 0.80 \pm 0.05$ leaves infested (Wilson & Room, 1983). Here, we set Type I and II error rates at 0.05, and $n_{\min}$ and $n_{\max}$ at 5 and 75 leaves, respectively. Sequential plans for spider mites using these parameters and Wald's and Iwao's procedures are summarized in Fig. 7.5.

In practice, a scout enters a field at weekly intervals during the period of highest risk and observes a leaf from each of five plants, and totals the number infested. If the total is outside the stop limits, then sampling stops and a decision is made to either do nothing or to apply an acaricide, depending on whether the lower or upper limit was exceeded. On the other hand, if the total is within the central "indecision zone," then additional leaves are inspected sequentially until a decision can be made. To limit the possibility that sampling might continue indefinitely, sampling is halted after a predetermined 75 leaves have been examined. If the total at that point is still in the indecision zone, it is recommended that the field be resampled in three days. For comparison, RSS = 64 leaves (via Eq. 7.6a) would be needed to estimate p when 0.8 with $h = 0.05$, $\alpha = 0.05$.

In this example, Wald's stop limits are slightly broader than Iwao's when sample sizes are below 25, but narrower when above (Fig. 7.5A). The OC curves indicate the two plans would have equivalent "no-act" profiles (Fig. 7.5B). Average sample number (ASN) curves with the two plans are equivalent at low infestation levels, but ASNs with Iwao's plan are greater than with Wald's plan when infestation rate is near or above threshold (Fig. 7.5C). Despite minor differences in performance, either plan would enable scouts to make treatment decisions with far fewer leaf inspections than would be required by a fixed-precision plan with equal type I error rate.

7.5 Practical considerations and future needs

The elements of IPM sampling discussed thus far should offer guidance in the design and use of sampling plans for individual pests on a one-by-one basis. In practice, though, most IPM situations involve more than one potential pest species. Furthermore, the pest fauna and flora is likely to change as the growing season progresses within any region, and from one geographic region to another. A practical difficulty for IPM scouts is that the optimal sample unit and sampling design for one species is not likely to be the same for all. Hence, compromises must be made if multiple pests are to be monitored simultaneously. Solutions are currently being found by practitioners through common sense and intuition. There is a need to develop an economic framework for optimizing sampling plans for multiple species situations, based on sampling costs and economic value of the information obtained.

IPM researchers and stakeholders working in some agricultural systems have invested to develop sampling programs that are practical and effective. However, Wearing (1988) surveyed IPM researchers and educators, and found that about 50 percent of respondents in Australia, New Zealand, Europe and the USA ranked "lack of simple monitoring methods" as the greatest technical barrier to IPM implementation. We are unaware of formal surveys that have assessed sampling practices in any IPM sector, but we suspect rigor and extent of adoption are greatest where the economic and environmental stakes are highest, but much room for improvement remains. Unfortunately, sampling requires labor that is costly in many parts of the world, so there is a universal need to find technologies that could reduce labor costs and improve efficiency. There is growing interest in remote sensing, forecasting models and global positioning systems to direct sampling resources to locations and times of greatest need. Additionally, web-based data entry, analysis and reporting procedures may ease adoption of rigorous sampling procedures, and permit retrospective analyses of their performance.

References

Alexander, C. J., Holland, J. H., Winder, L., Woolley, C. & Perry, J. (2005). Performance of sampling strategies in the presence of known spatial patterns. *Annals of Applied Biology*, **146**, 361–370.

Binns, M. R. (1994). Sequential sampling for classifying pest status. In *Handbook of Sampling Methods for Arthropods in Agriculture*, eds. L. P. Pedigo & G. D. Buntin, pp. 137–174. Boca Raton, FL: CRC Press.

Binns, M. R., Nyrop, J. P. & van der Werf, W. (2000). *Sampling and Monitoring in Crop Protection: The Theoretical Basis for Developing Practical Decision Guides*. Wallingford, UK: CABI Publishing.

Cochran, W. G. (1977). *Sampling Techniques*. New York: John Wiley.

Espino, L. A, Way, M. O. & Wilson, L. T. (2007). Determination of *Oebalus pugnax* (Hemiptera: Pentatomidae) spatial pattern in rice and development of visual sampling methods and population sampling plans. *Journal of Economic Entomology*, **101**, 216–225.

Fleischer, S. J., Blom, P. E. & Weisz, R. (1999). Sampling in precision IPM: when the objective is a map. *Phytopathology*, **89**, 1112–1118.

Hutchison, W. D. (1994). Sequential sampling to determine population density. In *Handbook of Sampling Methods for Arthropods in Agriculture*, eds. L. P. Pedigo & G. D. Buntin, pp. 207–243. Boca Raton, FL: CRC Press.

Iwao, S. (1975). A new method of sequential sampling to classify populations relative to a critical density. *Research in Population Ecology*, **16**, 281–288.

Karandinos, M. G. (1976). Optimum sample size and comments on some published formulae. *Bulletin of the Entomological Society of America*, **22**, 417–421.

Kogan, M. & Herzog, D. C. (eds.) (1980). *Sampling Methods in Soybean Entomology*. New York: Springer-Verlag.

Kuno, E. (1991). Verifying zero-infestation in pest control: a simple sequential test based on the succession of zero-samples. *Research in Population Ecology*, **33**, 29–32.

Lohr, S. L. (1999). *Sampling: Design and Analysis*. Pacific Grove, CA: Brooks/Cole Publishing.

Madden, L. V. & Hughes, G. (1999). Sampling for plant disease incidence. *Phytopathology*, **89**, 1088–1103.

Moon, R. D. (2007a). Estimation of Canada thistle areas along roadway right-of-ways. In *Radcliffe's IPM World Textbook*, eds. E. B. Radcliffe, W. D. Hutchison & R. E. Cancelado. St. Paul, MN: University of Minnesota. Available at http://ipmworld.umn.edu/textbook/moon1.htm.

Moon, R. D. (2007b). Detection of leafy spurge along roadway right-of-ways. In *Radcliffe's IPM World Textbook*, eds. E. B. Radcliffe, W. D. Hutchison & R. E. Cancelado. St. Paul, MN: University of Minnesota. Available at http://ipmworld.umn.edu/textbook/moon2.htm.

Naranjo, S. E. & Hutchison, W. D. (1997). Validation of arthropod sampling plans using a resampling approach: software and analysis. *American Entomologist*, **43**, 48–57.

Pedigo, L. P. & Buntin, G. D. (eds.) (1994). *Handbook of Sampling Methods for Arthropods in Agriculture*. Boca Raton, FL: CRC Press.

Perry, J. N. (1981). Taylor's power law for dependence of variance on mean in animal populations. *Applied Statistics*, **30**, 254–263.

Ruesink, W. G. (1980). Introduction to sampling theory. In *Sampling Methods in Soybean Entomology*, eds. M. Kogan & D. C. Herzog, pp. 61–78. New York: Springer-Verlag.

Southwood, T. R. E. & Henderson, P. A. (2000). *Ecological Methods*, 3rd edn. London: Blackwell Science.

Taylor, L. R. (1961). Aggregation, variance and the mean. *Nature*, **189**, 732–735.

Taylor, L. R., Woiwod, I. P. & Perry, J. N. (1978). The density-dependence of spatial behavior and the rarity of randomness. *Journal of Animal Ecology*, **47**, 383–406.

Venette, R. C., Moon, R. D. & Hutchison, W. D. (2002). Strategies and statistics of sampling for rare individuals. *Annual Review of Entomology*, **47**, 143–174.

Wearing, C. H. (1988). Evaluating the IPM implementation process. *Annual Review of Entomology*, **33**, 17–38.

Wilson, L. T. (1982). Development of an optimal monitoring program in cotton: emphasis on spider mites and *Heliothis* spp. *Entomophaga*, **27**, 45–50.

Wilson, L. T. (1994). Estimating abundance, impact, and interactions among arthropods in cotton agroecosystems. In *Handbook of Sampling Methods for Arthropods in Agriculture*, eds. L. P. Pedigo & G. D. Buntin, pp. 475–514. Boca Raton, FL: CRC Press.

Wilson, L. T. & Room, P. M. (1983). Clumping patterns of fruit and arthropods in cotton, with implications for binomial sampling. *Environmental Entomology*, **12**, 50–54.

Chapter 8

Application of aerobiology to IPM

Scott A. Isard, David A. Mortensen, Shelby J. Fleischer and Erick D. De Wolf

Insects, plant pathogens and weeds that move through the air create some of the most interesting pest management problems because their populations can increase dramatically, often with little warning and independent of factors that operate within fields (Irwin, 1999; Jeger, 1999). The advent of IPM programs has created an increased need to predict when, where and which pest populations are likely to grow rapidly and require control. Where dispersal is critical to the dynamics of populations, the realization of this demand requires information on the movements of pests into and out of agricultural fields and the degree to which fields within landscapes and regions are interconnected by these flows (Isard & Gage, 2001). Fundamental to this need is a solid understanding of aerobiology, the study of the biological and atmospheric factors that interact to govern aerial movements of biota (aerobiota) among geographic places (Aylor & Irwin, 1999).

Aerobiology, and dispersal research in general, is currently "on the move," in large part due to rapid advances in technologies for measuring and analyzing flows of organisms at relevant temporal and spatial scales (Gage *et al.*, 1999; Westbrook & Isard, 1999; Blackburn, 2006).[1] The renewed attention to issues of movement spans a wide range of pest and beneficial taxa that use air, water and/or land to change position on Earth for a multiplicity of reasons. Human-mediated dispersal is receiving much attention as well, although inadvertent pest movements in association with the globalization of commodity exchanges and human travel are extremely difficult to measure (National Research Council, 2002).

The success of IPM tactics may be limited by insufficient knowledge of aerobiology (Irwin & Isard, 1994). Being able to estimate the contribution of immigrants to a pest population, versus those derived from within-field development, and whether these immigrants are derived locally or from more distant sources, would dramatically improve our ability to manage pest density and resistance. Programs to introduce biological enemies would also benefit from improved predictions of aerial movements of these organisms. Technologies to bio-engineer crops and other organisms are advancing rapidly, but little is understood about how these engineered organisms respond to natural fluctuations of weather and their ability to spread through the air. This knowledge is necessary to meet

[1] See special sections in *Agricultural and Forest Meteorology*, 1999, vol. 97, no. 4; *Ecology*, 2003, vol. 84, no. 8; *Diversity and Distributions*, 2005, vol. 11, no. 1; and *Science*, 2006, vol. 313, no. 5788.

Integrated Pest Management, ed. Edward B. Radcliffe, William D. Hutchison and Rafael E. Cancelado. Published by Cambridge University Press.

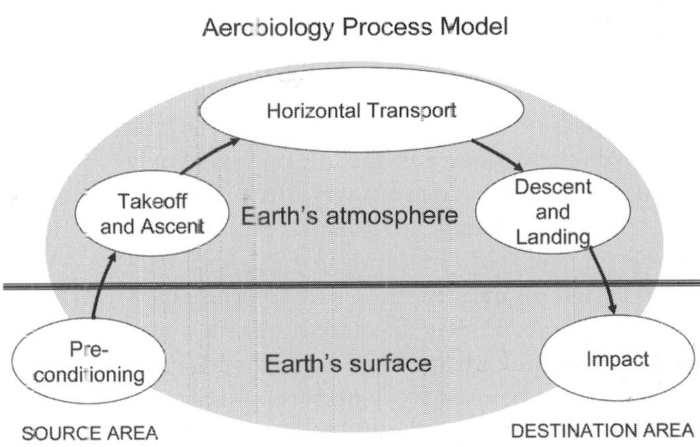

Fig. 8.1 The original version of the conceptual model was called the aerobiology pathway model and the stages were source, release, dispersion, deposition and impact. Figure adapted from Benninghoff & Edmonds (1972). The stages were renamed by Isard & Gage (2001) to extend the applicability of the conceptual model across a range of aerobiota from microorganisms to birds.

regulatory requirements for deploying genetically engineered organisms (Rissler & Mellon, 1996), and to design effective resistance management programs (Brown & Hovmoller, 2002). The potential risk from the airborne introductions of engineered animal and plant pathogens, pollen and seeds is also considerable (Wheelis *et al.*, 2002; Stack *et al.*, 2006). Whether these organisms are released purposefully or by mistake, the same aerobiological knowledge is required to understand and predict their spread.

8.1 The aerobiology process model and scales of movement

During the US Biophysical Program on Aerobiology in the 1970s, the systems approach began to be employed to focus research on the many interrelated factors that influence the movement of aerobiota including agricultural pests (Edmonds, 1979). From this collaborative effort, a conceptual model of the aerobiology process emerged (Fig. 8.1) which still serves as a foundation for aerobiological research today (Isard & Gage, 2001). The model holds that each organism that changes geographic location by moving through the air proceeds sequentially through five basic stages: (1) preconditioning, (2) takeoff and ascent, (3) horizontal transport, (4) descent and landing and (5) impact.

Knowledge of the atmospheric medium through which organisms move is fundamental to understanding the suite of ecological and environmental factors that they experience in each airborne life stage and consequently the impact they cause at their destination (Pedgley, 1982; Aylor, 2003). Although the aerobiology process model is scale independent, at any given time and place, the air is composed of a variety of atmospheric motion systems that are hierarchically nested in time and space (Stull, 1988). For example, a typical hurricane/midlatitude cyclone (macroscale) is composed of many thunderstorms/tornadoes (mesoscale) and a multitude of gusts/eddies (microscale). Organisms may encounter a number of these motion systems in a single journey. Consequently, the temporal and spatial scales relevant for organisms that move in the air are highly variable ranging from seconds to weeks and from plant parts to continents (Isard & Gage, 2001).

The biology and ecology of organisms that exploit atmospheric winds to disperse, within and among habitats, dictate the relevant temporal and spatial scales for particular individuals and populations and strongly influences their patterns of movement (Nathan, 2005). A taxonomy of movements useful for arthropods was constructed by Dingle (1996) based on the work of John S. Kennedy. Relating movement to life histories, Dingle grouped movement behaviors into three types: (1) those that are directed by a home range or are resource dependent, (2) those that are not directly responsive to resource or home range and (3) those that are not under the control of the organism. Movements that meet the second criteria are considered to be migratory. Although this classification does not apply to plant pathogens and weeds that disperse in the air, it is extremely useful for characterizing commonalities in migratory behaviors among arthropod pests and for identifying the spatial and temporal scales associated with their movements. Phenological, physiological and morphological adaptations for movement in the air expressed by plant pathogens and weeds also can be used to identify relevant scales of movement. Examples include the plant controlled traits that dictate terminal velocity, release height and abscission in seeds (Tackenberg *et al.*, 2003), liberation mechanisms in fungi (Gregory, 1973), and pigmentation in spores that influence survival to UV radiation exposure during aerial transport (Aylor, 2003). Price (1997, ch. 3) provides a synthesis of how the size and shape of organisms influence how they relate to physical forces and thus their movement behaviors, with examples that encompass organisms relevant for IPM applications. Development of these movement-adapted traits and behaviors among individuals in populations in source regions (preconditioning) are crucial to the initiation of movement and the status of the biota during the aerobiological process.

As a result of the structure of atmospheric motion systems and the biology of organisms that use the air to move, three scales of study are particularly relevant to the application of aerobiology in IPM: field, landscape and continental. At the field scale, the architecture of the plant canopy is shaped by interplant competition. The resulting canopy architecture strongly influences the temperature and moisture conditions as well as the small-scale motions within the atmospheric medium through which organisms move (Lowry & Lowry, 1989). Individual journeys occurring within a canopy or immediately above a field are typically shorter than 100 m and last a few minutes at most. An overwhelming proportion of passively transported organisms move only a few meters from their point of origin in a single transport event (Gregory, 1973). Because populations of pests that utilize crops for resources can build rapidly, and because growers typically only have realistic opportunities to manage fields, the individual field has been the spatial scale at which IPM is most often practiced.

While in many situations dispersal occurs at a field scale for much of a growing season, movements among landscape elements are often associated with behaviors related to colonization and overwintering. Management opportunities exist at these wider scales (National Research Council, 1996; Hunter, 2002) and in many circumstances aerobiology is a critical component of effective ecologically based management at scales larger than a single field. Pheromone disruption programs directed at insect pests, for example, should operate at scales that encompass the mate-searching area (Jones, 1998). Areawide efforts at pest management, programs aimed at altering the rate of geographic expansion (Tobin *et al.*, 2004) and eradication efforts directed against early-stage invasive pests require understanding of movement among host resources and especially between agricultural and surrounding landscapes (Ekbom, 2000). An understanding of movement biology is also key to developing effective strategies for using transgenic crops for areawide suppression of pest populations (Carriere *et al.*, 2003) and for successful pro-active deployment of resistance management strategies for transgenic crops (Sisterson *et al.*, 2005).

Local winds that blow across landscapes are influenced by the physical characteristics of the Earth's surface and the extent of the geographic assemblages of ecosystems that comprise the landscape (Oke, 1987). Relief in a landscape can induce mechanical winds while thermal circulations typically occur during calm atmospheric conditions when skies are clear and where there is spatial variation in the rate of surface heating. These landscape-scale winds provide opportunities for pests to move among diverse habitats (Nathan, 2005). Generally journeys of individuals span hundreds to thousands of meters and occur within a single diel period (Isard & Gage, 2001). To a large extent, landscape-scale movements of populations occur over relatively short time periods (hours to days), either because the life history of the species makes many individuals ready to move at the same time and/or because only a limited number of atmospheric motion systems facilitate movement to appropriate destinations (Pedgley, 1982). The resultant flows of organisms both among fields with similar crops and among diverse habitats within landscapes are often critical to the dynamics of pest populations in agricultural fields and provide opportunities for novel IPM practices.

At the continental scale, weather systems embedded in the global circulation govern the aerial movement of organisms (Pedgley *et al.*, 1995). These systems range from thunderstorms to large cyclones that traverse tens to thousands of kilometers over days to weeks. The progression of spring from the subtropics into mid-latitude continents (i.e. warm temperatures and longer days) triggers emergence and development of many plants and animals (Gage *et al.*, 1999). Large-scale advection of warm moist air from the subtropics at this time provides a mechanism for long-distance transport of organisms that are in dispersal-ready life stages from low latitudes to more poleward regions (Johnson, 1995). In summer, when the latitudinal temperature gradient is less pronounced, winds are weaker and opportunities for continental-scale biological movements are less frequent. However, opportunities for long-distance movement increase again in early autumn with the advent of tropical cyclones and hurricanes (Isard *et al.*, 2005). The incursion of cold polar air masses in autumn also stimulates the senescence of most crops and occurs at a time when many weedy species have mature seeds preconditioned for flight. At this time, enhanced latitudinal temperature gradients create weather systems with strong winds directed from the poles equatorward that are capable of transporting dispersal ready aerobiota from the middle latitudes to overwintering habitats in the subtropics (Johnson, 1995). Either because of their biology or due to the nesting of landscape and local-scale motion systems within large-scale weather patterns, or both, other organism such as seeds exploit continental-scale weather patterns to disperse over lesser distances. Sudden and often dramatic influxes of pests associated with these large-scale weather systems occur each spring in most midlatitude North American agricultural regions (Stakman & Harrar, 1957), and similar large-scale movements toward subtropical latitudes are facilitated by weather patterns during the fall (Showers *et al.*, 1993).

8.2 Application of aerobiology principles to IPM in the twenty-first century

Realizing the difficulties of making generalizations about the movement process because of the biological diversity of aerobiota and the relevance of events that occur at multiple and often nested temporal and spatial scales, a group of scientists and outreach specialists met in 1993 to create an aerobiology research focus (Isard, 1993; Isard & Gage, 2001). They developed a generic set of hypotheses pertaining to long-distance transport of biota during the middle three stages of the aerobiology process (Table 8.1). Although the hypotheses pertain to movement in the atmosphere, the participants acknowledged that long-distance movement can only be understood holistically if one understands what happens at each end of the movement process as well. Their vision was that researchers would evaluate the generic hypotheses across a wide range of both biological and meteorological systems. The resulting scientific principles would then provide the basis for applying aerobiology to IPM and ecosystem management in general.

The technologies for understanding and using aerobiology to enhance IPM decision support have advanced dramatically. Enabling technologies supported by the cyber infrastructure that link weather data and remotely sensed phenomena to pest movement, and enhance coordination of on-the-ground sampling of pests and/or hosts and real-time exchange of expert opinion by practitioners have made it possible to functionally integrate risk assessment across multiple disciplines and data sources to address complex pest management issues. The technologies that have enhanced observation, analysis and communication capabilities include: (1) Measurement, monitoring and diagnostics tools that allow rapid and more accurate spatially referenced data collection, (2) use of citizen scientists for expanding our biological observations networks over large

Table 8.1 General purpose hypotheses governing the atmospheric movement of biota

Maintenance of the movement process
(1) Long-distance movement is a one-way process
(2) Long-distance movement is a two-way process
  (2a) Return movement is ancillary to long-distance aerial transport
  (2b) Return movement reinforces the genetic control over long-distance transport
  (2c) Return movement is driven by existing environmental conditions

Components of the movement process
(A) Takeoff and ascent
(3) Takeoff and ascent into the atmosphere by organisms that move long distances is biologically mediated
  (3a) The phenological state that invokes initiation of ascent is genetically controlled
  (3b) Environmental preconditioning induces a physiological state that causes initiation of ascent
  (3c) Intraspecific/interspecific interactions influence the initiation of ascent
  (3d) Ascent may be influenced by aerodynamic properties
(4) Ascent by organisms into the atmosphere is influenced by environmental conditions
  (4a) Ascent is governed by convection within the lower atmosphere
  (4b) Thresholds of important atmospheric factors limit the tendency to take off and the success of ascent
  (4c) Ascent may be caused by hydrometeors
(B) Horizontal transport
(5) Organisms are concentrated within atmospheric layers during long-distance aerial movement
  (5a) Behavior and aerodynamic properties govern the vertical distribution of organisms during long-distance aerial movement
  (5b) Atmospheric factors govern the vertical distribution of organisms during long-distance aerial movement
(6) Horizontal transport of organisms within the atmosphere is predictable
  (6a) The duration and direction of movement are determined by the organism
  (6b) The duration and direction of movement are affected or driven by atmospheric processes
  (6c) The duration and direction of movement are influenced by environmental preconditioning
(C) Descent and landing
(7) Organisms actively descend from the atmosphere
  (7a) Environmental cues induce descent
  (7b) Physiological status govern descent
  (7c) Intraspecific/interspecific interactions influence the initiation of descent
(8) The descent of organisms from the atmosphere is governed by meteorological factors
  (8a) Descent is caused by hydrometeors
  (8b) Descent is caused by downdrafts
  (8c) Descent is caused by changes in stability/turbulence
  (8d) Descent is caused by gravity

Source: Taken from Isard (1993).

regions, (3) internet-based tools for data collection that enable immediate sharing of observations from around the world, (4) remote sensing instrumentation for monitoring important environmental variables (e.g. weather and land cover) that allow us to run models in near real time and to forecast change, (5) high-speed computing that enhances our capacity to add value to observations through modeling, (6) spatial analysis tools for higher levels of synthesis of processes that occur over space and time and (7) internet-based tools for rapid communication of commentary and guidelines for managing pest problems.

In the case studies that follow, we demonstrate the usefulness of applying the aerobiology process model in IPM. We describe novel types of pest management problems that operate at landscape to continental scales, and how advanced technologies can improve IPM decision making in the twenty-first century. To highlight the diversity of applications of aerobiology to IPM, we have chosen to focus on a combination of insect, plant pathogen and weed systems. The three case studies also represent different stages in the application of aerobiology to IPM. In the first, we related how aerobiology-based sampling methods have been used to increase our understanding of seed movement across landscapes and improve the effectiveness of management strategies for herbicide-resistant weeds in genetically modified crops. The second case study describes a network for monitoring migratory noctuid moths north of their overwintering range, which begins to provide early warnings and insights into the movement processes and seasonal patterns of this lepidopteran pest. In the final case study, we demonstrate the utility of coupling a system for forecasting the aerial spread of an important plant pathogen at the continental scale with an extensive monitoring system on a state-of-the-art information technology (IT) platform to provide real-time communications and IPM decision support.

8.3 Case studies

8.3.1 An aerobiology approach influencing herbicide resistance management plans

Horseweed (*Conyza canadensis*, syn. *Erigeron canadensis*) is a winter annual plant, native to North America. Its distribution is worldwide, though it is more abundant in temperate climates. It is found in a wide range of soil types and disturbance regimes from coarse sandy soils to organic soils and agricultural fields to roadsides and recently abandoned fields (Leroux *et al.*, 1996). Horseweed seedlings emerge from late August through October, forming rosettes that overwinter (Holm *et al.*, 1997), to emerge in late winter and early spring. The plant bolts in late spring, blooms in mid-July, and seed set occurs in early August. A mature plant can reach 2 m in height producing upwards of 200 000 seeds that are wind-borne with the aid of a pappus (Weaver, 2001).

In the summer of 1998, horseweed resistant to glyphosate herbicide was found infesting soybean fields along the Delaware–Virginia coast (VanGessel, 2001). It was the first weed in an annual crop to develop resistance to glyphosate herbicide. This event was particularly important as glyphosate tolerance was introduced in genetically modified soybean and made commercially available in 1996. The adoption rate for this transgenic crop is unprecedented. In the 2002 field season approximately 75% of USA soybean hectares were planted to glyphosate-tolerant cultivars and the use rate has now reached 87% of planted hectares (Chassy, 2005; Dauer *et al.*, 2006). Glyphosate-tolerant maize, alfalfa and cotton have since been commercialized and those crops have also experienced high adoption rates. Already, by 2005, 31% of the maize hectares in New York State had been planted to glyphosate-tolerant maize (National Agricultural Statistics Service, 2007). Repeated use of a high mortality practice selects for resistance to that practice. The selection pressure for resistance to glyphosate is high given that it is regularly used in most of the widely planted summer grain and fiber crops. Complete reliance on glyphosate for weed control goes against the fundamental tenet of IPM to promote the use of a diversity of tactics to avoid selection for adapted species. Preventing the appearance of new glyphosate-resistant biotypes and limiting the spatial extent and impact of existing resistant populations is a pressing IPM challenge with profound agronomic and economic implications.

While this discussion centers on resistant *C. canadensis* spread among fields, the underlying processes apply to many weedy species. In general, weed invasions are successful when the species can scatter seeds among heterogeneous environments, establish, adapt and then disperse again to colonize other areas (Sakai *et al.*, 2001). Wind dispersal provides species with an effective method of dispersing throughout a landscape. Some common agricultural weeds that are increasing on farmlands in the USA, including thistles (*Carduus* spp. and *Cirsium* spp.), dandelion (*Taraxacum officinale*), milkweed (*Asclepias syriaca*) and *C. canadensis*, use the atmosphere to move throughout landscapes and often long distances. Establishment in and adaptation to the varying cropping systems within which weed seeds land are critical next steps in the invasion process. Individuals that successfully establish themselves in new locations play an important role in the invasion because they are adapted to the destination habitat and act as seed sources for the next wave of the invasion. The increase in no-tillage hectares can be directly linked to adoption of glyphosate-tolerant crops and the result is a reduction in herbicide heterogeneity (Carpenter *et al.*, 2002). This aids in the establishment of resistant horseweed and the invasion of additional farms. Consequently, resistant horseweed provides an ideal subject for the study of long-distance dispersal across complex landscapes. Aerobiological knowledge gained through this research effort in turn provides a basis for developing management strategies for many of our most problematic weeds.

Seed dispersal is a dynamic process combining functions of plant and seed morphology, habitat characteristics (e.g. topography and rugosity), vector mediation and post-settlement dynamics (Cousens & Mortimer, 1995). Horseweed produces small seeds with an attached pappus. The terminal fall velocity (settling velocity) of a seed is dramatically reduced by the presence of a plume structure (Burrows, 1975). Anderson (1993) measured settling velocities of wind-dispersed seeds from 19 species of the family Asteraceae, and results for horseweed were the lowest of those tested. Combined, these characteristics indicate a high potential for wind dispersal of horseweed over considerable distances. To better understand just how far seeds could move, thus defining the interconnectedness of fields within an agricultural landscape, a series of experiments were initiated. Source patch populations were established and the source strength of each patch was quantified. Seed traps were positioned along radial transects from the source patches out to 500 m. The results of the study are striking (Dauer *et al.*, 2006). Small numbers of seeds were found near the ends of transects revealing that it was likely that they were moving beyond 500 m. This is five times greater than any previously reported dispersal distance for horseweed. Simulations conducted using a two-dimensional model and a range of realistic source strength values, indicated that it is highly likely that horseweed can disperse distances of 2–5 km. Not surprisingly, the dispersal distance was greatest in the direction of the prevailing wind direction. The results reveal that farm fields are much more highly interconnected with respect to weed management outcomes than previously thought. Prior to these aerobiology-based studies, herbicide resistance management and the development of weed infestations were considered field-specific processes.

To confirm this novel and important finding, Shields *et al.* (2006) deployed remotely controlled aircraft to sample seeds in the air above and downwind of source patches. Specifically, they were interested in whether or not seeds are able to leave the surface boundary layer of the atmosphere and attain an altitude where they are likely to be blown long distances. In this campaign, horseweed seeds were collected at the highest altitude sampled (110 m above the field). Concomitant sampling of seeds and measurements of wind speed, wind direction and turbulence immediately above the source patch provide the basis for aerobiological analyses. The greatest numbers of horseweed seeds were captured above the surface boundary layer, generally considered to be twice the height of the canopy in the field (Oke, 1987), during the mid afternoon. It is reasonable to assume that seeds which are carried upward through the surface layer during midday by thermals are mixed throughout the planetary boundary layer above by evening when turbulence in the lower atmosphere typically diminishes and the air becomes stable. In this situation, given a fall velocity for horseweed of 0.32 m/s (Dauer *et al.*, 2006), the seeds would require between 3 min and 1 h to settle out of the air. Assuming that the seeds were released near midday and landed 1 h after dusk, a light wind speed of 5 m/s (11 mph) would transport them between 75 and 150 km while a strong wind of 20 m/s (43 mph) would transport them over 550 km (Shields *et al.*, 2006).

The *Conyza* system and management problem provided an opportunity to apply aerobiology-based methods for sampling a pest organism and associated environment to enlighten our understanding of how to manage herbicide-resistant weeds in genetically modified crops. The fact that seed dispersal distances may stretch many kilometers is important in a number of ways. First, it demonstrates that plants possess the ability to move their propagules and novel genes much greater distances than previously thought. Such movement has important implications for the biology and ecology of the species. From an IPM perspective, such long-distance transport effectively decouples a farmer's management decision from the resulting weed control. A neighbor's field that contains a wind-dispersed weed will continue to act as a source for invasion into adjacent fields regardless of the management practices used to maximize control in adjacent fields. In the case of glyphosate-resistant weeds, the continued invasion could limit the options of farmers utilizing glyphosate-tolerant crops.

One solution is cooperation among farmers to slow the spread of this and other wind-dispersed species. Already such cooperation is evident at the level of the commercial applicator and consultant. In regions on the scale of commercial applicator districts, resistance management herbicide programs have been devised and deployed. Invariably, they involve either adding a pre-emergence treatment prior to the glyphosate application that targets horseweed or using an additional post-emergence herbicide with a different mode of action to control the glyphosate-resistant horseweed. Clearly, applying the knowledge of inter-field movement of horseweed seed to growing regions that have not witnessed the establishment of glyphosate-resistant horseweed populations would be the most prudent action in the long run. This is because managing resistant populations in ways mentioned above bring with it $US 20–37 per hectare increase in herbicide cost and the environmental impact of increased herbicide load. In a recent symposium addressing the subject of managing glyphosate-resistant weeds, Mortensen *et al.* (2007) argued that other areawide management practices should also be considered and could include establishing buffer regions around fields or farmsteads with resistant biotypes. Within the buffer, a more diverse cropping practice and herbicide program would be employed. For example, recently VanGessel *et al.* (2007) reported that winter cover crops like rye and hairy vetch significantly limit the establishment and fecundity of glyphosate-resistant *Conyza*.

We argue that detailed aerobiology research can be used to guide pesticide registration and agricultural policy as well as stewardship practices. For example, if spread of glyphosate-resistant biotypes occurs over large distances and over short time intervals, such information could help the US Environmental Protection Agency make decisions about future registrations of crops carrying this trait. For example, the impact of increasing the number of hectares on which glyphosate can be used by decreasing management practice diversity was explored using a spatially explicit modeling approach and with knowledge of the horseweed dispersal direction and distance (Mortensen *et al.*, 2003). The invasion speed of the resistant horseweed increased dramatically when it was assumed that the maize and alfalfa hectares in a Pennsylvania study region were planted to glyphosate-tolerant cultivars. Clearly herbicide management plans that target the individual field and grower will not contain the spread of problem weeds such as horseweed that move well beyond a farmstead in a single day.

8.3.2 An aerobiological approach influencing measurement of migratory lepidopterans

Insect species unable, or less able, to adapt to the harsh winters that cover most of the North American continent are eliminated or dramatically reduced in all or most of the region. Examples include lepidopterans in the family Noctuidae, e.g. the corn earworm (*Helicoverpa zea*), fall armyworm (*Spodoptera frugiperda*), black cutworm (*Agrotis ipsilon*), beet armyworm (*Spodoptera exigua*) and soybean looper (*Pseudoplusia includens*). These migratory lepidopterans annually reinvade their northern geographic range through long-distance aerial movements involving successive broods advancing northward (Holland *et al.*, 2006). The corn earworm, in the subfamily Heliothinidae, serves as a model where understanding its population dynamics requires an aerobiology knowledge base and should be considered at a continental scale.

The biology and ecology of corn earworm enable it to use both resource-directed and migratory dispersal to exploit dispersed host resources. Adaptations influencing this dispersal behavior can be placed in context of its evolutionary history. Prior to 1953, and for over 100 years, there were only one or two recognized taxa in this subfamily, all in *Heliothis* (*asaulta, armigera*, or *obsoleta*) with typically global distribution. A distinct New World taxon (*zea*) was recognized in the 1950s (Common, 1953; Todd, 1955), which Hardwick (1965) placed into a newly erected genus *Helicoverpa. Heliocoverpa*, considered the ancestral genus (Hardwick, 1965), retains much higher fecundity (hundreds to thousands of eggs) than the other major groups (*Heliothis* or *Schinia*), and a much wider host range. Derived groups have decreasing fecundity and increased specializations, such as the ability to oviposit within floral structures, which are tied to a more specific host range and perhaps more resource-directed dispersal. *Helicoverpa*, in contrast, widely disperses many eggs, and has retained a much wider host range, along with stronger migratory behaviors. This ancestral group is comprised primarily of tropical or warm-temperate species, with continuous development in tropical regions except where there is a dry season. Some members (*zea* and *armigera*) have limited diapause capabilities in higher latitudes; none do this well. Neither *zea* from the Americas nor *armigera* from the Old World overwinters above 40° latitude (Hardwick, 1965). Thus, *H. zea* annual reinvasions affect the entire North American continental interior extending into Canada. A similar process occurs with *armigera* in the Old World, and allozyme patterns suggest that *H. zea* represents a founding event from *H. armigera* (Mallet *et al.*, 1993).

Holland *et al.* (2006) suggest that insects are often facultative migrants, responding to existing or predicted changes in habitat quality to achieve "multi-generational bet-hedging." "Bet-hedging" for *H. zea* could involve portions of the population exhibiting local, resource-directed redistribution among nearby hosts in the landscape, while others exhibit long-range migratory flight. Landscape-scale redistribution is well documented and has been modeled as an ovipositional choice process weighted by distance among fields (Stinner *et al.*, 1986). Long-range migratory flight is also documented (Lingren *et al.*, 1994; Westbrook *et al.*, 1997). Probability of recaptured moths carrying a citrus-pollen marker, indicating migration, was modeled as a logistic function of the insect's flight trajectory, duration of flight the previous night, and local minimum air temperature (Westbrook *et al.*, 1997). Aerobiology analyses suggest that migratory moths utilize dynamic but definable pathways, and synoptic weather patterns correlate with corn earworm migration into southern (Lingren *et al.*, 1994; Westbrook *et al.*, 1997) and midwestern USA (Sandstrom *et al.*, 2005).

As with other migratory noctuids (e.g. black cutworm: see Sappington & Showers, 1991), corn earworm displays behavior that distinguishes migration from foraging, including a spiraling vertical ascent behavior at dusk (Lingren *et al.*, 1995) which carries them above the boundary layer to altitudes of >800 m where they often concentrate in nocturnal wind jets associated with temperature inversions (Wolf *et al.*, 1990; Beerwinkle *et al.*, 1995). Within these jets, moths often orient at an angle to the downwind direction, perhaps to influence migration direction (Wolf *et al.*, 1995), compensate for wind drift, or reduce fallout. Airborne radar indicated that this persistent and straightened-out movement enables earworms to move a few hundred kilometers in a single night (Wolf *et al.*, 1990). Such distinctive poleward migratory flights during spring have been documented (Lingren *et al.*, 1994), while there is indirect evidence for return equatorward fall flights (Gould *et al.*, 2002). Mark–release–recapture of black cutworm provide direct evidence for both poleward spring and equatorward fall movements (Showers *et al.*, 1989, 1993). Air parcel trajectory analyses and radiosonde measurements indicated that these moths move long distances at altitudes between 300 and 800 m (Showers, 1997). Black cutworm also shows an age dependency for migratory flights (Sappington & Showers, 1991) with flight initiation inhibited by specific weather conditions (Domino *et al.*, 1983).

Thus, corn earworm is well adapted to our ephemeral cropping systems, and achieves pest

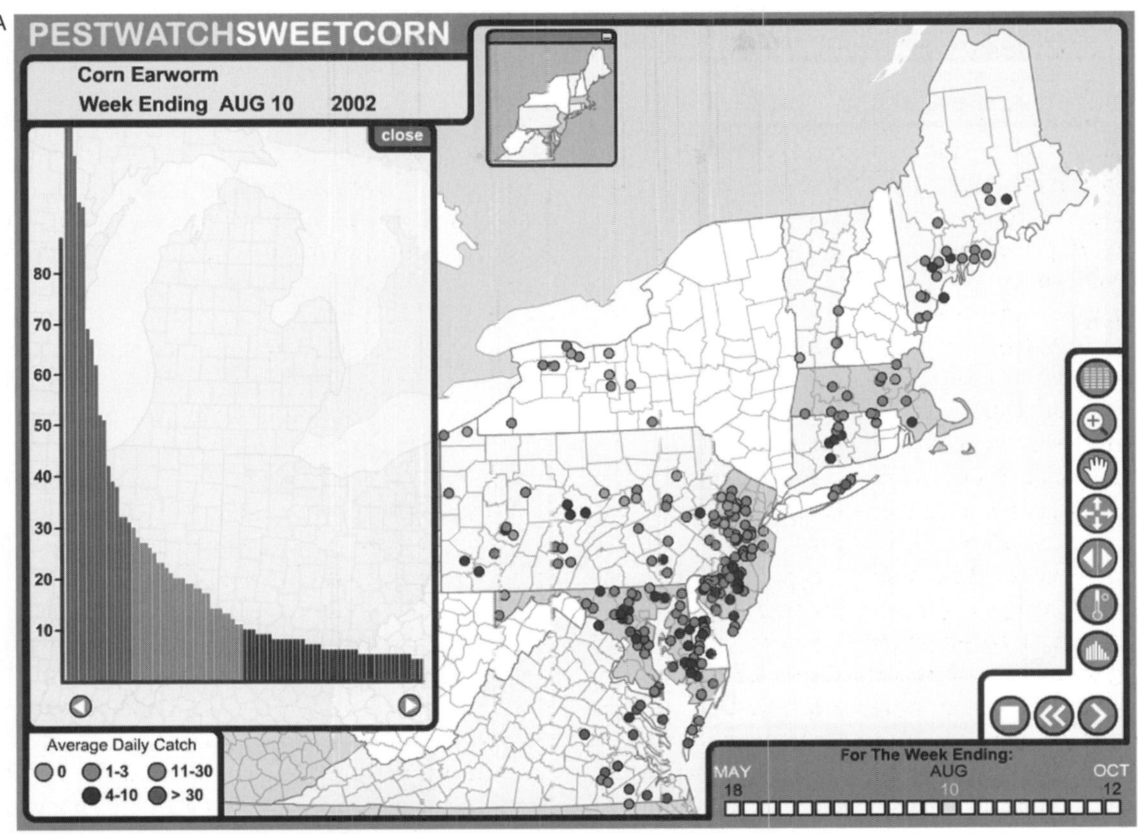

Fig. 8.2 Interactive map from PestWatch (Fleischer *et al.* 2007) showing captures of corn earworm for August 10 in the drought year of 2002 (A) compared to 2001 (B). Histogram insets show distribution of average daily capture for the current week among ~250 sites. Maps depicting changes over time may be viewed as animations while time-series from individual dates can be viewed as still frames at each sampling site. Geographic extents were expanded in 2007.

status in maize, cotton, tomatoes, snap beans, sorghum and soybeans. Larvae feed on flowers, fruits and seeds, boring quickly into reproductive tissue, causing direct damage to economically and nutritionally important plant parts. Management with insecticides or biological control is difficult once larvae access the interior of fruiting bodies. Coupled with a tendency to distribute eggs singly among fruiting structures, damage can become economically significant very quickly and can be disproportionately high relative to corn earworm density.

North of their overwintering range, dramatic influxes of corn earworm and other lepidopteran migrants can result in rapid population increase independent of factors that operate within fields. The timing and intensity of invasive, migratory flights can vary dramatically. For example (Fig. 8.2), in 2001, regional densities primarily from migrants into the northeastern USA had just begun to increase by approximately 10 August, whereas in the next year, 2002, much larger populations were evident throughout the region at the same time. Consequently, regional monitoring networks, a component of aerobiological approaches to IPM, have direct relevance. These include enhanced observation, analysis and communication capabilities. They can be used to rapidly and effectively visualize pest dynamics over wide geographic regions, at varied temporal and spatial scales, enabling us to discuss the migratory process as a comprehensive whole, and thus consider migratory processes when we consider management options. Aerobiological

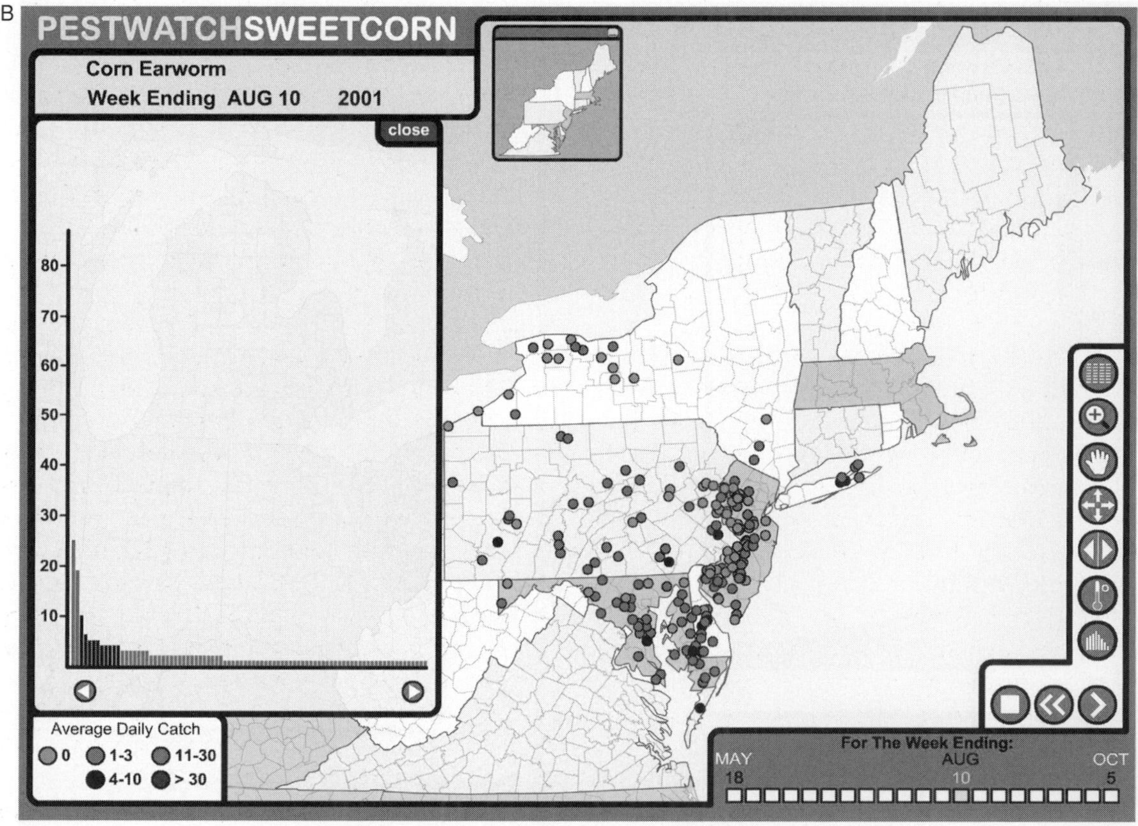

Fig. 8.2 (cont.)

structuring of data flows enables early warnings, both from historical knowledge gained through organized observations structured with information technologies, and from modeled forecasts that couple pest, host and meteorological processes.

Autonomous monitoring of corn earworm has occurred for decades for IPM of cotton, sweet corn or other vegetable crops. In the northeastern USA, almost entirely north of the overwintering range of the corn earworm, the migratory contribution to the density and dynamics of this pest made it critical to consider pest movement to improve IPM. A regional monitoring network was established across seven northeastern states in 1999 to provide advance warning about this migratory process (Fleischer *et al.*, 2007), and this network has recently expanded to approximately 500 sites. The maps influence insecticide spray decisions in vegetable crops.

8.3.3 An aerobiology based decision support system for managing soybean rust

In autumn 2004, the Asian strain of soybean rust (*Phakopsora pachyrhizi*), an exceptionally virulent fungus, was found infesting soybean fields in Louisiana (Schneider *et al.*, 2005). During the preceding years, the US Department of Agriculture (USDA) had prepared for the incursion of this pathogen by supporting grower, specialist and industry education, training, and surveillance programs, purchases of new equipment for diagnostic facilities, offshore fungicide evaluation trials, construction of risk assessments and searching breeding materials for novel sources of host plant resistance (Livingston *et al.*, 2004). The sense of urgency stemmed not only because *P. pachyrhizi* had demonstrated frequent long-distance aerial spread among all other major soybean growing areas worldwide, but also because yield losses from infected fields had been significant in each of these production regions (Miles *et al.*, 2003).

In 2003, shortly after soybean rust had blown from Africa across the South Atlantic Ocean to infect soybean fields in Brazil, Paraguay and Argentina, a number of research groups began developing models to assess the risk of *P. pachyrhizi* incursion into the USA, likely transport pathways, and when and where deposition of spores was most likely to occur (Isard *et al.*, 2005). At the onset of this program, the aerobiology process model (Fig. 8.1) was specified for the soybean rust system to identify relationships that needed to be incorporated into the transport models as well as to identify knowledge gaps that were used to guide field measurement programs in South America and the USA for evaluating aerobiology-related hypotheses (e.g. Isard *et al.*, 2006a). Output from the resulting Integrated Aerobiology Modeling System (IAMS) became the basis for the USDA, Economic Research Service assessment of the risk of a soybean rust incursion and its potential impact on USA agriculture (Livingston *et al.*, 2004).

IAMS simulations conducted immediately after discovery of soybean rust in the USA showed that airflows converging into Hurricane Ivan as it made landfall along the Gulf Coast had the potential to transport viable rust spores directly to the USA from the infected Rio Cauca source area in Colombia, South America. A model generated map delineating regions of spore deposition associated with the hurricane was provided to members of the USDA Animal and Plant Health Inspection Service (APHIS) Soybean Rust Rapid Response Team as they went into the field and was used successfully to scout for the pathogen (see Fig. 5 in Isard *et al.*, 2005).

Phakopsora pachyrhizi is an obligate parasite that requires green tissue to survive (Bromfield, 1984). In North America, it has primarily been found on two hosts, soybean and kudzu. Consequently, the geographic range of soybean rust during winter is restricted to areas along the Gulf Coast, and in the Caribbean basin, Mexico and Central America where either kudzu retains its foliage or soybeans are grown year round (Pivonia & Yang, 2004). To cause yield losses in major North American soybean production regions, *P. pachyrhizi* uridineospores must be blown from these overwintering areas into the continental interior between early May and August, when the crop is susceptible to the disease (Isard *et al.*, 2005).

During the 2004/2005 winter, APHIS provided funds to the IAMS team for the construction and operation of an IT platform to integrate soybean rust monitoring, database construction and communications to stakeholders (Isard *et al.*, 2006b). The team used the opportunity to integrate emerging technologies into a unique and highly functional, state-of-the-art cyber infrastructure. The level of cooperation among USDA agencies, state Departments of Agriculture, universities, industry and grower organizations, in support of the resulting USDA Soybean Rust Information System was unprecedented for an invasive agricultural pest in the USA, enabling the deployment of a pest information system with an exceptional level of utility and credibility. As a result of this success, government administrators, researchers, industry representatives and producers employed the same template the following year to launch a national Pest Information Platform for Extension and Education (PIPE) including soybean aphid along with soybean rust (Isard *et al.*, 2006b).

The IAMS is configured in a modular format to include all of the stages in the aerobiology process (Isard *et al.*, 2007). Host development and disease progression submodels, driven by weather data, are used to characterize source strength and distribution as well as colonization and disease progression at destinations. The IAMS also incorporates information on spore release and canopy escape in source areas, mortality due to exposure to solar radiation during atmospheric transport and both dry and wet deposition.

During the 2005 and 2006 growing seasons, the IAMS modeling team provided maps of output from daily simulations depicting deposition of spores on soybean and kudzu, host developmental stage and disease progression on these hosts for the subtropical and temperate regions of North America. Observations from an extensive network of "sentinel" soybean plots were used to define the geographic extent and severity of the disease in source areas. Because of the complexity of the biological and environmental interactions that are important to the soybean rust system and the paucity of knowledge about this

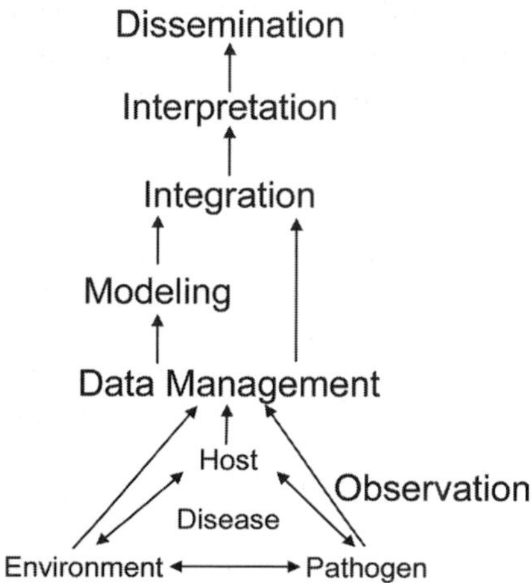

Fig. 8.3 The information technology structure of the Pest Information Platform for Extension and Education (PIPE) as designed for soybean rust. The observation component in the lower portion of the figure pertains to a plant pathogen and is different for insects and weeds.

disease in North America, one major limitation of the IAMS initially was the need to "nudge" the model. That is, to produce realistic simulations of pathogen transport and subsequent disease detection, model parameter values had to be continually adjusted and field observations of vegetative growth and changes in soybean rust distribution had to be continually fed into the IAMS. To achieve model adjustment in near real time and to provide a platform upon which to communicate interpretations of field observations and model output to users, the PIPE cyber infrastructure was developed.

The PIPE is people connected by computers using advanced IT to add value to field observations. The goals of the system are to enhance support for IPM decision making and provide information for documenting good management practices for crop insurance purposes. The IT structure of the PIPE is depicted in Fig. 8.3. Field observations of the pathogen, hosts and environment are channeled through standardized internet portals into a national database. This was accomplished by the developing protocols for field monitoring of the pathogen and its hosts and building a set of tools including PDA programs, internet forms and spreadsheet files for easy and rapid entry of observations into the national database. National Weather Service products are also downloaded daily and archived. These spatiotemporal-referenced data are immediately available to researchers throughout the country who are able to add value to the observations through modeling. Outputs from multiple models running on the platform are integrated with field observations of soybean rust into easy to read maps. Participants access information on the PIPE through two interrelated websites. A restricted access website provides a platform for extension specialists, researchers and administrators to view and interpret these maps. Extension specialists then use state-of-the-art IT tools to disseminate interpretations, management guidelines and other relevant materials to growers and industry agents through a public access website (Isard *et al.*, 2006b).

Realizing the success of the PIPE for providing useful information to support decision making for soybean rust IPM, the USDA Risk Management Agency (RMA), the Cooperative State Research Extension and Education Service (CSREES) and APHIS are developing a coordinated effort to maintain and expand the system capitalizing on the existing structure of regional IPM Centers and state extension specialists (Isard *et al.*, 2006b). The IPM Centers set national and regional pest priorities through interactions with industry and as a result of their leadership, the PIPE information technology structure was expanded in 2006 to include soybean aphid. Plans for future additions to the platform include viral diseases of dry beans, head scab of barley and lepidopteran pests of sweet corn. Even more critical to the success of the PIPE is the network of state extension specialists who coordinate the input of observations from monitoring networks into the national database and interpret field observations and model output to provide agricultural producers and industry with decision support for pest management and information for documenting their management activities.

8.4 Conclusions

Successful IPM implementation requires conceptual models that integrate knowledge of the biology, ecology and behavior of host–pest complexes

(Jeger, 1999). Where aerial dispersal of pests is an important component of the system, knowledge of the scale at which movements occur, and the flow structures within the atmospheric environment within which they take place, enable us to most effectively utilize ecologically based pest management. Understanding movement is absolutely essential for management that restricts geographic distribution of pests, alters genetic frequencies in populations, and thus potentially causes pesticide resistance. The *Conyza* case study demonstrates how aerobiology-based research can fill knowledge gaps and thus enhance understanding of movement across landscapes; this in turn, can improve the effectiveness of our management strategies for herbicide-resistant weeds in genetically modified crops.

In the corn earworm case study, IT tools are applied to spatially referenced monitoring of a pest for which migratory processes cause rapid increases in population independent of factors that operate within fields. The development of early warning systems based on regional monitoring networks such as PestWatch (Fleischer *et al.*, 2007), are a prerequisite for running and verifying aerobiology models that forecast impending movements. There are important feedback loops throughout this process. Efforts to visualize, describe and model movement of pests also reveal knowledge gaps and aerobiological methods can help direct future research priorities. Also, the information about pest density, dynamics and phenology at wider scales adds value to data collected at individual sites, thus increasing the effectiveness and efficiencies of scouting efforts.

The soybean rust example shows how an extensive biological observation network, coupled with mechanistic aerobiology transport models and state-of-the-art communications capabilities, all operating over large geographic scales can serve IPM. In 2006, the USDA Economic Research Service reported that many millions of USA soybean hectares that would otherwise have received unnecessary fungicide application for soybean rust in 2005 remained untreated for soybean rust in 2005 due to this application of aerobiology for IPM of soybean rust. In that year alone, the information disseminated through the USDA Soybean Rust Information System increased USA producers' profits between $US 11 to 299 million ($US 0.40 to 10.18 per hectare) at a cost that was only a fraction of this return (Roberts *et al.*, 2006).

References

Anderson, M. C. (1993) Diaspore morphology and seed dispersal in several wind-dispersed Asteraceae. *American Journal of Botany*, **80**, 487–492.

Aylor, D. E. (2003). Spread of plant disease on a continental scale: role of aerial dispersal of pathogens. *Ecology*, **84**, 1989–1997.

Aylor, D. E. & Irwin, M. E. (1999). Aerial dispersal of pests and pathogens: implications for integrated pest management. *Agricultural and Forest Meteorology*, **97**, 233–234.

Beerwinkle, K. R., Lopez, J. D. Jr., Schleider, P. G. & Lingren, P. D. (1995). Annual patterns of aerial insect densities at altitudes from 500 to 2400 meters in east-central Texas indicated by continuously-operating vertically-oriented radar. *Southwestern Entomology (Suppl.)*, **10**, 63–79.

Benninghoff, W. S. & Edmonds, R. L. (eds.) (1972). *Ecological Systems Approaches to Aerobiology. I. Identification of Component Elements and Their Functional Relationships*, United States/International Biophysical Program Aerobiology Program Handbook No. 2. Ann Arbor, MI: University of Michigan.

Blackburn, L. (2006). Tag team. *Science*, **313**, 780–781.

Bromfield, K. R., 1984. *Soybean Rust*, Monograph No. 11. St. Paul, MN: American Phytopathological Society Press.

Brown, J. K. M. & Hovmoller, M. S. (2002). Aerial dispersal of pathogens on the global and continental scales and its impact on plant disease. *Science*, **297**, 537–542.

Burrows, F. M., 1975. Calculation of the primary trajectories of dust seeds, spores and pollen in unsteady winds. *New Phytologist*, **75**, 389–403.

Carpenter, J., Felsot, A., Goode, T. *et al.* (2002). *Comparative Environmental Impacts of Biotechnology-Derived and Traditional Soybean, Corn and Cotton Crops*. Ames, IA: Council for Agricultural Science and Technology.

Carriere, Y., Ellers-Kirk, C., Sisterson, M. S. *et al.* (2003). Long-term regional suppression of pink bollworm by *Bacillus thuringiensis* cotton. *Proceedings of the National Academy of Sciences of the USA*, **100**, 1519–1523.

Chassy, B. (2005). *Crop Biotechnology and the Future of Food: A Scientific Assessment*. Ames, IA: Council for Agricultural Science and Technology.

Common, I. F. B. (1953). The Australian species of *Heliothis* (Lepidoptera, Noctuidae) and their pest status. *Australian Journal of Zoology*, **1**, 319.

Cousens, R. & Mortimer, A. M. (1995). *Dynamics of Weed Populations*. Cambridge, UK: Cambridge University Press.

Dauer, J. T., Mortensen, D. A. & VanGessel, M. (2006). Spatial and temporal dynamics governing long distance dispersal of *Conyza canadensis*. *Journal of Applied Ecology*, **44**, 105–114.

Dingle, H. (1996). *Migration: The Biology of Life on the Move*. Oxford, UK: Oxford University Press.

Domino, R. P., Showers, W. B., Taylor, E. & Shaw, R. H. (1983). Spring weather pattern associated with suspected black cutworm moth (Lepidoptera: Noctuidae) introduction into Iowa. *Environmental Entomology*, **12**, 1863–1871.

Ekbom, B. (2000). Interchanges of insects between agricultural and surrounding landscapes. In *Interchanges of Insects between Agricultural and Surrounding Landscapes*, eds. B. Ekbom, M. E. Irwin & Y. Robert, pp. 1–3. Dordrecht, Netherlands: Kluwer.

Edmonds, R. L. (ed.) (1979). *Aerobiology: The Ecological Systems Approach*. Stroudsburg, PA: Dowden, Hutchinson & Ross.

Fleischer, S. Payne, G., Kuhar, T. *et al.* (2007). Pestwatch sweetcorn pest monitoring system. In *Pestwatch*, eds. S. Fleischer *et al.* University Park, PA: Pennsylvania State University. Available at www.pestwatch.psu.edu/.

Gage, S. H., Isard, S. A. & Colunga-Garcia, M. (1999). Biological scales of motion for dispersal of biota. *Agricultural and Forest Meteorology*, **97**, 249–261.

Gould, F., Blair, N., Reid, M. *et al.* (2002). *Bacillus thuringiensis*-toxin resistance management: stable carbon isotope assessment of alternate host use by *Helicoverpa zea*. *Proceedings of the National Academy of Science of the USA*, **99**, 16 581–16 586.

Gregory, P. H (1973). *The Microbiology of the Atmosphere*. New York: John Wiley.

Hardwick, D. F. (1965). *The Corn Earworm Complex*, Memoir No. 40. Ottawa, Canada: Entomological Society of Canada.

Holland, R. A., Wikelski, M. & Wilcove, D. S. (2006). How and why do insects migrate? *Science*, **313**, 794–796.

Holm, L., Doll, J., Holm, E., Pancho, J. & Herberger, J. (1997). *Conyza canadensis* (L.) Cronq. (syn *Erigeron canadensis* L.). In *World Weeds: Natural Histories and Distribution*, pp. 226–235. New York: John Wiley.

Hunter, M. (2002). Landscape structure, habitat fragmentation, and the ecology of insects. *Agricultural and Forest Entomology*, **4**, 159–166.

Irwin, M. E. (1999). Implications of movement in developing and deploying integrated pest management strategies. *Agricultural and Forest Meteorology*, **97**, 235–248.

Irwin, M. E. & Isard, S. A. (1994). Movement and dispersal and IPM. In *Proceedings of the 2nd National Integrated Pest Management Symposium/Workshop*, pp. 136–137. Raleigh, NC: Experiment Station Committee on Organization and Policy, Pest Management Strategies Subcommittee and Extension Service IPM Task Force.

Isard, S. A. (ed.) (1993). *Alliance for Aerobiology Research Workshop Report*. Champaign, IL: Alliance for Aerobiology Research Workshop Writing Committee.

Isard, S. A. & Gage, S. H. (2001). *Flow of Life in the Atmosphere: An Airscape Approach to Understanding Invasive Organisms*. East Lansing, MI: Michigan State University Press.

Isard, S. A., Gage, S. H., Comtois, P. & Russo, J. M. (2005). Principles of aerobiology applied to soybean rust as an invasive species. *BioScience*, **55**, 851–861.

Isard, S. A., Dufault, N. S., Miles, M. R. *et al.* (2006a). The effect of solar irradiance on the mortality of *Phakopsora pachyrhizi* urediniospores. *Plant Disease*, **90**, 941–945.

Isard, S. A., Russo, J. M. & De Wolf, E. D. (2006b). *The Establishment of a National Pest Information Platform for Extension and Education*. Online, Plant Health Progress, doi:10. 1094/PHP-2006–0915-01-RV, available at www.ceal.psu.edu/Isard06.pdf.

Isard, S. A., Russo, J. M. & Ariatti, A. (2007). Aerial transport of soybean rust spores into the Ohio River Valley during September 2006. *Aerobiologia*, **23**, 271–281. DOI 10.1007/s10453-007-9073-z.

Jeger, M. J. (1999). Improved understanding of dispersal in crop pest and disease management: current status and future directions. *Agricultural and Forest Meteorology*, **97**, 331–349.

Johnson, S. J. (1995). Insect migration in North America: synoptic-scale transport in a highly seasonal environment. In *Insect Migration: Tracking Resources through Space and Time*, eds. V. A. Drake & A. G. Gatehouse, pp. 31–66. Cambridge, UK: Cambridge University Press.

Jones, O. T. (1998). Practical applications of pheromones and other semiochemicals. In *Insect Pheromones and Their Use in Pest Management*, eds. P. E. Howse, I. D. R. Stevens & O. T. Jones, pp. 263–351. London: Chapman & Hall.

Leroux, G. D., Benoit, D. L. & Banville, S. (1996). Effect of crop rotations on weed control, *Bidens cernua* and *Erigeron canadensis* populations, and carrot yields in organic soils. *Crop Protection*, **15**, 171–178.

Lingren, P. D., Westbrook, J. K., Bryant, V. M. Jr. *et al.* (1994). Origin of corn earworm (Lepidoptera: Noctuidae) migrants as determined by *Citrus* pollen markers and synoptic weather systems. *Environmental Entomology*, **23**, 562–570.

Lingren, P. D., Raulston, J. R., Popham, T. W. *et al.* (1995). Flight behavior of corn earworm (Lepidoptera: Noctuidae) moths under low wind speed conditions. *Environmental Entomology*, **24**, 851–860.

Livingston, M., Johansson, R., Daberkow, S. *et al.* (2004). *Economic and Policy Implications of Wind-Borne Entry of Asian Soybean Rust into the United States*. Electronic Outlook Report from the US Department of Agriculture, Economic Research Service, OCS-04D-02. Available at www.ers.usda.gov/publications/OCS/Apr04/OCS04D02/OCS04D02.pdf.

Lowry, W. P. & Lowry, P. P. II. (1989). *Fundamentals of Biometeorology: Interactions of Organisms and the Atmosphere*, vol. 1. McMinnville, OR: Peavine.

Mallet, J., Korman, A., Heckel, D. G. & King, P. (1993). Biochemical genetics of *Heliothis* and *Helicoverpa* (Lepidoptera: Noctuidae) and evidence for a founder event in *Helicoverpa zea*. *Annals of the Entomological Society of America*, **86**, 189–197.

Miles, M. R., Frederick, R. D. & Hartman, G. L. (2003). *Soybean Rust: Is the US Soybean Crop at Risk?* St. Paul, MN: American Phytopathological Society. APS Feature Story 06/2003, available at www.apsnet.org/online/feature/rust/.

Mortensen, D. A., Humston, R., Jones, B. & Dauer, J. (2003). A landscape analysis of an invasive composite. *Weed Science Society of America*, **43**, 24 (abstract).

Mortensen, D. A., Dauer, J. T. & Curran, W. S. (2007). Do ecological insights inform *Conyza* management? *Northeastern Weed Science Society Abstracts*, **61**, 117.

Nathan, R. (2005). Long-distance dispersal research: building a network of yellow brick roads. *Diversity and Distributions*, **11**, 125–130.

National Agricultural Statistics Service (2007). *Agricultural Chemical Use Database*. Washington, DC: US Department of Agriculture. Available at www.pestmanagement.info/nass/.

National Research Council (1996). *Ecologically Based Pest Management: New Solutions for a New Century*. Washington, DC: National Academy Press.

National Research Council (2002). *Predicting Invasions of Nonindigenous Plant and Plant Pests*. Washington, DC: National Academy Press.

Oke, T. R. (1987). *Boundary Layer Climates*. London: Routledge.

Pedgley, D. E. (1982). *Windborne Pests and Diseases: Meteorology of Airborne Organisms*. Chichester, UK: Ellis Horwood.

Pedgley, D. E., Reynolds, D. R. & Tatchell, G. M. (1995). Long-range insect migration in relation to climate and weather: Africa and Europe. In *Insect Migration: Tracking Resources through Space and Time*, eds. V. A. Drake & G. A. Gatehouse, pp. 3–29. Cambridge, UK: Cambridge University Press.

Pivonia, S. & Yang, X. B. (2004). Assessment of the potential year-round establishment of soybean rust throughout the world. *Plant Disease*, **88**, 523–529.

Price, P. (1997). *Insect Ecology*, 3rd edn. New York: John Wiley.

Rissler, J. & Mellon, M. (1996). *The Ecological Risks of Engineered Crops*. Cambridge, MA: MIT Press.

Roberts, M. J., Schimmelpfennig, D., Ashley, E. & Livingston, M. (2006). *The Value of Plant Disease Early-Warning Systems: A Case Study of USDA's Soybean Rust Coordinated Framework*, Economic Research Report No. 18. Washington, DC: US Department of Agriculture Economic Research Service. Available at www.ers.usda.gov/publications/err18/.

Sakai, A. K., Allendorf, F. W., Holt, J. S. *et al.* (2001). The population biology of invasive species. *Annual Review of Ecology and Systematics*, **32**, 305–332.

Sandstrom, M., Chagnon, D. & Flood, B. R. (2005). How weather and climate impact your pest management decisions. In *Vegetable Insect Management*, eds. R. Foster & B. Flood, pp. 23–29. Willoughby, OH: Meister.

Sappington, T. W. & Showers, W. B. (1991). Implications for migration of age-related variation in flight behavior of *Agrotis ipsilon* (Lepidoptera: Noctuidae). *Annals of the Entomological Society of America*, **84**, 360–365.

Schneider, R. W., Hollier, C. A., Whitam, H. K. *et al.* (2005). First report of soybean rust caused by *Phakopsora pachyrhizi* in the continental United States. *Plant Disease*, **89**, 774.

Shields, E. J., Dauer, J. T., VanGessel, M. J. & Neumann, G. (2006). Horseweed (*Conyza canadensis*) seed collected in the planetary boundary layer. *Weed Science*, **54**, 1063–1067.

Showers, W. B., 1997. Migratory ecology of the black cutworm. *Annual Review of Entomology*, **42**, 393–425.

Showers, W. B., Whitford, F., Smelser, R. B. *et al.* (1989). Direct evidence for meteorologically driven long-range dispersal of an economically important moth. *Ecology*, **70**, 987–992.

Showers, W. B., Keaster, A. J., Raulston, J. R. *et al.* (1993). Mechanism of southward migration of a noctuid moth (*Agrotis ipsilon* Hufnagel): a complete migrant. *Ecology*, **74**, 2303–2314.

Sisterson, M. S., Carriere, Y., Dennehy, T. J. & Tabashnik, B. E. (2005). Evolution of resistance to transgenic crops: interactions between insect movement and field distribution. *Journal of Economic Entomology*, **98**, 1751–1762.

Stack, J., Cardwell, K., Hammerschmidt, R. *et al.* (2006). The National Plant Diagnostic Network. *Plant Disease*, **90**, 128–136.

Stakman, E. C. & Harrar, J. C. (1957). *Principles of Plant Pathology*. New York: Ronald Press.

Stinner, R. E., Saks, M. & Dohse, L. (1986). Modeling of agricultural pest displacement. In *Insect Flight: Dispersal and Migration*, ed. W. Danthariarayana, pp. 235–241. Berlin, Germany: Springer-Verlag.

Stull, R. B. (1988). *An Introduction to Boundary Layer Meteorology*. Dordrecht, Netherlands: Kluwer.

Tackenberg, O., Poschlod, P. & Bonn, S. (2003). Assessment of wind dispersal potential in plant species. *Ecological Monographs*, **73**, 191–205.

Tobin, P. C., Sharov, A. A., Liebhold, A. A. *et al.* (2004). Management of the gypsy moth through a decision algorithm under the STS project. *American Entomologist*, **50**, 200–209.

Todd, E. L. (1955). The distribution and nomenclature of the corn earworm (Lepidoptera, Phalaenidae). *Journal of Economic Entomology*, **48**, 600–603.

VanGessel, M. J. (2001). Glyphosate-resistant horseweed from Delaware. *Weed Science*, **49**, 703–705.

VanGessel, M. J., Scott, B. A., Johnson, Q. R. & White, S. E. (2007). Horseweed response to no-till production systems. *Northeastern Weed Science Society Abstracts*, **61**, 116.

Weaver, S. E. (2001). The biology of Canadian weeds. 115. *Conyza canadensis*. *Canadian Journal of Plant Science*, **81**, 867–875.

Westbrook, J. K., Wolf, W. W., Lingren, P. D. *et al.* (1997). Early-season migratory flights of corn earworm (Lepidoptera: Noctuidae). *Environmental Entomology*, **26**, 12–20.

Westbrook, J. K. & Isard, S. A. (1999). Atmospheric scales of motion for dispersal of biota. *Agricultural and Forest Meteorology*, **97**, 263–274.

Wheelis, M., Casagrande, R. & Madden, L. V. (2002). Biological attack on agriculture: low-tech, high-impact bioterrorism. *BioScience*, **52**, 569–576.

Wolf, W. W., Westbrook, J. K., Raulston, J., Pair, S. D. & Hobbs, S. E. (1990). Recent airborne radar observations of migrant pests in the United States. *Philosophical Transactions of the Royal Society of London B*, **328**, 619–630.

Wolf, W. W., Westbrook, J. K., Pair, S. D., Raulston, J. R. & Lingren, P. D. (1995). Radar observations of orientation of noctuids migrating from corn fields in the Lower Rio Grande Valley. *Southwestern Entomology (Suppl.)*, **18**, 45–61.

Chapter 9

Introduction and augmentation of biological control agents

Robert J. O'Neil and John J. Obrycki

Natural enemies play key roles in pest management programs worldwide. Using natural enemies in pest management requires an understanding of their basic biology, how they impact pest population growth and how the environment and management system affect natural enemy dynamics and performance. Whereas there are abundant examples of using natural enemies in weed and plant pathogenic systems (VanDriesche & Bellows, 1996; Coombs *et al.*, 2004), in this chapter we focus on insect pests and their associated predatory, parasitic and pathogenic natural enemies. Although natural enemies that occur naturally in crops can provide substantial control and are the cornerstone of many pest management programs, we will focus on approaches that intentionally add natural enemies to affect pest control. Examples of manipulating the environment to increase the number or effectiveness of natural enemies, sometimes referred to as conservation biological control, are given in several chapters in the current volume or in VanDriesche & Bellows (1996), Barbosa (1998) and Bellows & Fisher (1999). In this chapter we review approaches to introduce or augment natural enemies, provide case histories to illustrate their use and suggest research needs to increase the use of biological control agents in pest management. General texts on biological control (e.g. Debach & Rosen, 1991; VanDriesche & Bellows, 1996; Bellows & Fisher, 1999) provide a broader overview of the methods and use of biological control in pest management systems.

9.1 Introducing new natural enemies

The history of natural enemy introductions to control introduced pests, sometimes referred to as "classical biological control," traces its origin to the successful use of vedalia beetle (*Rodolia cardinalis*), a predator coccinellid, to control the cottony cushion scale (*Icerya purchasi*) in California, USA in the 1880s (Debach & Rosen, 1991). That project identified many of the key tenets of classical biological control and was successfully replicated in most areas where cottony cushion scale had become a pest. The central idea behind the approach was that as an introduced insect (in this case from Australia), cottony cushion scale lacked the natural enemies in its new range that kept its population densities below economic levels in its homeland. Reuniting cottony cushion scale with its natural enemies would re-establish the natural enemy–prey dynamic, lower the pest population density and affect economic control. The steps to reunite so-called "exotic" natural enemies with an introduced pest include exploration in the pest's area of origin,

Integrated Pest Management, ed. Edward B. Radcliffe, William D. Hutchison and Rafael E. Cancelado. Published by Cambridge University Press.

Table 9.1 Examples of successful introductions of natural enemies as part of a classical biological control program

Pest (year)	Program costs (× $US 1000)	Benefit : cost (1st 10 years)
Sugarcane borer (1967) (*Diatraea* spp.)	150	88
Coffee mealybug (1939) (*Planococcus kenyae*)	75	202
Winter moth (1954) (*Operophthera brumata*)	150	15
Spotted alfalfa aphid (1958) (*Therioaphis maculata*)	900	44
Cottony cushion scale (1966) (*Icerya purchasi*)	2	39
Olive parlatoria scale (1951) (*Parlatoria oleae*)	228	42

Source: After Gutierrez *et al.* (1999).

quarantine examination for host specificity, release of the natural enemy into the environment and evaluation of the ecological and economic impacts of the introduction. Debach & Rosen (1991) estimated that classical biological control has resulted in the complete control of over 75 insect pests worldwide and substantial control of another 74 species. Since that publication additional successes have occurred in a variety of crop–pest systems. Accrued cost savings of successful projects can exceed many $US millions and obviate the need for further control interventions (Table 9.1). The environmental and health protections of classical biological control are less enumerated, but with often-significant reduction of insecticide use following successful biological control, they can be considerable.

Although classical biological control has had a long track record of success, a number of studies have reported deleterious effects on non-target populations, including increased parasitism and predation rates and associated declines (and reported extinctions) of non-target host populations (Follett & Duan, 2000). Concern over possible negative effects of releases has resulted in more stringent evaluations of potential non-target impacts (see below). As with any pest control technology, the introduction of natural enemies is not risk-free, and evaluation of the risks associated with introducing natural enemies need to balance the costs and benefits (both economic and ecological) of using natural enemies. Delfosse (2004) provides a framework to evaluate the relative risks of biological control (or other control technologies). Whereas earlier insect biological control programs paid scant attention to non-target impacts, current efforts include significantly more focus in this regard. (Non-target effects have been a concern for weed biological control for many years now.) We are encouraged that the history of biological control is replete with practitioners addressing challenges and modifying their approaches based on science and environmental concerns. We have every expectation that current and future practitioners will continue along these lines and expand the use of natural enemies in ever safer and more economic ways.

Below we use several case histories to describe the use of natural enemy introduction to control insect pests. We have selected these projects to illustrate that natural enemies can be successfully used in a variety of crop–pest systems, and that ultimately success depends on commitment predicated on basic knowledge of pest and natural enemy ecology and systematics.

9.1.1 Cassava mealybug

In the early 1970s farmers in Africa began noticing significant declines in cassava yields. Cassava, a crop introduced to Africa by Portuguese colonists in the 1500s, is originally from the Americas. Grown as a subsistence crop, cassava provides essential nutrition to people in the "cassava belt" – an area south of the Sahel, and approximately the size of the continental USA. The cause of the crop loss was identified as a mealybug new to Africa, the cassava mealybug (*Phenacoccus manihoti*). Beginning in 1977, scientists at the International Institute of Tropical Agriculture (IITA) began a classical biological control program. Initially, explorations focused on Central America and northern South America as this was identified as the center of diversity of *Phenacoccus*. However, parasitoids reared from mealybugs collected in this area failed to attack cassava mealybug. Subsequent taxonomic revision suggested cassava mealybug was not from this area, and that other locations needed to be surveyed. Collections were then made in more southern latitudes in South America, where cassava mealybug was subsequently located. In 1981, a parasitoid, *Apoanagyrus* (= *Epidinocaris*) *lopezi*, was collected from cassava mealybug in Paraguay, and following host testing in quarantine (in Europe), releases were made in Nigeria. The parasitoid quickly established, spread throughout most of the cassava belt, and caused a significant decline in cassava mealybug densities and associated damage (Neuenschwander & Herren, 1988). It has been estimated that the cassava mealybug biological control project saved subsistence farmers in Africa hundreds of millions of dollars and helped secure food reserves for over 200 million people.

The cassava mealybug project illustrates a number of points concerning classical biological control. First, the project shows how a fundamental understanding of the pest and natural enemy systematics and ecology is critical to success in importation biological control. While initial explorations focused on the area thought to be the most promising (because it was the center of diversity of the pest group), testing of natural enemies suggested additional areas should be explored. Exploration in the proper area led to the discovery of the key natural enemy and success in the project. The project also illustrates how using classical biological control is not limited by locale or the cropping system. Classical biological control had not been frequently attempted in subsistence crops and it was argued that resource-poor farmers in Africa would not benefit from the initial (*c.* $US 35 million) investment. Fortunately project personnel persevered and focused research to use biological control options for this introduced pest. The lead scientist on this project, Hans Herren, received the World Food Prize in 1995 for his leadership on the cassava management program. Finally, the success of the cassava mealybug led to work on other cassava pests, resulting in the biological control of the introduced cassava green mite (*Mononychellus tanajoa*) and later successes in other crops such as mangos and bananas.

9.1.2 Ash whitefly

Ash whitefly (*Siphoninus phillyreae*) was introduced into California in the late 1980s. A pest of trees, including ash and ornamental pear, ash whitefly infestations in urban centers resulted in early season defoliation of street trees and subsequent loss of aesthetic value, and the cooling effects provided by shade. At high-density whitefly populations, respiratory health risks were reported, and outdoor activities declined due to "clouds" of flying adults. A biological control program was started by the California Department of Food and Agriculture, and the University of California (Pickett & Pitcairn, 1999). Within three years, a parasitoid, *Encarsia inaron*, was imported from Italy and Israel and released throughout most of the affected areas of California. Immediate and significant declines of ash whitefly populations were noted, and the whitefly is no longer considered a pest in California. Economic analysis of the program suggests $US 200–300 million savings in tree replacement costs and associated aesthetic benefits. The ash whitefly program illustrates the use of biological control in urban settings as well as in protecting the aesthetic value of the urban landscape. The project benefited by having an experienced set of researchers familiar with biological control, who quickly developed a biological control option. The project also benefited from past taxonomic work on the parasitoid taxon, as *Encarsia* species have been the focus of considerable work in biological control, both in classical and augmentative approaches (see below).

9.1.3 Soybean aphid

Soybean aphid (*Aphis glycines*) was discovered infesting soybean fields in the Midwest USA in the summer of 2000. Having subsequently spread to 22 USA states and three Canadian provinces, the aphid directly reduces yield through feeding and is capable of transmitting a number of viruses. In outbreak years the pest has caused $US millions in damage, and an estimated 3 to 4 million hectares of soybeans have been treated with insecticides. Prior to the introduction of the soybean aphid, soybeans, particularly in the Midwest USA, were rarely treated with insecticides. The damage and associated control costs have made soybean aphid one of the most damaging insect pests of soybean production in the USA.

In response to the threat posed by the soybean aphid a group of entomologists from several midwestern land-grant universities and research institutions, and the US Department of Agriculture initiated a classical biological control program. Foreign exploration for soybean aphid natural enemies began in 2001 and to date, nearly 30 populations representing six to nine species of parasitoids from several areas of northeast China, Korea and Japan have been received and successfully established in quarantines in the USA. Host specificity has been initially evaluated for at least 21 Asian populations. One species, *Binodoxys communis*, a newly described braconid, has undergone regulatory review and was approved for release in 2007. At the start of the program additional studies were initiated to assess potential non-target impacts as part of a risk assessment of the biological control program. Included were in-country (Asia) field studies of host specificity of soybean aphid natural enemies, review of non-target impacts of previous aphid biological control projects, development of a baseline assessment of aphid diversity using suction trap collections and field sampling of selected native aphid species and their associated natural enemies. Finally, a region-wide sampling program has been initiated to determine the impact of released natural enemies, both on the soybean aphid as well as other, non-target aphid species.

The success of the soybean aphid project cannot be assessed at this time. However, it is important to note that the project began almost immediately following determination that the aphid was an introduced pest. Also, non-target impacts were of immediate concern, and studies were initiated and are an integral part of the research (Heimpel *et al.*, 2004). Unfortunately, for all too many insect pests, classical biological control is not attempted, with the rationalization that the crop–pest system is not amenable to success. This is particularly true for annual crop systems that are thought to be too ephemeral for classical biological control to "work." The soybean aphid project directly tests this perspective, and if successful will provide an additional case study to demonstrate that biological control in annual crop systems is a viable and important control option.

9.2 Augmentation

Compared to introduction, in which the goal is the permanent establishment of a natural enemy to reduce the density of a pest, augmentative biological control involves periodic releases to reduce pest densities; permanent establishment of the natural enemy is not expected. As a method of biological control, augmentative tactics can be organized within two overlapping categories: environmental manipulations and periodic releases. The goal of augmentation, increasing the effectiveness of a natural enemy, is achieved by periodic releases of natural enemies or by manipulating the environment to favor the natural enemy. Over the past 40 years, overviews of the methods and results of augmentative releases have been published numerous times (e.g. Ridgway & Vinson, 1977; Parrella *et al.*, 1999; van Lenteren & Bueno, 2003). In this chapter, we consider augmentation in a narrow perspective to include only periodic releases, and we highlight examples of successful augmentation programs, provide key references, outline methods and suggest areas for improvement. Readers interested in environmental manipulations (conservation biological control) can refer to several chapters in this book as well as VanDriesche & Bellows (1996), Barbosa (1998) and Bellows & Fisher (1999).

Augmentative methods can involve government sponsored areawide management programs, grower-based cooperatives or individuals making small-scale releases in gardens, greenhouses or urban landscapes. Regardless of the scale, augmentation biological control relies on the use of effective natural enemies, proper evaluation of efficacy and often integration with other control tactics (Ridgway & Vinson, 1977). Augmentative biological control has been used with semiochemicals for suppression of lepidopteran pests of cotton and tomatoes, chemical control for suppression of phytophagous mite pests in orchards and vineyards and plant resistance for management of whitefly pests of greenhouse-grown cucumbers. Methods for release and evaluation of natural enemies are still a work in progress for many systems, although substantial progress has been made in the refinement of release protocols and economic evaluation of efficacy for selected environments such as greenhouse pest management (van Lenteren, 2000), and releases of *Trichogramma* wasps for control of lepidopteran pests in several systems (Smith, 1996).

Augmentative releases can be inundative, when the natural enemy is released in high numbers to cause relatively rapid and direct mortality with no expectation of longer-term pest suppression. Examples of inundative programs include the use of entomopathogenic nematodes for suppression of the black vine weevil (*Otiorhynchus sulcatus*) in citrus, releases of high densities of *Trichogramma* targeting eggs of stalk boring lepidopteran pests of corn and sugarcane, and releases of immature Chrysopidae for reduction of homopteran pests (VanDriesche & Bellows, 1996; Barbosa, 1998). For many decades use of the bacterium *Bacillus thuringiensis* (*Bt*) was the most widely used pathogen for microbial control of lepidopteran pests based on an inundative strategy. Debates in the literature focused on questions of whether the use of *Bt* was a form of augmentative biological control or a biopesticide toxin delivery system. More recently, *Bt* genes have been incorporated into plants via genetic transformations, producing transgenic *Bt* corn, potato and cotton possessing a novel type of plant resistance (Shelton *et al.*, 2002).

The second type of augmentative release is inoculative, in which fewer numbers of natural enemies are released, with the expectation of longer-term (seasonal) pest suppression resulting from the offspring of the released individuals. The ultimate goal of these periodic releases is to increase densities of the natural enemy, based on the underlying assumption that greater numbers of natural enemies result in greater suppression of the pest species. For example, the release of the parasitic wasp *Encarsia formosa* for suppression of the greenhouse whitefly (*Trialeurodes vaporariorum*) is based on increasing the densities of this parasitoid over the growing period of greenhouse floriculture and vegetable production systems. Pupae of the predatory fly *Aphidoletes aphidimyza* are released for aphid suppression which is accomplished via the production of larvae which consume the target prey. Similarly, reproduction by the mealybug destroyer (*Cryptolaemus montrouzieri*) is required for the production of predatory larvae which contribute to sustained biological control of mealybugs in controlled environments.

Improvements in augmentation methodology include genetic selection of more efficient biotypes, pre-release conditioning of parasitoids through exposure to host materials, selection of appropriate natural enemy species for low or high pest densities and providing nourishment for released natural enemies. Several species of predatory mites have been selected for resistance against several pesticides commonly used for IPM in orchard systems. Studies demonstrated that individuals from releases of these resistant strains persisted in the environment when the pesticides were used in these systems. Pesticide-resistant strains of the western predatory mite (*Galendromus occidentalis*) are commercially available for biological control of spider mites in orchards and vineyards.

Improved release methods can protect natural enemies from environmental mortality factors. For example, *Trichogramma* wasps may be enclosed in small, waxed cardboard capsules (Trichocaps®) for efficient handling and to exclude predators. Each capsule contains approximately 500 parasitized Mediterranean flour moth (*Ephestia kuehniella*) eggs. Dormant immature *Trichogramma* within the Trichocaps can be stored at low temperatures until needed for field releases. Large-scale field releases, timed to the seasonal phenology of the target pest, can be made due to the mechanized production and long-term storage of

Trichocaps. This type of mass production and storage provides hundreds of thousands of Trichocaps for field releases against the European corn borer (*Ostrinia nubilalis*) in western Europe (Obrycki *et al.*, 1997).

Commercial production of natural enemies is the major source of natural enemies used in augmentation in Europe and North America. For augmentative releases to be a viable component of pest management systems, natural enemies must be consistently mass produced and available. During the past two decades, the number of commercially produced natural enemy species has increased to over 125 species. An online list of North American suppliers of beneficial organisms for augmentative biological control has been compiled by Hunter (1997). General quality control guidelines for several species have been developed by the International Organization for Biological Control (van Lenteren *et al.*, 2003). The biological parameters used to assess quality include the number alive in a container, adult size, longevity and fecundity, rates of emergence for natural enemies shipped as pupae, and estimates of rates of parasitism or predation. Quality control of commercially produced arthropod natural enemies remains a research focus within augmentative biological control (O'Neil *et al.*, 1998).

9.2.1 Augmentative biological control in greenhouses

Augmentation biological control is the foundation for many IPM systems for arthropod pests within the greenhouse environment (Parrella *et al.*, 1999; van Lenteren, 2000). Two cosmopolitan plant pests of a wide variety of crops produced in greenhouses, two-spotted spider mite (*Tetranychus urticae*) and greenhouse whitefly are effectively managed through augmentative releases of the predatory mite (*Phytoseiulus persimilis*) and *E. formosa*, respectively. In the Netherlands, over 90% of greenhouse-grown tomatoes, cucumbers and peppers are produced using IPM systems based on augmentation (see Chapter 27). These successful programs in greenhouses evolved from a long-term commitment and sustained interactions between researchers and producers. Faced with severe pest problems that could not be addressed economically with pesticides, as well as government policies that encouraged reduction of pesticide use, greenhouse plant production systems were developed based on the compatible use of natural enemies and other control tactics for the suppression of several arthropod pests (van Lenteren, 2000).

9.2.2 *Trichogramma* releases for suppression of stalk boring pests

In Latin America and China, several million hectares of corn, sugarcane, cotton and cereals are grown under IPM systems that use augmentative releases of egg parasitoids (*Trichogramma* spp. and *Telenomus remus*) for suppression of lepidopteran pests (VanDriesche & Bellows, 1996; van Lenteren & Bueno, 2003). Because of the relatively high cost of insecticides, several governmental agencies have become actively involved in the mass production and release of natural enemies. Recently, augmentative releases of *Trichogramma* have targeted the European corn borer in sweet corn and pepper systems in the eastern USA. Early season releases of *T. ostriniae* in sweet corn fields in New York produced season long parasitism of corn borer eggs (Hoffmann *et al.*, 2002). The efficacy of *Trichogramma* releases has typically been assessed by increased rates of egg parasitism and/or reduced levels of damage caused by the target pest. Smith (1996) summarized over 40 studies that examined the efficacy of releases of *Trichogramma* against lepidopteran pests; wide variation has been reported in the increase of egg parasitism and the percentage reduction in damage following releases.

9.2.3 Augmentative releases of insect pathogens

Viruses

During the 1970s Brazilian farmers began to use suspensions of diseased caterpillars to reduce densities of the velvet bean caterpillar (*Anticarsia gemmatalis*), a major lepidopteran pest of soybeans. These suspensions contained a nuclear polyhedrosis virus (AgNPV), applied as an inoculative microbial control, which spread to cause epizootics and suppressed populations of velvet bean caterpillar. Following a decade of research the Brazilian agricultural research organization Empresa Brasileira de Pesquisa Agropecuaria (EMBRAPA), formulated AgNPV as a wettable powder that was used by farmer cooperatives, individual soybean farmers and private companies. In 1982, AgNPV was applied to about 2000 hectares of soybeans; by 1993 this virus was applied to over 1 million hectares of soybeans in Brazil (Moscardi, 1999). The use of this virus has resulted in significant reductions in insecticide applied to soybeans and is one of the best examples of augmentative microbial control of an insect pest.

Bacteria

For over four decades, several subspecies of *Bacillus thuringiensis* (*Bt*) have been widely used in inundative biological control programs for suppression of lepidopteran, dipteran and coleopteran pests (Federici, 1999). *Bt* is commercially available and has been applied for suppression of pests in a wide range of systems (e.g. forests, organic vegetables and aquatic environments). *Bt* is used within an inundative strategy for microbial control and results in rapid mortality of larval stages with no reproduction in the field. The specificity and short residual properties of *Bt* have been viewed as either an advantage or disadvantage of the use of *Bt*. Federici (1999) summarizes several factors that have contributed to the successful augmentative use of *Bt*: the ability to mass produce and apply on a large scale, relatively rapid mortality of target pests, activity against a number of important pest species, and fewer non-target effects and negative environmental effects compared to many insecticides.

Nematodes

Several species and strains (geographic isolates) of entomopathogenic nematodes in the families Steinernematidae and Heterorhabditidae are commercially produced for augmentative releases against insect pests in soil or protected habitats (Kaya & Gaugler, 1993). Nematodes used in biological control contain mutualistic bacteria (*Xenorhabdus* spp.), which are injected into an insect host by the infective stages of these nematodes and rapidly kill the host. These nematodes are generally considered to have broad host ranges, but recent studies have shown that several species are restricted to particular habitats and may only attack specific host taxa within that habitat. Based on this new knowledge, previous failed attempts to use entomopathogenic nematodes in augmentative programs should be re-examined. Releases of entomopathogenic nematodes have provided biological control of numerous insect pests including the black vine weevil and citrus root weevil (*Diaprepes abbreviatus*) (Georgis *et al.*, 2006).

9.2.4 Perspectives on augmentation

Due to the repetitive nature of augmentative releases, comparisons with chemical control are inevitable. One of the recurring themes for augmentative releases is their relatively higher cost and lower effectiveness compared to the use of chemical insecticides (Collier & Van Steenwyk, 2004). These comparisons are based on the premise that augmentative releases can be substituted for insecticide applications, which cause immediate and close to 100% mortality. Using augmentative releases as part of an IPM program requires a change in perspective as natural enemies are not pesticides and in general they do not cause immediate mortality. The appropriate measures to determine the success of augmentative biological control should include relevant ecological and economic data from comparative field trials and an assessment of the numbers of growers using the approach, as well as the hectarage under augmentative biological control (van Lenteren, 2006). These measures will require new research foci and collaboration among producers of natural enemies, biological control specialists, extension educators and end-users. However, the case studies we have presented clearly show that such an effort will be worth it.

9.3 Conclusions

Adding natural enemies to control pests, whether within a program of classical biological control or augmenting commercially available species has a long track record of success. When done properly these methods provide economic control without substantial environmental damage. Success in adding natural enemies is often restricted not by the attributes of the crop–pest system or natural enemy involved, but by self-imposed limitations based on theoretical constructs, perceptions of what farmers want or limited experience using natural enemies. To expand the use of natural enemies in IPM will require a commitment to initiate research and extension programming when new pests are identified (as in the case of introduced pests) or as key elements of IPM programs for established or native pests. Critical research areas include increased efforts in natural enemy systematics, non-target research in the area of origin and development of protocols to maximize the effectiveness of released natural enemies. In augmentative programs, quality control and validation of efficacy are critical research areas.

For both types of biological control, extension programming is needed to provide end-users information on natural enemy identification, options for use and testing compatibility of natural enemies with other control tactics. With growing global trade, shifts in market demands and increased environmental awareness, using natural enemies will become an ever-increasing component of IPM programs.

References

Barbosa, P. (ed.) (1998). *Conservation Biological Control*. San Diego, CA: Academic Press.

Bellows, T. S. Jr. & Fisher, T. W. (eds.) (1999). *Handbook of Biological Control*. San Diego, CA: Academic Press.

Collier, T. & Van Steenwyk, R. (2004). A critical evaluation of augmentative biological control. *Biological Control*, **31**, 245–256.

Coombs, E. M., Clark, J. K., Piper, G. L. & Cofrancesco, A. F. Jr. (eds.) (2004). *Biological Control of Invasive Plants in the United States*. Corvallis, OR: Oregon State University Press.

Debach, P. & Rosen, D. (1991). *Biological Control by Natural Enemies*. New York: Cambridge University Press.

Delfosse, E. S. (2004). Introduction. In *Biological Control of Invasive Plants in the United States*, eds. E. M. Coombs, J. K. Clark, G. L. Piper & A. F. Cofrancesco, Jr., pp. 1–11. Corvallis, OR: Oregon State University Press.

Federici, B. A. (1999). A perspective on pathogens as biological control agents for insect pests. In *Handbook of Biological Control*, eds. T. S. Bellows, Jr. & T. W. Fisher, pp. 517–548. San Diego, CA: Academic Press.

Follett, P. A. & Duan, J. J. (eds.) (2000). *Nontarget Effects of Biological Control*. Norwell, MA: Kluwer.

Georgis, R., Koppenhofer, A. M., Lacey, L. A. *et al.* (2006). Successes and failures in the use of parasitic nematodes for pest control. *Biological Control*, **38**, 103–123.

Gutierrez, A. P., Caltagirone, L. E. & Meikle, W. (1999). Evaluation of results: economics of biological control. In *Handbook of Biological Control*, eds. T. S. Bellows, Jr. & T. W. Fisher, pp. 243–252. San Diego, CA: Academic Press.

Heimpel, G. E., Ragsdale, D. W., Venette, R. *et al.* (2004). Prospects for importation biological control of the soybean aphid: anticipating potential costs and benefits. *Annals of the Entomological Society of America*, **97**, 249–258.

Hoffmann, M. P., Wright, M. G., Pilcher, S. A. & Gardner, J. (2002). Inoculative releases of *Trichogramma ostriniae* for suppression of *Ostrinia nubilalis* (European corn borer) in sweet corn: field biology and population dynamics. *Biological Control*, **25**, 249–258.

Hunter, C. D. (1997). *Suppliers of Beneficial Organisms in North America*. Sacramento, CA: California Environmental Protection Agency, Department of Pesticide Regulation, Environmental Monitoring and Pest Management Branch. Available at www.cdpr.ca.gov/docs/pestmgt/ipminov/bensup.pdf.

Kaya, H. K. & Gaugler, R. (1993). Entomopathogenic nematodes. *Annual Review of Entomology*, **38**, 181–206.

Moscardi, F. (1999). Assessment of the application of baculoviruses for control of Lepidoptera. *Annual Review of Entomology*, **44**, 257–289.

Neuenschwander, P. & Herren, H. R. (1988). Biological control of the cassava mealybug, *Phenacoccus manihoti*, by the exotic parasitoid, *Epidinocaris lopezi*, in Africa. *Philosophical Transactions of the Royal Society of London B*, **318**, 319–333.

Parrella, M. P., Hansen, L. S. & van Lenteren, J. C. (1999). Glasshouse environments. In *Handbook of Biological Control*, eds. T. S. Bellows, Jr. & T. W. Fisher, pp. 819–839. San Diego, CA: Academic Press.

Pickett, C. H. & Pitcairn, M. J. (1999). Classical biological control of ash whitefly: factors contributing to its success in California. *Biological Control*, **14**, 143–158.

Obrycki, J. J., Lewis, L. C. & Orr, D. B. (1997). Augmentative releases of entomophagous species in annual cropping systems. *Biological Control*, **10**, 30–36.

O'Neil, R. J., Giles, K. L., Obrycki, J. J. *et al.* (1998). Evaluation of the quality of four commercially available natural enemies. *Biological Control*, **11**, 1–8.

Ridgway, R. L. & Vinson, S. B. (1977). *Biological Control by Augmentation of Natural Enemies*. New York: Plenum Press.

Shelton, A. M., Zhao, J.-Z. & Roush, R. T. (2002). Economic, ecological, food safety, and social consequences of the development of Bt transgenic plants. *Annual Review of Entomology*, **47**, 845–881.

Smith, S. M. (1996). Biological control with *Trichogramma*: advances, successes, and potential of their use. *Annual Review of Entomology*, **41**, 375–406.

VanDriesche, R. & Bellows, T. S. Jr. (1996). *Biological Control*. New York: Chapman & Hall.

van Lenteren, J. C. (2000). A greenhouse without pesticides: fact or fantasy? *Crop Protection*, **19**, 375–384.

van Lenteren, J. C. (2006). How not to evaluate augmentative biological control. *Biological Control*, **39**, 115–118.

van Lenteren, J. C. & Bueno, V. H. P. (2003). Augmentative biological control of arthropods in Latin America. *BioControl*, **48**, 121–139.

van Lenteren, J. C., Hale, A., Klapwijk, J. N., van Schelt, J. & Stenberg, S. (2003). Guidelines for quality control of commercially produced natural enemies. In *Quality Control and Production of Biological Control Agents: Theory and Testing*, ed. J. C. van Lenteren, pp. 265–304. Cambridge, MA: CABI Publishing.

Chapter 10

Crop diversification strategies for pest regulation in IPM systems

Miguel A. Altieri, Clara I. Nicholls and Luigi Ponti

Ninety-one percent of the 1500 million hectares of the worldwide cropland are mostly under annual crop monocultures of wheat, rice, maize, cotton and soybeans (Smith & McSorley, 2000). These systems represent an extreme form of simplification of nature's biodiversity, since monocultures, in addition to being genetically uniform and species-poor systems, advance at the expense of natural vegetation, a key landscape component that provides important ecological services to agriculture such as natural mechanisms of crop protection (Altieri, 1999). Since the onset of agricultural modernization, farmers and researchers have been faced with a main ecological dilemma arising from the homogenization of agricultural systems: an increased vulnerability of crops to insect pests and diseases, which can be devastating when infesting uniform crop, large-scale monocultures (Adams *et al.*, 1971; Altieri & Letourneau, 1982, 1984). The expansion of monocultures has decreased abundance and activity of natural enemies due to the removal of critical food resources and overwintering sites (Corbett & Rosenheim, 1996). With accelerating rates of habitat removal, the contribution to pest suppression by biocontrol agents using these habitats is declining and consequently agroecosystems are becoming increasingly vulnerable to pest outbreaks (e.g. Gurr *et al.*, 2004).

A key task for agroecologists is to understand the link between biodiversity reduction and pest incidence in modern agroecosystems in order to reverse such vulnerability by increasing functional diversity in agricultural landscapes. One of the most obvious advantages of diversification is a reduced risk of total crop failure due to pest infestations (Nicholls & Altieri, 2004). Research results from recent vineyard studies conducted in California illustrate ways in which biodiversity can contribute to the design of pest-stable agroecosystems by creating an appropriate ecological infrastructure within and around cropping systems.

10.1 Understanding pest vulnerability in monocultures

The spread of modern agriculture has resulted in tremendous changes in landscape diversity. There has been a consistent trend toward simplification that entails: (1) enlargement of fields, (2) aggregation of fields, (3) increase in the density of crop plants, (4) increase in the uniformity of crop population age structure and physical quality and (5) decrease in inter- and intraspecific diversity within the planted field. One of the main characteristics of the modern agricultural landscape is

Integrated Pest Management, ed. Edward B. Radcliffe, William D. Hutchison and Rafael E. Cancelado. Published by Cambridge University Press.

the large size and homogeneity of crop monocultures, which fragment the natural landscape. This can directly affect the abundance and diversity of natural enemies, as the larger the area under monoculture the lower the viability of a given population of beneficial fauna.

Moreover, monocultures do not constitute good environments for natural enemies (Andow, 1991). Such simple crop systems lack many of the resources such as refuge sites, pollen, nectar and alternative prey and hosts, that natural enemies need to feed, reproduce and thrive. To the pests, the monocrop is a dense and pure concentration of its basic food resource, so, of course, many insect herbivores boom in such fertilized, weeded and watered fields. For the natural enemies, such overly simplified cropping systems are less hospitable because natural enemies require more than prey/hosts to complete their life cycles. Many parasitoid adults, for instance, require pollen and nectar to sustain themselves while they search for hosts (Gurr *et al.*, 2004).

Normal cultural activities such as tillage, weeding, spraying and fertilization can have serious effects on farm insects. Insect pest outbreaks and/or resurgences following insecticide applications are common phenomena (Pimentel & Perkins, 1980). Pesticides either fail to control the target pests or create new pest problems. Development of resistance in insect pest populations is the main way in which pesticide use can lead to pest control failure. Pesticides can also foster pest outbreaks through the elimination of the target pest's natural enemies (Pimentel & Lehman, 1993). Research also suggests that the susceptibility of crops to insects may be affected by the application of chemical fertilizers. Studies show that increases in nitrogen rates dramatically increased aphid and mite numbers. In reviewing 50 years of research relating to crop nutrition and insect attack, Scriber (1984) found 135 studies showing increased damage and/or growth of leaf-chewing insects or mites in nitrogen-fertilized crops, versus fewer than 50 studies in which herbivore damage was reduced by normal fertilization regimes. Pest levels were highly correlated to increased levels of soluble nitrogen in leaf tissue, suggesting that high nitrogen inputs can precipitate high levels of herbivore damage in crops (Altieri & Nicholls, 2003).

Lately, research suggests that transgenic crops, now ubiquitous components of agricultural landscapes, may affect natural enemy species directly through inter-trophic-level effects of the toxin. The potential for *Bt* toxins moving through arthropod food chains poses serious implications for natural biocontrol in agricultural fields. Studies show that the *Bt* toxin can affect beneficial insect predators that feed on insect pests present in *Bt* crops (Hilbeck *et al.*, 1998). Inter-trophic-level effects of the *Bt* toxin raise serious concerns about the potential of the disruption of natural pest control (Altieri, 2000).

10.2 Biodiversity in agroecosystems: types and roles

Biodiversity refers to all species of plants, animals and microorganisms existing and interacting within an ecosystem, and which play important ecological functions such as pollination, organic matter decomposition, predation or parasitism of undesirable organisms and detoxification of noxious chemicals (Gliessman, 1998). These renewal processes and ecosystem services are largely biological; therefore their persistence depends upon maintenance of ecological diversity and integrity. When these natural services are lost due to biological simplification, the economic and environmental costs can be quite significant. Economically, in agriculture the burdens include the need to supply crops with costly external inputs, since agroecosystems deprived of basic regulating functional components lack the capacity to sponsor their own soil fertility and pest regulation (Conway & Pretty, 1991).

The biodiversity components of agroecosystems can be classified in relation to the role they play in the functioning of cropping systems. According to this, agricultural biodiversity can be grouped as follows (Altieri, 1994; Gliessman, 1998).

- *Productive biota*: crops, trees and animals chosen by farmers that play a determining role in the diversity and complexity of the agroecosystem.

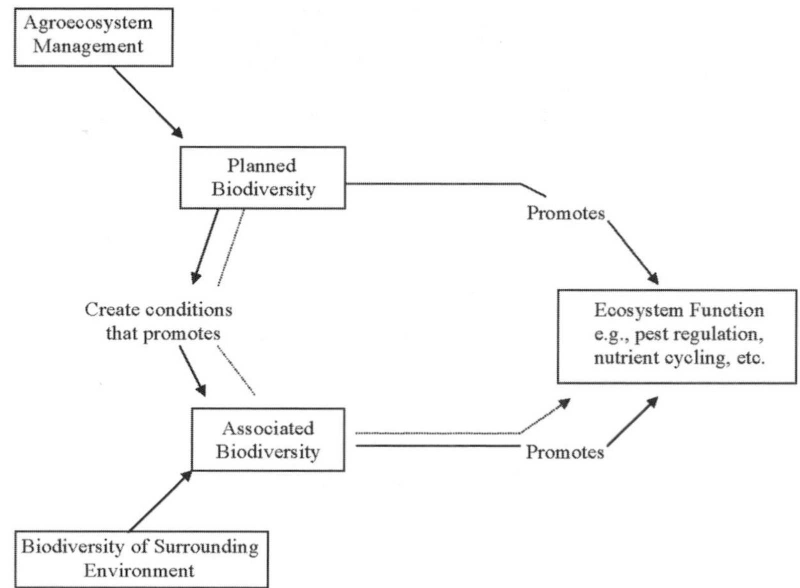

Fig. 10.1 The relationship between planned biodiversity (which the farmer determines, based on management of the agroecosystem) and associated biodiversity and how the two promote ecosystem function (modified from Vandermeer & Perefecto, 1995).

- *Resource biota*: organisms that contribute to productivity through pollination, biological control, decomposition, etc.
- *Destructive biota*: weeds, insects pests, microbial pathogens, etc., which farmers aim at reducing through cultural management.

The above categories of biodiversity can further be recognized as two distinct components (Vandermeer & Perefecto, 1995). The first component, *planned biodiversity*, includes the crops and livestock purposely included in the agroecosystem by the farmer, which will vary depending on the management inputs and crop spatial/temporal arrangements. The second component, *associated biodiversity*, includes all soil flora and fauna, herbivores, carnivores, decomposers, etc., that colonize the agroecosystem from surrounding environments and that will thrive in the agroecosystem depending on its management and structure. The relationship of both types of biodiversity components is illustrated in Fig. 10.1. Planned biodiversity has a direct function, as illustrated by the bold arrow connecting the planned biodiversity box with the ecosystem function box. Associated biodiversity also has a function, but it is mediated through planned biodiversity. Thus, planned biodiversity also has an indirect function, illustrated by the dotted arrow in the figure, which is realized through its influence on the associated biodiversity. For example, the trees in an agroforestry system create shade, which makes it possible to grow only sun-intolerant crops. So, the direct function of this second species (the trees) is to create shade. Yet along with the trees might come wasps that seek out the nectar in the tree's flowers. These wasps may in turn be the natural parasitoids of pests that normally attack understory crops. The wasps are part of the associated biodiversity. The trees, then create shade (direct function) and attract wasps (indirect function) (Vandermeer & Perefecto, 1995).

The optimal behavior of agroecosystems depends on the level of interactions between the various biotic and abiotic components. By assembling a functional biodiversity it is possible to initiate synergisms which subsidize agroecosystem processes by providing ecological services such as the activation of soil biology, the recycling of nutrients, the enhancement of beneficial arthropods and antagonists, and so on, all important components that determine the sustainability of agroecosystems (Nicholls *et al.*, 2000).

The key is to identify the type of biodiversity that it is desirable to maintain and/or enhance in order to carry out ecological services, and then to determine the best practices that will encourage the desired biodiversity components. There are many agricultural practices and designs that have the potential to enhance functional

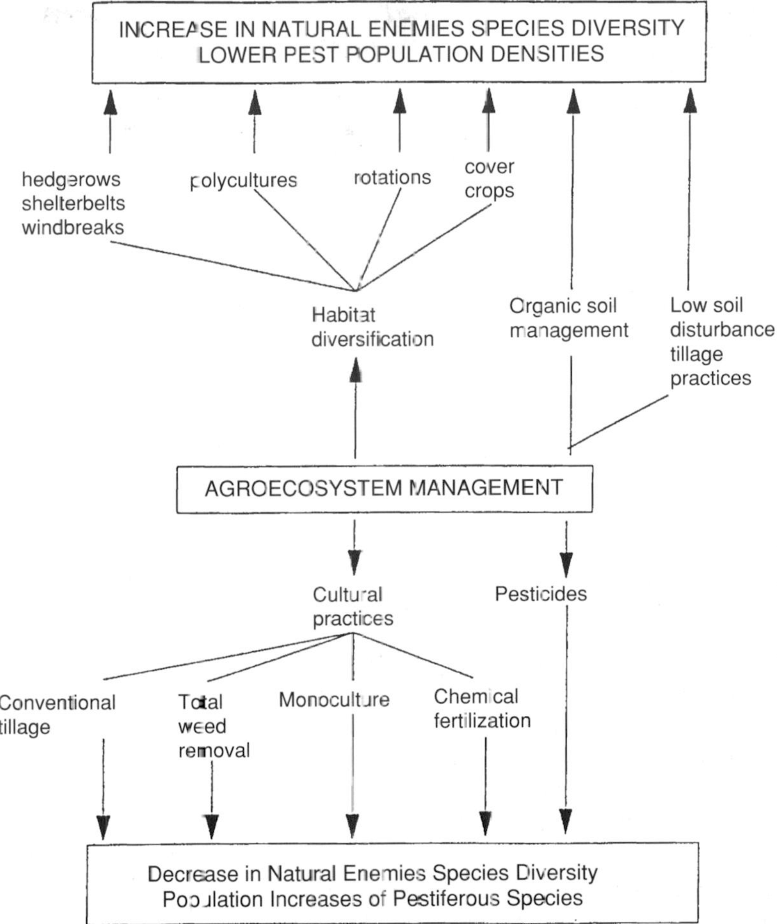

Fig. 10.2 Effects of agroecosystem management and associated cultural practices on biodiversity of natural enemies and abundance of insect pests.

biodiversity, and others that negatively affect it (Fig. 10.2). The idea is to apply the best management practices in order to enhance or regenerate the kind of biodiversity that can best subsidize the sustainability of agroecosystems by providing ecological services such as biological pest control, nutrient cycling, water and soil conservation, etc.

10.3 Diversified agroecosystems and pest management

Across the world, agroecosystems differ in age, diversity, structure and management. In fact, there is great variability in basic ecological and agronomic patterns among the various dominant agroecosystems. In general, agroecosystems that are more diverse, more permanent, isolated, and managed with low input technology (e.g. agroforestry systems, traditional polycultures) take fuller advantage of work usually done by ecological processes associated with higher biodiversity than highly simplified, input-driven and disturbed systems (e.g. modern vegetable monocultures and orchards).

All agroecosystems are dynamic and subjected to different levels of management so that the crop arrangements in time and space are continually changing in the face of biological, cultural, socioeconomic and environmental factors. Such landscape variations determine the degree of spatial and temporal heterogeneity characteristic of agricultural regions, which may or may not benefit the pest protection of particular agroecosystems. Thus, one of the main challenges facing agroecologists today is identifying the types of

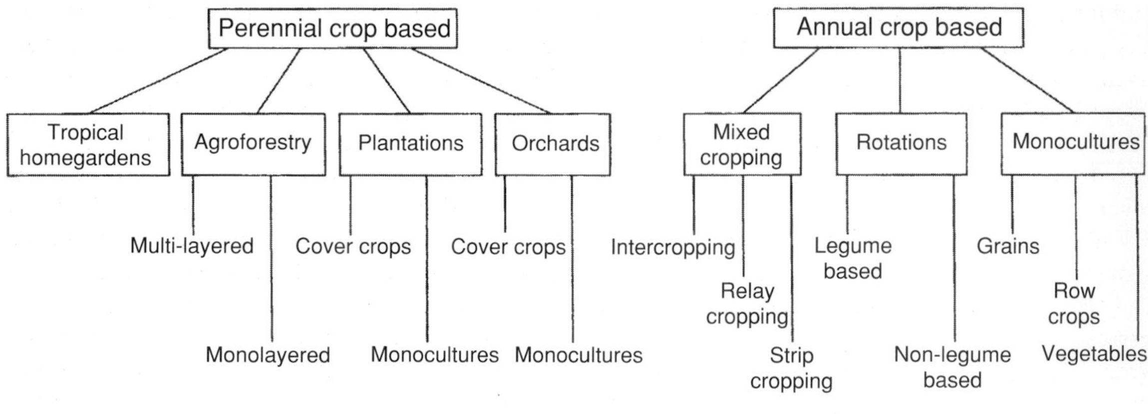

Fig. 10.3 A classification of dominant agricultural agroecosystems on a gradient of diversity and vulnerability to pest outbreak.

heterogeneity (either at the field or regional level) that will yield desirable agricultural results (i.e. pest regulation), given the unique environment and entomofauna of each area. This challenge can only be met by further analyzing the relationship between vegetation diversification and the population dynamics of herbivore species, in light of the diversity and complexity of site-specific agricultural systems. A hypothetical pattern in pest regulation according to agroecosystem temporal and spatial diversity is depicted in Fig. 10.3. According to this "increasing probability for pest buildup" gradient, agroecosystems on the left side of the gradient are more biodiverse, and tend to be more amenable to manipulation since polycultures already contain many of the key factors required by natural enemies. There are, however, habitat manipulations that can introduce appropriate diversity into the important (but biodiversity impoverished) grain, vegetable and row crop systems lying in the right half of Fig. 10.3.

Although herbivores vary widely in their response to crop distribution, abundance and dispersion, the majority of agroecological studies show that structural (i.e. spatial and temporal crop arrangement) and management (i.e. crop diversity, input levels, etc.) attributes of agroecosystems influence herbivore dynamics. Several of these attributes are related to biodiversity and most are amenable to management (i.e. crop sequences and associations, weed diversity, genetic diversity, etc.).

Diversified cropping systems, such as those based on intercropping and agroforestry or cover cropping of orchards, have been the target of much research recently. This interest is largely based on the emerging evidence that these systems are more stable and more resource conserving (Vandermeer, 1995). Many of these attributes are connected to the higher levels of functional biodiversity associated with complex farming systems. As diversity increases, so do opportunities for coexistence and beneficial interference between species that can enhance agroecosystem sustainability (Vandermeer, 1995). Diverse systems encourage complex food webs which entail more potential connections and interactions among members, and many alternative paths of energy and material flow through it. For this and other reasons a more complex community exhibits more stable production and less fluctuations in the numbers of undesirable organisms. Studies further suggest that the more diverse the agroecosystems and the longer this diversity remains undisturbed, the more internal links develop to promote greater insect stability. It is clear, however, that the stability of the insect community depends not only on its trophic diversity, but also on the actual density-dependence nature of the trophic levels (Southwood & Way, 1970). In other words, stability will depend on the precision of the response of any particular trophic link to an increase in the population at a lower level.

What is apparent is that functional characteristics of component species are as important as the total number of species in determining processes and services in ecosystems (Tilman, 1996). From an agroecosystem management point of view, the focus should be placed on enhancing a specific biodiversity component that plays a specific role, such as a plant that fixes nitrogen, provides cover for soil protection or harbors resources for natural enemies. In the case of farmers without major economic and resource limits and who can withstand a certain risk of crop failure, a crop rotation or a simple polyculture may be all it takes to achieve a desired level of stability. But in the case of resource-poor farmers, who can not tolerate crop failure, highly diverse cropping systems would probably be the best choice. The obvious reason is that the benefit of complex agroecosystems is low risk; if a species falls to disease, pest attack or weather, another species is available to fill the void and maintain full use of resources. Thus there are potential ecological benefits to having several species in an agroecosystem: compensatory growth, full use of resources and nutrients and pest protection (Ewel, 1999).

10.4 Diversity in traditional farming systems

The persistence of millions of hectares under traditional agriculture in the form of polycultures, agroforestry systems, etc. documents a successful indigenous agricultural biodiversification strategy for adapting to difficult environments (Altieri, 1999). These microcosms of traditional agriculture offer promising models for other areas as they promote biodiversity, thrive without agrochemicals and sustain year-round yields (Denevan, 1995). Traditional crop management practices used by many resource-poor farmers represent a rich resource for agroecologists interested in understanding the mechanisms at work in complex agroecosystems, especially the interactions between biodiversity and ecosystem function. Some agroecologists have recognized the virtues of diversified traditional agroecosystems whose sustainability lies in the complex ecological models they follow, quickly realizing that the prevalence of these systems is of key importance to peasants, as interactions between crops, animals and trees result in beneficial synergisms allowing agroecosystems to sponsor their own soil fertility, pest control and productivity (Altieri *et al.*, 1985; Reinjtes *et al.*, 1992). Considerable work was conducted on the biological mechanisms at play within intercropping systems which minimize crop losses due to insect pests, diseases and weeds (Altieri, 1994). The literature is full of examples of experiments documenting that diversified traditional cropping systems such as the prevalent maize–bean polyculture of the Latin American tropics exhibit reduced pest populations (Andow, 1991; Landis *et al.*, 2000; Altieri & Nicholls, 2004).

Researchers have shown that it is only when traditional systems are modernized, reducing their plant diversity, that herbivore abundance increases to pest levels, compounded by changes brought about by modern plant breeding and agronomy. In fact, although traditional farmers may be aware that insects can exert crop damage, they rarely consider them pests, as experienced by Morales *et al.* (2001) when studying traditional methods of pest control among the highland Maya of Guatemala. Influenced by Mayan attitudes, these Western scientists rapidly reformulated their research questions and rather than study how Mayan farmers control pest problems, they focused on why farmers do not have pest problems. This line of inquiry proved more productive as it allowed researchers to understand how farmers designed and managed pest-resilient cropping systems and to explore the mechanisms underlying agroecosystem health.

Many studies have transcended the research phase and have found applicability to control specific pests such as lepidopteran stem borers in Africa. Scientists at the International Center of Insect Physiology and Ecology (ICIPE) developed a habitat management system which uses plants in the borders of maize fields which act as trap crops (Napier grass and Sudan grass) attracting stem borer colonization away from maize (the push) and two plants intercropped with maize (molasses grass and silverleaf) that repel the stem borers (the pull) (Khan *et al.*, 2000). Border grasses also enhance the parasitization of stem borers by the wasp *Cotesia semamiae* and are important fodder plants. The leguminous silverleaf (*Desmodium uncinatum*) suppresses the parasitic weed *Striga* spp. by a factor of 40 when compared with maize monocrop. *Desmodium*'s nitrogen-fixing ability increases soil fertility; and it is an excellent forage. As an added bonus, sale of *Desmodium* seed is proving to be a new income-generating opportunity for women in the project areas. The push–pull system has been tested on over 450 farms in two districts of Kenya and has now been released for uptake by the national extension systems in East Africa. Participating farmers in the breadbasket of Trans Nzoia are reporting a 15–20% increase in maize yield. In the semi-arid Suba district – plagued by both stem borers and *Striga* – a substantial increase in milk yield has occurred in the last four years, with farmers now being able to support increased numbers of dairy cows on the fodder produced. When farmers plant maize together with the push–pull plants, a return of $US 2.30 for every dollar invested is made, as compared to only $US 1.40 obtained by planting maize as a monocrop (Khan *et al.*, 2000).

10.5 Plant diversity and insect pest incidence

An increasing body of literature documents that increased plant diversity in agroecosystems leads to pest population regulation by favoring the abundance and efficacy of associated natural enemies (Landis *et al.*, 2000). Research has shown that mixing certain plant species usually leads to density reductions of specialized herbivores. In a review of 150 published investigations Risch *et al.* (1983) found evidence to support the notion that specialized insect herbivores were less numerous in diverse systems (53% of 198 cases). In another comprehensive review of 209 published studies that deal with the effects of vegetation diversity in agroecosystems on herbivores arthropod species, Andow (1991) found that 52% of the 287 total herbivore species examined in these studies were less abundant in polycultures than in monocultures, while only 15.3% (44 species) exhibited higher densities in polycultures. In a more recent review of 287 cases, Helenius (1998) found that the reduction of monophagous pests was greater in perennial systems, and that the reduction of polyphagous pest numbers was less in perennial than in annual systems. Helenius (1998) concluded that monophagous (specialist) insects are more susceptible to crop diversity than polyphagous insects. He cautioned about the increased risk of pest attack if the dominant herbivore fauna in a given agroecosystem is polyphagous. In his analysis of available studies on crop–weed systems, Baliddawa (1985) found that 56% of pest reductions in weed diversified cropping systems were caused by natural enemies. In examining numerous studies testing the responses of pest and beneficial arthropods to plant diversification in cruciferous crops, Hooks & Johnson (2003) concluded that biological parameters of herbivores impacted by crop diversification were mainly related to the behavior of the insect studied. Mechanisms accounting for herbivore responses to plant mixtures include reduced colonization, reduced adult tenure time in the crop and oviposition interference.

The ecological theory relating to the benefits of mixed versus simple cropping systems revolves around two possible explanations of how insect pest populations attain higher levels in monoculture systems compared with diverse ones. The two hypotheses proposed by Root (1973) are:

The *enemies hypothesis* which argues that pest numbers are reduced in more diverse systems because the activity of natural enemies is enhanced by environmental opportunities prevalent in complex systems;

The *resource concentration hypothesis* argues that the presence of a more diverse flora has direct negative effects on the ability of the insect pest to find and utilize its host plant and also to remain in the crop habitat.

The resource concentration hypothesis predicts lower pest abundance in diverse communities because a specialist feeder is less likely to find its host plant due to the presence of confusing masking chemical stimuli, physical barriers to movement or other environmental effects such as shading. It will tend to remain in the intercrop for a shorter period of time simply because the probability of landing on a non-host plant is increased; it may have a lower survivorship and/or fecundity (Bach, 1980). The extent to which these factors operate will depend on the number of host plant species present and the relative preference of the pest for each, the absolute density and spatial arrangement of each host species and the interference effects from more host plants.

The enemies hypothesis attributes lower pest abundance in intercropped or more diverse systems to a higher density of predators and parasitoids (Bach, 1980). The greater density of natural enemies is caused by an improvement in conditions for their survival and reproduction, such as a greater temporal and spatial distribution of nectar and pollen sources, which can increase parasitoid reproductive potential and abundance of alternative host/prey when the pest species are scarce or at inappropriate stages (Risch, 1981). In theory, these factors can combine to provide more favorable conditions for natural enemies and thereby enhance their numbers and effectiveness as control agents.

10.6 Designing biodiverse pest-suppressive agroecosystems

In real situations, exploiting the complementarity and synergy that result from the various spatial and temporal polycultural combinations involves agroecosystem design and management and requires an understanding of the numerous relationships among soils, microorganisms, plants, insect herbivores and natural enemies.

Different options to diversify cropping systems are available depending on whether the current monoculture systems to be modified are based on annual or perennial crops. Diversification can also take place outside the farm, for example, in crop-field boundaries with windbreaks, shelterbelts and living fences, which can improve habitat for beneficial insects (Altieri & Letourneau, 1982). When done correctly, plant diversification creates a suitable ecological infrastructure within the agricultural landscape providing key resources and habitat for natural enemies. These resources must be integrated into the landscape in a way that is spatially and temporally favorable to natural enemies and practical for farmers to implement.

During the last decade we have applied biodiversification strategies to the design and management of pest suppressive organic vineyards in northern California. Results from some of our previously published studies (Nicholls *et al.*, 2000, 2001, 2005), are presented here in an effort to systematize the emerging lessons from our experience on biodiversity enhancement for ecologically based pest management in agroecosystems.

10.6.1 Vineyard studies

Our studies took advantage of an organic vineyard located in Mendocino County, California in which a 600-m corridor composed by 65 flowering species connected to a riparian forest cutting across the monoculture organic vineyard. This setting allowed for testing the idea whether such a corridor served as a biological highway for the movement and dispersal of natural enemies into the center of the vineyard (Nicholls *et al.*, 2001). We evaluated whether the corridor acted as a consistent, abundant and well-dispersed source of alternative food and habitat for a diverse community of generalist predators and parasitoids, allowing predator and parasitoid populations to develop in the area of influence of the corridor well in advance of vineyard pest populations. The corridor would serve as a conduit for the dispersion of predators and parasitoids within the vineyard, thus providing protection against insect pests within the area of influence of the corridor, by allowing distribution of natural enemies throughout the vineyard. As the vineyard also contained summer cover crops, we hypothesized that neutral insects (non-pestiferous herbivores) and pollen and nectar in the summer cover crops provide a constant and abundant supply of food

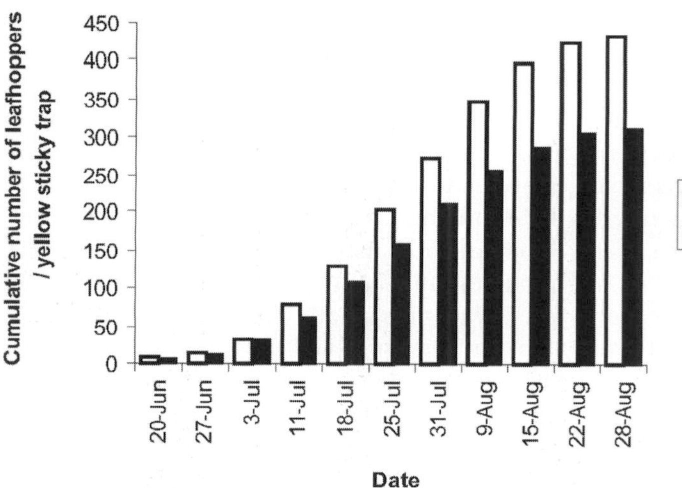

Fig. 10.4 Densities of adult *Erythroneura elegantula* leafhoppers in cover cropped and monoculture vineyards in Mendocino County, California, during the 1996 and 1997 growing seasons.

sources for natural enemies. This in turn decouples predators and parasitoids from a strict dependence on grape herbivores, allowing natural enemies to build up in the system, thereby keeping pest populations at acceptable levels.

We also conducted research at a 17-ha biodynamic vineyard located in Sonoma County, California. As part of a whole-farm biodiversity management strategy, a 0.5-ha island of flowering shrubs and herbs (*insectory*) was created at the center of the vineyard. This insectory was planted to provide flower resources from early April to late September to beneficial organisms, including natural enemies of grape insect pests (Nicholls *et al.*, 2005).

10.6.2 Enhancing within vineyard biodiversity with cover crops

Because most farmers either mow or plow under cover crops in the late spring, organic vineyards become virtual monocultures without floral diversity in early summer. Maintaining a green cover during the entire growing season is crucial to provide habitat and alternate food for natural enemies. An approach to achieve this is to sow summer cover crops that bloom early and throughout the season, thus providing a highly consistent, abundant and well-dispersed alternative food source, as well as microhabitats, for a diverse community of natural enemies (Nicholls *et al.*, 2000). Maintaining floral diversity throughout the growing season in the Mendocino vineyard in the form of summer cover crops of buckwheat and sunflower, substantially reduced the abundance of western grape leafhopper (*Erythroneura elegantula*) and western flower thrips (*Frankliniella occidentalis*) by allowing an early buildup of natural enemies. In two consecutive years (1996–1997), vineyard systems with flowering cover crops were characterized by lower densities of leafhopper nymphs and adults (Fig. 10.4). Thrips also exhibited reduced densities in vineyards with cover crops in both seasons (Nicholls *et al.*, 2000).

During both years, general predator populations on the vines were higher in the cover-cropped sections than in the monocultures. Generally, the populations were low early in the season and increased as prey became more numerous as the season progressed. Dominant predators included spiders, *Nabis* sp., *Orius* sp., *Geocoris* sp., Coccinellidae and *Chrysoperla* spp.

Although *Anagrus epos*, the most important leafhopper parasitoid wasp, achieved high numbers and inflicted noticeable mortality of grape leafhopper eggs, this impact was not substantial enough. Apparently the wasps encountered sufficient food resources in the cover crops, and few moved to the vines to search for leafhopper eggs. For this reason, cover crops were mowed every other row to force movement of *Anagrus* wasps and predators into the vines. Before mowing, leafhopper nymphal densities on vines were similar in the selected cover-cropped rows. One week after mowing, numbers of nymphs declined on vines

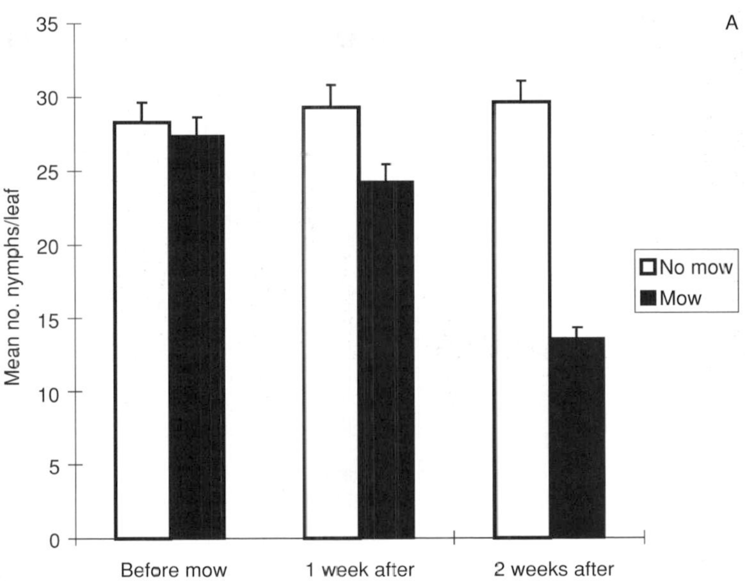

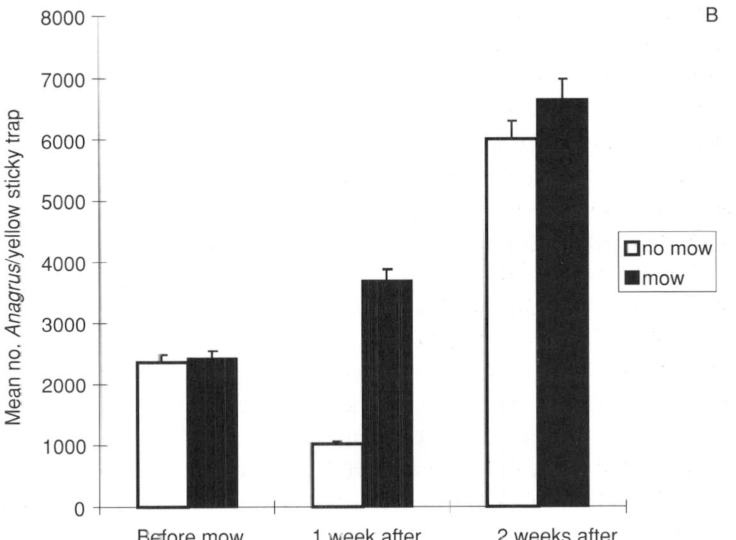

Fig. 10.5 (A) Effect of cover-crop mowing in vineyards on densities of leafhopper nymphs during the 1997 growing season in Mendocino County, California. (B) Effect of cover-crop mowing in vineyards on densities of *Anagrus epos* during the 1997 growing season in Mendocino County, California.

where the cover crop was mowed, coinciding with an increase in *Anagrus* densities in mowed cover-crop rows. During the second week, such nymphal decline was even more pronounced, coinciding with an increase in numbers of *Anagrus* wasps in the foliage (Fig. 10.5).

The mowing experiment suggests a direct ecological linkage, as the cutting of the cover crop vegetation forced the movement of *Anagrus* wasps and other predators harbored by the flowers, resulting in both years in a decline of leafhopper numbers on the vines adjacent to the mowed cover crops. Obviously, the timing of mowing must take place when eggs are present on the vine leaves in order to optimize the efficiency of arriving *Anagrus* wasps.

10.6.3 Corridor influences on population gradients of leafhoppers, thrips and associated natural enemies

Studies assessing the influence of adjacent vegetation or natural enemy refuges on pest dynamics

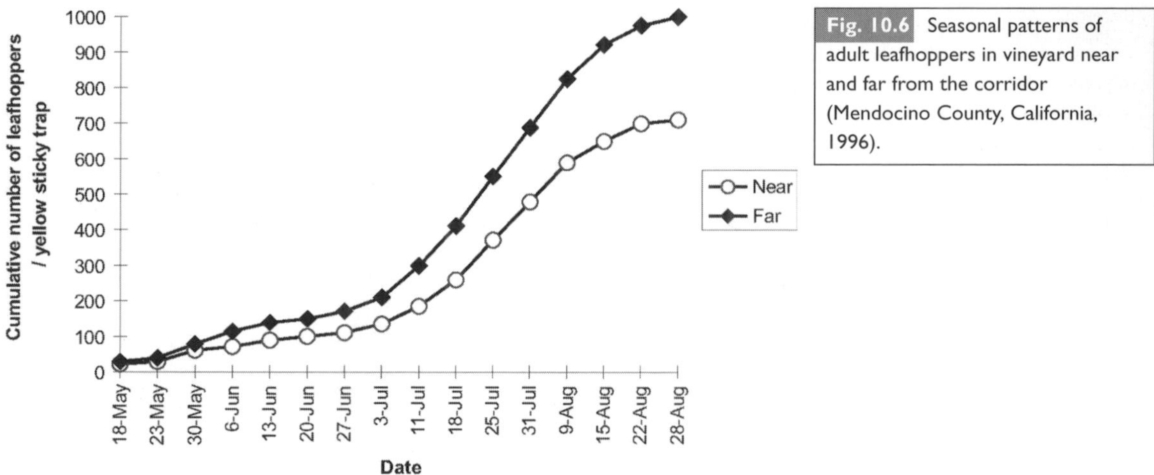

Fig. 10.6 Seasonal patterns of adult leafhoppers in vineyard near and far from the corridor (Mendocino County, California, 1996).

within vineyards show that, in the case of prune refuges, the effect is limited to only a few vine rows downwind, as the abundance of *A. epos* exhibited a gradual decline in vineyards with increasing distance from the refuge (Corbett & Rosenheim, 1996). This finding poses an important limitation to the use of prune trees, as the colonization of grapes by *A. epos* is limited to field borders, leaving the central rows of the vineyard void of biological control protection. The corridor connected to a riparian forest and cutting across the vineyard was established to overcome this limitation (Nicholls *et al.*, 2001).

The flowering sequence of the various plant species provided a continual source of pollen and nectar, as well as a rich and abundant supply of neutral insects for the various predator species, thus allowing the permanence and circulation of viable populations of key species within the corridor. In both years, adult leafhoppers exhibited a clear density gradient, with lowest numbers in vine rows near the corridor and increasing numbers towards the center of the field. The highest concentration of adult and nymph leafhoppers occurred after the first 20–25 rows (30–40 m) downwind from the corridor (Fig. 10.6). A similar population and distribution gradient was apparent for thrips. In both years, catches of leafhoppers and thrips were substantially higher in the central rows than in rows adjacent to the corridor.

The abundance and spatial distribution of generalist predators in the families Coccinellidae, Chrysopidae, Nabidae and Syrphidae were influenced by the presence of the corridor which channeled dispersal of the insects into adjacent vines (Fig. 10.7). Predator numbers were higher in the first 25 m adjacent to the corridor, which probably explains the reduction of leafhoppers and thrips observed in the first 25 m of vine rows near the corridor. The presence of the corridor was associated with the early vineyard colonization by *Anagrus* wasps, but this did not result in a net season-long prevalence in leafhopper egg parasitism rates in rows adjacent to the corridor. The proportion of eggs parasitized tended to be uniformly distributed across all rows in both blocks. Eggs in the center rows had slightly higher mean parasitization rates than eggs located in rows near the corridor, although differences were not statistically significant.

10.6.4 Creating flowering islands as a push–pull system for natural enemies in a Sonoma vineyard

Cover crops and corridors are all important practices to enhance insect biodiversity, but at times creating habitat on less productive parts of the farm to concentrate natural enemies may be a key strategy. This is the approach used at Benziger farm in Sonoma County, where a 0.25-ha island of flowering shrubs and herbs was created at the

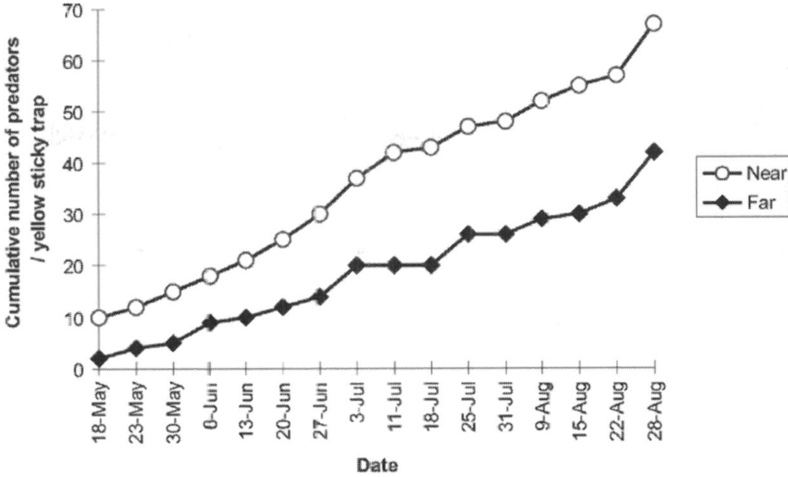

Fig. 10.7 Seasonal patterns of predator catches in vineyard as influenced by the presence or absence of forest edge and the corridor ($P < 0.05$; Mann–Whitney *U*-test) (Mendocino County, California, 1996).

center of the vineyard to act as a push–pull system for natural enemy species (Nicholls *et al.*, 2005).

The island and its mix of shrubs and herbs provides flower resources from early April to late September to a number of herbivore insects (pests, neutral non-pestiferous insects and pollinators) and associated natural enemies which build up in the habitat, later dispersing into the vineyard.

The island acts as a source of pollen, nectar and neutral insects which serve as alternate food to a variety of predators and parasites including *Anagrus* wasps. The island is dominated by neutral insects that forage on the various plants but also serve as food to natural enemies of thrips which slowly build up in numbers in the adjacent vineyard as the season progresses. Many natural enemies moved from this island insectory into the vineyard (up to 60 m). Responding to the abundance of habitat resources in the insectory, predators tended to decrease in abundance in vines 30 and 60 m away. *Orius* spp. reached significantly lower abundances in vines away from the insectory a trend that correlated with the densities of thrips which increased in vines far from the island. While the proportion of natural enemies in relation to the total number of insects caught in the traps remained relatively constant within the insectarium, their proportion increased from 1% to 10% and 13% in vines located 30 m and 60 m from the insectory respectively. *Orius* spp. and coccinellids are prevalent colonizers at the beginning of the season, but later syrphid flies and *Anagrus* wasps started dispersing from the insectarium into the vineyard (Fig. 10.8).

Table 10.1 Levels of leafhopper eggs parasitization by *Anagrus* wasps in the second and tenth vine rows from the insectory during the peak summer months (Sonoma County, California, 2004)

Date	Parasitization (%)	
	Second row	Tenth row
12 July	76.8	54.1
26 July	52.4	52.3
23 August	43.6	32.1

Parasitization of leafhopper eggs by *Anagrus* wasps was particularly high on the vines near the island (10 m from the island), with parasitization levels decreasing progressively towards the center of the vineyard away from the island (Table 10.1). It is possible that the presence of pollen and nectar in the island's flowers build up the populations of *Anagrus*, which moved from the island but confined their activity to nearby rows.

10.7 Conclusions

The instability of agroecosystems, manifested in the form of pest outbreaks, can be

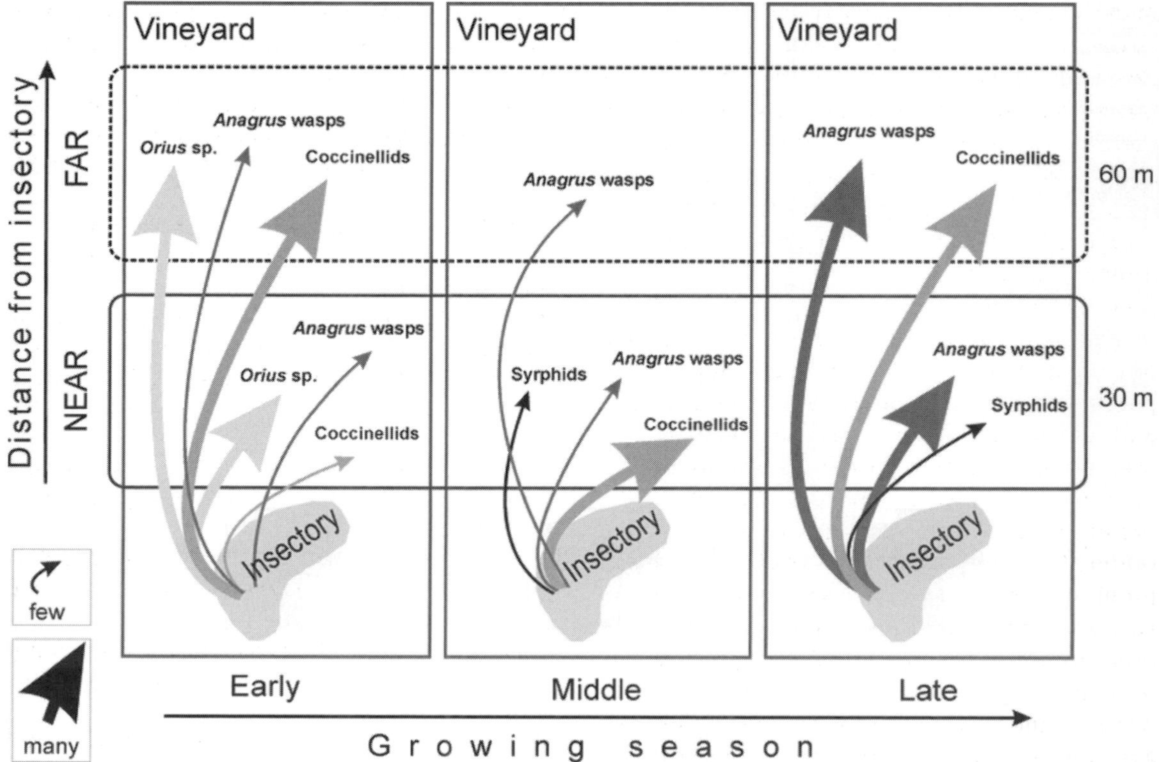

Fig. 10.8 Dispersal of *Anagrus* wasps and generalist predators from the island into the vineyard (Sonoma County, California, 2004).

increasingly linked to the expansion of crop monocultures (Altieri, 1994). Plant communities that are intensely modified to meet the special needs of humans become subject to heavy pest damage. The inherent self-regulation characteristics of natural communities are lost when humans modify such communities through the shattering of the fragile thread of community interactions. Agroecologists maintain that restoring the shattered elements of the community homeostasis through the addition or enhancement of biodiversity can repair this breakdown (Gliessman, 1998; Altieri, 1999).

A key strategy in sustainable agriculture is to reincorporate diversity into the agricultural landscape and manage it more effectively. Emergent ecological properties develop in diversified agroecosystems that allow the system to function in ways that maintain soil fertility, crop production and pest regulation. The role of agroecologists should be to encourage agricultural practices that increase the abundance and diversity of above- and below-ground organisms, which in turn provide key ecological services to agroecosystems.

The main approach in ecologically based pest management is to use management methods that increase agroecosystem diversity and complexity as a foundation for establishing beneficial interactions that keep pest populations in check (Altieri & Nicholls, 2004). This is particularly important in underdeveloped countries where resource-poor farmers have no access to sophisticated inputs and rely instead on the rich complex of predators and parasites associated with their mixed cropping systems for insect pest control. Reduction of plant diversity in such systems brought by modernization has the potential to disrupt natural pest control mechanisms, making farmers more dependent on pesticides.

When properly implemented, habitat management leads to establishment of the desired type of plant biodiversity and the ecological infrastructure necessary for attaining optimal natural enemy diversity and abundance. Habitat management may not always demand a radical change in farming as illustrated by the relative ease with which cover crops or corridors can be introduced into vineyard systems, providing a highly consistent, abundant and well-dispersed alternative food source, as well as microhabitats, for a diverse community of natural enemies, thus bringing biological control benefits to farmers (Landis *et al.*, 2000).

Long-term maintenance of diversity requires a management strategy that considers the design of environmentally sound agroecosystems above purely economic concerns. This is why several authors have repeatedly questioned whether the pest problems of modern agriculture can be ecologically alleviated within the context of the present capital-intensive structure of agriculture (Altieri & Nicholls, 2004; Gurr *et al.*, 2004). Many problems of modern agriculture are rooted within that structure and thus require consideration of major social change, land reform, redesign of machinery and ecological reorientation of research and extension to increase the possibilities of improved pest control through vegetation management. Whether the potential and spread of ecologically based pest management is realized will depend on policies, attitude changes on the part of researchers and policy makers, existence of markets for organic produce, and also the organization of farmer and consumer movements that demand a more healthy and viable agriculture.

References

Adams, M. W., Ellingbae, A. H. & Rossineau, E. C. (1971). Biological uniformity and disease epidemics. *BioScience*, **21**, 1067–1070.

Altieri, M. A. (1994). *Biodiversity and Pest Management in Agroecosystems*. New York: Haworth Press.

Altieri, M. A. (1999). The ecological role of biodiversity in agroecosystems. *Agriculture Ecosystems & Environment*, **74**, 19–31.

Altieri, M. A. (2000). The ecological impacts of transgenic crops on agroecosystem health. *Ecosystem Health*, **6**, 13–23.

Altieri, M. A. & Letourneau, D. K. (1982). Vegetation management and biological control in agroecosystems. *Crop Protection*, **1**, 405–430.

Altieri, M. A. & Letourneau, D. K. (1984). Vegetation diversity and outbreaks of insect pests. *CRC Critical Reviews in Plant Sciences*, **2**, 131–169.

Altieri, M. A. & Nicholls, C. I. (2003). Soil fertility management and insect pests: harmonizing soil and plant health in agroecosystems. *Soil and Tillage Research*, **72**, 203.

Altieri, M. A. & Nicholls, C. I. (2004). *Biodiversity and Pest Management in Agroecosystems*, 2nd edn. New York: Haworth Press.

Altieri, M. A., Wilson, R. C. & Schmidt, L. L. (1985). The effects of living mulches and weed cover on the dynamics of foliage–arthropod and soil–arthropod communities in 3 crop systems. *Crop Protection*, **4**, 201–213.

Andow, D. A. (1991). Vegetational diversity and arthropod population response. *Annual Review of Entomology*, **36**, 561–586.

Bach, C. E. (1980). Effects of plant diversity and time of colonization on an herbivore–plant interaction. *Oecologia*, **44**, 319–326.

Baliddawa, C. W. (1985). Plant species diversity and crop pest control: an analytical review. *Insect Science and Its Applications*, **6**, 479–487.

Conway, G. R. & Pretty, J. (1991). *Unwelcome Harvest: Agriculture and Pollution*. London: Earthscan.

Corbett, A. & Rosenheim, J. A. (1996). Impact of a natural enemy overwintering refuge and its interaction with the surrounding landscape. *Ecological Entomology*, **21**, 155–164.

Denevan, W. M. (1995). Prehistoric agricultural methods as models for sustainability. *Advances in Plant Pathology*, **11**, 21–43.

Ewel, J. J (1999). Natural systems as models for the design of sustainable systems of land use. *Agroforestry Systems*, **45**, 1–21.

Gliessman, S. R. (1998). *Agroecology: Ecological Processes in Sustainable Agriculture*. Chelsea, MI: Ann Arbor Press.

Gurr, G. M., Wratten, S. D. & Altieri, M. A. (2004). *Ecological Engineering for Pest Management: Advances in Habitat Manipulation for Arthropods*. Wallingford, UK: CABI Publishing.

Helenius, J. (1998). Enhancement of predation through within-field diversification. In *Enhancing Biological Control*, eds. E. Pickett & R. L. Bugg, pp. 121–160. Berkeley, CA: University of California Press.

Hilbeck, A., Baumgartner, M., Fried, P. M. & Bigler, F. (1998). Effects of transgenic *Bacillus thuringiensis* corn-fed prey on mortality and development time of immature *Chrysoperla carnea* (Neuroptera: Chrysopidae). *Environmental Entomology*, **27**, 480–487.

Hooks, C. R. R. & Johnson, M. W. (2003). Impact of agricultural diversification on the insect community of cruciferous crops. *Crop Protection*, **22**, 223–238.

Khan, Z. R., Pickett, J. A., van der Berg, J. & Woodcock, C. M. (2000). Exploiting chemical ecology and species diversity: stemborer and *Striga* control for maize in Africa. *Pest Management Science*, **56**, 1–6.

Landis, D. A., Wratten, S. D. & Gurr, G. M. (2000). Habitat management to conserve natural enemies of arthropod pests in agriculture. *Annual Review of Entomology*, **45**, 175–201.

Morales, H., Perfecto, I. & Ferguson, B. (2001). Traditional soil fertilization and its impacts on insect pest populations in corn. *Agriculture, Ecosystems and Environment*, **84**, 145–155.

Nicholls, C. I. & Altieri, M. A. (2004). Designing species-rich, pest-suppressive agroecosystems through habitat management. In *Agroecosystems Analysis*, eds. D. Rickerl & C. Francis, pp. 49–61. Madison, WI: American Society of Agronomy.

Nicholls, C. I., Parrella, M. P. & Altieri, M. A. (2000). Reducing the abundance of leafhoppers and thrips in a northern California organic vineyard through maintenance of full season floral diversity with summer cover crops. *Agricultural and Forest Entomology*, **2**, 107–113.

Nicholls, C. I., Parrella, M. & Altieri, M. A. (2001). The effects of a vegetational corridor on the abundance and dispersal of insect biodiversity within a northern California organic vineyard. *Landscape Ecology*, **16**, 133–146.

Nicholls, C. I., Ponti, L. & Altieri, M. A. (2005). Manipulating vineyard biodiversity for improved insect pest management: case studies from Northern California. *International Journal of Biodiversity Science and Management*, **1**, 191–203.

Pimentel, D. & Lehman, H. (1993). *The Pesticide Question*. New York: Chapman & Hall.

Pimentel, D. & Perkins, J. H. (1980). *Pest Control: Cultural and Environmental Aspects*, AAAS Selected Symposium No. 43. Boulder, CO: Westview Press.

Reinjtes, C., Haverkort, B. & Waters-Bayer, A. (1992). *Farming for the Future*. London: Macmillan.

Risch, S. J. (1981). Insect herbivore abundance in tropical monocultures and polycultures: an experimental test of two hypotheses. *Ecology*, **62**, 1325–1340.

Risch, S. J., Andow, D. & Altieri, M. A. (1983). Agroecosystem diversity and pest control: data, tentative conclusions, and new research directions. *Environmental Entomology*, **12**, 625–629.

Root, R. (1973). Organization of a plant–arthropod association in simple and diverse habitats: the fauna of collards (*Brassica oleracea*). *Ecological Monographs*, **43**, 95–124.

Scriber, J. M. (1984). Nitrogen nutrition of plants and insect invasion. In *Nitrogen in Crop Production*, ed. R. D. Hauck, pp. 441–460. Madison, WI: American Society of Agronomy.

Smith, H. A. & McSorley, R. (2000). Intercropping and pest management: a review of major concepts. *American Entomologist*, **46**, 154–161.

Southwood, T. R. E. & Way, M. J. (1970). Ecological background to pest management. In *Concepts of Pest Management*, eds. R. L. Rabb & F. E. Guthrie, pp. 6–29. Raleigh, NC: North Carolina State University.

Tilman, D. (1996). Biodiversity: population versus ecosystem stability. *Ecology*, **77**, 350–363.

Vandermeer, J. (1995). The ecological basis of alternative agriculture. *Annual Review of Ecology and Systematics*, **26**, 210–224.

Vandermeer, J. & Perefecto, I. (1995). *Breakfast of Biodiversity*. Oakland, CA: Food First Books.

Chapter 11

Manipulation of arthropod pathogens for IPM

Stephen P. Wraight and Ann E. Hajek

A great number and diversity of naturally occurring microorganisms are capable of causing disease in insects and other arthropods, and these pathogens have been manipulated for the purpose of pest control for more than 130 years (Lord, 2005). These efforts in applied invertebrate pathology have given rise to the field known today as microbial biological control, or simply microbial control, which is broadly defined as "that part of biological control concerned with controlling insects (or other organisms) by the use of microorganisms" (Onstad *et al.*, 2006). The term biological control does not have a universally accepted definition. The principal disagreement relates to whether or not the term should include control by non-organismal biological factors (e.g. toxic metabolites and other natural products). We believe that the term biological control should be restricted to use of living organisms, and in this context, the simplest definition of biological control is: *the use of living organisms to reduce damage by pests to tolerable levels*. This definition is similar to those presented by Van Driesche & Bellows (1996), Crump *et al.* (1999), Eilenberg *et al.* (2001) and Hajek (2004), all of which limit the definition to use of living organisms. Employing this definition, it follows that biological control agents are living organisms and microbial control agents are living microbes (including viruses, which some do not consider as living, and nematodes, whose inclusion will be explained below).

Most pathogenic microbes in general produce large numbers of propagules (e.g. bacterial and fungal spores and viral occlusion bodies) that function as the infectious units. In most cases, these propagules are resistant to environmental extremes of temperature and moisture and thus capable of surviving for extended periods outside the host. Indeed, the most important function of these propagules is survival in the environment for a period sufficient for successful transport (by myriad mechanisms) to new, susceptible hosts. Observations of this natural mode of action, especially when it culminated in impressive disease outbreaks, stimulated development of methods for mass production and formulation of these propagules for conventional broadcast applications against arthropod pests, an approach essentially equivalent to that employed in the use of synthetic chemical pesticides. In fact, the term microbial control, for much of its history, has been used primarily in reference to use of microbes in this way. These efforts worldwide have produced a number of notable successes, but far more commonly the result has been only partial control (pest population control or suppression insufficient to consistently reduce crop damage to acceptable levels). Consequently, as the result of

Integrated Pest Management, ed. Edward B. Radcliffe, William D. Hutchison and Rafael E. Cancelado. Published by Cambridge University Press.

competition from highly effective chemical pesticides, microbial control products have never comprised more than a very small percentage of the world pesticide market. This situation persists despite the increasing popular demand for environmentally safe pest control methods. Numerous authors have addressed this challenge in recent years (Lacey *et al.*, 2001; Lacey & Kaya, 2007; and see Chapter 13), and there is general agreement that in the future, microbial control agents will make their greatest contributions as components of IPM systems.

There are innumerable definitions of IPM; however, we here refer to those definitions that describe IPM systems as pest management systems based on coordinated use of multiple control agents (e.g. chemical toxins and biological control agents) and/or other control factors (e.g. host-plant resistance and cultural practices) to maintain pest populations at economically acceptable levels with minimal impacts on the environment. In such systems, control of any one pest may be achieved through the actions of a single agent or through the combined actions of multiple agents. IPM systems based primarily on biological agents or biologically derived agents (minimizing use of broad-spectrum synthetic chemical pesticides) may be referred to as biologically based (bio-based) or biologically intensive (bio-intensive) IPM systems. In natural systems, populations of insects that are well adapted to the prevailing environmental conditions are regulated primarily by various biotic factors including natural enemies (parasites, predators and pathogens). Virtually all arthropods interact with more than one natural enemy, and development of bio-based IPM systems is often viewed as an attempt to re-establish or replace a natural enemy complex that was lost when an arthropod moved into a new geographic region or new habitat.

11.1 Microorganisms as arthropod biocontrol agents

The pathogens that infect invertebrates are as diverse as those infecting vertebrates, and include all of the major, broadly defined groups of microorganisms: viruses, bacteria, fungi, algae and protozoa. Of these, however, only the viruses, bacteria and fungi (including microsporidia) are currently used for microbial control of arthropod pests. Arthropod-attacking nematodes, though more advanced than single-celled microorganisms, have historically been studied by invertebrate pathologists and are generally viewed as microbial control agents. The nematode species most commonly manipulated for biological control harbor symbiotic bacteria that are released into the body of the invaded host. Once in the host hemocoel, these bacteria act as virulent, lethal pathogens. Nematode reproduction subsequently takes place in the bacterium-colonized host cadaver.

Detailed descriptions of the myriad modes of action of these diverse microbial control agents are not provided here, but basics can be found in Hajek (2004) with more detailed descriptions in Boucias & Pendland (1998) and Gaugler (2002). Bacteria, viruses and microsporidia must be ingested to infect their hosts. In contrast, infections by the fungi most commonly used for microbial control are initiated externally, with fungal elements penetrating directly through the host body wall. The infective stages of nematodes also initiate infection by direct penetration, but invasion sites may be external or internal (e.g. through the gut wall). Nematodes also are highly mobile, moving through water or across moist surfaces to locate hosts, especially in soil. These differences in modes of infection influence the types of hosts that can be infected. For example, herbivorous arthropods that feed by piercing plants and sucking sap from the vascular tissues are not susceptible to the common bacterial and viral pathogens applied for biological control. These insects must therefore be targeted with fungal pathogens.

As will be discussed, one major difference among pathogens is whether they kill hosts by growing throughout the host (due to infection only) or through production of a toxin (toxinosis). It is difficult to generalize as to the ways in which each group of pathogens has been deployed for microbial biological control. Various pathogens from each group, for example, have been introduced into pest populations under the objectives of producing both long- and short-term pest control.

11.2 Microbial biological control strategies

Approaches for use of microbial control agents are commonly placed in three categories: (1) augmentation biological control, (2) classical biological control and (3) conservation biological control (Hajek, 2004) (Box 11.1).

Our principal emphasis in this chapter is with augmentation strategies and, usually, pathogens are applied for inundative control. Pathogens used for strategies other than inundative augmentation must have good potential to recycle in the host population. By recycling, we mean the ability of the pathogens initially infecting hosts to produce inoculum that infects more hosts and for this process to continue.

11.2.1 Types of pathogens used by strategy

Augmentation

Bacteria, viruses, fungi and nematodes comprise the major groups of pathogens that are mass-produced and commercially available for inundative control of arthropod pests. Pathogen-based pesticide products (generally referred to as biopesticides) are applied against many types of arthropod pests, but the major targets are insects in the Orders Lepidoptera, Coleoptera and Diptera. Numbers of products currently produced worldwide for microbial control are presented in Table 11.1 (and see Chapter 9). The microbial control agent that has been used most extensively worldwide is the bacterium *Bacillus thuringiensis* (*Bt*), with products worth $US 159.57 million at the end-user level in 2005 (Quinlan & Gill, 2006). This pathogen occurs as many varieties that produce virulent toxins specific to different groups of insects. Despite the considerable commercial success of *Bt*, its use as a biological control agent

Box 11.1 Definitions for categories of biological control

Type of biological control	Definition
Augmentation: inundative biological control	The use of living organisms to control pests when control is achieved exclusively by the organisms themselves that have been released.
Augmentation: inoculative biological control	The intentional release of a living organism as a biological control agent with the expectation that it will multiply and control the pest for an extended period, but not that it will do so permanently.
Classical biological control	The intentional introduction of an exotic biological control agent for permanent establishment and long-term pest control.
Conservation biological control	Modification of the environment or existing practices to protect and enhance specific natural enemies or other organisms to reduce the effect of pests.

Source: (Eilenberg *et al.*, 2001; Hajek, 2004).

Table 11.1 Major products available worldwide for microbial control of arthropod pests

Pathogen	Number of species in commercial products	Number of products with trade names[a]	Major hosts targeted
Viruses	22	68	Lepidoptera, Hymenoptera
Bacteria	4 (although including many varieties of *Bacillus thuringiensis*)	238	Lepidoptera, Diptera, Coleoptera
Fungi[b]	12	136	Diverse hosts
Nematodes	13	50	Diverse hosts

[a] In a few cases, products are mixes of different pathogens, and these products are listed for each pathogen.
[b] Including the microsporidian *Nosema locustae*. Recent molecular studies indicate that the microsporidia, formerly identified as protozoa or protists, are highly reduced fungi.
Source: Quinlan & Gill (2006).

as in the above definition has declined in recent years due to genetic engineering applications that will be discussed in a later section. Use of the numerous other agents listed in Box 11.1 has grown markedly in the past decade but is still constrained by many factors that will be reviewed herein.

Classical biological control
Classical biological control has been used only to a limited extent, but has resulted in excellent control in some systems (Hajek *et al.*, 2007). Pathogens that have been most successfully used for classical biological control are species with high epizootic potential that are active in stable habitats such as forests (Hajek *et al.*, 2007). The pathogen most frequently and successfully used for classical biological control has been a nudivirus that attacks palm rhinoceros beetles. An introduced nematode attacking the invasive pine woodwasp *Sirex noctilio* is considered the most important control agent in IPM programs that also include introductions of parasitoids and thinning of pine stands. A fungus accidentally introduced into North America from Japan is a key agent controlling the gypsy moth (*Lymantria dispar*), but inundative releases of a nucleopolyhedrovirus and *Bt* and mating disruption are also part of IPM programs to control this invasive forest defoliator (see Chapter 32).

Conservation biological control
This strategy has been investigated only sparingly with microbial control agents, but there have been notable successes that indicate considerable potential in agroecosystems. For example, a program that monitors naturally occurring fungal epizootics of cotton aphid (*Aphis gossypii*) in the southeastern USA prevents unnecessary insecticide applications that would kill beneficial predators and parasitoids (Steinkraus, 2007).

11.3 Attributes of microbial biocontrol agents

Though the microorganisms that have been developed for biological control are extraordinarily diverse, there are a number of general traits that characterize these control agents and greatly influence how they are manipulated for pest management.

11.3.1 Environmental safety

Most microbial control agents, like biological control agents in general, are considered environmentally safe or at least environmentally soft. Most have restricted host ranges, and even those species classified as generalists (e.g. the common entomopathogenic fungi *Beauveria bassiana* and *Metarhizium anisopliae*) comprise diverse strains that exhibit more restricted host ranges than the species as a whole. Some species of arthropod pathogens or strains within species produce broadly toxic or mutagenic metabolites (e.g. bacterial or fungal toxins); however, these compounds are not associated with dormant spores or other infectious units upon which most biopesticides are based, and in the environment, they are produced primarily within the infected host (during vegetative growth) and in quantities that are extremely small relative to the amounts of toxic materials typically released into the environment via conventional insecticide spray applications. Because they do not produce toxic residues, biopesticides typically have shorter post-application re-entry times and pre-harvest intervals than synthetic chemical pesticides, a factor that can translate into a strong economic advantage in labor-intensive fruit, vegetable and ornamental plant production systems.

11.3.2 Speed of action

With notable exceptions (toxigenic bacteria and nematodes), most microbial control agents have slow modes of action. Even following a successful inundative application, infected hosts do not succumb to infection for several to many days, or, in some cases, several weeks. During the disease incubation period, infected hosts may continue to damage crops and produce offspring at rates that support pest population growth. Pests infected with microbial control agents may also continue to vector virulent plant pathogens.

11.3.3 Natural epizootic potential and persistence

Many microbial control agents are capable of persisting or recycling in the environment (either as resistant spores or at low levels of host infection), and under favorable environmental conditions, some have extraordinary capacities for reproduction and dispersal and can develop rapidly to epizootic levels. This natural epizootic potential is a key factor in use of microbes in classical and inoculative biological control. In these systems, the great capacity of microbes to saturate a pest's habitat with infectious propagules can produce dramatically sudden pest population crashes (despite relatively slow action against the individual members of the population).

11.3.4 Host age-dependent susceptibility

Though there are numerous exceptions, most arthropod pathogens are more virulent against the immature (larval and nymphal) stages than the adults of their hosts. Within larval or nymphal instars, susceptibility also tends to decrease with increasing age. High virulence against immatures is an important trait of many biological control agents, because it confers the potential to compensate for slow action by preventing the pest from reaching larger stages that cause greater damage or from reaching reproductive maturity. Immatures of many pests also aggregate (e.g. at favored feeding sites on a host plant), making them vulnerable to pathogen epizootics. On the other hand, larvae of many important pests lead solitary lives in cryptic habitats, reducing susceptibility to pathogens capable of causing epizootics and to inundative or inoculative applications of pathogens. It should also be noted that few pathogens used for microbial control are highly virulent against the egg stage of their hosts.

11.3.5 Environmental sensitivity

Efficacy of many microbial control agents is highly dependent upon environmental conditions. Conditions of low or high temperature, low moisture and especially high insolation can severely limit both the initial and residual activity of microorganisms applied in inundation and inoculation biological control programs. Adverse abiotic conditions can also greatly diminish the natural epizootic potential of microbes released for inoculation or classical biological control.

11.3.6 Host contact

With few exceptions, lethal actions of microbial control agents are initiated only after infectious propagules (or infectious stages of nematodes) come into contact with the host. In most cases, microbial propagules are incapable of directed movement and therefore unable to actively search for susceptible hosts or avoid unfavorable environmental conditions; this trait is often cited as the reason many pathogenic microbes produce great numbers of propagules. Lack of mobility is especially significant when suitable hosts are present only in cryptic habitats or at low densities. In such cases, novel formulations (e.g. with baits or other attractants) or highly efficient, precisely targeted application methods may be required to achieve efficacy. This contrasts sharply with chemical control agents exhibiting vapor, translaminar, or systemic activity, for which less efficient application methods may suffice.

11.3.7 Mass production

Many microbial control agents can be mass-produced in vitro on industrial fermentation scales, and thus their populations can be manipulated more easily, rapidly and economically than those of most macrobiological control agents (i.e. insect parasitoids and predators). Inundation biological control is a major strategy in microbial control, whereas few parasitoids and predators are used in this way. On the other hand, some microbial control agents (especially obligate pathogens with high host specificity) are difficult or impossible to mass-produce and formulate, and some agents formulated as biopesticides have limited shelf-lives. Use of these agents for pest control may be limited to inoculative releases, conservation, or classical biological control approaches.

11.3.8 Product registration

Though microbial control agents in general have excellent environmental safety records, they are generally considered to pose greater health risks to humans and other non-target organisms than macrobiological control agents. Some arthropod pathogenic microbes are allergenic, and, as previously indicated, some have the potential to produce toxic metabolites. Mass production systems for microbial control agents also may be accidentally contaminated with other microbes that pose unknown risks. In many countries, therefore, microbial control agents mass-produced and formulated as biopesticides are required to undergo rigorous registration and quality control processes. Consequently, commercial development of these agents may be more costly than development of parasites or predators. As an exception, nematodes, once permitted into a country and cleared for release, are usually exempt from registration requirements.

11.3.9 Response to dose

Infectious pathogens exhibit dose–response regression lines with low slopes. The infectious units of pathogenic microbes generally act independently or largely independently in establishing lethal infections. One of the most significant effects of this independent action is to constrain the slope (regression coefficient) of the dose–response regression. Slopes of log dose-probit response regression lines of arthropod pathogens rarely exceed 2 and are often substantially <1. This contrasts sharply with chemical insecticides, whose molecules generally interact synergistically in causing host death and which typically exhibit much steeper dose–response regressions. It is noteworthy that insect pathogens whose primary mode of action is based on a toxin may also show similarly steep slopes.

11.3.10 Host resistance

It is often reported and generally perceived that use of microbial control agents poses a much lower risk than use of chemical control agents with respect to development of resistance. This is likely true, at least in cases where pathology is initiated by infection rather than toxinosis. Infection processes may involve myriad interacting and potentially redundant systems that make resistance development a difficult challenge for the host. However, no organism can survive without a capacity to resist lethal pathogens, be they toxic or infectious, and thus no relationship between a pathogen and its host can be considered constant. Economic issues have limited application of most microbial control agents at rates that would reproduce the levels of selective pressure generated by highly toxic chemical insecticides, and therefore the potential for pests to develop resistance to these control agents under field conditions remains largely unknown. Pathogens that kill their hosts with toxins can exhibit rapid lethal action comparable to synthetic chemical pesticides. Rapid action is generally a desirable characteristic of pest control agents; however, toxins applied extensively for pest control also have strong potential to select for resistance. To date, the only microbial control agents to which resistance has been documented in the field are toxigenic bacteria, including *Bt*, with resistance in only a few lepidopteran species (Glare & O'Callaghan, 2000) and *Bacillus sphaericus*, applied against mosquito larvae of the genus *Culex*. Researchers are actively working on resistance management strategies, which are similar to those for management of resistance to genetically engineered crops (see Chapter 19). Because of their unique modes of action, microbial control agents are also strong candidates for use in pesticide rotation programs designed to prevent development of pest resistance to conventional chemical pesticides.

11.3.11 Compatibility with other natural enemies

Microbial control agents are generally compatible with other natural enemies. Although use of microbial biological control agents accounts for a low percentage of pest control efforts worldwide, this is largely a cost and efficacy issue (see Chapter 13). There is strong public demand for environmentally safe pest control agents, and, despite previously mentioned difficulties, development of biologically based IPM systems combining multiple natural enemies and cultural practices to achieve acceptable pest control is an active area of research. Most microbial control agents, with their limited host ranges, are highly compatible with use of a broad range of arthropod predators and parasitoids. Even those pathogens characterized as generalists have proven safe to populations of beneficial insects and other non-target organisms when applied for microbial control (see Hokkanen & Hajek, 2003).

11.3.12 Compatibility with agrochemicals

Microbial control agents show highly variable levels of compatibility with agrochemicals. Most crops are treated with a broad range of chemical pesticides, including fungicides and bactericides for control of plant pathogenic microbes. These agents can be toxic to beneficial pathogens applied for arthropod pest control. On the other hand, arthropod pathogens are highly compatible with many chemical insecticides. Sublethal doses of some chemical insecticides have been shown to synergize the activity of arthropod pathogens through a variety of mechanisms.

11.3.13 Potential for increasing pathogen virulence

Arthropod pathogenic microbes, like most microorganisms, are highly amenable to a broad range of genetic manipulations, including: (1) simple selection of more virulent strains resulting from natural reproductive processes (e.g. transconjugant strains of *Bt*), (2) selection of strains genetically altered via exposure to mutagenic agents (e.g. selection of fungal strains altered via exposure to UV radiation), (3) selection of spontaneous mutants and (4) application of recombinant DNA technology (genetic engineering) to artificially modify the DNA of microbial control agents (e.g. engineering of insect viruses to produce insecticidal toxins). Pathogens can also be sources of genes (e.g. genetically modified crop plants expressing *Bt* toxins). According to our definition, however, use of transgenic plants cannot be considered microbial control because the microbial control agent is not employed as a living organism.

11.4 Manipulation of microbial control agents to maximize efficacy

Manipulation of arthropod pathogens to exploit positive attributes and overcome negative attributes is a major objective of microbial control research. Below, we will examine many of the difficulties associated with use of pathogens for pest control and describe methods currently in use or under development to improve the efficacy of microbial control.

11.4.1 Environmental safety/limited host range

Limited host range of pathogens is advantageous in that it minimizes environmental impacts. However, most agricultural crops are attacked by multiple, diverse pests, and there are economic constraints associated with targeting pests on an individual basis using highly host-specific control agents. Most successful microbial biological control programs have been based on use of highly virulent pathogens in crop systems with pest populations dominated by a single key pest. In the same light, however, such programs have proven highly vulnerable to abandonment in the face of needs to control other primary and secondary pests for which biological controls are not available. It is well documented, as well, that the usual alternative control option, broad-spectrum chemical insecticides, can severely disrupt naturally occurring or classical biological control systems, greatly exacerbating or even creating secondary pest problems.

This dilemma represents one of the most difficult challenges in all of biological control. Complex biologically based pest management systems employing multiple control agents against multiple pests are not only difficult to establish, but, once established, are quite unstable, especially in today's world where pest complexes are in constant flux due to pest and disease invasions.

In terms of microbial control, there have been a few efforts to produce biopesticide products based on more than one pathogen (e.g. multiple fungal pathogens) with the aim of expanding product host range, but costs of applying effective dosages of multiple biological control agents can be prohibitive. Much more effort has been directed toward incorporation of microbial control agents into broad-based IPM systems, especially systems employing predators, parasites and biorational (soft) chemicals.

11.4.2 Slow action and age-dependent susceptibility

Slow action is a common trait of biological control agents in general (both microbial and macrobial). Effective use of slow-acting agents for inundative and, in some cases, inoculative pest control often requires strategic timing of applications and targeting of specific pest life stages. This often involves targeting early-instar nymphs or larvae, that may be not only the stages most susceptible to infection. These can also be the primary targets in cases where adults have short preoviposition periods or where the later instars or adults are the most destructive to the crop. It is commonly recommended that inundative applications of microbial control agents be made during the early stages of pest population development (avoiding the need for rapid control of highly destructive populations). Few pests, however, develop as even-aged cohorts; even univoltine pests typically occur as mixed-age populations due to asynchronous oviposition or egg hatch. Inundation biological control of such populations with agents exhibiting host-age-specific virulence can be difficult and costly, requiring multiple applications, often at frequent intervals.

As previously noted, not all microbial control agents have slow modes of action. The bacteria associated with steinernematid and heterorhabditid nematodes usually kill the host within 48 hours. *Bacillus thuringiensis* and other *Bacillus* species produce virulent endotoxins that can kill highly susceptible hosts within hours after exposure. It is noteworthy, however, that these endotoxins are generally virulent in the absence of the living bacterial cell and microbes exhibiting primarily or exclusively toxic modes of action are perhaps more accurately defined as biologically derived or biorational chemical control agents than as biological control agents (the insecticidal toxin-producing actinomycete *Saccharopolyspora spinosa* is another excellent example). Toxins originating from arthropod pathogens and toxins from other organisms artificially associated with arthropod pathogens (e.g. insertion of genes for toxins specific to arthropods into pathogens) have great potential to play, or already play, a major role in pest control.

11.4.3 Epizootic potential

Exploiting the natural epizootic potential of arthropod pathogens is the principal objective for many microbial biological control efforts. Pathogens with the capacity to establish and cycle in a pest population are not necessarily constrained by slow modes of action, low regression coefficients, or high mass production costs. Such pathogens have potential to hold pest numbers below economic thresholds and provide great economic benefit. Outstanding examples of classical biological control include the introduction of a nudivirus into populations of coconut rhinoceros beetle (*Oryctes rhinoceros*) infesting palms on Pacific Islands (Huger, 2005) and the introduction (albeit apparently accidental) of the fungus *Entomophaga maimaiga* into gypsy moth populations in the USA (Hajek, 1999). Examples of inoculation biological control include the applications of the bacterium *Paenibacillus popilliae* and the fungus *Beauveria brongniartii* for long-term suppression of white grubs in turf and pasture lands (Klein, 1995).

Unfortunately, while classical and inoculation biological control represent the most efficient uses of arthropod pathogens, successes such as those cited above have proven difficult to achieve, especially in short-term agricultural crops. Cycling of pathogens in pest populations at rates sufficient to provide useful levels of pest control is, in most cases, extraordinarily dependent upon environmental conditions and characteristics of the host population (especially density). Environmental conditions most favorable to pathogen epizootics often do not coincide with periods in a cropping cycle when pests must be controlled, and buildup of pest numbers to a density supportive of pathogen epizootics may be intolerable (although high host densities are not always required for epizootics to occur). These problems also explain why manipulation of pathogen populations to initiate epizootics earlier in the season than occurs naturally has seen little success. With regard to such difficulties, it is important to note that great economic benefit may be provided by naturally occurring pathogens operating entirely without human intervention. These benefits are rarely quantified or even fully recognized, but potential exists for incorporating such natural pathogen cycles into IPM systems. In the southeastern USA a protocol has been developed whereby cotton aphid populations are monitored for prevalence of a naturally occurring fungal pathogen with extraordinarily high epizootic potential (*Neozygites fresenii*) (Steinkraus, 2007). If disease levels reach a threshold level before a specific time in the crop cycle, growers are advised to refrain from applying chemical insecticides, as the impending fungal epizootic can be relied upon to provide adequate control. In the sense that this reliance on *N. fresenii* limits applications of chemical insecticides and preserves populations of susceptible hosts for the fungus and other natural enemies, this approach can be considered a form of conservation biological control.

Despite the cost effectiveness of classical biological control where limited introductions are made with the goal of long-term control, there are some instances when high epizootic potential may be undesirable. This is primarily the case for pathogens capable of attacking a broad range of hosts (commonly referred to as generalist pathogens), thus posing potential risks to non-target organisms (usually other arthropods). In these cases, limited residual activity might be considered a positive attribute with respect to environmental safety. In addition, future development and use of genetically modified agents may involve mechanisms (e.g. programmed cell death) to prevent recycling or persistence in the environment.

11.4.4 Environmental requirements

Temperature and moisture

Temperature is one of the most difficult environmental constraints. Other than in greenhouses and other protected cultures, it is generally not possible to control temperature, and even in greenhouse situations, it may be too costly to moderate temperatures to levels optimal for activity of many pathogens. Temperatures that are substantially greater or less than optimum for a pathogen may directly inhibit development during infection or disease incubation. If the inhibitory temperature happens to be near optimal for the arthropod host, resulting enhanced activity of host immune systems may further disadvantage the pathogen. Fungi are particularly sensitive to ambient temperature and moisture conditions, because, unlike pathogens that infect per os (by mouth), most fungi remain exposed to the environment throughout much of the infection process.

Temperature may also indirectly affect pathogen success by influencing the rate of growth and development of insects and other arthropods. As insects develop through the larval instars to the adult stage, they tend to become less susceptible to infection by a broad range of microbes. In the case of fungal pathogens, the host is primarily infected via direct penetration of the body wall, and penetrating spores can be shed during molts. If warmer temperatures shorten intermolt periods, there is potential to markedly limit infection. Thus, rapid development promoted by high temperature can be a significant factor determining susceptibility. It is noteworthy, however, that because they do not molt, adults of some arthropods (especially those lacking heavily sclerotized cuticles) are markedly more susceptible to fungal infection than their immature stages.

Strains comprising many species of common microbial control agents exhibit a broad range of temperature optima, and screening for those most tolerant of temperature conditions in the target pest habitat may prove beneficial. There may be safety concerns, however, with respect to selection of pathogenic microbes exhibiting strong activity at temperatures approaching human body temperature (37 °C). In habitats characterized by temperature extremes, application of two complementary pathogens, one with high- and the other with low-temperature tolerance has been proposed (Inglis *et al.*, 1999).

In many agricultural systems, irrigation can be manipulated to provide at least some degree of evaporative cooling. Controlling irrigation to ensure constant plant hydration (maximal rates of transpiration) will also modulate temperature conditions within the phyllosphere, which is the principal habitat of many plant pests. Such manipulations of environmental or microenvironmental conditions as well as creation of new or additional habitat favorable to pathogen persistence, reproduction and transmission is a common objective of conservation biological control programs. For example, when windrows of alfalfa were left in Kentucky fields after cutting, alfalfa weevils aggregated in these warm, moist habitats, and fungal epizootics developed that controlled weevil populations (Brown, 1987).

Much of the discussion of temperature also applies to moisture and these factors are obviously closely interrelated. As the temperature of an air mass increases, its relative humidity decreases and drying capacity increases. Moisture is generally not a limiting factor for pathogens that are infectious per os (viruses, bacteria and microsporidia), but is essential for the germination and host infection processes of many arthropod-pathogenic fungi. Parasitic nematodes require free moisture when searching for hosts.

Manipulation of environmental moisture levels is potentially achievable via controlled irrigation. This is especially the case with the soil environment. However, moist air conditions resulting from irrigation are extremely transitory in the presence of significant air movement. In the field, manipulation of ambient relative humidity is most readily achieved under stagnant air conditions occurring at night (conditions that typically support dew formation). Greenhouses offer greater potential for environmental manipulations; however, this is generally more difficult than widely perceived. Greenhouses are typically ventilated throughout the day and early evening to counter the direct and residual effects of solar heating. Moreover, prolonged elevation of ambient moisture conditions in greenhouse crops increases the risk of crop losses due to plant pathogens. Nevertheless, favorable moisture conditions may result from specific crop production practices in greenhouses, and these have been shown to enhance efficacy of fungal pathogens. A prominent example involves application of a fungal pathogen into the favorable environment created when plants were covered with opaque shrouds to manipulate day length and control flowering (Hall, 1981).

Although high moisture levels are essential for fungal spore germination and host infection, many fungal pathogens are capable of infecting their hosts at rates largely independent of ambient moisture conditions. In these cases, the necessary moisture exists in the microhabitat, most commonly in humid boundary layers generated by moisture loss through the insect cuticle and spiracles, through the plant cuticle and stomata, or through a combination of these sources. In many cases, especially with respect to foliage-feeding insects with chewing mouthparts, moisture released from damaged plant tissues may contribute to creation of a zone of high moisture at the insect/plant interface.

An important aspect of fungal pathogen manipulation for biological control involves development of methods to enhance delivery of the infectious propagules into favorable microenvironments. Enhanced efficacy of fungal spores formulated in oils has been hypothesized to result, in part, from the capacity of these materials to carry the spores into recesses on the host body where humidity is higher (e.g. intersegmental folds) (Ibrahim *et al.*, 1999). Spraying of fungal conidia suspended in aqueous carriers with strong surfactants is another approach to achieving this objective. Potential also exists for developing pathogen formulations with materials that absorb moisture from the air (humectants) or otherwise reduce moisture loss following application (e.g. water in oil emulsions) (see Burges, 1998). However, these technologies have seen only limited commercial development.

Given a narrow window of favorable conditions for fungal pathogen activity, one potentially useful approach is the pretreatment of spores (by exposure to water, aqueous nutrient media, or high humidity) to initiate the germination process prior to application (see Burges, 1998). Conidia of some fungi are sensitive to rapid rehydration, and substantial viability can be preserved through preliminary exposure to high humidity (Moore *et al.*, 1997). Unfortunately, pretreatments may be difficult or impractical to implement on an operational scale.

Rain disseminates pathogens and thus has potential to promote epizootic development. With respect to fungal pathogens, rain also provides moisture needed for spore formation and survival. On the other hand, heavy or wind-driven rain may be disadvantageous if it washes propagules or insect cadavers serving as sources of inoculum from pest habitats. To counteract this effect, some biopesticide formulations have materials (stickers) added to improve rainfastness.

Solar radiation

Virtually all of the important arthropod pathogens are highly susceptible to solar radiation, especially the UVB portion of the spectrum (280–320 nm). Major research efforts have been undertaken and are continuing to identify UV screens with potential for incorporation into biopesticide formulations. Various materials including UV reflectors (e.g. zinc oxide, silicates, clays, starches and carbon) and UV absorbers (e.g. specialized dyes, lignins, stilbenes and melanins) have been tested. Many of these materials substantially increase pathogen survival under direct irradiation; however, this does not necessarily translate to greater efficacy under field conditions. This is primarily because direct sunlight is so rapidly lethal to most microbes (causing severe damage or death within a few hours) that half-life must be increased more than four- to five-fold to extend residual activity beyond one day. This level of protection is difficult to achieve without using large amounts of material, which can add substantially to product cost. Thick layers or capsules of protectant materials also affect the physical properties of formulations and may reduce efficacy by interfering with the host infection process. Nevertheless, development of efficient, economical sunscreen technologies continues to be an active area of research, and materials with UV protectant properties are included in many commercial biopesticide formulations.

Numerous alternative approaches to the application of pathogens are employed in which at least one objective is to minimize exposure to solar radiation. These include targeting pests in protected habitats, incorporating pathogens into granular formulations and applying pathogens during overcast weather conditions or at night. Persistence and thus potential efficacy of microbial control agents can be enhanced by targeting life stages of pests that inhabit niches shaded from solar radiation. This is an important advantage to targeting soil-dwelling stages of pests. However, the advantage of low radiation levels can be largely offset by presence of antagonistic soil microorganisms and by physical and economic constraints to achieving lethal titers of pathogens in soil. Pathogen propagules applied to soil surfaces tend to remain at or near the surface, and mechanical mixing of materials into the soil may be impractical and require large amounts of material. Greatest potential for such an approach may lie in potted nursery and greenhouse crops where pathogens can be premixed into potting substrates. It is noteworthy also that greenhouse glass and some plastic glazings block UVB and the shorter UVA wavelengths. Many entomopathogenic nematodes have the capacity to actively seek out prey in the soil and can be effective following simple drench application. Considerable research has also focused on modification of spray application technology to enhance coverage of leaf undersides or to penetrate dense foliage. Plant leaves absorb UV light, and the half-life of a pathogen may be many times greater when applied to the abaxial versus adaxial leaf surfaces. Pathogens may be applied to virtually any substrate likely to attract or harbor a specific pest; e.g. one method employed for introducing the *Oryctes* virus into rhinoceros beetle populations involved virus contamination of compost heaps of decaying palm leaves and sawdust where adults mate and beetle larvae develop (Huger, 2005).

Use of granular formulations, especially those incorporating bait materials, enables incorporation of high doses of microbes into large particles. Such particles may include thick coatings of UV-reflecting/absorbing materials and have greater potential to shield pathogens from solar radiation than formulations designed specifically to maximize dispersion (e.g. aqueous suspensions for foliar applications).

When practical, applying pathogens prior to cloudy weather conditions or in late afternoon or evening can also prolong pathogen survival. The latter approach may be most effective against pests that feed or are otherwise active throughout the night and may be particularly useful for inoculation or classical biological control programs.

11.4.5 Application

Because the infectious units of most arthropod pathogens are immobile and yet must come into direct contact with the target pest to effect control, their successful use in inundation and inoculation biological control programs is dependent upon development of effective delivery systems. Undersides of leaves, recesses in buds and flowers, areas beneath dense foliar canopies, tunnels in host plants and soil, etc. are favored habitats for many insects but are difficult to target with biopesticide applications.

Modification/reconfiguration of conventional pesticide or other application equipment

Targeting pests in cryptic habitats with spray applications requires adjustments to numerous parameters, including spray volume, trajectory, droplet size and turbulence. In some situations, especially where overwhelming doses are required (e.g. when targeting insect nests), effective application may be possible only with hand-held equipment. Formulation also may be a key factor; e.g. biopesticide penetration into the tunnels of leaf mining insects can be enhanced by formulation with strong surfactants, and dry powder formulations can be blown throughout the galleries of wood- or soil-inhabiting insects such as termites. Seed drill equipment has been adapted for applying granular formulations of pathogens at desired depths in soil.

Formulation of pathogens in food baits

These materials are usually targeted against pests that actively forage for food, with some species carrying materials back to nests. Examples have included incorporation of fungal pathogens into grain kernels for control of soil-dwelling caterpillars in Australia and into substrates baited with vegetable oils for control of fire ants in the USA. Strongly attractive baits may be especially useful in situations where large-scale broadcast applications are neither economically feasible nor environmentally acceptable.

Localized treatments

Pathogen preparations (usually highly concentrated) may be strategically placed across pathways or flyways in specific habitats where they are likely to be contacted by the pest. For example, the fungi *Beauveria brongniartii*, *B. bassiana* and *M. anisopliae* are cultured on fabric bands that are wrapped around the bole and major limbs of trees to intercept and inoculate ambulatory adult longhorned beetles.

Autodissemination and mechanical vectoring

Pests may be attracted by various means including food or pheromone baits to specially designed "traps" that direct them into an inoculation chamber provisioned with a pathogen preparation and then to an exit. The heavily inoculated pests then spread the inoculum to other individuals in the general population. This approach has been tested most extensively with fungal pathogens; the inoculum is often in the form of a dry conidial powder. A similar approach involves inoculation of non-target insects with heavy doses of a pathogen that is then delivered to pest habitats via foraging activities. Most commonly, this latter relatively new approach has involved inoculation of bees as they leave their hives or nests to forage for nectar and pollen. Initial tests have indicated potential for these vectors to deliver lethal doses of fungal pathogens to insect pests infesting flowers in field cages and greenhouses (e.g. Butt *et al.*, 1998). These approaches have obvious applicability for dissemination of pathogens in inoculation biological control programs. In Europe, extensive testing demonstrated that spray inoculation of adult chafers swarming at forest borders was an effective method of introducing a fungal pathogen into the larval populations developing in the soil in adjacent pastures. Epizootic spread ensued, reducing pest populations for many seasons (Keller *et al.*, 1997).

It should be noted that there are exceptions to this generalization of arthropod pathogens being non-motile. The most important of these with respect to microbial control are entomopathogenic nematodes that search for hosts in moist soil environments and actively attack them. In addition, insect pathogenic oomycete fungi produce motile spores called zoospores. Mycelia of the oomycete *Lagenidium giganteum* introduced into mosquito breeding pools produce zoospores that seek out and infect mosquito larvae.

Noteworthy also is the fact that some arthropod pathogenic microbes exhibit insecticidal vapor action and some may have systemic activity as endophytes. *Muscodor albus*, a recently discovered fungal endophyte of the cinnamon tree, produces a mixture of volatile organic compounds (alcohols, esters, ketones, acids and lipids) lethal against a broad range of plant and animal pathogens and against some insect pests (Lacey & Neven, 2006). Possibilities for using this fungus (or mixtures of volatiles that mimic those generated by the fungus) to control insect pests of stored products are under investigation. While the common entomopathogenic fungi *B. bassiana* and *M. anisopliae* also have been reported as either endophytes or rhizosphere colonizers of crop plants, the ability of these species when living in these specialized habitats to significantly impact pest populations and thus protect plants from pest attack requires further study.

11.4.6 Mass production efficiency

Advances in in vitro mass production technologies continue to improve the economics for many microbial control agents. Numerous bacteria and fungi are mass producible in large-scale fermentation systems. Many obligate pathogens, especially the viruses, are still produced in vivo, which can be labor intensive and therefore costly. Costs of many microbial control agents developed for inundation or inoculation biological control are competitive with synthetic chemical insecticides on a per-application basis at label rates. However, on a per-application basis, microbial control agents are usually less efficacious than chemical insecticides (at least when pesticide resistance is not an issue). The synergistic action of toxin molecules, reflected in high dose–response regression coefficients, enables formulation of chemical insecticides at many times the rate of active ingredient required to produce a high rate of pest mortality; this greatly increases the efficacy, consistency and reliability of these control agents. In contrast, the low regression coefficients associated with many microbial control agents make it economically infeasible to formulate products with a dose sufficient to achieve this result. As a consequence, effective use of these agents may be dependent upon multiple, frequent applications, which add greatly to cost. Competition from highly effective synthetic chemical insecticides remains one of the most important reasons for the limited success of microbials in the overall pesticide market (see Chapter 13).

Another economic factor relates to shelf-life and storage requirements. As one would expect, preparations whose active ingredients are living organisms are not nearly as stable as chemical insecticides. Shelf-lives of most agents do not extend beyond a single growing season when stored at room or warehouse temperatures. Refrigeration can markedly increase the usable life of many products, but cold storage increases overall cost.

Numerous approaches have been devised to improve the economics of pathogen production, formulation and storage. One of the most prominent involves local production by grower organizations. Fungal pathogens for control of spittlebugs have been produced for many years by large sugarcane cooperatives in Brazil. Fungi are cultured on an inexpensive grain-based substrate (usually rice) and used in a single season, circumventing storage stability problems. Another program developed in Brazil has involved mass production of the velvetbean caterpillar (*Anticarsia gemmatilis*) nucleopolyhedrovirus in caterpillar populations established in large outdoor plots. Following virus applications, diseased caterpillars are simply collected and frozen. The larvae are then macerated and sprayed in growers' fields. Systems such as this are applicable even at the level of individual farmers, who can harvest infected larvae from their production fields and apply the collected material directly back into the fields. Such approaches, which eliminate costs of mass production facility operation/maintenance and product purification and formulation, can be very economical, although there are quality control/regulatory issues that must be addressed. Efforts are still continuing toward in vitro production of viral pathogens in tissue culture. Efficient production systems have been developed in terms of numbers of infectious units producible per unit volume of cell culture; however, problems persist with respect to high costs of tissue culture ingredients and maintenance of virulence. Viruses, in general, tend to lose virulence during passage in cell culture; this is an advantage in production of vaccines, but an obvious disadvantage in production of biological control agents.

11.4.7 Registration requirements

Requirements of registration of microbial control agents apply almost exclusively to pathogens developed into commercial products for inundation and inoculation biological control. Exotic (non-indigenous) microbial control agents (and exotic biological control agents in general) are considered to pose greater environmental risks than indigenous pathogens. Thus, registration of exotic pathogens for biological control purposes can be more difficult, time consuming and costly than registration of indigenous pathogens. Introduction of exotic agents into a country or region for classical biological control may require a rigorous importation approval process, but registration in the sense of obtaining a periodically renewable license to produce and sell a quality-controlled product is not an issue.

11.4.8 Interactions among pest control agents

In reality, natural enemy species do not act in isolation; multiple natural enemies usually compete for hosts, and agrochemicals are commonly an important part of pest management systems. One of the most important characteristics of arthropod pathogens is their capacity to operate in conjunction with a broad range of other pest control agents (both chemical and biological). This attribute underscores the great potential for use of pathogens in IPM systems. Results of interactions between two or more control agents targeted against a single host species vary. The major categories of interactions include *independent action*, *synergism* and *antagonism*.

In the context of IPM, the term *independent action* is used to describe joint actions of two or more control agents resulting in impacts on a pest population that would be expected if each agent were acting independently (exhibiting independent, uncorrelated joint action). Mathematically, such effects are expressed by the combination of independent probabilities (Robertson *et al.*, 2007), and are commonly referred to in the ecological literature as additive effects (not to be confused with simple algebraic additivity (see Wraight, 2003)).

A *synergistic* or positive interaction is defined as an effect greater than that predicted by independent action. The mechanisms underlying synergistic interactions generally involve a non-lethal effect of one agent that increases the susceptibility of the host to another agent, or more generally, increases the probability of mortality from another agent.

An *antagonistic* or negative interaction is defined as an effect less than that predicted by independent action. Such interactions commonly occur when one agent directly interferes with the action of another agent. An example would be infection of an insect parasitoid or predator by a microbial control agent.

Successful manipulation of microorganisms as components of multiple-control-agent IPM systems requires knowledge of these biological interactions. In particular, pest managers attempt to minimize antagonistic interactions and enhance synergistic interactions. However, antagonistic interactions do not always yield an overall decrease in control; combined use of two or more control agents can provide an increase in control and can be very useful even if the interaction involves some level of antagonism. Of prime importance is the final level of control and crop protection. Here, we will discuss general trends in interactions of entomopathogens with other natural enemies and entomopathogens with agrochemicals.

11.4.9 Compatibility with other biological control agents

In virtually all systems, pest species are under pressure from multiple natural enemies. A complex of natural enemies associated with a pest species is termed a guild, and interactions among these natural enemies are referred to as intraguild interactions. In multiple pest systems, interguild interactions among natural enemies are also possible. Microbial pathogens, including entomopathogenic nematodes, and insect parasitoids and predators interact in a variety of ways with varied results. Here, we emphasize positive interactions but also mention important antagonistic interactions. For a more detailed account of different types of interactions, we refer you to Wraight (2003).

PHYSIOLOGY/IMMUNITY

In some cases, infection by a pathogen may interfere with the normal physiology of the host to such a degree that it becomes more susceptible to attack by other natural enemies. Some fungal and bacterial pathogens produce toxins that inhibit the cellular immune response of their hosts (Boucias & Pendland, 1998), and any animal with a compromised immune system is vulnerable to attack by a broad range of natural enemies. Direct or indirect effects of microbial toxins on host physiology may underlie the synergism reported between *Bt* and fungal pathogens (e.g. Wraight & Ramos, 2005). Both microbial and macrobial control agents also produce effects that result in a breach of the host's first line of defense, the internal and external body walls. Scarab beetle grubs infected with bacteria have been found more susceptible to attack by nematodes, and it has been hypothesized that these pathogens weaken the guts of their hosts and make them more susceptible to penetration by nematodes (Thurston *et al.*, 1994). Injuries inflicted by predators and parasitoids also predispose insects to infection by various pathogens.

HOST BEHAVIOR

Infection by pathogens often changes host behavior. Most notably, heavily infected arthropods may become sluggish. Insects that ingest sublethal doses of toxin-producing microbes such as *Bt* may also exhibit reduced responsiveness for a period of time prior to recovery. It is well documented that such behavioral changes can reduce the capacity of insects to escape or repel predators or parasites (see Wraight, 2003). On the other hand, foraging activities of predators and parasites may make pests more active, increasing the chances they will come into contact with pathogen propagules (e.g. see Roy & Pell, 2000). These effects have obvious implications for IPM, especially considering that many pathogens are slow acting, and their insect hosts continue to feed and reproduce for several to many days before succumbing to infection.

HOST DEVELOPMENT

Pathogen infections may also interfere with host development, and as in cases of host behavioral changes, these effects can greatly influence susceptibility or vulnerability to attack by other natural enemies. For example, infections that slow development of the immature stages of a host may greatly increase the window of time during which it is susceptible to stage- or instar-specific predators or parasitoids. A number of studies have reported synergistic effects of *Bt* treatments on the success of predators and parasitoids that attack insect larvae. Microsporidia are generally not highly virulent against their hosts, and few species have been developed as biopesticides. Many, however, have pronounced sublethal effects on their hosts (especially reduced growth and fecundity), which, despite their low virulence, make them potentially useful agents for integrated biological control.

POPULATION DYNAMICS

Predator and parasite populations generally increase more slowly than pest populations, and their effective use in biological control often depends on their capacity to become active during the early stages of pest population development (to prevent the pest population from reaching damaging levels). Thus, any agent that selectively contributes toward slowing the rate of pest population increase has the potential to enhance the efficacy of these biological control agents. Applications of microbial control agents, for example, may reduce pest population growth sufficiently to enable parasitoid and predator populations to reach effective levels. In terms of pest or crop damage control, the ultimate effect may be synergism, even if the control agents do not directly interact. Such indirect interactions often result from a partitioning of natural enemy activity against different life stages of the host. For example, effects of baculoviruses and parasitoids are often concentrated on different host instars. Pathogens, parasitoids and predators can also concentrate on different host stages based on the habitat they occupy (e.g. soil versus foliage). Natural enemies also partition their efforts by host population density. Pathogens have limited capabilities for directed host searching, and thus are often not effective in controlling low-density pest populations. However, due to their combined capacities for commercial mass production and rapid epizootic increase, pathogens have advantages over parasitoids and predators in high-density host populations. Microbial control agents can be useful against outbreak populations of pests that have escaped control by other natural enemies. Among the most prominent examples are applications of insect viruses to control hymenopteran and lepidopteran pests in forests.

PATHOGEN DISPERSAL

Both parasitoids and predators are known to aid in dispersal of pathogens, and this interaction has been investigated as a means to spread pathogens within and among host populations. For example, aphid predators contaminated with fungal pathogens have been shown to vector the pathogens into uninfected pest populations (see Roy & Pell, 2000).

ANTI-ANTAGONISM

While antagonistic interactions act to decrease the effects of natural enemies acting at the same time, the term anti-antagonism was coined to describe interactions where natural enemies avoid each other with an end result of greater impact on host populations. Following host invasion, parasitoids generally develop more slowly than pathogens, and may fail to develop before the host succumbs to the microbial infection. Studies have shown that many parasitoids can identify and avoid or only minimally utilize pathogen-infected hosts. Predators do not face the same risks as parasitoids in attacking hosts infected by pathogens; however, some insect predators have been observed to reject pathogen-killed hosts. Such behaviors result in increased reproduction by the two interacting natural enemies and are highly beneficial from the perspective of biological control. IPM systems have been designed with the aim of minimizing negative interactions between natural enemies and pathogens. For example, for integrated biological control of greenhouse whitefly (*Trialeurodes vaporariorum*) in glasshouses, it has been recommended that the fungus *Aschersonia aleyrodis* be applied 7 days before release of the parasitoid *Encarsia formosa* (Fransen & van Lenteren, 1994). With this timing, parasitoids can detect and avoid infected whiteflies and seek out suitable hosts, which maximizes survival of their offspring. Partitioning of natural enemy activity described in the above section on population dynamics result in an effect comparable to anti-antagonism.

Compatibility with agrochemicals

Among the most prominent and extensively investigated negative interactions between agrochemicals and microbial control agents are those between fungicides and entomopathogenic fungi. Many broad-spectrum fungicides are highly toxic to these beneficial fungi, including such commonly used materials as carbamate, dithiocarbamate, dicarboxamide and triazole compounds. However, even the strongest incompatibilities observed under in vitro laboratory conditions are not consistently expressed in the field environment. General unpredictability of results suggests that combined use of arthropod pathogenic fungi and fungicides requires evaluation on a case-by-case basis (Majchrowicz & Poprawski, 1993). This conclusion can generally be extended to virtually any microbial control agent–agrochemical combination. In some cases, problems can be attributed to toxic materials such as organic solvents used in the formulation of agrochemicals. More consistent findings of negative interactions have been observed in soils beneath fungicide-treated crops. This has been attributed to weathering of fungicides from plant foliage and their accumulation in the soil.

Attempts to overcome these problems have focused primarily on manipulation of microbial control applications to minimize direct interactions between arthropod pathogens and antagonistic agrochemicals. Research has demonstrated that even the most incompatible fungicides such as benomyl (a carbamate) and mancozeb (an ethylene bisdithiocarbamate) can be used in conjunction with inundative foliar applications of entomopathogenic fungi if the fungicides and fungi are applied asynchronously (Gardner *et al.*, 1984; Jaros-Su *et al.*, 1999). Much more difficult are situations where classical, conservation or inoculation biocontrol systems are jeopardized by fungicide applications. At present, other than selection of compatible fungicides, the most practical solution involves planning fungicide rotations so that the least antagonistic materials are applied during the most critical periods of activity of insect pathogenic fungi. This is especially important with respect to development of entomophthoralean (Zygomycetes) epizootics, as laboratory and field studies have shown these fungi to be very sensitive to fungicides.

While many chemicals are toxic to arthropod pathogens and have the potential to significantly interfere with microbial control, many others are highly compatible, and some are synergistic. Synergistic chemicals may act by either increasing host susceptibility or increasing pathogen virulence. Though there are many exceptions, chemical insecticides are generally more compatible with microbial control agents than fungicides. In many cases insecticides and pathogens can be tank mixed, and there is a long history of efforts to employ low doses of insecticides to synergize the activity of arthropod pathogens. Recommendations for use of *B. bassiana* for control of Colorado potato beetle in the Soviet Union and Eastern Europe during the 1970s and 1980s, for example, called for combination with a 1/4 dose of a chemical insecticide. It should be noted, however, that many early reports of synergism between fungal pathogens and various chemical insecticides were not supported by rigorous statistical analyses. More recent work has indicated synergism of fungal pathogens by the chloronicotinyl insecticide imidacloprid. Termites treated with imidacloprid ceased normal grooming behavior (Boucias *et al.*, 1996), treated weevil larvae exhibited reduced movement in soil (Quintela & McCoy, 1998) and intoxicated potato beetle larvae ceased feeding (Furlong & Groden, 2001). Each of these effects was identified as the likely mechanism underlying enhanced activity of fungal pathogens. Neonicotinoids disrupt grooming and evasion by white grubs in soil leading to synergism with entomopathogenic nematodes (Koppenhöfer *et al.*, 2002). Synergism also has been reported between the insecticide spinosad and the fungus *M. anisopliae* (Ericsson *et al.*, 2007), as well as between many chemical insecticides and baculoviruses (see McCutchen & Flexner, 1999).

Activity of arthropod pathogens may also be enhanced by a broad range of inorganic and organic chemicals not generally viewed as agrochemicals but with great potential to act as beneficial ingredients in biopesticide formulations. For example, non-insecticidal fluorescent brighteners have been found to synergize the activity of several different types of insect viruses (e.g. Shapiro, 2000), glycerol and other materials with humectant and/or nutrient qualities have been shown to enhance efficacy of fungal pathogens, and organosilicone surfactants and other powerful wetting agents can promote penetration of bacterial spores and toxins into cryptic pest habitats (e.g. tunnels of leaf-mining insects).

11.4.10 Genetic manipulations to enhance microbial control

Most microbe species comprise many strains having differing levels of virulence (i.e. ability to invade a host and cause disease. In development of microbial control agents, the first step usually involves selection of a highly virulent strain for use against the pest of concern. Researchers have ventured beyond this to genetic manipulations in the laboratory to enhance virulence for virtually all groups of pathogens; however, it is only among the bacteria that genetically modified strains are available for control.

Many of the myriad *Bt* toxins are active only against a few pests. The toxin genes are often present on plasmids that can be transferred between bacterial cells via the natural process of conjugation (Sanchis, 2000), and this process has been used to construct strains of *Bt* with novel combinations of toxins (for example, strains have been developed that produce toxins effective against both lepidopteran and coleopteran pests). Desired combinations of genes cannot always be achieved through conjugation, but there now exist numerous artificial means for achieving gene transfer. Manipulations via these methods, broadly referred to as genetic engineering, have enabled transfers of genetic material among extraordinarily diverse organisms, and one of the most significant applications of this technology has been insertion of genes from arthropod-pathogenic bacteria into plants. Tobacco plants were engineered to express a *Bt* toxin in 1877, and since then, the principal interest in genetic-transformation of this pathogen has shifted from *Bt* cells to *Bt*-engineered crops. Insertion of multiple *Bt* genes (known as gene pyramiding or gene stacking) has become a prime strategy for protecting plants from multiple pest species and also for managing *Bt* resistance (see Chapter 19). Major crops such as maize and cotton engineered to produce *Bt* toxins have been so successful that conventional spray application of *Bt* for control of herbivorous insects has decreased markedly in recent years. However, *Bt* spores and toxins continue to be applied against some pests and for some purposes (e.g. against mosquito larvae and in organic agriculture).

Another type of genetic manipulation that was developed and resulted in a commercial product involved moving *Bt* toxin genes into the genome of another bacterium. A *Bt* toxin gene was expressed in *Pseudomonas fluorescens* cells, which were then killed; this process led to enhanced UV protection for the toxins within the dead bacterial cells and, since the cells were killed, there was no possibility that genetically engineered bacteria would persist in the environment.

11.5 Conclusions

A great diversity of pathogenic microorganisms and nematodes has been developed for microbial biological control of insect and other arthropod pests. These control agents have many characteristics that determine their capacities to provide reliable pest control, and these characteristics must be taken into account when designing effective microbial control methodologies and strategies. In this chapter we have outlined what we consider to be the most important traits of microbial control agents, including positive attributes of environmental safety, epizootic potential, mass production capacity, compatibility with other natural enemies and negative attributes of slow lethal action, immobility of infectious propagules, environmental sensitivity, and low dose–response regression coefficients. We have explained how these various attributes affect microbial control efficacy and described many of the novel methods and strategies that have been devised to take advantage of positive attributes and minimize, circumvent or compensate for negative attributes. In view of the general incapacity of most microbial control agents to consistently provide economically acceptable levels of pest control as stand-alone agents, we conclude that their greatest potential lies in their use as components of carefully conceived and fully executed IPM systems.

References

Boucias, D. G. & Pendland, J. C. (1998). *Principles of Insect Pathology*. Boston, MA: Kluwer.

Boucias, D. G., Stokes, C., Storey, G. & Pendland, J. C. (1996). The effects of imidacloprid on the termite *Reticulitermes flavipes* and its interaction with the mycopathogen *Beauveria bassiana*. *Pflanzenschutz Nachrichten Bayer*, **49**, 231–238.

Brown, G. C. (1987). Modeling. In *Epizootiology of Insect Diseases*, eds. J. R. Fuxa & Y. Tanada, pp. 43–68. New York: John Wiley.

Burges, H. D. (ed.) (1998). *Formulation of Microbial Biopesticides: Beneficial Microorganisms, Nematodes, and Seed Treatments*. Dordrecht, Netherlands: Kluwer.

Butt, T. M., Carreck, N. L., Ibrahim, L. & Williams, I. H. (1998). Honey bee-mediated infection of pollen beetle (*Meligethes aeneus* Fab.) by the insect-pathogenic fungus *Metarhizium anisopliae*. *Biocontrol Science and Technology*, **8**, 533–538.

Crump, N. S., Cother, E. J. & Ash, G. J. (1999). Clarifying the nomenclature in microbial weed control. *Biocontrol Science and Technology*, **9**, 89–97.

Eilenberg, J., Hajek, A. & Lomer, C. (2001). Suggestions for unifying the terminology in biological control. *BioControl*, **46**, 387–400.

Ericsson, J. D., Kabaluk, J. T., Goettel, M. S. & Myers, J. H. (2007). Spinosad interacts synergistically with the insect pathogen *Metarhizium anisopliae* against the exotic wireworms *Agriotes lineatus* and *Agriotes obscurus* (Coleoptera: Elateridae). *Journal of Economic Entomology*, **100**, 31–38.

Fransen, J. J. & van Lenteren, J. C. (1994). Survival of the parasitoid *Encarsia formosa* after treatment of parasitized whitefly larvae with fungal spores of *Aschersonia aleyrodis*. *Entomologia Experimentalis et Applicata*, **71**, 235–243.

Furlong, M. J. & Groden, E. (2001). Evaluation of synergistic interactions between the Colorado potato beetle (Coleoptera: Chrysomelidae) pathogen *Beauveria bassiana* and the insecticides, imidacloprid, and cyromazine. *Journal of Economic Entomology*, **94**, 344–356.

Gardner, W. A., Oetting, R. D. & Storey, G. K. (1984). Scheduling of *Verticillium lecanii* and benomyl applications to maintain aphid (Homoptera: Aphididae) control on chrysanthemums in greenhouses. *Journal of Economic Entomology*, **77**, 514–518.

Gaugler, R. A. (ed.) (2002). *Entomopathogenic Nematology*. New York: CABI Publishing.

Glare, T. R. & O'Callaghan, M. (2000). Bacillus thuringiensis: *Biology, Ecology and Safety*. Chichester, UK: John Wiley.

Hajek, A. E. (1999). Pathology and epizootiology of the lepidoptera-specific mycopathogen *Entomophaga maimaiga*. *Microbiology and Molecular Biology Reviews*, **63**, 814–835.

Hajek, A. E. (2004). *Natural Enemies: An Introduction to Biological Control*. Cambridge, UK: Cambridge University Press.

Hajek, A. E., McManus, M. L. & Delalibera Júnior, I. (2007). A review of introductions of pathogens and nematodes for classical biological control of insects and mites. *Biological Control*, **41**, 4–13.

Hall, R. A. (1981). The fungus *Verticillium lecanii* as a microbial insecticide against aphids and scales. In *Microbial Control of Pests and Plant Diseases: 1970–1980*, ed. H. D. Burges. pp. 483–498. New York: Academic Press.

Hokkanen, H. M. T. & Hajek, A. E. (2003). *Environmental Impacts of Microbial Insecticides*. Dordrecht, Netherlands: Kluwer.

Huger, A. (2005). The *Oryctes* virus: its detection, identification and implementation in biological control of the coconut palm rhinoceros beetle, *Oryctes rhinoceros* (Coleoptera: Scarabaeidae). *Journal of Invertebrate Pathology*, **89**, 78–84.

Ibrahim, L., Butt, T. M., Beckett, A. & Clark, S. J. (1999). The germination of oil-formulated conidia of the insect pathogen, *Metarhizium anisopliae*. *Mycological Research*, **103**, 901–907.

Inglis, G. D., Duke, G. M., Kawchuk, L. M. & Goettel, M. S. (1999). Influence of oscillating temperatures on the competitive infection and colonization of the migratory grasshopper by *Beauveria bassiana* and *Metarhizium flavoviride*. *Biological Control*, **14**, 111–120.

Jaros-Su, J., Groden, E. & Zhang, J. (1999). Effects of selected fungicides and the timing of fungicide application on *Beauveria bassiana*-induced mortality of the Colorado potato beetle (Coleoptera: Chrysomelidae). *Biological Control*, **15**, 259–269.

Keller, S., Schweizer, C., Keller, E. & Brenner, H. (1997). Control of white grubs (*Melolontha melolontha* L.) by treating adults with the fungus *Beauveria brogniarta*. *Biocontrol Science and Technology*, **7**, 105–116.

Klein, M. G. (1995). Microbial control of turfgrass insects. In *Handbook of Turfgrass Insect Pests*, eds. R. L. Brandenburg & M. G. Villani, pp. 95–100. Lanham, MD: Entomological Society of America.

Koppenhöfer, A. M., Cowles, R. S., Cowles, E. A., Fuzy, E. M. & Baumgartner, L. (2002). Comparison of neonicotinoid insectides as synergists for entomopathogenic nematodes. *Biological Control*, **24**, 90–97.

Lacey, L. A. & Kaya, H. K. (eds.) (2007). *Field Manual of Techniques in Invertebrate Pathology: Application and Evaluation of Pathogens for Control of Insects and Other Invertebrate Pests*, 2nd edn. Dordrecht, Netherlands: Springer-Verlag.

Lacey, L. A. & Neven, L. G. (2006). The potential of the fungus, *Muscodor albus*, as a microbial control agent of potato tuber moth (Lepidoptera: Gelechiidae) in stored potatoes. *Journal of Invertebrate Pathology*, **91**, 195–198.

Lacey, L. A., Frutos, R., Kaya, H. K. & Vail, P. (2001). Insect pathogens as biological control agents: do they have a future? *Biological Control*, **21**, 230–248.

Lord, J. C. (2005). From Metchnikoff to Monsanto and beyond: the path of microbial control. *Journal of Invertebrate Pathology*, **89**, 19–29.

Majchrowicz, I. & Poprawski, T. J. (1993). Effects *in vitro* of nine fungicides on growth of entomopathogenic fungi. *Biocontrol Science and Technology*, **3**, 321–336.

McCutchen, B. F. & Flexner, L. (1999). Joint action of baculoviruses and other control agents. In *Biopesticides: Use and Delivery*, eds. F. R. Hall & J. J. Menn, pp. 341–355. Totowa, NJ: Humana Press.

Moore, D., Langewald, J. & Obognon, F. (1997). Effects of rehydration on the conidial viability of *Metarhizium flavoviride* mycopesticide formulations. *Biocontrol Science and Technology*, **7**, 87–94.

Onstad, D. W., Fuxa, J. R., Humber, R. A. *et al.* (2006). *An Abridged Glossary of Terms Used in Invertebrate Pathology*. Knoxville, TN: Society for Invertebrate Pathology. Available at www.sipweb.org/glossary

Quinlan, R. & Gill, A. (2006). *The World Market for Microbial Biopesticides*. Wallingford, UK: CPL Business Consultants, CAB International Centre.

Quintela, E. D. & McCoy, C. W. (1998). Conidial attachment of *Metarhizium anisopliae* and *Beauveria bassiana* to the larval cuticle of *Diaprepes abbreviatus* (Coleoptera: Curculionidae) treated with imidacloprid. *Journal of Invertebrate Pathology*, **72**, 220–230.

Robertson, J. L., Russell, R. M., Preisler, H. K. & Savin, N. E. (2007). *Bioassays with Arthropods*, 2nd edn. Boca Raton, FL: CRC Press.

Roy, H. E. & Pell, J. K. (2000). Interactions between entomopathogenic fungi and other natural enemies: implications for biological control. *Biocontrol Science and Technology*, **10**, 737–752.

Sanchis, V. (2000). Biotechnological improvement of *Bacillus thuringiensis* for agricultural control of insect pests: benefits and ecological implications. In *Entomopathogenic Bacteria: From Laboratory to Field Application*, eds. J.-F. Charles, A. Delécluse & C. Nielsen-LeRoux, pp. 441–459. Dordrecht, Netherlands: Kluwer.

Shapiro, M. (2000). Enhancement in activity of homologous and heterologous baculoviruses infectious to beet armyworm (Lepidoptera: Noctuidae) by an optical brightener. *Journal of Economic Entomology*, **93**, 572–576.

Steinkraus, D. C. (2007). Documentation of naturally occuring pathogens and their impact in agroecosystems. In *Field Manual of Techniques in Invertebrate Pathology: Application and Evaluation of Pathogens for Control of Insects and Other Invertebrate Pests*, 2nd edn, eds. L. A. Lacey & H. K. Kaya, pp. 267–281. Dordrecht, Netherlands: Springer-Verlag.

Thurston, G. S., Kaya, H. K. & Gaugler, R. (1994). Characterizing the enhanced susceptibility of milky disease-infected scarabaeid grubs to entomopathogenic nematodes. *Biological Control*, **4**, 67–73.

Van Driesche, R. G. & Bellows, T. S. Jr. (1996). *Biological Control*. New York: Chapman & Hall.

Wraight, S. P. (2003). Synergism between insect pathogens and entomophagous insects, and its potential to enhance biological control efficacy. In *Predators and Parasitoids*, eds. O. Koul & G. S. Dhaliwal, pp. 139–161. London: Taylor & Francis.

Wraight, S. P. & Ramos, M. E. (2005). Synergistic interaction between *Beauveria bassiana*- and *Bacillus thuringiensis tenebrionis*-based biopesticides applied against field populations of Colorado potato beetle. *Journal of Invertebrate Pathology*, **90**, 139–150.

Chapter 12

Integrating conservation biological control into IPM systems

Mary M. Gardiner, Anna K. Fiedler, Alejandro C. Costamagna and Douglas A. Landis

Most agricultural production systems harbor many species of herbivorous arthropods capable of damaging crops. However, the vast majority of these species do not reach damaging levels. In this chapter we explore the role of predators and parasitoids in suppressing pest abundance and damage. In particular, we focus on factors that influence the abundance of beneficial arthropods in agricultural landscapes. Finally, we address ways to manage these systems to increase the effectiveness of beneficial arthropods.

There are three primary means by which managers influence biological control of insects. *Importation* of natural enemies against pests of exotic origin is sometimes referred to as classical biological control, while *augmentation* is the rearing and release of natural enemies already present to increase their effectiveness. *Conservation* of natural enemies involves improving conditions for existing natural enemies by reducing factors which interfere with natural enemies or increasing access to resources that they require to be successful (Ehler, 1998). Habitat management is considered a subset of conservation practices that focus on manipulating habitats within agricultural landscapes to provide resources to enhance natural enemies (Landis *et al.*, 2000).

Managing agricultural landscapes to improve biological control relies on a detailed understanding of factors that influence both pest and natural enemy abundance (Fig. 12.1). We begin by examining landscape processes that influence pests and beneficial insects at larger spatial scales. Next we focus on processes that influence these organisms and their interactions at local scales. Finally, we detail steps that pest managers may take to alter agricultural habitats and landscapes to favor natural enemies in IPM systems. Throughout, we attempt to show how both temporal and spatial factors influence the outcome of pest–enemy interactions.

12.1 Landscape processes and IPM

Agricultural landscapes consist of a mosaic of crop and non-crop habitats. The diversity and abundance of both pests and natural enemies depends on the large-scale structure of these landscapes. Several patterns have emerged that illustrate the impact of landscape structure on pests and beneficial insects. First, the diversity and abundance of predators and parasitoids often increases as landscape complexity and the proportion of non-crop habitat increase. Second, several studies suggest that there may be thresholds in landscape structure below which the search efficiency and ability

Integrated Pest Management, ed. Edward B. Radcliffe, William D. Hutchison and Rafael E. Cancelado. Published by Cambridge University Press.

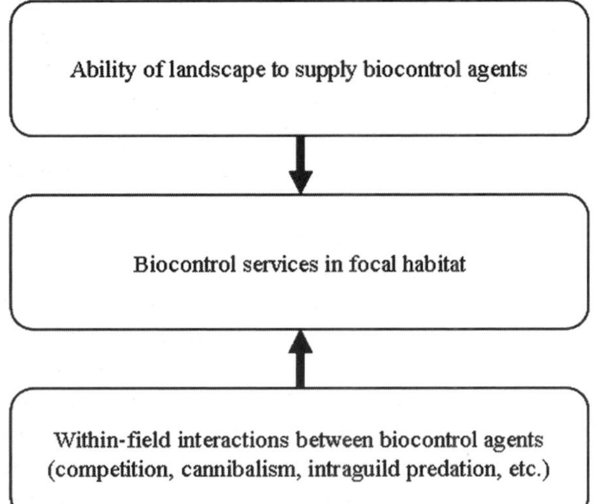

Fig. 12.1 Hierarchy of factors influencing biological control. The landscape constrains the diversity and abundance of natural enemies available for biological control in a focal habitat. Within the field interactions mediate the strength of their combined impact on pests.

of natural enemies to aggregate and control pests is diminished. Finally, landscape characteristics may not influence all species equally, or at the same scale.

12.1.1 Landscape complexity

The diversity of habitats within a landscape can greatly affect communities of herbivores and their natural enemies within an agricultural crop (Marino & Landis, 1996; Östman, 2002; Schmidt & Tscharntke, 2005; Tscharntke *et al.*, 2005). A majority of studies found a decrease in herbivore density and damage as the proportion of non-crop habitat in a landscape increased. For example, den Belder *et al.* (2002) found the density of onion thrips (*Thrips tabaci*) decreased with an increase in woodlot area in the landscape. This was attributed to trees and shrubs restricting the dispersal of onion thrips between fields. In addition, Thies *et al.* (2003) examined herbivory and parasitism of rape pollen beetle (*Meligethes aeneus*) in oilseed rape landscapes that varied in structural complexity. They found decreased plant damage and increased larval parasitism in structurally complex landscapes.

The presence of non-crop habitats may increase natural enemy abundance by providing resources such as alternative prey or overwintering habitat. While pest abundance and herbivory generally decline with structural complexity, the abundance and diversity of predators and parasitoids often increases as non-crop habitat increases. For example, Elliott *et al.* (1999) used aerial photographs to demonstrate that increases in uncultivated land and habitat patchiness were both associated with increasing abundance and richness in predator communities in wheat fields. Similarly, Bommarco (1998) found that fecundity of the generalist predatory ground beetle *Pterostichus cuperus* was greater in areas with small fields, higher perimeter-to-area ratios and a higher coverage of perennial crops compared to less spatially complex areas.

While many beneficial insects increase with the proportion of non-crop habitat and overall complexity of a landscape, these responses vary by taxa. Colunga-Garcia *et al.* (1997) studied changes in species assemblages of predatory lady beetles in response to landscapes with different mosaics of crop (alfalfa, corn, wheat) and non-crop (deciduous forest, field succession) habitats. Although the deciduous forest habitat had the highest lady beetle species richness, there were exceptions. Seven-spotted lady beetle (*Coccinella septempunctata*) was the dominant species in the overall landscape and was equally abundant in all sites, while the pink lady beetle (*Coleomegilla maculata*) was more abundant when corn was present in the landscape.

There is also temporal variation in the response of natural enemies to landscape structure. Menalled *et al.* (2003) examined parasitism of armyworm (*Pseudaletia unipunctata*) by two braconid parasitoids, *Glyptapanteles militaris* and *Meteorus communis*, across five years in simple and complex landscapes. The simple landscape had 29% non-crop habitat, and the complex landscape had 41% non-crop habitat and smaller crop fields. Although percentage parasitism in the complex landscape was equal to or higher than the simple landscape in four of the five years of the study, distribution of the parasitoids varied. *Glyptapanteles militaris* was most abundant in the simple landscape during the last two years of the study while *M. communis* was the dominant species in the complex landscape during the first three years of the study.

12.1.2 Habitat thresholds

Several studies suggest thresholds in landscape structure below which the search efficiency and ability of natural enemies to aggregate and control pests is diminished. Simulation experiments have shown that search success of natural enemies declined when suitable habitat fell below 20% (With & King, 1999). This observation has also been documented in the field; Thies & Tscharntke (1999) found that parasitism rates declined in agricultural landscapes when the non-crop area fell below 20%. Thies *et al.* (2003) found that in simple landscapes, parasitism of rape pollen beetle declined below threshold levels needed for successful biological control.

The impact of landscape structure is dependent not only on the total amount of suitable habitat within a landscape but also on the arrangement of crop and non-crop habitat fragments. Some species readily disperse from non-crop habitats into crop fields while others do not. Herbivorous pests and their natural enemies all vary in their capacity for dispersal. Species that are able to fly several kilometers may survive in a landscape with a few isolated fragments of suitable habitat while species with limited dispersal may not. Therefore, the arrangement and size of suitable habitat patches within a landscape are likely to have species-specific impacts on biocontrol in agricultural crops.

12.1.3 Edge effects

Natural enemy populations may build up in field borders and move into a crop when pest populations begin to build. However, habitat edges may also impede the movement of arthropods into crops by acting as physical barriers (Wratten *et al.*, 2003) or by providing a more stable and suitable habitat than the crop (Fortin & Mauffette, 2001). A substantial amount of research has been conducted to understand the role of field edges on biological control of pests by natural enemies. Overall, these data show that measuring the activity and richness of predatory populations in field boundaries can, but doesn't always, accurately predict their potential impact on herbivores in neighboring crops (Hunter, 2002). In some cases, field edges and neighboring habitats contribute to within-field natural enemy assemblages. For example, Alomar *et al.* (2002) found that the abundance of greenhouse whitefly (*Trialeurodes vaporariorum*) and the predatory bugs *Dicyphus tamaninii* and *Macrolophus caliginosus* was highest in outer tomato rows close to surrounding non-crop vegetation. This indicates that the presence of resources such as winter refuges in nearby habitats increased predator colonization of tomato. In other cases, the presence of non-crop habitat surrounding agricultural fields either had no effect on natural enemy density in the crop or had a positive effect on herbivore abundance. For example, Bugg *et al.* (1987) found that weedy strips at the edge of alfalfa fields contained a high density of spiders; however, spiders did not disperse from strips into adjacent crop fields. The authors suspected that spiders had no stimulus to disperse because weedy borders were more hospitable than crops.

12.1.4 Functional spatial scale and trophic level

Landscape complexity may not influence all species equally, or at the same scale. Thies *et al.* (2003) measured the functional spatial scale at which rape pollen beetle and its parasitoids were affected by landscape structure. They tested the effects of landscape on trophic-level interactions using simple (<3% non-crop habitat) to complex (>50% non-crop habitat) landscapes, analyzing them at eight spatial scales (concentric circles 0.5–6 km in diameter). They found that herbivory and parasitism were best correlated with percent non-crop area using a spatial scale of 1.5 km. At all spatial scales the predictive power of non-crop area was higher for herbivory than for parasitism. Schmidt *et al.* (2005) used pitfall traps to collect spider species in wheat fields located in landscapes that varied in their proportion of non-crop habitat. The proportion of non-crop habitat did not affect spider abundance. Conversely, spider species richness in wheat fields rose with the proportion of non-crop habitat in the surrounding landscape. The most common family of spiders collected were wolf spiders (Lycosidae). Ballooning species were poorly represented overall, suggesting that ballooning species may respond to a larger spatial scale of landscape complexity than walking species (Schmidt *et al.*, 2005).

12.1.5 Impacts of landscape variables depend on farming practice

A potential complication of studying the effects of landscape complexity in agricultural systems is the effect that production practices may have on arthropod communities. Weibull *et al.* (2003) compared the effect of farm management (conventional farming versus organic) and landscape features on species richness in agricultural fields and pastures. They found that the species richness of several taxonomic groups did not differ between organic and conventional farms. However, the effects of landscape structure were more pronounced on conventional farms where species richness of plants, butterflies and carabids increased with either large (25 km^2) or small-scale (400 m^2) landscape heterogeneity. There were no significant correlations between species richness and landscape variables among organic farms (Weibull *et al.*, 2003). Östman *et al.* (2001) measured the impact of natural enemies on the population development of the bird cherry-oat aphid (*Rhopalosiphum padi*) on conventional and organic farms. Here the impact of natural enemies on establishment of the aphid was greater on organic than conventional farms. Regardless of farming practices, early season establishment of bird cherry-oat aphid was lower in landscapes with abundant field margins and perennial crops (Östman *et al.*, 2001). These studies illustrate that both production practices occurring at the local scale and the surrounding landscape can affect the diversity and impact of arthropods in farm fields.

12.2 Local processes and IPM

Although potential pests and biological control agents occur in a regional pool, local processes impact their interactions and the outcome of biological control at the field level. Some of these local processes include negative impacts of pesticide application and cultural practices such as soil cultivation on natural enemy populations (Croft, 1990; DeBach & Rosen, 1991; Barbosa, 1998; Stark & Banks, 2003; Desneux *et al.*, 2007). Here, we focus on several phenomena that mediate the effect of natural enemies at the local scale and have received relatively less attention, including timing between natural enemy and pest arrival and interactions within multiple enemy assemblages.

12.2.1 Early-season pest suppression

The difference in timing between the arrival of prey and their natural enemies into a crop often determines the outcome of their interaction. Generalist natural enemies have the potential to be present in crops early in the season, before pests arrive and experience population growth. However, such early season predation usually goes unnoticed due to the relatively low numbers of natural enemies required to suppress initial pest populations. Using a theoretical model, Chang & Kareiva (1999) demonstrated that early predation by generalist predators can reduce pest populations as much as later immigration of more specialized natural enemies. Landis & van der Werf (1997) showed that the assemblage of generalist predators present in sugar beet fields early in the season significantly reduced aphid abundance and the impact of the aphid-vectored viruses. Using predator exclusion cages, several studies have demonstrated strong suppressive effects of generalist predators on soybean aphid (*Aphis glycines*) early in the season, both during outbreak and non-outbreak aphid years (Fox *et al.*, 2004; Costamagna & Landis, 2006; Desneux *et al.*, 2006; Costamagna *et al.*, 2007). In all these studies, predator populations had densities of only a few individuals per square meter, indicating that a relatively low number of generalists are capable of suppressing soybean aphid.

12.2.2 Interactions within multiple enemy assemblages

Natural enemy assemblages in agroecosystems are typically composed of multiple species, including both generalists and specialists (Symondson *et al.*, 2002). Within these diverse assemblages, positive interactions such as predator facilitation and negative interactions such as predator interference, cannibalism, predator avoidance behavior and intraguild predation commonly occur and can modify the level of herbivore pest suppression

(Roland & Embree, 1995; Losey & Denno, 1998; Sih *et al.*, 1998; Snyder & Wise, 1999; Prasad & Snyder, 2006a).

12.2.3 Positive interactions

Synergistic interactions, such as predator facilitation, have been demonstrated to increase aphid control. For example, significantly greater suppression of pea aphid (*Acyrthosiphon pisum*) resulted from combining seven-spotted lady beetle, a foliar predator that increases aphid dropping behavior, and the ground beetle *Harpalus pennsylvanicus*, a predator that attacks aphids on the ground or lower part of the plant. Thus, the ground beetle consumes significantly more aphids in the presence of the lady beetle, which maintains its rate of attack regardless of ground beetle presence. The result is synergistic aphid suppression with the two predators combined (Losey & Denno, 1998). Another example of synergistic effects was found between introduced parasitoids of the winter moth (*Operophtera brumata*) and rove beetles predators. Rove beetles attacked proportionally fewer parasitized than unparasitized winter moth pupae, resulting in increased moth mortality relative to control by either parasitoids or rove beetles alone (Roland & Embree, 1995).

12.2.4 Negative interactions

Although positive interactions between natural enemies occur, it is more common that they engage in negative interactions, resulting in a variety of consequences for prey suppression. In multiple natural enemy assemblages, the presence of generalist natural enemies often leads to predatory interactions between members of the same guild, a phenomenon termed intraguild predation (IGP) (Polis *et al.*, 1989). Intraguild predation may involve direct effects when species interact by eating each other and indirect effects when species compete for a shared resource (Polis *et al.*, 1989) (Fig. 12.2). Several reviews have shown that IGP can cause disruption of biocontrol (Rosenheim *et al.*, 1995; Brodeur & Rosenheim, 2000; Müller & Brodeur, 2002; Rosenheim & Harmon, 2006). Impacts of these interactions, however, vary from biocontrol enhancement to disruption, and may even have no effect on herbivore populations.

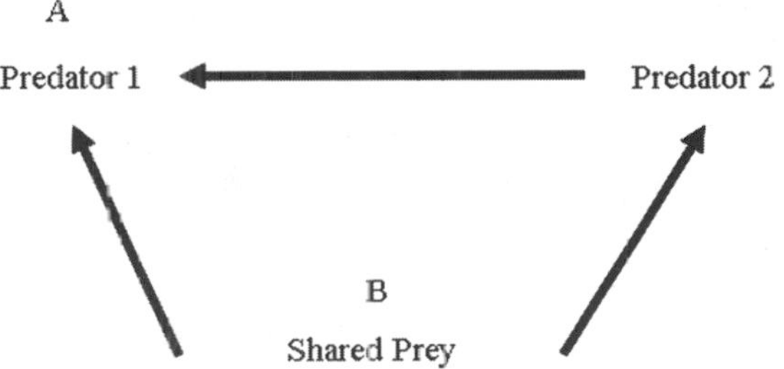

Fig. 12.2 Intraguild predation may result in both (A) direct effects (reduction in the population of an intraguild prey (Predator 1) due to consumption by an intraguild predator (Predator 2) and (B) indirect effects (increase in the population of a shared prey due to the reduction in the population of the intraguild prey by the intraguild predator). Arrows indicate the direction of energy flow.

There are multiple examples of IGP that do not result in biological control disruption. Field studies with the braconid *Lysiphlebus testaceipes* in combination with convergent lady beetle (*Hippodamia convergens*) showed that despite high predation on parasitoid mummies, suppression of cotton aphid (*Aphis gossypii*) was enhanced in the presence of the lady beetle (Colfer & Rosenheim, 2001). Similarly, strong suppression of soybean aphid by multicolored Asian lady beetle (*Harmonia axyridis*) compensated for IGP of *L. testaceipes* (Costamagna *et al.*, 2007), the predatory midge *Aphidoletes aphidimyza* and the common green lacewing (*Chrysoperla carnea*) (Gardiner & Landis, 2007). Both Colfer & Rosenheim (2001) and Costamagna *et al.* (2007) found that despite IGP, predation by the intraguild predator on a shared prey resulted in increased crop biomass and yield, termed a trophic cascade. Lang (2003) showed that despite strong IGP, winter wheat aphid populations were reduced to a greater extent by carabid beetles and spiders (Lycosidae, Linyphiidae) together than in plots where only one group of predators was present. Finally, Snyder *et al.* (2006) found increased suppression in collards with two pests together, green peach aphid (*Myzus persicae*) and cabbage aphid (*Brevicoryne brassicae*) and groups of natural enemies including predators and parasitoids. They showed that multiple predator species groups affected greater aphid suppression than single predator species and found no evidence of IGP among predator groups. The authors concluded that predator diversity *per se* resulted in increased predator suppression in this system. In each of these systems, intraguild predators exert such strong suppression on pests that IGP does not disrupt biological control.

In contrast, higher-order predators and intraguild predators that do not cause strong pest suppression have been shown to disrupt biological control. A well-documented case is disruption of cotton aphid biological control by the assassin bug *Zelus renardii* (Rosenheim, 2001). By feeding on the green lacewing, a generalist predator that also consumes cotton aphid, the assassin bug reduced overall aphid control. Another example includes the release of squash bug (*Anasa tristis*) from predation by damsel bugs (*Nabis* sp.) due to IGP by wolf spiders (Lycosidae), which resulted in decreased squash yield (Snyder & Wise, 2001). In these studies, IGP released herbivore populations by removing the predator most important in overall suppression of the pest.

12.2.5 Indirect interactions

The presence and availability of alternative prey may affect the efficacy of herbivore suppression by natural enemies. Significant reductions in abundance of cowpea aphid (*Aphis craccivora*) by multicolored Asian lady beetle and damsel bugs yielded increased parasitism of pea aphid by the aphelinid parasitoid *Aphidius ervi*. This resulted in synergistic suppression of pea aphid by the three natural enemies combined (Cardinale *et al.*, 2003). In contrast, although carabid activity density increased significantly near beetle banks, there was not a significant effect on fly egg predation rates near banks compared with more distant locations (Prasad & Snyder, 2006a). This failure to increase predation rates was attributed to two factors. First, the large ground beetle *Pterosticus melanarius* acted as an intraguild predator on small ground beetles that are more effective at attacking fly eggs, and second, the presence of aphids acted as alternative prey and diverted small predator attention from fly eggs. However, further studies in the same system revealed that increased abundance of alternative prey also relaxed IGP. Field experiments showed that increased aphid densities restored fly egg predation rates to the same level as in the absence of the main intraguild predator, *P. melanarius* (Prasad & Snyder, 2006b).

12.2.6 Role of key natural enemies

In a study on green peach aphid suppression on potato, Straub & Snyder (2006) found that predator identity dominated the effect of multiple natural enemies on this aphid pest. Specifically, they found that *per capita* predation rates of different predator groups better predicted the level of aphid suppression than predator diversity, facilitation or interference. These results suggest that biological control efforts that target key predators of the pest involved, thus promoting the "right" diversity, rather than an increase of natural enemy diversity *per se*, are more likely to be effective (Landis *et al.*, 2000; Straub & Snyder, 2006).

12.3 Managing agricultural landscapes to increase pest suppression

Given an understanding of landscape and local processes affecting pest suppression, pest managers may wish to modify production practices to enhance natural enemy populations. Selective pesticide use is an accepted technique which may decrease natural enemy mortality (Croft, 1990; Ruberson *et al.*, 1998; Johnson & Tabashnik, 1999; Mansfield *et al.*, 2006). Altered cultural practices, such as no-till production (Witmer *et al.*, 2003) and strip cropping (Hossain *et al.*, 2002; Weiser *et al.*, 2003), may also decrease natural enemy mortality. In addition to refuge from disturbance, natural enemies frequently live longer and are more fecund when provided access to shelter, overwintering sites, alternate hosts, prey, and nectar and pollen (Ehler, 1998; Landis *et al.*, 2000). Habitat management involves establishing plants or plant communities within managed systems that provide a limiting resource to natural enemies (Wilkinson & Landis, 2005). Theoretically, adding habitat strips to a landscape will increase natural enemy populations available for biological control (Fig. 12.3). Below we discuss the resources that are frequently absent from agricultural landscapes and their effect on the presence of natural

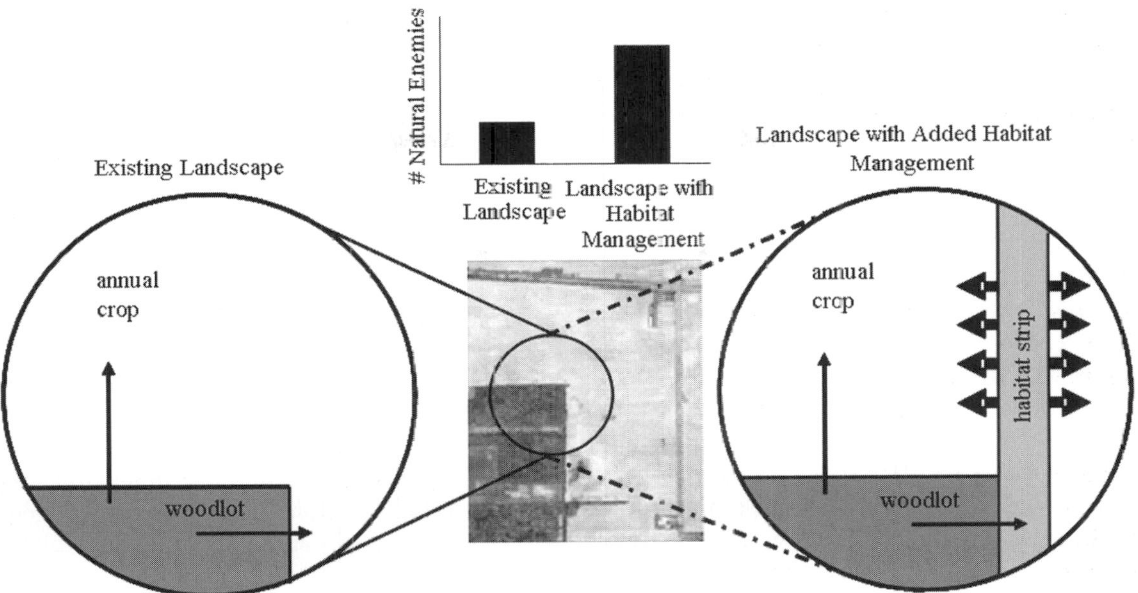

Fig. 12.3 Non-crop habitats are a source of natural enemies for biological control in annual crop fields. While natural enemies will colonize crop fields directly from non-crop habitats, providing a habitat strip may increase their abundance. Habitat strips provide natural enemies with ready access to shelter and food resources including pollen, nectar and alternative prey. Theoretically, these resources will increase natural enemy populations in the strip, resulting in greater dispersal into annual crop fields. Arrow width indicates the magnitude of natural enemy population dispersal.

enemies, as well as methods to increase habitat suitability for natural enemies from field to landscape scales.

12.3.1 Overwintering sites

Many natural enemy taxa require undisturbed sites near crop fields to increase overwintering success. Pickett *et al.* (2004) found that two aphelinid parasitoids, *Eretmocerus eremicus* and *Encarsia* spp., moved from overwintering refuges into cantaloupe and cotton crops adjacent to the refuge, where they parasitized sweet potato whitefly (*Bemisia tabaci*). They also found that refuges harbored greater numbers of sweet potato whitefly than aphelinid parasitoids early in the growing season. This result illustrates the potential of an overwintering site or refuge strip to harbor crop pests in addition to natural enemies. Lemke & Poehling (2002) examined the effect of "weed strips" within crop fields on spider abundance. They found that spiders emigrate from overwintering sites in the weed strips into the crop field in spring. Varchola & Dunn (2001) studied the effect of grassy and woody refuge strips on ground beetle diversity and abundance in both refuge strips and bordering corn fields. They found ground beetle diversity and abundance were highest in corn fields bordered by woody hedges early in the growing season, indicating that these strips provided suitable overwintering sites for ground beetles. Finally, Pywell *et al.* (2005) examined the quality of overwintering sites, measured by habitat type and habitat age. They found that habitat type had a greater impact on natural enemy abundance than habitat age, with grassy field margins containing fewer ground beetle, rove beetle and spider predators than woody hedgerows, despite age differences. These studies provide evidence that both predators and parasitoids respond to overwintering habitat on relatively small scales, including crops adjacent to overwintering habitat.

12.3.2 Alternate hosts and prey

Parasitoids and predators may also require alternate hosts or prey to complete their life cycle, or when primary hosts or prey are not available (DeBach & Rosen, 1991; Menalled *et al.*, 1999; van Emden, 2003). When alternate hosts live on specific plant species, plants that commonly harbor these species may be planted near crops to increase pest control nearby (e.g. Corbett & Rosenheim, 1996). In other cases, predators and parasitoids may find prey or host alternatives in a variety of non-crop habitats. For example, the braconid parasitoid *Diaeretiella rapae* is capable of surviving on 19 aphid host species (Pike *et al.*, 1999). Many predators are generalist feeders, and the presence of a variety of prey will have a positive effect on population growth of generalist predators (DeBach & Rosen, 1991). In agricultural landscapes, alternate hosts and prey often occur in uncultivated land adjacent to crop fields.

A greater diversity of alternate hosts and prey may be available in landscapes with more non-crop habitat. Menalled *et al.* (1999) examined parasitoid diversity and parasitism rates in simple landscapes primarily composed of cropland versus complex landscapes that contained cropland and successional non-crop habitats. They found no difference between two of three paired simple and complex landscapes, but did find an increase in both parasitoid diversity and rate of parasitism in a third landscape comparison. Similarly, Elliott *et al.* (2002) found that aphid predator diversity was affected by landscape complexity. These studies indicate that the availability of multiple alternate hosts or prey may be greater in more complex landscapes, increasing natural enemy abundance.

12.3.3 Nectar and pollen

Access to nectar and pollen has been shown to significantly increase parasitoid longevity in laboratory studies (Dyer & Landis, 1996; Baggen *et al.*, 1999). In addition to feeding on prey, predators feed on pollen (Harmon *et al.*, 2000) which can increase fecundity (Hickman & Wratten, 1996) and may be required for egg maturation (Ehler, 1998). Some species also use plant resources such as phloem fluids to supplement their diet (Eubanks & Denno, 1999).

Field studies indicate that access to nectar and pollen may be vital to natural enemy survival and reproduction. Winkler *et al.* (2006) found large differences in parasitism rates of the ichneumonid parasitoid *Diadegma semiclausum* when it was released into field cages. Parasitoids with access to nectar and hosts parasitized 96 times as many hosts as those with access to hosts only. Multiple studies have found increased natural enemy abundance in the presence of flowering plants (Colley & Luna, 2000; Frank & Shrewsbury, 2004; Lee & Heimpel, 2005; Rebek *et al.*, 2005; Forehand *et al.*, 2006; Pontin *et al.*, 2006). Fiedler & Landis (2007a) compared native Michigan plant species with several exotic species that were commonly used for habitat management in the past. They found that natural enemy abundance on flowering native plants was equal to or greater than abundance on flowering exotic species. The most attractive plant of the 51 plant species tested was the native boneset *Eupatorium maculatum*, with an average of 199 natural enemies/m^2 during full bloom. In addition, they found that floral area per square meter explained 20% of the variability in natural enemy abundance on native plants (Fiedler & Landis, 2007b).

12.3.4 Shelter

Disturbances including cultivation and pesticide use are frequent within fields in agricultural systems. Refuge from these disturbances can be promoted outside of the crop field, and may decrease natural enemy mortality rates (Gurr *et al.*, 1998). Lee *et al.* (2001) examined the effects of non-crop refuge strips on carabid populations in corn with and without pesticide application. In pesticide-treated areas, they found greater ground beetle abundance in refuge strips than in corn-planted control strips. As the growing season progressed, ground beetles were also more abundant in pesticide-treated crops near refuge strips. These results indicate that refuge strips containing perennial flowering plants, legumes and grasses benefit ground beetle populations exposed to in-field disturbances.

12.3.5 The role of habitat management in IPM at landscape scales

Research on habitat management has focused primarily on maximizing the abundance and temporal duration of local natural enemy resources (Idris & Grafius, 1995; Corbett & Rosenheim, 1996; Patt *et al.*, 1997; Freeman-Long *et al.*, 1998; Baggen *et al.*, 1999; Nicholls *et al.*, 2000). Far less attention has been paid to the ability of the landscape to supply insects that could benefit from plant-provided resources and no studies have addressed the interactive effects of arthropod community structure, resource supply and the realized ecosystem service of pest suppression. Elucidating these interactions will be an important next step in habitat management research.

12.4 Conclusions

- Agricultural landscapes are a mosaic of crop and non-crop habitats that support unique pools of insect pests and their natural enemies.
- The relative abundance and diversity of these species depends on the diversity of habitats within the landscape as well as habitat patch size, arrangement and connectivity.
- Understanding the relative ability of a landscape to provide biocontrol agents is a critical first step in implementing an IPM strategy.
- At local scales, the timing of pest and natural enemy arrival into the crop, enemy interactions such as intraguild predation and the occurrence of key natural enemies all influence the outcome of pest enemy interactions.
- By understanding the resources natural enemies need in order to be effective, managers can manipulate local cropping systems and the agricultural landscapes in which they are embedded to enhance biological control.

References

Alomar, O., Goula, M. & Albajes, R. (2002). Colonization of tomato fields by predatory mirid bugs (Hemiptera: Heteroptera) in northern Spain. *Agriculture Ecosystems and Environment*, **89**, 105–115.

Baggen, L. R., Gurr, G. M. & Meats, A. (1999). Flowers in tri-trophic systems: mechanisms allowing selective exploitation by insect natural enemies for conservation biological control. *Entomologia Experimentalis et Applicata*, **91**, 155–161.

Barbosa, P. (1998). *Conservation Biological Control*. San Diego, CA: Academic Press.

Bommarco, R. (1998). Reproduction and energy reserves of a predatory carabid beetle relative to agroecosystem complexity. *Ecological Applications*, **8**, 846–853.

Brodeur, J. & Rosenheim, J. A. (2000). Intraguild interactions in aphid parasitoids. *Entomologia Experimentalis et Applicata*, **97**, 93–108.

Bugg, R. L., Ehler, L. E. & Wilson, L. T. (1987). Effect of common knotweed (*Polygonum aviculare*) on abundance and efficiency of insect predators of crop pests. *Hilgardia*, **55**, 1–51.

Cardinale, B. J., Harvey, C. T., Gross, K. & Ives, A. R. (2003). Biodiversity and biocontrol: emergent impacts of a multi-enemy assemblage on pest suppression and crop yield in an agroecosystem. *Ecology Letters*, **6**, 857–865.

Chang, G. C. & Kareiva, P. (1999). The case of indigenous generalists in biological control. In *Theoretical Approaches to Biological Control*, ed. H. V. Cornell, pp. 103–115. Cambridge, UK: Cambridge University Press.

Colfer, R. G. & Rosenheim, J. A. (2001). Predation on immature parasitoids and its impact on aphid suppression. *Oecologia*, **126**, 292–304.

Colley, M. R. & Luna, J. M. (2000). Relative attractiveness of potential beneficial insectary plants to aphidophagous hoverflies (Diptera: Syrphidae). *Environmental Entomology*, **29**, 1054–1059.

Colunga-Garcia, M., Gage, S. H. & Landis, D. A. (1997). Response of an assemblage of Coccinellidae (Coleoptera) to a diverse agricultural landscape. *Environmental Entomology*, **26**, 797–804.

Corbett, A. & Rosenheim, J. A. (1996). Impact of a natural enemy overwintering refuge and its interaction with the surrounding landscape. *Ecological Entomology*, **21**, 155–164.

Costamagna, A. C. & Landis, D. A. (2006). Predators exert top–down control of soybean aphid across a gradient of agricultural management systems. *Ecological Applications*, **16**, 1619–1628.

Costamagna, A. C., Landis, D. A. & DiFonzo, C. D. (2007). Suppression of *A. glycines* by generalist predators results in a trophic cascade in soybean. *Ecological Applications*, **17**, 441–451.

Croft, B. A. (1990). *Arthropod Biological Control Agents and Pesticides*. New York: John Wiley.

DeBach, P. & Rosen, D. (1991). *Biological Control by Natural Enemies*. Cambridge, UK: Cambridge University Press.

den Belder, E., Elderson, J., van den Brink, W. J. & Schelling, G. (2002). Effect of woodlots on thrips density in leek fields: a landscape analysis. *Agriculture Ecosystems and Environment*, **91**, 139–145.

Desneux, N., O'Neil, R. J. & Yoo, H. J. S. (2006). Suppression of population growth of the soybean aphid, *Aphis glycines* Matsumura, by predators: the identification of a key predator and the effects of prey dispersion, predator abundance, and temperature. *Environmental Entomology*, **35**, 1342–1349.

Desneux, N., Decourtye, A. & Delpuech, J. M. (2007). The sublethal effects of pesticides on beneficial arthropods. *Annual Review of Entomology*, **52**, 81–106.

Dyer, L. E. & Landis, D. A. (1996). Effects of habitat, temperature, and sugar availability on longevity of *Eriborus terebrans* (Hymenoptera: Ichneumonidae). *Environmental Entomology*, **25**, 1192–1201.

Ehler, L. E. (1998). Conservation biological control: past, present, and future. In *Conservation Biological Control*, ed. P. Barbosa, pp. 1–8. San Diego, CA: Academic Press.

Elliott, N. C., Kieckhefer, R. W., Lee, J. H. & French, B. W. (1999). Influence of within-field and landscape factors on aphid predator populations in wheat. *Landscape Ecology*, **14**, 239–252.

Elliott, N. C., Kieckhefer, R. W., Michels, G. J. & Giles, K. L. (2002). Predator abundance in alfalfa fields in relation to aphids, within-field vegetation, and landscape matrix. *Environmental Entomology*, **31**, 253–260.

Eubanks, M. D. & Denno, R. F. (1999). The ecological consequences of variation in plants and prey for an omnivorous insect. *Ecology*, **80**, 1253–1266.

Fiedler, A. K. & Landis, D. A. (2007a). Attractiveness of Michigan native plants to arthropod natural enemies and herbivores. *Environmental Entomology*, **36**, 751–776.

Fiedler, A. K. & Landis, D. A. (2007b). Plant characteristics associated with natural enemy attractiveness to Michigan native plants. *Environmental Entomology*, **36**, 871–877.

Forehand, L. M., Orr, D. B. & Linker, H. M. (2006). Evaluation of a commercially available beneficial insect habitat for management of Lepidoptera pests. *Journal of Economic Entomology*, **99**, 641–647.

Fortin, M. & Mauffette, Y. (2001). Forest edge effects on the biological performance of the forest tent caterpillar (Lepidoptera: Lasiocampidae) in sugar maple stands. *Ecoscience*, **8**, 164–172.

Fox, T. B., Landis, D. A., Cardoso, F. F. & DiFonzo, C. D. (2004). Predators suppress *Aphis glycines* Matsumura population growth in soybean. *Environmental Entomology*, **33**, 608–618.

Frank, S. D. & Shrewsbury, P. M. (2004). Effect of conservation strips on the abundance and distribution of natural enemies and predation of *Agrotis ipsilon* (Lepidoptera: Noctuidae) on golf course fairways. *Environmental Entomology*, **33**, 1662–1672.

Freeman-Long, R. F., Corbett, A., Lamb, C. *et al.* (1998). Beneficial insects move from flowering plants to nearby crops. *California Agriculture*, **52**, 23–26.

Gardiner, M. M. & Landis, D. A. (2007). Impact of intraguild predation by adult *Harmonia axyridis* (Coleoptera: Coccinellidae) on *Aphis glycines* (Hemiptera: Aphididae) biological control in cage studies. *Biological Control*, **40**, 386–395.

Gurr, G. M., van Emden, H. F. & Wratten, S. D. (1998). Habitat manipulation and natural enemy efficiency: implications for the control of pests. In *Conservation Biological Control*, ed. P. Barbosa, pp. 155–183. San Diego, CA: Academic Press.

Harmon, J. P., Ives, A. R., Losey, J. E., Olson, A. C. & Rauwald, K. S. (2000). *Coleomegilla maculata* (Coleoptera: Coccinellidae) predation on pea aphids promoted by proximity to dandelions. *Oecologia*, **125**, 543–548.

Hickman, J. M. & Wratten, S. D. (1996). Use of *Phacelia tanacetifolia* strips to enhance biological control of aphids by hoverfly larvae in cereal fields. *Journal of Economic Entomology*, **89**, 832–840.

Hossain, Z., Gurr, G. M., Wratten, S. D. & Raman, A. (2002). Habitat manipulation in lucerne *Medicago sativa*: arthropod population dynamics in harvested and "refuge" crop strips. *Journal of Applied Ecology*, **39**, 445–454.

Hunter, M. D. (2002). Landscape structure, habitat fragmentation, and the ecology of insects. *Agricultural and Forest Entomology*, **4**, 159–166.

Idris, A. B. & Grafius, E. (1995). Wildflowers as nectar sources for *Diadegma insulare* (Hymenoptera: Ichneumonidae), a parasitoid of diamondback moth (Lepidoptera: Yponomeutidae). *Environmental Entomology*, **24**, 1726–1735.

Johnson, M. W. & Tabashnik, B. E. (1999). Enhanced biological control through pesticide selectivity. In *Handbook of Biological Control*, eds. T. S. Bellows Jr. & T. W. Fisher, pp. 297–317. San Diego, CA: Academic Press.

Landis, D. A. & van der Werf, W. (1997). Early-season predation impacts the establishment of aphids and spread of beet yellows virus in sugar beet. *Entomophaga*, **42**, 499–516.

Landis, D. A., Wratten, S. D. & Gurr, G. M. (2000). Habitat management to conserve natural enemies of arthropod pests in agriculture. *Annual Review of Entomology*, **45**, 175–201.

Lang, A. (2003). Intraguild interference and biocontrol effects of generalist predators in a winter wheat field. *Oecologia*, **134**, 144–153.

Lee, J. C. & Heimpel, G. E. (2005). Impact of flowering buckwheat on Lepidopteran cabbage pests and their parasitoids at two spatial scales. *Biological Control*, **34**, 290–301.

Lee, J. C., Menalled, F. B. & Landis, D. A. (2001). Refuge habitats modify impact of insecticide disturbance on carabid beetle communities. *Journal of Applied Ecology*, **38**, 472–483.

Lemke, A. & Poehling, H. M. (2002). Sown weed strips in cereal fields: overwintering site and "source" habitat for *Oedothorax apicatus* (Blackwall) and *Erigone atra* (Blackwall) (Araneae: Erigonidae). *Agriculture Ecosystems and Environment*, **90**, 67–80.

Losey, J. E. & Denno, R. F. (1998). Positive predator–predator interactions: enhanced predation rates and synergistic suppression of aphid populations. *Ecology*, **79**, 2143–2152.

Mansfield, S., Dillon, M. L. & Whitehouse, M. E. A. (2006). Are arthropod communities in cotton really disrupted? An assessment of insecticide regimes and evaluation of the beneficial disruption index. *Agriculture Ecosystems and Environment*, **113**, 326–335.

Marino, P. C. & Landis, D. A. (1996). Effect of landscape structure on parasitoid diversity and parasitism in agroecosystems. *Ecological Applications*, **6**, 276–284.

Menalled, F. D., Marino, P. C., Gage, S. H. & Landis, D. A. (1999). Does agricultural landscape structure affect parasitism and parasitoid diversity? *Ecological Applications*, **9**, 634–641.

Menalled, F. D., Costamagna, A. C., Marino, P. C. & Landis, D. A. (2003). Temporal variation in the response of parasitoids to agricultural landscape structure. *Agriculture Ecosystems and Environment*, **96**, 29–35.

Müller, C. B. & Brodeur, J. (2002). Intraguild predation in biological control and conservation biology. *Biological Control*, **25**, 216–223.

Nicholls, C. I., Parrella, M. & Altieri, M. A. (2000). Reducing the abundance of leafhoppers and thrips in a northern California organic vineyard through maintenance of full season floral diversity with summer cover crops. *Agricultural and Forest Entomology*, **2**, 107–113.

Östman, O. (2002). Distribution of bird cherry-oat aphids (*Rhopalosiphum padi* (L.)) in relation to landscape and farming practices. *Agriculture Ecosystems and Environment*, **93**, 67–71.

Östman, O., Ekbom, B. & Bengtsson, J. (2001). Landscape heterogeneity and farming practice influence biological control. *Basic and Applied Ecology*, **2**, 365–371.

Patt, J. M., Hamilton, G. C. & Lashcmb, J. H. (1997). Impact of strip-insectary intercropping with flowers on conservation biological control of the Colorado potato beetle. *Advances in Horticultural Science*, **11**, 175–181.

Pickett, C. H., Roltsch, W. & Corbett, A. (2004). The role of a rubidium marked natural enemy refuge in the establishment and movement of *Bemisia* parasitoids. *International Journal of Pest Management*, **50**, 183–191.

Pike, K. S., Stary, P., Miller, T. *et al.* (1999). Host range and habitats of the aphid parasitoid *Diaeretiella rapae* (Hymenoptera: Aphidiidae) in Washington state. *Environmental Entomology*, **28**, 61–71.

Polis, G. A., Myers, C. A. & Holt, R. D. (1989). The ecology and evolution of intraguild predation: potential competitors that eat each other. *Annual Review of Ecology and Systematics*, **20**, 297–330.

Pontin D. R., Wade, M. R., Kehrli, P. & Wratten, S. D. (2006). Attractiveness of single and multiple species flower patches to beneficial insects in agroecosystems. *Annals of Applied Biology*, **148**, 39–47.

Prasad, R. P. & Snyder, W. E. (2006a). Polyphagy complicates conservation biological control that targets generalist predators. *Journal of Applied Ecology*, **43**, 343–352.

Prasad, R. P. & Snyder, W. E. (2006b). Diverse trait-mediated interactions in a multi-predator, multi-prey community. *Ecology*, **87**, 1131–1137.

Pywell, R. F., James, K. L., Herbert, I. *et al.* (2005). Determinants of overwintering habitat quality for beetles and spiders on arable farmland. *Biological Conservation*, **123**, 79–90.

Rebek, E. J., Sadof, C. S. & Hanks, L. M. (2005). Manipulating the abundance of natural enemies in ornamental landscapes with floral resource plants. *Biological Control*, **33**, 203–216.

Roland, J. & Embree, D. G. (1995). Biological control of the winter moth. *Annual Review of Entomology*, **40**, 475–492.

Rosenheim, J. A. (2001). Source–sink dynamics for a generalist insect predator in habitats with strong higher-order predation. *Ecological Monographs*, **71**, 93–116.

Rosenheim, J. A. & Harmon, J. P. (2006). The influence of intraguild predation on the suppression of a shared prey population: an empirical reassessment. In *Trophic and Guild Interactions in Biological Control*, ed. G. Boivin, pp. 1–20. Dordrecht, Netherlands: Springer-Verlag.

Rosenheim, J. A., Kaya, H. K., Ehler, L. E., Marois, J. J. & Jaffee, B. A. (1995). Intraguild predation among biological control agents: theory and evidence. *Biological Control*, **5**, 303–335.

Ruberson, J. R., Nemoto, H. & Hirose, Y. (1998). Pesticides and conservation of natural enemies in pest management. In *Conservation Biological Control*, ed. P. Barbosa, pp. 207–220. San Diego, CA: Academic Press.

Schmidt, M. H. & Tscharntke, T. (2005). Landscape context of sheetweb spider (Araneae: Linyphiidae) abundance in cereal fields. *Journal of Biogeography*, **32**, 467–473.

Schmidt, M. H., Roschewitz, I., Thies, C. & Tscharntke, T. (2005). Differential effects of landscape and management on diversity and density of ground-dwelling farmland spiders. *Journal of Applied Ecology*, **42**, 281–287.

Sih, A., Englund, G. & Wooster, D. (1998). Emergent impacts of multiple predators on prey. *Trends in Ecology and Evolution*, **13**, 350–355.

Snyder, W. E. & Wise, D. H. (1999). Predator interference and the establishment of generalist predator populations for biocontrol. *Biological Control*, **15**, 283–292.

Snyder, W. E. & Wise, D. H. (2001). Contrasting trophic cascades generated by a community of generalist predators. *Ecology*, **82**, 1571–1583.

Snyder, W. E., Snyder, G. B., Finke, D. L. & Straub, C. S. (2006). Predator biodiversity strengthens herbivore suppression. *Ecology Letters*, **9**, 789–796.

Stark, J. D. & Banks, J. E. (2003). Population-level effects of pesticides and other toxicants on arthropods. *Annual Review of Entomology*, **48**, 505–519.

Straub, C. S. & Snyder, W. E. (2006). Species identity dominates the relationship between predator biodiversity and herbivore suppression. *Ecology*, **87**, 277–282.

Symondson, W. O. C., Sunderland, K. D. & Greenstone, M. H. (2002). Can generalist predators be effective biocontrol agents? *Annual Review of Entomology*, **47**, 561–594.

Thies, C., Steffan-Dewenter, I. & Tscharntke, T. (2003). Effects of landscape context on herbivory and parasitism at different spatial scales. *Oikos*, **101**, 18–25.

Thies, C. & Tscharntke, T. (1999). Landscape structure and biological control in agroecosystems. *Science*, **285**, 893–895.

Tscharntke, T., Klein, A. M., Kruess, A., Steffan-Dewenter, I. & Thies, C. (2005). Landscape perspectives on agricultural intensification and biodiversity – ecosystem service management. *Ecology Letters*, **8**, 857–874.

van Emden, H. F. (2003). Conservation biological control: from theory to practice. In *Proceedings of the International Symposium on Biological Control of Arthropods*, Honolulu, HI, January 14–18, 2002, ed. R. Van Driesch, pp. 199–208. Morgantown, WV: US Department of Agriculture Forest Service.

Varchola, J. M. & Dunn, J. P. (2001). Influence of hedgerow and grassy field borders on ground beetle (Coleoptera: Carabidae) activity in fields of corn. *Agriculture Ecosystems and Environment*, **83**, 153–163.

Weibull, A. C., Östman, O. & Granqvist, A. (2003). Species richness in agroecosystems: the effect of landscape, habitat and farm management. *Biodiversity and Conservation*, **12**, 1335–1355.

Weiser, L. A., Obrycki, J. J. & Giles, K. L. (2003). Within-field manipulation of potato leafhopper (Homoptera: Cicadellidae) and insect predator populations using an uncut alfalfa strip. *Journal of Economic Entomology*, **96**, 1184–1192.

Wilkinson, T. K. & Landis, D. A. (2005). Habitat diversification in biological control: the role of plant resources. In *Plant-Provided Food for Carnivorous Insects*, eds. F. L. Wackers, P. C. J. van Rijn & J. Bruin, pp. 305–325. Cambridge, UK: Cambridge University Press.

Winkler, K., Wackers, F., Bukovinszkine-Kiss, G. & van Lenteren, J. (2006). Sugar resources are vital for *Diadegma semiclausum* fecundity under field conditions. *Basic and Applied Ecology*, **7**, 133–140.

With, K. A. & King, A. W. (1999). Extinction thresholds for species in fractal landscapes. *Conservation Biology*, **13**, 314–326.

Witmer, J. E., Hough-Goldstein, J. A. & Pesek, J. D. (2003). Ground-dwelling and foliar arthropods in four cropping systems. *Environmental Entomology*, **32**, 366–376.

Wratten, S. D., Bowie, M. H., Hickman, J. M. *et al.* (2003). Field boundaries as barriers to movement of hoverflies (Diptera: Syrphidae) in cultivated land. *Oecologia*, **134**, 605–611.

Chapter 13

Barriers to adoption of biological control agents and biological pesticides

Pamela G. Marrone

According to the Biopesticide, Pollution, Prevention Division (BPPD) at the US Environmental Protection Agency (EPA), which registers biological pesticides for sale, biological pesticides or biopesticides are certain types of pesticides derived from such natural materials as animals, plants, bacteria and certain minerals. Biopesticides fall into three major classes: microbial pesticides, plant-incorporated-protectants (PIPs) or biochemical pesticides.

- Microbial pesticides consist of a microorganism (e.g. a bacterium, fungus, virus or protozoan) as the active ingredient. The most widely used microbial pesticides are subspecies and strains of *Bacillus thuringiensis* (*Bt*).
- Plant-incorporated-protectants (PIPs) are pesticidal substances that plants produce from genetic material that has been added to the plant. For example, scientists can take the gene for the *Bt* pesticidal protein, and introduce the gene into the plant's own genetic material. Then the plant, instead of the *Bt* bacterium, manufactures the substance that kills the pest.
- Biochemical pesticides are naturally occurring substances that control pests by non-toxic mechanisms. Conventional pesticides, by contrast, are generally synthetic materials that directly kill or inactivate the pest. Biochemical pesticides include substances such as insect sex pheromones that interfere with mating, as well as various scented plant extracts that attract insect pests to traps (Environmental Protection Agency, 2007).

This chapter will focus only on microbial and biochemical pesticides because microbial and biochemical pesticides continue to have challenges with adoption on a large scale, whereas PIPs are widely adopted on millions of hectares in US agriculture. Biological control agents (BCAs) are living organisms that infect and kill the pest or pathogen or are predators of pests and weeds. For purposes of this chapter, BCAs are macroorganisms like predators and parasites. Microorganisms that some may consider as BCAs (e.g. *Bacillus subtilis* for controlling plant diseases) are considered (microbial) biopesticides.

13.1 Overview of the biopesticide market compared to synthetic chemical market

The chemical pesticide market is flat to declining and was estimated at between $US 2700 million (Venkataraman *et al.*, 2006) and $US 33 600 million (Agrow, 2006a) in 2005 (Tables 13.1 and 13.2). In 2005, increased sales of insecticides and

Integrated Pest Management, ed. Edward B. Radcliffe, William D. Hutchison and Rafael E. Cancelado. Published by Cambridge University Press.

Table 13.1 Global pesticide market by region, 2005

Region	Sales ($US millions)
North America	8 600
Europe	8 740
Asia	8 160
Latin America	5 650
Rest of world	2 450
Total	33 660

Source: Allan Woodburn Associates/Cropnosis.

Table 13.2 Global pesticide market by product, 2005

Pesticide	Sales ($US millions)
Herbicides	15 400
Insecticide	8 800
Fungicides	7 700
Others	1 600
Total	33 500

Source: Allan Woodburn Associates/Cropnosis.

fungicides in the USA offset the decline in herbicide sales, resulting in a 2% increase in agrochemical sales to $US 6515 million, according to Phillips McDougall and CropLife America data (Agrow, 2006b) (Table 13.3). This analysis includes biotechnology product sales, which rose by 6.1% to $US 3909 million in 2005, bringing the total market for crop protection chemicals and traits to $US 10 424 million.

In contrast, biopesticides today represent about 2.4% of the overall pesticide market ($US 512 million), are growing quickly and are expected to grow to about 4.2% of the global agrochemical market by 2010 (more than $US 1000 million). Key developments expected in the coming years are more research and development in biopesticides, the continuing increase in genetically modified crops, the application of IPM concepts and an increase in organic food sales (Fig. 13.1). Orchard crops hold the largest share of biopesticide use at 55% (Venkataraman *et al.*, 2006).

The demand for biopesticides is rising steadily in all parts of the world. This is because of increased public awareness of the environment,

Table 13.3 USA agrochemical sales, 2005 ($US millions)

Crop	Herbicides	Insecticides	Fungicides	Other	Total
Maize	1239	338	6	2	1585
Fruit/vegetables	369	486	393	50	1298
Potatoes[a]	34	84	49	10	177
Grapevines[a]	40	35	53	6	134
Pome fruit[a]	30	57	36	0	123
Others[a]	265	310	255	34	864
Soybeans	811	3	60	0	874
Cotton	269	178	6	267	720
Cereals	390	42	40	5	477
Rice	115	8	25	3	151
Sugar beet	70	13	11	1	95
Sunflower	40	8	0	0	48
Sugarcane	32	6	0	2	40
Oilseed rape	12	2	0	0	14
Others	778	225	175	35	1213
Total	4125	1309	716	415	6515

[a] Use included in totals for fruit/vegetables.
Source: Phillips McDougall/Crop Life America.

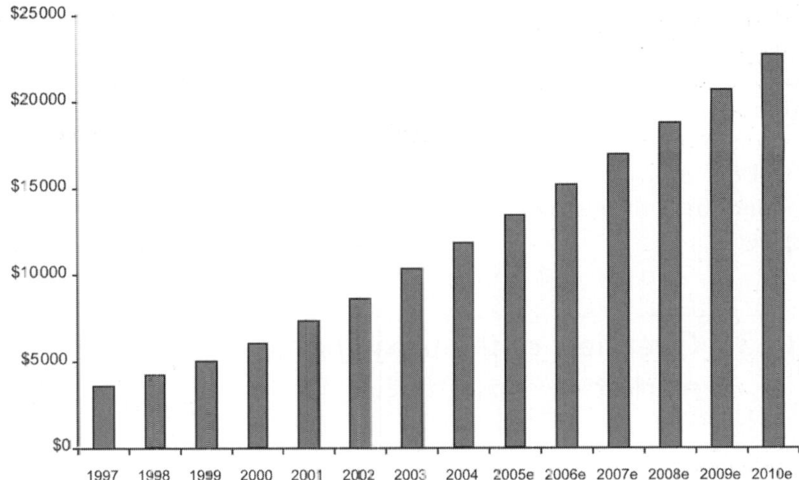

Fig. 13.1 Organic food sales in the USA, estimate. (Data from *Nutrition Business Journal*, 2006.)

and the pollution potential and health hazards related to many conventional pesticides (worker safety, bird toxicity, air pollution, and surface and groundwater contamination). The issues are most acute at the urban/housing and agricultural interface, where the rapid growth of housing into rural areas creates clashes of farmers, environmental groups and residents. Northern European countries (Denmark, Sweden, the Netherlands) have legislated a 50% reduction in on-farm chemical pesticide use. Many countries pay farmers large subsidies to farm organically – sustainable and organic farming are codified into the EU Common Agricultural Policy. Many Asian countries have banned classes of toxic chemicals.

13.2 Microbial biopesticide market and growth

The world market for microbial biopesticides was worth $US 268 million at end-user level in 2005 (Quinlan & Gill, 2006) (Table 13.4).

This represents a 15% increase on an estimate made by Michel Guillon, President of the International Biocontrol Manufacturers' Association, in 2003. Despite the apparent growth, these figures indicate that microbial biopesticide sales remain at less than 1% of the total global pesticide market. Products based on *Bt* dominate the market, taking a 60% share. This proportion is significantly down on estimates made in the 1990s, which

Table 13.4 Estimated sales of microbial and nematode-based biopesticides ($US millions)

	North America	Asia and Europe	Australasia	Africa and Latin America	Middle East	Total
Total *Bt*[a]	72.75	26.77	43.12	14.45	2.48	159.57
Other bacteria	15.20	2.60	11.58	0.65	0.10	30.13
Viruses	1.03	5.45	5.33	2.89	0.55	15.25
Fungi	11.76	2.17	11.52	22.16	0.49	48.10
Protozoa	0.05	0.00	0.02	0.00	0.00	0.07
Nematodes	8.25	6.00	0.37	0.03	0.00	14.65
Total	109.04	42.99	71.94	40.18	3.62	267.77

[a] Products based on *Bacillus thuringiensis* serotypes.

routinely, and consistently, estimated the market share taken by *Bt* products as 80–90% (Quinlan & Gill, 2006). This share is reduced by competition from new, safer pesticides derived from fermentation products, such as insecticides marketed by Dow Agrosciences based on spinosad compounds from the microorganism *Saccharopolyspora spinosa*.

13.3 Overview of the urban/home market

Globally, the number of chemical pesticides for the home and garden market has decreased due to regulatory action. In Europe and Canada, most home and garden pesticides were removed from the market. In many cities in the USA and Canada (e.g. New York, San Francisco, Toronto), chemical pesticides are used only as a last resort or not at all. Postings of pesticide applications are required. The increase in organic gardening and the use of natural garden products parallels the organic food market in growth. According to the National Gardening Association (2002), of the $US 2000 million spent by consumers on insecticides, 10% are natural and less toxic materials (2004) and are growing at 20% per year. Fredonia Market Research Group says that in 2006 organic lawn and garden products will be $US 400 million (includes fertilizers) and that they are growing at 9.4% per year (Fredonia Group, 2006). According to the National Gardening Association (2004), 54 million of 90 million households applied pesticides and fertilizers on their lawns, 28 million use both natural and chemicals and 6 million use naturals only.

The 2004 survey from the National Gardening Association and *Organic Gardening Magazine* found that, while only 5% of households in the USA now use all-organic methods in their yards, some 21% said they would definitely or probably do so in the future. Nearly two dozen states, including New York and Wisconsin, now require public notification when pesticides are being applied by professionals, according to Beyond Pesticides (2008), a Washington, DC advocacy group. At least 13 towns in the USA, including Lawrence (Kansas) and Chatham (New Jersey) have pesticide-free parks, and 33 states and several hundred school districts have laws or policies designed to minimize children's exposure to pesticides. In 2005, New York City passed legislation requiring the city to phase out acutely toxic pesticides on city-owned or leased property and make commercial landscapers give neighbors notice before spraying certain pesticides. Marblehead (Massachusetts) recently converted 6000 ha of athletics fields to organic care. Even Walt Disney World has reduced its use of traditional pesticides by 70% since the 1990s and is using all-natural composts in some areas of the park (Bounds & Brat, 2006).

13.4 Advantages of biopesticides

Because steady advances were made in the 1990s and 2000s in microbial and biochemical research and in formulation technology, today's biopesticides are much improved over biopesticides from earlier eras. The advantages of biopesticides are driving the increased use in farming, landscaping and home gardening:

- When used in IPM systems, the efficacy of biopesticides can be equal or better than conventional products, especially for "minor" crops like fruits, vegetables, nuts and flowers.
- Biopesticides provide greater margins of safety for applicators, farm workers and rural neighbors and have much shorter field restricted-entry intervals, which makes it easier for farmers to complete essential agronomic practices on a timely basis and schedule harvest operations.
- Biopesticides generally affect only the target pest and closely related species. They pose little or no risk to many if not all non-target organisms including birds, fish, beneficial insects and mammals.
- Many of today's biopesticides are biodegradable, resulting in essentially no risk to surface and groundwater. Biopesticides also generally have low volatile organic chemicals (VOC) content and can be used to reduce the air pollution caused by high-VOC chemicals (e.g. by fumigants in the San Joaquin Valley in California).

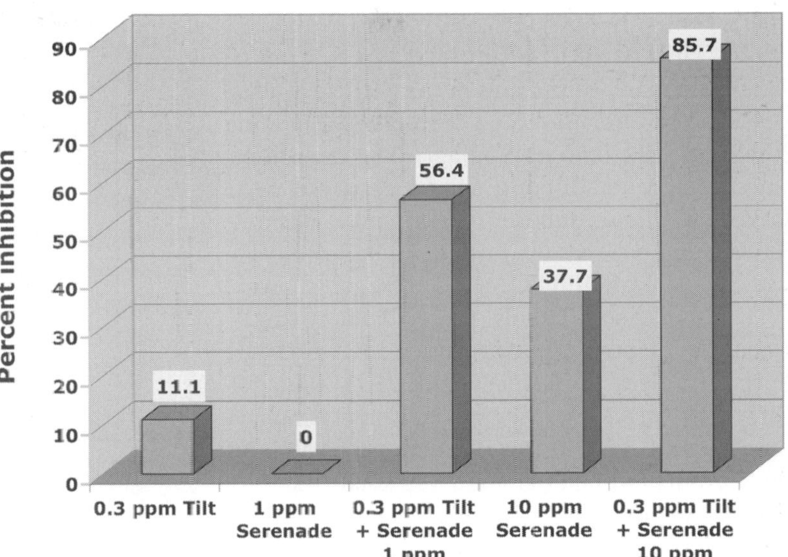

Fig. 13.2 Synergistic action of a biopesticide (*Bacillus subtilis* QST713, Serenade®) and a chemical pesticide (Tilt) in controlling *Mycosphaerella fijiensis* ascospores collected from farms resistant to propiconazole. (Data from Monreri Project, Teresa Arroyo, Costa Rica, courtesy of AgraQuest, Inc.)

- Biopesticides typically have a lower chance for the development of resistance to pests than single-site chemicals because of their complex mode of action. Biopesticides are excellent resistance management tools when used alone or in combinations with chemicals as tank mixes and rotations. Figure 13.2 shows an example of synergistic action of a biopesticide when combined with a chemical pesticide. Strains of black sigatoka (*Mycosphaerella fijiensis*) resistant to propiconazole were effectively controlled when combined with Serenade®.
- Biopesticides typically have no pre-harvest interval and very short restricted-entry intervals, allowing the grower to harvest the crop immediately after spraying. This is particularly important for export crops now that produce is shipped globally and is subject to international maximum residue levels.
- Biopesticides are produced by environmentally friendly and sustainable production processes. Microbial biopesticides are produced by fermentation using readily available biomass (agricultural raw materials) such as soy flour and corn starch. Waste from fermentation processes is often applied back to farms as fertilizer.
- Biopesticides can be, and many are, approved for use in organic farming, the fastest growing segment of the food industry.

Of all types of growers, those with the largest operations tend to be the most avid users of biopesticides. Millions of hectares of cropland receive at least one application of biopesticides each year. Growers who incorporate biopesticides into their programs are typically among the more progressive and entrepreneurial growers in their markets. Growers who use biopesticides do so because they have needs that include but extend beyond efficacy, needs relating to resistance management, residue management, harvest flexibility, maintaining beneficial populations, and worker and environmental safety. Growers who use biopesticides would only do so if they saw a tangible return on investment. However, as the data presented below show, there are still significant issues with the perceived value and efficacy of biopesticides.

13.5 Barriers to adoption

Below are the main barriers to adoption based on the author's personal experience after starting three biopesticide companies and commercializing more than eight biopesticide products and survey data summarized in a subsequent section.

13.5.1 Highly competitive, capital intensive marketplace

There are many companies in the pesticide industry chasing after the same customers, including the multi-$US 1000 million agrichemical companies, Monsanto, Dupont, Syngenta, Bayer, BASF and Dow. The second tier of companies with revenues up to $US1000 million includes Arysta, Advan, Makhteshim, FMC, Cerexagri, Gowan, Valent and others. Finally, there are many small–medium biopesticide companies including the largest, Valent Biosciences and Certis USA, followed by AgraQuest, Suterra, Bioworks, Prophyta and several others. The plethora of companies and products makes it hard for small biopesticide companies to stand out among the others and it requires millions of dollars to properly conduct field trials, to organize farm demonstrations and to develop marketing programs. Most small biopesticide companies are undercapitalized and often skimp on field development and customer education.

13.5.2 Risk-averse customer: no reason to change

Growers and their gatekeepers and key influencers (distributors and pest control advisors: PCAs) have become accustomed to affordable chemicals that generally perform up to expectations. When evaluating a pesticide the questions a grower or PCA asks:

- Does it work (as well or better than existing chemical pesticides)?
- What does it cost?
- What other benefits are there?
- Is it safer?

Growers will try a new biopesticide product and compare it to their existing pest management programs in demonstration trials on their farms. Conducting these demonstrations is the best, if not the only, way to gain adoption. The grower will typically try some on a small number of hectares the first season and then on some more the second and adopt it fully during the third season (Fig. 13.3). Some early adopters will try a new product on more hectares the first season, but mainstream growers typically follow a three-season adoption curve for biopesticides.

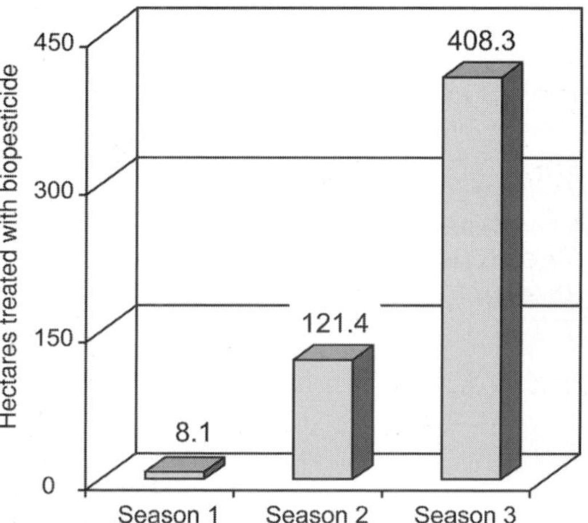

Fig. 13.3 Typical farmer adoption curve (total ha in IPM in example program).

Distributors and PCAs typically will be the gatekeepers for the grower, deciding which products the grower will try or not. To the PCA, it is risky to change to a biopesticide because a failure means he could lose his job and the grower could lose his crop. Thus the adoption time is gated by the willingness of the PCA to allow access to the grower and the number of years and demonstrations that will convince the PCA there is a benefit and low risk to switch. University Extension researchers will also test pesticide products and provide their recommendations. Therefore, adoption can be faster the more buy-in these researchers have. This buy-in is possible by providing research dollars to test the product and by developing relationships with these researchers through scientific forums.

13.5.3 Complex selling channel

In order for a biopesticide company to make a sale, it must go through traditional selling channels. Figure 13.4 shows the sales channel pathway and participants. The sales chain in agrichemicals starts at the manufacturer who typically has a sales force and technical support that:

- Motivates the distributor to sell the product to the farmer (push) via margin, other incentives

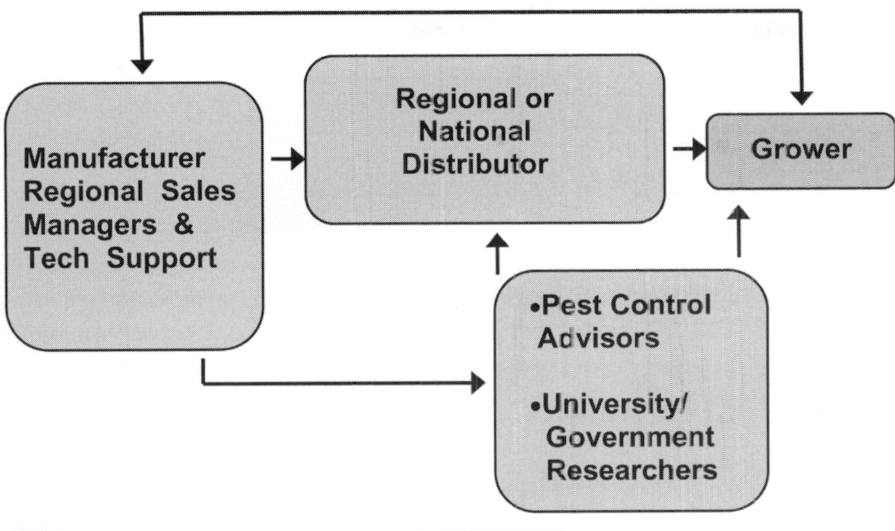

Fig. 13.4 Typical sales channel for a pesticide in the USA.

and having a product that provides value to the grower compared to competing products.

- Motivates the pest control advisor (PCA) to test, demonstrate and sell the product to the grower. Typically, the PCA is employed by the distributor (and gets a commission of pesticide sales), although there are some independent PCAs who are hired by large growers.
- Develops relationships with university researchers who test products and provide technical and product recommendations to the grower.
- Accesses the grower directly (pull through to distributor) and conducts on-farm demonstrations showing the value of the product in the grower's program compared to his or her standard program.

Most biopesticide companies choose to set up their own sales force, which is expensive and challenging. This business is very relationship-driven and the best salespeople with strong relationships with all participants in the sales channel are already working at other pesticide companies. Most biopesticide companies do not have the capital to spend the millions of dollars it takes to develop a professional sales force with the best individuals hired away from other companies. Hence, many biopesticides languish in the marketplace, not getting the sales and marketing support needed to drive adoption. Biopesticides are not always perceived in a positive light. Many surveys have been conducted on this topic and are summarized and discussed in the next section.

13.6 Perception of biopesticides

Obviously the small size of the biopesticide market compared to the chemical pesticide market indicates that despite the benefits biopesticides bring, there are reasons why their usage is not more widespread. The Biopesticide Industry Alliance (BPIA), a non-profit trade group of small biopesticide companies formed in 2000, conducted a survey in 2003 to determine the attitudes of customers toward biopesticides (Donaldson, 2003). The survey was designed to investigate the degree of awareness of biological pesticides, profile respondent's personal experience with biologicals, ascertain current perception of biopesticide technology and probe for specific reasons why biopesticides are used or not used by target segments. The survey was conducted by telephone by a market research firm and 198 persons were interviewed.

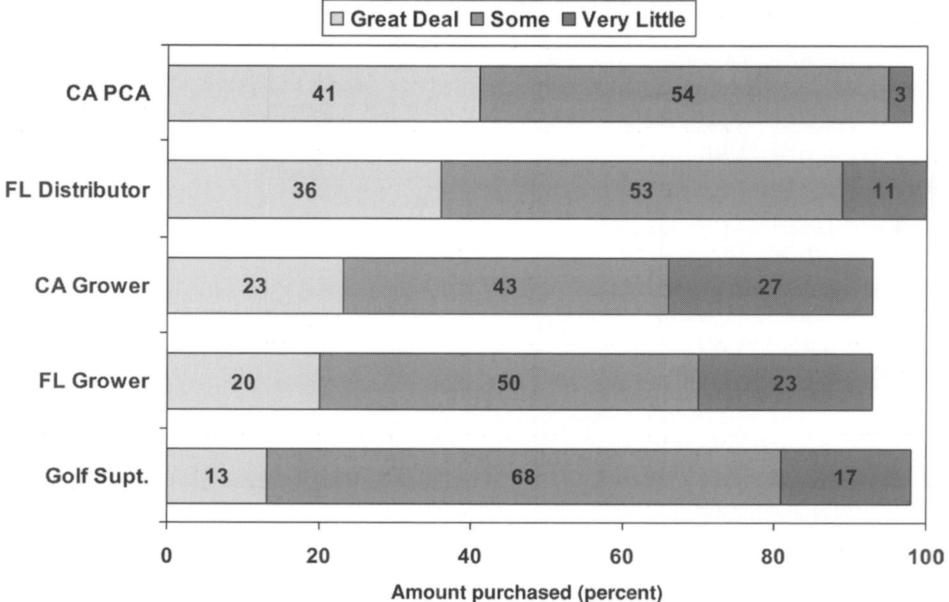

Fig. 13.5 Usage of biopesticides. Key: CA PCA, 70 pest control advisors in California; FL Distributor, 29 distributors in Florida; CA Grower, 30 growers in California; FL growers in Florida; Golf Supt., 40 turf and ornamental superintendents.

For the agriculture audience, the survey included 70 California pest control advisors (from a master list provided by the state of California) and 30 growers in California and 28 distributors and 30 growers in Florida. The market focus was specialty (high value) agriculture (trees, nuts, vines, vegetables). Meister Publishing provided lists of Florida distributors, Florida growers and California growers. Also participating were 40 turf and ornamental superintendents randomly selected from the Golf Course Superintendents Directory from the ten largest turf and ornamental states. The sampling procedure targeted key cross-sectional groups, but was not large enough for statistical analyses. Awareness of biopesticides was highest at the PCA/distributor level and lowest at the grower level. Awareness in the golf course industry is lower than the agricultural industry. To increase usage of biopesticides, increased awareness initiatives are needed at end-user (customer) level. Figure 13.5 shows the respondents' usage of biopesticides.

The results indicated a higher past usage by PCAs/Florida distributors than growers. Growers may be using them without knowing what they are. Use by the golf course industry was very low, relative to ever used. Figure 13.6 shows user perceptions of biopesticides.

The perception of biopesticide usefulness was generally positive by all groups. Growers' and golf superintendents' perception generally was more variable than that of PCAs and Florida distributors. The issue could be lack of awareness and limited understanding of biopesticide utility at the end-user level. The survey also asked what the main reasons were why biopesticides were used or not. Table 13.5 summarizes user responses as to why biopesticides were used.

Environmental safety profile was the highest ranked reason for using biopesticides. Efficacy was rather low as a reason to use biopesticides. The implication is that perceived efficacy is low, but products are used based on environmental and safety attributes. Table 13.6 shows the reasons why biopesticides were not used by respondents in this BPIA survey.

Perceived lack of efficacy is mainly an issue at the PCA/Florida distributor level. Growers admitted they don't know about biopesticides – lack of awareness. The perception by most groups is that the cost/efficacy ratio is not balanced.

Table 13.5 Reasons why biopesticides are used (response percent)

Reason volunteered	California PCA	Florida distributors	California growers	Florida growers	Golf superintendents
Environmentally safe	30	46	43	30	75
Operator safety	23	29	20	23	15
Product efficacy	17	29	17	13	5
More natural/safe	10	11	10	13	15
Public perception	9	–	3	7	13
Organic farming	13	–	–	3	–

Source: BPIA survey, 2003.

Table 13.6 Reasons why biopesticides are not used (response percent)

Reason volunteered	California PCAs	Florida distributors	California growers	Florida growers	Golf superintendents
Not as effective	44	50	27	13	30
Higher cost	46	36	23	23	30
Lack of awareness	14	29	47	47	43
Old habits	4	–	13	10	10
Lack of research	6	7	–	7	15

Source: BPIA survey, 2003.

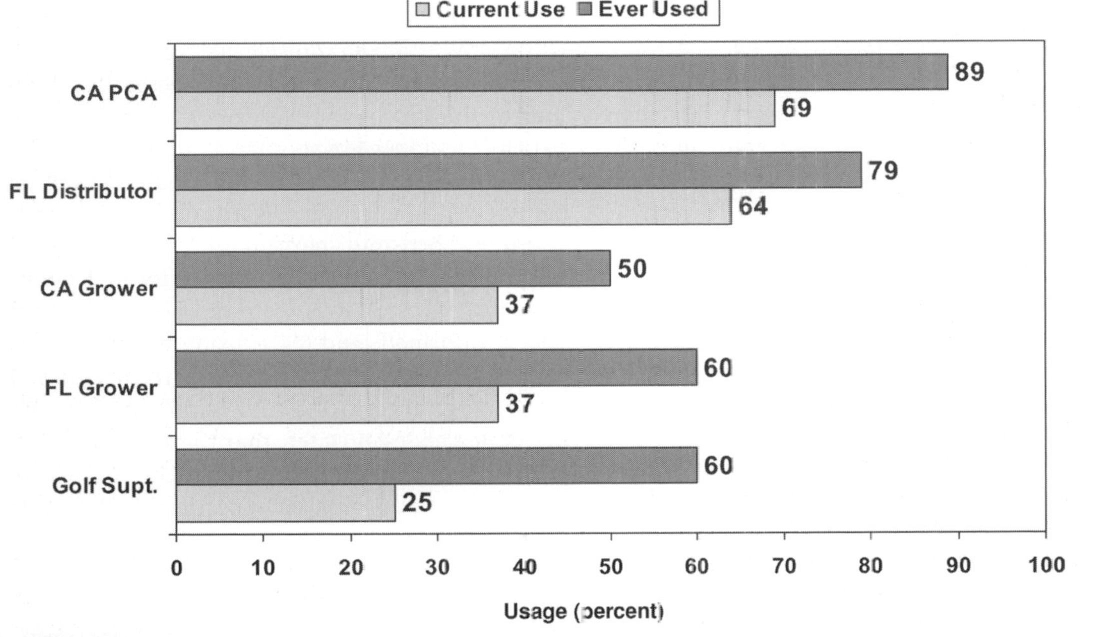

Fig. 13.6 Usage of biopesticides.

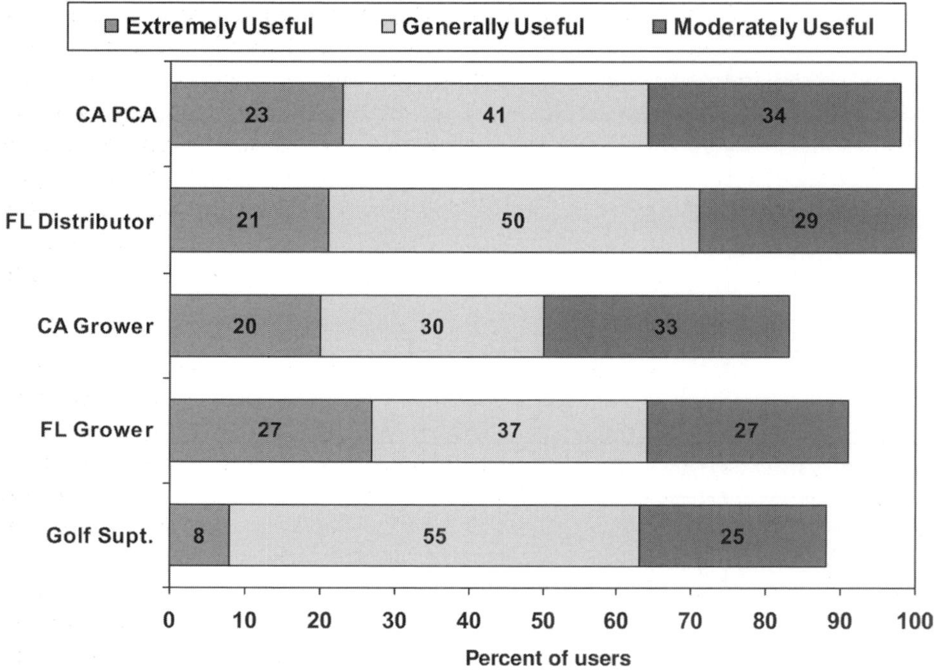

Fig. 13.7 User perceptions of biopesticides. Key as for Fig. 13.5.

Awareness of biopesticide use and overall value decreased significantly at the grower level. This provides opportunities to increase emphasis on "demand creation" for biopesticides at the end-user level.

The conclusion from this survey is that there is a large gap between the awareness and perceived value of biopesticides in the golf course segment versus agriculture segments. This may be due to lack of available products for golf courses. Cost/value ratios for biopesticides were perceived to be unfavorable. While there are many examples of successful, effective biopesticides used today, the overall perception is not consistent with the potential value these products can provide.

The Biocontrol Network (www.biocontrol.ca), a group of Canadian universities, government agencies, pesticide manufacturers and end-users dedicated to increasing the research and usage of biopesticides, also conducted a survey about attitudes and barriers to biocontrols (includes beneficial insects such as predators and parasites) (Cuddeford, 2006). In winter 2005, the Biocontrol Network and World Wildlife Fund–Canada (WWF) conducted two separate surveys on public attitudes to biocontrol. The Network surveyed 1000 randomly selected members of the public, while WWF–Canada interviewed a small group of opinion leaders in the agri-food sector. The findings of the Biocontrol Network's survey are statistically valid within standard parameters, while WWF's survey, because of the small sample, is best viewed as a pointer towards needs for further research. Forty-nine percent of the respondents reported being somewhat well informed about biological control, 47% were only poorly or very poorly informed, and 3% considered themselves very well informed. Sixty-three percent believed that biological control would be preferable to the use of pesticides, 60% felt that there was less risk of food being contaminated with the use of biological controls compared to pesticides. Sixty-four percent reported that they would buy food where pests had been controlled with biological agents rather than pesticides. Seventy-three percent were in favor of using beneficial insects to control pests, as compared to 25% for pesticides and 41% for genetically modified organisms. Forty-six percent reported feeling wary of eating food if "beneficial microbes" were used to control pests. Eighty-seven percent wanted food to be labeled indicating that biocontrol agents had been used to control pests.

The strength of support for biocontrol was positively associated with number of years of education. Regional differences of opinion were sometimes significant. For example, the proportion of respondents who considered themselves well informed on biocontrol ranged from a low of 34% in Quebec to a high of 76% in British Columbia while the percentage who supported use of biological agents rather than pesticides varied from 55% in the Maritimes to 81% in Quebec. The strongest finding from the WWF survey was that efficacy, cost and availability are viewed as the major barriers to wider adoption of biological control products. There was strong agreement that the greatest benefit of biological control products is their softer environmental footprint. Perceptions of the future of biological control were on the whole positive but, for many, qualified by the need to address the barriers mentioned. A number of trends were identified by respondents as supporting the view that increased adoption of biological pest control products is likely: the growing perception of conventional chemical pesticides as damaging to health and the environment, an aging and health-conscious generation of "boomers," and growing support for products marketed as "natural" or "organic." The scenario of older chemical pesticides going off the market, via withdrawal and other means, was also perceived as a positive opportunity for biocontrol agents (Cuddeford, 2006).

Like the BPIA survey, these Canadian surveys indicate that there is more work needed for improved adoption of biopesticides and biocontrol agents. Education of the benefits and value is key and will be discussed in a later section.

13.7 Recommendations to increase usage

The survey data discussed in the previous section clearly show that the perception of the efficacy of biopesticides and general lack of awareness are significant impediments to increased usage. The EPA has a committee of stakeholders on the Pesticide Program Dialogue Committee (PPDC) that meets on a regular basis to help the EPA and US Department of Agriculture (USDA) implement reduced risk alternatives that meet Food Quality Protection Act of 1996 transition goals. At the May 2002 PPDC meeting, the group discussed the barriers to adoption of biopesticides and ways to increase adoption. The top ideas were:

- Lack of research support for biopesticides, especially how to refine their use to improve effectiveness.
- Lack of demonstration programs to show users that biopesticides fit into IPM and other pest management programs. Users need training on what to expect from the product.
- Lack of realistic performance and cost criteria for biopesticides and lack of incentives to encourage growers and researchers to judge biopesticide performance and value fairly, especially in programs with conventional products.
- Users are skeptical about biopesticides, there is lack of brand name recognition and users expect "silver bullets."
- Users need early experience with biopesticides; registration decisions are too slow for biopesticides.

A workshop was sponsored by the BPIA and National Foundation for IPM Education at the 4th National IPM Symposium/Workshop (US Department of Agriculture, 2003). Below is a summary of the discussion and recommendations by the workshop attendees (over 100 people). The discussion conclusions are consistent with the other surveys discussed previously.

With PCAs, conduct grower demonstrations on a small scale to start adoption

- PCAs are a main driving force in California agriculture as to the use of biopesticides.
- PCAs are concerned they will lose their client (the grower) if they use a biopesticide, or any other product, and it does not work.
- Put the PCA and the grower in the same room to agree to try a little on a small scale to see if it works.

- Compare the program with the biopesticide to the standard grower program. Progressively increase hectarage as performance is verified.

US Department of Agriculture IR-4 program
The USDA Interregional Research Project No. 4 (IR-4 program) is involved in making sure that pesticides are registered for use on minor crops. The USDA IR-4 program has provided grant funding to key influencers such as land-grant university extension specialists to demonstrate with the end-user the performance of biopesticides in realistic programs with conventional pesticides. More funding for this valuable grant program would significantly advance perception and adoption.

Show profit of the program with the biopesticide (return on investment) to the grower and PCA
When you can show that a program with a biopesticide nets the grower equal or more profit, it is more likely that growers will start using biopesticides.

Improve image and efficacy of biopesticides by implementing the BPIA's certification standards for efficacy, quality and registration
The plethora of small companies with unregistered products hurt the legitimate biopesticide industry. Certification will shift the image away from these "snake oils."

Public funding must be available for developing new kinds of monitoring for the pests
It is important to look at the basic biology and research is needed at an ecological level. For example: understanding the peach twig borer biology in stone fruit led to monitoring that enabled the use of *Bt* as replacements for organophosphate insecticides.

Promote biopesticides in resistance management programs, residue management programs (especially for exported produce) and IPM programs to preserve beneficials
Much of the new pesticide chemistry is single site mode of action, e.g. strobilurin fungicides and imidacloprid insecticides. Funding should be obtained (from USDA's IR-4) for resistance management programs. To show the benefits of biopesticides as preservers of beneficials, include measurements of beneficials in IR-4 grant applications.

Educate growers and researchers as to what biopesticides are
Many growers and researchers are unclear on the definition of a biopesticide. For example, growers often don't think they are using a pesticide when they are using pheromones. They think abamectin, spinosad and strobilurins are biopesticides. Conversely, researchers try to use biopesticides as chemicals.

Provide incentives for use of biopesticide
Subsidies, lower mill taxes (California) and ecolabels were suggested as possible incentives.

Develop promotional and marketing agreements between biopesticide companies and large agrichemical companies, promoting resistance management
Adoption could be jump-started if companies worked together to combine their products in IPM programs to increase yield and quality above existing programs.

Through a consortium of USDA, Small Business Innovation Research (SBIR) and EPA, fund turn key commercialization projects as they do in Europe
A North American example of partnership between government and private industry to commercialize a pesticide is the Bipesco cost-914 project to commercialize *Beauveria brongniartii*.

Find a champion outside of the industry to champion biopesticides
USDA's IR-4 program is such a champion but it has only a few hundred thousand dollars of funding each year to support biopesticides.

Decrease the reference to "biopesticides," and promote them as solutions to pest problems
Biopesticide references may prevent a grower with bad experiences in the past from using them.

13.8 Additional ways to address adoption of biopesticides

The survey data confirmed what the BPIA members already knew about the perception of efficacy of biopesticides. As a condition of membership and to insure that its members uphold the highest standards of product stewardship, BPIA supports basic principles that demonstrate commitment to customers and key influencers (see: www.biopesticideindustryalliance.org/html/operprinciples.htm):

- BPIA-supported products are legally registered by a recognized regulatory agency in the USA, Canada, EU or other countries where they have been demonstrated to meet the required efficacy and safety criteria necessary to receive approval.
- BPIA-supported products are expected to have stewardship policies that maintain product integrity and appropriate complaint management policies related to product efficacy and performance.
- BPIA member products are expected to be supported by scientifically valid tests that confirm the product's performance claims. Such claims provide commercially recognized product efficacy levels in the target geography with minimal performance variability over multiple years of use by the end-user.
- BPIA-supported products should be recognized by peer groups and end-user customers as having demonstrated commercially valid efficacy and product stewardship claims over multiple growing seasons.

These operating principles are being developed into standards that would result in a "seal of approval" so that any distributor, end-user and key influencer would recognize that a commercial biopesticide product adhered to the highest standards of efficacy, quality and product stewardship. Survey data indicate that these standards would help the negative perceptions of biopesticides and increase market adoption. At the very least, adoption of standards would distinguish products that are registered by a regulatory agency versus unregulated products that are comprised of exempt materials listed on EPA 25-B product list and other materials such as compost teas that do not have to adhere to government quality and efficacy standards.

The standards are still in draft form (see Box 13.1) but the concepts and principles will remain the same, even if the specifics change.

An unpublished survey conducted by a team at the University of California, Graduate School of Management (supervised by the author) targeted medium to large California growers and PCAs. These customers claimed that biopesticide input companies should place a heavy emphasis on education in order to establish trust and sustainable use of the product. They indicated that the most successful companies are very careful in targeting specific markets, either by crop, pest or disease. In turn, companies should be very clear about the protection and value being provided to the grower. It is important to tell customers how to use products – precisely which sprays you will replace and when.

On-farm demonstrations are critical and biopesticide companies should obtain more efficacy data (stand-alone and in programs – rotation/tank mix) and find ways to better shepherd university trials (although resources are a constraint).

Where do biopesticides shine?

- Used for pesticide resistance management
- In rotations and alternations – tank mixes
- Early season – low pest pressure
- Late season – short pre-harvest interval
- During critical field events and to save labor costs – short re-entry intervals get you back in the field

Often, researchers report biopesticide results with a "glass half-empty" viewpoint. By way of illustration, the following is a summary of a biopesticide in trial with a chemical pesticide (Box 13.2) (example from Michael Braverman, Biopesticide Director, IR-4 program; see Braverman, 2005).

Thus it is important to recognize that stand-alone data do not reflect the potential value of biopesticide, and biopesticides should be tested in programs reflective of grower practices.

Box 13.1 Proposed standards for biopesticide standards (BPIA)

Certification criteria requirements	Documentation requirements
(1) Product registration by EPA or Ag Canada	(1) Stamped, approved label and letter
(2) Active ingredient(s) consistent with EPA definition of "biopesticide"	(2) EPA or other regulatory agency documents
(3) Process description from company indicating plan for maintaining quality integrity in normal trade channel	(3) Explanation of corporate stewardship program. Signed documents from company representatives (President, Sales/Marketing, R&D)
(4) Verification that product has been sold in typical trade channel represented by label claims for minimum of two full seasons	(4) Documented invoices (modified to maintain confidentiality, if necessary) (see Documentation Requirement Number 6)
(5) Achievement of product efficacy levels in target geography for targeted use	(5) Documentation of scientifically valid efficacy trials and claims representing: • >85% efficacy, as compared to typical chemical standard (standard used should be explained), OR • >90% achievement of indicated response expectation indicated on advertising collateral
(6) Testimonials from peer groups, attesting to quality claims, specific efficacy claims and overall satisfaction with product use	(6) Signed testimonials from each of the following relating to quality, consistency and efficacy: • at least five independent end-users; also include proof of purchase • at least three university co-operators • at least two trade channel representatives; also include proof of purchase
(7) Achievement of field trial variability within +/−25% of collateral or advertising claims	(7) Documentation of scientifically valid efficacy trials and claims representing statistical significance relating to promotional claims (Note: no penalty for exceeding claims, but excessive variability through over formulation is not desirable)
(8) Product complaint history demonstrating product's commercial use meeting grower expectations	(8) Documentation showing complaints represent less than 5% of product sold (volume basis) over the previous two-year period

Box 13.2 Summary of results from trial comparing biopesticide with conventional chemical insecticide

Treatment	Control
Conventional insecticide alone	83%
Conventional insecticide – biopesticide rotation	79%
Control	0%

"There was no advantage to rotating the conventional treatment with a biopesticide." However, viewing the trial with a "glass half-full" the trial would be reported as follows:

The level of control was maintained while gaining the following:

- Resistance management
- Re-entry period – flexibility with labor
- Reduced time to harvest – multiple harvest
- Residue management

Biopesticide companies do not have the dollars of the big companies to conduct trials. Marketable yield is the most important measure of performance; however, pest/disease level is usually the measure in performance trials. Biopesticides often increase marketable yield (and results often can be as good as or better than with chemicals) but trials may have more diseases/pests than in trials with conventional chemicals. Chemicals fail, yet the perception is they always work. Excitement often accompanies the introduction of new conventional chemistry, but often there is less enthusiasm over a new biopesticide even when has good efficacy.

13.10 Conclusions

By combining performance and safety, biopesticides offer benefits not generally realized with conventional pesticides. These benefits include efficacy while providing customers the flexibility of:

- minimum application restrictions,
- residue management,
- resistance management, and
- human and environmental safety.

Consumers are driving growth of organic food and food produced with reduced-risk pesticides and fewer pesticides. Due to government regulations around the world, older, more toxic chemical pesticides are being removed from the market and there is increased emphasis on worker safety and pesticide exposure and attention to pesticide contamination of air and water. In addition, the global production and shipment of food products has resulted in an emphasis on maximum residue levels on food at time of shipment and entry into the receiving country. These trends are driving the global growth of biopesticides at rates much faster than the mature, slow growth chemical pesticide market. Despite these trends and continued fast biopesticide growth, perception of biopesticide products and lack of awareness remain the largest barriers to increased adoption. Key influencers (e.g. university extension specialists and PCAs) and gatekeepers (e.g. distributors) may have decided opinions about the efficacy of biopesticides, or lack thereof, and end-users (growers, consumers and superintendents) generally may not understand their benefits and value, how they work and how to use them. This provides a good opportunity to policy makers and the industry to finds ways to educate and increase awareness. For example, the IR-4 program, funded by the EPA, has developed a searchable biopesticide database (Environment Protection Agency, 2007). For the first time, an end-user or key influencer can find information about biopesticides all in one place.

It is a good first step. It is time to move from listing and discussing the barriers to adoption, which are now well known and focus on breaking down these barriers. These include:

- Increased federal funding through the IR-4 program for biopesticide efficacy trials and demonstration programs.
- BPIA standards and seal of approval for biopesticide products.
- Increased applied research at land-grant universities integrating biopesticides with chemicals and with other biopesticides.
- Increased effort by biopesticide companies to educate their customers and specifically position their products in the marketplace and selling through experienced channel partners.
- Increased education of end-users by biopesticide companies, extension specialists and other key influencers (USDA, EPA, California Department of Pesticide Regulation).

References

Agrow (2006a). Global agrichemical market flat in 2005. *Agrow: World Crop Protection News*, February 24, 2006, **490**, 15.

Agrow (2006b). US agrochemical sales up 2% in 2005. *Agrow: World Crop Protection News*, October 6, 2006, **505**, 14.

Beyond Pesticides (2008). Home page. Available at www.beyondpesticides.org.

Bounds, G. & Brat, I. (2006). Turf Wars: armed with chicken manure, organic lawn product makers are battling for your backyard. *Wall Street Journal*, April 15, 2006, P1. Available at http://online.wsj.com/article/SB114505266809026471.html.

Braverman, M. (2005). Why use biopesticides in IPM programs? In *Proceedings of 2005 National IPM Symposium*. Available at www.ipmcenters.org/ipmsymposium/sessions/51_Braverman.pdf.

Cuddeford, V. (ed.) (2006). Surveys gauge attitude of Canadian public. *Biocontrol Files: Canada's Bulletin on Ecological Pest Management*, January 2006, **5**, 8. Available at http://biocontrol.ca/pdf/Bio5EN.pdf.

Donaldson, M. (2003). Results of a customer survey on attitudes towards biopesticides. In *Proceedings of 2003 National IPM Symposium*. Available at www.cipm.info/symposium/getsession.cfm?sessionID=47.

Environmental Protection Agency (2007). *What Are Biopesticides*? Washington, DC: US Environmental Protection Agency. Available at www.epa.gov/pesticides/biopesticides/whatarebiopesticides.htm.

Fredonia Group (2006). Focus on lawn and garden consumables. *Fredonia Focus Reports, Consumer Goods*, August 1, 2006. Available at www.fredonia.edu/business/ama/home.htm. Cleveland, OH: The Freedonia Group.

Guillon, M. (2003). President's Address. Available at www.ibma.ch/pdf/20041028%20Presentation.

National Gardening Association (2002). *2002 National Gardening Survey*. South Burlington, VT: National Gardening Association.

National Gardening Association (2004). *Environmental Lawn and Garden Survey*. South Burlington, VT: National Gardening Association and *Organic Gardening Magazine*.

Quinlan, R. & Gill, A. (2006). *The World Market for Microbial Biopesticides, Overview Volume*. Wallingford, UK: CPL Business Consultants.

US Department of Agriculture (2003). Symposium: Barriers to the adoption of biocontrol agents and biological pesticide. In *Proceedings of the 4th National IPM Symposium/Workshop*. Washington, DC: US Department of Agriculture. Available at www.cipm.info/symposium/getsession.cfm?sessionID=47.

Venkataraman, N. S., Parija, T. K., Panneerselvam, D., Govindanayagi, P. & Geetha, K. (2006). *The New Biopesticide Market*. Wellesley, MA: Business Communications Company Inc. (BCC) Research Corporation. Available at www.bccresearch.com/chm/CHM029B.asp.

Chapter 14

Integrating pesticides with biotic and biological control for arthropod pest management

Richard A. Weinzierl

Insecticides and acaricides are used commonly for pest suppression in agriculture, forestry and public health. The National Research Council (2000) and others have documented their value in protecting crops, livestock and humans from injury by insects and other arthropods. However, adverse effects of pesticide use can include the killing of non-target organisms, contamination of water supplies and persistence of unwanted residues on foods and animal feed. The chronic health effects of even extremely low concentrations of pesticides in food and water remain under debate today just as they were when the National Research Council (1993) reported on pesticides in children's diets in the early 1990s. Resistance to one or more pesticides has evolved in populations of over 500 insect and mite species (Clark & Yamaguchi, 2002), rendering many of those pesticides ineffective against the resistant populations. For these and other reasons, supplementing or replacing pesticides with non-chemical control tactics, including biological control, is a goal in many crop and livestock production systems.

As biological control programs that rely on parasites, predators and pathogens are developed, they rarely are so robust as to provide all the pest suppression needed in an entire cropping system for the lifespan of the crop. Even if such robust biological control is attainable over time in specific crops or situations, initial efforts that target only one or a few pests often leave a need for pesticides to manage other species if their populations exceed economic thresholds. Insecticides used under such circumstances may be (and often are) toxic to organisms used for biological control (Newsom *et al.*, 1976; Hull & Beers, 1985; Croft, 1990), and the effectiveness of the biocontrol program is thereby threatened. For this reason, effective integration of biological and chemical controls is essential. Previous works by Stern *et al.* (1959), Stern & van den Bosch (1959), Smith & Hagen (1959), Gonzalez & Wilson (1982) and Hoy (1989) illustrate that the need for such integration has long been a key issue. Integrating biological and chemical control is, after all, the foundational idea in the "integrated control" concept proposed by Stern *et al.* (1959) now known as IPM.

The integration of chemical and biological control tactics can take many forms, depending on the characteristics of the biological control agent and its host, the nature of the cropping system or habitat, the persistence and toxicity of the pesticides to be used and the methods by which pesticides are applied. Approaches to integration are presented and discussed in this chapter under four headings: (1) reducing pesticide use; (2) using selective pesticides and applying pesticides selectively; (3) modifying biological control agents to survive

Integrated Pest Management, ed. Edward B. Radcliffe, William D. Hutchison and Rafael E. Cancelado. Published by Cambridge University Press.

pesticide applications; and (4) combining biological control agents and pesticides for increased effectiveness. The chapter closes with a brief discussion of the current status of efforts to integrate the use of chemical pesticides and natural enemies in arthropod pest management.

In this chapter, the term *biotic control* is used to describe the naturally occurring effects of predators, parasites and pathogens; *biological control* is used to describe human-managed efforts to preserve or use natural enemies. The terms blur a bit when IPM includes well-planned efforts to conserve natural enemies by specific habitat management practices, pesticide selection and other actions. For purposes of illustrating these concepts, the scope of the discussion is limited to crops and livestock.

14.1 Reducing pesticide use

Reducing the use of pesticides is an obvious way to reduce their negative impacts on biotic and biological control organisms. Clearly, pesticides should not be applied unnecessarily, and ending unneeded, ineffective and unprofitable uses in any crop or setting would save money, reduce environmental risks and allow survival of natural enemies. It would be incorrect, however, to assume that all pesticide applications are unnecessary and could be ended without dramatic increases in losses to pests (National Research Council, 2000). What steps, then, are realistic for reducing the use of pesticides and allowing greater success for biological control? The main categories of actions are: (1) designing production systems that minimize the need for pesticides; (2) identifying natural enemies as beneficial species; (3) developing and using thresholds and models that accurately reflect control needs and account for the impacts of natural enemies; and (4) delivering educational programs to farmers, landscapers, greenhouse managers, pest management consultants and the public.

14.1.1 Production systems that rely less on pesticides

Approaches to reducing pesticide use incorporate such practices as planting resistant cultivars, practicing effective crop rotations, altering time of planting to avoid pest presence, using physical barriers to exclude pests and maintaining habitats that favor natural enemy survival. All of these practices reduce the need for pesticides that might also kill biotic or biological control agents. For example, where application of broad-spectrum insecticides for control of onion thrips (*Thrips tabaci*) in cabbage would interfere with biological control of lepidopterans such as cabbage looper (*Trichoplusia ni*) and diamondback moth (*Plutella xylostella*), choosing thrips-resistant cultivars (Egel *et al.*, 2007) might avoid the need for insecticide use and unwanted side effects. Similarly, planting sorghum cultivars resistant to sorghum midge (*Stenodiplosis sorghicola*) and/or greenbug (*Schizaphis graminum*) reduces the likelihood that insecticides will be needed for the control of these pests (Teetes, 1994). Soil applications of insecticides to protect cucurbits (cucumbers, squash, melons and pumpkins) from seedcorn maggot (*Delia platura*) likely reduce populations of beneficial predaceous carabid and staphylinid beetles (Hassan, 1969; Brust *et al.*, 1985; Curtis & Horne, 1995), but where these crops are direct-seeded into cool, wet soils, seedcorn maggot control often is necessary. Using transplants instead of direct-seeding reduces the risk of damage by seedcorn maggot and usually avoids the need for insecticidal control. Delaying planting of cucurbits would likewise escape injury by seedcorn maggot. Floating row covers and pest-proof screening exclude a wide range of insect pests in outdoor and greenhouse production of horticultural crops, reducing the need to apply pesticides that might kill biological control agents. Because many natural enemies benefit from stable habitats, maintaining ground covers, standing crops and crop residues (instead of practicing clean tillage) generally results in greater numbers of predators, parasites and pathogens. Additionally, maintaining plants that provide nectar and pollen for adults of parasitic wasps increases their retention and survival (Landis *et al.*, 2000).

14.1.2 Identifying natural enemies

Recognizing natural enemies is an obvious basic requirement for minimizing pesticide use. Although certain predators, e.g. lady beetles and praying mantids, may be widely recognizable, many farmers and gardeners do not recognize adults of parasitic insects (Hymenoptera and Diptera). Likewise, larvae of lacewings and lady beetles and adults and immatures of carabid and staphylinid beetles, syrphid flies and predaceous Hemiptera are common but often not identified by farmers and gardeners. If a failure to identify them as beneficial organisms leads to killing them with insecticides, many other biological control agents often are killed as well.

14.1.3 Improving and using pest thresholds and predictive models

Refinement of economic thresholds and development of models to predict changes in pest densities based on field-level population sampling may be among the most important steps for better integration of biological and chemical control in agriculture. Insect management guidelines that include pest density thresholds for insecticide applications often mention natural enemies and suggest that pesticide application may not be needed if they are present (Steffey & Gray, 2007). However, few such recommendations available for growers or consultants give specifics on the number of natural enemies needed to provide enough biotic or biological control to prevent crop losses that would exceed the cost of a pesticide application. Establishing and communicating clear guidelines on the numbers of natural enemies needed to control pest populations of varying densities would allow a grower or consultant to reject a decision that the pest population exceeds the economic injury level and decide instead to allow biotic or biological control to occur (avoiding unneeded expense and the non-target impacts of pesticide application).

Examples of information specific enough to really guide decision making include assessments of the numbers of eggs of common green lacewing (*Chrysoperla carnea*) needed if subsequent larvae are to prevent economic losses to *Heliothis* spp. in cotton (Ridgway & Jones, 1969), the impacts of specific numbers of the parasitoids *Bathyplectes cucurlionis* and *Tetrastichus incertus* on alfalfa weevil (*Hypera postica*) (Davis, 1974 and Horn, 1971, respectively), necessary ratios of common green lacewing for suppression of apple aphid (*Aphis pomi*) in apples (Niemczyk *et al.*, 1974), and the number of convergent lady beetles (*Hippodamia convergens*) per sweep necessary for adequate control of pea aphid (*Acyrthosiphon pisum*) in alfalfa (Hagen & McMurtry, 1979). Tamaki *et al.* (1974) used the term "predator power" to describe the idea that different natural enemies kill different numbers of aphids and therefore have greater or lesser pest management impact. Polyphagous hemipteran predators such as nabids, anthocorids and geocorids were given a predator power rating of 1; immature coccinellids, syrphids and chrysopids were rated 4 and large adult coccinellids were rated 8. Combined predator power ratings for populations sampled in an individual field were calculated to predict aphid suppression (Tamaki *et al.*, 1974). Tamaki & Long (1978) further developed a predator efficacy model that included temperatures and functional responses to predict impacts of inundative releases. Tamaki and colleagues focused on the impacts of inundative releases and did not suggest modifying pest thresholds or decisions on pesticide applications based on natural enemy densities. However, Naranjo & Hagler (1998) assessed impacts of heteropteran predators for just such a purpose.

Adjusting thresholds for aphid control based on parasitism (Fernandez *et al.*, 1998; Giles *et al.*, 2003) or infection by fungal pathogens (Hollingsworth *et al.*, 1995; Conway *et al.*, 2006) has been proposed for cereal crops and cotton. Hoffman *et al.* (1990, 1991) proposed incorporating impacts of *Trichogramma* spp. on *Helicoverpa* (= *Heliothis*) *zea* to modify thresholds for egg counts in processing tomatoes. Gonzalez & Wilson (1982) took a season-long approach to this idea and suggested that economic thresholds should consider the food webs that exist in cropping systems, particularly cotton. Nyrop & van der Werf (1994) summarized methods to monitor natural enemies for the prediction and assessment of biological control.

Brown (1997) outlined two categories of models designed to incorporate the impacts of natural enemies in estimates of economic thresholds, one for generalist natural enemies with population densities not tied closely to the dynamics of a given pest and a second for specialist natural enemies whose population densities are coupled tightly with the pest's population dynamics. Musser *et al.* (2006) suggested that natural enemy densities and their projected impacts should be incorporated into thresholds and that thresholds should differ for individual pesticides according to the negative impacts of those pesticides on natural enemies. This approach incorporates and expands upon ideas expressed well over a decade ago by Kovach *et al.* (1992) and Higley & Wintersteen (1992).

Despite the research presented in professional journals, relatively few pest management recommendations that reach growers incorporate assessments of biotic control. Weires (1980), in a fact sheet written for apple growers, summarized recommendations for control of European red mite (*Panonychus ulmi*) based on numbers of European red mites and their predators on a per-leaf basis. Possible actions recommended for mite management included applying a miticide, waiting to resample, or concluding that biotic control is likely. If growers or pest management consultants are to incorporate counts of natural enemies into their decisions on the need for insecticide application (and thereby integrate biotic and biological control with pesticides), specific guidelines such as these are essential. For example, for growers to weigh the impacts of lady beetle predation on soybean aphid (*Aphis glycines*) and determine whether biotic control will suppress aphid densities enough to prevent economic losses, detailed recommendations based on predator–prey population dynamics are needed for a range of population densities and environmental conditions.

14.1.4 Educating growers and consumers

Providing growers and consumers with the information needed to make educated decisions is a constant need in guiding insecticide use. Numerous guides to IPM, natural enemy identification and the practice of biological control provide recommendations on when pesticides are needed (and which ones are most effective or appropriate) and when alternatives – including no control at all – may be better choices. Nonetheless, educational programs that discourage unnecessary pesticide applications face strong competition when marketing efforts of pesticide companies strive for maximum sales. Marketing efforts that use biased presentations of research data remain as much of a concern as they were when Turpin & York (1981) wrote "Insect management and the pesticide syndrome." For example, in Illinois, despite conservative thresholds for soybean aphid control (i.e. 250 aphids per plant) (Ragsdale *et al.*, 2007), soybean growers tell of pesticide sales representatives who recommend insecticide application at much lower densities, using the rationale that a few aphids today will increase to exceed the threshold soon. They ignore the impacts of biotic control that are conservatively incorporated in the recommended threshold and they urge that "the sooner the better" is the rule to follow for insecticide application. It is indeed an understatement that education on the benefits of not spraying is an ongoing need.

14.2 Using selective pesticides and applying pesticides selectively

Where insecticides must be used in pest management but their integration with biotic and biological control also is a goal, the use of selective insecticides or selective application methods commonly is advised (Newsom *et al.*, 1976; Weinzierl & Henn, 1991). More than a half-century ago, Ripper *et al.* (1951) defined selective insecticides as "insecticides toxic to the pests but not toxic to the beneficial insects." They noted that beneficial insects that survive spray treatments (with selective insecticides) exert: (1) an immediate effect when they "mop up the surviving individuals of the pest population and thus increase the apparent effectiveness of the insecticide application," and (2) a delayed effect when "(they) or their progeny act on the surviving pest population and prevent the start of the rapid build-up of the pest population which is so often observed when conventional insecticides are used." They considered selective insecticides to include the systemic organophosphate schraden, as well as bait formulations of DDT encapsulated in hemicelluloses that were poisonous only to phytophagous insects that consumed the bait. Similarly, Stern & van den Bosch (1959) and Smith & Hagen (1959) considered the organophosphate systox to be selective because when used as a systemic its toxicity to key natural enemies was low. More recently, the concept of selectivity has been discussed under two broad headings: (1) physiological selectivity, in which specific pests and natural enemies differ in susceptibility to individual broad-spectrum toxins or in which the mode of action of the pesticide is specific to certain groups of insects; and (2) ecological selectivity, which depends largely on either spatial or temporal separation of the insecticide's effects and the occurrence of key natural enemies (Theiling & Croft, 1988; Croft, 1990).

14.2.1 Physiological selectivity

The physiological selectivity of pesticides is discussed in detail by Theiling & Croft (1988) and Croft (1990). Their SELECTV database summarizes results from studies of more than 12 000 insecticide–natural enemy combinations. The authors calculated selectivity ratios to reflect the relative toxicity (based on LD_{50}s, LC_{50}s and similar measures) of selected pesticides to natural enemies in comparison with their hosts or prey (the target pests). Hundreds of additional reports on the toxicity of individual insecticides to specific natural enemies have been reported since the SELECTV database was first published. Representative examples of studies included in the database and those that have been published more recently include works by Hassan & Oomen (1985), Hagley & Laing (1989), Boyd & Boethel (1998), Medina *et al.* (2003) and Tillman & Mullinix (2004).

When data from multiple trials examining multiple pesticides are analyzed, assessments of the relative toxicity of available pesticides can guide growers' selections of the least toxic broad-spectrum insecticides where such compounds must be used. For example, choosing insecticides for control of direct pests of apples, including codling moth (*Cydia pomonella*), plum curculio (*Conotrachelus nenuphar*) and apple maggot (*Rhagoletis pomonella*), has for many years included consideration of the toxicity of those insecticides to predaceous mites that control European red mite. The organophosphates azinphosmethyl and phosmet became standards in apple pest management because predator species developed resistance to them, rendering them less disruptive to biotic control than most other organophosphates, carbamates, pyrethroids and neonicotinoids (Croft & Meyer, 1973; Rytter & Travis, 2006). Hoy (1995) suggested that information on the toxicity of new pesticides to natural enemies important in specific cropping systems be required for US Environmental Protection Agency (EPA) registration of those pesticides. This information could be included in a database available to pesticide users, crop consultants and IPM advisers. One problem for such a database is that selectivity of broad-spectrum insecticides and miticides to specific natural enemies may differ significantly over time and among geographical regions as these organisms respond to selection pressure (Croft & Meyer, 1973; Croft, 1990).

In more recent years, physiological selectivity based on taxon-specific modes of action has provided greater options for integrating pesticides and biological control. Examples of insecticides that are selective based on mode of action include microbial insecticides and synthetic compounds. Among the microbials (some of which also may be considered as biological control agents themselves) are insecticides derived from subspecies of the soil-borne bacterium *Bacillus thuringiensis* (*Bt*); different products are toxic only to larvae of Lepidoptera, certain lower Diptera, or a few species of Coleoptera (Maagd *et al.*, 2001). Other microbial insecticides contain taxon-specific viruses, fungi or microsporidia. Selective synthetic insecticides include compounds primarily toxic to Lepidoptera (tebufenozide, methoxyfenozide, diflubenzuron, novaluron), Hemiptera/Homoptera (buprofezin, kinoprene and pyriproxyfen) or Diptera (cyromazine) (Medina *et al.*, 2003; Stark *et al.*, 2004; Insecticide Resistance Action Committee, 2006). Cloyd & Dickinson's (2006) summary of the toxicity of buprofezin, pyriproxyfen, flonicamid, acetamiprid, dinotefuron and clothianidin to a parasite and a predator of citrus mealybug (*Planococcus citri*) illustrates the value of the selectivity of taxon-specific pesticides for a greenhouse system. Selective miticides that are low in toxicity to predaceous mites and certain other predators in orchards include chlofentezine, acequinocyl and hexythiazox. Miticides that are somewhat more toxic to predaceous mites but are not toxic to predaceous insects include bifenazate, spirodiclofen and fenpyroximate (Rytter & Travis,

2006). Additionally, pheromone- or kairomone-based management using mass trapping, mating disruption, or attract-and-kill concepts offers physiological selectivity and little or no direct mortality to natural enemies.

It is important to note that insecticides considered to be selective, including lepidopteran or hemipteran/homopteran growth regulators, may be toxic to natural enemies in taxa other than those targeted by the insecticide's primary mode of action. For example, Rothwangi *et al.*, (2004) found that growth regulators thought to narrowly target hemipterans/homopterans were toxic to the mealybug parasite *Leptomastix dactylopii*. Conversely, even broad-spectrum insecticides and miticides may be considered at least somewhat selective within a biological control program that relies on pathogens as the organisms employed by humans (Barbara & Buss, 2005; Ericsson *et al.*, 2007).

Finally, transgenic crops that produce *Bt* toxins might be deemed selective pesticides. While the toxins they produce directly kill only certain taxa, the extremely high level of control they provide, often in a high percentage of crop area in a given region, has obvious ecological impacts on natural enemies of target pests. Their sustainable use as selective insecticides is dependent on steps that provide ecological selectivity as discussed later in this chapter.

Assessing the toxicity of specific pesticides against specific natural enemies in laboratory tests has been the most common approach to identifying physiological selectivity (Theiling & Croft, 1988; Croft, 1990), but Stark & Banks (2003) suggest that data gained from such bioassays need to be interpreted carefully. They propose that population-level effects of pesticides must be measured to understand the impacts of physiological selectivity. Recent examples of studies that have assessed physiological selectivity (perhaps in combination with ecological selectivity) at the population level include those by Bernard *et al.* (2004), Furlong *et al.* (2004), Wilkinson *et al.* (2004) and Desneux *et al.* (2005).

14.2.2 Ecological selectivity

As noted previously, ecological selectivity results from the separation of a pesticide's effects from the occurrence of susceptible natural enemies. Either time or space may be the separating factor.

Where broad-spectrum pesticides are used, integrating biological and chemical control may depend on timing of pesticide applications in comparison with natural enemy occurrence, rates of application and persistence of the pesticides applied. Clausen (1956) noted that with the introduction of DDT and other persistent insecticides, the impacts of pesticide applications on natural enemies were dramatically greater than they had been when less persistent botanical insecticides were the primary or only chemicals used. Although pesticides currently in use are far less persistent than DDT and other organochlorines that Clausen referenced, even moderate persistence means that if time is to provide ecological selectivity, the broad-spectrum insecticide usually must be applied *after* natural enemies have exerted the majority of their potential benefit in a given crop season. For example in the midwestern USA, conserving natural enemies of diamondback moth while still using insecticides for lepidopteran control is a goal in production of cabbage, broccoli and related crops. To accomplish this goal, growers are urged to use microbial insecticides containing *Bt* when infestations of lepidopteran larvae exceed thresholds prior to heading (Weinzierl & Cloyd, 2007). This approach provides adequate control before heading and allows natural enemies of diamondback moth, cabbage looper and imported cabbageworm (*Pieris rapae*) to survive and attack larvae not killed by *Bt* applications. Pyrethroids are used as needed after heading, especially for control of cabbage looper. Pyrethroids are the most effective insecticides against this pest, especially if middle or late instars are present, and high levels of control are needed at this time to prevent damage to and contamination of heads. Withholding pyrethroid use earlier in crop development allows biotic control agents to suppress diamondback moth, resulting in temporal integration of biotic and chemical control.

The scale of spatial separation required to provide ecological selectivity for broad-spectrum pesticides may range from millimeters to decimeters. The following examples all represent selective applications of pesticides in ways that allow survival of natural enemies.

- Systemic seed treatments and in-furrow applications of systemic insecticides kill insects in direct contact with the treated seed or soil and those that feed on young plants, but natural enemies outside the treated furrow or on the surface of the treated crop are not directly affected (though consumption of prey or parasitism of hosts feeding on plant tissue may result in indirect poisoning). Banded applications of insecticides (applied to only a narrow strip of soil surrounding the row) used to protect against root-feeding insects such as *Diabrotica* spp. allow survival of carabids and other soil-inhabiting predators outside the treated area.
- Spot treatment of two-spotted spider mite (*Tetranychus urticae*) infestations in soybeans – usually treating field edges where mites build up first as they move into fields from surrounding vegetation – instead of treating entire fields allows survival of lady beetles and other predators and parasites that provide biotic control of soybean aphid.
- Alternate-row applications of insecticides in apples and other tree fruit and nut crops (where sprays are applied from only one side of each row of trees, resulting in less than complete coverage) allow survival of predators of European red mites on the leeward side of trees, and redistribution of these predators within trees provides some level of continuous survival and predation. Hull & Beers (1985) described additional practices for achieving ecological selectivity in orchard systems.
- Where pteromalid parasites (Hymenoptera: Pteromalidae) are used in biological control of housefly (*Musca domestica*) or stablefly (*Stomoxys calcitrans*), insecticides can be used as residual sprays to walls of livestock buildings to kill adult flies without interfering significantly with parasite survival (Jones & Weinzierl, 1997). Although adult flies are poisoned when they rest on treated surfaces, these parasites attack fly pupae in manure, and their host-finding flights do not bring them into contact with spray residues. Insecticides applied as manure sprays or as aerosols would cause more mortality to these parasites, but they are separated in space from the effects of surface residual sprays.

14.2.3 Refuges

Finally, a unique form of ecological selectivity resulting from spatial separation of natural enemies from pesticides occurs when refuges are left untreated to allow pest survival. Leaving refuges of untreated crops is recommended almost exclusively for the purpose of managing pesticide resistance, and the only widespread use of the practice to date has been in managing (slowing the development of) pest resistance to *Bt* toxins in transgenic plants (Ostlie *et al.*, 1997). Transgenic crops are regulated in the USA as pesticides and the planting of non-*Bt* refuges is required as a condition of their EPA registration.

In general, neither the lepidopteran-specific *Bt* toxins (e.g. Cry1Ab and Cry1F in corn and Cry1Ac in cotton) nor the corn-produced *Bt* toxins that kill corn rootworm larvae (Cry3Bb1, Cry34Ab1 and Cry35Ab1) have been found to be directly toxic to predators or parasites of key pests in controlled laboratory studies or limited-duration field surveys (Shelton *et al.*, 2002; Al-Deeb & Wilde, 2003; Al-Deeb *et al.*, 2003). However, Andow & Hilbeck (2004), Lövei & Arpaia (2005) and Naranjo *et al.* (2005) argue that the methods used in many assessments of impacts of transgenic crops on natural enemies have been inadequate and that long-term ecological impacts are possible, if not likely. There is no argument that transgenic crops offer near-total control of several target pests and that hosts and prey for specialist natural enemies of those pests are absent in "treated" fields (fields planted exclusively to a *Bt* crop). Refuges therefore serve not only in resistance management as a necessary habitat for pests that are susceptible to *Bt* toxins, but also as habitat where biotic and biological control agents might survive. Integrating chemical control in the form of transgenic crops with biotic and biological control is dependent on the use of refuges.

Considered in total, physiological selectivity and ecological selectivity can be exploited to varying degrees in different crops or settings to gain the benefits of pesticides and still allow survival of biotic control agents and maintenance of a meaningful biological control program. Given the opportunity to use selective chemicals or selective approaches to their application, one might ask why non-selective chemicals and application practices are used so widely despite their negative impacts on biological control. Although part of the answer lies in the need to better educate growers and other pesticide users about assessing the need for control, the benefits of biological control and the negative impacts of pesticide use, other factors also lead to broad-spectrum pesticide use. Where more than one pest organism may be targeted by a single control method, such as the application of a broad-spectrum insecticide to control codling moth, apple maggot and white apple leafhopper (*Typhlocyba pomaria*) in apples, applying three separate selective insecticides, all of which might be less toxic to many natural enemies, may represent three times the dollar cost. If they are present at levels that warrant control, killing several pests with a single insecticide application appeals even to conscientious growers.

14.3 Modifying biological control agents to survive pesticide applications

Many insecticides are more toxic to parasites and predators than they are to pests attacked by these natural enemies (Croft, 1990). However, resistance to pesticides can develop in natural enemy populations just as it does in pests. Pielou & Glasser (1952) reported on selection of DDT resistance over 50 years ago in the parasitic wasp *Macrocentrus ancylivorus*. Resistance can evolve in natural enemies in field conditions, or laboratory selection may be used to produce resistant populations for release in the field. Increased resistance in natural enemies renders broad-spectrum insecticides at least somewhat selective and therefore allows their use in conjunction with biotic or biological control.

Field selection resulting from exposure to repeated insecticide or miticide applications has led to evolution of resistance to one or more pesticides in several predaceous or parasitic arthropods, including *Metaseiulus occidentalis*, *Typhlodromus pyri*, *Neoseiulus* (= *Amblyseius*) *fallacis*, *Coleomegilla maculata*, *Bracon mellitor* and *M. ancylivorus* (Croft & Brown, 1975). Croft & Meyer (1973) observed that azinphosmethyl resistance in *N. fallacis* increased 300-fold in a Michigan apple orchard treated five to seven times per year over a 4-year period. Field selection of resistance in natural enemies is a major cause of differing observations of physiological selectivity for specific pest and pesticide combinations over time or among locations, as represented in the data summarized by Croft (1990).

Laboratory selection programs for increasing pesticide resistance in natural enemies have been used most extensively to develop resistant strains of predaceous mites that are important in the suppression of tetranychid mites in deciduous tree fruits and nuts (Hoy, 1985). Selection programs comprised of both field and laboratory components have produced strains with resistance to multiple pesticides (Hoy, 1989; Whitten & Hoy, 1999). Although initial efforts centered on phytoseiid mites, laboratory selection has also been used to develop resistant strains of insects such as *C. carnea* (Grafton-Cardwell & Hoy, 1986), *Aphytis melinus* (Rosenheim & Hoy, 1988) and *Trioxis pallidus* (Hoy & Cave, 1988).

In addition to standard selection practices for development of pesticide-resistant natural enemies, molecular genetic techniques now make it possible to transfer resistance genes among species. Li & Hoy (1996) and Hoy (2000) used maternal microinjection to transform the predaceous mite *M. occidentalis*. Currently, rules for the release and use of transgenic natural enemies have not been established, and none are in use (Hoy, 2006).

14.4 Combining biological controls and pesticides for increased effectiveness

Many natural enemies of arthropods are other arthropods – predators and parasites – and many of those are killed when broad-spectrum insecticides are applied (with exceptions as noted above in the discussion of selectivity). However, many insect pathogens – viruses, bacteria, fungi, microsporidia and nematodes – are not harmed by the use of most insecticides or acaricides. Where pathogens are used in classical biological control or in augmentation as microbial insecticides, conventional insecticides may be applied as well if needed, and in many instances there is no direct mortality to the pathogens. For example, codling moth granulosis virus may be used in orchards where conventional insecticides also are applied, and it infects and kills codling moth larvae not controlled by the conventional treatments (although virus buildup may be limited by low host densities). Insect-pathogenic fungi (e.g. *Beauveria bassiana*) or nematodes (e.g. *Steinernema* spp. and *Heterorhabditis* spp.) may be used along with many soil insecticides to increase the range of pest species controlled. In some instances, insecticides and pathogens may act synergistically. For example, Quintella & McCoy (1998) found that imidacloprid and two entomopathogenic fungi acted synergistically against the root weevil *Diaprepes abbreviatus*. Barbara & Buss (2005) found that using insecticides and parasitic nematodes together gave increased control of mole crickets (*Scapteriscus* spp.). Brinkman & Gardner (2001), Furlong & Groden (2001) and Ericsson *et al.* (2007) reported similar findings of insecticide and pathogen synergism. Combining pathogens and chemical pesticides does not always result in additive or synergistic effects. For example, James & Elzen (2001) reported antagonism between *B. bassiana* and imidacloprid when combined for control of silverleaf whitefly (*Bemisia argentifolii*).

14.5 Conclusions

The integration of biotic and biological controls with chemical pesticides has presented challenges for several decades, at least since the rapid increase in use of synthetic insecticides that began in the 1950s. Methods to accomplish this integration, as outlined in this chapter – reducing pesticide use, increasing the selectivity of pesticide use, modifying natural enemies to survive pesticide exposures, and combining compatible pesticides and natural enemies – are yet to be employed as fully as possible, or even as fully as practical. Hoy (1989) suggested that a reordering of priorities was needed if biological control is to be integrated into agricultural IPM systems, and that chemical control should not be viewed as the central component around which other tactics must fit. Selective modes of action of some new pesticides provide opportunities that were not available in the past, and transgenic plants produce toxins that are selective as well. Nonetheless, a balanced, integrated approach to arthropod pest management seems little closer to reality than it was nearly two decades ago when Hoy (1989) suggested the need for new priorities. Transgenic crops seem to have become the new application method for insecticides, but integrating their use with natural enemies seems far beyond the vision of the pest control mindset that again seems to govern current agriculture. Challenges still loom large for true integration in modern-day IPM.

References

Al-Deeb, M. A. & Wilde, G. E. (2003). Effect of *Bt* corn expressing the Cry3Bb1 toxin for corn rootworm control on aboveground nontarget arthropods. *Environmental Entomology*, **32**, 1164–1170.

Al-Deeb, M. A., Wilde, G. E., Blair, J. M. & Todd, T. C. (2003). Effect of *Bt* corn for corn rootworm control on nontarget soil microarthropods and nematodes. *Environmental Entomology*, **32**, 859–865.

Andow, D. A. & Hilbeck, A. (2004). Science-based risk assessment for nontarget effects of transgenic crops. *BioScience*, **54**, 637–649.

Barbara, K. A. & Buss, E. A. (2005). Integration of insect parasitic nematodes (Rhabditida: Steinernematidae) with insecticides for control of pest mole crickets (Orthoptera: Gryllotalpidae: *Scapteriscus* spp.). *Journal of Economic Entomology*, **98**, 689–693.

Bernard, M. B., Horne, P. A. & Hoffman, A. A. (2004). Developing an ecotoxicological testing standard for predatory mites in Australia: acute and sublethal effects of fungicides on *Euseius victoriensis* and *Galendromus occidentalis* (Acarina: Phytoseiidae). *Journal of Economic Entomology*, **97**, 891–899.

Boyd, M. L. & Boethel, D. J. (1998). Susceptibility of predaceous hemipteran species to selected insecticides used on soybeans in Louisiana. *Journal of Economic Entomology*, **91**, 401–409.

Brinkman, M. A. & Gardner, W. A. (2001). Use of diatomaceous earth and entomopathogen combinations against the red imported fire ant (Hymenoptera: Formicidae). *Florida Entomologist*, **84**, 740–741.

Brown, G. C. (1997). Simple models of natural enemy action and economic thresholds. *American Entomologist*, **43**, 117–124.

Brust, G. E., Stinner, B. R. & McCartney, D. A. (1985). Tillage and soil insecticide effects on predator-black cutworm (Lepidoptera: Noctuidae) interactions in corn agroecosystems. *Journal of Economic Entomology*, **78**, 1389–1392.

Clark, J. M. & Yamaguchi, I. (2002). Scope and status of pesticide resistance. In *Agrochemical Resistance: Extent, Mechanism, and Detection*, eds. J. M. Clark & I. Yamaguchi, pp. 1–22. Washington, DC: American Chemical Society.

Clausen, C. P. (1956). *Biological Control of Insect Pests in the Continental* United States, Technical Bulletin No. 1139. Washington, DC: US Department of Agriculture.

Cloyd, R. A. & Dickinson, A. (2006). Effect of insecticides on mealybug destroyer (Coleoptera: Coccinellidae) and parasitoid *Leptomastix dactylopii* (Hymenoptera: Encyrtidae), natural enemies of citrus mealybug (Homoptera: Pseudococcidae). *Journal of Economic Entomology*, **99**, 1596–1604.

Conway, H. E., Steinkraus, D. C., Ruberson, J. R. & Kring, T. J. (2006). Experimental treatment threshold for cotton aphid (Homoptera: Aphididae) using natural enemies in cotton. *Journal of Entomological Science*, **41**, 361–373.

Croft, B. A. (1990). *Arthropod Biological Control Agents and Pesticides*. New York: John Wiley.

Croft, B. A. & Brown, A. W. A. (1975). Responses of arthropod natural enemies to insecticides. *Annual Review of Entomology*, **20**, 285–334.

Croft, B. A. & Meyer, R. H. (1973). Carbamate and organophosphorus resistance patterns in populations of *Amblyseius fallacis*. *Environmental Entomology*, **2**, 691–695.

Curtis, J. E. & Horne, P. A. (1995). Effect of chlorpyrifos and cypermethrin applications on non-target invertebrates in a conservation-tillage crop. *Australian Journal of Entomology*, **34**, 229–231.

Davis, D. W. (1974). Parasite–prey ratios among alfalfa weevil larvae in northern Utah. *Environmental Entomology*, **3**, 1031–1032.

Desneux, N., Fauvergue, X., Dechaume-Moncharmont, F. *et al.* (2005). *Diaterella rapae* limits *Myzus persicae* populations after application of deltamethrin in oilseed rape. *Journal of Economic Entomology*, **98**, 9–17.

Egel, D., Lam, F., Foster, R. *et al.* (2007). *Midwest Vegetable Production Guide for Commercial Growers, 2007*. West Lafayette, IN: Purdue University.

Ericsson, J. D., Kabaluk, J. T., Goettel, M. S. & Myers, J. H. (2007). Spinosad interacts synergistically with the insect pathogen *Metarhizium anisopliae* against the exotic wireworms *Agriotes lineatus* and *Agriotes obscurus* (Coleoptera: Elateridae). *Journal of Economic Entomology*, **100**, 31–38.

Fernandez, O. A., Wright, R. J. & Mayo, Z. B. (1998). Parasitism of greenbugs (Homoptera: Aphididae) by *Lysiphlebus testaceipes* (Hymenoptera: Braconidae) in grain sorghum: implications for augmentative biological control. *Journal of Economic Entomology*, **91**, 1315–1319.

Furlong, M. J. & Groden, E. (2001). Evaluation of synergistic interactions between Colorado potato beetle (Coleoptera: Chrysomelidae) pathogen *Beauveria bassiana* and the insecticides imidacloprid and cyromazine. *Journal of Economic Entomology*, **94**, 344–356.

Furlong, M. J., Zu-Hua, S., Yin-Quan, L. *et al.* (2004). Experimental analysis of the influence of pest management practices on the efficacy of an endemic arthropod natural enemy complex of the diamondback moth. *Journal of Economic Entomology*, **97**, 1814–1827.

Giles, K. L., Jones, D. B., Royer, T. A., Elliott, N. C. & Kindler, S. D. (2003). Development of a sampling plan in winter wheat that estimates cereal aphid parasitism levels and predicts population suppression. *Journal of Economic Entomology*, **96**, 975–982.

Gonzalez, D. & Wilson, L. T. (1982). A food-web approach to economic thresholds: a sequence of pests/predaceous arthropods on California cotton. *Entomophaga*, **27**, 31–43.

Grafton-Cardwell, E. E. & Hoy, M. A. (1986). Genetic improvement of common green lacewing, *Chrysoperla carnea* (Stephens) (Neuroptera: Chrysopidae): selection for carbaryl resistance. *Environmental Entomology*, **15**, 1130–1136.

Hagen, K. S. & McMurtry, J. A. (1979). Natural enemies and predator–prey ratios. In *Biological Control and Insect Management*, Bulletin No. 1191, eds. D. W. Davis, S. C. Hoyt, J. A. McMurtry & M. T. Aliniazee, pp. 28–40. Riverside, CA: University of California, Agricultural Experiment Station.

Hagley, E. A. C. & Laing, J. E. (1989). Effect of pesticides on parasitism of artificially distributed eggs of the codling moth, *Cydia pomonella* (Lepidoptera: Tortricidae), by *Trichogramma* spp. (Hymenoptera: Trichogrammatidae). *Proceedings of the Entomological Society of Ontario*, **120**, 25–33.

Hassan, S. A. (1969). Observations on the effect of insecticides on coleopterous predators of *Erioischia brassicae* (Diptera: Anthomyiidae). *Entomologia Experimentalis et Applicata*, **12**, 157–168.

Hassan, S. A. & Oomen, P. A. (1985). Testing the side effects of pesticides on beneficial organisms by OILP working party. In *Biological Pest Control: The Glasshouse Experience*, eds. N. W. Hussey & N. Scopes, pp. 145–152. Ithaca, NY: Cornell University Press.

Higley, L. G. & Wintersteen, W. K. (1992). A novel approach to environmental risk assessment of pesticides as a basis for incorporating environmental costs into economic injury levels. *American Entomologist*, **38** 34–39.

Hoffman, M. P., Wilson, L. T., Zalom, F. G. & Hilton, R. J (1990). Parasitism of *Heliothis zea* (Lepidoptera: Noctuidae) eggs: effect on pest management decision rules for processing tomatoes in the Sacramento Valley of California. *Environmental Entomology*, **19**, 753–763.

Hoffman, M. P., Wilson, L. T., Zalom, F. G. & Hilton, R. J. (1991). Dynamic sequential sampling plan for *Helicoverpa zea* (Lepidoptera: Noctuidae) eggs in processing tomatoes: parasitism and temporal patterns. *Environmental Entomology*, **20**, 1005–1012.

Hollingsworth, R. G., Steinkraus, D. C. & McNew, R. W. (1995). Sampling to predict fungal epizootics in cotton aphids (Homoptera: Aphididae). *Environmental Entomology*, **24**, 1414–1421.

Horn, D. J. (1971). The relationship between a parasite, *Tetrastichus incertus* (Hymenoptera: Eulophidae), and its host the alfalfa weevil, *Hypera postica* (Coleoptera: Curculionidae), in New York. *Canadian Entomologist*, **103**, 83–94.

Hoy, M. A. (1985). Recent advances in genetics and genetic improvement of the Phytoseiidae. *Annual Review of Entomology*, **30**, 345–370.

Hoy, M. A. (1989). Integrating biological control into agricultural IPM systems: reordering priorities. In *Proceedings, National Integrated Pest Management Symposium/Workshop*, ed. E. H. Glass, pp. 41–57. Geneva, NY: New York State Agricultural Experiment Station, Cornell University.

Hoy, M. A. (1995). Multitactic resistance management: an approach that is long overdue? *Florida Entomologist*, **78**, 443–451.

Hoy, M. A. (2000). Transgenic arthropods for pest management programs: risks and realities. *Experimental and Applied Acarology*, **24**, 463–495.

Hoy, M. A. (2006). Evaluating potential risks of transgenic arthropods for pest management programmes. In *Status and Risk Assessment of the Use of Transgenic Arthropods in Plant Protection*, Proceedings of the FAO/IAEA Programme of Nuclear Techniques in Food and Agriculture, International Atomic Energy Association TECDOC-1483, pp. 121–144. Rome, Italy: Food and Agriculture Organization of the United Nations.

Hoy, M. A. & Cave, F. E. (1988). Guthion-resistant strain of the walnut aphid parasite. *California Agriculture*, **42**(4), 4–6.

Hull, L. A. & Beers, E. H. (1985). Ecological selectivity: modifying chemical control practices to preserve natural enemies. In *Biological Control in Agricultural IPM Systems*, eds. M. A. Hoy & D. C. Herzog, pp. 103–122. Orlando, FL: Academic Press.

Insecticide Resistance Action Committee (2006). *IRAC Mode of Action Classification Version 5.2.* Available at www.irac-nline.org/documents/IRAC_MoA_Classification_v5.2.pdf.

James, R. R. & Elzen, G. W. (2001). Antagonism between *Beauveria bassiana* and imidacloprid when combined for *Bemisia argentifolii* (Homoptera: Aleyrodidae) control. *Journal of Economic Entomology*, **94**, 357–361.

Jones, C. J. & Weinzierl, R. A. (1997). Geographical and temporal variation in pteromalid (Hymenoptera: Pteromalidae) parasitism of stable fly and house fly (Diptera: Muscidae) pupae collected from cattle feedlots in Illinois. *Environmental Entomology*, **26**, 421–432.

Kovach, J., Petzold, C., Degni, J. & Tette, J. (1992). *A Method to Measure the Environmental Impact of Pesticides*, New York Food and Life Sciences Bulletin No. 139. Geneva, NY: Cornell University New York State Agricultural Experiment Station Available at http://ecommons.library.cornell.edu/handle/1813/5203.

Landis, D. A., Wrattan, S. D. & Gurr, G. M. (2000). Habitat management to conserve natural enemies of arthropod pests in agriculture. *Annual Review of Entomology*, **45**, 175–201.

Lövei, G. L. & Arpaia, S. (2005). The impact of transgenic plants on natural enemies: a critical review of laboratory studies. *Entomologica Experimentalis et Applicata*, **114**, 1–14.

Li, J. & Hoy, M. A. (1996). Adaptability and efficacy of transgenic and wild-type *Metaseiulus occidentalis* (Acari: Phytoseiidae) compared as part of a risk assessment. *Experimental and Applied Acarology*, **20**, 563–573.

Maagd, R. A., Bravo, A. & Crickmore, N. (2001). How *Bacillus thuringiensis* has evolved specific toxins to colonize the insect world. *Trends in Genetics*, **17**, 193–199.

Medina, P., Smagghe, G., Budia, F., Tirry, L. & Viñuela, E. (2003). Toxicity and absorption of azadirachtin, diflubenzuron, pyriproxyfen, and tebufenozide after topical application in predatory larvae of *Chrysoperla carnea* (Neuroptera: Chrysopidae). *Environmental Entomology*, **32**, 196–203.

Musser, F. E., Nyrop, J. P. & Shelton, A. M. (2006). Integrating biological and chemical control in decision making: European corn borer (Lepidoptera: Crambidae) control in sweet corn as an example. *Journal of Economic Entomology*, **99**, 1538–1549.

Naranjo, S. E. & Hagler, J. R. (1998). Characterizing and estimating the impact of heteropteran predation. In *Predatory Heteroptera: Their Ecology and Use in Biological Control*, Thomas Say Symposium Proceedings, eds. M. Coli & J. Ruberson, pp. 170–197. Lanham, MD: Entomological Society of America.

Naranjo, S. E., Head, G. P. & Dively, G. P. (2005). Field studies assessing arthropod non-target effects in *Bt* transgenic crops: Introduction. *Environmental Entomology*, **34**, 1178–1180.

National Research Council (1993). *Pesticides in the Diets of Infants and Children*. Washington, DC: National Academy of Sciences, National Research Council, Board on Agriculture and Board on Environmental Studies and Toxicology, Committee on Pesticides in the Diets of Infants and Children.

National Research Council (2000). *The Future Role of Pesticides in US Agriculture*. Washington, DC: National Academy of Sciences, National Research Council, Board on Agriculture and Natural Resources and Board on Environmental Studies and Toxicology, Commission on Life Sciences, Committee on the Future Role of Pesticides in US Agriculture.

Newsom, L. D., Smith, R. F. & Whitcomb, W. H. (1976). Selective pesticides and selective use of pesticides. In *Theory and Practice of Biological Control*, eds. C. Huffaker & P. Messenger, pp. 565–591. New York: Academic Press.

Niemczyk, E., Olszak, R., Miszczayk, M. & Bakowski, G. (1974). *Effectiveness of Some Predaceous Insects in the Control of Phytophagous Mites and Aphids on Apple Trees*, Annual Report. Skierniewice, Poland: Research Institute of Pomology.

Nyrop, J. P. & van der Werf, W. (1994). Sampling to predict and monitor biological control. In *Handbook of Sampling Methods for Arthropod Pests in Agriculture*, eds. L. Pedigo & D. G. Buntin. Boca Raton, FL: CRC Press.

Ostlie, K. R., Hutchinson, W. D. & Hellmich, R. L. (eds.) (1997). *Bt Corn and European Corn Borer*, NCR Publication No. 602. St Paul, MN: University of Minnesota.

Pielou, D. P. & Glasser, R. F. (1952). Selection for DDT resistance in a beneficial insect parasite. *Science*, **115**, 117–118.

Quintela, E. D. & McCoy, C. W. (1998). Synergistic effect of imidacloprid and two entomopathogenic fungi on the behavior and survival of larvae of *Diaprepes abbreviatus* (Coleoptera: Curculionidae) in soil. *Journal of Economic Entomology*, **91**, 110–122.

Ragsdale, D. W., McCornack, B. P., Venette, R. C. *et al.* (2007). Economic threshold for soybean aphid (Hemiptera: Aphididae). *Journal of Economic Entomology*, **100**, 1258–1267.

Ridgway, R. L. & Jones, S. L. (1969). Inundative release of *Chrysopa carnea* for control of *Heliothis* in cotton. *Journal of Economic Entomology*, **62**, 177–180.

Ripper, W. E., Greenslade, R. M. & Hartley, G. S. (1951). Selective insecticides and biological control. *Journal of Economic Entomology*, **44**, 448–459.

Rosenheim, J. A. & Hoy, M. A. (1988). Genetic improvement of a parasitoid biological control agent: artificial selection for insecticide resistance in *Aphytis melinus* (Hymenoptera: Aphelinidae). *Journal of Economic Entomology*, **81**, 1161–1173.

Rothwangi, K. B., Cloyd, R. A. & Wiedenmann, R. N. (2004). Effects of growth regulators on citrus mealybug parasitoid *Leptomastix dactylopii* (Hymenoptera: Encyrtidae). *Journal of Economic Entomology*, **97**, 1239–1244.

Rytter, J. L. & Travis, J. W. (2006). *Pennsylvania Tree Fruit Production Guide, 2006–07* edn. University Park, PA: Pennsylvania State University. Available at http://tfpg.cas.psu.edu/.

Shelton, A. M., Zhao, J. Z. & Roush, R. T. (2002). Economic, ecological, food safety, and social consequences of the deployment of *Bt* transgenic plants. *Annual Review of Entomology*, **47**, 845–881.

Smith, R. F. & Hagen, K. S. (1959). Impact of commercial insecticide treatments. *Hilgardia*, **29**, 131–154.

Stark, J. D. & Banks, J. E. (2003). Population-level effects of pesticides and other toxicants on arthropods. *Annual Review of Entomology*, **48**, 505–519.

Stark, J. D., Vargas, R. & Miller, N. (2004). Toxicity of spinosad in protein bait to three economically important tephritid fruit fly species (Diptera: Tephritidae) and their parasitoids (Hymenoptera: Braconidae). *Journal of Economic Entomology*, **97**, 911–915.

Steffey, K. L. & Gray, M. E. (2007). Insect pest management for field and forage crops. In *2007 Illinois Agricultural Pest Management Handbook*, ed. S. Bissonnette, pp. 1–19. Urbana, IL: University of Illinois Extension.

Stern, V. M. & van den Bosch, R. (1959). Field experiments on the effects of insecticides. *Hilgardia*, **29**, 103–130.

Stern, V. M., Smith, R. F., Van Den Bosch, R. & Hagen, K. S. (1959). The integrated control concept. *Hilgardia*, **29**, 81–101.

Tamaki, G. & Long, G. E. (1978). Predator complex of the green peach aphid on sugar beets: expansion of the predator power and efficacy model. *Environmental Entomology*, **7**, 835–842.

Tamaki, G., McGuire, J. V. & Turner, J. E. (1974). Predator power and efficacy: a model to evaluate their impact. *Environmental Entomology*, **3**, 625–630.

Teetes, G. L. (1994). Adjusting crop management recommendations for insect-resistant crop varieties. *Journal of Agricultural Entomology*, **11**, 191–200.

Theiling, K. M. & Croft, B. A. (1988). Pesticide side-effects on arthropod natural enemies: a database summary. *Agriculture, Ecosystems and Environment*, **21**, 191–218.

Tillman, P. G. & Mullinix, B. G. Jr. (2004). Comparison of susceptibility of pest *Euschistus servus* and predator *Podisus maculiventris* (Heteroptera: Pentatomidae) to selected insecticides. *Journal of Economic Entomology*, **97**, 800–806.

Turpin, F. T. & A. C. York. (1981). Insect management and the pesticide syndrome. *Environmental Entomology*, **10**, 567–572.

Weinzierl, R. & Cloyd, C. (2007). Insect pest management for commercial vegetable crops. In *2007 Illinois Agricultural Pest Management Handbook*, ed. S. Bissonnette, pp. 199–235. Urbana, IL: University. of Illinois Extension.

Weinzierl, R. & Henn, T. (1991). *Alternatives in Insect Management: Biological and Biorational Approaches*, North Central Regional Publication No. 401. Urbana, IL: University of Illinois Extension. Available at www.ag.uiuc.edu/~vista/abstracts/aaltinsec.html.

Weires, R. W. (1980). *Predatory Phytoseiid Mite*, Amblyseius fallacis (Garman), Tree Fruit IPM Insect Identification Sheet No. 7. Geneva, NY: Cornell University New York State Agricultural Experiment Station.

Whitten, M. J. & Hoy, M. A. (1999). Genetic improvement and other genetic considerations for improving the efficacy and success rate of biological control. In *Principles and Applications of Biological Control*, eds. T. S. Bellows Jr. & T. W. Fisher, pp. 271–296. San Diego, CA: Academic Press.

Wilkinson, T. K., Landis, D. A. & Gut, L. J. (2004). Parasitism of obliquebanded leafroller (Lepidoptera: Tortricidae) in commercially managed Michigan apple orchards. *Journal of Economic Entomology*, **97**, 1524–1530.

Chapter 15

Pesticide resistance management

Casey W. Hoy

Good pesticide resistance management is just good IPM. Managing pesticide resistance means using the chemistry with enough restraint, and enough understanding of its role in an agroecosystem, to sustain that use. One interpretation is to maximize the number of applications that result in close to 100% control. To achieve this goal means optimizing the trade-off between the selection pressure exerted by use of the chemistry and the control benefits in terms of rapid reduction of pest population density. Too much population reduction too quickly leads quickly to resistance, whereas too much restriction of use denies economically justified control. The hope is that the chosen pattern of use provides many more effective applications than any alternative pattern, ideally an infinite number of effective applications. Models of resistance development, however, typically predict a finite number of effective applications, followed by a rapid increase in resistance gene frequency and rather sudden control failure. Experience in agroecosystems has generally been consistent with model predictions, for example resistance within a few generations to a long list of insecticides in Colorado potato beetle (*Leptinotarsa decemlineata*) on Long Island (Forgash, 1985; Mota-Sanchez *et al.*, 2006).

Agroecosystems include both people and the land, and their functioning needs to be considered in light of this combination and at large spatial extents and long time-frames. Pesticide resistance management programs that have taken a comprehensive and cooperative approach to maximizing the number of effective applications have been in place for about 20 years. A comprehensive strategic and theoretical analysis appeared in a National Research Council study (Glass, 1986) at a time when the first comprehensive and nationwide resistance management program was beginning in Australia (Forrester & Cahill, 1987), although pesticide resistance has been recognized and studied for a much longer period of time (e.g. Georghiou, 1972). So far, what has been sustained is an industrial cycle of innovation, therapeutic product use and product obsolescence because of resistance, all based on users who are heavily dependent on pesticides as economically valuable, therapeutic solutions to pest problems. One might ask, however, if sustaining reliance on a singular albeit profitable therapeutic solution to high pest population density is the goal of IPM (see Chapter 1).

An alternative interpretation of pesticide resistance management would be to prevent it completely. Meeting this objective could be considerably more difficult than maximizing the number of effective applications. It would require finding trade-offs that lead to evolutionary dead ends

Integrated Pest Management, ed. Edward B. Radcliffe, William D. Hutchison and Rafael E. Cancelado. Published by Cambridge University Press.

for resistant insects, as sometimes documented in natural systems, rather than the ultimate success that resistant insects typically have achieved in pesticide treated environments. Alternatively, it could push insect adaptation in other directions, such as adaptation to non-crop hosts. The impact of pesticide chemistry and use pattern on the full range of life history traits, trophic relationships and other control tactics would need to be explored and understood in detail. We currently tend to understand little of this biology and ecology compared with our knowledge of a use pattern that kills pests.

A pesticide use pattern that truly could be sustained indefinitely may not be geared to providing quick and complete control of pests, but rather may be geared towards shifting their population dynamics and host plant relationships in much more subtle ways, adding to existing ecosystem services rather than replacing them. Such shifts are theoretically the role of a single tactic in IPM, to be combined with many other similar tactics to prevent economic injury. But in practice, this level of integration of chemical controls is rarely achieved and is not considered to be pesticide resistance management. The economic benefits of such a use pattern may be too diffuse to support the level of industrial activity needed to produce and distribute the pesticide in the first place. Nevertheless, use patterns that avoid resistance entirely would be more consistent with the theory of IPM, as discussed in Chapter 1, and should be the ultimate goal of pesticide resistance management as well whether or not it is currently achievable with available chemistries, technologies for their use and the agroecosystems in which this use takes place.

This chapter will outline first the biology and ecology of pesticide resistance and second the extent to which resistance management has improved the resilience of agroecosystems. Insecticide resistance in crop production will be the focus, although many of the principles could apply to other commodities and public health as well as weeds and pathogens. Although the institutional responses to resistance have become well organized and entrenched, opportunities for improving the resilience of agroecosystems may still be available.

15.1 Genotypes and phenotypes in individuals, populations and ecosystems

15.1.1 Evolutionary underpinnings of pesticide resistance

Herbivores have been dealing with chemical defenses in host plants, a basis of host plant resistance, for a long time. Although plant defenses are ubiquitous, a number of trade-offs act to keep toxin concentrations very heterogeneous in natural systems at spatial scales from within leaves to across continents (Berenbaum, 1995; Hoy *et al.*, 1998). Furthermore, variation in environmental conditions can lead to variation in chemical defenses in plants (Herms & Mattson, 1992). Physiological adaptation to plant toxins does occur in natural systems, particularly in specialist herbivores for which the ultimate physiological adaptation is complete immunity and storage or sequestering of the toxic compounds from the host plant to provide chemical defense to the herbivore itself. In these cases, however, the chemical defense can still play a role in protection from other herbivores and microbes. Expression of naturally occurring plant chemical defenses is shaped by multiple herbivore guilds, leading to complex and varying chemical profiles and responses (Adler & Kittelson, 2004). In most cases, however, stability in plant defense is afforded by the heterogeneous distributions resulting from trade-offs, including trade-offs between resistance to multiple threats such as pathogens and herbivores (Felton & Korth, 2000). Theories behind these trade-offs generally require a cost to the plant for production of the chemical defense. Initial attempts to quantify these costs suggested that they are small or non-existent (Bergelson & Purrington, 1996). More recent meta-analysis, however, indicates that resistance costs to plants are much more common than previously estimated and accrue from both the direct cost of resource allocation within plants and ecological interactions between plants and their environment (Koricheva, 2002; Strauss *et al.*, 2002).

Two kinds of responses have been documented in insect herbivores to heterogeneous distributions of naturally occurring toxins in plants.

Physiological adaptation has been documented, but often constrained to an apparently stable pattern of defense and adaptation between the plants and insects (e.g. Berenbaum *et al.*, 1986; Berenbaum & Zangerl, 1998; Nitao, 1995). Far more common, however, are behavioral adaptations to plant toxins that shift feeding away from higher concentration and toward lower concentration (Hoy *et al.*, 1998). The concentrations and spatial scales of pesticide treated and untreated areas typically are not intended to allow behavioral responses to result in survival of the pests, despite the fact that avoidance of treated areas can often accomplish the objectives of IPM. The typical adaptive response of pest populations in managed systems is adaptation by physiological resistance (Mota-Sanchez *et al.*, 2002). Despite long-standing recognition (Berenbaum, 1995) of these differences between deployment of defensive chemistry in agriculture and in nature, and associated problems including resistance to pesticides, we persist in using pesticides very differently from the way chemical defenses have evolved in plants to a relatively stable arrangement.

15.1.2 Mechanisms and genes for pesticide resistance

Although many investigations of insecticide resistance have focused on finding the gene (singular) responsible, a wide range of genetic mechanisms responsible for reduced penetration or absorption, detoxification, sequestering or target site insensitivity have been identified for many if not most insecticide chemistries, and these have been reviewed previously (Georghiou, 1972; Denholm & Rowland, 1992; Brown, 1996). These mechanisms are not mutually exclusive and frequently are found in combination, often with uncharacterized modifiers or genes of "minor effect." The modifier genes may contribute only small amounts of variation in tolerance, but their potential correlation with other traits or contribution to continued adaptation, reduction of fitness effects, etc. should not be overlooked. Fitness effects in particular are typically presented as if static, but may remain under selection in a resistance management program that relies on them in repeated cycles of selection. Fitness effects also may not be extensive enough to provide sufficient suppression of resistance gene frequencies in a practical pest management program. For example, rotating between pyrethroids and other chemistries in one of the first comprehensive resistance management programs in Australian cotton resulted in a "sawtooth pattern" of resistance frequency but with a long-term incline toward greater and greater frequencies of pyrethroid resistance (Forrester & Bird, 1996).

15.1.3 Phenology, dispersal and other life history characteristics and the spatial and temporal scales of resistance

When resistance occurs it typically arises at a particular spatial and temporal scale, which depends on the life history characteristics of the pest, particularly the spatial extent of gene flow, and the pesticide use pattern. Cases of resistance can arise over areas from individual fields or farms to entire regions. Following initial occurrence it will typically expand geographically, but the rate of expansion will be dependent upon dispersal patterns of the pest at multiple scales. The temporal scales of resistance evolution are determined by such factors as the selection pressure, gene flow, stability of the resistance, presence of cross-resistance patterns and associated pesticide use, number of mechanisms and other correlated traits (Georghiou, 1986).

15.2 Navigating the everywhere/always – nowhere/never continuum

In general, resistance management means a reduction in use of the pesticide from what could theoretically be beneficial in the short term, i.e. if resistance was not of concern. A restriction in use can come in one or more of three ways: the concentration used, the area treated or the frequency of application. A simple set of rules to restrict pesticide use can be devised but the result of using them will be very complex due to variable spatial arrangements, time-frames and pest population dynamics. In practice, however, these are connected in ways that need to be considered

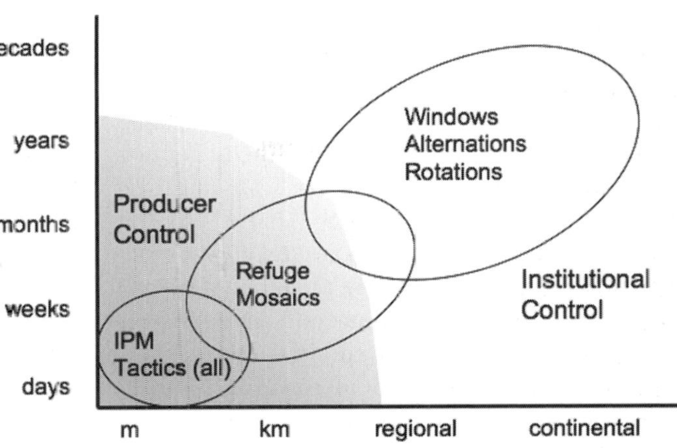

Fig. 15.1 Spatial and temporal scales of common resistance management tactics. Producers dominate the decisions and implementation within the shaded area whereas institutions dominate the management of tactics outside of the shaded area.

carefully and matched with the spatial and temporal scales at which resistance could develop (Fig. 15.1). For example, temporal alternations or rotations operated independently within small contiguous areas like individual farms amount to spatial mosaics over larger areas like farm landscapes. Refuge areas are likely to be moved from season to season, to avoid continued buildup of pest populations in a particular area. The sequence of treatments within a particular refuge area, therefore, scales up to an alternation or rotation over longer time-frames. The larger the area treated, the less likely will the timing be optimal for all pests species within the area, given variation in microclimate, oviposition and development. But the smaller the area treated, the less likely exposure will be confined to just the pests in the treated area as movement across the treated area boundaries is more likely to occur. Weathering of residues and dilution of concentrations by plant growth begin immediately after an application, so the concentration of pesticide is constantly in flux. Despite the need to keep resistance management programs simple, the situation in the field is likely to remain complex.

15.2.1 Prevention or cure?

Managing resistance can take the form of preventing it from occurring in the first place or decreasing the frequency of already resistant pests in a population. Theoretical studies suggest that prevention will only be possible if the preventive strategy is in place when resistance gene frequencies are very low (Tabashnik & Croft, 1982), but at this point the genetics of resistance are usually unknown or at best a guess. Prevention is generally preferred, but reliance on pesticides for economically beneficial, rapid and thorough population reduction results in use patterns that make resistance very likely, along with a hope that it will be unstable and manageable. Selectively reducing the frequency of resistant pests or increasing the frequency of susceptible pests requires some behavioral or biochemical means of selection.

Human pathogen resistance to antibiotics has provided both theoretical and empirical results that have been borrowed in developing approaches for managing pesticide resistance. In particular the question of low versus high dose, or the attention given to eradication of the pathogen with each use, has been applied to resistance management for agricultural pests. The focus on leaving no survivors leads to choices for how aggressively to attack the pest population, using single or multiple pesticides simultaneously. The more aggressive high dose and pesticide mixture approaches both maximize selection pressure and minimize initial numbers of resistant organisms, potentially increasing the effectiveness of a spatial refuge (Caprio, 1998).

The specific tactics available to manage resistance in pest populations have been reviewed comprehensively (Roush, 1989; Denholm & Rowland, 1992). In addition to variations on restricted use of a given pesticide in space or time, they include the use of pesticides, or mixtures of multiple pesticides and/or synergists, that are known to be effective for a particular resistant population of pests. The stated goal in these cases, however, has been to delay rather than prevent resistance. Testing these various strategies is difficult because they take place at the landscape scale and over relatively long periods of time. One of the few large-scale field studies undertaken took advantage of isolated pockets along the coast of Mexico to examine mosaics, rotations and single insecticides (i.e. the first stage of an alternation. There, all of these strategies led to elevated levels of resistance in mosquitos within one year (Hemingway *et al.*, 1997). The opportunities for prevention of resistance deserve some additional attention, despite the obvious change in the entire approach to pesticide use that would be required.

15.2.2 Preserving susceptible pests

Options for improving the survival of susceptible pests in a treated environment include reducing the concentration of pesticides, and moving from always and everywhere toward never and nowhere on the use continuum. In practice, strategies have focused on some variation on either windows in time or refuges in space, or both, where pesticides are not used at all. An additional and rarely considered possibility is the use of concentrations that are too low to kill even susceptible insects. Such a strategy relies on effects other than direct mortality, such as behavioral responses or altered development times, to protect against pest damage.

Spatial heterogeneity in pesticide concentrations typically has taken the form of high doses in treated areas, meant to kill insects that are heterozygous for resistance, and completely untreated refuge areas to support populations of susceptible insects (see Chapter 19 for specific examples with transgenic crops). The untreated areas can occur naturally, particularly for polyphagous pests with suitable abundance of alternate hosts. In the absence of a naturally occurring refuge, a design for a refuge arrangement can be challenging. The hope is that enough susceptible insects will disperse into treated areas to ensure production of heterozygotes from any resistant pests present, which will then be killed by the insecticide. The corollary is a hope that resistant pests do not move into the refuge and interbreed there, creating more homozygotes. Knowing or estimating gene flow relative to scale of treated and untreated areas can permit estimation of a size for treated areas that prevents the accumulation of resistant genotypes, based on a number of assumptions including a single gene for resistance and stable fitness effects (Lenormand & Raymond, 1998). In fact, the assumptions needed for an effective refuge are fairly restrictive (Carrière & Tabashnik, 2001). Lots of things can go wrong including insufficient gene flow or non-random mating (Caprio, 2001).

The common assumption regarding low pesticide concentrations is that they could permit survival of heterozygotes and make resistance effectively dominant (see Chapter 19). Certainly exposure to a low and uniform concentration of a pesticide, a concentration that selectively kills susceptible insects and permits survival of resistant ones, will have a very predictable result. Low concentrations, however, do not have to be spatially or temporally uniform, and adaptation in an insect need not be entirely physiological. Insects can respond either behaviorally or physiologically to low and spatially heterogeneous concentrations of pesticides (Hoy *et al.*, 1998), and behavioral responses tend to take place at concentrations well below those required to kill even susceptible insects. When a pesticide is applied to a surface, particularly a target like a crop canopy with rather complex architecture, the state of current application technology guarantees that it will be heterogeneous at a scale to which insects can respond behaviorally. Breakdown by UV light and weathering and dilution by growth occur immediately, leading to a very ephemeral and patchy deposit on plant surfaces, and even within plant tissues for systemic pesticides. Understanding how insects respond and adapt behaviorally as well as physiologically can be very important.

Low and heterogeneous concentrations of pesticides can select for *susceptibility* rather than resistance. Georghiou (1972) proposed that avoidance and tolerance of a pesticide should be negatively correlated in insect populations, i.e. the most behaviorally responsive, hypersensitive insects should also be the most susceptible. A similar trade-off is hypothesized for mammalian herbivores between tolerance and avoidance (Iason

& Villalba, 2006). Using simple models with a single gene for resistance and a single gene for avoidance, simulation results indicate that avoidance can result in very stable behavioral adaptation rather than physiological resistance (Gould, 1984). Particularly when behavioral responses are involved, however, single-gene assumptions are not likely to be valid. Using a variety of experimental techniques and examining a wide variety of arthropod species, a range of correlations between behavioral responsiveness and physiological tolerance have been recorded from negative to neutral to positive (Hoy *et al.*, 1998). Empirical results describing the outcome of selection in a heterogeneously treated environment are very rare but intriguing results have been observed. For example, susceptible aphids survived better than moderately resistant aphids on treated plants presumably by moving to untreated parts of the plant canopy (Ffrench-Constant *et al.*, 1988).

Our work on diamondback moth (*Plutella xylostella*) using quantitative genetics methodology has demonstrated significant genetic variation in both tolerance and behavioral responsiveness to permethrin, and negative genetic correlations between these traits (Head *et al.*, 1995a, b, c; Jallow & Hoy, 2005, 2006). We recently conducted a selection experiment (Jallow & Hoy, 2007) in greenhouse cages in which all life stages were exposed to one of three treatment regimes: untreated control, relatively uniform high concentration of permethrin (applied to the entire plant to runoff) and a relatively heterogeneous low concentration (applied only to the center leaves). Not surprisingly, the uniform high concentration resulted in a substantial increase in LC_{50} within 20 generations. However, the heterogeneous low dose resulted in a significant decrease in LC_{50} after 20 generations. Although it may take longer because of generally lower heritability for behavioral traits and indirect selection on behavior, it is possible to increase susceptibility to an insecticide in a treated environment. Furthermore, the behavioral response should be self-reinforcing in this case, as more susceptible diamondback moths were also more behaviorally responsive. The behavioral response resulted in avoidance of the treated parts of the plant, which were also the harvestable parts. The genetic correlation can also be positive when toxin concentrations are high (Hoy & Head, 1995), in which case behavior could select indirectly for physiological resistance. Pest genetics and toxin distribution must be carefully matched to push selection in a direction that is advantageous to agriculture.

Restricting pesticide use over time results in susceptible pests surviving either within or among generations. Control is typically most beneficial at critical times in the development of a crop, providing a focus or window in time when the pesticide can be used with the greatest benefit and least selection pressure. Logistical constraints to implementing windows and refuges can be very challenging (Forrester, 1990). Not the least among these challenges is that the restricted use of the pesticide often provides less immediate benefit to the producer than if used everywhere, constantly and at high concentration, i.e. attempted eradication rather than IPM.

15.2.3 Destroying resistant pests

If resistance is to be cured, then some means of weeding out the resistant individuals in the population is needed. Options include very high doses, mixtures that overwhelm pests that are resistant to only part of the mixture, synergists that inhibit or overcome a particular resistance mechanism, mixtures with a negative cross-resistance pattern or some other means of selection based on the fitness effects of resistance (Roush, 1989; Denholm & Rowland, 1992). There is no guarantee that such mixtures or synergists exist for a given case of resistance.

Barring insecticidal means of suppressing already resistant pests, the remaining means are the same suite of mortality factors that were available without the pesticide, and perhaps fitness effects of resistance. If these other mortality factors were sufficient for control in a given production system then pesticides would not be entirely necessary and probably would not have been used enough to result in resistance in the first place. Insect natural enemies and prey other than pest species tend to fare even more poorly than pests in treated environments. Pathogens and other mortality factors often depend on high pest population density. Thus, the reliance on therapeutic control tends to be self-reinforcing.

15.3 Agroecosystem-level responses to pesticide resistance

Taken over the longer term and considering the social system accompanying pesticide use, pesticide resistance management fits a typical pattern of interplay between natural and social systems seen elsewhere in natural resource management, referred to as the resource management pathology (Holling *et al.*, 2002). An immediate response to a serious challenge results in rapid and visible results, but sets in place a social system that precludes alternatives and further adaptation. Essentially the response and associated infrastructure that arises in social institutions (government organizations such as the US Environmental Protection Agency, industry organizations such as the Insecticide Resistance Action Committee (IRAC), universities and consortia such as the European Network for the Management of Arthropod Resistance to Insecticides and Acaricides) lock a system in place (the pesticide product cycle) that constrains the options for a lasting solution. A common assumption for environmental management that is driven by industrial interests underlies this syndrome, that nature can be controlled with sufficient engineering and management control, including new product development in this case. This is despite the recognition over 20 years ago by the National Research Council's Committee on Strategies for the Management of Pesticide Resistant Pest Populations (Glass, 1986) that "Control of pest populations by combining in cycles the use of old and novel chemical pesticides, as they become available, is unlikely to be a viable long-term strategy." "Victory" in this form of "combat" against pesticide resistance doesn't appear to be any more feasible than winning the coevolutionary arms race for a plant species (Kareiva, 1999).

The resistance management strategies recommended by IRAC include a healthy dose of IPM, suggesting as usual that good IPM will eliminate the concern about resistance. But in practice, new insecticides are still released with a marketing strategy based on providing very clear and saleable benefits of the product, i.e. use the product only when a pest population has exceeded threshold (or prophylactically as in the case of systemic insecticides such as neonicotinoids applied at planting) and then achieve as close to 100% control as possible. Transgenic crops are following the same path, with the high dose (and impressive results) strategy taken as a given despite documented violation of key resistance management assumptions and severe practical difficulty with implementation especially in an IPM context (Bates *et al.*, 2005). Pesticide resistance is perceived by industry to be a part of doing business, a motivator for better product stewardship, and therefore a contributor to sustainable agriculture (Urech *et al.*, 1997). These perceived benefits, however, rest upon the assumption that pesticides must continue in their current role in agriculture. For example, implementation of the 1996 US Food Quality Protection Act has required a very careful and deliberate search for effective replacements for any pesticides removed from commercial use.

The technical community takes an expected role in pesticide resistance: scientific inquiry, skepticism, debate and thorough analysis of the topic from many angles. This debate has played out in the scientific literature over virtually every aspect of resistance management. The debate itself leads to a certain amount of resistance to change in the approach taken by industry and government, which will be influenced by political pressure in the absence of complete agreement in the scientific community. These debates continue (for good reason, they're part of good science) despite early recognition that one can already put into practice a simple and useful strategy based on existing knowledge that will gain most of the benefit of a more thoroughly researched approach (Roush, 1989), particularly if the strategy maintains the ability to change and adapt (Forrester, 1990).

Are we moving toward greater resilience in the face of the adaptive capability of pests or trading resilience for a series of short-term solutions to this long-term problem? In general, experience with environmental management leads to a number of relevant questions that could guide an evaluation of where we are headed (Berkes *et al.*, 2003). As cycles of pesticide product release and resistance have been repeated, is there a social and

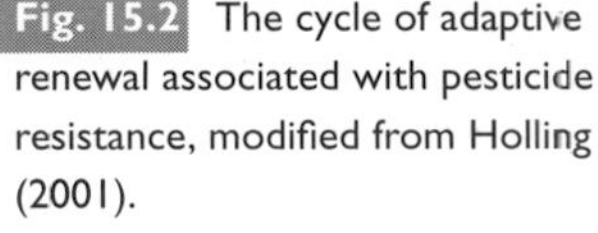

Fig. 15.2 The cycle of adaptive renewal associated with pesticide resistance, modified from Holling (2001).

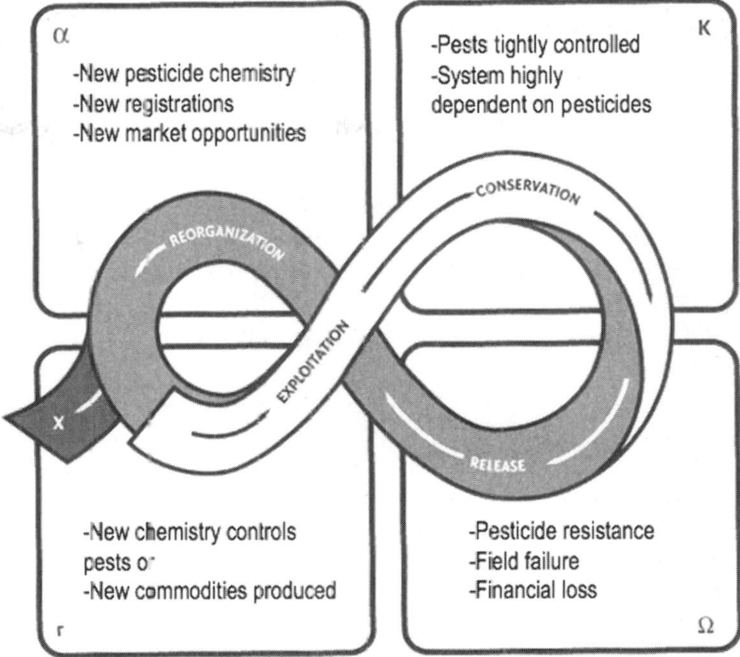

ecological memory accumulating that is improving the capacity of agroecosystems to adapt to rapid change (the "back-loop in the cycle of ecological and social change caused by resistance"; (see Fig. 15.2)? As expected, resistance still occurs, e.g. Colorado potato beetle has recently become resistant to imidacloprid on Long Island, NY, with cross-resistance to the other neonicotinoids and possibly to spinosad (Mota-Sanchez *et al.*, 2006), despite extensive experience with resistance in this pest. The institutional response is that it was monitored closely and documented quickly and farmers can now move on to the next chemistry. The system we have in place around resistance is one in which the resistance problems occur locally and progress to regional and national or international levels, whereas the solutions (the next product in line) flow in the opposite direction (Fig. 15.3).

Four properties have been suggested to lead to sustainability and resilience (Folke *et al.*, 2003), and these will be considered in light of current approaches to resistance management:

(1) *Learning to live with change and uncertainty* Resistance management systems currently in place do a good job of living with change and uncertainty in terms of assuming that resistance will occur and taking a proactive stance, but do a poor job of living with change and uncertainty in their assumption that resistance will be manageable (identifiable single recessive or incompletely dominant genes, fitness costs or negative cross-resistance patterns, etc.) and that replacement chemistry will be available when needed.

(2) *Nurturing diversity for reorganization and renewal* The cycle of resistance and new chemistry introduction encourages innovation on an industrial scale. Competition against the first product of a very effective new chemistry, however, can be very difficult and may suppress innovation even more than pesticide regulation. Industry, regulatory agencies and university or government research organizations have become very well organized to respond to changes in pesticide product cycles. The monitoring and educational programs needed to implement industry-wide resistance management programs are being developed more quickly and consistently, and have been very important in the response to increasing resistance problems (McCaffery & Nauen, 2006). Resistance management strategies are

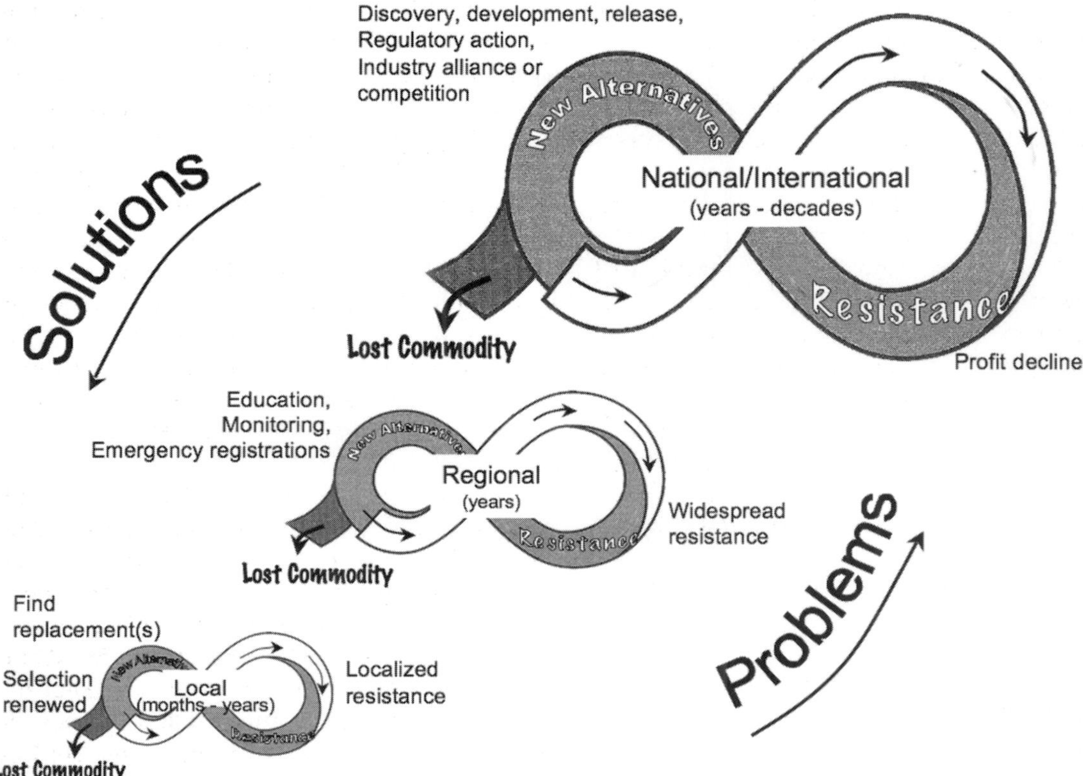

Fig. 15.3 The panarchy of the adaptive renewal cycle currently associated with pesticide resistance. Although resistance occurs at a local scale, the problems scale upwards to the national and international scale, whereas solutions flow from innovation in multinational corporations down to local scales.

heavily reliant upon cooperation among individual producers and in many cases this cooperation has been successfully gained. At more local scales, however, agroecosystems that are reliant enough on pesticides to make resistance management an important issue tend to be homogeneous biologically, having natural capital suppressed and replaced by pesticides. As a result, when resistance occurs a shift to more biologically based pest management tends to be very difficult. When no replacement chemistry is available, the commodity may simply be lost, placing severe economic stress on producers who generally are not gaining flexibility. Strategies that require refuges, however, provide some opportunity to introduce increased biological diversity within agroecosystems. Management of biological diversity in refuges, rather than just production of susceptible insects, is an opportunity that deserves further research.

(3) *Combining different types of knowledge for learning* Experience with pesticide resistance management has accumulated at the institutional level. Institutional learning appears to have prepared industry, research and educational and regulatory agencies to address similar issues with the next new chemistry. However, this institutional level of organization tends to suppress rather than enhance innovation and inventiveness on the part of individual producers or users of pesticides, and even inhibit new players in the pesticide industry. The free market forces at play encourage simply using or even overusing pesticides for the sake of consistency in production, leaving little learning among users other than to follow the instructions of extension and industry in use of the next product in the cycle when resistance occurs.

(4) *Creating opportunity for self-organization* Pesticide resistance management efforts have to some extent achieved a self-organizing community with an ability to quickly adapt at the institutional level, with coordination by government agencies such as the US Environmental Protection Agency, the industry-based IRAC, commodity organizations like the US Cotton Board, universities and associated grower outreach programs. However, little self-organization is encouraged at more local scales given the suppression of human and natural capital described above.

Chemistries and uses that foster and improve natural capital, rather than replacing it, may be the only means of re-engaging the producer and creating opportunities for self-organization at local scales, relieving some of the dependence on solutions from larger scales and creating more resilient agroecosystems.

15.4 Conclusions

Much of the current focus on resistance management stems from the need to respond to recalcitrant problems in a highly evolved and efficient system of agriculture that is very dependent upon external inputs like pesticides. The scientific and technical community is expected to respond with answers and solutions. We can either proactively prevent such problems by redesigning agroecosystems based on our current science and technology or accept the system as is and continue responding. Given that pesticides will continue to be an important management tactic for the foreseeable future, finding ways to create evolutionary dead ends for resistant pests deserves much more attention. Because most major companies are shifting from development of chemistries to development of traits, the advances in this approach could play out in crop and animal genetics (see Chapter 19) rather than in pesticide application. Unfortunately, genetic engineering currently is continuing along the same high-dose, high-kill, results-that-sell path as pesticides have been on, based at least as much on the business need to compete with pesticides as on the scientific arguments for the high dose-refuge strategy of resistance management. Although no cases of resistance to *Bacillus thuringiensis* (*Bt*) crops have emerged so far, the selection pressure they impose on pest populations is such that physiological resistance would be the likely pest response, eventually. A social and ecological system with a cycle of product innovation and obsolescence that takes years to decades is well established, at least in more developed countries, and should be able to continue on its own. The resistance management side of this system should require less support from public sources for research and regulation over time, and the private sector would prefer the reduction as well (Thompson & Head, 2001), at least in regulation.

Although pesticide resistance has been around for almost a century and the science to understand it began at the first observations, we've really been responding as a scientific, technical and agricultural community in a coordinated and serious fashion to pesticide resistance for about two decades. What appeared to be an exponential increase in reported cases of resistance in the early 1980s has been an approximately linear increase since, with most of the new reported cases repeating previously reported cases in new locations (Mota-Sanchez *et al.*, 2002). Farmers continue to rely on pesticides for economic survival, resistance continues to result and the agroecosystems that produce resistance continue to have fewer farmers and lower prospects for economic survival among those who remain. A new direction described in the 2006 strategic plan of the International Rice Research Institute focuses on the well-being of rice farmers, rather than strictly on rice production. This perceptive approach recognizes that farmers can think long term and use sustainable practices when they have sufficient relief from the economic pressures of their production system. Desperation measures taken by farmers who do not have that freedom, on the other hand, tend to be self-reinforcing, perpetuating a cycle of unsustainable approaches and declining resilience. This same "vicious cycle" is at play in more developed countries with more industrial approaches to agriculture. With dependence on pesticides that exert complete control comes a self-reinforcing cycle of resistance and dependence on new pesticides.

Shifting the emphasis of resistance management to self-organizing systems at local scales is a difficult challenge but at least two opportunities deserve more consideration:

(1) *Engage the creativity of producers* Resistance monitoring should be in the hands of the user community. Some progress has been made with offering simple test kits to producers but more attention and more simple monitoring tools would be worthwhile. If agricultural producers could gain more flexibility in use patterns within limits that prevent environmental, health and resistance risk, we could benefit from their creativity and innovation, but current pest management systems do not encourage this kind of innovation. The extension service can help with sharing information more rapidly and widely on what works at the farm level.

(2) *Better imitate the diversity in toxins found in natural systems* Heterogeneity in toxins and toxin concentrations at multiple spatial scales is a foundation of the relative stability seen in natural systems. Pesticide application remains imperfect enough to inherently provide diversity in concentrations. Beyond the considerable attention given to a high dose and refuge strategy for transgenic crops, however, the rational design of spatial heterogeneity in toxin distributions has received far less attention than the attempt to avoid it with uniform high concentrations. Behavioral adaptation in arthropod pests can be a means of stabilizing crop defense rather than a loss of control, if accompanied by shifts in the location of feeding damage that decrease economic losses or shifts in life history traits that enhance the activity of natural enemies. Perhaps the focus of discovery should also be intensified for new chemistries or families of toxicants with inherent negative cross-resistance patterns. New application technology may provide opportunities to create more diverse toxin mixture distributions with chemical applications. For example, a relatively new double nozzle (Chapple *et al.*, 1996, 1997) expresses small pesticide solution droplets into a stream of larger plain water droplets that carry the pesticide droplets into a plant canopy with improved efficiency of delivery. Droplets from multiple solutions could theoretically be injected into the same water carrier droplet stream to generate complex deposits with individual droplets of different chemistries. If this heterogeneity can incorporate negative cross-resistance patterns then it could further stabilize pest adaptive responses (Pittendrigh & Gaffney, 2001; Pittendrigh *et al.*, 2004).

The final conclusion on this subject remains unchanged over the past two decades. Pesticide resistance management will be enhanced more by the multiple tactics (especially non-chemical) and their integration described in the rest of this book than in continued refinement of resistance management strategies in agroecosystems that do not yet live up to the preventive potential of true IPM. Defensive chemistry could still be part of this potential but with a rather different role than the silver bullet it currently represents. The design capability needed to fit new chemistry into this potential role will take commitment, painstaking research and detailed ecological understanding. Meanwhile, pesticide resistance management programs that focus on maintaining the impressive benefits of complete control with pesticides will maintain both reliance and dependence on them.

References

Adler, L. S. & Kittelson, P. M. (2004). Variation in *Lupinus arboreus* alkaloid profiles and relationships with multiple herbivores. *Biochemical Systematics and Ecology*, **32**, 371–390.

Bates, S. L., Zhao, J. Z., Roush, R. T. & Shelton, A. M. (2005). Insect resistance management in GM crops, past, present and future. *Nature Biotechnology*, **23**, 57–62.

Berenbaum, M. R. (1995). The chemistry of defense: theory and practice. *Proceedings of the National Academy of Sciences of the USA*, **92**, 2–8.

Berenbaum, M. R. & Zangerl, A. R. (1998). Chemical phenotype matching between a plant and its insect herbivore. *Proceedings of the National Academy of Sciences of the USA*, **95**, 13 743–13 748.

Berenbaum, M. R., Zangerl, A. R. & Nitao, J. K. (1986). Constraints on chemical coevolution, wild parsnips and the parsnip webworm. *Evolution*, **40**, 1215–1228.

Bergelson, J. & Purrington C. B. (1996). Surveying costs of resistance in plants. *American Naturalist*, **148**, 536–558.

Berkes, F., Colding, J. & Folke, C. (eds.) (2003). *Navigating Social–Ecological Systems*. Cambridge, UK: Cambridge University Press.

Brown, T. (1996). *Molecular Genetics and Evolution of Pesticide Resistance*. Washington, DC: American Chemical Society.

Caprio, M. (1998). Evaluating resistance management strategies for multiple toxins in the presence of external refuges. *Journal of Economic Entomology*, **91**, 1021–1031.

Caprio, M. (2001). Source–sink dynamics between transgenic and non-transgenic habitats and their role in the evolution of resistance. *Journal of Economic Entomology*, **94**, 698–705.

Carrière, Y. & Tabashnik, B. (2001). Reversing insect adaptation to transgenic insecticidal plants. *Proceedings of the Royal Society of London B*, **268**, 1475–1480.

Chapple, A. C., Downer, R. A., Wolf T. M., Taylor, R. A. J. & Hall, F. R. (1996). The application of biological pesticides: limitations and a practical solution. *Entomophaga*, **41**, 465–474.

Chapple, A. C., Wolf, T. M., Downer, R. A., Taylor, R. A. J. & Hall, F. R. (1997). Use of nozzle-induced air-entrainment to reduce active ingredient requirements for pest control. *Crop Protection*, **16**, 323–330.

Denholm, I. & Rowland, B. W. (1992). Tactics for managing pesticide resistance in arthropods, theory and practice. *Annual Review of Entomology*, **37**, 91–112.

Felton, G. W. & Korth, K. L. (2000). Trade-offs between pathogen and herbivore resistance. *Current Opinion in Plant Biology*, **3**, 309–314.

Ffrench-Constant, R. H., Clark, S. J. & Devonsire, A. L. (1988). Effect of decline of insecticide residues on selection for insecticide resistance in *Myzus persicae* Sulzer (Hemiptera: Aphididae). *Bulletin of Entomological Research*, **78**, 19–30.

Folke, C., Colding, J. & Berkes, F. (2003). Synthesis: building resilience and adaptive capacity in social-ecological systems. In *Navigating Social–Ecological Systems*, eds. F. Berkes, J. Colding & C. Folke, pp. 352–387. Cambridge, UK: Cambridge University Press.

Forgash, A. J. (1985). Insecticide resistance in the Colorado potato beetle. In *Proceedings of 17th International Congress of Entomology*, Research Bulletin No. 704, eds. D. N. Ferro & R. H. Voss, pp. 33–52. Amherst, MA: Massachusetts Agricultural Experiment Station, University of Massachusetts.

Forrester, N. W. (1990). Designing, implementing and servicing an insecticide resistance management strategy. *Pesticide Science*, **28**, 167–179.

Forrester, N. W. & Bird, L. J. (1996). The need for adaption to change in insecticide resistance management strategies: the Australian experience. In *Molecular Genetics and Evolution of Pesticide Resistance*, ed. T. Brown, pp. 160–168. Washington, DC: American Chemical Society.

Forrester, N. W. & Cahill, M. (1987). Management of insecticide resistance in *Heliothis armigera* (Hübner) in Australia. In *Combating Resistance to Xenobiotics: Biological and Chemical Approaches*, eds. M. Ford, B. P. S. Khambay, D. W. Holloman & R. M. Sawicki, pp. 127–137. Chichester, UK: Ellis Horwood.

Georghiou, G. P. (1972). The evolution of resistance to pesticides. *Annual Review of Ecology and Systematics*, **3**, 133–168.

Georghiou, G. P. (1986). The magnitude of the resistance problem. In *Pesticide Resistance: Strategies and Tactics for Management*, pp. 14–43. Washington, DC: National Academy Press.

Glass, E. H. (ed.) (1986). *Pesticide Resistance: Strategies and Tactics for Management*. Washington, DC: Committee on Strategies for the Management of Pesticide Resistant Pest Populations, National Academy Press.

Gould, F. (1984). Role of behavior in the evolution of insect adaptation to insecticides and resistant host plants *Bulletin of the Entomological Society of America*, **30**, 34–41.

Head, G., Hoy, C. W. & Hall, F. R. (1995a). The quantitative genetics of behavioral and physiological response to a pyrethroid in diamondback moth. *Journal of Economic Entomology*, **88**, 447–453.

Head, G., Hoy, C. W. & Hall, F. R. (1995b). Permethrin droplets influence larval *Plutella xylostella* (Lepidoptera: Plutellidae) movement. *Pesticide Science*, **45**, 271–278.

Head, G., Hoy, C. W. & Hall, F. R. (1995c). Effects of direct and indirect selection on behavioral response to permethrin in larval diamondback moths. *Journal of Economic Entomology*, **88**, 461–469.

Hemingway, J., Penilla R. P., Rodriguez, A. D. *et al.* (1997). Resistance management strategies in malaria vector mosquito control: a large-scale field trial in southern Mexico. *Pesticide Science*, **51**, 375–382.

Herms, D. A. & Mattson, W. J. (1992). The dilemma of plants, to grow or defend. *Quarterly Review of Biology*, **67**, 283–335.

Holling, C. S. (2001). Understanding the complexity of economic, ecological, and social systems. *Ecosystems*, **4**, 390–405.

Holling, C. S., Gunderson, L. H. & Ludwig, D. (2002). In quest of a theory of adaptive change. In *Panarchy: Understanding Transformations in Human and Natural Systems*, eds. L. H. Gunderson & C. S. Holling, pp. 3–22. Washington, DC: Island Press.

Hoy, C. W. (1999). Colorado potato beetle resistance management strategies for transgenic potatoes. *American Journal of Potato Research*, **76**, 215–219.

Hoy, C. W. & Head, G. (1995). Correlation between behavioral and physiological responses to transgenic potatoes containing *Bacillus thuringiensis* δ-endotoxin in *Leptinotarsa decemlineata* Say (Coleoptera: Chrysomelidae). *Journal of Economic Entomology*, **88**, 480–486.

Hoy, C. W., Head, G. & Hall, F. R. (1998). Spatial heterogeneity and insect adaptation to toxins. *Annual Review of Entomology*, **43**, 571–594.

Iason, G. R. & Villalba, J. J. (2006). Behavioral strategies of mammal herbivores against plant secondary metabolites, the avoidance–tolerance continuum. *Journal of Chemical Ecology*, **32**, 1115–1132.

Jallow, M. F. A. & Hoy, C. W. (2005). Phenotypic variation in adult behavioral response and offspring fitness in *Plutella xylostella* (Lepidoptera: Plutellidae) in response to permethrin. *Journal of Economic Entomology*, **98**, 2195–2202.

Jallow, M. & Hoy, C. W. (2006). Quantitative genetics of adult behavioral response and larval physiological tolerance to permethrin in diamondback moth *Plutella xylostella* (Lepidoptera: Plutellidae). *Journal of Economic Entomology*, **99**, 1388–1395.

Jallow, M. & Hoy, C. W. (2007). Indirect selection for increased susceptibility to permethrin in diamondback moth (*Plutella xylostella*) (Lepidoptera: Plutellidae). *Journal of Economic Entomology*, **100**, 526–533.

Kareiva, P. (1999). Coevolutionary arms races: is victory possible? *Proceedings of the National Academy of Sciences of the USA*, **96**, 8–10.

Koricheva, J. (2002). Meta-analysis of sources of variation in fitness costs of plant antiherbivore defenses. *Ecology*, **83**, 176–190.

Lenormand, T. & Raymond M. (1998). Resistance management, the stable zone strategy. *Proceedings of the Royal Society of London B*, **265**, 1985–1990.

McCaffery, A. & Nauen, R. (2006). The Insecticide Resistance Action Committee (IRAC), public responsibility and enlightened industrial self-interest. *Outlooks on Pest Management*, **17**, 11–14.

Mota-Sanchez, D., Bills, P. S. & Whalon, M. E. (2002). Arthropod resistance to pesticides. In *Pesticides in Agriculture and the Environment*, ed. W. Wheeler, pp. 241–272. New York: Marcel Dekker,

Mota-Sanchez, D., Hollingworth, R. M., Grafius, E. J. & Moyer, D. D. (2006). Resistance and cross-resistance to neonicotinoid insecticides and spinosad in the Colorado potato beetle, *Leptinotarsa decemlineata* (Say) (Coleoptera: Chrysomelidae). *Pest Management Science*, **62**, 30–37.

Nitao, J. K. (1995). Evolutionary stability of swallowtail adaptations to plant toxins. In *Swallowtail Butterflies, their Ecology and Evolution*, eds. J. M. Scriber, Y. Tsubaki & R. C. Lederhouse, pp. 39–52. Gainsville, FL: Scientific Publishers Inc.

Pittendrigh, B. R. & Gaffney, P. J. (2001). Pesticide resistance: can we make it a renewable resource? *Journal of Theoretical Biology*, **211**, 365–375.

Pittendrigh, B. R., Gaffney, P. J., Huesing, J. E. *et al.* (2004). "Active" refuges can inhibit the evolution of resistance in insects towards transgenic insect-resistant plants. *Journal of Theoretical Biology*, **231**, 461–474.

Roush, R. T. (1989). Designing resistance management programs: how can you choose? *Pesticide Science*, **26**, 423–441.

Strauss, S. Y., Rudgers, J. A., Lau, J. A. & Irwin, R. E. (2002). Direct and ecological costs of resistance to herbivory. *Trends in Ecology and Evolution*, **17**, 278–285.

Tabashnik, B. E. & Croft, B. (1982). Managing pesticide resistance in crop–arthropod complexes: interactions between biological and operational factors. *Environmental Entomology*, **11**, 1137–1144.

Thompson, G. D. & Head, G. (2001). Implications of regulating insect resistance management. *American Entomologist*, **47**, 6–10.

Urech, P. A., Staub, T. & Voss, G. (1997). Resistance as a concomitant of modern crop protection. *Pesticide Science*, **51**, 227–234.

Chapter 16

Assessing environmental risks of pesticides

Paul C. Jepson

For synthetic chemicals to pose an environmental risk, a complex sequence of events must unfold that result in a toxic compound reaching a site of action within a susceptible organism with resultant impacts on behavior, fitness or survival. For most chemicals, other than pesticides, toxicity to any organism within the environment, including humans, is an unintended consequence of their use. Pesticides are however among a group of compounds that are synthesized and utilized in such a way that they exhibit direct toxicity to particular organisms, specifically those organisms that humans define as pathogens or pests. Humans compete directly with these organisms for the harvestable yield of crop plants and using poisons to remove them and protect the food and fiber supply has been considered an acceptable, even desirable, activity for centuries, despite evidence that adverse effects on non-target organisms can occur (Devine & Furlong, 2007; Kogan & Jepson, 2007).

16.1 Defining environmental risk

Environmental risk is normally expressed as the probability that a defined adverse impact or endpoint will occur within a particular organism as a result of exposure to an environmental stressor (Suter, 1993). Throughout this chapter, I will use the term risk broadly when referring to different kinds of pesticide environmental impact. In some cases, the risks that I refer to could be defined more narrowly as the estimate of probability derived from the product of dose or exposure to the toxin and the susceptibility of the organism to it. This narrow definition becomes problematic, however, when referring to ecological impacts that may unfold in populations and communities generations after the original toxic effect, and away from the location where this impact occurred. This is a complex and difficult field to describe fully, but I have attempted to point out where the conceptual and methodological challenges lie when these arise.

Although risk to humans is very much a part of the environmental risk spectrum for pesticides, the role of this chapter is to focus on non-human organisms, and to address ecological risks as an important subset of the problems that pesticides pose. Despite the fact that most of the regulatory procedures surrounding pesticides are designed to address risks to humans, it is important to recognize that the assessment and management of human risks alone will not result in adequate protection of living organisms and the environment. There are a number of reasons why non-human organisms may be more sensitive to pesticides than humans, and therefore why ecological risk assessment is a critical component of pesticide regulation and management. These reasons

Integrated Pest Management, ed. Edward B. Radcliffe, William D. Hutchison and Rafael E. Cancelado. Published by Cambridge University Press.

include (adapted to be more relevant to pesticides from Suter, 1993):

- The existence of unique routes of non-human chemical exposure to pesticides including the use of water as a respiratory medium, oral cleaning of the body, and root uptake by plants.
- The higher susceptibility of certain non-human species to pesticides because of specific toxicological modes of action (e.g. herbicide toxicity to plants that are relatives of weed species or insecticide toxicity to beneficial insects), the presence of unique biochemical or physiological pathways for effects within the organism (e.g. eggshell synthesis in birds), or higher metabolic rates that increase rates of exposure for a given body size.
- Specific features of size, structure, distribution or behavior (including feeding) that confer greater rates of exposure and uptake, including total immersion in potentially contaminated environmental media such as soil and water, exploitation of disproportionately contaminated diet sources, or possession of membranes or body parts that are particularly permeable to pesticides.
- The occurrence of mechanisms of action at higher levels of biological organization that are not relevant to humans, including certain population and community impacts that arise through disruption to food webs (e.g. secondary pest outbreaks or pest resurgence that result from disproportionate toxicity to predatory and parasitic species) or disruption to ecological services, including pollination and nutrient cycling.
- The close coupling of non-human organisms to their environments which confers greater susceptibility to secondary and indirect effects that impact critical food or habitat resources.

The degree of environmental risk associated with a particular pesticide will be a function of the amount of the active ingredient that is applied, the location and placement of the compound, its partitioning, breakdown and transport in the environment, and its toxicity to the organisms that are exposed to it (van der Werf, 1996). These risks are unfortunately widespread and often severe because of the nature of pesticide use and the properties of the compounds themselves.

Environmental contamination occurs because pesticides are delivered to the sites of treatment by very inefficient processes including spray application, that result in only a tiny fraction of the applied material reaching the intended target organism, and a large proportion of the applied chemical being effectively wasted (Graham-Bryce, 1977; Matthews, 2000). Pesticides may also cause contamination as a result of spillage via accidental release or because of deliberate misuse or abuse. Following release, pesticide physicochemical properties then enable the active ingredients to partition between air, soil, water and the biota (Hornsby *et al.*, 1996; Mackay, 2001) where they must persist long enough to expose and affect the pest, diseases or weeds that they target. This persistence, however, enables pesticides to enter untargeted environmental compartments where their mobility enables them to travel beyond the site of treatment into ground and surface waters (Gilliom *et al.*, 2006) and in some cases to globally distant marine and terrestrial ecosystems (Ueno *et al.*, 2003; Kelly *et al.*, 2007).

The characteristics of widespread and inefficient use patterns, accidental spillages, deliberate misuse, environmental partitioning, mobility and persistence result inevitably in the exposure of non-pest species to pesticides and their toxic breakdown products. Given that the toxicological modes of action of pesticides are rarely specific to target pest species alone, non-target organisms that pose no threat to agricultural yields or public health may also be susceptible to them and succumb to toxic impacts as a result of being exposed (McGlaughlin & Mineau, 1995; Nimmo & McEwen, 1998).

This chapter explores the assessment of pesticide environmental impact once these compounds are marketed and in commercial use. Although environmental risks are evaluated prior to regulatory approval in a number of jurisdictions internationally, the standard of these assessments varies widely, as does the degree of post-release monitoring and evaluation of the regulatory standards that are set. There are also some regions of the world where little if any environmental risk assessment is undertaken prior to the approval of pesticides for use, and still others where regulatory oversight of the pesticide market is virtually absent. The pesticide label is intended to provide guidance on whether or not a chemical can be used legally in a certain setting, and if the chemical is used as instructed on the label, there is at least an expectation of a low likelihood of adverse environmental harm. Although the regulatory system that operates prior to marketing can work effectively, there is evidence that environmental impacts do in fact occur even as a result of widespread, legal pesticide application. It is not the role of this chapter to review critically the effectiveness of the pesticide regulatory system: it simply begins from the premise that environmental impacts are associated with pesticide use, no matter how comprehensive the local regulatory system may be, and that assessment and understanding of these impacts will contribute to the adoption of sustainable IPM systems, and the avoidance of environmental harm.

16.2 Evidence for pesticide-related environmental risks

Although many literature sources cite the most severe environmental impacts of pesticides, there are very few if any comprehensive summaries of the broader evidence for pesticide impacts on the environment. This gives the impression that the realization of pesticide environmental risks is exceptional and unusual, rather than widespread and normal, which is more probably the case (Devine & Furlong, 2007).

The following review provides a critical analysis of the most readily available sources of information concerning pesticide environmental risks. It provides guidance on the assembly of these data for particular chemicals to enable an assessment of the potential for environmental harm in a given set of circumstances. Data that pertain to the exact set of local conditions and potential uses are rarely available, and thus when making these assessments, it is extremely important to draw data from as many sources as possible. It is also important to identify, for each data source, the similarities and differences between the environmental and ecological contexts presented and the context that prompted the assessment in the first place. As a whole, the assembled data set from these disparate sources is more likely to equate with impacts across a large scale than localized impacts.

16.2.1 Pesticide incident monitoring and reporting

Wildlife deaths that result from exposure to pesticides are reported in some countries and reveal that incidents involving terrestrial wildlife are still a common occurrence (de Snoo *et al.*, 1999). The majority of incidents are caused by deliberate abuse and misuse of pesticides rather than approved use, but this may reflect the fact that vertebrate cadavers are less concealed and more likely to be detected at frequently visited, open sites, where deliberate abuse occurs, rather than within crop canopies on private land, where legal uses take place. In the UK in 2005, 103 wildlife mortality incidents were attributed to pesticides, the majority of which were caused by organophosphate and carbamate insecticides and various rodenticides (Barnett *et al.*, 2006). Spillages during manufacturing, storage or transport may also provide documentation of ecological hazards, or the risk that these may occur. Rice crop losses in Liaoning Province in China in 1997 provided evidence of herbicide contamination in irrigation water from the Tiaozi and Zhaosutai Rivers that was eventually attributed to accidental spillage upstream (Li *et al.*, 2007). Acts of warfare and looting at a pesticide store in Somalia in 1988 caused contamination by organochlorine and organophoshate pesticides (Lambert, 1997). Five years later, reptiles were absent where surface residues exceeded 10 ppm, they avoided zones contaminated at levels above 1 ppm and reptile species richness was reduced in the valley where the spillage occurred. Additionally surface residues were still lethal to certain reptiles and amphibians and pesticide residue burdens were detectable in ground lizards and in both well and rainpool-inhabiting frogs.

16.2.2 Biological monitoring and surveillance

There is a long history of biological monitoring programs in agricultural systems, particularly in Western Europe, many of which have detected changes, mainly declines, in the abundance and diversity of multiple taxa (Robinson & Sutherland, 2002; Jepson, 2007a). Pesticides may have played a role in these declines, but unless monitoring programs are designed to explicitly address specific mechanisms, they can rarely be used to unambiguously determine cause (Noon, 2003). For example, declines in UK farmland birds have been particularly severe (Ormerod & Watkinson, 2000), but although changes in food quality or quantity on the farm, partly driven by herbicide and insecticide use, have been cited as possible influences (Benton *et al.*, 2002; Newton, 2004), many other factors including climate and land use change have also played a role. A unique and historical example of the use of the biological monitoring data is the retrospective analysis of eggshell thickness in peregrine falcons (*Falco peregrinus*) from preserved collections in the UK. This analysis revealed the abrupt onset of eggshell thinning in 1947, coinciding with the introduction and widespread use of DDT and its gradual recovery following the banning of this insecticide (Radcliffe, 1967, 1993). More targeted, often shorter-term monitoring investigations may also be used to obtain evidence of pesticide impacts. For example, a 49-site survey by Good & Giller (1991) demonstrated reduced species richness in Staphylinidae (Coleoptera) within fields treated with the organophosphate pesticide dimethoate.

16.2.3 Chemical monitoring and surveillance

Environmental monitoring of pesticide residues is more recent, and less widespread than biological monitoring. Contamination of fresh water as a result of runoff, leaching, droplet and vapor drift and over-spraying is still a very serious problem, and large-scale monitoring in the USA revealed that more than half of the streams sampled had pesticide concentrations that exceeded benchmark concentrations for aquatic life (Gilliom *et al.*, 2006). Pesticides also volatilize at the site of application and may move downwind and deposit at cooler locations a significant distance away from the site of application. Monitoring of summertime atmospheric transport of non-persistent insecticides, fungicides and herbicides in California's Central Valley has shown that air and surface water concentrations in the Valley, and the Sequoia National Park which lies downwind, are directly proportional to levels of use of these compounds (LeNoir *et al.*, 1999). Pesticides were also detected in the rain and snow precipitating in the same mountain range (McConnell *et al.*, 1998). Global monitoring of persistent organochlorine pesticides in skipjack tuna (*Katsuwonus pelamis*) tissue has found residues in all of the fish sampled from a number of oceanic systems (Ueno *et al.*, 2003), and these compounds have also been shown to accumulate in terrestrial food webs, including birds and mammals in the Arctic (Kelly *et al.*, 2007). Ecotoxicologists are still developing methods to understand the effects of complex pesticide mixtures (e.g. Posthuma *et al.*, 2002), and our understanding of the possible effects of long-term, multiple, low-dose exposures is still very limited. Despite these constraints, chemical monitoring data are vital in enabling mapping of contamination, demonstration of the potential for exposure and identification of compounds of potential concern for risk assessment and risk management purposes.

16.2.4 Risk estimates derived from chemical monitoring

Monitoring data can only be translated into estimates of ecological impacts if they are supported by measurements of effects or by risk assessment procedures that relate chemical exposure estimates to specific biological endpoints. Streams and drainage canals in the Yakima River Basin in Washington State, USA with the highest and most potentially toxic pesticide concentrations tended to have the highest numbers of pollution-tolerant benthic aquatic invertebrates (Fuhrer *et al.*, 2004), implying that more sensitive species have been lost as a result of pesticide toxicity. Organophosphorous residues and depressed cholinesterase activity in amphibian tissues from the California Sierra Nevada, USA, were associated with sites where there was poor to moderate amphibian population status (Sparling *et al.*, 2001). The inferential power inherent within the chemical monitoring data may however be poor, and it is not always possible to assert causality if there are no control comparisons, or if there are other stressors that confound the data.

16.2.5 Published field experiments

There is a limited literature from field experiments undertaken in open natural or agricultural systems that contains evidence concerning the impacts of individual and multiple pesticides on a wide array of non-target taxa. At best, properly designed experiments engender a high degree of statistical and inferential power. With the rigorous statement of hypotheses, effective and realistic design combined with careful pesticide application procedures and efficient sampling of organisms, they may provide powerful evidence of the nature, level and duration of pesticide impacts.

Experiments in open freshwater ecosystems

Although investigations made in natural still (lentic) and flowing (lotic) surface waters may not be laid out as formal experiments, they may carry a high degree of inferential power if they adhere to rigorous procedures for design, layout and analysis. There are however some substantial methodological challenges. It may be difficult to quantify the chemical exposure of the stream flora and fauna because it is often not possible to account for and measure all the sources of pesticide contamination, and because invertebrates and particularly fish are highly mobile in lotic systems. Without mechanistic hypotheses concerning potential impacts, it may also be impossible to define the most appropriate ecological endpoints to measure, and many studies simply record abundance and diversity of readily sampled taxa. The sites themselves may not lend themselves to selection of control or reference areas that are unaffected by pesticide exposure, and matched sites that are otherwise identical in terms of hydrology, riparian habitat characteristics, pond and stream chemical and physical conditions; other potential stressors and surrounding land uses may not be available.

Leonard *et al.* (2000) investigated storm runoff of the organochlorine insecticide endosulfan that could potentially lead to fish kills in the Namoi River, Australia. They sought evidence for impacts in macroinvertebrate communities because these occurred in high densities and were relatively sedentary, compared with fish. Population densities were reduced at downstream sites with 10–25-fold higher pesticide concentrations compared with multiple reference sites in upstream reaches. The weight of the evidence from this study pointed to endosulfan exposure being responsible for these effects. The authors were able to make this assertion because they employed a relatively symmetrical BACI (Before–After–Control–Impact) sampling design that included several control sites, thereby avoiding the low statistical and inferential power associated with asymmetrical studies that include only a single reference or control location (Underwood, 1994). They also had mechanistic evidence for pesticide effects on key taxa that were depleted in the contaminated sites because they had previously demonstrated that riverine pesticide concentrations exceeded the 48-h LC_{50}s for mayfly (Ephemeroptera) and caddisfly (Trichoptera) nymphs from the same location (Leonard *et al.*, 1999).

Thiere & Schulz (2004) found large differences in both turbidity and pesticide exposure between agricultural and upstream reaches of the Lourens River in South Africa. They argued that either or both of these may be responsible for differences in macroinvertebrate community structure between the two sites that left only the more water-quality-insensitive taxa in agricultural reaches. The requirements for high statistical and inferential power are far better satisfied when treatments can be allocated to separate, matched replicate aquatic habitats at a range of treatment concentrations; these conditions occur in the case of temporary pond invertebrate communities exposed to pesticide drift during locust control operations in the Sahelian zone of West Africa.

Lahr *et al.* (2000) detected negative direct (toxicological) effects of fenitrothion, diflubenzuron, deltamethrin and bendiocarb on cladocerans, fairy shrimps and backswimmers, and also found indirect effects, particularly patterns of superabundance, that could be explained by reduced predation or competition in treated ponds. Only 12 of the 80 taxa in the ponds that they studied, however, satisfied the requirements for statistical analysis, and the fluctuating populations and low abundance of these taxa meant that inferences about effects were drawn from a combination of statistics, graphical analysis and expert judgment, combined with evidence for direct toxicity derived from laboratory experiments (Lahr *et al.*, 2001).

Experiments in open freshwater agroecosystems

Some of the most sophisticated understanding of pesticide impacts in aquatic systems have come from IPM studies in rice agriculture, as opposed to ecotoxicological investigations in natural water bodies. Combined observational and experimental studies in Indonesian rice demonstrated for example, that carbofuran and monocrotophos applications caused pest resurgence by suppressing predators and their alternative food supplies early in the season (Settle *et al.*, 1996). This investigation, and others in rice, are among the only research studies of pesticide impacts to address the importance of trophic linkages and their disruption in sprayed systems; in this case the linkages among organic matter, detritivores, plankton feeders and generalist predators that result in herbivore (pest) suppression early in the season by generalist natural enemies that do not rely solely upon pest herbivores for their survival. Deltamethrin applications to rice have been found to increase rice herbivores (mainly Delphacidae) by 4 million/ha per sampling date, and decrease natural enemies by 1 million/ha per sampling date, significantly reducing plant-borne invertebrate food web length through the removal of certain spider, coccinellid and heteropteran predators (Schoenly *et al.*, 1994). The result of this food web simplification and associated disorganization of the natural enemy community is greatly increased unpredictability in pest population dynamics (Cohen *et al.*, 1994).

Experiments in open fields and farms: impacts on invertebrates

Terrestrial agriculture is very poorly supported by investigations of pesticide impacts on agriculturally relevant scales. Investigators tend to be lured by the simplicity and logistic tractability of replicated small plot experiments and lose sight of the fact that the small scale of these investigations is trivial in comparison with the scale upon which the population processes of non-target organisms ensue (Jepson, 1989, 1993, 2007b); hence, effects may be routinely unrecorded or underestimated. Perhaps because they are not confined solely to an agricultural matrix, ecotoxicological investigations of the side effects of disease vector and plague pest pesticides tend to be undertaken on scales that are ecologically relevant. Malaise traps and fallout funnels beneath sprayed trees captured invertebrates and demonstrated reductions in Diptera and Hymenoptera after aerial and ground-based endosulfan treatments against tsetse fly (*Glossina* spp.) (Everts *et al.*, 1983). All pool-dwelling fish were also killed. Pitfall traps and Malaise traps in 3–15-ha tsetse fly treatments with synthetic pyrethroids also exhibited reductions in multiple invertebrate taxa, including a ctenizid spider that was selected as an indicator taxon (Everts *et al.*, 1985). Barrier treatments over 5–20-km^2 areas against locusts using the insect growth regulator diflubenzuron enabled sweep net comparisons between invertebrates in 50-m-wide sprayed and 600-m-wide unsprayed strips (Tingle, 1996). Assessments were made of 300 species, distributed between 120 families and 17 orders, and revealed significant reductions in Lepidoptera and Acrididae. Investigations on this scale enable the potential of unsprayed areas as refuges for non-target taxa to be determined, and also the ecological selectivity of barrier spraying, which biases exposure towards mobile animals that walk through the treated area and intercept sprayed strips (Jepson & Sherratt, 1996).

Experimental regimes may also be undertaken on a large scale, but there are trade-offs to address between replication and plot size. Peveling *et al.* (1999a) argued that replicated 16-ha plots were too small to enable assessment of effects on flying invertebrates, but larger 400-ha plots could not be replicated in their investigation. They demonstrated >75% reductions for more than 3 months in Collembola, Formicidae, Carabidae and Tenebrionidae in fenitrothion-treated plots. They found successively more limited impacts with a fenitrothion–esfenvalerate mixture and with the insect growth regulator triflumuron. Impacts of triflumuron on non-target Lepidoptera may, however, have been among the most significant and damaging effects detected, because 70% of these species were endemic to the area of Madagascar where this investigation took place. Peveling *et al.* (1999b) also explored the use of presence/absence assessments in replicated, intermediate-sized plots (50 ha) in order to develop a more rapid risk assessment program.

They again demonstrated that fenitrothion had significant impacts on non-target invertebrates, with 75% of Carabidae, Tenebrionidae, Formicidae and Ephydridae reduced significantly in the pesticide treatment.

In temperate systems, one of the only investigations to address realism in spatial and temporal scales is the Boxworth Study, which investigated the impacts of different spray regimes allocated to contiguous blocks of whole fields in the UK (Greig-Smith *et al.*, 1992). When whole fields were treated with organophosphate insecticides in a high input regime, certain ground beetles (Carabidae) became locally extinct (Burn, 1992). Jepson & Thacker (1990) demonstrated experimentally that Carabidae reinvade organophosphate pesticide-treated areas as a result of random dispersal behavior, once residual toxicity has ameliorated after 7–9 days (Unal & Jepson, 1991), but that this process can take up to 30 days at distances of 100 m into treated plots. Within-field, replicated experiments on unrealistically small scales will therefore not detect significant impacts on dispersive taxa because these will invade rapidly from untreated control areas. Modeling (Sherratt & Jepson, 1993) demonstrated that delayed reinvasion by Carabidae of the larger, organophosphate-treated whole fields in sprayed systems could result in the local extinctions detected by Burn (1992), and they also predicted that these effects could result in locally increased pest density (also termed pest resurgence). This was subsequently confirmed experimentally by Duffield *et al.* (1996), who demonstrated a wave of aphid pest and collembolan population outbreaks in advance of the reinvasion front of epigeal predatory species that were temporarily impacted by spray applications.

Large or realistically scaled experiments tend to reveal the long-term effects of broad-spectrum insecticides that are not detected within smaller-scale investigations. Our overall understanding of the effects of broad-spectrum insecticides in agroecosystems might be transformed if experiments were conducted on scales that equated better with the scales of commercial treatment and the scales over which the population processes of non-target organisms takes place. At risk however is the low statistical power inherent in experiments with low levels of replication. Experimental research into the ecological impacts of GM crops expressing insecticidal genes compared with conventional insecticides has demonstrated that smaller, replicated plots may offer an effective alternative to large plots in certain commodities such as cotton, and that statistical power may be further strengthened by conducting multi-year investigations (Naranjo, 2005). Prasifka *et al.* (2005) recommended that small plots should be avoided altogether for investigating pesticide and GM crop ecological impacts in corn, and that intermediate-sized plots could be isolated from one another by cultivated strips to enable effect size to be more accurately measured.

Experiments in open fields and farms: impacts on vertebrates

Assessments of pesticide impacts on vertebrate wildlife also comprise a relatively limited international literature, with impacts on birds being the best represented. Plague pest and vector management campaigns in Africa may constitute some of the largest scale wildlife exposures that occur, but data concerning impacts on lizards, birds and small mammals are relatively sparse. A non-replicated set of 800-ha treatments, with paired controls, revealed that a low dosage of a fenitrothion–esfenvalerate mixture significantly reduced the abundance of the endemic lizard *Chalarodon madagascarensis* by 39%, possibly through direct mortality of juvenile lizards (Peveling & Nagel, 2001). There were similar but non-significant reductions in a triflumuron treatment. The same lizard species, with another species, *Mabuya elegans*, and the lesser hedgehog tenrec (*Echinops telfairi*), are termite predators and the potential for food chain perturbations caused by pesticides was demonstrated when all three species were reduced following locust control barrier treatments over 45 km^2 with fipronil that caused large reductions in the harvester termite *Coarctotermes clepsydra* (Peveling *et al.*, 2003). This compound has also been shown, however, to be directly toxic to lizards and both direct and indirect effects may underlie observed effects (Peveling & Demba, 2003).

Mineau (2002) reviewed 181 field investigations of cholinesterase-inhibiting pesticide effects on birds in the literature, encompassing field and pasture crops, forestry, orchards, mosquito control, forage crop or turf treatments and exposure via drinking (i.e. from puddles, irrigation equipment, or crop leaf whorls). Of these, no effects were detected in 67 studies, sublethal effects only were found in 23, lethal effects were detected in 60, and 31 exhibited mass bird mortality. Mineau also demonstrated by probabilistic modeling that pesticide application rate, intrinsic toxicity, relative dermal toxicity and possibly volatility were all critical factors in explaining the effects exhibited in these investigations. These models contributed to predictions of the widespread lethal risk of pesticides to birds from insecticide use in the USA (Mineau & Whiteside, 2006). This analysis revealed that the crops most susceptible to bird mortality in the USA were corn and cotton.

16.2.6 Published laboratory and outdoor enclosed-system experiments

Laboratory bioassay data may provide essential mechanistic underpinning to field investigations and evidence for potential environmental risks. They certainly have a degree of explanatory power, and lie at the heart of the complex risk assessment procedures employed by regulatory agencies. The bioassay data of greatest explanatory power are those that reveal no effects in a particular group of organisms, at least in the concentration ranges encountered in the field. If a pesticide is simply not toxic, then mechanisms of toxicity can not be recruited in the analysis of any observed effects in the field. When toxicity is revealed, the likelihood that this will be translated into actual environmental impacts, and the relationship between these possible impacts and the relative toxicities of different compounds is far more complex and challenging. Evidence of toxicity from laboratory-based bioassays therefore provides evidence of the potential for harm, and provides the justification for further inquiry and analysis. This section critically reviews data and analyses from laboratory and highly controlled field enclosure experiments to provide a basis for interpreting the value of this information in the evaluation of environmental risks.

Aquatic toxicologists use tanks, artificial pools or ditches, termed microcosms and larger mesocosms, to simulate aquatic habitats. These are amenable to experimental manipulation and are thought to offer more realism than laboratory-based, single-species bioassays. The systems used are enormously diverse, as are the flora and fauna selected and the endpoints that are recorded. The most frequently studied taxa are phytoplankton and zooplankton, macrophytes (vascular plants and loose filamentous algae), macroinvertebrates and fish (Brock & Buddle, 1994). The chosen endpoints are also enormously diverse, and include community composition, chlorophyll *a* content, biomass, abundance, similarity, diversity and spatial distribution. Results reveal primary effects, which many argue are likely to equate to those that may occur in the real world, and secondary effects and recovery dynamics which are also affected by the construction and characteristics of the experimental system. In an indoor system, designed to simulate drainage ditches, Brock *et al.* (1992a) demonstrated effects of chlorpyrifos on amphipods, insects and isopods, and also revealed the importance of macrophytes in determining water and sediment concentrations of the pesticide. Secondary effects in the same system were also affected by the presence of macrophytes, and included trophic effects on primary producers, herbivores and other functional groups, that resulted from loss of arthropods (Brock *et al.*, 1992b). Functional impacts were also detected, particularly effects on community metabolism, signaled by decreases in dissolved oxygen and pH, and increases in alkalinity and conductivity (Brock *et al.*, 1993). There is a large and critical literature concerning the degree to which data from mesocosms can contribute to regulatory toxicology (e.g. Shaw & Kennedy, 1996; Maund *et al.*, 1997). Artificial pond systems may not relate closely to any natural system (e.g. Williams *et al.*, 2002); however, a recent review has found consistent patterns of response, relative to laboratory-based single-species toxicity tests which revealed that the ecological effects threshold is normally about ten times the maximum concentrations that are determined to be acceptable in the laboratory tests (van Wijngaarden *et al.*, 2005).

The use of laboratory-based bioassay data in the evaluation of pesticide risks to non-target invertebrates can be equally problematic, although these data are available for hundreds of compounds and thousands of invertebrate species (e.g. Croft, 1990). In the case of these organisms so-called semi-field data, from artificial enclosures, although useful, are only available for a small number of pesticides and taxa (Jepson, 1993). Stark *et al.* (1995) used several methods for extrapolating laboratory test data to predict field effects against a number of predatory and parasitic invertebrates and found that these gave variable predictions of the compatibility of these compounds with IPM. Although laboratory test data are widely used to compare pesticides and determine their compatibility with IPM, in reality, these data must be considered alongside information concerning exposure rate, chemical fate and invertebrate behavior if they are to contribute to predictions of either short-term or long-term ecological effects (Jepson, 1989, 1993, 2007b). Differences in key life history variables determine the population trajectories of affected populations after initial toxic effects have taken place, further questioning the basis for using toxicological data alone in the evaluation of the potential risks that compounds pose (Stark *et al.*, 2004).

One way of overcoming the requirement for large plot sizes to compensate for the dispersed populations and mobility of small mammals, is to enclose small populations within replicated barriers. Edge *et al.* (1996) demonstrated a dose-response by vole (*Microtus canicaudus*) populations in a series of replicated enclosures, when they were exposed to a single treatment with azinphos-methyl. They argued that the short-term nature of the impact could reflect a rapid population response by the voles, and that effects might be severe when voles were exposed to multiple treatments. Enclosure techniques in this case, may have presented the only effective method for obtaining data concerning pesticide side effects.

16.3 Conclusions

This chapter has outlined the sources of data and approaches that are used for the evaluation of pesticide environmental risks. It has argued implicitly that evidence for potential effects can be drawn from a number of sources, relying most heavily on those data that reveal impacts under conditions of field use. This is an evidence-based approach, which requires critical evaluation of the data derived from each of the sources that are listed.

Table 16.1 provides a summary of the main constraints and uses associated with data derived from the sources of evidence for pesticide environmental impacts outlined in this chapter. In a particular location, data sets from these sources may be assembled for specific compounds or closely related materials, but these must be qualified by an assessment of their relevance to local conditions. These data may also be used to aid interpretation of observations or experiments, and also to assist end-users in ranking pesticides for their potential to inflict environmental harm. Some materials may accumulate a positive "environmental profile" because for example, chemical monitoring for these specific materials does not detect them, and because widespread use does not generate evidence for potential impacts from either monitoring data or experiments. Other materials may warrant far greater caution, because they have been repeatedly detected in concentrations that may cause adverse effects, and also because well-designed monitoring programs and experiments with these compounds have sufficient inferential power to attribute cause.

In summarizing some of the key challenges associated with some of these data sources, the chapter has revealed the conceptual challenges faced by organizations and individuals that might wish to develop more quantitative risk assessments. The chapter has not referred to simple environmental impact indices that are advanced by some as useful tools in IPM decision making or evaluation, because in the author's opinion, none of these to date has been adequately validated for use in real-world decision making. Similarly, it has not referred to the complex risk assessment procedures used by regulatory agencies, which must undertake this process for hundreds of compounds and use data sets that are usually built from laboratory bioassays and chemical fate and behavior data, prior to extensive field use. The author argues that there is a role for the characterization of compounds, based upon the properties

Table 16.1 Summary of the main constraints and uses associated with data derived from the sources of evidence for pesticide environmental impacts outlined in this chapter

Source of evidence	Major constraints	Main uses for data	Comments
Pesticide incident reporting	Biased towards cases where wildlife cadavers are readily detectable	• Evaluation of status and trends in wildlife poisoning as contributory factor to population status of threatened or endangered species • Early warning of unexpected hazards associated with specific compounds in commercial use • Detection and prosecution of illegal use and abuse, and monitoring of compliance with the law	Only a limited number of incident reporting schemes internationally, and requires high quality forensic and chemical analysis facilities, combined with an alert public; may be aggregated to a national or international scale
Biological monitoring and surveillance	• Normally not possible to definitively attribute cause, or to tease out the contribution of pesticide mortality to population status • Not effective in early warning	• Evaluation of status and trends in wildlife populations associated with agriculture	Monitoring and surveillance are rarely sensitive enough to detect anthropogenic changes in floral or faunal biodiversity, or specific effects associated with pesticides; successful programs are built upon clearly stated hypotheses and rigorous design, including control areas
Chemical monitoring and surveillance	Expensive, logistically and methodologically complex, and temporal and spatial resolution normally poor	• Measurement of off-crop movement and environmental distribution of pesticides • Contributes to accurate environmental concentration estimates for risk assessment	When available, these data are extremely informative and helpful in developing risk assessments; when sufficiently localized and detailed, they can be used to backtrack to particular uses that may then be addressed by IPM programs

		• Very effective early warning tool for potential ecological effects • Evaluation of the benefits of IPM programs and regulatory instruments	
Risk estimates from chemical monitoring	Risk estimates based upon toxicological data are not readily extrapolated to the field, even when based upon accurate chemical monitoring; the ecological context and ecological processes must also be taken into consideration	• Mapping and identification of places where impacts may be unacceptably high • Support for regulatory instruments, including the setting of limits • Very effective in developing early warning for potential ecological effects from chemical monitoring data • Evaluation of the benefits of IPM programs and regulatory instruments, particularly when followed up with field measurements	Most effective when follow-up assessments are possible *in situ*, following risk assessments that identify the possibility that adverse impacts may be occurring
Published field experiments	Highly context specific, and difficult to extrapolate to larger scales and other locations until a body of complementary experimental evidence has been assembled	• Testing of hypotheses that address specific mechanisms of environmental harm • Development of reduced-risk IPM regimes • Verification or further ecological exploration of initial risk assessments	Data from multiple experiments may be assembled within databases, or employed within meta-analysis of environmental impacts
Published laboratory and outdoor, enclosed system experiments	Exposure pathways and pesticide concentrations may not be sufficiently relevant to field conditions for simple assessments of environmental risk	• Essential data for the derivation of risk estimates when combined with environmental concentrations • Evidence of low, or no toxicity for certain materials is of great value in IPM program design	Toxicological datasets are extremely useful when handled and interpreted appropriately; particularly data that demonstrate no effects

and effects that they reveal in widespread field use, where chemicals confront ecological communities and processes and exert impacts that we must respond to and manage. IPM researchers and practitioners should not be inhibited or discouraged from assembling these data sets and discussing them with stakeholders as an additional element in the complex decision support process that underlies IPM.

References

Barnett, E. A., Fletcher, M. R., Hunter, K. & Sharp, E. A. (2006). *Pesticide Poisoning of Animals in 2005*. London: Department of Environment and Rural Affairs, Pesticide Safety Division. Available at www.pesticides.gov.uk/uploadedfiles/Web_Assets/PSD/finalPPA2005report.pdf

Benton, T. G., Bryant, D. M., Cole, L. & Crick, H. Q. P. (2002). Linking agricultural practices to insect and bird populations: a historical study over three decades. *Journal of Applied Ecology*, **39**, 673–687.

Brock, T. C. M. & Buddle, B. J. (1994). On the choice of structural parameters and endpoints to indicate responses of freshwater ecosystems to pesticide stress. In *Freshwater Field Tests for Hazard Assessment of Chemicals*, eds. I. R. Hill, F. Heimbach, P. Leeuwangh & P. Matthiessen, pp. 19–56. Boca Raton, FL: CRC Press.

Brock, T. C. M., Crum, S. J. H., van Wijngaarden, R. *et al.* (1992a). Fate and effects of the insecticide Dursban 4E in indoor *Elodea*-dominated and macrophyte-free freshwater model ecosystems. I. Fate and primary effects of the active ingredient chlorpyrifos. *Archives of Environmental Contamination and Toxicology*, **23**, 69–84.

Brock, T. C. M., van den Bogaert, M., Bos, A.R. *et al.* (1992b). Fate and effects of the insecticide Dursban 4E in indoor *Elodea*-dominated and macrophyte-free freshwater model ecosystems. II. Secondary effects on community structure. *Archives of Environmental Contamination and Toxicology*, **23**, 391–409.

Brock, T. C. M., Vet, J. J. R. M., Kerkhofs, M. J. J. *et al.* (1993). Fate and effects of the insecticide Dursban 4E in indoor *Elodea*-dominated and macrophyte-free freshwater model ecosystems. III. Aspects of ecosystem functioning. *Archives of Environmental Contamination and Toxicology*, **25**, 160–169.

Burn, A. J. (1992). Interactions between cereal pests and their predators and parasites. In *Pesticides and the Environment: The Boxworth Study*, eds. P. Greig-Smith, G. H. Frampton & A. Hardy, pp. 110–131. London: Her Majesty's Stationery Office (HMSO).

Cohen, J. E., Schoenly, K., Heong, K. L. *et al.* (1994). A food web approach to evaluating the effect of insecticide spraying on insect pest population dynamics in a Philippine irrigated rice ecosystem. *Journal of Applied Ecology*, **31**, 747–763.

Croft, B. A. (1990). *Arthropod Biological Control Agents and Pesticides*. New York: Wiley Interscience.

de Snoo, G. R., Scheidegger, N. M. I. & de Jong, F. M. W. (1999). Vertebrate wildlife incidents with pesticides: a European survey. *Pesticide Science*, **55**, 47–54.

Edge, W. D., Carey, R. L., Wolff, J. O., Ganio, L. M. & Manning, T. (1996). Effects of Guthion 2S on *Microtus canicaudus*: a risk assessment validation. *Journal of Applied Ecology*, **33**, 269–278.

Everts, J. W., Van Frankenhuyzen, K., Roman, B. *et al.* (1983). Observations on side effects of endosulfan used to control tsetse in a settlement area in connection with a campaign against human sleeping sickness in Ivory Coast. *Tropical Pest Management*, **29**, 177–182.

Everts, J. W., Kortenhoff, B. A., Hoogland, H. *et al.* (1985). Effects on non-target terrestrial arthropods of synthetic pyrethroids used for the control of tsetse fly (*Glossina* spp.) in settlement areas of the southern Ivory Coast, Africa. *Archives of Environmental Contamination and Toxicology*, **14**, 641–650.

Devine, D. J. & Furlong, M. J. (2007). Insecticide use: contexts and ecological consequences. *Agriculture and Human Values*, **24**, 281–306.

Duffield, S. J., Jepson, P. C., Wratten, S. D. & Sotherton, N. W. (1996). Spatial changes in invertebrate predation pressure in winter wheat following treatment with dimethoate. *Entomologia Experimentalis et Applicata*, **78**, 9–17.

Fuhrer, G. J., Morace, J. L., Johnson, H. M. *et al.* (2004). *Water Quality in the Yakima River Basin, Washington, 1999–2000*, US Geological Survey Circular No. 1237. Reston, VA: US Department of the Interior.

Gilliom, R. J., Barbash, J. E., Crawford, C. G. *et al.* (2006). *The Quality of our Nation's Waters: Pesticides in the Nation's Streams and Ground Water, 1992–2001*, US Geological Survey Circular No. 1291. Reston, VA: US Department of the Interior.

Good, J. A. & Giller, P. S. (1991). The effect of cereal and grass management on staphylinid (Coleoptera) assemblages in South-East Ireland. *Journal of Applied Ecology*, **28**, 810–826.

Graham-Bryce, I. J. (1977). Crop protection: a consideration of the effectiveness and disadvantages of current methods and of the scope for improvement. *Philosophical Transactions of the Royal Society of London B*, **281**, 163–179.

Greig-Smith, P., Frampton, G. H. & Hardy, A. (eds) (1992). *Pesticides and the Environment: The Boxworth Study*. London: Her Majesty's Stationery Office (HMSO).

Hornsby, A. G., Wauchope, R. D. & Herner, A. E. (1996). *Pesticide Properties in the Environment*. New York: Springer-Verlag.

Jepson, P. C. (ed.) (1989). *Pesticides and Non-Target Invertebrates*. Wimborne, UK: Intercept.

Jepson, P. C. (1993). Insects, spiders and mites. In *Handbook of Ecotoxicology*, ed. P. Calow, pp. 299–325. Oxford, UK: Blackwell Scientific Publications.

Jepson, P. C. (2007a). Challenges to the design and implementation of effective monitoring for GM crop impacts: lessons from conventional agriculture. In *Genetically Modified Organisms in Crop Production and their Effects on the Environment: Methodologies for Monitoring and the Way Ahead*, eds. K. Ghosh & P. C. Jepson, pp. 34–57. Rome, Italy: Food and Agriculture Organization of the United Nations.

Jepson, P. C. (2007b). Ecotoxicology and IPM. In *Perspectives in Ecological Theory and Integrated Pest Management* eds. M. Kogan & P. C. Jepson, pp. 522–551. Cambridge UK: Cambridge University Press.

Jepson, P. C. & Sherratt, T. N. (1996). The dimensions of space and time in the assessment of ecotoxicological risks. In *ECOtoxicology: Ecological Dimensions*, eds. D. J. Baird, P. W. Grieg-Smith & P. E. T. Douben, pp. 44–54. London: Chapman & Hall.

Jepson, P. C. & Thacker, J. R. M. (1990). Analysis of the spatial component of pesticide side-effects on non-target invertebrate populations and its relevance to hazard analysis. *Functional Ecology*, **4**, 349–358.

Kelly, B. C., Ikonomou, M. G., Blair, J. D., Morin, A. E. & Gobas, F. A. P. C. (2007). Food-web specific biomagnification of persistent organic pollutants. *Science*, **317**, 236–239.

Kogan, M. & Jepson, P. C. (2007). Ecology, sustainable development and IPM. In *Perspectives in Ecological Theory and Integrated Pest Management*, eds. M. Kogan & P. C. Jepson, pp. 1–44. Cambridge, UK: Cambridge University Press.

Lahr, J., Diallo, A. O., Gadji, B. *et al.* (2000). Ecological effects of experimental insecticide applications on invertebrates in Sahelian temporary ponds. *Environmental Toxicology and Chemistry*, **19**, 1278–1289.

Lahr, J., Badji, A., Marquenie, S. *et al.* (2001). Acute toxicity of locust insecticides to two indigenous invertebrates from Sahelian ponds. *Ecotoxicology and Environmental Safety*, **48**, 66–75.

Lambert, M. R. K. (1997). Environmental effects of heavy spillage from a destroyed pesticide store near Hargeisa (Somaliland) assessed during the dry season, using reptiles and amphibians as bioindicators. *Archives of Environmental Contamination and Toxicology*, **32**, 80–93.

LeNoir, J. S., McConnell, L. L., Fellers, G. M., Cahill, T. M. & Seiber, J. N. (1999). Summertime transport of current-use pesticides from California's Central Valley to the Sierra Nevada mountain range, USA. *Environmental Toxicology and Chemistry*, **18**, 2715–2722.

Leonard, A. W., Hyne, R. V., Lim, R. P. & Chapman, J. C. (1999). Effect of endosulfan runoff from cotton fields on macroinvertebrates in the Namoi River. *Ecotoxicology and Environmental Safety*, **42**, 125–134.

Leonard, A. W., Hyne, R. V., Lim, R. P., Pablo, F. & van den Brink, P. J. (2000). Riverine endosulfan concentrations in the Namoi River, Australia: link to cotton field runoff and macroinvertabrate population densities. *Environmental Toxicology and Chemistry*, **19**, 1540–1551.

Li, Q., Luo, Y., Song, J. & Wu, L (2007). Risk assessment of atrazine polluted farmland and drinking water: a case study. *Bulletin of Environmental Contamination and Toxicology*, **78**, 187–190.

Maund, S. J., Sherratt, T. N., Strickland, T. *et al.* (1997). Ecological considerations in pesticide risk assessment for aquatic systems. *Pesticide Science*, **49**, 185–190.

McConnell, L. L., LeNoir, J. S., Datta, S. & Seiber, J. N. (1998). Wet deposition of current use pesticides in the Sierra Nevada mountain range, California, USA. *Environmental Toxicology and Chemistry*, **17**, 1908–1916.

McGlaughlin, A. & Mineau, P (1995). The impact of agricultural practices on biodiversity. *Agriculture, Ecosystems and the Environment*, **55**, 201–212.

Mackay, D. (2001). *Multimedia Environmental Models: The Fugacity Approach*. Boca Raton, FL: CRC Press.

Matthews, G. A. (2000). *Pesticide Application Methods*. Oxford, UK: Blackwell Science.

Mineau, P. (2002). Estimating the probability of bird mortality from pesticide sprays on the basis of the field study record. *Environmental Toxicology and Chemistry*, **21**, 1497–1506.

Mineau, P. & Whiteside, M. (2006). Lethal risk to birds from insecticide use in the United States: a spatial and temporal analysis. *Environmental Toxicology and Chemistry*, **25**, 1214–1222.

Naranjo, S. E. (2005). Long-term assessment of the effects of transgenic *Bt* cotton on the abundance of non-target arthropod natural enemies. *Environmental Entomology*, **34**, 1193–1210.

Newton, I. (2004). The recent declines of farmland bird populations in Britain: an appraisal of causal factors and conservation actions. *Ibis*, **146**, 579–600.

Nimmo, D. R. & McEwen, L. C. (1998). Pesticides. In *Handbook of Ecotoxicology*, vol. 2, ed. P. Calow, pp. 619–667. Oxford, UK: Blackwell Scientific Publications.

Noon, B. R. (2003). Conceptual issues in monitoring ecological resources. In *Monitoring Ecosystems: Interdisciplinary Approaches for Evaluating Ecoregional Initiatives*, eds. D. E. Busch & J. C. Trexler, pp. 27–72. Washington, DC: Island Press.

Ormerod, S. J. & Watkinson, A. R. (2000). Special profile: birds and agriculture. *Journal of Applied Ecology*, **37**, 699–705.

Peveling, R. & Demba, S. A. (2003). Toxicity and pathogenicity of *Metarhizium anisopliae* var. *acridium* (Deuteromycotina, Hyphomycetes) and fipronil to the fringe-toed lizard *Acathodactylus dumerili* (Squamata: Lacertidae). *Environmental Toxicology and Chemistry*, **22**, 1437–1447.

Peveling, R. & Nagel, P. (2001). Locust and Tsetse fly control in Africa: does wildlife pay the bill for animal health and food security? In *Pesticides and Wildlife*, ed. J. J. Johnston, pp. 82–108. Washington, DC: American Chemical Society.

Peveling, R., Rafanomezantsoa, J.-J., Razafinirina, R., Tovonkery, R. & Zafimaniry, G. (1999a). Environmental impact of the locust control agents fenitrothion, fenitrothion–esfenvalerate and triflumuron on terrestrial arthropods in Madagascar. *Crop Protection*, **18**, 659–676.

Peveling, R., Attignon, S., Langewald, J. & Oouambama, Z. (1999b). An assessment of the impact of biological and chemical grasshopper control agents on ground-dwelling arthropods in Niger, based on presence/absence sampling. *Crop Protection*, **18**, 323–339.

Peveling, R., McWilliam, A. N., Nagel, P. *et al.* (2003). Impact of locust control on harvester termites and endemic vertebrate predators in Madagascar. *Journal of Applied Ecology*, **40**, 729–741.

Posthuma, L., Suter, G. W. & Traas, T. P. (2002). *Species Sensitivity Distributions in Ecotoxicology*. Boca Raton, FL: Lewis Publishers.

Prasifka, J. R., Hellmich, R. L., Dively, G. P. & Lewis, L. C. (2005). Assessing the effects of pest management on nontarget arthropods: the influence of plot size and isolation. *Environmental Entomology*, **34**, 1181–1192.

Radcliffe, D. A. (1967). Decrease in eggshell weight in certain birds of prey. *Nature*, **215**, 208–210.

Radcliffe, D. A. (1993). *The Peregrine Falcon*, 2nd edn. London: Poyser.

Robinson, R. A. & Sutherland, W. J. (2002). Post-war changes in arable farming and biodiversity in Great Britain. *Journal of Applied Ecology*, **39**, 157–176.

Schoenly, K. G., Cohen, J. E., Heong, K. L. *et al.* (1994). Quantifying the impact of insecticides on food web structure of rice-arthropod populations in a Philippine farmer's irrigated field: a case study. In *Food Webs: Integration of Patterns and Dynamics*, eds. G. A. Polis, G. A. & K. Wisemiller, pp. 343–351. London: Chapman & Hall.

Shaw, J. L. & Kennedy, J. H. (1996). The use of aquatic field mesocosm studies in risk assessment. *Environmental Toxicology and Chemistry*, **15**, 605–607.

Settle, W. H., Ariawan, H., Astuti, E. T. *et al.* (1996). Managing tropical rice pests through conservation of generalist natural enemies and alternative prey. *Ecology*, **77**, 1975–1988.

Sherratt, T. N. & Jepson, P. C. (1993). A metapopulation approach to modelling the long-term impact of pesticides on invertebrates. *Journal of Applied Ecology*, **30**, 696–705.

Sparling, D. W., Feller, G. M. & McConnell, L. L. (2001). Pesticides and amphibian population declines in California, USA. *Environmental Toxicology and Chemistry*, **20**, 1591–1595.

Stark, J. D., Jepson, P. C. & Mayer, D. F. (1995). Limitations to use of topical toxicity data for predictions of pesticide side-effects in the field. *Journal of Economic Entomology*, **88**, 1081–1088.

Stark, J. D., Banks, J. E. & Vargas, R. (2004). How risky is risk assessment: the role that life history strategies play in susceptibility of species to stress. *Proceedings of the National Academy of Sciences of the USA*, **101**, 732–736.

Suter, G. W. (1993). *Ecological Risk Assessment*. Albany, GA: Lewis Publishers.

Thiere, G. & Schulz, R. (2004). Runoff-related agricultural impact in relation to macroinvertebrate communities of the Lourens River, South Africa. *Water Research*, **38**, 3092–3102.

Tingle, C. C. D. (1996). Sprayed barriers of diflubenzuron for control of the migratory locust (*Locusta migratoria capito* (Sauss.)) [Orthoptera: Acrididae] in Madagascar: short-term impact on relative abundance of terrestrial non-target invertebrates. *Crop Protection*, **15**, 579–592.

Ueno, D., Takahashi, S., Tanaka, H. *et al.* (2003). Global pollution monitoring of PCBs and organochlorine pesticides using skipjack tuna as a bioindicator. *Archives of Environmental Contamination and Toxicology*, **45**, 378–389.

Unal, G. & Jepson, P. C. (1991). The toxicity of aphicide residues to beneficial invertebrates in cereal crops. *Annals of Applied Biology*, **118**, 493–502.

Underwood, A. J. (1994). On beyond BACI: sampling designs that might reliably detect environmental disturbances. *Ecological Applications*, **4**, 3–15.

van der Werf, H. M. G. (1996). Assessing the impact of pesticides on the environment. *Agriculture, Ecosystems and the Environment*, **60**, 81–96.

van Wijngaarden, R. P. A., Brock, T. C. M. & van den Brink, P. J. (2005). Threshold levels for effects of insecticides in freshwater ecosystems: a review. *Ecotoxicology*, **14**, 355–380.

Williams, P., Whitfield, M., Biggs, J. *et al.* (2002). How realistic are outdoor microcosms? A comparison of the biota of microcosms and natural ponds. *Environmental Toxicology and Chemistry*, **21**, 143–150.

Chapter 17

Assessing pesticide risks to humans: putting science into practice

Brian Hughes, Larry G. Olsen and Fred Whitford

IPM has long been a staple of decision making in agricultural settings. In recent years, the concept of IPM has been widely utilized in non-agricultural pest management programs. While the foundation of IPM has been to better manage pests, in many aspects, IPM has come to symbolize the management of pesticide risks. Thus, it seems that IPM is both a pest management and an integrated pesticide management program. IPM practitioners today use the IPM principles to predict and manage the risks that both pests and pesticides pose to people and the environment (see Chapter 37). Here we will address how the risks of pesticides are evaluated under USA federal statutes and how mitigation measures can reduce risk.

17.1 Regulatory framework

Many of the older pesticides were metals (e.g. arsenic or copper) or compounds derived from plants (e.g. nicotine). Concerns were expressed that some of these substances could end up in processed food leading to food supply contamination. In 1906, the USA Congress passed the Food and Drugs Act to address the issue of adulterated products entering the food supply. This Act while general in language was quite powerful in that it became unlawful to manufacture a food product if it contained any unwanted or deleterious ingredients that could be injurious to human health. Congress would add to and clarify the Food and Drug Act over the years to update the law to meet growing concerns over food safety.

A landmark event occurred in 1938 when the US Congress passed the Food, Drug, and Cosmetic Act which set the benchmark as it related to pesticides in food and setting general standards for reducing the risks of exogenous chemicals in and on the food supply. The basic foundation to the law was that federal agencies would have to set tolerances for pesticides used to produce food and fiber. The tolerance is the permissible residue level for pesticides in raw agricultural products or processed food. In 1954, the Miller Pesticide Amendment listed procedures for setting these safety limits for pesticide residues on raw agricultural commodities. Government, specifically the US (Food and Drug Administration) (FDA), was now controlling the levels and risk.

Another federal law in the USA that tried to define risk was the Food Additives Amendment to the Food, Drug, and Cosmetic Act in 1958. What would become one of the most controversial aspects of the law was the Delaney clause. This clause prohibited food additives to be found in processed food if they were shown to induce cancer in humans or in laboratory animals. Pesticides were considered food additives to processed foods. There arose an inconsistency in these laws

Integrated Pest Management, ed. Edward B. Radcliffe, William D. Hutchison and Rafael E. Cancelado. Published by Cambridge University Press.

which granted on one hand a tolerance for a pesticide on raw agricultural commodities based on a risk/benefit analysis versus a no tolerance for the pesticide that might cause cancer if found in processed food.

The next major regulatory reform was a sweeping piece of legislation known as Food Quality Protection Act (FQPA) of 1996 (Food Quality Protection Act of 1996 § 1, 7 U.S.C. § 101). The law eliminated the inconsistency posed by the Delaney clause by excluding pesticides from the definition of food additives. It created a single standard for pesticides in both raw and processed foods. It greatly strengthened risk evaluation by considering combined exposures resulting from diet, drinking water and residential uses instead of evaluating each risk separately. With the FQPA, the exposures from each of these routes are added and compared to the toxicity data as the means of determining risk. The only exception is that occupational risks currently are still evaluated separately.

Beyond setting tolerances for aggregate exposures, FQPA required the US Environmental Protection Agency (EPA) to consider cumulative effects of classes of pesticides or those having the same mechanism of toxicity. Thus, exposures over multiple active ingredients will be used in some cases to determine if those exposures impact human health.

Another hallmark of FQPA was the incorporation of an additional safety factor for infants and children. The safety factor was intended to protect subpopulations that had special sensitivities as reported by the National Academy of Sciences Committee on Pesticide Residues in the Diets of Infants and Children in 1993 (National Academy of Sciences, 1993). FQPA requires that up to an additional ten-fold safety factor be incorporated into the risk assessment process to take into account pre- and postnatal toxicity shown in laboratory animals.

17.2 Risk assessment

The EPA has an established risk assessment process by which pesticides are evaluated to determine their level of risks to human health and the environment. The risk assessment process has three overarching components: toxicity assessment/hazard identification, exposure assessment and risk characterization. The end result of risk assessment is the fourth step called risk management where EPA evaluates the risk(s), and determines whether additional steps need to be undertaken to reduce the risks to acceptable levels.

17.2.1 Toxicity assessment/hazard identification

This assessment requires a variety of controlled experiments using laboratory animals which are used to determine the biological consequences of exposure to that pesticide. These tests are conducted by the product manufacturer or registrant and involve subjecting animals to a range of doses, through various routes and durations to determine the toxicological profile of the chemical. A key question to answer is at what level a pesticide produces no adverse effects.

Acute toxicity

Acute toxicity studies are used to mimic short-term exposures to relatively high levels of the pesticide. These studies help to determine the adverse effects when the pesticide is taken by an oral, inhalation or dermal route. These are the initial studies that provide the underlying data that are used to develop the hazard assessment of a pesticide. These data help identify what problems arise out of an exposure duration of a day or less. They serve as one of the benchmarks that mimic occupational exposures to a pesticide. These studies provide information to address how occupational exposures resulting from early re-entry into fields, mixing/loading or application measure against levels that produce adverse effects such as skin irritation, blindness and death. They form the basis for later evaluations to determine what personal protective equipment will be required on the pesticide label. On the basis of these acute studies, pesticides are then classified by signal words as Danger (highly toxic), Warning (moderately toxic) and Caution (slightly toxic) (Table 17.1).

Chronic toxicity

Pesticide exposures occur over longer periods of time especially if one considers dietary, including

Table 17.1 Classification of pesticide toxicity

Category	Signal word	Oral LD_{50} (mg/kg)	Dermal LD_{50} (mg/kg)	Inhalation LC_{50} (mg/l)
I	Danger	<50	<200	<0.2
II	Warning	50–500	200–2000	0.2–2
III	Caution	500–5000	2000–20 000	2–20
IV	Caution	>5000	>20 000	>20

Source: Adapted from 40 Code of US Government Federal Regulations Part 156.62 (US Federal Government, 2003).

drinking water, or residential contributions. Therefore, EPA requires the manufacturer to conduct a series of studies that measure the adverse impacts of subchronic (e.g. 90 days) studies and lifelong studies that mimic exposure over a lifetime (e.g. chronic studies). Subchronic studies involve exposing multiple animal species for up to 90 days through the same routes – oral, dermal and inhalation – in a similar manner as the acute studies. Chronic studies are used to determine the effects of a chemical to prolonged and repeated exposures covering much of the laboratory animal's lifespan. Chronic studies provide the data that addresses the role of the pesticide to cause cancer, liver disorders, etc.

Reproductive effects

Reproductive studies are an important set of studies that allow one to address the potential impact on the developing fetus and the mother. These studies determine if the pesticide has any deleterious effects on the ability of adult animals to conceive and reproduce or the offspring to grow and develop.

Mutagenicity

The last battery of tests that help to describe the toxicological properties of a pesticide is one that measures its mutagenetic potential. This testing is usually conducted in vitro with the pesticide and/or its metabolites to determine if it induces mutations in bacterial or mammalian DNA. The data obtained in these studies are reviewed with the other studies that are designed to measure the pesticide's oncogenicity and teratogenicity potential.

Lastly, a number of toxicokinetic studies define how the body handles the pesticide regarding absorption (oral, dermal or through inhalation), distribution (where does the pesticide go in the body), metabolism (how is the parent chemical altered) and excretion (how is the chemical eliminated).

The preceding tests are those basic to the understanding of the toxicology of a pesticide. They are conducted according to published guidelines by the EPA. In some cases EPA will ask the manufacturer to conduct additional tests when the agency believes that additional toxicological data is needed to better describe at what levels adverse effects of a pesticide are observed.

Laboratory studies that describe the toxicologic properties are used in lieu of using human subjects. Recently, the EPA outlined various protections for subjects in human research involved in pesticide testing. These protections included a ban on using sensitive subpopulations such as children or women of reproductive age and enhanced oversight of human studies conducted to support registration (US Federal Government, 2006). The hesitancy of using human subjects involve ethical considerations such as weighing the risk to the participants to the benefits they gain as a result of conducting the research as well as methodological considerations such as how much confidence to place on the low number of volunteers used in such studies.

Another direct or indirect look at human health effects associated with pesticide exposure is through the use of epidemiological studies. Epidemiology is an observational science that looks at events within a population to see if there are strong or weak associations between a population's exposure to a pesticide and an adverse effect within that population. Two basic studies are used to determine whether a pesticide exposure can be linked to a health impact.

A cohort study is conducted by selecting a group of people with similar exposures to a pesticide versus a control group of individuals who do not have the exposure. The point of demarcation is whether one is exposed or not. The epidemiologists will collect health information over time and into the future from the subjects to determine whether the incidence of disease in the exposed group is significantly different than the disease in the unexposed control group. The other classic epidemiology study is called a case-control study. In this method, an epidemiologist will start out with individuals with a disease and those without the disease. The scientist then goes back in time to gather detailed histories from each subject including diet, smoking and chemical exposures. In this study, attempts are made to link any exposures that might have resulted in higher disease rates.

17.2.2 Exposure assessment

From the toxicology studies one can determine at what levels adverse effects occur in laboratory animals. However, these numbers in themselves have little meaning without knowing the full extent and magnitude of the exposure. Thus, an additional step of assessing exposure is needed to address the risk that a pesticide might pose. Exposure assessment asks the questions: how are humans exposed, how frequent is the exposure and what is the amount of exposure? While toxicology studies can be controlled, the answers to these questions asked in exposure studies are rather complex because they must include the variations associated with human behavior. In general, two assessments are conducted. A general exposure assessment is made to describe intake levels from pesticides through diet, drinking water and use around and in the home. The second is an occupational assessment where one evaluates agricultural workers that either contact pesticides in the field post-application or while performing tasks using pesticides.

General exposure

Diet represents the primary source of exposure for most pesticides. The nature and extent of pesticide exposure through food and meat varies with the type of food, the pesticides used to produce those products, and the amount of food ingested. The basic estimate of dietary exposure is represented by the following equation:

$$\text{Pesticide intake in diet} = \sum(\text{Concentration of residue on each food} \times \text{Amount of each food consumed})$$

The food consumption data and pesticide residue information used in the equation is derived from multiple sources. The simplest evaluation is when EPA uses the legal tolerance levels that are set for the pesticide. Tolerances are the upper legal amount of a pesticide allowed on each crop (e.g. maize, spinach, tomato). These are called worst-case scenarios in that seldom are tolerance levels found on food products. It is an overestimation of the amount on food products. In other instances, data on pesticide residues that are actually on the food supply can be obtained from the US Department of Agriculture (USDA) Pesticide Data Program (PDP) (US Department of Agriculture, 2005). PDP is a national pesticide database program that has been in existence since 1991. Each year this program publishes a detailed report of their findings of how much pesticide is actually measured on food. Each year, new priorities are established that direct which foods are to be examined in greater detail. For instance, foods and juices consumed by infants and children have been given a high priority by the USDA. The data provides a wealth of residue (e.g. exposure) information for risk assessment calculations, such as ranges, means and lower limits of detection.

Even the PDP can overestimate levels of pesticides consumed especially in those products that are cleaned and cooked. The Total Diet Study (TDS), popularly called the market basket study, is an ongoing FDA program that determines levels of various pesticides, contaminants and nutrients in foods after they are prepared for consumption. A unique aspect of the TDS is that foods are prepared as they would be consumed (table-ready) prior to analysis. It allows for the pesticide to be washed off the vegetables and fruits as well as taking into account the destructive forces of heat in breaking down the pesticide. While it provides realistic estimates of how much pesticide is actually consumed, a single study can cost $US 1 million to conduct.

The amount of pesticide residue on the food is but half of the equation. Obviously, the more one eats of a specific crop, the more potential there is for greater consumption of pesticides when it is found on that food. Food consumption data is critical in the calculation of the amount of pesticide residue consumed in the diet. The USDA is the primary agency assigned the responsibility for collecting food consumption data for the USA population. Since 1989, the consumption data has been collected annually through a program called the Continuing Survey of Food Intake by Individuals. Personal interviews with volunteers determine what the subject consumed within a 24-hour period. As expected, the American diet has drastically changed over the past decades. Within the last 20 years, increased emphasis is also given to the unique dietary habits of children since their proportionally greater intake of fruit and fruit juices may lead to increased intakes of pesticides used on these commodities. Since 2001, the USDA partnered with the National Health and Nutrition Examination Survey (NHANES) has worked to improve and conduct the collection, assessment and dissemination of food consumption and related data of Americans. This current survey is called the What We Eat In America (WWEIA) survey. The data collected is based on such descriptions as pizza. The pizza is then broken down into the various raw product components such as tomato, maize, green pepper, etc. The researchers can then break down each product into an amount consumed.

The EPA considers drinking water to be an exposure pathway for certain pesticides and it is considered part of the aggregate assessment. The difficulty in assessing pesticides in drinking water is due to seasonal variations in the levels of contamination found in surface and groundwater. Various agencies monitor drinking water and have provided the data to the EPA. The most notable of these is the US Geological Survey's National Water Quality Assessment Program. Generally speaking, the amount of pesticide exposure is rather small as compared to the amount consumed in the diet. In other situations, the EPA will use various exposure models to predict the amount of estimated exposure of a pesticide in drinking water.

Residential and non-occupational exposures may be among the most difficult estimates to predict; scientists need to examine, for example, exposure from homes, school and daycare centers. Use of pesticides in and around the home, on turf or on pets provides multiple points of exposure. For example, the general population also is exposed through how a homeowner manages pests in and around the home. The indoor environment provides opportunity for exposure from insect control like foggers and crack and crevice treatments. Also, flea and tick treatments are often applied to pets and transferred to humans. Sanitizers and disinfectants are also commonly used in the home.

Indoor assessments are highly complicated due in part to the pesticides used, their application methods and human populations exposed. There are a number of factors that must be used when evaluating and calculating residential exposures:

(1) *Residential building factors* There are many types of residential construction: mobile homes, houses and apartments. Each varies with regard to the type of construction materials and the heating, air conditioning and ventilation requirements.
(2) *Demographic factors* Each age group differs with respect to their residential exposures. Infants and toddlers have greater inhalation rates for their size.
(3) *Human activity patterns* Infants and toddlers are more likely to crawl on floors near areas where a pesticide has been applied and concentrations are the greatest (Fenske *et al.*, 1990). They also spend a greater amount of their lives indoors compared to working adults who spend significant amount of time away from home. The young and the elderly may be considered more highly exposed due to the lack of mobility. Adults may also handle and apply pesticides, particularly to lawns and gardens. Children on the other hand may play in these

areas incurring greater exposure to the skin and they may even eat treated foliage.

The outdoor environment also has its unique exposures. Outdoors, weed and insect control products may be applied to lawns, gardens, trees and sidewalks to control pest problems around the house. Algaecides are used in swimming pools, paint and wood preservatives to protect decking, termiticides to protect structures and rodenticides to control mice and rats. Each use allows for exposures through use or human interactions in the environment.

Turf applications are a major outside source of pesticide residue to the homeowner including the handling and application of pesticides. Foliar residues are a potential source of dermal pesticide exposure. Residues that can be transferred from foliage to humans are termed dislodgeable residues. The type and duration of human activity determine the residue concentrations transferred during normal human contact with treated surfaces. Dislodgeable residue studies are taken from foliage immediately after the applied residues have dried and for several days after to determine the decay or dissipation rate of the pesticide.

Occupational exposure

The EPA is required under the Code of Federal Regulation, Title 40 (40 CFR 158.202) to establish minimum re-entry times before agricultural workers can re-enter agricultural fields after a pesticide has been applied. Beyond requiring toxicity data, the EPA requires studies involving dissipation of foliar pesticide and residue exposure to farm workers from post-application entry. These studies have been outlined by the EPA in their Occupational and Residential Exposure Guidelines.

In 1994, the Agricultural Re-entry Task Force was organized to develop a generic agricultural re-entry exposure database intended to address agricultural post-application/re-entry data requirements. Passive whole-body dosimetry and inhalation exposure monitoring are the standard methodologies used to determine exposure levels. The EPA determined that exposure to pesticide handlers is more a function of the job and application technique rather than the specific product being applied.

Generic re-entry exposures were calculated using a set of crop groupings. Exposures are considered similar in these crop groupings if the plants are similar in structure, height and appearance. Thus, the plants were grouped according to characteristics they shared with each other (e.g. height, leaf structure, etc.) and agronomic activities involved (e.g. harvesting, thinning, pruning, scouting, etc.). Thus, maize would be a tall row crop which includes sorghum and sunflower as crops with similar physical characteristics.

Exposures are also defined based on the types of activity involved in crop production. This information can be obtained from crop profiles. Crop profiles are descriptions of crop production and pest management practices compiled on a state basis for specific commodities and include worker activities that occur during the growing season. (Crop profiles can be found at: www.ncipmc.org/profiles/index.cfm.) These profiles provide a tool that the EPA can use to identify low to high risk activities. Figure 17.1 illustrates the types of information available from crop profiles.

Intuitively, exposure might be more if the activity required lots of contact when the plants are in the later growth stages (e.g. harvesting) as compared with activities when the plants are smaller and there is less plant surface area to come into contact. The end result is to predict potential exposures based on duration and activity of the workforce. For instance, workers involved in seed maize production can be divided into eight distinct activities and durations and ranked with their relative exposure potentials (Table 17.2).

The important question is how much residue would dislodge itself from the crop onto the person walking through or working in a field. True exposure to the person is not how much is on the plant at the time of application, but how much of the residue remaining over time can be rubbed off the plant onto the person while performing various activities. Residue dissipation studies are conducted taking foliar samples and determining the dislodgeable foliar residue. Dislodgeable foliar residues (DFRs) are often described as residues present on the surface of a leaf that are available for transfer from the leaf onto skin or clothing.

Foliar samples are taken prior to the day of application, directly after application (4 or

Table 17.2 Maize seed activities, duration and relative exposure potentials

Activity	Duration (weeks)	Exposure potential
Planting	5	Low
Isolation	6	Very low
Irrigation	14	Very low
Pest scouting	12–16	Moderate to high
Rogueing	6	Moderate
Detasseling	6	Moderate to high
Certification of detasseling	6	Moderate to high
Phytosanitary inspection	8	Moderate to high

Source: L. G. Olsen (personal communication).

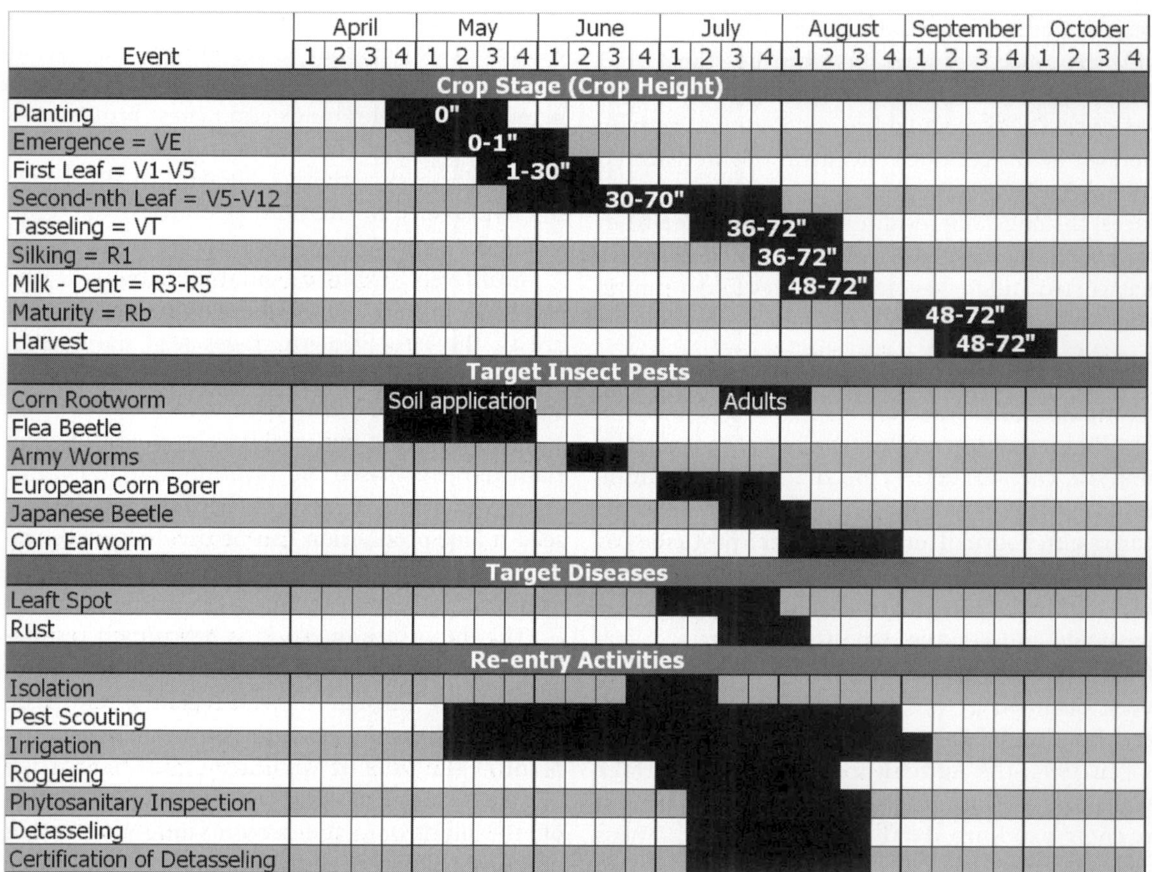

Fig. 17.1 Timeline indicating when events and activities occur in the production of maize seed (L. G. Olsen, personal communication).

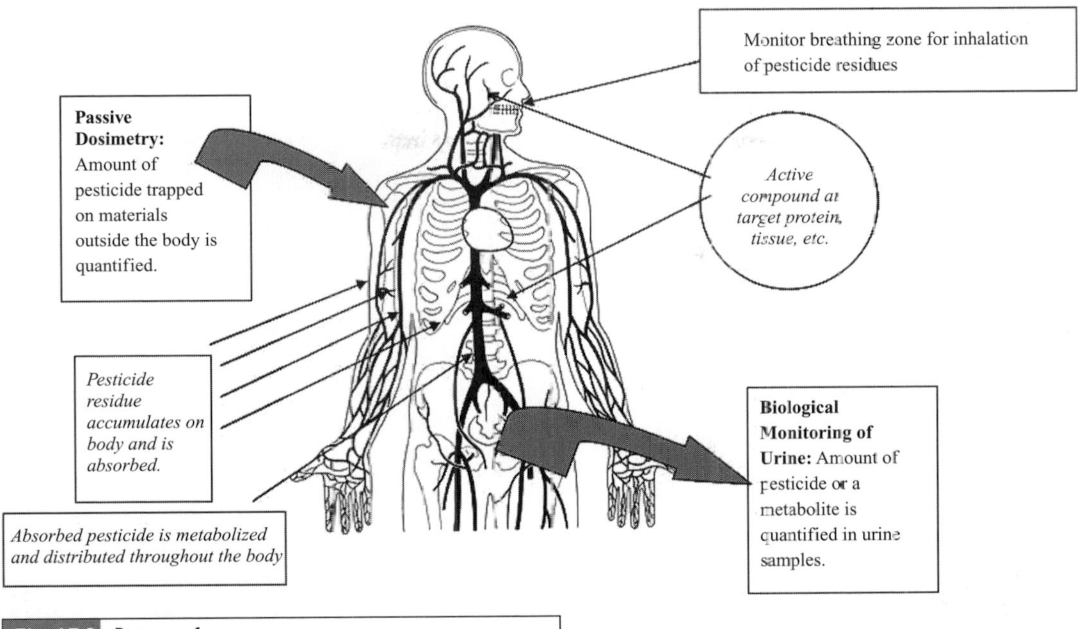

Fig. 17.2 Routes of exposure.

12 hours as appropriate) and at 1, 2, 4, 7, 10, 14, 21, 28 and 35 days after application. Foliar samples consist typically of a leaf punch that can accurately measure the recommended 400 cm^2 of leaf surface to be processed and analyzed. The samples are taken randomly within each field and at varying heights in the crop.

The processing of the plant material requires the leaf surface to be washed with an appropriate chemical to remove dislodgeable residues. The testing process requires the leaf tissue to be processed within 4 hours. The results are reported in μg of residue/cm^2 of plant material.

The data from the dislodgeable foliar residue analysis is used to calculate the transfer coefficient (TC). The TC is the amount of foliage that a worker comes into contact within 1 hour and is calculated by the following equation.

$$\text{Transfer coefficient (TC)} = \text{Dermal exposure } (\mu g/hr)/\text{DFR}(\mu g/cm^2)$$

Transfer coefficients are dependent on the type of crop, application method and worker activity in the field. The application of this TC is to calculate the dose or dermal exposure of an individual.

$$\text{Dose (mg/kg per day)} = (TC)(DFR)(AT)(AB)/(BW)$$

where

TC = transfer coefficient in cm^2/h
DFR = dislodgeable foliar residue in μg/cm^2
AT = activity time in h
AB = dermal absorption (fraction absorbed through the skin)
BW = body weight in kg.

The result is a dose in milligrams/kg body weight which can be compared to the appropriate RfD (see discussion under "Threshold effects" in Section 17.2.3).

Dermal exposure from post-pesticide application is assessed by estimating two main points of body entry: inhalation and dermal contact. Estimates of dermal exposure often use the patch testing, whole body dosimetry, and hand rinse and face wipes. The external estimate of exposure is also known as passive dosimetry although some have used biological monitoring to estimate absorbed dose (Fig. 17.2).

Patch testing involves placing 10–12 absorbent cloth patches on the outside of clothing on the chest, back, upper arm, forearm, thigh and lower leg. Each patch covers an approximate area of 100 cm^2. Residues are trapped and collected on the cloth. At the end of a predetermined exposure period, the patches are analyzed for the presence of the pesticide. The resulting data collected for each patch is in μg of pesticide residue/cm^2 of patch material. These results of how much pesticide was collected per cm^2 are then extrapolated to determine how much pesticide would have been collected on the entire outside surface of the body. For example, a pesticide applicator is found to have 0.1 μg/cm^2 from patches placed on the chest. This 0.1 μg/cm^2 value would be multiplied by 3454 cm^2 which is the average size of a male chest. Thus, it would be predicted that 345.4 μg of the pesticide would be expected to be found on the front chest.

The EPA will also accept a similar test method known as whole body dosimetry. Whole-body dosimeters generally consist of an inner layer such as long underwear garments covered with long pants and a long-sleeved shirt of an absorbent material (Fig. 17.3). These clothes are worn throughout the exposure period while the person is involved in conducting field tasks such as planting, hoeing, weeding and harvesting. The inner dosimeter or undergarments are sectioned into upper and lower leg, upper and lower arm, and front and rear torso. Each piece of the dosimeter is analyzed separately and then are added together to obtain a total amount of residue. The outside layer also is analyzed for the amount of transferable residue. Often, these studies can show that the outer layer proves very difficult for many pesticides to move through onto the inner garments.

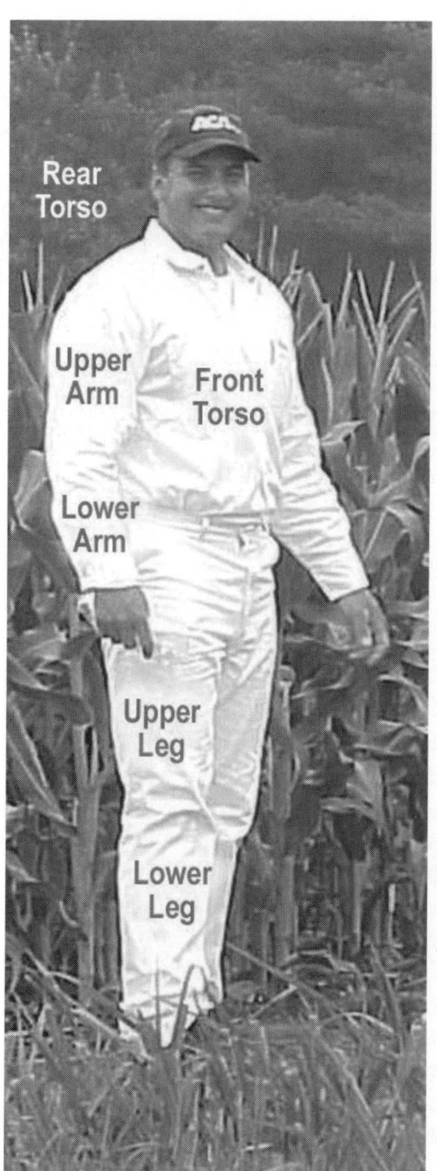

Fig. 17.3 Whole-body outer dosimeters.

The hands are often the part of the body which receives the majority of the pesticide exposure to the whole body. To estimate this amount of exposure on the skin surface, the subject is asked to wash their hands after they have performed their various jobs for a specified period of time. After the allotted time has expired, workers are asked to wash their hands with a solution known to remove the product from the surface of the skin. The wash solution is collected and analyzed. Gloves too can be used to capture this data in much the same way that whole-body dosimeters are used. Areas of the face and neck are assessed by wiping the area with a moist absorbent material and then submitting the material for analysis.

Personal air samplers are used to estimate the amount of pesticide inhaled by workers. A portable battery powered monitoring pump is clipped to the belt and a tube is run up the back and clipped to the collar of each worker. Inserted into the tube near the collar is an absorbent filter that traps airborne pesticides. The filter is then analyzed to determine the amount of pesticide per liter of airflow.

For mixer/loaders and applicators, industry and the regulatory agencies cooperated in establishing the Pesticide Handlers Exposure Database (PHED) in the early 1990s to gather data for occupational risk assessments for these activities. Companies contributed their exposure data for the common benefit of all registrants. PHED was then used by the regulatory agencies and registrants to satisfy the data requirements. PHED satisfied the need for occupational exposure data during the 1990s. However, PHED is incomplete since many exposure scenarios are not covered and today there are new handling systems and formulations that are not covered in PHED.

Similar to the Agricultural Re-entry Task Force, industry and the EPA formed the Agricultural Handlers Exposure Task Force to gather a generic database to determine exposures to pesticide mixer, loader and applicators in support of pesticide registration.

17.2.3 Risk characterization

The integration of data from the toxicity studies compared to the estimates of exposure is a detailed process called risk characterization. Risk assessors conducting these evaluations often describe risks in numerical terms. This risk characterization is necessary to determine if actions are required to increase the margin of safety to the general population or agricultural workers from pesticide use. Therefore, the greater the extent to which the exposure is below an adverse effect level, then that product is less risky. Concerns are raised when the estimates of exposure are above levels known to have produced a recognizable symptom.

Making conclusions around a single line – exposures below known toxicity levels are "safe" – are in themselves risky statements. There are many reasons why one has concerns about simple models such as these. There are variables that are hard to define. For instance, might there be differences in response between laboratory animals and people? Might individuals have different sensitivities compared to others resulting in different responses like between children and adults? Thus, risk characterization errs on the side of safety. The EPA scientists take into account the uncertainties in the data and the unanswered questions by reducing the toxicological line that separates one level that creates an adverse effect from the one that doesn't. For instance, the toxicological data indicates that at 100 mg/kg of body weight the pesticide does not produce an observable effect. Instead, the EPA might divide the level by a factor of 100 to derive a level of limited concern at 1 mg/kg. The exposure level is then compared to the lower level which has the built-in 100 × safety factor. The threshold for safety in this example is increased by 100-fold. The risk characterization adequately takes into account the quality of the data in terms of its variability and uncertainty and how each is accounted for in the assessment.

Threshold effects

A threshold effect is one where certain levels produce an adverse effect, while lower levels do not. For threshold effects only, the EPA determines a Reference Dose (RfD) or acceptable daily intakes in which the agency can reasonably anticipate the level found in food is unlikely to harm the consumer. RfDs are based on oral intakes for various periods of exposure (acute, subchronic and chronic) in different species of laboratory animals. For example, one may establish a No Observable Adverse Effect Level (NOAEL) is 10 mg/kg per day based on a 2-year (chronic) dietary study in rats. To establish an RfD, the EPA will divide the NOAEL by various safety factors. The safety factors include 10× based on the uncertainty for interspecies extrapolation from animals to humans or another 10× based on the variability in the human population. Additional safety or modifying factors may be incorporated into the RfD including 3× to 10× based on database sufficiency or as directed by FQPA where children and infants need additional protection. Dietary and drinking water evaluations depend on the RfDs as comparison values. For instance, the NOAEL from the laboratory animals is 10 mg/kg per day. With the uncertainties inherent in the risk characterization, the assessor could divide by a safety factor of 300 (10× for interspecies, 10× for human variability and 3× for children and infants). Thus, whether a pesticide is characterized as risky depends on whether the exposure that is estimated goes above 0.003 mg/kg per day.

RfD = NOAEL/Safety factors

In the assessment of occupational risk, a Margin of Exposure (MOE) is developed by dividing the NOAEL by the exposure estimate. The difference between the two values is obviously a percentage. The EPA uses the guideline of a MOE of 100 or greater as a margin of safety to assure that worker exposures will not result in any adverse health effects. The acceptable margin of safety is developed similar to that of the RfD safety factor approach: 10× to account for the uncertainty in extrapolating from effects seen in animal studies to humans and 10× to account for the variable responses within the human population to pesticides.

Non-threshold (carcinogenic) effect

It is believed that cancers and some chronic diseases can result from accumulation of adverse effects resulting from low-level exposures over a period of time and may not be evident until some time after the exposures have occurred. In other words, these diseases do not operate in terms of a threshold creating the problem. The EPA assesses the effects of pesticides differently to calculate the risk of exposure and the development of cancer. The medical community and the EPA make the assumption that any exposure can be accompanied by an increase risk of cancer. Therefore, a risk of 1 in 1 million indicates that if everyone in a population of 1 million were similarly exposed to a chemical, there may be one additional cancer case in that population. The level of acceptable risk to the general population is usually 1 in 1 million but may vary as a policy decision between various state and federal agencies and the affected population.

Dose–response curves from animal studies are used to estimate carcinogenic risk. The lower end of the dose response curve is extrapolated back to zero using various mathematical models. Since this mathematical construct is an estimate, a dose–response curve that is at the upper 95th confidence level is used for risk assessment purposes. The use of upper 95th percentile indicates that the true dose–response curve will be represented 95% of the time. Thus, at a risk of 1 in 1 million there is a 95% certainty that at most only one additional cancer case in a population will result and may be zero.

The slope of the dose-response curves at the 95% confidence level represents cancer potency. This is called the q^* with units of reciprocal mg/kg body weight per day or (mg/kg body weight)$^{-1}$. Risk is a unitless value calculated by finding the dose absorbed multiplied by the q^*.

Aggregate risk

In the past, risk characteristics were often made separately for dietary, residential, water intake and occupation. FQPA requires that the EPA calculate the total dose from pesticides through all routes of exposure. In other words, the EPA must take all of the exposures from food, water and residential use for a specific pesticide to determine how that level compares to some toxicological endpoint.

For dietary risk, the total dose is delivered through residues on a wide variety of fruits and vegetables all through the oral route. Therefore, the evaluation of dietary risk and toxicology endpoints derived from an animal study involving oral administration of the pesticide is straightforward. The process for determining aggregate risk is more complicated because other activities and routes of exposure are summed. For instance, a child's risk is dietary and also residential which involves exposure to treated surfaces in the home and turf outside the home. So the aggregate to a child involves oral exposure not only to foods but also from hand-to-mouth behaviors, inhalation and dermal exposures in and around the home.

The aggregate must also be specific to the duration of the exposure. If an exposure is acute, then only data obtained from acute exposure studies should be used in the assessment. Second, the studies must match with respect to their route of exposure; if the exposure is oral, dermal or inhalation then the route of exposure in the animal study should match. Should the data not be available for the specific route of exposure then the data can be used to extrapolate an absorbed dose. For instance, where exposure is on the skin and no reliable dermal study can be found, then it can be assumed that a certain percentage of the dose is absorbed through the skin.

In this way the risks are calculated, then summed as follows: Dietary (i.e. eating an apple):

$$I_a = C_a \times R_a$$

where

C_a = consumption rate (in g/kg body weight per day)
R_a = residue (μg of pesticide/grams of food consumed)
I_a = intake (μg of pesticide/kg of body weight per day).

17.2.4 Risk management

The goal of the risk assessment process is to provide a mechanism to manage pesticide risks. During the risk management process the EPA determines the risk and mitigation measures needed to reduce food, water, worker and environmental risks to acceptable levels. The worst-case scenario for pesticides that do not meet that standard would be not to register that active ingredient, or the cancellation of the right of the manufacturer to sell their product in the market. Often this is not a good solution for either the producer or the consumer particularly when there are few alternate options to control pests safely and economically. One option is for the registrant to generate more realistic numbers for input into the risk assessment process.

For food residues that exceed the RfD, the EPA can take action to reduce the allowable uses on different commodities. With each use removed or denied, the theoretical exposure to that pesticide is reduced. Furthermore, the EPA can also limit the number of applications of a pesticide or limit the amount per application to ensure such levels are met. Pre-harvest intervals can also be lengthened to allow time for residue dissipation in the field.

For pest control in the home, many insecticides are now in the form of gels and baits. Application techniques have also changed so that pesticides applied indoors via wide area applications are now crack and crevice treatments which reduce the amount of pesticide and the potential for exposures. In many states, laws prevent the pesticide from being used in or around institutional settings like schools and daycare centers when buildings or grounds are being occupied. Pesticide use can also be made less risky by repackaging or reformulating products or by adding dyes or bitterants.

Workers may be exposed upon re-entering fields after pesticide application. The EPA can manage the risks to these workers by limiting the number of applications or reducing the rate of application or increasing the Restricted Entry Interval (REI). Consider the case in Table 17.3 where the various MOEs for azinphos methyl have been calculated for almonds and apples. The MOEs have been calculated using the labeled rate and the particular TC for each activity listed. Based on the former REIs the MOEs are less than 100. To manage the risks the REIs have been lengthened to where the calculated MOEs meet the target level of 100.

For those mixing/loading or applying pesticides, personal protective equipment (PPE) is required by label. EPA has instituted worker protection standards to protect workers by requiring PPE, central notification, availability of sanitary facilities and posting requirements restricting entry after pesticide application.

Risk management decisions may also take into account the availability of other control methods. For example, the availability of other chemistries where the toxicity is lower can result in a greater MOE. Newer chemistries can include the use of neonicotinoids or insect growth regulators. Plant incorporated protectants such as the *Bacillus thuringiensis* Bt maize can lower the number of pesticide applications and the problems with timing of application.

17.3 Using pesticide risk information to make IPM decisions

The EPA and the manufacturers identify the level of risks posed by pesticides not just as an exercise for product registration for the marketplace, but to provide safety information. The identification of the product's toxicological characteristics is just as much a part of the decision making

Table 17.3 Margins of exposure (MOEs) for post-application agricultural re-entry activities

Crop (maximum label rate per application)	MOEs	Transfer coefficients for each activity
Almonds (2.0 lbs ai/A)	2 day REL: MOE = 3 for irrigating and scouting 14 day REL: N/A (no hand thinning) 28 day PHI: MOE = 3 for poling mummy nuts and pruning; REL where the MOE reaches 100: 71 days for irrigating, scouting & hand weeding, 104 days for poling & pruning	TC = 400 cm^2/h for irrigating, scouting and hand weeding TC = 2500 cm^2/h for poling and pruning
Apples, crab apples (1.5 lbs ai/A)	2 day REL: MOE = 23 for propping; MOE = 2 for irrigation and scouting; MOE = 1 for pruning, tying and training 14 day REL: MOE = 1 for hand thinning 14/21 day PHI: MOE = 2 for hand harvesting. REL where the MOE reaches 100: 32 days for propping; 79 days for irrigating, scouting and weeding; 102 days for hand harvesting, hand thinning, pruning, tying and training	TC = 100 cm^2/h for propping TC = 1000 cm^2/h for irrigating, scouting and weeding TC = 3000 cm^2/h for hand harvesting and thinning, pruning, tying and training

Source: Data from Environmental Protectioin Agency (2001).

process for growers as is using insect thresholds, disease forecast models, product efficacy and scouting programs.

Registration of a new active ingredient involves years of research and millions of dollars to bring the risk side into balance with the benefits side. This is important because the risk assessment process provides pesticide applicators reduced risk alternatives to reach IPM goals. IPM professionals see themselves as environmental problem-solving consultants who use pesticides more judiciously than in the past with the goal of minimizing the occurrence of pests using the lowest risk combination of tools appropriate to the individual situation. To reach these goals, IPM professionals are challenged to develop new approaches to manage insects, rodents, plant diseases and weeds. No matter how diligent the IPM practitioner is in the use of alternative management strategies, they will find that pesticides (conventional or organic) are an important tool in their IPM program.

An IPM practitioner can use the information developed for the EPA registration process to identify products that pose less risk to people and environment. The following criteria are useful in the selection of a pesticide product for use in an IPM program.

(1) *What is the signal word on a pesticide label?* The relative acute toxicity of a pesticide product is reflected on the label by one of three signal words: Danger – most toxic, Warning – moderately toxic, Caution – least toxic. The signal word can also reflect the product's nonlethal effects such as skin and eye irritation. Products in the Caution category are preferred for IPM programs when a pesticide is required.

(2) *Is the product classified for restricted use or general use?* Compounds that are classified by EPA as restricted-use products (RUP) are analogous to prescription drugs that must be administered by a trained person such as a doctor or nurse. On the other hand, general-use pesticides (everything not a RUP) are like over-the-counter medicines that any person can purchase and use. A pesticide product is classified as an RUP due to concerns about the potential for harm if the product is not used properly. RUPs are only sold to and used by persons with certification issued by the state agency which is responsible for pesticide regulation. Although general-use pesticides are generally less risky than the RUPs, both can cause harm if not used according to label directions. RUPs are not used in IPM programs if alternatives exist.

(3) *Is the product a reduced-risk pesticide?* The EPA has created a category of pesticides that pose fewer risks when compared to other products currently in the marketplace for the control of similar pests. To obtain EPA's reduced-risk classification manufacturers must submit substantive data. Claims must be supported by evidence of reduced toxicity to humans or to other non-target organisms and/or improved environmental fate and transport. In return, EPA expedites the registration process for these products.

(4) *Does the product have acute and/or chronic effects?* Manufacturers produce a Material Safety Data Sheet (MSDS) for every pesticide active ingredient. The MSDS provides a great deal of information about the hazards of a product including acute and chronic effects, carcinogenicity and reproductive/developmental defects. It is important to remember that most MSDS are produced for the concentrated product in the jug or bag. Therefore the MSDS serves as a hazard communication document mainly for occupational use and the risks when the product is diluted are typically lower than for the concentrated form in the bag or jug.

Selecting reduced risk products in concert with limiting exposure through sound application methods is essential to sound IPM practices. The product label in essence is the ultimate risk management document used to minimize any risks associated with pesticide use and still allow users to reap its benefits. It is more than a list of directions on how to control specific pests – product costs, pests controlled, application rates and crops it can be used on. It also carries a tremendous amount of information based on the human health, wildlife protection and environmental testing process put in place by the EPA. For instance, it provides valuable information on preharvest application requirements that must be followed to prevent unacceptable residues on raw and processed foods. Another example is the re-entry interval that prevents individuals from returning to fields after a pesticide application to protect themselves from unacceptable exposures.

17.4 Conclusions

Many IPM professionals are providing what customers want, service that maximizes pest control while minimizing pesticide exposure to the environment and to applicators, farmers and clients. IPM represents a major step forward for growers, licensed pesticide applicators and the firms for which they work. It is the natural progression for growers and companies focused upon risk management as well as pest management. As governmental agencies and customers struggle with the positive and negative aspects of pesticides in every day life, IPM provides a "middle path" for a holistic and integrated approach to manage pest problems coupled with responsible pesticide use. Clients appreciate and often are willing to pay a premium for a program managed by a professional trained and skilled in the tenets of IPM. Pest management service providers understand that the best IPM programs include more narrowly focused selection and application of reduced risk pesticides, and better trained, more observant and more skilled advisors.

References

Fenske, R. A., Black, K. G., Elkner, K. P. *et al.* (1990). Potential exposure and health risks of infants following indoor residential pesticide applications. *American Journal of Public Health*, **80**, 689–693.

National Academy of Sciences (1993). *Pesticides in the Diets of Infants and Children*. Washington, DC: National Academies Press.

US Department of Agriculture (2005). *Pesticide Data Program: Annual Summary Calendar Year 2005*. Washington, DC: US Department of Agriculture, Agricultural Marketing Service.

Environmental Protection Agency (2001). *Interim Reregistration Eligibility Decision for Azinphos-Methyl Case No. 0235*, October 2001. Washington, DC: US Environmental Protection Agency, Office of Prevention, Pesticides and Toxic Substances.

US Federal Government (2003). 156.62 Toxicity category. *Code of Federal Regulation*, Title 40 CFR Chapter 1 (7–1-03 edition). Available at http://a257.g.akamaitech.net/7/257/2422/08aug20031600/edocket.access.gpo.gov/cfr_2003/julqtr/pdf/40cfr156.62.pdf.

US Federal Government (2006). Final rule. Protections for subjects in human research. February 6, 2006. *Federal Register: US Environmental Protection Agency (EPA)*, **71**, 6137–6176.

Whitford, F. (2002). *The Complete Book of Pesticide Management: Science, Regulation, Stewardship, and Communication*. New York: Wiley Interscience.

Whitford, F., Pike, D., Burroughs, F. *et al.* (2006). *The Pesticide Marketplace: Discovering and Developing New Products*, Purdue Pesticide Programs No. PPP-71. West Lafayette, IN: Purdue University.

Chapter 18

Advances in breeding for host plant resistance

C. Michael Smith

Production of crop plants with heritable arthropod resistance traits has been recognized for more than 100 years as a sound approach to crop protection (Painter, 1951; Smith, 2005). Hundreds of arthropod-resistant crops are grown globally and represent the results of long-standing cooperative efforts of entomologists and plant breeders. These crops significantly improve world food production, increase producer profits and contribute to reduced insecticide use and residues in food crops (Smith, 2004).

It is essential to determine the inheritance of arthropod resistance genes. Plant breeders do so by observing progeny segregating from crosses between resistant and susceptible parents to determine the mode of inheritance and action of the resistance gene or genes. Breeding methods such as mass selection, pure line selection, recurrent selection, backcross breeding and pedigree breeding are often used to incorporate arthropod resistance genes into cultivars of such crops as maize, rapeseed, rice, wheat, potato, cotton and alfalfa (Smith, 2005). The focus of this chapter is on how the inheritance of resistance has been determined for the development of these crops and how new methods have been adapted in twentieth- and twenty-first-century plant breeding to select for arthropod resistance genes.

18.1 Inheritance of resistance

Khush & Brar (1991) and Gatehouse *et al.* (1994) have prepared extensive reviews on the inheritance of arthropod resistance in food and fiber crops. The following subsections summarize the number of genes involved in resistance and the mode of inheritance of these genes to pests of fruit, forage, oilseed and vegetable crops, as well as the major cereal crops of maize, rice, sorghum and wheat. Specific reference citations in this section are listed in the supplemental bibliography.

18.1.1 Fruit, forage, oilseed and vegetable crops

Resistance in apple, lettuce, peach and raspberry to several species of aphids is controlled by the action of one or two genes inherited as dominant traits. Crosses between cultivated potato and different *Solanum* species have also shown that glandular trichome-based resistance to the green peach aphid (*Myzus persicae*) is a partially dominant trait controlled by one or two dominant genes. Similarly, resistance to the potato tuber moth (*Phthorimaea operculella*) is controlled by a small number of major genes.

Integrated Pest Management, ed. Edward B. Radcliffe, William D. Hutchison and Rafael E. Cancelado. Published by Cambridge University Press.

Resistance in mung bean (*Vigna radiata*) to the Azuki bean weevil (*Callosobruchus chinensis*) and the cowpea weevil (*Callosobruchus maculatus*) as well as cowpea (*Vigna unguiculata*) resistance to cowpea aphid (*Aphis craccivora*) are all also controlled by single genes inherited as dominant traits. The phytomelanin layer in *Helianthus* species resistant to sunflower moth (*Homoeosoma electellum*) is also inherited as a dominant trait controlled by a single gene. Resistance in common bean (*Phaseolus vulgaris*) to Mexican bean weevil (*Zabrotes subfasciatus*) is controlled by a toxic seed protein (arcelin) and is also inherited as a dominant trait. However, resistance to a bruchiid (*Acanthoscelides obtectus*) also derived from a *Phaseolus* accession, is controlled by two genes, inherited as recessive traits. Resistance in tomato to the spider mite *Tetranychus evansi* is also controlled by the action of one gene that segregates as a recessive trait.

Resistance in soybean to several species of defoliating Lepidoptera is linked to quantitative trait loci (QTLs). Similarly, resistance in common bean and lima bean to a leafhopper (*Empoasca kraemeri*) and the inheritance of hooked trichomes, an *Empoasca* resistance mechanism, are linked to QTLs. In the forage legumes alfalfa (*Medicago sativa*) and sweetclover (*Melilotus infesta*) single dominant genes control resistance to pea aphid (*Acyrthosiphon pisum*) and sweetclover aphid (*Therioaphis riehmi*). In contrast, alfalfa resistance to spotted alfalfa aphid (*Therioaphis maculata*) is controlled by QTLs.

18.1.2 Maize

Multiple genes linked to QTLs control resistance in maize to larval defoliation and stalk damage inflicted by several species of Lepidoptera. These include corn earworm (*Helicoverpa zea*), European corn borer (*Ostrinia nubilalis*), stem borer (*Sesamia nonagrioides*), spotted stem borer (*Chilo partellus*), southwestern corn borer (*Diatraea grandiosella*) and sugarcane borer (*D. saccharalis*). Depending on the pest, gene action involves epistatic as well as additive-dominance effects. Different genes condition resistance to both first and second generations of the European corn borer, but some genes condition resistance to both generations.

18.1.3 Sorghum

Sorghum resistance to greenbug (*Schizaphis graminum*) was originally found to be controlled by a single gene inherited as a partially dominant trait. QTL analyses have more recently been used to document resistance to different greenbug biotypes as well as resistance to the sorghum midge (*Stenodiplosis sorghicola*). In some genotypes of sorghum resistant to the sorghum shootfly (*Atherigona soccata*) resistance is controlled by QTLs and also inherited as a partially dominant trait. In other genotypes expressing resistance based on leaf trichomes, resistance is expressed as a recessive trait conditioned by a single gene. Two genes inherited as recessive traits control resistance in sorghum to head bug (*Eurystylus oldi*). Resistance to another pentatomid, *Calocoris angustatus*, is inherited as a partially dominant trait controlled by both additive and nonadditive gene action.

18.1.4 Rice

Many genes have been identified in rice or related wild relatives for resistance to a complex of pests including brown planthopper (*Nilaparvata lugens*), green rice leafhopper (*Nephotettix cincticeps*), green leafhopper (*Nephotettix virescens*), Asian rice gall midge (*Orseolia oryzae*), whitebacked planthopper (*Sogatella furcifera*) and zigzag leafhopper (*Recilia dorsalis*) (Table 18.1).

Four genes control resistance to green rice leafhopper, and QTLs for each are located on four different chromosomes. Eight genes control expression of resistance in rice to green leafhopper. Six are inherited as dominant traits and two genes are inherited as recessive traits and linked to QTLs. Four genes inherited as dominant traits and one gene inherited as a recessive trait control whitebacked planthopper resistance in rice, while three different dominant genes condition resistance to zigzag leafhopper.

In the most extensively studied system, 13 genes control resistance to brown planthopper. Of these, six genes are inherited as dominant traits and seven genes are inherited as recessive traits. QTLs associated with antixenosis and tolerance resistance to brown planthopper have been identified. Rice gall midge resistance is also an

Table 18.1 Genes in rice and related wild relatives controlling resistance to leafhoppers, planthoppers and the rice gall midge

Insect	Resistance genes[a]	References[b]
Green rice leafhopper *Nephotettix cincticeps*	*Grh1, 2, 3, 4*	Kobayashi *et al.*, 1980; Saka *et al.*, 1997; Fukuta *et al.*, 1998; Yazawa *et al.*, 1998; Tamura *et al.*, 1999; Wang *et al.*, 2003
Green leafhopper *Nephotettix virescens*	*Glh1, 2, 3, 5, 6, 7* *glh4, 8*	Athwal & Pathak, 1971; Siwi & Khush, 1977; Rezaul Kamin & Pathak, 1982; Pathak & Khan, 1994; Wang *et al.*, 2004
Brown planthopper *Nilaparvata lugens*	*Bph1, 3, 6, 9, 10, 13* *bph2, 4, 5, 7, 8, 11, 12*	Athwal & Pathak, 1971; Lakashminarayana & Khush, 1977; Ikeda & Kaneda, 1981; Kabir & Khush, 1988; Ishii *et al.*, 1994; Kawaguchi *et al.*, 2001; Renganayaki *et al.*, 2002
Rice gall midge *Orseolia oryzae*	*Gm1, 2, 4, 5, 6, 7, 8, 9* *gm3*	Satyanarayanaiah & Reddi, 1972; Sastry & Praska Rao, 1973; Chaudhary *et al.*, 1986; Sahu & Sahu, 1989; Tomar and Prasad, 1992; Srivastava *et al.*, 1993; Yang *et al.*, 1997; Kumar & Sahu, 1998; Kumar *et al.*, 2000a, b; Katiyar *et al.*, 2001; Shrivastava *et al.*, 2003
Zigzag leafhopper *Recilia dorsalis*	*Zlh1, 2, 3*	Angeles *et al.*, 1986
Whitebacked planthopper *Sogatella furcifera*	*Wbph1, 2, 3, 5* *wbph4*	Sidhu *et al.*, 1979; Angeles *et al.*, 1981; Hernandez & Khush, 1981; Wu & Khush, 1985

[a] Uppercase – inherited as a dominant trait; lowercase – inherited as a recessive trait.
[b] References provided in the supplemental online bibliography (Smith, 2007).

extensively studied gene-for-gene system in arthropod resistance. Eight different genes inherited as dominant trait genes and one monogenic recessive gene have been documented as controlling gall midge resistance in India and China. In-depth discussions of rice–arthropod gene-for-gene interactions involving Asian rice gall midge and brown planthopper are provided in Chapter 12 of Smith (2005).

18.1.5 Wheat

Genes from barley, rye and wheat wild relatives have been transferred into bread wheat to provide resistance to numerous arthropod pests (see review by Berzonsky *et al.*, 2003). The most extensively studied pest is Hessian fly (*Mayetiola destructor*), which began infesting wheat in the USA in the first decade of the twentieth century. More than 25 rye or wheat genes control Hessian fly resistance and all but one (*h4*) are inherited as dominant or partially dominant traits (Table 18.2). The deployment of these resistance genes in response to Hessian fly biotypes is also discussed in Chapter 12 of Smith (2005).

Seven genes expressing resistance to five biotypes of the greenbug have been characterized. *Gb2*, *Gb3*, *Gb5*, *Gb6*, *Gbx*, *Gby* and *Gbz* are all inherited as single dominant traits. Similarly, ten genes in wheat and two genes in barley confer resistance to Russian wheat aphid (*Diuraphis noxia*). All but one, the recessive gene *dn3*, are inherited as dominant traits. Resistance to the wheat curl mite (*Aceria tosichella*), a vector of wheat streak mosaic virus, is also controlled by four different genes from rye or various wheat wild relatives. All are inherited as dominant traits.

Table 18.2 Genes in wheat expressing resistance to arthropod pests

Arthropod	Resistance genes[a]	References[b]
Wheat curl mite *Aceria tosichilla*	*Cm1, 2, 3, 4*	Thomas & Conner, 1986; Schlegel & Kynast, 1987; Whelan & Hart, 1988; Whelan & Thomas, 1989; Chen *et al.*, 1996; Cox *et al.*, 1999; Malik *et al.*, 2003
Russian wheat aphid *Diuraphis noxia*	*Dn1, 2, 4, 5, 6, 7, 8, 9, x, dn3, Rdn1, Rdn2*	du Toit, 1987, 1988, 1989; Nkongolo *et al.*, 1989, 1991a, b; Harvey & Martin, 1990; Marais & du Toit, 1993; Marais *et al.*, 1994, 1998; Mornhinweg *et al.*, 1995, 2002; Saidi & Quick, 1996; Ma *et al.*, 1998; Zhang *et al.*, 1998; Liu, 2001; Liu *et al.*, 2002
Greenbug *Schizaphis graminum*	*Gb2, 3, 5, 6, x, y, z,*	Livers & Harvey, 1969; Sebesta & Wood, 1978; Harvey *et al.*, 1980; Joppa *et al.*, 1980; Hollenhorst & Joppa, 1983; Tyler *et al.*, 1987; Porter *et al.*, 1994; Weng & Lazar, 2002; Boyko *et al.*, 2004; Zhu *et al.*, 2004
Hessian fly *Mayetiola destructor*	*H1, 2, 3, 5, 6, 7, 8, 9, 10, 11, 12, 13, 14, 15, 16, 17, 18, 19, 20, 21, 22, 23, 24, 25, 26, 28, 29, 30, h4*	Stebbins *et al.*, 1982, 1983; Maas *et al.*, 1987; Ratcliffe & Hatchett, 1997; Martin-Sanchez *et al.*, 2003

[a] Uppercase – inherited as a dominant trait; lowercase – inherited as a recessive trait.
[b] References provided in the supplemental online bibliography (Smith, 2007).

18.2 Resistance gene clusters

Several of the maize QTLs mentioned above that play major roles in resistance to the European corn borer, the southwestern corn borer and the sugarcane borer, occur on maize chromosomes 2, 5, 7 and 9. The major QTLs for production of maysin and apimaysin, flavonoid allelochemicals controlling resistance to feeding by the corn earworm, also occur on chromosomes 5 and 9. The relationships between the QTLs for the different types of resistance on chromosomes 5 and 9 have not been investigated.

Genes for resistance to arthropod pests occur in clusters in maize, rice and wheat. In rice, four brown planthopper resistance genes map to a block on chromosome 12 and two genes map to a second cluster on chromosome 3.

In wheat, five genes controlling Russian wheat aphid resistance are located in a cluster on the short arm of wheat chromosome 7D, and five genes for resistance to greenbug are clustered on the distal portion of the long arm of chromosome 7D (Fig. 18.1). More recently, QTL on both the long and short arm of wheat chromosome 7D have been shown to play a role in resistance to Russian wheat aphid and greenbug (Castro *et al.*, 2004).

18.3 Resistance gene mapping

High-density genetic maps of nearly all major agricultural crops such as barley, maize, potato, rye, sorghum, soybean, tomato and wheat have been developed (see review by Smith, 2005), and molecular markers in many of these crops are linked to genes expressing resistance to several major arthropod pests (see review by Yencho *et al.*, 2000). Using these resources, genetic mapping techniques allow comparison of a plant phenotype and a plant genotype. Plant genotypes are

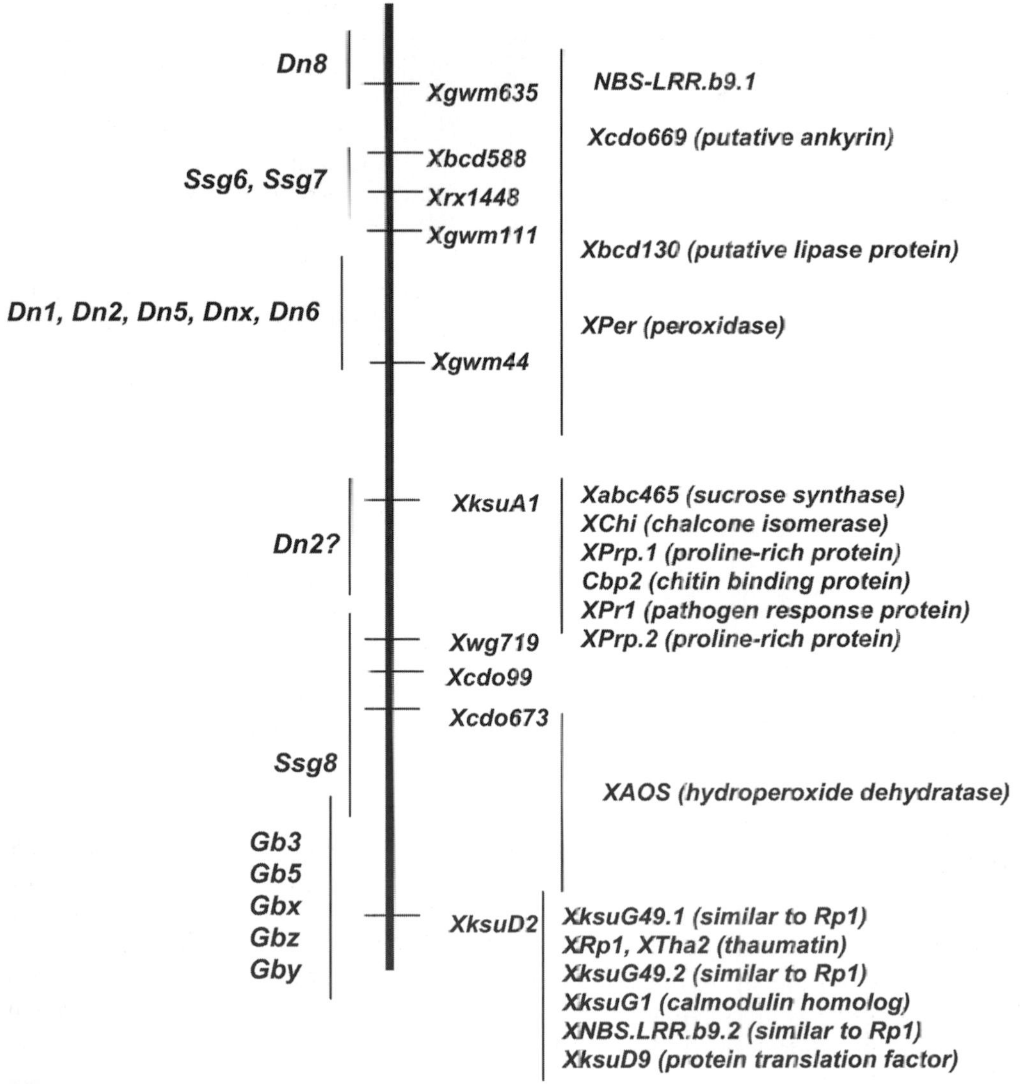

Fig. 18.1 Aphid resistance gene loci in sorghum and wheat on Triticeae homoeologous chromosome group 7. Left side of chromosome: *Dn* – single dominant genes from wheat for *Diuraphis noxia* resistance, *Gb* – single dominant genes from wheat for *Schizaphis graminum* resistance *Ssg* – major restriction fragment length polymorphism (RFLP) loci from sorghum controlling *S. graminum* resistance. Right side of chromosome: RFLP loci spanned on far right by disease defense response loci or resistance gene analog loci (as indicated). Positions of loci are not ordered. For complete description and discussion see Smith (2005).

determined after the amplification of plant DNA with multiple molecular markers of known chromosome location, and estimates of the genetic linkage between the resistance gene and specific markers are then determined. Many molecular markers linked to single resistance genes inherited as dominant traits have been identified. Multiple markers linked to groups of QTLs controlling resistance have also been identified. The marker-assisted selection of plants based on genotype, before phenotypic resistance is determined, is becoming more common.

Linkage between resistance genes and molecular markers varies greatly. They may be completely linked, where no crossing-over occurs between the gene and the marker during meiosis, or incompletely linked, with crossing-over between the two. Genes and markers may have no linkage, because they are located on separate chromosomes or far from one another on the same chromosome. Estimates of the recombination between a resistance gene and a linked marker are measured as the recombination frequency, which is measured in segregating plant populations by pairing the phenotype and genotype of each progeny and analyzing these data with computer software such as Mapmaker or Mapmaker/QTL (Lincoln *et al.*, 1993).

To estimate recombination frequency, DNA is collected from tissues of resistant and susceptible parent plants and 100–200 F_2 plants or plants from 100–200 F_2-derived F_3 families of known resistance or susceptibility. Molecular markers from multiple chromosome locations are screened to identify those producing polymorphisms between the DNA of the parents, and if parent DNA banding pattern polymorphisms exist, the loci of a marker is said to be informative of the resistance gene location (Fig. 18.2). Two DNA samples, one from several highly resistant and one from several highly susceptible plants, are then amplified with the informative marker. If parent polymorphisms exist between the bulked resistant and susceptible DNA samples, the marker is putatively linked to the gene, DNA of all F_2 plants or F_3 families is amplified, and phenotype and genotype data are subjected to Mapmaker analysis.

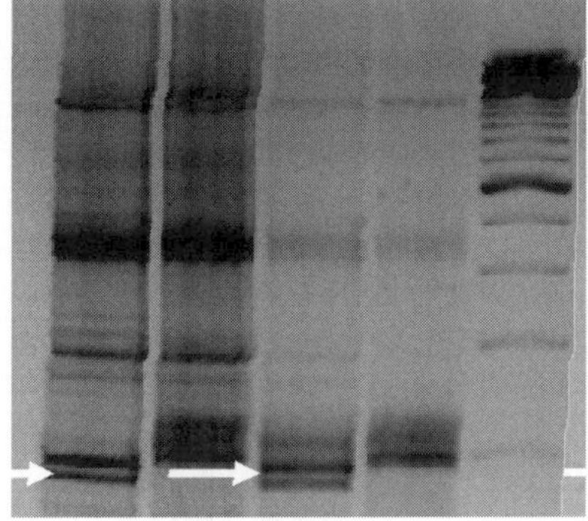

Fig. 18.2 Banding patterns of DNA from leaves of wheat plants lines amplified with a simple sequence repeat (SSR) marker of known wheat chromosome location. R, resistant phenotype parent; S, susceptible phenotype parent; L, 100 base pair DNA reference ladder. Arrow, putative resistance-specific DNA band. Amplification products were electrophoresed in a 3% agarose gel stained with ethidium bromide (from Flinn, 2000).

18.4 Molecular markers

The use of molecular markers has advantages compared to morphological markers. Some molecular markers behave in a co-dominant manner to detect heterozygotes in segregating progeny when morphological markers detect dominant or recessive traits. In addition, the allelic variation detected by molecular markers is considerably greater than that detected by morphological markers, and molecular markers are unaffected by environmental affects. Molecular markers used to determine arthropod resistance gene location include restriction fragment length polymorphism (RFLP) markers, random amplified polymorphic (RAPD) markers, amplified fragment length polymorphism (AFLP) markers and simple sequence repeats (SSRs) or microsatellite markers.

RFLP markers detect differences between genotypes when restriction enzymes cut genomic DNA to yield variable-sized DNA fragments that are then separated by electrophoresis. Digested DNA is transferred to a nylon membrane (Southern blotting) and the membrane is probed with a radioactive labeled, single-stranded DNA probe of known chromosome location. Membrane-bound DNA is denatured by heat and the probe sequences bind to complementary sites in the restriction digest. Unbound probe is removed by washing and the dried membrane is exposed to x-ray film and photographically developed as an autoradiogram. Binding between probe and membrane-bound DNA provides information about the location of a resistant gene in the form of different (polymorphic) autoradiogram DNA banding patterns between two genotypes. When restriction sites in the vicinity of a gene are compared between genotypes, one genotype may have the site, while the other does not. If differences exist, they are referred to as polymorphisms between the two genotypes.

As indicated above, RFLP markers detect heterozygotes and have been used to map arthropod resistance gene loci in numerous crops (Yencho *et al.*, 2000). The disadvantages of RFLP linkage analysis include the 7 to 10 day time period required to complete an analysis and the use of radioactive isotopes.

Polymerase chain reaction (PCR) primers from known chromosome locations are reacted with template DNA, amplified in a thermal cycler, and the amplification products are electrophoresed to identify primers yielding polymorphic banding patterns between resistant and susceptible plants. Compared to RFLP hybridization, PCR amplification is many times faster and does not require the use of radioactive materials. Several types of PCR primers have been used to identify plant resistance genes.

The use of AFLP markers involves digesting DNA with different restriction enzymes and annealing restriction enzyme adaptors to the restriction products. Restriction digests are preselected by PCR amplification with general restriction enzymes attached to unique oligonucleotide primers. Preselected PCR products are then selectively amplified using specific oligonucleotide primers, amplified fragments are separated by electrophoresis and gel images are converted to autoradiograms. AFLP markers have been used to successfully map arthropod resistance genes in apple, rice and wheat.

Simple sequence repeat (SSR) or microsatellite PCR primers are two to five base dinucleotide repeats widely distributed in eukaryotic DNA. Microsatellite primers generate high levels of polymorphism, detect patterns of co-dominant inheritance and have been used to map several arthropod resistance genes in rice and wheat (Yencho *et al.*, 2000). Comparison of RFLP, AFLP and SSR markers in cultivated and wild soybean revealed a high correlation between the three marker types (Powell *et al.*, 1996).

Linkage between QTLs and marker loci is based on the relation between the phenotypic expression of several minor resistance genes and molecular markers at multiple loci. QTL analyses determine which loci explain the greatest amount of phenotypic variation for a biochemical or biophysical character controlling resistance. QTLs are linked to arthropod resistance genes in barley, maize, potato, rice, sorghum, soybean, tomato and wheat (see reviews by Yencho *et al.*, 2000 and Smith, 2005).

18.5 Molecular marker-assisted selection

Cost and labor savings have been documented for the use of a molecular marker-assisted selection system based on SSR markers linked to genes for nematode resistance in soybean and barley (Mudge *et al.*, 1997; Kretshmer *et al.*, 1997) and for disease resistance in rice (Hittalmani *et al.*, 2000; Toenniessen *et al.*, 2003). In arthropod resistance research, the use of maker-assisted selection continues to increase as a strategy for single-gene resistance inherited as a dominant trait. An online bibliography provides additional references on marker-assisted selection (Smith, 2007).

For QTLs in general, these markers are yet to be used to their fullest extent in marker-assisted selection (see reviews of Babu *et al.*, 2004 and Food and Agriculture Organization, 2003). The same situation exists for arthropod resistance studies. In a study to determine transfer of soybean QTLs for Lepidoptera resistance into cultivars, Narvel *et al.* (2001) found very few resistant genotypes with multiple QTLs from different soybean linkage groups, and suggested marker-assisted selection to introgress QTLs for resistance into elite germplasm.

Results of research with maize QTLs linked to Lepidoptera resistance further illustrate the limited use of QTLs in marker-assisted selection. Comparisons of marker-assisted selection and phenotypic selection of maize resistance to southwestern cornborer and sugarcane borer by Groh *et al.* (1998), Bohn *et al.* (2001) and Willcox *et al.* (2002) demonstrated that the efficiency of both methods was similar, and suggested that phenotypic selection is more favorable, due to reduced costs. Environmental variation also affects the use of QTLs for *Diatrea* resistance marker-assisted selection (Groh *et al.*, 1998). Thus, as noted in these above studies, the effective use of marker-assisted selection for *Diatrea* resistance will depend on major marker-assisted selection cost reductions with this technique, the development of more QTLs that explain more variance for resistance and the expression of QTLs over a broad range of environments.

18.6 Deploying insect resistance genes

As reviewed above, resistance inherited as either single genes (monogenic) or QTLs (polygenic) has been deployed in many different insect pest management systems. This has occurred in spite of the fact that monogenic resistance often results in the development of virulent biotypes of insects that are unaffected by the original plant resistance gene. Comparatively, polygenic resistance is often considered more stable than monogenic resistance and is not readily overcome by resistance-breaking arthropod biotypes. The polygenic nature of maize resistance to lepidopteran larvae is a good example of this relationship. Nevertheless, the majority of the insect-resistant cultivars, including transgenic plants possessing insecticidal proteins, contain single-gene resistance. This is because plants containing single traits are easier to score and the population size necessary to study the inheritance of resistance is smaller than for the evaluation of polygenic resistance.

The practice of releasing a resistant cultivar containing a single major gene, planting it until it becomes ineffective and making additional sequential releases of other major genes is quite common in rice and wheat insect pest resistance breeding programs. Pyramiding (incorporation of more than one resistance gene) of two or more major genes in one cultivar, although time consuming, increases longevity of resistance genes and has been used successfully to protect rice cultivars with brown planthopper resistance. In the case of greenbug resistance in wheat, pyramiding provides no additional protection over that provided by single resistance genes released sequentially. Finally, the development and deployment of multiline cultivars composed of different combinations of major and minor resistance genes has been used for rice resistance to brown plant hopper and green rice leafhopper, wheat resistance to Hessian fly and sorghum resistance to sorghum midge.

18.7 Conclusions

Insect-resistant crops will continue to play a very important role in world sustainable agricultural systems. It is very likely that the benefits of insect-resistant crops will become more acute as world climate change increases and food needs and food availability become more uncertain, especially in the underdeveloped and developing countries of the semi-tropics. Although conventional and transgenic resistance breeding efforts have made major strides to improve maize, rice and wheat during the past century, the important food crops of the semi-arid tropics crops (sorghum, millet, pigeon pea and chickpea) remain in need of identification and deployment of increased amounts of insect resistance. Refinement and increased use of marker-assisted selection techniques should be encouraged in order to accelerate the rate and accuracy of breeding of all crop plants for insect resistance. The continued evolution of virulent biotypes dictates the need for identification of new sources of resistance and heightens need for marker-assisted selection systems to identify and track these genes.

Genomic technologies have opened completely new avenues of research in plant resistance to insects over the last decade. Genomic microarrays of several crop plants (barley, maize, rice, tomato, wheat) are beginning to provide critical information about the identity of resistance genes, their chromosome location and the gene products mediating the function of resistance genes. The sequencing of the genomes of rice and the model plant *Arabidopsis* are providing insights into plant insect resistance gene structure, function and location. The soon to be completed sequencing of maize and the model legume *Medicago truncatulata* will provide additional insights as well. As more plant genomes are sequenced, existing and new information about resistance gene synteny can be used to make foresighted decisions about the design and breeding of insect resistant crop plants. Knowledge about the diversity of resistance genes will permit breeding of crop cultivars with resistance genes of diverse sequence and function that will help delay the development of resistance-breaking insect biotypes.

Sequence information afforded by resistance gene analogs in many crop plants will also allow plant resistance researchers to use this *in silico* resource to determine more in-depth information about location of candidate resistance genes and the biochemical and biophysical gene products mediating their function. The ultimate goal of resistance gene expression studies, genomic studies, and marker-assisted selection systems should be to identify plant genes that can be cloned and used to transform crop plants for durable insect resistance.

References

Angeles, E. R., Khush, G. S. & Heinrichs, E. A. (1981). New genes for resistance to whitebacked planthopper in rice. *Crop Science*, **21**, 47–50.

Angeles, E. R., Khush, G. S. & Heinrichs E. A. (1986). Inheritance of resistance to planthoppers and leafhoppers in rice. In *Rice Genetics, Proceedings of the International Rice Genetics Symposium*, May 27–31, 1985, pp. 537–549. Manila, Philippines: Island Publishing Company.

Athwal, D. S. & Pathak, M. D. (1971). Genetics of resistance to rice insects. In *Rice Breeding*, pp. 375–386. Los Baños, Philippines: International Rice Research Institute.

Babu, R., Nair, S. K., Prasanna, B. M. & Gupta, H. S. (2004). Integrating marker-assisted selection in crop breeding: prospects and challenges. *Current Science*, **87**, 607–619.

Berzonsky, W. A., Ding, H., Haley, S. D. *et al.* (2003). Breeding wheat for resistance to insects. *Plant Breeding Review*, **22**, 221–296.

Bohn, M., Groh, S., Khairallah, M. M. *et al.* (2001). Re-evaluation of the prospects of marker-assisted selection for improving insect resistance against *Diatraea* spp. in tropical maize by cross validation and independent validation. *Theoretical and Applied Genetics*, **103**, 1059–1067.

Boyko, E. V., Starkey, S. R. & Smith, C. M. (2004). Molecular genetic mapping of *Gby*, a new greenbug resistance gene in bread wheat. *Theoretical and Applied Genetics*, **109**, 1230–1236.

Castro, A. M., Vasicek, A., Ellerbrook, C. *et al.* (2004). Mapping quantitative trait loci in wheat for resistance against greenbug and Russian wheat aphid. *Plant Breeding*, **123**, 229–332.

Chaudhary, B. P., Srivastava, P. S., Shrivastava, M. N. & Khush, G. S. (1986). Inheritance of resistance to gall midge in some cultivars of rice. In *Rice Genetics*, pp. 523–528. Los Baños, Philippines: International Rice Research Institute.

Chen, Q., Conner, R. L. & Laroche, A. (1996). Molecular characterization of *Haynaldia villosa* chromatin in wheat lines carrying resistance to wheat curl mite colonization. *Theoretical and Applied Genetics*, **93**, 679–684.

Cox, T. S., Bockus, W. W., Gill, B. S. *et al.* (1999). Registration of KS96WGRC40 hard red winter wheat germplasm resistant to wheat curl mite, stagonospora leaf blotch, and septoria leaf blotch. *Crop Science*, **39**, 597.

du Toit, F. (1987). Resistance in wheat (*Triticum aestivum*) to *Diuraphis noxia* (Hemiptera: Aphididae). *Cereal Research Communications*, **15**, 175–179.

du Toit, F. (1988). Another source of Russian wheat aphid (*Diuraphis noxia*) resistance in *Triticum aestivum*. *Cereal Research Communications*, **16**, 105–106.

du Toit, F. (1989). Inheritance of resistance in two *Triticum aestivum* lines to Russian wheat aphid (Homoptera: Aphididae). *Journal of Economic Entomology*, **82**, 1251–12153.

Food and Agriculture Organization (2003). *Molecular Marker Assisted Selection as a Potential Tool for Genetic Improvement of Crops, Forest Trees, Livestock and Fish in Developing Countries*. FAO Electronic Forum on Biotechnology in Food and Agriculture. Available at www.fao.org/biotech/C10doc.htm.

Flinn, M. (2000). A molecular marker linked to tolerance in *Aegilops tauschii* Accession 1675 to greenbug (Homoptera: Aphididae). Manahattan, KS: M.S. Thesis, Kansas State University.

Fukuta, Y., Tamura, K., Hirae, M. & Oya, S. (1998). Genetic analysis of resistance to green rice leafhopper (*Nephotettix cincticeps* Uhler) in rice parental line, Norin-PL6, using RFLP markers. *Breeding Science*, **48**, 243–249.

Gatehouse, A. M. R., Boulter, D. & Hilder, V. A. (1994). Potential of plant-derived genes in the genetic manipulation of crops for insect resistance. *Plant Genetic Manipulation for Crop Protection*, **7**, 155–181.

Groh, S., Gonzalez-deLeon, D., Khairallah M. M. *et al.* (1998). QTL mapping in tropical maize. III. Genomic regions for resistance to *Diatraea* spp. and associated traits in two RIL populations. *Crop Science*, **38**, 1062–1072.

Harvey, T.L. & Martin, T. J. (1990). Resistance to Russian wheat aphid, *Diuraphis noxia*, in wheat (*Triticum aestivum*). *Cereal Research Communications*, **18**, 127–129.

Harvey, T. L., Martin, T. J. & Livers, R. W. (1980). Resistance to biotype C greenbug in synthetic hexaploid wheats derived from *Triticum tauschii*. *Journal of Economic Entomology*, **73**, 387–389.

Hernandez, J. E. & Khush, G. S. (1981). Genetics of resistance to whitebacked planthopper in some rice (*Oryza sativa* L.) varieties. *Oryza*, **18**, 44–50.

Hittalmani, S., Parco, A., Mew, T. W., Zeigler, R. S. & Huang, N. (2000). Fine mapping and DNA marker-assisted pyramiding of the three major genes for blast resistance in rice. *Theoretical and Applied Genetics*, **100**, 1121–1128.

Hollenhorst, M. M. & Joppa, L. R. (1983). Chromosomal location of genes for resistance to greenbug in 'Largo' and 'Amigo' wheats. *Crop Science*, **23**, 91–93.

Ikeda, R. & Kaneda, C. (1981). Genetic analysis of resistance to brown planthopper, *Nilaparvata lugens* Stål, in rice. *Japan Journal of Breeding*, **31**, 279–285.

Ishii, T., Brar, D. S., Multani, D. S. & Khush, G. S. (1994). Molecular tagging of genes for brown planthopper resistance and earliness introgressed from *Oryza australiensis* into cultivated rice, *O. sativa*. *Genome*, **37**, 217–221.

Joppa, L. R., Timian, R. G. & Williams, N. D. (1980). Inheritance of resistance to greenbug toxicity in an amphiploid of *Triticum turgidum/T. tauschi*. *Crop Science*, **20**, 343–344.

Kabir, M. A. & Khush, G. S. (1988). Genetic analysis of resistance to brown planthopper in rice (*Oryza sativa* L.). *Plant Breeding*, **100**, 54–598.

Katiyar, S. K., Tan, Y., Huang, B. *et al.* (2001). Molecular mapping of gene *Gm-6(t)* which confers resistance against four biotypes of Asian rice gall midge in China. *Theoretical and Applied Genetics*, **103**, 953–961.

Kawaguchi, M., Murata, K., Ishii, T. *et al.* (2001). Assignment of a brown planthopper (*Nilaparvata lugens* Stål) resistance gene *bph4* to the rice chromosome 6. *Breeding Science*, **51**, 13–18.

Khush, G. S. & Brar, D. S. (1991). Genetics of resistance to insects in crop plants. *Advances in Agronomy*, **45**, 223–274.

Kobayashi, A., Kaneda, C., Ikeda, R. & Ikehashi, H. (1980). Inheritance of resistance to green rice leafhopper, *Nephotettix cincticeps*, in rice. *Japan Journal of Plant Breeding*, **30** (Suppl. 1), 56–57.

Kretshmer, J. M., Chalmers, K. J., Manning, S. *et al.* (1997). RFLP mapping of the *Ha2* cereal cyst nematode resistance gene in barley. *Theoretical and Applied Genetics*, **94**, 1060–1064.

Kumar, A. & Sahu, R. K. (1998). Genetic analysis for gall midge resistance: a reconsideration. *Rice Genetics Newsletter*, **15**, 142–143.

Kumar, A., Shrivasta, M. N. & Shukla, B. C. (2000a). Genetic analysis of gall midge (*Orseolia oryzae* Wood Mason) biotype 1 resistance in the rice cultivar RP 2333-156-8. *Oryza*, **37**, 79–80.

Kumar, A., Bhandarkar, S., Pophlay, D. J. & Shrivasta, M. N. (2000b). A new gene for gall midge resistance in rice accession Jhitpiti. *Rice Genetics Newsletter*, **17**, 83–84.

Lakashminarayana, A. & Khush, G. S. (1977). New genes for resistance to the brown planthopper in rice. *Crop Science*, **17**, 96–100.

Lincoln, S. E., Daly, M. J. & Lander, E. S. (1993). *MAPMAKER/EXP Version 3.0., A Tutorial and Reference Manual*, 3rd edn. Cambridge, MA: a Whitehead Institute for Biomedical Research.

Liu, X. (2001). Molecular mapping of wheat genes expressing resistance to the Russian wheat aphid, *Diuraphis noxia* (Mordvilko) (Homoptera: Aphididae). Manhattan, KS: Ph.D. dissertation, Kansas State University.

Liu, X. M., Smith, C. M. & Gill, B. S. (2002). Mapping of microsatellite markers linked to the *Dn4* and *Dn6* genes expressing Russian wheat aphid resistance in wheat. *Theoretical and Applied Genetics*, **104**, 1042–1048.

Livers, R. W. & Harvey, T. L. (1969). Greenbug resistance in rye. *Journal of Economic Entomology*, **62**, 1368–1370.

Ma, Z.-Q., Saidi, A., Quick, J. S. & Lapitan, N. L. V. (1998). Genetic mapping of Russian wheat aphid resistance genes *Dn2* and *Dn4* in wheat. *Genome*, **41**, 303–306.

Maas, F. B. III, Patterson, F. L., Foster, J. E. & Hatchett, J. H. (1987). Expression and inheritance of resistance of 'Marquillo' wheat to Hessian fly biotype D. *Crop Science*, **27**, 49–52.

Malik, R., Smith, C. M., Harvey, T. L. & Brown-Guedira, G. L. (2003). Genetic mapping of wheat curl mite resistance genes *Cmc3* and *Cmc4* in common wheat. *Crop Science*, **43**, 644–650.

Marais, G. F. & du Toit, F. A. (1993). A monosomic analysis of Russian wheat aphid resistance in the common wheat PI294994. *Plant Breeding*, **111**, 246–248.

Marais, G. F., Horn, M. & du Toit, F. A. (1994). Intergeneric transfer (rye to wheat) of a gene(s) for Russian wheat aphid resistance. *Plant Breeding*, **113**, 265–271.

Marais, G. F., Wessels, W. G. & Horn, M. (1998). Association of a stem rust resistance gene (Sr45) and two Russian wheat aphid resistance genes (*Dn5* and *Dn7*) with mapped structural loci in common wheat. *South African Journal of Plant and Soil*, **15**, 67–71.

Martin-Sanchez, J. A., Gomez-Colmenarejo, M., Del Moral, J. *et al.* (2003). A new Hessian fly resistance gene (H30) transferred from the wild grass *Aegilops triuncialis* to hexaploid wheat. *Theoretical and Applied Genetics*, **106**, 1248–55.

Mornhinweg, D. W., Porter, D. R. & Webster, J. A. (1995). Inheritance of Russian wheat aphid resistance in spring barley. *Crop Science*, **35**, 1368–1371.

Mornhinweg, D. W., Porter, D. R. & Webster, J. A. (2002). Inheritance of Russian wheat aphid resistance in spring barley germplasm line STARS-9577B. *Crop Science*, **42**, 1891–1893.

Mudge, J., Cregan, P. B., Kenworthy, J. P. *et al.* (1997). Two microsatellite markers that flank the major soybean cyst nematode resistance locus. *Crop Science*, **37**, 1611–1615.

Narvel, J. A., Walker, D. R., Rector, B. G. *et al.* (2001). A retrospective DNA marker assessment of the development of insect resistant soybean. *Crop Science*, **41**, 1931–1939.

Nkongolo, K. K., Quick, J. S., Meyers, W. L. & Peairs, F. B. (1989). Russian wheat aphid resistance of wheat, rye, and triticale in greenhouse tests. *Cereal Research Communications*, **17**, 227–232.

Nkongolo, K. K., Quick, J. S., Limin, A. E. & Fowler, D. B. (1991a). Sources and inheritance of resistance to Russian wheat aphid in *Triticum* species amphiploids and *Triticum tauschii*. *Canadian Journal of Plant Science*, **71**, 703–708.

Nkongolo, K. K., Quick, J. S., Peairs, F. B. & Meyer W. L. (1991b). Inheritance of resistance of PI 373129 wheat to the Russian wheat aphid. *Crop Science*, **31**, 905–906.

Painter, R. H. (1951). *Insect Resistance in Crop Plants.* Lawrence, KS: University of Kansas Press.

Pathak, M. D. & Khan, Z. R. (1994). *Insect Pests of Rice.* Los Baños, Philippines: International Rice Research Institute.

Porter, D. R., Webster, J. A. & Friebe, B. (1994). Inheritance of greenbug biotype G resistance in wheat. *Crop Science*, **34**, 625–628.

Powell, W., Morgante, M., Andre, C. *et al.* (1996). The comparison of RFLP, RAPD, AFLP and SSR (microsatellite) markers for germplasm analysis. *Journal of Molecular Breeding*, **2**, 225–238.

Ratcliffe, R. H. & J. H. Hatchett (1997). Biology and genetics of the Hessian fly and resistance in wheat. In *New Developments in Entomology*, ed. K. Bondari, pp. 47–56. Trivandrum, India: Research Signpost, Scientific Information Guild.

Renganayaki, K., Fritz, A. K., Sadasivam, S. *et al.* (2002). Mapping and progress toward map-based cloning of brown planthopper biotype-4 resistance gene introgressed *Oryza officinalis* into cultivated rice, *O. sativa*. *Crop Science*, **42**, 2112–2117.

Rezaul Kamin, A. N. M. & Pathak, M. D. (1982). New genes for resistance to green leafhopper, *Nephotettix virescens* (Distant) in rice, *Oryza sativa* L. *Crop Protection*, **1**, 483–490.

Sahu, V. N. & Sahu, R. K. (1989). Inheritance and linkage relationships of gall midge resistance with purple leaf, apiculus and scent in rice. *Oryza*, **26**, 79–83.

Saidi, A. & Quick, J. S. (1996). Inheritance and allelic relationships among Russian wheat aphid resistance genes in winter wheat. *Crop Science*, **36**, 256–258.

Saka, N., Toyama, T., Tuji, T., Nakamae, H. & Izawa, T. (1997). Fine mapping of green ricehopper resistant gene *Grh-3 (t)* and screening of *Grh-3 (t)* among green ricehopper resistant and green leafhopper resistant cultivars in rice. *Breeding Science*, **47** (Suppl. 1), 55 (in Japanese).

Sastry, M. V. S. & Prakasa Rao, P. S. (1973). Inheritance of resistance to rice gall midge *Pachydiplosis oryzae* Wood Mason. *Current Science*, **42**, 652–653.

Satyanarayanaiah, K. & Reddi, M. V. (1972). Inheritance of resistance to insect gall midge (*Pachydiplosis oryzae*, Wood Mason) in rice. *Andhra Agriculture Journal*, **19**, 1–8.

Schlegel, R. & Kynast, R. (1987). Confirmation of 1A/1R wheat–rye chromosome translocation in the wheat variety 'Amigo'. *Plant Breeding*, **98**, 57–60.

Sebesta, E. E. & Wood, Jr., E. A. (1978). Transfer of greenbug resistance from rye to wheat with x-rays. *Agronomy Abstracts*, **1978**, 61–62.

Shrivastava, M. N., Kumar, A., Bhandarkar, S., Shukla, B. C. & Agrawal, K. C. (2003). A new gene for resistance in rice to Asian rice gall midge (*Orseolia oryzae* Wood

Mason) biotype 1 population at Raipur, India. *Euphytica*, **130**,143–145.

Sidhu, G. S., Khush, G. S. & Medrano, F. G. (1979). A dominant gene in rice for resistance to whitebacked planthopper and its relationship to other plant characteristics. *Euphytica*, **28**, 227–232.

Siwi, B. H. & Khush, G. S. (1977). New genes for resistance to the green leafhopper in rice. *Crop Science*, **17**, 17–20.

Smith, C. M. (2004). Plant resistance against pests: issues and strategies. In *Integrated Pest Management: Potential, Constraints and Challenges*, eds. O. Koul, G. S. Dhaliwal & G. W. Cuperus, pp. 147–167. Wallingford, UK: CABI Publishing.

Smith, C. M. (2005). *Plant Resistance to Arthropods: Molecular and Conventional Approaches*. Dordrecht, Netherlands: Springer-Verlag.

Smith, C. M. (2007). *Advances in Breeding for Host Plant Resistance: Supplemental Bibliography*. Available at http://ipmworld/umn.edu/smith.htm.

Srivastava, M. N., Kumar, A., Shrivastava, S. K. & Sahu, R. K. (1993). A new gene for resistance to rice gall midge in rice variety Abhaya. *Rice Genetics Newsletter*, **10**, 79–80.

Stebbins, N. B., Patterson, F. L. & Gallun, R. L. (1983). Inheritance of resistance of PI94587 wheat to biotypes B and D of Hessian fly. *Crop Science*, **23**, 251–253.

Stebbins, N. B., Patterson, F. L. & Gallun, R. L. (1982). Interrelationships among wheat genes H3, H6, H9, and Hl0 for Hessian fly resistance. *Crop Science*, **22**, 1029–1032.

Tamura, K., Fukuta, Y., Hirae, M. *et al.* (1999). Mapping of the *Grh1* locus for green rice leafhopper resistance in rice using RFLP markers. *Breeding Science*, **49**, 11–14.

Thomas, J. B. & Conner, R. L. (1986). Resistance to colonization by the wheat curl mite in *Aegilops squarrosa* and its inheritance after transfer to common wheat. *Crop Science*, **26**, 527–530.

Toenniessen, G. H., O'Toole, J. C. & DeVries, J. (2003). Advances in plant biotechnology and its adoption in developing countries. *Current Opinion in Plant Biology*, **8**, 191–198.

Tomar, J. B. & Prasad, S. C. (1992). Genetic analysis of resistance to gall midge (*Orseolia oryzae* Wood Mason) in rice. *Plant Breeding*, **109**, 159–167.

Tyler, J. M., Webster, J. A. & Merkle, O. G. (1987). Designations for genes in wheat germplasm conferring greenbug resistance. *Crop Science*, **27**, 526–527.

Wang, C., Yasui, H., Yoshimura, A., Zhai, H. & Wan, J. (2003). Green rice leafhopper resistance gene transferred through backcrossing and CAPs marker assisted selection. *Chinese Agricultural Science*, **2**, 13–18.

Wang, C., Yasui, H., Yoshimura, A., Zhai, H. & Wan, J. (2004). Inheritance and QTL mapping of antibiosis to green leafhopper in rice. *Crop Science*, **44**, 389–393.

Weng, Y. & Lazar, M. D. (2002). Amplified fragment length polymorphism- and simple sequence repeat-based molecular tagging and mapping of greenbug resistance gene *Gb3* in wheat. *Plant Breeding*, **121**, 218–223.

Whelan, E. D. P. & Hart, G. E. (1988). A spontaneous translocation that transfers wheat curl mite resistance from decaploid *Agropyron elongatum* to common wheat. *Genome*, **30**, 289–292.

Whelan, E. D. P. & Thomas, J. B. (1989). Chromosomal location in common wheat of a gene (*Cmc*1) from *Aegilops squarrosa* that conditions resistance to colonization by the wheat curl mite. *Genome*, **32**, 1033–1036.

Willcox, M. C., Khairallah, M. M., Bergvinson, D. *et al.* (2002). Selection for resistance to southwestern corn borer using marker-assisted selection and conventional backcrossing. *Crop Science*, **42**, 1516–1528.

Wu, C. F. & Khush, G. S. (1985). A new dominant gene for resistance to whitebacked planthopper in rice. *Crop Science*, **25**, 505–509.

Yang, D., Parco, A., Nandi, S. *et al.* (1997). Construction of a bacterial artificial chromosome (BAC) library and identification of overlapping BAC clones with chromosome 4-specific RFLP markers in rice. *Theoretical and Applied Genetics*, **95**, 1147–1154.

Yazawa, S., Yasui, H., Yoshimura, A. & Iwata, N. (1998). RFLP mapping of genes for resistance to green rice leafhopper (*Nephotettix cincticeps* Uhler) in rice cultivar DV85 using near isogenic lines. *Science Bulletin of the Faculty of Agriculture at Kyushu University*, **52**, 169–175.

Yencho, G. C., Cohen, M. B. & Byrne, P. F. (2000). Applications of tagging and mapping insect resistance loci in plants. *Annual Review of Entomology*, **45**, 393–422.

Zhang Y., Quick, J. S. & Liu, S. (1998). Genetic variation in PI 294994 wheat for resistance to Russian wheat aphid. *Crop Science*, **38**, 527–530.

Zhu, L. C., Smith, C. M., Fritz, A., Boyko, E. V. & Flinn, M. B. (2004). Genetic analysis and molecular mapping of a wheat gene conferring tolerance to the greenbug (*Schizaphis graminum* Rondani). *Theoretical and Applied Genetics*, **109**, 289–293.

Chapter 19

Resistance management to transgenic insecticidal plants

Anthony M. Shelton and Jian-Zhou Zhao

Among the biological concerns expressed about the use of transgenic insecticidal plants are the potential for their genes to spread to wild and cultivated crops, have deleterious effects on non-target organisms, and for insects to evolve resistance to the toxins(s) expressed in the plant (Shelton *et al.*, 2002). These same concerns apply to many other forms of pest management, but transgenic plants have come under more scrutiny than most other strategies. For a broad review of the risks and benefits of insecticidal transgenic plants, the reader is referred to Sanvido *et al.* (2007). The focus of this chapter is solely on the risks of insects developing resistance to transgenic insecticidal plants and how such risks can be reduced. There is a long history of literature on resistance to conventional insecticides and to traditionally bred plants that are resistant to insects, and they provide a solid foundation for understanding the evolution and management of insect resistance to transgenic insecticidal plants.

A commonly used definition of resistance is that it is a genetic change in response to selection by toxicants that may impair control in the field (Sawicki, 1987). Such changes could result from physiological or behavioral adaptation, although many more examples are from the former. Resistance to traditional synthetic pesticides has become one of the major driving forces altering the development of IPM programs worldwide (Shelton & Roush, 2000). There are over 500 species of arthropods that have developed strains resistant to one or more of the five principal classes of insecticides (Georghiou & Lagunes-Tejeda, 1991), and that list continues to grow (http://whalonlab.msu.edu/rpmnews/). There are now well-documented cases of insects having developed resistance not only to synthetic insecticides, but also to pathogens including bacteria, fungi, viruses and nematodes (Shelton & Roush, 2000). Resistance has recently been found to spinosad and indoxacarb in the diamondback moth (*Plutella xylostella*) after less than 3 years of use (Zhao *et al.*, 2006). This insect has a long history of rapidly becoming resistant to most insecticides used intensively against it, and is a harbinger of potential problems with other insects. This is especially relevant because it is the only insect that has developed high levels of resistance to proteins from the bacterium *Bacillus thuringiensis* (*Bt*), in the field, although this was to foliar sprays of *Bt*.

Resistance is not one-dimensional. Not only do organisms have various physiological, biochemical and behavioral methods of developing resistance, but resistance has spatial and temporal components (Croft, 1990). Insects in one field may be susceptible to an insecticide while insects in

Integrated Pest Management, ed. Edward B. Radcliffe, William D. Hutchison and Rafael E. Cancelado. Published by Cambridge University Press.

a nearby field may have developed high levels of resistance to that same insecticide (Shelton *et al.*, 2006). Likewise, insects that were once resistant to an insecticide may regain susceptibility if that material is not used for some length of time, especially if fitness costs (e.g. lower fecundity) to the insect are associated with resistance (Croft, 1990). It is the variability in the spatial and temporal aspects of resistance to a particular insecticide (or class of insecticide) that can be exploited and allow the insecticide to be part of an overall insecticide resistance management (IRM) program. Reducing the selection pressure to a particular insecticide, and hence allowing it to be part of an IRM program, will delay the evolution of resistance and this should be the goal of any IRM strategy. To achieve this end will require an understanding of the genetics of resistance, monitoring the frequency of resistance alleles, and finding other tactics that will reduce the use of a particular insecticide while still providing adequate management of the insect population. Croft (1990) provides a list of examples of IRM programs for arthropods that have been "more successful" or "less successful." Although more recent examples could be added to each category, the important message is that resistance can be managed and that the more successful programs have done so by: increased understanding of resistance mechanisms and enhanced resistance monitoring, increased cooperation of growers and government agencies and integration of IRM into an overall IPM program.

19.1 Insect-resistant plants

Host plant resistance should be a foundation of IPM. There are over 100 insect-resistant crop cultivars grown in the USA and probably twice that many worldwide (Smith, 1989, ch. 18). The majority of these cultivars are field crops and the main target pests are sucking insects (aphids and leafhoppers). The benefits of genetically incorporated insect control can be substantial and include increased security to the grower, decreased use of insecticides and the potential to enhance biological control through conservation of natural enemies. However, like any technology directed toward insect management, there is the risk that insects will evolve methods of overcoming the plant's defenses. An important historical example illustrates this well: Hessian fly (*Mayetiola destructor*), a significant pest of wheat in the USA (see Gould, 1986). At least 16 wheat genes that confer resistance to this insect have been identified and a number of resistant wheat cultivars have been planted. However, their effectiveness has been diminished by Hessian fly biotypes that have developed resistance to the once-resistant plants. In many respects this is similar to the situation with traditional insecticides noted above.

Considerable efforts have been made to breed resistance to Lepidoptera in many crops, but there have been few commercial successes outside of increasing the concentration of the cyclic hydroxamic acid DIMBOA in some cereal crops, most notably maize (Smith, 1989). DIMBOA affects the development of some insects and thus is classified as a form of antibiosis using Painter's (1951) classification. DIMBOA has activity against some Lepidoptera as well as some sucking pests and can reduce the level of damage by the European corn borer (*Ostrinia nubilalis*) during the early growth stages of the maize. However, for protection of the ear, other strategies (most notably insecticide sprays) are used. No high level of host plant resistance to Lepidoptera or Coleoptera, orders that contain the most destructive insect pests of crops worldwide, has been developed and commercialized through conventional breeding methods.

With the advent of biotechnology, breeders were no longer limited to genes from plants that could be used to produce insect-resistant plants, even to Lepidoptera and Coleoptera. The bacterium *B. thuringiensis* (*Bt*) offered some unique opportunities since different strains of most *Bts* contain varying combinations of insecticidal crystal proteins (ICPs), and different ICPs are toxic to different groups of insects (Tabashnik, 1994). Insecticidal products containing *Bt* subspecies were first commercialized in France in the late 1930s and dozens of *Bt* products have been sprayed on crops since. *Bt* genes that express specific ICPs were first introduced into tobacco plants for control of Lepidoptera in 1987 but more effective plants that used synthetic genes modeled on those from *Bt*, but designed (plant codon optimized) to be more compatible with plant expression, were introduced a few years later (see review by Shelton *et al.*, 2002). *Bt* plants displayed the first high level of host plant resistance to major lepidopteran pests and to some coleopteran pests. Of the $US 8100 million spent annually on all insecticides worldwide, it has been estimated that nearly $US 2700 million could be substituted with *Bt* biotechnology products for Lepidoptera (Krattiger, 1997). In 2007, 42.1 million ha of plants (maize and cotton) that express *Bt* proteins were utilized (James, 2007). Thus, *Bt*, which was once a fairly minor insecticide (<2% of the worldwide insecticide market), has now become a major insecticide. With the recent registration in the USA of *Bt* maize to control the corn rootworm complex, which has control costs estimated at $US 1000 million annually (Metcalf & Metcalf, 1993), use of *Bt* for insect control will increase rapidly. Besides maize and cotton, other crops are being developed for commercialization including rice, eggplant, cauliflower and cabbage, canola, potato, tobacco, tomato, apples, soybean and peanut.

19.2 Insecticide resistance management concepts for *Bt* crops: genes, promoters and deployment

As the regulatory agency responsible for commercial use of *Bt* crops in the USA, the Environmental Protection Agency (EPA) made a decision, in response largely to organic growers, that susceptibility to *Bt* was a public good and should be preserved. Although the same argument could be made for maintaining susceptibility to other products used by the organic community and conventional growers (e.g. pyrethrum and pyrethroids), the *Bt* argument had political clout and was adopted. Thus, EPA, industry and the farming community began to develop IRM plans for *Bt* crops. In the decade leading up to the first commercial release of *Bt* plants, several deployment tactics designed to delay resistance were proposed and debated. These strategies built on a considerable body of theoretical and empirical work on resistance to conventional insecticides (Bates *et al.*, 2005a). The strategies can be broken down into three elements: the toxins, the promoters that drive toxin expression and the deployment of the plants in the field. The overall goal is to use these three elements to reduce pest populations while maintaining susceptible alleles for the *Bt* toxin in the insect population. It is useful to discuss these strategies because they continue to shape IRM strategies for insecticidal plants.

19.2.1 Expression of *Bt* toxins at a moderate level to allow a proportion of susceptible insects to survive

The idea behind this concept is that low doses of *Bt* endotoxins retard growth rates of susceptible genotypes and allow natural enemies or other events to further reduce pest populations. Caprio *et al.*, (2000) and Bates *et al.* (2005a) point out many problems with this concept including the unacceptable and/or unpredictable control in crops that might cause growers to spray other insecticides, the temporal asynchrony that may develop between the emergence of susceptible and resistant adults, and the variable efficacy of *Bt* at lower doses due to environmental influences. Models indicated that moderate expression provides some delay in resistance, but the delay is small compared to other tactics.

19.2.2 Expression of toxins at a level high enough to ensure that individuals heterozygous for resistance are killed

Because the initial occurrence of individuals homozygous for resistance (RR) is likely to be so rare that it can essentially be ignored, the rate of resistance evolution is driven primarily by the frequency and survival of heterozygotes (RS) (Roush, 1997a). Thus, from an IRM perspective, a dose that is high enough to cause mortality to heterozygotes is preferred. From an IPM perspective, a high dose will also ensure that crop damage is minimized. A central issue of the high-dose strategy is determining what constitutes a sufficiently high dose and whether that dose will be expressed at all times in all parts of the plant subject to insect attack. The general recommendation is that it should be 25-fold the toxin concentration needed to kill 99% of susceptible individuals when heterozygotes are not available for testing, but others suggest higher concentrations (Caprio *et al.*, 2000). Additionally, there is concern that what is a high dose for one insect species may be lower for another species.

The concept of the high dose of *Bt* expressed in the plant can be contrasted with what occurs when a plant is sprayed with a *Bt* product. Assume a sprayer is properly calibrated and the concentration of the *Bt* in the sprayer is sufficient to kill all the larvae (e.g. 100 ppm). However, what actually lands on the plant will be variable. For example, 100 ppm may be the dose that lands on the upper leaf surface of the top leaf but what lands on the lower leaf surface of the top leaf will be much lower. Additionally, the dose that lands on the upper and lower surfaces of a leaf in the middle and lower parts of the plant canopy will be substantially lower. Thus, the insect population on the plant is subjected to a mosaic of doses, many of which are likely to be sublethal to heterozygotes in the population. As mentioned above, it is the survival of the heterozygotes that drives resistance evolution and this can explain why diamondback moth resistance to a *Bt* protein developed faster with foliar sprays than when the insects were exposed to high dose *Bt* plants (Roush, 1997a).

19.2.3 Deployment of *Bt* varieties expressing different toxins simultaneously in a mosaic, in a rotation, sequentially (e.g. use one toxin until control failure occurs), or by incorporating both toxins in a single variety (i.e. pyramided)

The first *Bt* plants were deployed with a single toxin, and as more companies with different toxins have entered the market there is increased concern about the need to have a cross-company IRM strategy. Models (Roush, 1997b) and empirical data (Zhao *et al.*, 2003) indicate that the use of two plant varieties, each expressing one different *Bt* toxin, when grown in the same area (i.e. a mosaic pattern) is the least effective method of delaying resistance. Thus, caution should be warranted if multiple varieties expressing different single toxins are grown concurrently in the field, because this will result in a landscape-level mosaic. Avoiding this situation is a challenge to IRM and regulatory agencies. Pyramiding two (or more) toxins, each with a different binding site to reduce the likelihood of cross-resistance, into one variety appears to be the best way to delay resistance (Roush, 1997b; Zhao *et al.*, 2003). Unlike resistance to a single toxin, it is not the mortality of heterozygous larvae that is the most influential factor, but the mortality of the susceptible homozygotes (SS genotype). If both toxins are expressed at high levels, the frequency of individuals with complete resistance to two toxins will be far rarer than those resistant to one. Pyramiding toxins is also favored over rotations of toxins on a temporal scale because, in many cases, resistance to *Bt* toxins can remain stable or decrease only slowly in the absence of selection pressure. For species with only moderate susceptibility to a given *Bt* toxin, such as *Helicoverpa* spp., pyramided plants can also provide superior control. Pyramided toxins may also require smaller refuge sizes (see below), and there have been suggestions that regulatory agencies should actively promote their registration (Zhao *et al.*, 2003, 2005).

19.2.4 Expression of *Bt* toxins in plants only at certain times or in certain parts through the use of temporal, tissue-specific or chemically inducible promoter

Current *Bt* plants express the toxin continuously because the plant promoters are always turned on. This is analogous to continuous application of a foliar insecticide (although the "coverage" would be different). The use of selective expression (either in time or on different plant parts) through promoters is possible, but has been limited by the available technology. However, it may provide a "within-plant" refuge that might be easy for growers to use for IRM and be more in line with the philosophy of economic thresholds, i.e. treating only when the insect reaches a certain density that would cause economic loss. For example, in many crops the plant part to be protected develops later in the season (e.g. maize ear or cotton boll)

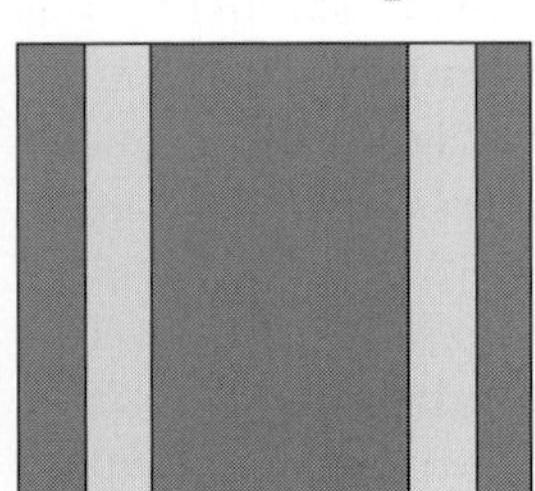

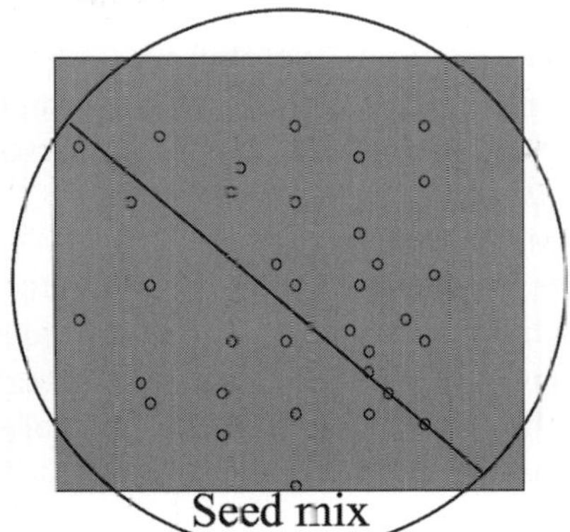

Fig. 19.1 Examples of structured refuge options for *Bt* transgenic plants. Light-colored areas represent non-*Bt* refuge plants (diagrams courtesy of S. Matten, US Environmental Protection Agency).

and plants can withstand tissue defoliation in the earlier part of the season without reduction in the size of the marketable part. It is possible that plants could be bred that would have the capacity to express a toxin but specific promoters would be required that can be turned on so that the toxin would be expressed only when needed. Thus, prior to the toxin expression, susceptible alleles in the insect population would be maintained in an early-season within-plant refuge. The potential of this strategy has been examined using diamondback moth and *Bt* broccoli (Bates *et al.*, 2005b; Cao *et al.*, 2006) and it is clear that use of such promoters would have to be tailored to an individual crop–pest system. In the case of tissue-specific promoters, larvae that move easily between toxic and non-toxic plant structures may negate the benefits of selective *Bt* expression, so its usefulness would also depend on the specific crop–pest system. However, from an IRM standpoint, the use of temporal or tissue-specific promoters in conjunction with pyramided toxins could greatly aid in IRM of insecticidal plants.

19.2.5 Provision of non-toxic plants to maintain susceptible insects within the field in seed mixtures or external to the transgenic crop in refuges

The use of refuges combined with a high dose of the toxin (i.e. the high-dose/refuge strategy) is presently the EPA-mandated strategy for *Bt* corn and cotton in the USA, and several other countries have adopted this strategy. The refuge is composed of non-transgenic plants that will generate enough susceptible insects to dilute resistant alleles during mating. An example of some structured refuge options are shown in Fig. 19.1, but there are other options such as "community refuges" in which one non-*Bt* field may serve as the refuge for several other fields. The idea behind a refuge is that if resistance is a recessive trait (as has been the case so far), then any resistant homozygotes (RR genotype individuals) that do develop and mate with SS individuals will produce offspring that are RS and therefore susceptible to the *Bt* plants that were developed to express a dose sufficient to kill the RS genotype. It has been suggested that refuges should generate 500 individuals (SS and RS genotypes) for each resistant individual (RR) that may arise from the transgenic plant (EPA, 1998). The method of planting the refuge and the size of the refuge has long been a subject of debate (Bates *et al.*, 2005a).

Seed mixtures of transgenic and non-transgenic plants once seemed an attractive method of creating a refuge because they required little effort on the part of growers to implement IRM. They were also expected to favor random mating between susceptible and resistant insects. Subsequent studies have shown, however, that for some pests, interplant movement by larvae would render this strategy less effective. Larvae may actively avoid feeding on *Bt* plants, or larvae developing on non-*Bt* plants may move to toxic plants and die, thus reducing the effective size of the refuge. Seed mixture is not an EPA-approved option for refuges; however, it has been of interest for farmers, research scientists and industry. In field tests using the diamondback moth and transgenic broccoli, Shelton *et al.* (2000) found that unsprayed refuges external to the *Bt* crop were more effective for conserving susceptible alleles and reducing the overall number of resistant insects on transgenic plants compared to mixed or sprayed refuges. However, care must be taken to ensure the insects in the refuge are managed from both an economic and IRM standpoint. Managing insects in the refuge, while conserving *Bt* susceptible alleles, continues to be a challenge.

19.3 Challenges for IRM and *Bt* crops

Early simulation models suggested that pest populations could evolve resistance in as little as 1–2 years under worse case conditions (Gould *et al.*, 1997; Roush, 1997a; Tabashnik, 1997; International Life Sciences Institute & Health and Environmental Sciences Institute, 1998). However, despite the large-scale planting of *Bt* crops (42.1 million ha. in 2007: James, 2007), up to 2006 no cases of resistance to *Bt* in the field had been observed (Matten & Reynolds, 2003; Tabashnik *et al.*, 2003). However, in 2007 it was documented that unexpected performance failures of TC1507 maize to fall armyworm (*Spodoptera frugiperda*) observed in 2006 in Puerto Rico were due to Cry1F-resistant fall armyworm (Matten, 2007) and the product was taken off the market in Puerto Rico. Perhaps the overall success of the current IPM program is due to the foresight of the high-dose/refuge strategy, which is the only currently approved program in the USA, but it is impossible to tell. Several sets of challenges remain to IRM for *Bt* crops.

19.3.1 Biological assumptions for IRM

There are three biological assumptions of the high-dose/refuge strategy. The biological assumptions are: (1) the dose of toxin expressed in the plant is high enough that the SS and RS genotypes will be killed (this is usually stated as resistance is recessive or functionally so on the plant); (2) resistance alleles are rare (e.g. $p < 10^{-3}$), so that there will be few homozygous (RR) survivors ($p^2 \ll 10^{-6}$); and (3) resistant and susceptible insects will mate more or less randomly. Because there have not been any cases of insects having developed high levels of resistance to *Bt* proteins (except for diamondback moth in the field and cabbage looper *Trichoplusia ni* in greenhouse, both of which are not major pests to the currently commercialized *Bt* plants and developed resistance from foliar sprays of *Bt*, not *Bt* expressed in plants), these assumptions are hard to evaluate (Caprio *et al.*, 2000). Nevertheless, alleles conferring resistance to *Bt* cotton have been found in field populations of the pink bollworm (*Pectinophora gossypiella*) (Tabashnik *et al.*, 2000) in the USA and cotton bollworm (*Helicoverpa armigera*) in Australia (Akhurst *et al.*, 2003). Similar studies conducted in *Bt* maize (Andow *et al.*, 2000; Bourguet *et al.*, 2003; Stodola *et al.*, 2006) suggest that the frequencies of resistant alleles are rare and would enable successful implementation of *Bt* maize for the major Lepidoptera in several countries.

However, a note of caution is needed in regard to creating plants that will meet the first assumption. In some situations, a *Bt* plant that delivers a high dose of toxin to one species may be only moderately toxic to another. Hence the high-dose/refuge strategy may not be appropriate as a one-size-fits-all IRM tool for *Bt* crops with multiple pests (Bates *et al.*, 2005a).

19.3.2 Operational requirements for IRM: refuges

The value of refuges for reducing the frequency of resistant alleles in a population of insects exposed to *Bt* plants has been demonstrated experimentally in the greenhouse (Tang *et al.*, 2001) and in the field (Shelton *et al.*, 2000) using the diamondback moth/*Bt* broccoli system. Prior to such empirical data, models had indicated the usefulness of refuges and they have been required for USA registration of *Bt* crops since early on in the registration process. Debate over the appropriate size, placement and management of refuges is arguably the most contentious issue surrounding the high-dose/refuge tactic (Bates *et al.*, 2005a). Refuges need to be large enough that they can generate sufficient susceptible alleles in the insect population yet produce a marketable crop, since growers are unlikely to plant them otherwise. Furthermore, refuges should be placed so they will encourage random mating of the different insect genotypes, but often insufficient data on movement patterns of the insects are available to guide refuge placement. Likewise, the availability of alternative hosts such as weeds needs to be considered when determining the size of refuges since such "unstructured refuges" may play a major role in conserving susceptible alleles. Also, within an area the percentage of land grown to a *Bt* crop will initially be small and thus the area grown to non-*Bt* varieties will serve as a refuge. This situation will change as more growers adopt *Bt* crops and this may require revisiting the refuge requirements periodically. Several types of refuge strategies are available (Matten & Reynolds, 2003) and include different sizes (from 4% to 50%) depending on the crop and whether the refuge is sprayed (spraying must be done with a non-*Bt* insecticide but spraying of the refuge will lower the populations of *Bt*-susceptible individuals in the landscape), and different locations, depending on whether a non-*Bt* crop can serve as a refuge for several surrounding *Bt* fields (i.e. a "community refuge"). Additional complexities can occur when an insect attacks multiple crops (such as corn earworm [*Helicoverpa zea*] on maize and cotton in the southeastern USA) and could be exposed to the same toxin in each crop. In this case, the refuge size for maize is 50%, a figure that would likely limit adoption of *Bt* maize. Ultimately, landscape studies that incorporate movement of insects between different plantings and wild hosts are needed for an IRM program to be carried out on a regional basis (Kennedy & Storer, 2000).

Although regulators may require refuges, legitimate questions arise about whether growers are adopting such practices. If one examines existing data, it appears that the majority of USA growers are complying with their obligations of planting refuges for IRM, although the data differ depending on the source (Bates *et al.*, 2005a). In China and India where millions of hectares of *Bt* cotton are grown on small holdings, planting refuges is far more problematic.

19.3.3 Operational requirements for IRM: monitoring

Resistance monitoring for *Bt* plants has been part of the registration requirements since 1996 when the first *Bt* plant was registered (no other pesticide has a similar requirement). Each company that sells *Bt* plants must conduct an annual resistance-monitoring program that requires field collection of insects, laboratory bioassays and reporting the results back to the EPA. The goal of resistance monitoring is to detect resistance (significant changes in susceptibility) before widespread field failure so that modifications to the IRM plan can be made to ensure the longevity of the product. A number of monitoring techniques for resistance detection have been proposed and include: grower reports of unexpected damage; systematic field surveys of *Bt* maize; discriminating concentration assays; the F_2 screen; sentinel plots of *Bt* maize and isolation of resistance genes from laboratory selected colonies for development of molecular diagnostic techniques. The most common method is the discriminating concentration assay by which one looks for increases in the percent survival at a toxin concentration derived from baseline susceptibility of previously unexposed populations (Hawthorne *et al.*, 2002). Its major limitation is its inability to detect the development of resistance early. When allele frequencies are low (i.e. prior

Table 19.1 Features and limitations of three monitoring methods

Monitoring method	Features	Limitations[a]
Diagnostic dose	Field validated	Insensitive to highly recessive R alleles
	Precise and accurate	Measurement of R allele frequencies in the range (10^{-5}–10^{-2}) is impossible without extremely large sample sizes
	Relatively inexpensive	Consumes high amounts of purified toxin
	Rapid turn around	Indirect calculation of R allele frequency
	Historical data set allows future comparisons	
	Infrastructure in place	
F_2 screen	Detects recessive alleles at low frequencies	Large sample size needed
	Direct measurement of R allele frequency and does not require a resistant strain of the target insect	Very labor and space intensive
	Realistic test dosage used in assays	Precision, accuracy, and sensitivity unvalidated and some assumptions (e.g. females mated once and inbreeding depression is absent) are unchecked
	Allows direct recovery of R alleles	Unsuitable for polygenic resistance
In-field screen	Realistic field-based exposure	Time sensitive and labor intensive
	Large number of insects assayed	Insensitive to recessive alleles
	Multiple pest species screened	Indirect calculation of R allele frequency
		Most variable and unquantifiable dosage
		Precision, accuracy, and sensitivity not validated

[a] R alleles are those that confer increased tolerance or resistance to an insecticide.
Source: Modified from Caprio *et al.* (2000) and Hawthorne *et al.* (2002).

to selection pressure), resistance alleles are most frequently found as heterozygotes. Because these alleles will be missed by the diagnostic dose methods, the measurement of allele frequencies in the range (10^{-5}–10^{-2}) is impossible without extremely large sample sizes. A modification of this technique is the larval growth inhibition assay in which a sublethal dose of the toxin is used.

Each monitoring method has unique benefits and liabilities, as listed in Table 19.1. One method that has received a large amount of interest is the F_2 screen which focuses on the detection of recessive and partially recessive resistance alleles assay (Andow & Alstad, 1998). While this assay is a very interesting approach for evaluating initial gene frequency, it is clearly impractical for routine monitoring (Caprio *et al.*, 2000). However, it may be a very useful tool for capturing resistant alleles in a population and thereby lead to further studies on the genetics and molecular basis of resistance. Another approach that requires less time and effort is to plant small plots of transgenic and non-transgenic plants and sample for resistant individuals, which are then tested against a diagnostic dose. This method is commonly called an in-field screen and has been described by

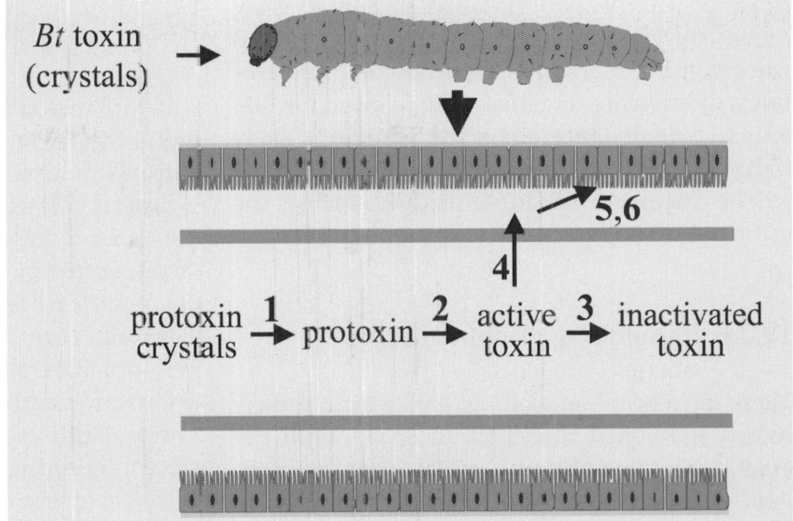

Fig. 19.2 Mode of pathogenesis of *Bt* toxins (diagram courtesy of P. Wang, Cornell University).

Venette and colleagues (Venette *et al.*, 2000, 2002). While this method has some advantages because it is easy to use, it suffers from the same limitations as mentioned above, but with the additional problem of generating false positives from non-expressing "*Bt* plants" (off-types). The holy grail of monitoring for resistance involves molecular techniques, but this has eluded researchers so far (see Section 19.3.4).

19.3.4 Mechanisms for *Bt* resistance and the potential to use genetic and molecular methods for monitoring

The mode of pathogenesis for *Bt* in insects is complex and follows several steps (Fig. 19.2). At any one of these steps, changes may occur that will influence pathogenesis. There have been several species of insects that have been selected in the laboratory for resistance to *Bt* toxins (Ferré & Van Rie, 2002), and these have been utilized for studying the mechanisms of resistance. The following list of potential mechanisms for *Bt* resistance is provided by Wang *et al.*, (2007).

(1) Alteration of midgut proteases which are critically involved in solubilization and proteolytic processing of Cry proteins in the insect midgut.
(2) Modification of midgut binding sites for *Bt* toxins, which results in reduced binding of the *Bt* toxins to the midgut targets.
(3) Retention of the *Bt* toxin by the midgut peritrophic membrane.
(4) Aggregation of *Bt* toxin proteins by the midgut esterase.
(5) Elevated melanization activity of the hemolymph and midgut cells.
(6) Increased rate of repair or replacement of affected epithelial cells, and possibly increased antioxidation activities in *Bt*-resistant insects.

These hypotheses were developed and tested largely with laboratory-selected resistant populations and the majority of studies have focused on midgut binding sites. The putative midgut receptors for Cry1A include the midgut cell membrane proteins cadherin, aminopeptidase N, alkaline phosphatase and a 252-kDa high-molecular-weight protein, and midgut glycolipids (Wang *et al.*, 2007). Several studies have focused on cadherin but in Cry1Ac-resistant diamondback moth strains originated from field-developed resistant populations from two geographical regions, the reduced binding of Cry1A toxins is not conferred by cadherin gene mutations (Baxter *et al.*, 2005). The holy grail for monitoring resistance is the use of molecular markers, but this has been elusive since different resistant populations of the same insect species from different origins or selected in different laboratories may have distinctly different mechanisms of resistance to the same *Bt* Cry toxin (Bates *et al.*, 2005a; Wang *et al.*, 2007).

However, the benefits of using molecular techniques for monitoring resistance are large and work continues in this important effort. Resistance monitoring remains a critical need for IRM to insecticidal plants and in the future it is likely that molecular methods for resistance detection will be combined with other methods such as traditional bioassays.

19.3.5 Remedial action if resistance occurs

There have been no documented cases of insects having developed resistance to *Bt* plants in the field (except for the case of fall armyworm in Puerto Rico: Matten, 2007), and this is quite remarkable (Bates *et al.*, 2005a) and greatly exceeds the length of time that typically passes in the field before resistance is first documented with most conventional neurotoxic pesticides (McCaffrey, 1998). However, caution is warranted and plans should be put into place if resistance is detected. According to the EPA (Matten & Reynolds, 2003), several actions will be implemented if resistance is detected in the course of the annual resistance monitoring programs. If resistance is detected in the random population sampling assays (but is not yet detectable in the field), a number of steps will be taken prior to any remedial action. These include: (1) confirmation that the resistance is heritable; (2) confirmation that the resistance will be observed in the field (i.e. on live *Bt* maize plants); (3) determination of the nature of resistance (dominant, recessive); (4) estimation of the frequency of the resistance allele; (5) analysis of whether the resistance allele distribution is increasing; (6) determination of the geographic extent of the resistance allele distribution; and (7) design of a remedial action plan if the resistance allele distribution is spreading. The purpose of the remedial action plan is to contain or slow the spread of resistant populations and it contains several steps including: mandatory reporting to the EPA of suspected resistance; use of alternative control measures in the affected areas; investigation of the causes of resistance; increased resistance monitoring and suspension of *Bt* crop sales in the affected areas.

19.4 Conclusions

Since its introduction in the USA in 1996, susceptibility to *Bt* has not measurably decreased nor has confirmed field resistance been detected in insects affecting maize or cotton, except for the isolated case of fall armyworm in Puerto Rico reported by Matten (2007). This includes all the monitored pests and *Bt* maize toxins: Cry1Ab (European corn borer, corn earworm, southwestern corn borer [*Diatraea grandiosella*], fall armyworm) and Cry1F (European corn borer, corn earworm, southwestern corn borer). Similarly, resistance monitoring reports since 1996 indicate that no decrease in susceptibility to Cry1Ac in *Bt* maize has been observed in tobacco budworm (*Heliothis virescens*), cotton bollworm (*Helicoverpa zea*) and pink bollworm (S. Matten, EPA, personal communication); however, there is a controversial report (Tabashnik *et al.*, 2008) claiming that resistance in *H. zea* has occurred. The reasons for this may be varied and include: the wisdom of the high-dose/refuge strategy and its adoption by growers; the abundance of non-structured refuges (non-*Bt* weed hosts and other host crops); or perhaps our present inability to detect resistance. In the case of the latter, it is important to continue refining the monitoring strategies in order to detect resistance before field failure.

Large challenges will be faced in countries that adopt insecticidal plants, but lack a well-regulated resistance management strategy. In these situations growers may not be able or willing to plant refuges and areawide resistance monitoring programs may not be performed. This situation could lead to an early loss of this valuable technology. To overcome these situations, modifications will be needed. For example, models (Roush, 1997b) have indicated that a 30–40% refuge for single gene plants equals a 5–10% refuge for dual-gene plants (i.e. plants expressing two dissimilar toxins simultaneously, commonly referred to as pyramided plants). Greenhouse tests have confirmed the improved durability of pyramided plants compared to single gene plants (Zhao *et al.*, 2003). Such pyramided cotton plants are currently registered in the USA and Australia and efforts are under way to develop pyramided *Bt* maize in the USA and crucifers for use against the diamondback moth in India (Srinivasin *et al.*, 2005).

Bt plants are the first type of host plant resistance that has displayed high efficacy against Lepidoptera and Coleoptera. As with the example of Hessian fly noted at the beginning of this chapter, host plant resistance can break down and it is likely that over time the present *Bt* plants will lose some of their effectiveness. While the effectiveness can be restored by the use of another *Bt* Cry protein, it is important to preserve the use of each for the longest time possible. *Bt* plants have provided great economic benefits to growers and the environment (Shelton *et al.*, 2002; Brookes & Barfoot, 2006; Sanvido *et al.*, 2007), and IRM strategies are needed to preserve their benefits. *Bt* insecticidal plants are only the first wave of genetically engineered plants for insect management, and lessons learned from them can be applied to future technologies.

References

Akhurst, R. J., James, W., Bird, L. J. & Beard, C. (2003). Resistance to the Cry1Ac δ-endotoxin of *Bacillus thuringiensis* in the cotton bollworm, *Helicoverpa armigera* (Lepidoptera: Noctuidae). *Journal of Economic Entomology*, **96**, 1290–1299.

Andow, D. A. & Alstad, D. N. (1998). F_2 screen for rare resistance alleles. *Journal of Economic Entomology*, **91**, 572–578.

Andow, D. A., Olson, D. M., Hellmich, R. L., Alstad, D. N. & Hutchison, W. D. (2000). Frequency of resistance to *Bacillus thuringiensis* toxin Cry1Ab in an Iowa population of European corn borer (Lepidoptera: Crambidae). *Journal of Economic Entomology*, **93**, 26–30.

Bates, S. L., Zhao, J.-Z., Roush, R. T. & Shelton, A. M. (2005a). Insect resistance management in GM crops: past, present and future. *Nature Biotechnology*, **23**, 57–62.

Bates, S. L., Cao, J., Zhao, J.-Z. *et al.* (2005b). Evaluation of a chemically inducible promoter for developing a within-plant refuge for resistance management. *Journal of Economic Entomology*, **98**, 2188–2194.

Baxter, S. W., Zhao, J.-Z., Gahan, L. J. *et al.* (2005). Novel genetic basis of field-evolved resistance to *Bt* toxins in *Plutella xylostella*. *Insect Molecular Biology*, **14**, 327–334.

Bourguet, D., Chaufaux, J., Seguin, M. *et al.* (2003). Frequency of alleles conferring resistance to *Bt* maize in French and US corn belt populations of the European corn borer, *Ostrinia nubilalis*. *Theoretical and Applied Genetics*, **106**, 1225–1233.

Brookes, G. & Barfoot, P. (2006). Global impact of biotech crops: socio-economic and environmental effects in the first ten years of commercial use. *AgBioForum*, **9**, 139–151.

Cao, J., Bates, S. L., Zhao, J.-Z., Shelton, A. M. & Earle, E. D. (2006). *Bt* protein production, signal transduction and insect control in chemically inducible PR-1aAb broccoli plants. *Plant Cell Reports*, **25**, 554–560.

Caprio, M. A., Summerford, D. V. & Simms, S. R. (2000). Evaluating transgenic plants for suitability in pest and resistance management programs. In *Field Manual of Techniques in Invertebrate Pathology*, eds. L. A. Lacey & H. K. Kaya, pp. 805–828. Dordrecht, Netherlands: Kluwer.

Croft, B. (1990). Developing a philosophy and program of pesticide resistance management. In *Pesticide Resistance in Arthropods*, eds. R. T. Roush & B. E. Tabashnik, pp. 277–296. New York: Chapman & Hall.

Environmental Protection Agency (1998). *FIFRA Scientific Advisory Panel, Subpanel on* Bacillus thuringiensis (Bt) *Plant Pesticides and Resistance Management, Transmittal of the Final Report*, Docket Number OPP #00231. Washington, DC: US Environmental Protection Agency.

Ferré, J. & Van Rie, J (2002). Biochemistry and genetics of insect resistance to *Bacillus thuringiensis*. *Annual Review of Entomology*, **47**, 501–533.

Georghiou, G. P. & Lagunes-Tejeda, A. (1991). *The Occurrence of Resistance to Pesticides in Arthropods*. Rome, Italy: Food and Agriculture Organization.

Gould, F. (1986). Simulation models for predicting durability of insect-resistant germ plasm: Hessian fly-resistant winter wheat. *Environmental Entomology*, **15**, 11–23.

Gould, F., Anderson, A., Jones, A. *et al.* (1997). Initial frequency of alleles for resistance to *Bacillus thuringiensis* toxins in field populations of *Heliothis virescens*. *Proceedings of the National Academy of Science of the USA*, **94**, 3519–3523.

Hawthorne, D. B., Sigfried, B., Shelton, A. & Hellmich, R. (2002). *Monitoring for Resistance Alleles: A Report from an Advisory Panel on Insect Resistance Monitoring Methods for* Bt *Corn*, Agricultural Biotechnology Stewardship Committee Report. Washington, DC: Biotechnology Industry Organization.

International Life Sciences Institute & Health and Environmental Sciences Institute (ILSI HESI) (1998). *An Evaluation of Insect Resistance Management in* Bt *Field Corn: A Science-Based Framework for Risk Assessment and Risk Management*, Report of an Expert Panel. Washington, DC: ILSI Press.

James, C. (2007). *Global Status of Commercialized Transgenic Crops*, ISAAA Brief No. 37. Ithaca, NY: International Service for the Acquisition of Agri-biotech Applications.

Kennedy, G. G. & Storer, N. P. (2000). Life systems of polyphagous arthropod pests in temporally unstable cropping systems. *Annual Review of Entomology*, **45**, 467–493.

Krattiger, A. F. (1997). *Insect Resistance in Crops: A Case Study of Bacillus thuringiensis* (Bt) *and its Transfer to Developing Countries*, ISAAA Brief No. 2. Ithaca, NY: International Service for the Acquisition of Agri-biotech Applications.

Matten, S. R. & Reynolds, A. H. (2003). Current resistance management requirements for *Bt* cotton in the United States. *Journal of New Seeds*, **5**, 137–178.

Matten, S. R. (2007). Review of Dow AgroScience's (and Pioneer Hibred's) submission (dated 12 July 2007) regarding fall armyworm resistance to the Cry1F protein expressed in TC1507 Herculex® I insect protection maize in Puerto Rico (EPA registrations 68467-2 and 29964-3). Memorandum from S. R. Matten, USEPA/OPP/BPPD to M. Mendelsohn, USEPA/OPP/BPPD, 24 August 2007.

McCaffrey, A. R. (1998). Resistance to insecticides in heliothine Lepidoptera: a global view. *Philosophical Transactions of the Royal Society of London, B*, **353**, 1735–1750.

Metcalf, R. L. & Metcalf, R. A. (1993). *Destructive and Useful Insects: Their Habits and Control*, 5th edn. New York: McGraw-Hill.

Painter, R. H. (1951). *Insect Resistance to Crop Plants*. Lawrence, KS: University of Kansas Press.

Roush, R. T. (1997a). Managing resistance to transgenic crops. In *Advances in Insect Control*, eds. N. Carozzi & M. Koziel, pp. 271–294. London: Taylor & Francis.

Roush, R. T. (1997b). *Bt*-transgenic crops: just another pretty insecticide or a chance for a new start in resistance management? *Pesticide Science*, **51**, 328–334.

Sanvido, O., Romeis, J. & Bigler, F. (2007). Ecological impacts of genetically modified crops: ten years of field research and commercial cultivation. *Advances in Biochemical Engineering and Biotechnology*, **107**, 235–278.

Sawicki, R. M. (1987). Definition, detection and documentation of insecticide resistance. In *Combating Resistance to Xenobiotics: Biological and Chemical Approaches*, eds. M. G. Ford, D. W. Holloman, B. P. S. Khambay & R. M. Sawicki, pp. 105–117. Chichester, UK: Ellis Horwood.

Shelton, A. M. & Roush, R. T. (2000). Resistance to insect pathogens and strategies to manage resistance. In *Field Manual of Techniques in Invertebrate Pathology*, eds. L. A. Lacey & H. K. Kaya, pp. 829–846. Dordrecht, Netherlands: Kluwer.

Shelton, A. M., Tang, J. D., Roush, R. T., Metz, T. D. & Earle, E. D. (2000). Field tests on managing resistance to *Bt*-engineered plants. *Nature Biotechnology*, **18**, 339–342.

Shelton, A. M., Zhao, J.-Z. & Roush, R. T. (2002). Economic, ecological, food safety, and social consequences of the deployment of *Bt* transgenic plants. *Annual Review of Entomology*, **47**, 845–881.

Shelton, A. M., Zhao, J.-Z., Nault, B. A. *et al.*(2006). Patterns of insecticide resistance in onion thrips, *Thrips tabaci*, in onion fields in New York. *Journal of Economic Entomology*, **99**, 1798–1804.

Smith, C. M. (1989). *Plant Resistance to Insects*. New York: John Wiley.

Srinivasin, R., Talekar, N. S. & Dhawan, V. (2005). Transgenic plants with dual *Bt* gene: an innovation initiative for sustainable management of *Brassica* pests. *Current Science*, **88**, 1877–1879.

Stodola, T. J., Andow, D. A., Hyden, A. R. *et al.* (2006). Frequency of resistance to *Bacillus thuringiensis* toxin Cry1Ab in southern United States Corn Belt population of European corn borer (Lepidoptera: Crambidae). *Journal of Economic Entomology*, **99**, 502–507.

Tabashnik, B. E. (1994). Evolution of resistance to *Bacillus thuringiensis*. *Annual Review of Entomology*, **39**, 47–79.

Tabashnik, B. E. (1997). Seeking the root of insect resistance to transgenic plants. *Proceedings of the National Academy of Science of the USA*, **94**, 3488–3490.

Tabashnik, B. E., Patin, A. L., Dennehy, T. J. *et al.* (2000). Frequency of resistance to *Bacillus thuringiensis* in field populations of pink bollworm. *Proceedings of the National Academy of Science of the USA*, **97**, 12980–12984.

Tabashnik, B. E., Carriere, Y., Dennehy, T. J. *et al.* (2003). Insect resistance to transgenic *Bt* crops: lessons from the laboratory and field. *Journal of Economic Entomology*, **96**, 1031–1038.

Tabashnik, B. E., Gassmann, A. J., Crowder, D. W. & Carrière, Y. (2008). Insect resistance to *Bt* crops: evidence versus theory. *Nature Biotechnology*, **26**, 199–202.

Tang, J. D., Collins, H. L., Metz, T. D. *et al.* (2001). Greenhouse tests on resistance management of *Bt* transgenic plants using refuge strategies. *Journal of Economic Entomology*, **94**, 240–247.

Venette, R. C., Hutchison, W. D. & Andow, D. A. (2000). An in-field screen for early detection of insect resistance in transgenic crops: practical and statistical considerations. *Journal of Economic Entomology*, **93**, 1055–1064.

Venette, R. C., Moon, R. D. & Hutchison, W. D. (2002). Strategies and statistics of sampling for rare individuals. *Annual Review of Entomology*, **47**, 143–174.

Wang, P., Zhao, J.-Z., Rodrigo-Simón, A. *et al.* (2007). Mechanism of resistance to *Bacillus thuringiensis* toxin Cry1Ac in a greenhouse population of the cabbage looper, *Trichoplusia ni*. *Applied and Environmental Microbiology*, **73**, 1199–1207.

Zhao, J.-Z., Cao, J., Li, Y. *et al.* (2003). Plants expressing two *Bacillus thuringiensis* toxins delay insect resistance compared to single toxins used sequentially or in a mosaic. *Nature Biotechnology*, **21**, 1493–1497.

Zhao, J.-Z., Cao, J., Collins, H. L. *et al.* (2005). Concurrent use of transgenic plants expressing a single and two *Bt* genes speeds insect adaptation to pyramided plants. *Proceedings of the National Academy of Science of the USA*, **102**, 8426–8430.

Zhao, J.-Z., Collins, H. L., Li, Y. *et al.* (2006). Monitoring of diamondback moth resistance to spinosad, indoxacarb and emamectin benzoate. *Journal of Economic Entomology*, **99**, 176–181.

Chapter 20

Role of biotechnology in sustainable agriculture

Jarrad R. Prasifka, Richard L. Hellmich and Michael J. Weiss

A basic concept of sustainable agriculture includes using resources in a way that does not deplete or permanently damage systems used for plant and animal production. In early history, humans survived as hunter–gatherers and perhaps less than 1% of biomass could be used as food (Diamond, 1997). As a result, most resources in the environment were not likely to be used directly by humans. The limited availability of food also restricted population growth, helping to make the hunter–gatherer way of life sustainable. In contrast, domestication of crops and animals for food has greatly increased edible biomass, leading to dramatic population growth and the possibility that production of adequate food will lead to long-term damage to agricultural systems.

The high productivity of twenty-first-century agriculture is the cumulative result of periods of change called agricultural revolutions. Another revolution based on biotechnology is arguably under way. Some have called the biotechnology-based changes in agriculture the "gene revolution" because they follow the green revolution of the twentieth century, during which high-yielding crop varieties and other changes in production were spread to developing nations. The use of biotechnology in agriculture includes well-publicized techniques such as production of genetically modified (GM; alternatively called transgenic or genetically engineered [GE]) plants and animals, but also less controversial techniques (Herdt, 2006). For example, biotechnology may be used to improve or supplement conventional agricultural methods, such as when marker-assisted selection is employed to enhance traditional breeding of crops.

It is worth noting that each time agricultural methods advance, new problems related to sustainability may be resolved and created (Evans, 2003). For example, in the twentieth century the development of new synthetic insecticides delivered effective and long-lasting control of insect pests (Casida & Quistad, 1998). However, the adverse effects from uncontrolled pesticide use were brought to public attention by the book *Silent Spring* (Carson, 1962). Since the 1960s, increased regulation has considerably reduced the threat of environmental and agricultural problems stemming from overuse of pesticides. Therefore, it seems reasonable to predict that a biotech revolution will have both positive and negative effects on sustainability, and the degree to which either aspect dominates will be based on the choices society makes regarding how to use biotechnology in agriculture.

Along with changes to agriculture and society over the last century, the concept of sustainability has been popularized and expanded. Broader definitions of sustainable agriculture reveal that the concept suggested above (using resources in a

Integrated Pest Management, ed. Edward B. Radcliffe, William D. Hutchison and Rafael E. Cancelado. Published by Cambridge University Press.

way that does not deplete or permanently damage agricultural systems) may be too simple. One representative definition suggests that sustainable agriculture "enhances environmental quality and the resource base on which agriculture depends; provides for basic human food and fiber needs; is economically viable; and enhances the quality of life for farmers and society as a whole" (American Society of Agronomy, 1989). Understanding more complex definitions can be aided by considering three common components associated with sustainable agriculture: (1) economic, (2) environmental and (3) social or community effects (Lyson, 2002). Though most agricultural practices will impact more than one of these three components, the categories are useful to organize thinking about sustainability and to emphasize the broad nature of sustainable agriculture.

20.1 Potential of biotechnology to enhance sustainability

Many specific issues relate to agricultural sustainability, but it can be argued that there are two basic challenges (Schaller, 1993). First, agriculture must be profitable for those producing plant- and animal-based food for the rest of the world. Second, agriculture must be able to produce sufficient food (quantity and quality) to support a growing global population projected to exceed 9000 million before the year 2050. However, distinctions between applications of biotechnology that address profitability and production may not be very useful for two reasons. First, because revenue from agriculture is a basic incentive for farmers to remain involved in agriculture, profit and production are related. Second, many applications of biotechnology would clearly influence both profitability and production to some degree.

The most serious threats to agricultural profitability and production are limitations or excesses of basic resources required by plants and animals (water, heat, nutrients). Even in relatively wealthy industrialized countries like the USA, the short-term impacts of drought and long-term prospects of depleted groundwater are serious agricultural and environmental problems. One approach to increase yields under drought conditions and perhaps reduce water use is the development of crops with increased drought tolerance. Genetic engineering has been used to produce drought tolerance for many major crops including rice, wheat, maize (corn) and soybean. Marker-assisted selection and genetic engineering have been used to produce crops tolerant to other stresses including high salt levels, flooding and extreme temperatures. Stress tolerance incorporated into elite crop varieties would not only increase yields in some areas, but allow the expansion of agriculture into areas currently unfit for production of certain crops. Because resistance to multiple plant stresses may be controlled by expression of a single protein, biotechnology should make breeding multiple stress tolerant plants faster and more effective than previously possible.

Complementary biotechnology approaches also are being used to increase the efficiency of agricultural production. Increasing crops' ability to effectively use nitrogen would decrease fertilizer costs in industrialized countries and help maintain water quality by reducing the amount of nitrogen added to crops (and later leaching into groundwater). In developing nations, improved crop nitrogen use efficiency would increase yields for many farmers in developing nations who may be unable to afford synthetic fertilizer. Other biotechnology applications include modification of the nutrient content of agricultural products. Perhaps the best-known example is the beta-carotene-enriched Golden Rice, which could reduce vitamin A deficiency and save thousands of lives annually (Stein *et al.*, 2006). Many other promising examples of biofortification (nutrient enrichment through genetic engineering or conventional breeding) of crops exist, highlighting the potential to combat malnutrition using foods that are more nutrient-rich rather than simply requiring greater amounts and more types of food. Crop nutrient enrichment is also under way for livestock production, enhancing the nutritional value of crop residues fed to farm animals.

The development of alternatives to petroleum-based fuels is one of the best-known biotechnology projects. Currently most farmers are dependent on diesel and gasoline to power agricultural equipment. This makes them reliant on a

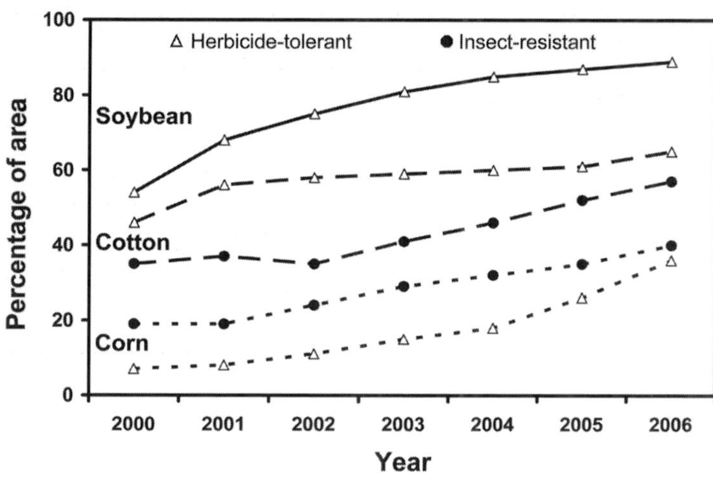

Fig. 20.1 Adoption of genetically modified crops in the USA, 2000–2006. Data from annual National Agricultural Statistics Service farm operator surveys summarized by the Economic Research Service (2006). Areas planted with varieties that are "stacked" (with herbicide-tolerance and insect-resistance traits) are represented in both lines for cotton and corn (maize).

resource that is (1) non-renewable, (2) environmentally detrimental and (3) subject to price fluctuations arguably manipulated by petroleum-exporting countries. The substitution of biologically based fuels (biofuels) such as ethanol or biodiesel may help to insulate farmers from price increases or price instability, and provide an additional source of revenue if maize, soybean or other crops are used to produce biofuels. Biotechnology is being used to more effectively produce ethanol from cellulose by the use of GM yeasts and bacteria. Similarly, genetic engineering is helping create plants that yield greater energy returns than currently available varieties. Applications of biotechnology also may allow fuels to be produced from by-products of agriculture otherwise considered waste. The benefits to the environment may increase as methods and technology related to biofuels advance. Non-food crops, including native perennial grasses, may offer the benefits of biofuels produced from maize or soybean, but with further advantages of reduced fertilizer, pesticide and energy inputs and helping to mitigate carbon dioxide emissions.

20.2 Biotechnology-based pest management and sustainability

While pest management is only one of many aspects of agriculture, and genetic engineering only one of several tools of biotechnology, transgenic management of crop pests has been the most commercially successful application of agricultural biotechnology. Herbicide-tolerance, insect-resistance and virus-resistance traits are currently available in maize, cotton, soybean, canola (oilseed rape), beets, rice, squash, papaya and alfalfa. Argentina, Brazil, Canada, China and India are among the top adopters of GM crops, though their combined GM production areas trail the leading producer of transgenic crops, the USA, which planted an estimated 54.6 million ha of the 102 million ha of global GM crops in 2006 (James, 2006).

The most successful combinations of crops and traits in the USA include insect resistance and herbicide tolerance in maize and cotton and herbicide tolerance in soybean. These traits have been commercially available since the mid-1990s with steadily increasing adoption from 2000 to 2006 (Fig. 20.1). As a result, much of the cost–benefit research on biotechnology relates to insect-resistance and herbicide-tolerance traits in maize, cotton and soybean, which have also attracted the greatest amount of scrutiny by critics of GM crops. Though some issues do not fit neatly within a single component of sustainability, the following sections further discuss what is known regarding the impact of pest management on economic, environmental and social concerns. Brookes & Barfoot (2006) provide an overview of global economic and environmental impacts of GM crops with less specific information on pest management.

Table 20.1 Estimated impacts on yield, costs and overall profitability of GM insect-resistant cotton

		Percent change = [(*Bt*/conventional) – 1] × 100					
Country	Years	Yield (kg/ha)	Seed cost	Insecticide cost[a]	Labor[c]	Profit	Reference
China	1999–2001	+19	+95	−67	−18	+340[b]	Pray *et al.*, 2002
India	2002–2003	+53	+8	−2[c]	–	+54	Kambhampati *et al.*, 2006
Mexico	1997–1998	+11	+165	−77	–	+12	Traxler *et al.*, 2003
South Africa	1998–2000	+64	+89	−58	+2	+198	Bennett *et al.*, 2006

[a] Dashes (–) indicated data not presented or collected for a study.
[b] Non-*Bt* cotton farmers produced an overall loss during this period.
[c] Costs of seed and insecticides combined.

20.2.1 Economic profitability

To make biotechnology-derived agriculture profitable, a combination of increased crop yield, quality or cost savings must be sufficient to offset any additional or premium costs associated with purchasing of the biotechnology-derived product. This premium for purchasing transgenic crop seed is commonly referred to as a technology fee. For transgenic pest management, a farmer is less likely to make up for the added cost when the targeted pests (insects, weeds or pathogens) are absent or only present in low numbers, or the price of the agricultural commodity is low.

In the USA, transgenic maize varieties expressing insect-active toxins derived from the soil bacterium *Bacillus thuringiensis* (*Bt*) help illustrate the sometimes complex economics of biotechnology. The first varieties of *Bt* maize were primarily intended to control the European corn borer (*Ostrinia nubilalis*). During 1998 and 1999, low maize prices and low European corn borer populations combined to make planting *Bt* maize an economic disadvantage (Carpenter & Gianessi, 2001). However, 1998 and 1999 were exceptionally poor economic conditions for producing *Bt* maize; analysis including more typical conditions for the USA (Sankula, 2006) show increased profitability for *Bt*-maize farmers. Research on lepidopteran-active *Bt* maize in Spain (Demont & Tollens, 2004) and the Philippines (Yorobe & Quicoy, 2006) also suggests farmers gain from using transgenic insect control. The overall economic benefits from reduction of insect damage and costs associated with insecticidal control (scouting, insecticide, application) are changing as new hybrids express additional *Bt* toxins. "Stacks," adding Cry3Bb1 or Cry34/35Ab1 toxins, are used to protect maize from both European corn borers and corn rootworms (*Diabrotica* spp.) (Rice, 2004). Similarly, the use of two or more complementary *Bt* toxins in "pyramids" should enhance the economic value of *Bt* maize by improving toxicity to broader groups of lepidopteran maize pests; future adoption rates for multiple pests will also depend upon the degree to which technology fees also increase.

The other widely adopted transgenic insect-resistant crop, *Bt* cotton, also has economic benefits for control of lepidopteran pests. Gains may be produced by large reductions in pest damage (leading to increased yield) or expenses associated with insecticide applications, as shown for farmers in Argentina (Qaim & de Janvry, 2005), China (Pray *et al.*, 2002), India (Kambhampati *et al.*, 2006), Mexico (Traxler *et al.*, 2003), South Africa (Bennett *et al.*, 2006) and the USA (Cattaneo *et al.*, 2006). Though the added costs of transgenic seed are considerable, a combination of benefits related to yield and production costs can combine to far exceed technology fees (Table 20.1).

The adoption and profitability of herbicide-tolerant crops present an equally interesting case. Though transgenic herbicide-tolerant crops are planted on approximately three times the area of *Bt* maize and cotton combined (James, 2006), markedly less information on the economic benefits of herbicide-tolerant crops is available. For soybean, the most widely grown herbicide-tolerant crop, economic benefits have been shown in the USA (Heatherly *et al.*, 2002) and Argentina (Qaim & Traxler, 2005). An economic analysis of transgenic glyphosate-resistant sugar beets in the USA also showed benefits from increased yield, quality and potential to decrease herbicide costs (Kniss *et al.*, 2004). Herbicide-tolerant canola in Canada also appears to present an overall economic benefit to farmers (Stringam *et al.*, 2003). However, some have suggested that convenience may better explain the broad and rapid farmer adoption of herbicide-tolerant crops (Economic Research Service, 2002; Stringam *et al.*, 2003). It is also possible that grower surveys used in some studies of biotech crops do not reflect some types of economic gains (e.g. reduced labor) from growing herbicide-tolerant varieties. As noted above for *Bt* maize, profitability of all biotechnology-derived crops may depend on several factors including differences among years (Kambhampati *et al.*, 2006), locations (Heatherly *et al.*, 2002) or farmer education (Yang *et al.*, 2005).

20.2.2 Environmental impact

Production of conventional or biotechnology-derived crops may impact agricultural fields and the surrounding environment in many different ways. Below the possible effects of GM and other biotech-derived crops are summarized with regard to (1) species abundance and diversity, (2) sustainability of pest management and (3) overall environmental health.

Effects on species abundance and diversity

Potential unintended effects of biotech crops on species abundance and diversity are often referred to as non-target effects. For GM insect-resistant crops, non-targets include any species other than the pests that an insecticidal trait is intended to control. The effects of biotechnology-derived crops on non-target species have been examined in hundreds of laboratory and field experiments. For *Bt* crops, the toxins generally impact only a few species closely related to target pests. Though this may effectively eliminate certain pests within a field, additional impacts on abundance and diversity are mostly limited to other species reliant on target pests, such as host-specific parasitoids. As a result, insect control with *Bt* crops should have far less impact on non-target species than conventional (broad-spectrum) insecticides.

Compared to conventional insecticide use, *Bt* crops conserve non-target species leading to greater arthropod abundance or diversity (Dively, 2005; Torres & Ruberson, 2005; Cattaneo *et al.*, 2006) and better biological control of pests not susceptible to *Bt* toxins (Naranjo, 2005) (Fig. 20.2). Because many beneficial arthropods move between cropping systems (Prasifka *et al.*, 2004a, b), conservation of non-target species in *Bt* fields also could improve biological pest control in nearby (non-transgenic) crops. Some research has shown unexpected adverse effects of *Bt* crops on non-target insects (e.g. Monarch butterfly larvae in Losey *et al.*, 1999; predatory lacewings in Hilbeck *et al.*, 1998), but such studies generally have been shown to be misleading or scientifically flawed (Hellmich *et al.*, 2001; Romeis *et al.*, 2004).

Impacts on plant biodiversity also have been considered. Because the use of GM and other herbicide-tolerant crops facilitates the use of herbicides, the abundance and diversity of weeds and weed seeds within agricultural systems will be reduced, leading to fewer herbivorous insects and birds (Chamberlain *et al.*, 2007). However, such an effect is more accurately caused by very effective weed control rather than biotechnology-derived crops. Another concern suggests that introduction of transgenic crops has reduced the diversity (among elite lines) within crop species (Gepts & Papa, 2003), though research on cotton and soybean varieties in the USA suggests introduction of transgenic varieties produced little or no impact on genetic diversity (Bowman *et al.*, 2003; Sneller, 2003). Further, hundreds of public-sector collections of germplasm from cultivated crops and their wild relatives exist for the purpose of preserving diversity (e.g. the National Genetic Resources Program in the USA).

Sustainability of pest management

The largest threat to sustainability for insect-resistant and herbicide-tolerant crops is the widespread evolution of resistant pest populations. As with conventional pesticide use,

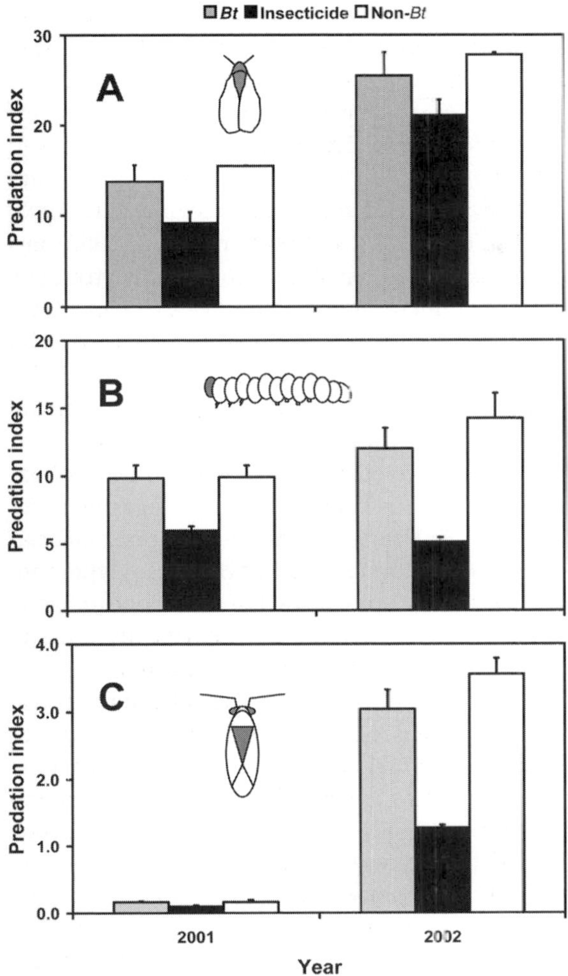

Fig. 20.2 Predation indices for cotton pests (A) *Bemisia tabaci* (Gennadius), (B) *Pectinophora gossypiella* (Saunders) and (C) *Lygus hesperus* Knight in *Bt* cotton and non-*Bt* cotton with or without insecticides. The predation index is a summed product of predator densities and predation frequency (from gut content immunoassays). (Figure modified from Naranjo, (2005.)

increasing reliance on a single GM trait to control insect or weed pests increases the likelihood that resistant genotypes will spread.

In the USA and other countries that produce insect-resistant *Bt* crops, steps to delay resistance evolution in target pests are organized into insect resistance management (IRM) plans. Such IRM plans outline mandatory actions for farmers and seed companies, including the planting of non-*Bt* refuges (Environmental Protection Agency, 2001). Refuges provide susceptible insects to mate with any resistant individuals emerging from *Bt* crops, resulting in hybrid progeny that cannot survive on insect-resistant plants. Evidence from several years of resistance monitoring in *Bt* cotton suggests the combination of effective resistance and careful management (i.e. the high-dose/structured refuge strategy: Environmental Protection Agency, 2001) has effectively delayed resistance and provided a means of sustainable management of insect pests (Tabashnik *et al.*, 2003, 2005). In China, *Bt* cotton may have improved sustainability of insecticide-based management; reductions in use of common insecticides appear to have lowered the levels of resistance in cotton bollworm (*Helicoverpa armigera*) (Wu *et al.*, 2005). There are legitimate concerns that in developing countries IRM may be more difficult. In particular, large numbers of small farms and less communication between farmers and advisors could result in ineffective use of non-*Bt* refuges. In such cases, the best solution may be to develop crops that utilize two or more toxins for which different adaptive mechanisms are required (i.e. pyramids). It appears this type of multiple-toxin strategy could effectively delay resistance with fewer or smaller planted refuges (Zhao *et al.*, 2003).

Unlike GM insect resistance, sustainability of transgenic herbicide-tolerant crops is not preserved by mandatory resistance management plans. The lack of a systematic plan to delay resistance may have contributed to the spread of glyphosate-resistant weeds (Owen & Zelaya, 2005; Sandermann, 2006). However, development of resistant weeds in herbicide-tolerant crops is not a necessary result of using a biotech approach to weed control, but of an unsustainable over-reliance on a single combination of herbicide and herbicide-tolerant crop. To prevent increases in weed resistance to glyphosate and other herbicides, increasing the duration of crop and herbicide rotations should be useful. Conventional and biotechnology-derived resistance to herbicides other than glyphosate indicate producing diverse herbicide-tolerance traits for crops is scientifically realistic (Duke, 2005), but farmers may be hesitant to adopt a more complex (though perhaps more effective and sustainable) weed management system. The cost to commercialize new products also may be unattractive to agricultural biotechnology companies (Devine, 2005).

Overall environmental health

In terms of environmental quality, the largest potential benefit related to pest management may come from significant reductions in the quantity of pesticides used in agriculture. Transgenic crops with resistance to insects, herbicides and plant pathogens may allow reductions in the use of pesticides whose toxic effects are a concern for humans and other vertebrate animals through acute or chronic exposure.

The level of pesticide reduction possible through biotechnology is largely dependant on crop and pest combinations. For example, without transgenic control of lepidopteran pests, cotton farmers have relied on intensive use of broad-spectrum insecticides. Since commercial use of *Bt* cotton began, control of lepidopteran pests has been accomplished with remarkable reductions in pesticide use by farmers in Australia (Knox *et al.*, 2006), China (Pray *et al.*, 2002), India (Kambhampati *et al.*, 2006), South Africa (Morse *et al.*, 2006) and the USA (Cattaneo *et al.*, 2006). However, reductions in insecticide use may allow pests previously controlled by regular spraying to become more common. Dramatic reductions in pesticide use (90%) also have been recorded in China for transgenic rice varieties that include *Bt* or a modified cowpea trypsin inhibitor (Huang *et al.*, 2005).

The effects of GM insect-resistance on insecticide use in maize are less clear. The first *Bt*-maize varieties primarily targeted the European corn borer. In the USA, insecticides are not frequently used to control *O. nubilalis* in field maize (see Shelton *et al.*, 2002), meaning only modest reductions in insecticide use might be possible. However, in other areas *Bt* maize has provided significant environmental benefits. Control of the Asian corn borer (*Ostrinia furnacalis*) in the Philippines has reduced insecticide use by half (Yorobe & Quicoy, 2006). Reduced insecticide use is possible in the USA for *Bt* sweet corn, which receives more insecticide applications per unit area than maize produced for grain (Shelton *et al.*, 2002). Also, the use of multiple-toxin stacks and pyramids to control other insect pests should expand the potential to reduce insecticide use with *Bt* maize. For example, crop rotation previously used to control corn rootworms is becoming both less effective in midwestern USA; corn rootworms have adapted to defeat a 2-year crop rotation by laying eggs on crops other than maize and exhibiting extended diapause (Levine *et al.*, 1992; Rondon & Gray, 2004). Rotation of maize with soybean has also become less economically attractive because of increasing maize prices, causing more farmers to plant maize in consecutive years. In this instance, coleopteran-active *Bt* maize could prove an environmentally favorable substitute for soil insecticides.

Herbicide-tolerant GM crops also impact the environment, in part, through changes in pesticide use. Although glyphosate-resistant soybean fields received more total herbicides, glyphosate was used as a substitute for considerably more toxic herbicides (Qaim & Traxler, 2005). Similarly, the use of glyphosate in midwestern USA has increased following the introduction of transgenic glyphosate-tolerant maize and soybean. Although overall a small increase in herbicide use in soybeans appears to be due to glyphosate-tolerant soybeans, these increases may have a net environmental benefit by reducing the use of other, more persistent herbicides (Economic Research Service, 2002). In glyphosate-resistant cotton in the USA, although herbicide-tolerant varieties appeared to receive more herbicide applications, no statistically significant change could be detected (Cattaneo *et al.*, 2006). The estimated effects of some GM insect-resistant and herbicide-tolerant crops on pesticide use are shown in Table 20.2, although data on insecticide use expressed as kilograms active ingredient (a.i.) or number of insecticide applications may not provide the best measure of environmental impacts (see Section 20.3.1, "*Bt* cotton in South Africa").

Beyond possible benefits from changes in chemical weed control, herbicide-tolerant GM crops appear to have allowed for reduced use of mechanical weed control using tillage (Ammann, 2005; Young, 2006). Increased adoption of reduced- or no-tillage agriculture is beneficial to overall environmental health and agricultural sustainability by conserving water, soil and fuel. Consequently, even increases in herbicide use because

Table 20.2 Estimated impacts of some herbicide-tolerant and insect-resistant crops on pesticide use

Crop	Trait[b]	Country	Percent change = [(BT/conventional − 1] × 100[a] Kg active ingredient	Number of applications	Reference
Canola	HT	Canada	−51%	–	Brimner *et al.*, 2005
Cotton	HT	USA	–	+40%	Cattaneo *et al.*, 2006
Cotton	IR	Argentina	−61%	−48%	Qaim & de Janvry, 2005
		China	−67%	–	Pray *et al.*, 2002
		Mexico	–	−39%	Traxler *et al.*, 2003
		South Africa	−53%	–	Morse *et al.*, 2006
		USA	–	−40%	Cattaneo *et al.*, 2006
Soybean	HT	Argentina	–	+17%	Qaim & Traxler, 2005

[a] Not all changes are statistically significant.
[b] IR, insect resistance; HT, herbicide tolerance.

of herbicide-tolerant crops may produce an overall positive effect on the environment.

Little research is available on the environmental effects of biotechnology-derived crops with resistance to plant pathogens. However, there are likely to be significant reductions of insecticides for pathogens transmitted by insects (e.g. papaya ringspot virus: Gonsalves *et al.*, 2007; references in Gaba *et al.*, 2004; but see Gatch & Munkvold, 2002). Environmental benefits also seem likely from resistance to plant pathogens in cases where pesticides are currently the only effective treatment.

20.2.3 Social impacts

The social impacts of biotechnology-based pest management will be most direct for farmers and others who live in or near farming communities. Whether impacts on these communities are positive or not depends on whether biotechnology effectively addresses social needs, which may differ between industrialized and developing nations.

In industrialized nations like the USA, one critical need is to preserve farming as a basic lifestyle or form of employment. Over less than a century, the USA farm population has declined from over 34% to less than 2% of the total population. Even though crop yields have continued to increase, most income in farming households is now derived from non-farm sources (Lobao & Meyer, 2001). Because preservation of family farms is one key to maintaining rural communities and quality of life (Lyson & Welsh, 2005), increasing the income that farming households derive directly from farming could help to sustain rural communities. Biotechnology and its applications to pest management may help by giving farmers more choices for management of crop pests. More importantly, most of the economic gains produced by transgenic insect-resistant or herbicide-tolerant crops are retained by farmers (Falck-Zepeda *et al.*, 2000). How significant this contribution is to preserving family farms and whether gains will be stable over the long term is difficult to predict.

In contrast, agriculture in developing countries requires more people to be directly involved in crop and animal production on small farms. Though significant, non-farm income provides a minority of income for most farming households (Food and Agriculture Organization, 1998), meaning stable yields are a key to both short- and long-term survival. Substitution of transgenic insect-control for conventional insecticides may have benefits for the health of farming families and communities (see Section 20.3.1, "*Bt* cotton in South Africa"), while herbicide-tolerant crops may help to improve weed control while conserving soil and water. However, because some crops and developing nations may not provide attractive markets to private biotechnology companies, strong public-sector involvement appears necessary to ensure that developing nations benefit from developments in biotechnology (Pingali & Raney, 2005).

20.3 Case studies: biotechnology and pest management

It is important to incorporate a broad view of how a technology may impact sustainable agriculture. In part, this is true because sustainability includes economic, environmental and social components. Positive and negative effects on each component of sustainability combine to determine the overall effect of a technology. However, even within a single component, complex relationships may exist. For example, the use of herbicide-tolerant crops may maintain soil quality by reducing mechanical weed control (and erosion), but might also lead to increased problems with insect pests previously controlled by tilling the soil. In such a case, the environmental value of soil conservation would need to be assessed against a potential increase in insecticide use to control a newly created pest problem. The case studies below highlight the results and limitations of large-scale attempts to measure the environmental effects of transgenic pest management in Africa and Europe.

20.3.1 *Bt* cotton in South Africa

The effects of changing pest management from a traditional insecticide-based regime to a program relying largely on biotechnology-derived cotton was studied over several years in the Republic of South Africa (Bennett *et al.*, 2003, 2006; Morse *et al.*, 2006). The impact of genetically engineered cotton expressing the *Bt* toxin Cry1Ac was evaluated, with particular emphasis given to the environmental and economic impacts. Though it can be suggested that *Bt* cotton does not address the adequacy of the world food supply, this argument is misleading; cotton is consumed directly (when used as cottonseed oil) and indirectly (when used as feed for cattle) by humans. More importantly, the use of cotton as a cash crop (i.e. crops grown for money rather than direct consumption) influences the ability of cotton farmers to buy food and remain profitably involved in agriculture. The studies specifically target economically disadvantaged farmers to explore the validity of concerns that GM crops could be inappropriate for use in developing countries. In an attempt to improve upon previous research, data on a large number of smallholder (small-scale) farmers were obtained from the records of a commercial seed supplier, Vunisa, which acted as the sole seed supplier and purchaser of cotton in the area. However, the validity of information provided by the company was checked against data collected by other researchers and surveys with area cotton farmers.

The economic analysis (Bennett *et al.*, 2006) showed large financial gains for smallholders growing *Bt* cotton. The cost of inputs, including insecticides used for bollworms (targets of the Cry1Ac toxin), insecticides used for other pests and labor required for pesticide applications were substantially reduced for growers of *Bt* cotton. Though growing *Bt* cotton required more labor for weed control (one of two years) and harvesting (all three years), labor cost increases resulted from picking a larger cotton crop. Overall savings from reduced insecticide use and increased revenue from higher yields exceeded additional labor and seed costs associated with growing *Bt* cotton. To put the economic gains into context, the economic advantage realized per hectare by *Bt* cotton farmers was 387–715 South African Rand ($US 70–130), or approximately two to four months of relatively well-paid labor for one worker. It also appeared that gains from growing *Bt* were maintained during an unusually wet year when conventional cotton was produced at a financial loss. Further, farmers with less land received an equal or greater benefit compared to farmers with larger fields.

Environmental impacts of *Bt* cotton production were linked to overall reductions in insecticide use (Morse *et al.*, 2006). However, data on insecticide use expressed as changes in active ingredient used can be misleading because the toxicity and persistence of any two insecticides used to control the same pest may be very different. As a result, additional methods were used to estimate environmental impacts of insecticide use. Over three years, decreases in insecticide use (kg a.i.) for *Bt* cotton farmers were 53%, 50% and 63%. Similar significant reductions in environmental impact were found by using methods that emphasized effects on mammals (Biocide Index) or broader groups of organisms (Environmental Impact Quotient).

Alhough Bennett *et al.* (2003) come closest to examining the social impacts of South African *Bt* cotton production, Bennett *et al.* (2006) and Morse *et al.* (2006) also make significant points regarding effects of *Bt* cotton on the community. First, while the yields of *Bt* cotton producers were variable, the increases in yield and revenue seem to make smallholders better able to tolerate price fluctuations. Second, reductions in pesticide applications may have been particularly beneficial to women and children, who help with insecticide applications; accidental insecticide poisonings in the area declined considerably over the course of the study (Bennett *et al.*, 2003). On balance, *Bt* cotton seems likely to improve the economic resilience and quality of life for South African farmers. However, Morse *et al.* (2006) note that the *Bt* cotton production will not be a cure-all for area farmers, and that reliance on a single company for credit, seeds, pesticides and a market for their crops makes smallholders particularly vulnerable.

20.3.2 Herbicide-tolerant crops in the United Kingdom

The possible effects of GM herbicide-tolerant crops on the environment were evaluated over three years in sugar beet, maize and canola fields in the United Kingdom (UK). These trials, often referred to as the Farm-Scale Evaluations (FSE), used split fields with farmers' conventional weed management on one half (planted to a non-GM variety) and herbicides applied to the second half (planted with a GM herbicide-tolerant cultivar). Herbicides used in the herbicide-tolerant crops included glufosinate-ammonium (maize and canola) and glyphosate (sugar beet). The environmental impacts assessed focused on potential changes in farmland biodiversity, included the abundance of weeds and arthropods. Because the FSE included controlled trials with GM crops on an almost unprecedented scale (>60 fields per crop), these trials have been among the most discussed field research on genetically modified crops.[1]

The results indicate the production of herbicide-tolerant sugar beet and canola reduced the abundance of butterflies, bees (sugar beet only), weeds, weed seeds and seed-feeding beetles. In herbicide-tolerant maize, no significant reductions were found for bees or butterflies and increases in the abundance of dicotyledonous weeds, weed seeds and seed-feeding beetles were seen. For herbicide-tolerant beet and maize, the abundance of springtails (many of which feed on decaying plant matter) was increased. Research on a subset of the fields in the FSE examined the effects of management on birds, finding significantly fewer granivorous birds in the herbicide-tolerant maize fields (Chamberlain *et al.*, 2007).

Interestingly, the likelihood of adverse environmental impacts caused by increasing use of herbicide-tolerant crops in the UK are most noted by those not involved in the FSE research. In fact, the FSE researchers are careful to note that the experiments examined the effects of changes in herbicide use rather than effects directly caused by genetic modification of the crops (Firbank *et al.*, 2003). Accordingly, Chamberlain *et al.* (2007) clarify that the differences in bird abundance and diversity only occurred after herbicide applications in maize. Other researchers have correctly noted the difficulty in determining clear cause-and-effect relationships for the FSE (Andow, 2003). Perhaps the most significant problem in assessing the likely impact of herbicide-tolerant crops on agriculture in the UK is that economic and social impacts were not assessed, and some possible environmental benefits were not included. Consequently, the research does not allow an overall evaluation of how herbicide-tolerant crops might affect agricultural sustainability.

20.4 Conclusions

Biotechnology has the potential to reduce the severity of many problems posed by an expanding population and limited or degraded resources. Agriculture enhanced by new technologies may be capable of producing an adequate supply of more

[1] Most of the results from the FSE were published concurrently in a 2003 issue of the *Philosophical Transactions of the Royal Society of London Series B, Biological Sciences*, vol. 351 no. 1342.

nutritious foods as well as biologically based fuels that use marginal land and fewer resources. With regard to pest management, biotechnology can provide improved control of pests, generally with reduced stresses on agricultural and surrounding environments.

Despite the exciting potential of biotechnology, it should not be considered a panacea for sustainability. Genetic engineering has been unable to fulfill the projections of benefits forecast in previous years. In part, benefits of biotechnology from the private sector are certainly constrained by the need to maximize profits. Further, the tools of biotechnology can be used in ways that decrease sustainability. However, there are good examples of competitive public-sector biotech products (Pray *et al.*, 2002), successful public–private partnerships (Gonsalves *et al.*, 2007) and responsible use of agricultural biotechnology (Tabashnik *et al.*, 2005) that illustrate the potential of biotechnology to benefit economic, environmental and social components of sustainability.

References

American Society of Agronomy (1989). Decision reached on sustainable ag. Madison, WI: American Society of Agronomy, *Agronomy News* (January).

Ammann, K. (2005). Effects of biotechnology on biodiversity: herbicide-tolerant and insect-resistant GM crops. *Trends in Biotechnology*, **23**, 388–394.

Andow, D. A. (2003). UK farm-scale evaluations of transgenic herbicide-tolerant crops. *Nature Biotechnology*, **21**, 1453–1454.

Bennett, R., Buthelezi, T. J., Ismael, Y. & Morse, S. (2003). *Bt* cotton, pesticides, labour and health: a case study of smallholder farmers in the Makhathini Flats, Republic of South Africa. *Outlook on Agriculture*, **32**, 123–128.

Bennett, R., Morse, S. & Ismael, Y. (2006). The economic impact of genetically modified cotton on South African smallholders: yield, profit and health effects. *Journal of Development Studies*, **42**, 662–677.

Bowman, D. T., May, O. L. & Creech, J. B. (2003). Genetic uniformity of the US upland cotton crop since the introduction of transgenic cottons. *Crop Science*, **43**, 515–518.

Brimner, T. A., Gallivan, G. J. & Stephenson, G. R. (2005). Influence of herbicide-resistant canola on the environmental impact of weed management. *Pest Management Science*, **61**, 47–52.

Brookes, G. & Barfoot, P. (2006). GM crops: the global economic and environmental impact – the first nine years 1996–2004. *AgBioForum*, **8**, 187–196.

Carpenter, J. E. & Gianessi, L. P. (2001). *Agricultural Biotechnology: Updated Benefit Estimates*. Washington, DC: National Center for Food and Agricultural Policy. Available at www.ncfap.org/reports/biotech/updatedbenefits.pdf.

Carson, R. (1962). *Silent Spring*. Boston, MA: Houghton Miffin.

Casida, J. E. & Quistad, G. B. (1998). Golden age of insecticide research: past, present, or future? *Annual Review of Entomology*, **43**, 1–16.

Cattaneo, M. G., Yafuso, C., Schmidt, C. *et al.* (2006). Farm-scale evaluation of the impacts of transgenic cotton on biodiversity, pesticide use, and yield. *Proceedings of the National Academy of Sciences of the USA*, **103**, 7571–7576.

Chamberlain, D. E., Freeman, S. N. & Vickery, J. A. (2007). The effects of GMHT crops on bird abundance in arable fields in the UK. *Agriculture, Ecosystems and Environment*, **118**, 350–356.

Demont, M. & Tollens, E. (2004). First impact of biotechnology in the EU: *Bt* maize adoption in Spain. *Annals of Applied Biology*, **145**, 197–207.

Devine, M. D. (2005). Why are there not more herbicide-tolerant crops? *Pest Management Science*, **61**, 312–317.

Diamond, J. M. (1997). *Guns, Germs, and Steel: The Fates of Human Societies*. New York: W. W. Norton.

Dively, G. P. (2005). Impact of transgenic VIP3A × Cry1Ab lepidopteran-resistant field corn on the nontarget arthropod community. *Environmental Entomology*, **34**, 1267–1291.

Duke, S. O. (2005). Taking stock of herbicide-resistant crops ten years after introduction. *Pest Management Science*, **61**, 211–218.

Economic Research Service (2002). *Adoption of Bioengineered Crops*, Agricultural Economic Report No. 810. Washington, DC: US Department of Agriculture. Available at www.ers.usda.gov/publications/aer810/aer810.pdf.

Economic Research Service (2006). *Adoption of Genetically-Engineered Crops in the US*. Washington, DC: US Department of Agriculture. Available at www.ers.usda.gov/Data/BiotechCrops/alltables.xls.

Environmental Protection Agency (2001). Insect resistance management. In *Biopesticides Registration Action Document* Bacillus thuringiensis *Plant-Incorporated Protectants*. Washington, DC: US Environmental Protection Agency. Available at www.epa.gov/pesticides/biopesticides/pips/bt_brad2/4-irm.pdf.

Evans, L. T. (2003). Agricultural intensification and sustainability. *Outlook on Agriculture*, **32**, 83–89.

Food and Agriculture Organization (1998). Part III: Rural non-farm income in developing countries. In *The State of Food and Agriculture 1998*, FaO Agriculture Series no. **31**, pp. 227–248. Rome, Italy: Food and Agriculture Organization of the United Nations.

Falck-Zepeda, J. B., Traxler, G. & Nelson, R. G. (2000). Surplus distribution from the introduction of a biotechnology innovation. *American Journal of Agricultural Economics*, **82**, 360–369.

Firbank, L. G., Heard, M. S., Woiwod, I. P. *et al.* (2003). An introduction to the farm-scale evaluations of genetically modified herbicide-tolerant crops. *Journal of Applied Ecology*, **40**, 2–16.

Gaba, V., Zelcer, A. & Gal-On, A. (2004). Cucurbit biotechnology: the importance of virus resistance. *In Vitro Cellular and Developmental Biology – Plant*, **40**, 346–358.

Gatch, E.W. & Munkvold, G.P. (2002). Fungal species composition in maize stalks in relation to European corn borer injury and transgenic insect protection. *Plant Disease*, **86**, 1156–1162.

Gepts, P. & Papa, R. (2003). Possible effects of trans(gene) flow from crops to the genetic diversity from landraces and wild relatives. *Environmental Biosafety Research*, **2**, 89–113.

Gonsalves, C., Lee, D. R. & Gonsalves, D. (2007). The adoption of genetically modified papaya in Hawaii and its implications for developing countries. *Journal of Development Studies*, **43**, 177–191.

Heatherly, L. G., Elmore, C. D. & Spurlock, S. R. (2002). Weed management systems for conventional and glyphosate-resistant soybean with and without irrigation. *Agronomy Journal*, **94**, 1419–1428.

Hellmich, R. L., Siegfried, B. D., Sears, M. K. *et al.* (2001). Monarch larvae sensitivity to *Bacillus thuringiensis*-purified proteins and pollen. *Proceedings of the National Academy of Sciences of the USA*, **98**, 11 925–11 930.

Herdt, R. W. (2006). Biotechnology in agriculture. *Annual Review of Environmental Resources*, **31**, 265–295.

Hilbeck, A., Baumgartner, M., Fried, P. M. & Bigler, F. (1998). Effects of transgenic *Bacillus thuringiensis* corn-fed prey on mortality and development time of immature *Chrysoperla carnea* (Neuroptera: Chrysopidae). *Environmental Entomology*, **27**, 480–487.

Huang, J. K., Hu, R. F., Rozelle, S. & Pray, C. (2005). Insect-resistant GM rice in farmers' fields assessing productivity and health effects in China. *Science*, **308**, 688–690.

James, C. (2006). *Global Status of Commercialized Biotech/GM Crops*: 2006, ISAAA Brief No. 35. Ithaca, NY: International Service for the Acquisition of Agri-Biotech Applications.

Kambhampati, U., Morse, S., Bennett, R. & Ismael, Y. (2006). Farm-level performance of genetically modified cotton: A frontier analysis of cotton production in Maharashtra. *Outlook on Agriculture*, **35**, 291–297.

Kniss, A. R., Wilson, R. G., Martin, A. R., Burgener, P. A. & Feuz, D. M. (2004). Economic evaluation of glyphosate-resistant and conventional sugar beet. *Weed Technology*, **18**, 388–396.

Knox, O. G. G., Constable, G. A., Pyke, B. & Gupta, V. V. S. R. (2006). Environmental impact of conventional and Bt insecticidal cotton expressing one and two Cry genes in Australia. *Australian Journal of Agricultural Research*, **57**, 501–509.

Levine, E., Oloumi-Sadeghi, H. & Fisher, J.R. (1992). Discovery of multiyear diapause in Illinois and South Dakota northern corn rootworm (Coleoptera: Chrysomelidae) eggs and incidence of the prolonged diapause trait in Illinois. *Journal of Economic Entomology*, **85**, 262–267.

Lobao, L. & Meyer, K. (2001). The great agricultural transition: crisis, change, and social consequences of twentieth century US farming. *Annual Review of Sociology*, **27**, 103–124.

Losey, J. E., Rayor, L. S. & Carter, M. E. (1999). Transgenic pollen harms monarch larvae. *Nature*, **399**, 214.

Lyson, T. A. (2002). Advanced agricultural biotechnologies and sustainable agriculture. *Trends in Biotechnology*, **20**, 193–196.

Lyson, T. A. & Welsh, R. (2005). Agricultural industrialization, anticorporate farming laws, and rural community welfare. *Environment and Planning A*, **37**, 1479–1491.

Morse, S., Bennett, R. & Ismael, Y. (2006). Environmental impact of genetically modified cotton in South Africa. *Agriculture, Ecosystems and Environment*, **117**, 277–289.

Naranjo, S. E. (2005). Long-term assessment of the effects of transgenic *Bt* cotton on the function of the natural enemy community. *Environmental Entomology*, **34**, 1211–1223.

Owen, M. D. K. & Zelaya, I. A. (2005). Herbicide-resistant crops and weed resistance to herbicides. *Pest Management Science*, **61**, 301–311.

Pingali, P. & Raney, T. (2005). *From the Green Revolution to the Gene Revolution: How Will The Poor Fare?*, ESA Working Paper No. 05-09. Rome, Italy: Food and Agriculture Organization of the United Nations. Available at ftp://ftp.fao.org/docrep/fao/008/af276e/af276e00.pdf.

Prasifka, J. R., Heinz, K. M. & Minzenmayer, R. R. (2004a). Relationships of landscape, prey and agronomic variables to the abundance of generalist predators in cotton (*Gossypium hirsutum*) fields. *Landscape Ecology*, **19**, 709–717.

Prasifka, J. R., Heinz, K. M. & Winemiller, K. O. (2004b). Crop colonization, feeding, and reproduction by the predatory beetle, *Hippodamia convergens*, as indicated by stable carbon isotope analysis. *Ecological Entomology*, **29**, 226–233.

Pray, C. E., Huang, J., Hu, R. & Rozelle, S. (2002). Five years of *Bt* cotton in China: the benefits continue. *Plant Journal*, **31**, 423–430.

Qaim, M. & de Janvry, A. (2005). *Bt* cotton and pesticide use in Argentina: economic and environmental effects. *Environment and Development Economics*, **10**, 179–200.

Qaim, M. & Traxler, G. (2005). Roundup Ready soybeans in Argentina: farm level and aggregate welfare effects. *Agricultural Economics*, **32**, 73–86.

Rice, M. E. (2004). Transgenic rootworm corn: assessing potential agronomic, economic, and environmental benefits. *Plant Health Progress* doi:10.1094/PHP-2004–0301-01-RV, available at www.plantmanagementnetwork.org/pub/php/review/2004/rootworm/.

Romeis, J., Dutton, A. & Bigler, F. (2004). *Bacillus thuringiensis* toxin (Cry1Ab) has no direct effect on larvae of the green lacewing *Chrysoperla carnea* (Stephens) (Neuroptera: Chrysopidae). *Journal of Insect Physiology*, **50**, 175–183.

Rondon, S. I. & Gray, M. E. (2004). Ovarian development and ovipositional preference of the western corn rootworm (Coleoptera: Chrysomelidae) variant in East Central Illinois. *Journal of Economic Entomology*, **97**, 390–396.

Sandermann, H. (2006). Plant biotechnology: ecological case studies on herbicide resistance. *Trends in Plant Science*, **11**, 324–328.

Sankula, S. (2006). *A 2006 Update of Impacts on US Agriculture of Biotechnology-Derived Crops Planted in 2005: Executive Summary*. Washington, DC: National Center for Food and Agricultural Policy. Available at www.ncfap.org/whatwedo/pdf/2005biotechExecSummary.pdf/.

Schaller, N. (1993). The concept of agricultural sustainability. *Agriculture, Ecosystems and Environment*, **46**, 89–97.

Shelton, A. M., Zhao, J. Z. & Roush, R. T. (2002). Economic, ecological, food safety, and social consequences of the deployment of Bt transgenic plants. *Annual Review of Entomology*, **47**, 845–881.

Sneller, C. H. (2003). Impact of transgenic genotypes and subdivision on diversity within elite North American soybean germplasm. *Crop Science*, **43**, 409–414.

Stein, A. J., Sachdev, H. P. S. & Qaim, M. (2006). Potential impact and cost-effectiveness of Golden Rice. *Nature Biotechnology*, **24**, 1200–1201.

Stringam, G. R., Ripley, V. L., Love, H. K. & Mitchell, A. (2003). Transgenic herbicide tolerant canola: the Canadian experience. *Crop Science*, **43**, 1590–1593.

Tabashnik, B. E., Carrière, Y., Dennehy, T. J. *et al.* (2003). Insect resistance to transgenic *Bt* crops: lessons from the laboratory and field. *Journal of Economic Entomology*, **96**, 1031–1038.

Tabashnik, B. E., Dennehy, T. J. & Carrière, Y. (2005). Delayed resistance to transgenic cotton in pink bollworm. *Proceedings of the National Academy of Sciences of the USA*, **102**, 15 389–15 393.

Torres, J. B. & Ruberson, J. R. (2005). Canopy- and ground-dwelling predatory arthropods in commercial *Bt* and non-*Bt* cotton fields: patterns and mechanisms. *Environmental Entomology*, **34**, 1242–1256.

Traxler, G., Godoy-Avila, S., Falck-Zepeda, J. & Espinoza-Arellano, J. (2003). Transgenic cotton in Mexico: a case study of the Comarca Lagunera. In *The Economic and Environmental Impacts of Agbiotech: a Global Perspective*, ed. N. Kalaitzandonakes, pp. 183–202. Norwell, MA: Kluwer-Plenum.

Wu, K., Mu, W., Liang, G. & Guo, Y. (2005). Regional reversion of insecticide resistance in *Helicoverpa armigera* (Lepidoptera: Noctuidae) is associated with the use of *Bt* cotton in northern China. *Pest Management Science*, **61**, 491–498.

Yang, P. Y., Li, K. W., Shi, S. B. *et al.* (2005). Impacts of transgenic *Bt* cotton and integrated pest management education on smallholder cotton farmers. *International Journal of Pest Management*, **51**, 231–244.

Yorobe, J. M. & Quicoy, C. B. (2006). Economic impact of *Bt* corn in the Philippines. *Philippine Agricultural Scientist*, **89**, 258–267.

Young, B. G. (2006). Changes in herbicide use patterns and production practices resulting from glyphosate-resistant crops. *Weed Technology*, **20**, 301–307.

Zhao, J., Cao, J., Li, Y. *et al.* (2003). Transgenic plants expressing two *Bacillus thuringiensis* toxins delay insect resistance evolution. *Nature Biotechnology*, **21**, 1493–1497.

Chapter 21

Use of pheromones in IPM

Thomas C. Baker

During the past 68 years that have elapsed since the identification of the first insect pheromone (Butenandt, 1959) there has been a bourgeoning of basic and applied research that has resulted in an amazingly diverse and effective use of pheromones in IPM. Other behavior-modifying semiochemicals have had more limited success on a commercial level, although much research is continuing to try to find new ways to make such chemicals as host plant volatiles more useful in IPM settings as attractants or deterrents.

Although pheromones have established themselves in IPM systems, most end-users and even applied researchers do not realize how much work goes into identifying and optimizing pheromone blends so that they become highly species-specific and optimally attractive to the target species so that they can be used to the best effect. Research to determine the most effective pheromone blend compositions and dispenser dosages for monitoring and detection typically takes five to ten years to complete. Optimizing trap design targeting particular species can take several more years. Sometimes effective pheromones cannot be elucidated at all despite decades of intensive effort.

Delivery of effective commercial products presents another hurdle. For instance, applied pheromone researchers may spend years establishing that a particular mating disruption system is highly effective at disrupting mating and reducing crop damage after conducting experiments in which disruptant dispenser dosages and deployment densities have been varied. A mating disruption system may work successfully from a biological standpoint, but at the commercial level it may be too costly or not sufficiently user-friendly compared to standard practices, and the system will have been judged to "fail" at this level. It is important to distinguish between failure in the commercial arena versus biological failure of the mating disruption system itself.

21.1 Monitoring established populations

The most widespread use of pheromones has been for monitoring endemic pest species' adult populations. Comprehensive apple IPM programs that were initiated in New York State and Michigan during the early 1970s were based on good, species-specific monitoring traps for the complex of tortricid moth pests that cause direct and indirect damage to fruit. Monitoring of leafroller pests coupled with computer assisted degree-day models (Riedl & Croft, 1974; Riedl *et al.*, 1976) allowed sprays to be timed for optimum efficacy against eggs and first instars on such key pests as the codling moth (*Cydia pomonella*) and

Integrated Pest Management, ed. Edward B. Radcliffe, William D. Hutchison and Rafael E. Cancelado. Published by Cambridge University Press.

oriental fruit moth (*Grapholita molesta*). Spray-or-no-spray decisions based on abundance of adults in monitoring trap grids were also made possible by effective, standardized monitoring traps and deployment schemes developed against some pests such as the codling moth (Madsen, 1981). During the mid-1970s to late 1980s, insecticide applications were reduced by more than 50% in New York State, Michigan and the Pacific Northwest due to monitoring programs that improved decision making about the need to spray insecticides (Madsen, 1981) as well as their timing. Such programs have become even more refined over the years, with concomitant further reductions in insecticide applications being documented (Agnello *et al.*, 1994). On other crops in the USA as well as around the world, pheromone monitoring traps have asserted themselves in IPM programs as essential elements to the success of these programs. Their use is accepted as an integral and routine part of IPM. The chemical composition of thousands of pheromones can be readily accessed on the websites "Pherobase" (www.pherobase.com) and "Pherolist" (www.nysaes.cornell.edu/pheronet/). Many companies can be accessed on the web that provide pheromone monitoring trap kits for hundreds of pests.

21.2 Detection and survey programs for invasive species

Pheromone traps have played a large role in detecting influxes of adult pest species from one region to another and also even from non-crop areas into crop areas. Survey programs involving grids of pheromone traps are used so routinely to report and track the yearly arrival of migrating adult populations of insects such as the black cutworm (*Agrotis ipsilon*) in the Midwest (Showers *et al.* 1989a, b) or the spread of expanding populations such as the gypsy moth (*Lymantria dispar*) (Elkinton & Cardé, 1981) that it has become an essential part of our arsenal of tools for tracking and ameliorating pest-movement-related threats.

One example of a successful use of pheromone traps in detecting invasive species is the pink bollworm (*Pectinophora gossypiella*). Every summer since the 1970s, the California Department of Food and Agriculture (CDFA) has monitored an extensive grid of pheromone traps to protect against the pink bollworm becoming established in the Central Valley of California. The grid density averages approximately one trap per 250 ha, but it extends at various densities throughout the Central Valley (Baker *et al.*, 1990). When a male moth is discovered in a trap, an airdrop of sterile moths, reared at the US Department of Agriculture Animal and Plant Health Inspection Service (USDA-APHIS) sterile pink bollworm moth-rearing facility in Phoenix, is implemented. The plane is guided to the exact field by GPS coordinates. Sterile moths are dyed pink by the diet they ingest at the rearing facility, and so the many sterile males captured in the pheromone traps are easily distinguished from non-dyed males that have been blown in on weather systems from southern desert valleys.

The presence of a non-dyed male in a trap is assumed to be indicative of fertile, wild females being present in the same area, and that the threat of an increasing endemic population of pink bollworms becoming established is real. Airdrops of sterile moths are aimed at creating a ratio of at least 60 : 1 pink : normal-colored (sterile : wild) male captures in the pheromone traps. When the 60 : 1 ratio cannot be attained, either by too many wild males present or too few sterile individuals being produced at a particular time by the sterile moth-rearing facility, the CDFA then applies pheromone mating disruption. Monitoring and sanitation of this exotic pest under this program has provided a significant savings to growers and to the environment, allowing cotton to continue to be grown in central California with minimal insecticide load and maximum profits.

21.3 Mass trapping

The once-discounted technique of mass trapping using either male-produced or female-produced pheromones has become a newly appreciated, highly effective, environmentally friendly and relatively inexpensive means of suppressing populations of certain pest species whose pheromone communication systems and bionomical characteristics make them susceptible to this approach.

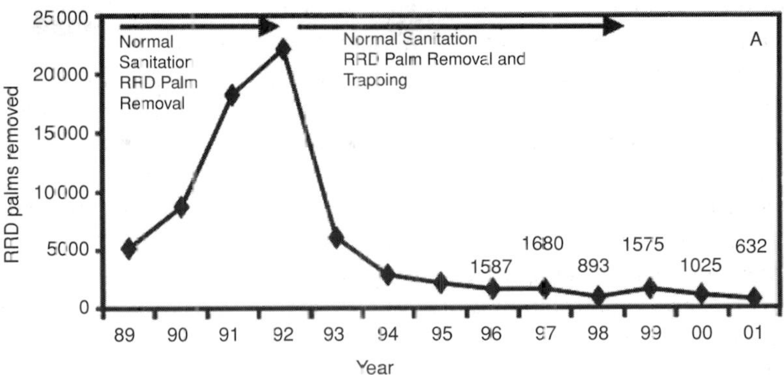

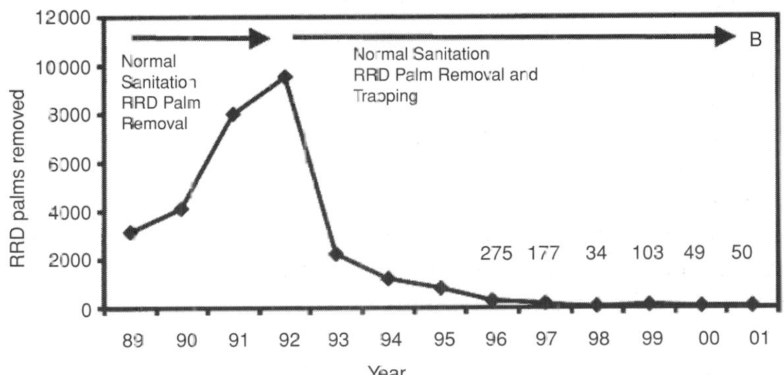

Fig. 21.1 Incidence of American palm weevil-vectored red ring disease in oil palms in Costa Rica, before and after mass trapping of the weevils, as measured by the number of diseased trees that had to be removed each year from two plantations, A and B, from 1989 to 2001. Plantation A comprised 6514 ha and Plantation B totaled 8719 ha. From 1989 to 1992, before the implementation of mass trapping of the weevils, disease incidence steadily rose. During this period only normal sanitation and diseased palm tree removal were used to try to control the weevils. After 1992, when pheromone mass trapping was added to the program, the incidence of red ring disease declined dramatically (from Oehlschlager *et al.*, 2002).

In this technique after much experimentation, traps are deployed at densities on the crop plants that have proven to attract and capture sufficiently large numbers of insects such that reduced damage to the crop occurs.

One example of successful mass trapping using pheromone traps concerns the American palm weevil (*Rhynchophorus palmarum*). This species is a highly damaging pest of oil and coconut palms in Central and South America. A related species, *R. ferrugineus*, is a major pest of oil and other palms in the Middle East. Larvae of *R. palmarum* cause direct damage when they bore into the trunks of the trees, but they also are a vector of red ring disease, which is caused by a nematode, *Rhadinaphelenchus cocophilus*. In the early 1990s, losses of trees either due to the weevil itself or to red ring disease (which necessitates removal of the infected trees) often routinely reached 15%.

In 1991 following successful preliminary experiments, plantation-wide mass trapping of the American palm weevil was undertaken on two major oil palm plantations in Costa Rica, one comprising 6514 ha and the other comprising 8719 ha (Oehlschlager *et al.*, 2002). Four-liter bucket traps baited with the male-produced sex pheromone plus insecticide-laced sugarcane (Chinchilla & Oehlschlager, 1992; Oehlschlager *et al.*, 1992, 1993) were deployed at chest level at a density of only one trap per 6.6 ha. The results in the two plantations were virtually identical (Fig. 21.1). Before the mass-trapping program was implemented, normal sanitation procedures involving removal of red ring disease-infected palms had failed to reduce the incidence of this disease. However, once mass trapping was added to the system, after one year the incidence of diseased trees needing to be removed dropped by nearly 80%, from about 10 000 diseased trees removed to 2000 (Oehlschlager *et al.*, 2002) (Fig. 21.1). During that first year (1992–1993) more than 200 000 weevils were trapped. In each successive year, the incidence of diseased trees declined until in 2001 only 50 diseased trees needed to be removed through sanitation (Oehlschlager *et al.*, 2002) (Fig. 21.1).

In subsequent commercial development of the *R. palmarum* mass-trapping system, approximately

25 000 ha of palm plantings were estimated to be under mass-trapping control yearly at the start of this century in Central and South America (A. C. Oehlschlager, personal communication). The density of traps used has typically been only one trap per 7 ha. In addition to the ability of this male-based pheromone to attract females, the success of this mass-trapping system (Oehlschlager *et al.*, 2002) is due first to the fact that although the weevils are highly damaging and cause high levels of tree mortality on a per-weevil basis, they are present in relatively small numbers. Second, the weevils have a long adult life, and so the steady capture of moderate numbers of weevils throughout the year can remove a large proportion of a generation and have a significant impact on population growth. Third, the adults are strong flyers, which, in conjunction with the highly attractive pheromone blend, allows the pheromone traps to be widely spaced (Oehlschlager *et al.*, 2002).

These same characteristics have come into play in other highly successful and effective commercial mass-trapping systems against other tropical weevil pest species. For *R. ferrugineus*, more than 35 000 ha of palm plantings are treated with mass trapping every year in the Middle East. For another pest of palm, coconut rhinocerous beetle (*Oryctes rhinoceros*), more than 50 000 ha yearly are under a mass-trapping program in the Middle East (A. C. Oehlschlager, personal communication). An estimated 10 000 ha of commercially grown bananas in the American tropics is under mass-trapping programs every year against the banana weevil, *Cosmopolites sordidus* (Germar) (A. C. Oehlschlager, personal communication).

In the early 1980s a "boll weevil eradication" program was initiated in the southeastern USA that relied extensively on a boll weevil lure and trap system developed over many years (Hardee *et al.*, 1967a, b; Tumlinson *et al.* 1968, 1969, 1970, 1971). In the early years, the traps served mainly as detection and monitoring tools and guided decision making on insecticide spraying. If trap captures exceeded an average of 0.1 weevils per trap, insecticides were applied. However, if capture levels were lower than this, mass trapping alone was considered adequate for population suppression. In the later years, the traps operated in a mass-trapping mode because populations had been reduced during the preceding years and weevil captures rarely exceeded 0.1 per trap. The scale of this effort was immense; in 1988 alone approximately 590 000 traps were deployed and more than 8.25 million pheromone dispensers were used in these southeastern states (Ridgway *et al.*, 1990).

The impact of the program was significant (Table 21.1), in terms of allowing cotton production to once again expand and flourish, in terms of the reduction in insecticides applied per hectare (savings of $US 69–74/ha in Virginia, North Carolina and South Carolina) and in terms of the increased yield value of the harvested cotton per hectare (increase of $US 85.25/ha) (Ridgway *et al.* 1990).

21.4 Mating disruption

Mating disruption involves dispensing relatively large amounts of sex pheromone over crop hectarage and suppressing males' abilities to locate females for mating. It used to be assumed that most, if not all, females would need to remain unmated in order for mating disruption to be effective, but current thinking has shifted, with evidence showing that females' ability to mate merely needs to be impaired such that their first and second matings are delayed, not prevented. Since the introduction of the first commercial pheromone mating disruptant in the world in 1976 against the pink bollworm on cotton, use of the mating disruption technique has grown slowly but steadily. Worldwide, over the past several years nearly 400 000 ha of various agricultural crops and forests have been under commercial mating disruption targeting a wide variety of insect pests. Examples of two of many such successful mating disruption programs are given below. Many other examples exist worldwide, in addition to these from the USA.

21.4.1 Successful mating disruption in IPM using commercial dispensers

In the early 1990s, apple and pear growers in California and the Pacific Northwest adopted a Codling Moth Areawide Management Program (CAMP) that relied on mating disruption for controlling the codling moth (*Cydia pomonella*). One overall goal of this study was to achieve

Table 21.1 Boll weevil captures in successive years of the expanded boll weevil eradication zones in North Carolina and South Carolina, USA

Year	Hectares	No. of fields	Percentage of fields capturing indicated numbers of weevils		
			0	1–5	>5
1983					
N.C.	6 600	800	0	1	99
S.C.	21 240	2 000	2	5	93
1984					
N.C.	9 200	1 000	21	35	44
S.C.	32 000	2 300	22	25	53
1985					
N.C.	9 400	1 000	90	7	3
S.C.	37 000	4 300	81	12	7
1986					
N.C.	8 600	1 000	>99.9	0	0
S.C.	34 400	4 200	97.3	2.3	0.4

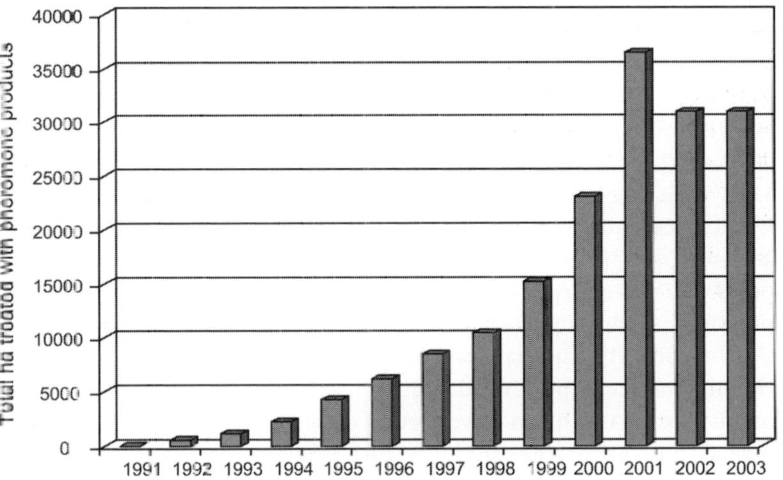

Fig. 21.2 The number of hectares treated each year with codling moth commercial mating disruption products in Washington State, USA from 1990 to 2002 (Brunner *et al.*, 2001; years 2001 and 2002 provided by Brunner, personal communication). In 1995 the CAMP program (Codling Moth Areawide Management Project) began with subsidized mating disruption applications provided to growers. The three CAMP sites in Washington State comprised a constant 760 hectares during the five years of the program (1995–1999).

an 80% or greater reduction of the use of broad-spectrum conventional insecticides by the end of the five-year program. A subsidy was provided to all participating growers of $US 125/ha for the first three years of the project to help defray the cost ($US 275/ha) of the mating disruption treatments (Brunner *et al.*, 2001). For the final two years, growers had to pay the full cost of the treatments themselves. Standard high-dose codling moth monitoring traps were used to assess trap-capture reductions and damage assessments were made for all blocks within the CAMP project and compared with fruit from non-CAMP-participating grower blocks.

The number of hectares of apples under mating disruption in Washington State increased steadily during this project as word of successful codling moth population suppression spread (Fig. 21.2). Growers statewide continued to use this technique for years after the subsidies had disappeared, indicating that they were satisfied with the population suppression and their economic

Table 21.2 Pink bollworm damage to cotton bolls during successive years of the Parker Valley Mating Disruption Project, Arizona, USA

	Larvae per 100 bolls				
	1989	1990	1991	1992	1993
9 July	–	0.3	–	0.19	0
16 July	–	0.6	0.6	0.09	–
23 July	3.6	1.4	0.03	0.95	–
30 July	7.7	2.7	0.09	0.61	0
6 Aug	17.9	1.5	0.03	0	0
13 Aug	25.9	5.9	0.03	2.45	0
27 Aug	36.4	20.6	1.6	1.19	0
3 Sept	34.5	10.9	1.9	1.6	0
10 Sept	21.6	10.4	3.7	1.78	0
17 Sept	28.4	33.3	6.6	–	0
Total bolls sampled/yr	23 847	31 630	21 675	25 603	22 852

balance sheets. Participants in the CAMP program achieved a 75% reduction in insecticide applications while reducing damage to unprecedented levels (Brunner *et al.*, 2001). Damage at harvest fell from 0.8% the year before the program started to 0.55% during year 1, 0.2% during year 2 and to between 0.01% and 0.03% during year 4. These levels were accomplished during 1998 with one-half the density of dispensers per hectare, because this was the first year without a cost subsidy from the program. Also, secondary pests did not arise with the reduced insecticide pressure, as had initially been feared would occur. On the contrary, comparison plots under conventional practice experienced higher levels of secondary pests and often lower levels of beneficial insects and mites than did the CAMP program plots (Brunner *et al.*, 2001).

In the Parker Valley of Arizona with over 10 000 ha of cotton, growers in the late 1980s mandated an areawide mating disruption program due to this pest's high level of resistance to insecticides (Staten *et al.*, 1997). As in the codling moth CAMP program, damage diminished year by year in the Parker Valley with continued areawide application of pheromone mating disruption formulations. Whereas the percentage of infested bolls out of the tens of thousands that were sampled each year was more than 25% during mid-August of year 1, during the same August period in year 2 (1990) damage was only 5.9% (Staten *et al.*, 1997) (Table 21.2). At the mid-August point of year 3 damage was only 0.03% and by 1993, out of more than 22 000 bolls sampled season-long, not a single infested boll was found (0% damage) (Table 21.2). In contrast, the central Arizona average infestation rate for conventionally insecticide-treated hectarage in 1993 was 9–10% by mid-August (Staten *et al.*, 1997).

21.4.2 Behavioral mechanisms underlying mating disruption success

There is evidence from many studies that what has now been more aptly termed "competition" (Cardé, 1990; Cardé & Minks, 1995) but had previously been called "confusion" or "false trail-following," does in fact occur in mating-disruption-treated fields, especially those receiving discrete point sources of dispensers such as hollow fibers or ropes. It is unknown, but we regard it as unlikely, that male moths in fields treated with sprayable microcapsules, creating a nearly uniform fog of pheromone from the closely spaced, weak point source emitters, would be subject to the competition mechanism. The majority of species that have been tested for their responses to uniform clouds of pheromone quickly habituate when released in the pheromone fog, and cease upwind flight after only 1or 2 seconds, reverting to cross-wind casting within the cloud (Kennedy *et al.*, 1981; Baker, 1985; Justus & Cardé, 2002).

Cardé & Minks (1995) and Cardé *et al.* (1997) hypothesized that combinations of mechanisms will likely be operating in concert to various degrees, depending on the type of pheromone emission by the dispensers of a particular mating disruption formulation, to reduce males' abilities to respond to pheromone plumes from their females. For dispensers acting in a competitive mode, when males are being "confused" and flying upwind in the plumes of synthetic pheromone emitted by hollow fibers (Haynes *et al.*, 1986; Miller *et al.*, 1990) or Shin-Etsu ropes (Cardé *et al*, 1997; Stelinski *et al.* 2004, 2005), they are receiving high-concentration contacts with the strands of pheromone in those plumes, and habituation of the olfactory pathways is occurring as a result of the male remaining in upwind flight and continuing to maintain contact with those strong pheromone strands (Baker *et al*, 1998). Habituation that results in reduced upwind flight in response to Shin-Etsu ropes has been observed in pink bollworm males by Cardé *et al.* (1997), and Stelinski *et al.* (2003a, b) have found evidence of long-lasting adaptation of the peripheral pheromone receptors in tortricids. Stelinski *et al.* (2004, 2005) documented that males of four tortricid orchard pest species are attracted in various degrees to individual rope dispensers in the field but often do not proceed all the way to the dispenser.

Miller *et al.* (2006a, b) developed an intriguing new mathematical foundation for judging the degree to which competition (attraction) and non-competitive (habituation) mechanisms contribute to the efficacy of a particular mating disruptant formulation and dispenser deployment density. As formulations use increasingly widely spaced, high emission rate strategies for emitting pheromones, optimal attraction using the blends most closely approximating the natural female blend will become necessary to prolong the time a male spends locked onto the plume dosing himself with high amounts of pheromone in the plume strands to become habituated. At the other extreme, sprayable microcapsule formulations will likely depend very little, if at all, on the attraction–competition mechanism, relying almost exclusively on habituation or plume camouflage (Cardé, 1990; Cardé & Minks, 1995). Regardless of the type of formulation, it is expected that all should be able to take advantage of the advanced period of sexual responsiveness that males exhibit before females begin to emit pheromone.

21.4.3 How mating disruption may suppress population growth

It had routinely been assumed that successful mating disruption can only occur if the majority of females in a population are prevented from mating after the application of a mating disruption formulation. In reality, the females' ability to obtain their first or second matings merely needs to be impaired and delayed. In all but a handful of the huge number of mating disruption field trials that have been conducted over the years, disruption of mating with freely flying females has not been directly assessed. The use of tethered females or clipped-wing females placed on "mating tables" or at stations deployed throughout the disruption plot and then dissected for the presence (mated) or absence (unmated) of spermatophores does not assess what is really happening to feral females that have the ability to fly freely throughout the area. The tethered female technique gives the illusion of being a robust, real-world assessment of mating disruption efficacy, and although it is one of many good indicators, it is especially deficient when the insects distribute themselves unevenly within the habitat due to environmental factors such as heat, humidity and wind. Unless the researcher knows ahead of time the locations where adults typically are most densely clumped and can tether the females there, the ability of the disruptant formulation to keep males from finding females will be overestimated.

Knight (1997) introduced the concept of delayed mating of females within mating disruption plots, based on the relatively high proportion of codling moth females that he found had mated in mating disruption orchards, yet the formulation successfully reduced damage. Earlier studies on the oriental fruit moth that monitored the mating success of freely flying females had suggested that something other than elimination of mating was operating. For the highly successful and grower-accepted Shin-Etsu rope formulation, Rice & Kirsch (1990) found that in plot after plot treated with mating disruptant, females' abilities to mate at least once in disruption-treated plots was suppressed at most by 50% relative to check plots during a flight. The suppression of mating by the disruptant was often as little as 15–20%, despite the reduction of fruit damage in these plots to acceptable levels comparable to those using standard insecticide regimes (Rice & Kirsch, 1990). Thousands of females were captured in terpinyl acetate bait pails and analyzed for the presence or absence of spermatophores in their bursae copulatrices in both the disruptant-treated plots and the check plots. Vickers *et al.* (1985) and Vickers (1990) reported similar results for oriental fruit moth feral female mating success in Australian rope-treated mating disruption plots (23% mated females in mating disruption plots versus 90% in check plots).

Delayed mating was directly confirmed in studies on European corn borer (*Ostrinia nubilalis*) using very high-release-rate metered semiochemical timed release system (MSTRS^{TM}: www.mstrs.com) dispensers (Fadamiro *et al.*, 1999). Analyses were made of the bursae copulatrices of more than 2400 feral females that were captured by hand-netting during the daylight hours as they were flushed from their grassy aggregation areas. During each of the two summer flights, 100% of the females eventually became mated despite the application of high-release-rate, low-point-source density dispensers (Fadamiro *et al.*, 1999). During the first flight about 50% of the females were virgin for the first few days of the flight in the mating disruption plots, but the mating success of females in these plots eventually reached 100% during the ensuing weeks as the flight proceeded. However, females attained this 100%-mated status more slowly in the mating disruption plots than in the check plots, in which females were 100% mated beginning at day 1. Analysis of the number of matings by (number of spermatophores in) European corn borer females showed that throughout the entire flight females captured in the disruption plots were attaining first and second matings at a significantly lower rate than those from the check plots (Fadamiro *et al.*, 1999). The mating disruptant was impairing the ability of females to attract and mate with males on a constant, daily basis but it did not completely eliminate mating. The application of this MSTRS formulation has subsequently been shown to reduce damage to corn by an average of 50–70% in various trials (T. C. Baker, unpublished data). Thus, as demonstrated in the oriental fruit moth (Rice & Kirsch, 1990) and codling moth studies (Knight, 1997), mating disruption success does not require keeping the population of females virgin, but rather just needs to impede females' ability to attract males and retard the dates at which they achieve their first or even second matings. Retarding the dates at which first or second matings occur achieved significantly affects fecundity in the European corn borer and codling moth (Knight, 1997; Fadamiro and Baker, 1999).

21.4.4 Methods to assess the efficacy of mating disruption

Of primary importance to growers and to companies marketing mating disruption products is the ability of a formulation to reduce crop damage to acceptable levels. In this context, "successful" mating disruption means assessing crop damage in pheromone-treated plots versus untreated check plots and finding that damage in mating disruption plots is lower. This process is problematic, but considered by many to be a significant outcome. The assessment is relatively straightforward, but it is essential that plots be large enough to reduce the probability that significant numbers of gravid females can fly in from nearby untreated plots and confound the damage data in the pheromone-treated plots.

Being an indirect measure of actual mating disruption efficacy, damage assessment evaluates the end result of many processes that are of interest to those operating in commercial IPM and agronomic arenas. Conclusions that mating disruption was "successful" in the context of suppressing damage and being cost-effective can be arrived at without knowing exactly to what degree the formulation affected the behavior of males to reduce mating by freely flying feral females.

Considered over many years to be the ultimate test of mating disruption efficacy, tethering females either on a thread or by clipping their wings and placing them on open arenas so that they cannot move from the location at which they are placed has been a good tool in assessing mating disruption efficacy (see Evendon *et al.*, 1999a b), but this is only one of several indirect measures

If the adult moths in the natural population reside in a more clumped distribution in some preferred substructure of the vegetative habitat, placing females on artificial stations outside of each of these clumps will overestimate the efficacy of the disruption formulation with regard to preventing mating of feral females. The formulation in effect will only be assessed for its ability to prevent the tethered females from attracting males such that they leave their aggregation sites. This long-distance attraction will be easier to disrupt than will be the disruption of males within the same clumps that also contain females.

Another limitation to this technique is that once a female mates with the first male arriving at her station, she emits no more pheromone and the ability to assess disruption of communication using that female ends. The data are binomial, with no opportunity for a graded assessment from individual females as to how many males the female could have attracted had it been calling for the entire activity period that night and not been mated.

In another technique, a few virgin females are placed in small screen cages containing sugar water source to keep the females alive and hydrated (see Cardé *et al.*, 1977). The cage is situated within a sticky trap, and so males that are attracted to the calling virgin females become ensnared before reaching the females. Males that do manage to land on the cage containing the females without getting trapped will still not be able to mate with any female in the cage. We feel that this technique has many advantages over other indirect measurement techniques such as the use of tethered females.

First, as with tethered females, the caged females are emitting their natural blend at its natural emission rate. However, unlike tethered females, the caged females in traps provide a graded assessment of their ability to attract males. If males are able to get close enough to the calling females that they can be trapped, they certainly will have mated if given that opportunity at such close range. The final step, mating, is an unnecessary one to examine because if the disruptant was not able to prevent a male's long-distance orientation to the female's plume, certainly it will not be sufficient to stop the male's orientation to the female over the last 10–20 cm or so.

The most widely used technique for assessing the disruption of mate-finding communication by disruptant formulations is the use of standard pheromone monitoring traps containing synthetic pheromone lures (see Rice & Kirsch, 1990; Knight, 1997; Staten *et al.*, 1997; Baker *et al.*, 1998; Fadamiro *et al.*, 1999). This technique can be as informative as the use of caged calling females if the lure that is used has been shown previously in untreated check plots to be able to attract equivalent numbers of males as do caged calling females.

The advantage of assessing trap capture reduction is that it provides a robust, graded data set from these continuously emitting sources. The disruption formulation is challenged throughout the attraction period each night or evening for its ability to continuously suppress the ability of males to locate "females." Using monitoring traps baited with a good synthetic lure is also easier and less problematic than is the use of live, calling females.

21.4.5 Pheromone component blend composition in the disruptant formulation

Minks & Cardé (1988) and Cardé & Minks (1995) reviewed results from many key sex pheromone communication disruption experiments and concluded that for a given species, the synthetic blend compositions and ratios most closely mimicking the natural blend for that species should be the most effective mating disruptants at a given dose per hectare because they will be able to make use of more of the mechanisms (see above) that result in disruption than will suboptimal, partial blends or off-ratios. Although the majority of field experimental evidence supports this conclusion, this does not mean that one or more of the more minor components, due to their more subtle effects, might be able to be eliminated from the final formulated product and still retain sufficient efficacy. There are, however, a few apparent exceptions to this general rule (see Evendon *et al.*, 1999a, b, c, 2000).

21.4.6 Some limitations to mating disruption

A primary obstacle to the use of pheromone mating disruption on any crop is that it must be applied no later than the start of the first adult flight period, that is, before a grower knows for sure that there is even going to be a pest problem that season. The requirement of this "up-front" investment in pheromone puts mating disruption at a disadvantage compared to curative pest management tools that provide a wait-and-see option (see also Chapter 4). For example, insecticide applications can be made after a preliminary assessment of the population density during the first flight of adults or even after oviposition when larval damage can be assessed. Pheromone mating disruption cannot be used curatively in that manner.

Charmillot (1990) summed up lessons he and his co-workers learned concerning when it is that mating disruption becomes less effective for codling moth population suppression. His lessons regarding codling moth are applicable to mating disruption efforts on other pests as well. First, mating disruption works best when applied on an areawide basis and is not advisable on very small areas (less than 1 ha). The crop borders represent vulnerable edges for immigration of mated females from adjoining untreated crop areas, and they also serve as a zone that concentrates females along the borders when there are no nearby planting of the same crop. Due to geometry, small hectarage accentuates the problem because the edge-to-area ratio becomes greater. Also, the use of more widely spaced dispensers requires that borders be given special attention to reduce the presence of pheromone-free clean-air "holes" along the borders. This can be accomplished either by decreasing the spacing between dispensers, or else through the use of different dispenser technology along borders, such as sprayable microcapsules.

Except for a few small groups or pairs of species, pheromones are extremely species-specific. The replacement of broad-spectrum insecticides in some IPM systems with mating disruption that targets only one species in a pest complex, might lead to increases in the populations of species that had been only secondary pests before mating disruption was used (see Rice & Kirsch, 1990). No pheromone disruption formulation has been created thus far that functions effectively as a "broad-spectrum" formulation. Evendon *et al.* (1999b, c) experimented with single blend formulations targeting obliquebanded leafroller (*Choristoneura rosaceana*) and threelined leafroller (*Pandemis limitata*) with some success. Perhaps there are other situations where effective multi-species blend formulations can be developed, although there are still none that have attained commercial success.

Pheromone mating disruption formulations have to this point been expensive compared with curative applications of insecticides. Although growers of many crops are now well aware that mating disruption "works" as a pest management tool, pheromones' expense plus the up-front nature of the cost has put them at a disadvantage relative to curative pest management tools. The active ingredient, the pheromone itself, is the most expensive part of a formulation. Costs per gram of even the least expensive pheromone components are approximately $US 1.00, and many other more expensive major pheromone components cost $US 3 to 20 per gram. Formulating the active ingredient into specialized dispensers adds to the cost, but nevertheless, if cheaper organic synthetic routes to the active ingredients can be developed, costs of the final formulated products could be reduced substantially.

21.5 Conclusions

Insect pheromone-related technologies for monitoring endemic pest populations, detecting invasive species, mass trapping for population suppression and mating disruption have had a relatively recent history of development in IPM compared to biological control and insecticide technologies. New progress in the application of pheromones in IPM is being made in many areas, including the knowledge that mass trapping can be a highly effective and economically beneficial use of these behavior-modifying chemicals. Novel lure-and-trap technologies continue to be developed for new pest species as they come on the scene in various regions of the world. New insights are also being made regarding ways to determine empirically the modes of action of mating disruption formulations, and the acceptance of the mating disruption technique by growers and government agencies has continued to grow in recent years. It remains to be seen whether other behavior-modifying chemicals such as host plant volatiles can become as widely used as pheromones for insect IPM in field situations where pheromones have been an integral part of insect IPM programs for approximately 35 years.

References

Agnello, A. M., Kovach, J., Nyrop, J. P. *et al.* (1994). Extension and evaluation of a simplified monitoring program in New York apples. *American Entomologist*, **40**, 37–49.

Baker, T. C. (1985). Chemical control of behaviour. In *Comprehensive Insect Physiology, Biochemistry, and Pharmacology*, eds. G. A. Kerkut & L. S. Gilbert, pp. 621–672. Oxford, UK: Pergamon Press.

Baker, T. C., Fadamiro, H. Y. & Cossé, A. A. (1998). Moth uses fine tuning for odor resolution. *Nature*, **393**, 530.

Baker, T. C., Staten, R. T. & Flint, H. M. (1990). Use of pink bollworm pheromone in the southwestern United States. In *Behavior-Modifying Chemicals for Insect Management*, eds. R. L. Ridgway, R. M. Silverstein & M. N. Inscoe, pp. 417–436. New York: Marcel Dekker.

Brunner, J., Welter, S., Calkins, C. *et al.* (2001). Mating disruption of codling moth: a perspective from the Western United States. *IOBC/WPRS Bulletin*, **25**, 207–215.

Butenandt, A. (1959). Wirkstoffe des Insektenveiches. *Naturwissenschaften*, **46**, 461–471.

Cardé, R. T. (1990). Principles of mating disruption. In *Behavior-Modifying Chemicals for Insect Management*, eds. R. L. Ridgway, R. M. Silverstein & M. N. Inscoe, pp. 47–71. New York: Marcel Dekker.

Cardé, R. T. & Minks, A. K. (1995). Control of moth pests by mating disruption: successes and constraints. *Annual Review of Entomology*, **40**, 559–585.

Cardé, R. T., Baker, T. C. & Castrovillo, P. J. (1977). Disruption of sexual communication in *Laspeyresia pomonella* (codling moth), *Grapholitha molesta* (oriental fruit moth) and *G. prunivora* (lesser appleworm) with hollow fiber attractant sources. *Entomologia Experimentalis et Applicata* **22**, 280–288.

Cardé, R. T., Mafra-Neto, A., Staten, R. T. & Kuenen, L. P. S. (1997). Understanding mating disruption in the pink bollworm moth. *IOBC/WPRS Bulletin*, **20**, 191–201.

Charmillot, P. J. (1990). Mating disruption technique to control codling moth in western Switzerland. In *Behavior-Modifying Chemicals for Insect Management*, eds. R. L Ridgway, R. M. Silverstein & M. N. Inscoe, pp. 165–182. New York: Marcel Dekker.

Chinchilla, C. M. & Oehlschlager, A. C. (1992). Capture of *Rhynchophorus palmarum* in traps baited with the male-produced aggregation pheromone. *ASD Oil Palm Papers*, **5**, 1–8.

Elkinton, J. S. & Cardé, R. T. (1981). The use of pheromone traps to monitor distribution and population trends of the gypsy moth. In *Management of Insect Pests with Semiochemicals*, ed. E. R Mitchell, pp. 41–55. New York: Plenum Press.

Evendon, M. L., Judd, G. J. R. & Borden, J. H. (1999a). Pheromone-mediated mating disruption of *Choristoneura rosaceana*: is the most attractive blend really the most effective? *Entomologia Experimentalis et Applicata*, **90**, 37–47.

Evendon, M. L., Judd, G. J. R. & Borden, J. H. (1999b). Mating disruption of two sympatric, orchard-inhabiting tortricids, *Choristoneura rosaceana* and *Pandemis limitata* (Lepidoptera: Tortricidae), with pheromone components of both species' blends. *Journal of Economic Entomology*, **92**, 380–390.

Evendon, M. L., Judd, G. J. R. & Borden, J. H. (1999c). Simultaneous disruption of pheromone communication in *Choristoneura rosaceana* and *Pandemis limitata* with pheromone and antagonist blends. *Journal of Chemical Ecology*, **25**, 501–517.

Evendon, M. L., Judd, G. J. R. & Borden, J. H. (2000). Investigations of mechanisms of pheromone communication disruption of *Choristoneura rosaceana* (Harris) in a wind tunnel. *Journal of Insect Behavior*, **13**, 499–510.

Fadamiro, H. Y. & Baker, T. C. (1999). Reproductive performance and longevity of female European corn borer, *Ostrinia nubilalis*: effects of multiple mating, delay in mating, and adult feeding. *Journal of Insect Physiology*, **45**, 385–392.

Fadamiro, H. Y., Cossé, A. A. & Baker, T. C. (1999). Mating disruption of European corn borer, *Ostrinia nubilalis*, by using two types of sex pheromone dispensers deployed in grassy aggregation sites in Iowa cornfields. *Journal of Asia-Pacific Entomology*, **2**, 121–132.

Hardee, D. D., Mitchell, E. B. & Huddleston, P. M. (1967a). Field studies of sex attraction in the boll weevil. *Journal of Economic Entomology*, **60**, 1221–1224.

Hardee, D. D., Mitchell, E. B. & Huddleston, P. M. (1967b). Procedure for bioassaying the sex attractant of the boll weevil. *Journal of Economic Entomology*, **60**, 169–171.

Haynes, K. F., Li, W. G. & Baker, T. C. (1986). Control of pink bollworm moth (Lepidoptera: Gelechiidae) with insecticides and pheromones (attracticide): lethal and sublethal effects. *Journal of Economic Entomology*, **79**, 1466–1471.

Justus, K. A. & Cardé, R. T. (2002). Flight behaviour of two moths, *Cadra cautella* and *Pectinophora gossypiella*, in homogeneous clouds of pheromone. *Physiological Entomology*, **27**, 67–75.

Kennedy, J. S., Ludlow, A. R. & Sanders, C. J. (1981). Guidance of flying male moths by wind-borne sex pheromone. *Physiological Entomology*, **6**, 395–412.

Knight, A. L. (1997). Delay of mating of codling moth in pheromone disrupted orchards. *IOBC/WPRS Bulletin*, **20**, 203–206.

Madsen, H. F. (1981). Monitoring codling moth populations in British Columbia apple orchards. In *Management of Insect Pests with Semiochemicals*, ed. E. R. Mitchell, pp. 57–62. New York: Plenum Press.

Miller, E., Staten, R. T., Nowell, C. & Gourd, J. (1990). Pink bollworm (Lepidoptera: Gelechiidae): point source density and its relationship to efficacy in attracticide formulations of gossyplure. *Journal of Economic Entomology*, **83**, 1321–1325.

Miller, J. R., Gut, L. J., de Lame, F. M. & Stelinski, L. L. (2006a). Differentiation of competitive vs. non-competitive mechansisms mediating disruption of moth sexual communication by point sources of sex pheromone. I. Theory. *Journal of Chemical Ecology*, **32**, 2089–2114.

Miller, J. R., Gut, L. J., de Lame, F. M. & Stelinski, L. L. (2006b). Differentiation of competitive vs. non-competitive mechanisms mediating disruption of moth sexual communication by point sources of sex pheromone. II. Case studies. *Journal of Chemical Ecology*, **32**, 2115–2144.

Minks, A. K. & Cardé, R. T. (1988). Disruption of pheromone communication in moths: is the natural blend really most efficacious? *Entomologia Experimentalis et Applicata*, **49**, 25–36.

Oehlschlager, A.C., Pierce, H.D., Morgan, B. *et al.* (1992). Chirality and field testing of Rhynchophorol, the aggregation pheromone of the American palm weevil. *Naturwissenschaften (Berlin)*, **79**, 134–135.

Oehlschlager, A. C., Chinchilla, C. M., Gonzales, L. M. *et al.* (1993). Development of a pheromone-based trapping system for *Rhynchophorus palmarum* (Coleoptera: Curculionidae). *Journal of Economic Entomology*, **86**, 1381–1392.

Oehlschlager, A. C., Chinchilla, C., Castillo, G. & Gonzalez, L. (2002). Control of red ring disease by mass trapping of *Rhynchophorus palmarum* (Coleoptera: Curculionidae). *Florida Entomologist*, **85**, 507–513.

Rice, R. E. & Kirsch, P. (1990). Mating disruption of oriental fruit moth in the United States. In *Behavior-Modifying Chemicals for Insect Management*, eds. R. L. Ridgway, R. M. Silverstein & M. N. Inscoe, pp. 193–211. New York: Marcel Dekker.

Ridgway, R. L., Inscoe, M. N. & Dickerson, W.A. 1990. Role of the boll weevil pheromone in pest management. In *Behavior-Modifying Chemicals for Insect Management*, eds. R. L. Ridgway, R. M. Silverstein & M.N. Inscoe, pp. 437–471. New York: Marcel Dekker.

Riedl, H. & Croft, B. A. (1974). A study of pheromone trap catches in relation to codling moth (Lepidoptera: Olethreutidae) damage. *Canadian Entomologist*, **112**, 655–663.

Riedl, H., Croft, B. A. & Howitt, A. G. (1976). Forecasting codling moth phenology based on pheromone trap catches and physiological-time models. *Canadian Entomologist*, **108**, 449–460.

Showers, W. B., Smelser, R. B., Keaster, A. J. *et al.* (1989a). Recapture of marked black cutworm (Lepidoptera: Noactuidae) males after long-range transport. *Environmental Entomology*, **18**, 447–458.

Showers, W. B., Whitford, F., Smelser, R. B. *et al.* (1989b). Direct evidence for meteorologically driven long-range dispersals of an economically important moth. *Ecology*, **70**, 987–992.

Staten, R. T., El-Lissy, O. & Antilla, L. (1997). Successful area-wide program to control pink bollworm by mating disruption. In *Insect Pheromone Research: New Directions*, eds. R. T. Cardé & A. K. Minks, pp. 383–396. New York: Chapman & Hall.

Stelinski, L. L., Miller, J. R. & Gut, L. J. (2003a). Presence of long-lasting peripheral adaptation in the obliquebanded leafroller, *Choristoneura rosaceana* and the absence of such adaptation in the redbanded leafroller, *Argyrotaenia velutinana*. *Journal of Chemical Ecology*, **29**, 405–423.

Stelinski, L. L., Gut, L. J. & Miller, J. R (2003b). Concentration of air-borne pheromone required for long-lasting peripheral adaptation in the obliquebanded leafroller, *Choristoneura rosaceana*. *Physiological Entomology*, **28**, 97–107.

Stelinski, L. L., Gut, L J., Pierzchala, A. V. & Miller, J. R. (2004). Field observations quantifying attraction of four tortricid moth species to high-dosage pheromone rope dispensers in untreated and pheromone-treated apple orchards. *Entomologia Experimentalis et Applicata*, **113**, 187–196.

Stelinski, L. L., Gut, L. J., Epstein, D. & Miller, J. R. (2005). Attraction of four tortricid moth species to high dosage pheromone rope dispensers: observations implicating false plume following as an important factor in mating disruption. *IOBC/WPRS Bulletin*, **28**, 313–317.

Tumlinson, J. H., Hardee, D. D., Minyard, J. P. *et al.* (1968). Boll weevil sex attractant: isolation studies. *Journal of Economic Entomology*, **61**, 470–474.

Tumlinson, J. H., Hardee, D. D., Gueldner, R. C. *et al.* (1969). Sex pheromones produced by male boll weevil: Isolation, identification, and synthesis. *Science*, **166**, 1010–1012.

Tumlinson, J. H., Gueldner, R. C., Hardee, D. D. *et al.* (1970). The boll weevil sex attractant. In *Chemicals Controlling Insect Behavior*, ed. M. Beroza, pp. 41–59. New York: Academic Press.

Tumlinson, J. H., Gueldner, R. C., Hardee, D. D. *et al.* (1971). Identification and synthesis of the four compounds comprising the boll weevil sex attractant. *Journal of Organic Chemistry*, **36**, 2616–2621.

Vickers, R. A. (1990). Oriental fruit moth in Australia and Canada. In *Behavior-Modifying Chemicals for Insect Management*, eds. R. L. Ridgway, R. M. Silverstein & M. N. Inscoe, pp. 183–192. New York: Marcel Dekker.

Vickers, R. A., Rothschild, G. H. L. & Jones, E. L. (1985). Control of the oriental fruit moth, *Cydia molesta* (Busck) (Lepidoptera:Tortricidae), at a district level by mating disruption with synthetic female pheromone. *Bulletin of Entomological Research*, **75**, 625–634.

Chapter 22

Insect endocrinology and hormone-based pest control products in IPM

Daniel Doucet, Michel Cusson and Arthur Retnakaran

IPM methods were developed largely in response to the negative consequences of the intensive use of broad-spectrum pesticides in the early to mid twentieth century (Kogan, 1998). These insecticides, belonging to the carbamate, organophosphate and organochlorine families, have unintended side effects such as environmental persistence, bioaccumulation, development of resistance among target pests, toxicity to non-target species (especially natural enemies) and human health risks. While IPM focuses mainly on preventative tactics (e.g. crop rotation) rather than remedial ones, synthetic chemical insecticides are still very much needed to achieve effective control in many agricultural systems.

The study of insect physiology has been driven, in no small part, by the need for safe alternatives to broad-spectrum insecticides. Theoretically at least, digestion, excretion, neuronal communication, metabolism and other physiological processes all comprise "insect-specific" components that are vulnerable and could be targeted by synthetic molecules. To this day, however, IPM-compatible pest control products that target the insect endocrine system far outnumber those targeting other systems. In particular, hormone mimics that control development have enjoyed not only wide appeal but also many commercial successes, and additional control products targeting hormone production and function are currently under development. In this chapter we provide an overview of (1) insect endocrinology, (2) existing control products that mimic ecdysone and juvenile hormone (JH) action and (3) possible development of disruption control strategies based on novel endocrine functions that are likely to generate new IPM tools in the future.

22.1 Basics of hormone biochemistry and biology

The term "hormone" was first coined by Ernest Starling a hundred years ago to define any chemical messenger, secreted by an organ, which travels through the bloodstream to affect the physiology of another, distant organ (reviewed in Henderson, 2005). At that time, hormones were known to be important mediators of vertebrate physiology, but Kopec (1917) soon demonstrated that similar secretions existed in insects, with his isolation of a brain factor promoting molting in a lepidopteran larva.

Insect hormones fall into four classes based on their chemical structures: (1) peptide and protein hormones, (2) biogenic amines, (3) prostaglandins and (4) terpenoid lipids. Peptide hormones are chains of amino acids usually shorter than 20–30 residues. Longer chains are traditionally referred

Integrated Pest Management, ed. Edward B. Radcliffe, William D. Hutchison and Rafael E. Cancelado. Published by Cambridge University Press.

to as proteins. This is by far the most abundant class of insect hormones, with several hundred different peptides isolated to date, from various species. Biogenic amines are derivatives of amino acids and are involved in signal transduction. In insects, octopamine and tyramine are two such compounds that regulate important aspects of locomotor and non-locomotor behavior, circadian rhythms and stress response (Gole & Downer, 1979; Fussnecker *et al.*, 2006). Prostaglandins are oxygenated metabolites of arachidonic acid and play diverse roles including mediation of cellular immunity and release of egg-laying behavior (Stanley, 2006). Terpenoid hormones are lipid molecules constructed from the basic hydrocarbon unit 2-methyl-1,3-butadiene, also called isoprene. In insects, two important subclasses exist: the juvenile hormones, which are derivatives of linear chains of three isoprene units (i.e. sesquiterpenes) and the ecdysteroids, which consist of a tetracyclic cholestane ring system derived from cholesterol, which is elaborated from smaller isoprenoids. Both hormones act on the timing and nature of molting in all insects. The most successful hormone-based insecticides either mimic or inhibit the activity of these hormones and are presented in greater detail in Sections 22.4 and 22.5.

22.2 Overview of the insect endocrine system

Hormone secretion in insects is accomplished by a limited number of glands, organs and tissues, and, as observed in other animal groups, the central nervous system (CNS) plays an overarching role. The CNS integrates sensory information and translates it into nervous or hormonal outputs that bring about the physiological, behavioral and developmental processes necessary for survival and reproduction (Nijhout, 1994). The endocrine control of these processes by the CNS can be either direct or indirect. A good example of direct control is the induction of diapause in silkworm (*Bombyx mori*) by the diapause hormone (DH), released by secretory neurons from the subesophageal ganglia (Hasegawa, 1957; Sato *et al.*, 1993). More frequently, however, the CNS modulates the secretions of other endocrine glands by releasing inhibitory or stimulatory (tropic) neurohormones. This hierarchical organization between hormones and neurohormones can be highly complex and include endocrine feedback loops. Such is the case for insect ecdysis, where the proper unfolding of this innate behavior is ensured by a tightly controlled spatial and temporal release of at least half a dozen hormones (Kim *et al.*, 2006). While a thorough review of insect hormones (and their sites of production) would be beyond the scope of this chapter, a survey of some of the more important ones was presented by Doucet *et al.* (2007a).

22.2.1 CNS and associated neurohemal organs

The CNS is composed of two major endocrine centers: the brain–retrocerebral complex (RC) and the perisympathetic organs (PSOs) of the ventral nerve cord. The RC is a bipartite structure, posterior to the brain, composed of the corpora allata (CA) and corpora cardiaca (CC). Clusters of neurosecretory cells are located in various parts of the insect brain (e.g. medial, lateral and ventral) and the majority release their secretions via the CC (Nijhout, 1994). Some brain neurosecretory cells instead use the CA as a neurohemal organ (e.g. for the secretion of prothoracicotropic hormone, PTTH), while fewer still release their secretions distally, for example in the vicinity of the proctodeum (hindgut) (e.g. the proctodeal nerves in tobacco hornworm [*Manduca sexta*] are the release sites of eclosion hormone [EH], which is synthesized by the ventromedial neurosecretory cells in the brain (Truman & Copenhaver, 1989). The glandular portions of the CA and the CC also secrete their own hormones besides those from the neurosecretory cells: the CA secrete juvenile hormone (JH) (see Section 22.4) while the CC contain endocrine cells that produce peptides such as the adipokinetic hormone (AKH). PSOs are segmentally distributed neurohemal organs that are functionally close to the CC (Nijhout, 1994). However, recent mass spectrometry analysis of PSO extracts indicate that they release a cocktail of neuropeptides distinct from those produced by the CC (Predel *et al.*, 1999, 2000).

22.2.2 Other endocrine glands

While the CNS is the most structurally and functionally complex hormone production site in insects, non-neural peripheral secretory organs contribute in important ways to the regulation of physiological processes. For example, a few highly specialized glands such as the prothoracic glands (PGs), the Inka cells in the epitracheal glands and the epiproctodeal glands (EPGs) have confirmed or suspected roles in endocrine regulation of molting. PGs are located in the prothoracic segments of immature insects and are devoted to the secretion of the hormone ecdysone. The Inka cells come into play later during the molting cycle, at the time of ecdysis. These cells are specialized in the secretion of two peptides regulating ecdysis-related behaviors: the pre-ecdysis triggering hormone (PETH) and the ecdysis triggering hormone (ETH) (Žitňan *et al.*, 2003). The number, morphology and distribution of Inka cells within the body can be quite variable among insects of different orders, but they are always in close association with the epitracheal glands, near the spiracles. EPGs are secretory structures that have been described in *Manduca*. They are present as a pair of multinucleated cells located at the junction of the hindgut and the rectum, in close contact with the proctodeal nerve (Davis *et al.*, 2003). EPGs synthesize a myoinhibitory-like peptide (MIP-like I) that is speculated to shut off ecdysone production by the PGs, at the end of each molt (Davis *et al.*, 2003).

Non-specialized endocrine tissues include the gut, the fat body and ovaries. Midgut endocrine cells have features of typical secretory cells (e.g. abundant secretory granules, clear cytoplasm: Brown *et al.*, 1985; Neves *et al.*, 2003) and have been found to release peptides involved in midgut contraction and other aspects of digestive physiology. Hormones secreted by these cells include allatostatins and allatostatin-like peptides (AS-like) (Davey *et al.*, 2005) and crustacean cardioactive peptide (CCAP) (Sakai *et al.*, 2004). Interestingly, both AS and CCAP were originally discovered in functions unrelated to invertebrate midgut physiology: as an inhibitor of JH synthesis by the CA (Stay & Tobe, 2007) and CCAP as a peptide regulating heartbeat in the crab and ecdysis in insects (Žitňan & Adams, 2005). Fat body cells from the desert locust (*Schistocerca gregaria*) have been shown to synthesize neuroparsins (Claeys *et al.*, 2003). Very little is known about the function of these neuroparsins, but their expression is stage- and sex-dependent, and is regulated by JH and 20-hydroxyecdysone (Claeys *et al.*, 2006). Rachinsky *et al.* (2006) also detected low levels of allatotropin, a JH biosynthesis stimulating peptide, in the fat body of *Manduca sexta*. Finally, the ovaries of many insects constitute a source of ecdysteroids that play important roles in reproductive physiology. These steroids either stimulate the transcription of yolk protein precursor genes in females (e.g. vitellogenin, lipophorin: Raikhel *et al.*, 2005) or presumably assist in cuticle formation during the development of the embryo (Lagueux *et al.*, 1984).

22.3 Ecdysone and ecdysone agonists

22.3.1 Ecdysone functions

The elucidation of the structure of ecdysone (Butenandt & Karlson, 1954) was the major catalyst for research on the physiology of the molting hormone and its role in insect metamorphosis. We now know that the biologically active form of ecdysone is 20-hydroxyecdysone (20E), which acts as the ligand of a heterodimeric receptor system consisting of two nuclear receptors, the ecdysone receptor (EcR) and ultraspiracle (USP). USP is an allosteric effector for ligand binding by the EcR. 20E binds to the ligand-binding domain of the EcR subunit of the EcR–USP dimer (EcR complex). In turn, this complex binds to a specific sequence on the DNA, designated as the "ecdysone response element" (EcRE), located upstream from the gene that is to be activated or repressed (Nordeen *et al.*, 1998). Binding of the 20E-liganded EcR complex to an EcRE causes transactivation of gene transcription. Conversely, the unliganded complex causes repression. The degree of binding affinity appears to correlate with activity. The bipolar transcriptional ability is probably aided by corepressors and coactivators, as in vertebrate steroid hormone receptor systems, but details are lacking at present (King-Jones & Thummel, 2005).

22.3.2 Molting regulation by 20E

The major function of 20E is the regulation of the molting process in insects. In the Lepidoptera, this hormone appears as a single peak during the middle of each larval stadium, except for the last one where there are two peaks; the earlier, smaller peak is the commitment peak which triggers the reprogramming towards pupal development, followed by the larger molt peak. The presence of high titers of JH during the larval 20E peak defines the molt as larval whereas the absence of JH during the commitment peak results in pupal transformation, which involves the turning off of larval genes and turning on of pupal genes. The physiological and molecular basis of the switch has been elegantly elucidated by Riddiford and her associates showing for instance that the *Broad Complex* gene (*BR-C*) encodes the transcription factor that specifies the pupal cuticle (Riddiford *et al.*, 2003). Thus, 20E can both repress and activate genes, and its effects can be modulated by JH.

22.3.3 Steroidal and non-steroidal analogs of ecdysone

Various plant ecdysteroids (phytoecdysones) and synthetic steroids have been assessed for their ability to act as agonists or antagonists of ecdysone and for their insecticidal activity, but with limited success. Besides being difficult to synthesize, steroids are large molecules lacking contact activity and are prone to oxidative breakdown. Over 200 plant ecdysteroids have been studied and none of them has shown high potential as control products (Dinan, 2001). Similarly, several synthetic steroids were tested for biological activity, but none proved very effective (Robbins *et al.*, 1970). A few steroid analogs such as cucurbitacins and brassinosteroids were tested for antagonistic activity but, again, the effects were weak (Charrois *et al.*, 1996; Dinan *et al.*, 1997). More recently, however, non-steroidal ecdysone agonists such as tetrahydroquinoline compounds showed some activity, especially against mosquitoes (Palli *et al.*, 2005). By far the most effective compounds among the non-steroidal ecdysone agonists are the diacylhydrazines, serendipitously discovered in the Rohm and Haas (RH) laboratories at Spring House, PA (Hsu, 1991). Initial optimization following discovery of the chemistry led to a first promising candidate, RH-5849, but soon three other compounds displaying greater activity were synthesized: (1) RH-5992 (tebufenozide) was active against several lepidopterans and was registered under the commercial names Mimic®, Confirm® and Romdan®; (2) RH-2485 (methoxyfenozide) was several times more active than RH-5992 on lepidopterans, and was marketed under the names Intrepid®, Runner®, Prodigy® and Falcon®; (3) RH-0345 or halofenozide was active against coleopterans in addition to lepidopterans and was registered under the name Mach 2®. Rohm and Haas since sold all these compounds to Dow Agrosciences, Indianapolis, IN. More recently Nippon Kayaku, Saitane and Sankyo, Ibaraki, from Japan, have come up with a new diacylhydrazine which they have named ANS-118/CM-001 or chromafenozide, which is active against lepidopterans and is registered under the names Matric® and Killat® (Nakagawa, 2005) (Fig. 22.1).

Diacylhydrazines are functionally similar to 20E and bind to the EcR receptor complex as ligands (Nakagawa, 2005). While they mimic the natural hormone in triggering the molting process, the latter is never completed. Treated larvae show all the initial signs of molting, such as feeding cessation, head-capsule slippage (with the new head capsule remaining untanned), and loosening of the cuticle due to apolysis. However, the similarity between 20E action and diacylhydrazines stops at this juncture. There is no ecdysis, resumption of feeding, or sclerotization and darkening of the new head capsule that normally follow clearance of 20E. Rather, the larva is in a state of developmental arrest or suspended animation and eventually dies of starvation and desiccation.

22.3.4 Diacylhydrazines as pest-control products

The non-steroidal ecdysone agonists have been tested on a variety of insects with varying degrees of success. The activity of different diacylhydrazines on a representative list of insects was presented by Doucet *et al.* (2007b). The Mimic® formulation of RH-5992 (tebufenozide) works very well against the spruce budworm, a serious pest of the balsam fir (*Abies balsamea*) and white spruce (*Picea glauca*) in the boreal forests of North America (Retnakaran *et al.*, 1997; Cadogan *et al.*, 1998, 2005).

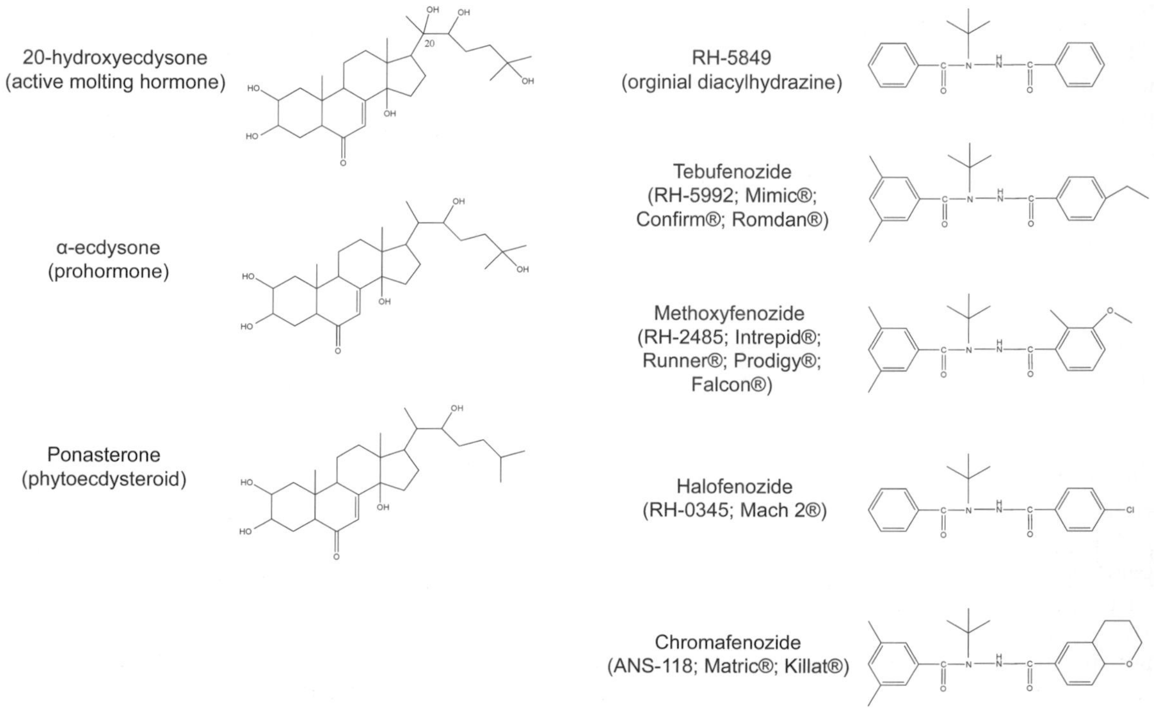

20-hydroxyecdysone
(active molting hormone)

α-ecdysone
(prohormone)

Ponasterone
(phytoecdysteroid)

RH-5849
(orginial diacylhydrazine)

Tebufenozide
(RH-5992; Mimic®; Confirm®; Romdan®)

Methoxyfenozide
(RH-2485; Intrepid®; Runner®; Prodigy®; Falcon®)

Halofenozide
(RH-0345; Mach 2®)

Chromafenozide
(ANS-118; Matric®; Killat®)

Fig. 22.1 Chemical structures of insect molting hormones, a phytoecdysteroid and five of the most promising diacylhydrazine ecdysone agonists.

Laboratory studies show that methoxyfenozide is about ten times more active than tebufenozide (Sundaram *et al.*, 1998). The grape berrymoth (*Lobesia botrana*) is very sensitive to methoxyfenozide, which shows excellent potential for control of this pest. Older larvae are more susceptible than younger ones, and treated adults display reduced fecundity and fertility (Sáenz-de-Cabezón Irigaray *et al.*, 2005).

In some instances the molt-inducing activity of diacylhydrazines is very weak, and thus this class of compounds cannot be retained as an effective control solution. The obliquebanded leafroller (*Choristoneura rosaceana*) shows relatively low susceptibility to tebufenozide and methoxyfenozide (Ahmad *et al.*, 2002). In the case of the white-marked tussock moth (*Orgyia leucostigma*), treatment with tebufenozide induces head capsule slippage but the larva, after remaining quiescent for a week to 10 days, becomes active and molts into the next instar (Retnakaran *et al.*, 2003). *In vitro* and *in vivo* data tend to indicate that the potency of diacylhydrazines is related to absorption and excretion rates in a given species (Retnakaran *et al.*, 2001; Smagghe *et al.*, 2001). Other cases of low susceptibility include the codling moth (*Cydia pomonella*) (Sauphanor & Bouvier, 1995), and the green-headed leafroller (*Planotortrix octo*) from New Zealand (Wearing, 1998), to name a few. This only goes to show that these compounds, while very effective on some species, have their own limitations and are not a general cure-all.

22.3.5 Ecdysone agonists as candidates for IPM

Over the years pest management methods have progressively evolved towards an ecologically based systems approach, combining biological, cultural, physical and chemical tools in a way that minimizes economic, health and environmental risks. Such an IPM approach is a dynamic process constantly aiming at maximizing target specificity and minimizing environmental side effects. Ecdysone agonists have so far been shown to be insect-specific and, among the various agonists, some are far more active on one group of insects than others. In general, their mammalian toxicity is very low; the acute oral toxicity for

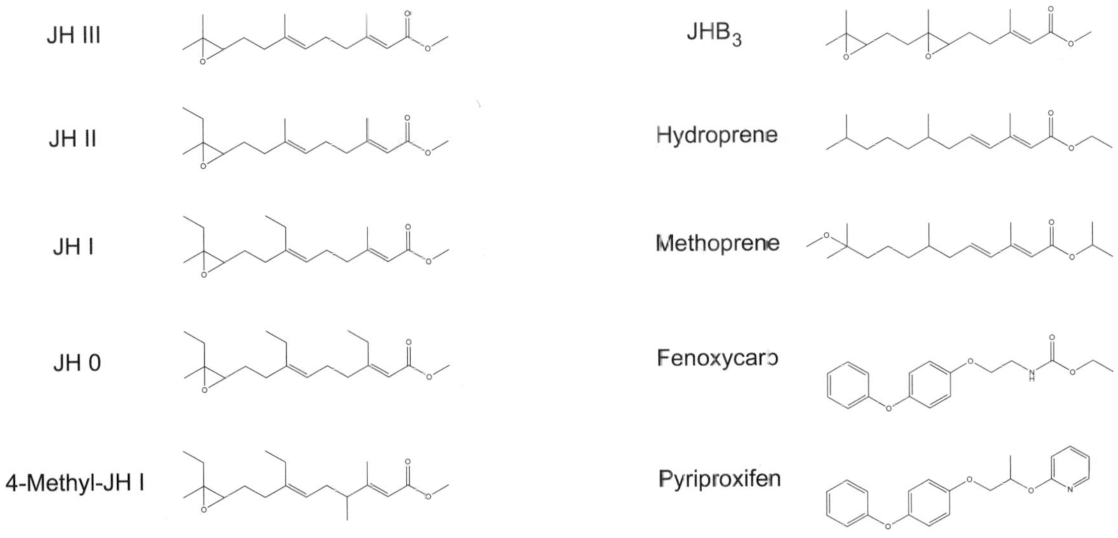

Fig. 22.2 Structures of juvenile hormones (JHs) produced by insects and of four JH analogs commercially available.

rat and mouse is >5000 mg/kg for tebufenozide, methoxyfenozide and chromafenozide, and >2850 mg/kg for halofenozide. These compounds have no detectable effects on reproduction, and are negative in the Ames mutation assay (Dhadialla *et al.*, 2005). These non-steroidal ecdysone agonists are also relatively safe for the environment and do not have any negative impact on forest-litter-dwelling organisms such as earthworms and Collembola (Addison, 1996). Macroinvertebrates in freshwater ponds are unaffected by tebufenozide (Kreutzweiser *et al.*, 1994). Tebufenozide and methoxyfenozide were shown to have little effect on the bumblebee (*Bombus terrestris*) (Mommaerts *et al.*, 2006). A summary of safety to parasites, predators and pollinators is provided by the National Registration Authority of Australia (Anonymous, 2002). Tebufenozide was also found to have little impact on a generalist predator, the common green lacewing (*Chrysoperla carnea*) (Medina *et al.*, 2003). The persistence, breakdown and catabolism of tebufenozide have been well studied. In conifer needles and forest litter some persistence was observed, but well within tolerance levels (Sundaram *et al.*, 1996). Thus the environmental safety and narrow spectrum of activity of these non-steroidal ecdysone agonists make them a valuable addition to the arsenal of candidates for IPM.

22.4 Juvenile hormones

22.4.1 Chemical nature, biosynthesis and functions

The juvenile hormones (JHs) form a family of lipophilic, sesquiterpenoid molecules with epoxide and methyl-ester functionalities. All are derived from the mevalonate pathway intermediate, farnesyl diphosphate (FPP), or from one of its ethyl-branched homologs. The majority of insects produce only one chemical form of JH, JH III (C-16), but the Lepidoptera produce four additional, ethyl-substituted JHs (JH 0 [C-19], JH I [C-18], JH II [C-17] and 4-methyl JH I [C-19]) and the Diptera produce a bis-epoxy form of JH III (JH-B3) (Fig. 22.2). These hormones are all produced *de novo* by the CA. JH biosynthesis begins with the condensation of three units of acetyl-CoA (for JH III and JH-B3) or two units of acetyl-CoA and one of propionyl CoA (lepidopteran JHs), leading to the formation of the isoprene (C-5) and homo-isoprene (C-6) building blocks of FPP and ethyl-substituted FPPs. Unlike the enzymatic steps that are responsible for FPP formation, which are common to most living organisms, those that convert FPP into JH are specific to insects.

Although JH plays multiple roles in insects, it owes its name to its juvenilizing effects during larval molts: the presence of elevated JH titers during ecdysone secretion represses the expression of metamorphic genes, thus maintaining the insect in a juvenile (larval) state. Shortly after the final larval molt, JH biosynthesis ceases and a JH-specific catabolic enzyme known as JH esterase (JHE) is secreted, which leads to a rapid decline in the JH titer. Under these conditions, ecdysone triggers the metamorphic program leading to the pupal molt. In adult insects, JH is a gonadotropin and is best known for its stimulatory effects on vitellogenesis, inducing the production of vitellogenin by the fat body and/or its uptake by developing oocytes. In males, JH has been implicated in the production of accessory sex gland secretions and in control of courtship behavior. JH has also been shown to be involved in regulation of various other processes, including embryogenesis, migration, cast differentiation, polyphenism and reproductive diapause (Cusson, 2004). Although many of the roles played by JH have been well characterized, its mode of action at the molecular level remains unclear. Several proteins have been tentatively identified as JH receptors, but conclusive evidence regarding their role as receptors is still lacking (see Palli & Cusson, 2007, for a more complete account of recent work in this area). The presence of a JH response element (JHRE) in the promoter region of the JH-responsive JHE gene suggests that JH can act through a nuclear receptor. Some nuclear proteins do, indeed, bind to this JHRE, but binding requires that the proteins be dephosphorylated, a process that appears to be induced by JH (Kethidi *et al.*, 2006).

22.4.2 Juvenile hormone analogs and IPM

Following the initial isolation of JH, Caroll Williams (1956) predicted the dawn of JH-based insecticides. It was surmised that synthetic JH-like molecules would fatally interfere with JH functions and that insects would not likely develop resistance against such compounds (Williams, 1967). Many JH analogs (molecules with JH effects, with or without a JH-like terpenoid structure) were subsequently designed, synthesized and assayed for insecticidal activity (Sláma *et al.*, 1974). Some of them (e.g. methoprene, hydroprene, fenoxycarb, pyriproxifen, diofenolan; Fig. 22.2) were found to be effective against certain pests and have since enjoyed commercial success, particularly for the control of insects that are injurious in the adult stage (e.g. mosquitoes, fleas, whiteflies, etc.), but their efficacy against phytophagous larval insects (e.g. caterpillars) has often proven to be limited, largely because the analogs interfere with metamorphosis, once larval feeding has ended (see Dhadialla *et al.*, 1998, 2005 for reviews).

22.4.3 Anti-JHs

It has long been recognized that a strategy involving the induction of precocious metamorphosis through the inhibition of JH biosynthesis would be better suited to the control of immature phytophagous insects than one involving the disruption of metamorphosis with JH analogs (Cusson & Palli, 2000). Although a number of naturally occurring and synthetic inhibitors of JH biosynthesis have been evaluated for their ability to trigger precocious metamorphosis, none has yet been developed commercially. The "precocenes," isolated from the bedding plant (*Ageratum houstonianum*) originally generated much hope and enthusiasm as they caused premature metamorphosis in the milkweed bug (*Oncopeltus fasciatus*). However, these compounds proved to be ineffective against holometabolous insects. Inhibitors of mevalonate pathway enzymes known to have hypocholesterolemic activity in mammals were also tested on insects; examples include the fungal metabolite compactin, an inhibitor of HMG-CoA reductase (Monger *et al.*, 1982), and the fluorinated mevalonate analog, fluoromevalonate, an inhibitor of enzymes involved in the processing of mevalonate (Quistad *et al.*, 1981), both of which were found to induce precocious metamorphosis in lepidopteran larvae, but required high doses and/or repeated applications. Allylic alcohol derivatives of dimethylallyl diphosphate, the C-5 chain initiator for FPP production by FPP synthase (FPPS), provided similar results (Quistad *et al.*, 1985). Inhibitors of later, insect-specific steps of JH biosynthesis, such as formation of the methyl ester moiety and epoxidation, were also examined for their ability to block JH biosynthesis and induce precocious metamorphosis. Brevioxime, a compound with a sesquiterpene-like structure isolated from the fungus *Penicillium brevicompactum*, was shown to inhibit *in vitro* JH biosynthesis by the CA of migratory locus (*Locusta migratoria*) (Castillo *et al.*, 1998). Similarly, the synthetic 1,5-disubstituted imidazole, KK-42, an inhibitor of the P450-linked enzyme that epoxidizes the JH precursor methyl farnesoate (MF), inhibited in vitro JH biosynthesis by the CA of the migratory locust (Castillo *et al.*, 1998) and the Pacific beetle cockroach (*Diploptera punctata*) (Pratt *et al.*, 1990). Again, concentrations of the compounds required to achieve inhibition were relatively high.

Recent developments in the area of insect genomics have led to the cloning and characterization of enzymes specific to JH biosynthesis, including JH acid methyl transferase from the silkworm (Shinoda & Itoyama, 2003) and MF epoxydase from Pacific beetle cockroach (Helvig *et al.*, 2004). The ability to produce these enzymes in heterologous expression systems should lead to substantial improvements in our capacity to assess their three-dimensional structures and design potent and highly specific inhibitors that could be used as anti-JH insecticides. In addition to the above proteins, several mevalonate pathway-specific enzymes have been cloned from various species of insects. Although these may not provide the most suitable target sites for insecticide development, given that they are found in most living organisms, the lepidopteran homolog of at least one of them, FPPS, appears to display significant structural singularities believed to be instrumental in the biosynthesis of the ethyl-substituted JHs (Cusson *et al.*, 2006; Sen *et al.*, 2007); this enzyme may thus prove to be a suitable target for the design of lepidoptera-specific inhibitiors (see Palli & Cusson, 2007 for a more in-depth review).

Identification of a JH receptor is perhaps the most promising avenue for the development of highly effective anti-JH compounds. As indicated above, however, this receptor has so far remained elusive, despite sustained and continuing research efforts. Once a receptor has been isolated, it should become possible to design cell-based high-throughput assays for the screening of potential JH antagonists that are effective at the target tissue level (Minakuchi & Riddiford, 2006; Palli & Cusson, 2007).

22.5 Conclusion

Insecticides targeting endocrine functions have a young history, and it remains to be seen at what rate new products will be successfully brought on the market. Agonists and antagonists of peptidic hormones are currently being investigated, such as cyclic backbone peptides inhibiting the action of pheromone biosynthesis activating neuropeptide (PBAN) (Altstein, 2004). Other suitable targets for peptidomimetics include receptors of PTTH and ecdysis-related peptides (Palli & Cusson, 2007). An important factor in the development of new insecticidal compounds will be the identification of new target sites ranging from hormone receptors to hormone biosynthetic and degradative enzymes. Target site identification has been greatly helped by the complete sequencing of insect genomes, including species of economic importance, e.g. the honeybee (*Apis mellifera*) (Honeybee Genome Sequencing Consortium, 2006); and the silkworm (Xia *et al.*, 2004 [*Bombyx mori* Biology Analysis Group]) as well as vectors of human diseases, e.g. the malaria mosquito (*Anopheles gambiae*) (Holt *et al.*, 2002). Thus the full repertoire of peptidic hormones, and receptors for both peptidic and non-peptidic hormones, can eventually be known for a given sequenced insect genome by mining for genes with endocrine functions. This is exemplified by the discovery of 56 G-protein-coupled receptors (GPCRs) for neurohormones and biogenic amines in the honeybee genome (Hauser *et al.*, 2006). Similarly, *Drosophila* genomics has made possible the identification of the suite of enzymes controlling ecdysone biosynthesis in this species (Gilbert & Warren, 2005). With large-scale DNA sequencing becoming more affordable, additional fully sequenced insect genomes should become available in the near future, including those of agricultural pests.

This large-scale gene identification effort will feed into the strategies currently used in agrochemistry to synthesize and screen compounds against target sites (i.e. proteins). High-throughput instrumentation now allows the screening of 500 to 1000 compounds per day using in vitro assays on a given target (Allenza & Eldridge, 2007). In vitro assays, while allowing a higher throughput than whole-insect assays, are still trial-and-error operations. A target protein needs to be expressed in sufficient quantities and in the proper conformation to retain its in vivo functions in an in vitro context. Furthermore, some hormone receptors require two or more subunits to be functional (e.g. the ecdysone receptor), thus increasing the complexity and cost of some in vitro assays (Allenza &

Eldridge, 2007). To avoid these pitfalls, the discovery of novel hormone-based insecticidal compounds will require an intimate knowledge of the target at the molecular, cellular and whole-organism levels.

IPM specialists will likely witness the introduction of additional, endocrine-based control products in the foreseeable future. New approaches in the way endocrine disruption is delivered are also in development, for instance by using genetically modified plants or microorganisms that interfere with hormone action (Palli & Cusson, 2007). By exploiting the diversity and specificity of insect hormone systems, these new application methods should bring even more environmentally attractive control options for IPM.

References

Addison, J. A. (1996). Safety testing of tebufenozide, a new molt-inducing insecticide, for effects on nontarget forest soil invertebrates. *Ecotoxicology and Environmental Safety*, **33**, 55–61.

Ahmad, M., Hollingworth, R. M. & Wise, J. C. (2002). Broad-spectrum insecticide resistance in obliquebanded leafroller *Choristoneura rosaceana* (Lepidoptera: Tortricidae) from Michigan. *Pest Management Science*, **58**, 834–838.

Allenza, P. & Eldridge, R. (2007). High-throughput screening and insect genomics for new insecticide leads. In *Insecticide Design Using Advanced Technologies*, eds. I. Ishaaya, R. Nauen & A. R. Horowitz, pp. 67–86. Berlin, Germany: Springer-Verlag.

Altstein, M. (2004). Novel insect control agents based on neuropeptide antagonists: the PK/PBAN family as a case study. *Journal of Molecular Neuroscience*, **22**, 147–157.

Anonymous (2002). *Evaluation of the New Active Methoxyfenozide in the Product PRODIGY 240 SC Insecticide.* Canberra, Australia: National Registration Authority for Agricultural and Veterinary Chemicals. Available at www.apvma.gov.au/publications/downloads/prsmeth.pdf/.

Bombyx mori Biology Analysis Group, Xia, Q., Zhou, Z., Lu, C. *et al.* (2004). A draft sequence for the genome of the domesticated silkworm (*Bombyx mori*). *Science*, **306**, 1937–1940.

Brown, M. R., Raikhel, A. S. & Lea, A. O. (1985). Ultrastructure of midgut endocrine cells in the adult mosquito, *Aedes aegypti*. *Tissue and Cell*, **17**, 709–721.

Butenandt, A. & Karlson, P. (1954). Über die Isolierung eines Metamorphosen-Hormons der Insekten in kristallisierter Form. *Zeitschrift fur Naturforschung*, **9B**, 389–391.

Cadogan, B. L., Thompson, D. G., Retnakaran, A. *et al.* (1998). Deposition of aerially applied tebufenozide (RH-5992) on balsam fir (*Abies balsamea*) and its control of spruce budworm (*Choristoneura fumiferana* [Clem.]). *Pesticide Science*, **53**, 80–90.

Cadogan, B. L., Scharbach, R. D., Knowles, K. R. & Krause, R. E. (2005). Efficacy evaluation of a reduced dosage of tebufenozide applied aerially to control spruce budworm (*Choristoneura fumiferana*). *Crop Protection*, **24**, 557–563.

Castillo, M., Moya, P., Couillaud, F., Garcerá, M. D. & Martinez-Pardo, R. (1998). A heterocyclic oxime from a fungus with anti-juvenile hormone activity. *Archives of Insect Biochemistry and Physiology*, **37**, 287–294.

Charrois, G. J. R., Mao, H. & Kaufman, W. R. (1996). Impact on salivary gland degeneration by putative ecdysteroid antagonists and agonists in the ixodid tick *Amblyomma hebraeum*. *Pesticide Biochemistry and Physiology*, **55**, 140–149.

Claeys, I., Simonet, G., Van Loy, T., De Loof, A. & Vanden Broeck, J. (2003). cDNA cloning and transcript distribution of two novel members of the neuroparsin family in the desert locust, *Schistocerca gregaria*. *Insect Molecular Biology*, **12**, 473–481.

Claeys, I., Breugelmans, B., Simonet, G. *et al.* (2006). Regulation of *Schistocerca gregaria* neuroparsin transcript levels by juvenile hormone and 20-hydroxyecdysone. *Archives of Insect Biochemistry and Physiology*, **62**, 107–115.

Cusson, M. (2004). Juvenile hormone. In *Encyclopedia of Entomology*, ed. J. L. Capinera, pp. 1228–1230. Dordrecht, Netherlands: Kluwer.

Cusson, M. & Palli, S. R. (2000). Can juvenile hormone research help rejuvenate integrated pest management? *Canadian Entomologist*, **132**, 263–280.

Cusson, M., Béliveau, C., Sen, S. E. *et al.* (2006). Characterization and tissue-specific expression of two lepidopteran farnesyl diphosphate synthase homologs: implications for the biosynthesis of ethyl-substituted juvenile hormones. *Proteins*, **65**, 742–758.

Davey, M., Duve, H., Thorpe, A. & East, P. (2005). Helicostatins: brain-gut peptides of the moth, *Helicoverpa armigera* (Lepidoptera: Noctuidae). *Archives of Insect Biochemistry and Physiology*, **58**, 1–16.

Davis, N. T., Blackburn, M. B., Golubeva, E. G. & Hildebrand, J. G. (2003). Localization of myoinhibitory peptide immunoreactivity in *Manduca sexta* and *Bombyx mori*, with indications that the peptide has a role in molting and ecdysis. *Journal of Experimental Biology*, **206**, 1449–1460.

Dhadialla, T. S., Carlson, G. R. & Le, D. P. (1998). New insecticides with ecdysteroidal and juvenile hormone activity. *Annual Review of Entomology*, **43**, 545–569.

Dhadialla, T. S., Retnakaran, A. & Smagghe, G. (2005). Insect growth- and development-disrupting insecticides. In *Comprehensive Molecular Insect Science*, vol. 6, eds. L. I. Gilbert, K. Iatrou & S. S. Gill, pp. 55–117. St Louis, MO: Elsevier.

Dinan, L. (2001). Phytoecdysteroids: biological aspects. *Phytochemistry*, **57**, 325–339.

Dinan, L., Whiting, P., Girault, J. P. *et al.* (1997). Cucurbitacins are insect steroid hormone antagonists acting at the ecdysteroid receptor. *Biochemical Journal*, **327**, 643–650.

Doucet, D., Cusson, M. & Retnakaran, A. (2007a). Insect endocrinology and hormone-based pest control products in IPM. Table 1. In *Radcliffe's IPM World Textbook*, eds. E. B. Radcliffe, W. D. Hutchison and R. E. Cancelado. Available at http://ipmworld.umn.edu/textbook/doucet1.htm/.

Doucet, D., Cusson, M. & Retnakaran, A. (2007b). Insect endocrinology and hormone-based pest control products in IPM. Table 2. In *Radcliffe's IPM World Textbook*, eds. E. B. Radcliffe, W. D. Hutchison and R. E. Cancelado. Available at http://ipmworld.umn.edu/textbook/doucet2.htm.

Fussnecker, B. L., Smith, B. H. & Mustard, J. A. (2006). Octopamine and tyramine influence the behavioral profile of locomotor activity in the honeybee (*Apis mellifera*). *Journal of Insect Physiology*, **52**, 1083– 1092.

Gilbert, L. I. & Warren, J. T. (2005). A molecular genetic approach to the biosynthesis of the insect steroid molting hormone. *Vitamins and Hormones*, **73**, 32–59.

Gole, J. W. D. & Downer, R. G. H. (1979). Elevation of adenosine 3′, 5′-monophosphate by octopamine in fat body of the American cockroach *Periplaneta americana* L. *Comparative Biochemistry and Physiology*, **64C**, 223–226.

Hasegawa, K. (1957). The diapause hormone of the silkworm, *Bombyx mori*. *Nature*, **179**, 1300–1301.

Hauser, F., Cazzamali, G., Williamson, M., Blenau, W. & Grimmelikhuijzen, C. J. (2006). A review of neurohormone GPCRs present in the fruitfly *Drosophila melanogaster* and the honeybee *Apis mellifera*. *Progress in Neurobiology*, **80**, 1–19.

Helvig, C., Koener, J. F., Unnithan, G. C. & Feyereisen, R. (2004). CYP15A1, the cytochrome P450 that catalyzes epoxidation of methyl farnesoate to juvenile hormone III in cockroach corpora allata. *Proceedings of the National Academy of Science of the USA*, **101**, 4024–4029.

Henderson J. (2005). Ernest Starling and "hormones": an historical commentary. *Journal of Endocrinology*, **184**, 5–10.

Holt, R. A., Subramanian, G. M., Halpern, A *et al.* (2002). The genome sequence of the malaria mosquito *Anopheles gambiae*. *Science*, **298**, 129–149.

Honeybee Genome Sequencing Consortium (2006). Insights into social insects from the genome of the honeybee *Apis mellifera*. *Nature*, **443**, 931–949.

Hsu, A. C.-T. (1991). 1,2-Diacyl-1-alkyl-hydrazines: a novel class of insect growth regulators. In *Synthesis and Chemistry of Agrochemicals II*, eds. D. R. Baker, J. G. Fenyes & W. K. Moberg, pp. 478–490. Washington, DC: American Chemical Society.

Kethidi, D. R., Li, Y. & Palli, S. R. (2006). Protein kinase C phosphorylation blocks juvenile hormone action. *Molecular and Cellular Endocrinology*, **247**, 127–134.

Kim, Y. J., Žitňan, D., Cho, K. H. *et al.* (2006). Central peptidergic ensembles associated with organization of an innate behavior. *Proceedings of the National Academy of Science of the USA*, **103**, 14 211–14 216.

King-Jones, K. & Thummel, C. S. (2005). Nuclear receptors: a perspective from *Drosophila*. *Nature Reviews Genetics*, **6**, 311–323.

Kogan, M. (1998). Integrated pest management: historical perspectives and contemporary developments. *Annual Review of Entomology*, **43**, 243–270.

Kopec, S. (1917). Experiments on metamorphosis of insects. *Bulletin of International Academy Cracov B*, pp. 57–60.

Kreutzweiser, D. P., Capell, S. S., Wainio-Keizer, K. L. & Eichenber, D.C. (1994). Toxicity of new molt-inducing insecticide (RH-5992) to aquatic macroinvertibrates. *Ecotoxicology and Environmental Safety*, **28**, 14–24.

Lagueux, M., Hoffmann, J. A., Goltzené, F. *et al.* (1984). Ecdysteroids in ovaries and embryos of *Locusta migratoria*. In *Biosynthesis Metabolism and Mode of Action of Invertebrate Hormones*, eds. J. A. Hoffmann & M. Porchet, pp 168–180. Heidelberg, Germany: Springer-Verlag.

Medina, P., Smagghe, G., Budia, F., Tirry, L. & Vinuela, E. (2003). Toxicity and absorption of azadirachtin, diflubenzuron, pyriproxyfen, and tebufenozide after topical application in predatory larvae of *Chrysoperla carnea* (Neuroptera: Chrysopidae). *Environmental Entomology*, **32**, 196–203.

Minakuchi, C. & Riddiford, L. M. (2006). Insect juvenile hormone action as a potential target of pest management. *Journal of Pesticide Science*, **31**, 77–84.

Mommaerts, V., Sterk, G. & Smagghe, G. (2006). Bumblebees can be used in combination with juvenile hormone analogues and ecdysone agonists. *Ecotoxicology*, **15**, 513–521.

Monger, D. J., Lim, W. A., Kezdy, F. J. & Law, J. H. (1982). Compactin inhibits insect HMG-CoA reductase and juvenile hormone biosynthesis. *Biochemical and Biophysical Research Communications*, **105**, 1374–1380.

Nakagawa, Y. (2005). Nonsteroidal ecdysone agonists. *Vitamins and Hormones*, **73**, 131–173.

Neves, C. A., Gitirana, L. B. & Serrao, J. E. (2003). Ultrastructure of the midgut endocrine cells in *Melipona quadrifasciata anthidioides* (Hymenoptera, Apidae). *Brazilian Journal of Biology*, **63**, 683–690.

Nijhout, H. F. (ed.) (1994). *Insect Hormones*. Princeton, NJ: Princeton University Press.

Nordeen, S. K., Ogden, C. A., Taraseviciene, L. & Lieberman, B. A. (1998). Extreme position dependence of a canonical hormone response element. *Molecular Endocrinology*, **12**, 891–898.

Palli, S. R. & Cusson, M. (2007). Future insecticides targeting genes involved in the regulation of molting and metamorphosis. In *Insecticides Design Using Advanced Technologies*, eds. I. Ishaaya, R. Nauen & A. R. Horowitz, pp. 105–134. Berlin, Germany: Springer-Verlag.

Palli, S. R., Tice, C. M., Margam, V. M. & Clark, A. M. (2005). Biochemical mode of action and differential activity of new ecdysone agonists against mosquitoes and moths. *Archives of Insect Biochemistry and Physiology*, **58**, 234–242.

Pratt, G. E., Kuwano, E., Farnsworth, D. E. & Feyereisen, R. (1990). Structure/activity studies on 1, 5-disubstituted imidazoles as inhibitors of juvenile hormone biosynthesis in aisolated corpora allata of the cockroach *Diploptera punctata*. *Pesticide Biochemistry and Physiology*, **38**, 223–230.

Predel, R., Eckert, M. & Holman, G. M. (1999). The unique neuropeptide pattern in abdominal perisympathetic organs of insects. *Annals of the New York Academy of Science*, **897**, 282–290.

Predel, R., Kellner R., Baggerman. G., Steinmetzer, T. & Schoofs, L. (2000). Identification of novel periviscerokinins from single neurohaemal release sites in insects: MS/MS fragmentation complemented by Edman degradation. *European Journal of Biochemistry*, **267**, 3869–3873.

Quistad, G. B., Cerf, D. C., Schooley, D. A. & Staal, G. B. (1981). Fluoromevalonate acts as an inhibitor of insect juvenile hormone biosynthesis. *Nature*, **289**, 176–177.

Quistad, G. B., Cerf, D. C., Kramer, S. J., Bergot, B. J. & Schooley, D. A. (1985). Design of novel insect anti juvenile hormones: allylic alcohol derivatives. *Journal of Agricultural Food Chemistry*, **33**, 47–50.

Rachinsky, A., Mizoguchi, A., Srinivasan, A. & Ramaswamy, S. B. (2006). Allatotropin-like peptide in *Heliothis virescens*: tissue localization and quantification. *Archives of Insect Biochemistry and Physiology*, **62**, 11–25.

Raikhel, A. S., Brown, M. B. & Belles, X. (2005). Hormonal control of reproductive processes. In *Comprehensive Molecular Insect Science*, vol. 3, eds. L. I. Gilbert, K. Iatrou & S. S. Gill, pp. 433–491. Amsterdam, Netherlands: Elsevier.

Retnakaran, A., Smith, W. L., Tomkins, W. L. *et al.* (1997). Effect of RH-5992, a nonsteroidal ecdysone agonist, on the spruce budworm, *Choristoneura fumiferana* (Lepidoptera: Tortricidae): laboratory, greenhouse and ground spray trials. *Canadian Entomologist*, **129**, 871–885.

Retnakaran, A., Gelbic, I., Sundaram, M. *et al.* (2001). Mode of action of the ecdysone agonist tebufenozide (RH-5992), and an exclusion mechanism to explain resistance to it. *Pest Management Science*, **57**, 951–957.

Retnakaran, A., Krell, P., Feng, Q. & Arif, B. (2003). Ecdysone agonists: mechanism and importance in controlling insect pests of agriculture and forestry. *Archives of Insect Biochemistry and Physiology*, **54**, 187–199.

Riddiford, L. M., Hiruma, K., Zhou, X. & Nelson, C.A. (2003). Insights into the molecular basis of the hormonal control of molting and metamorphosis from *Manduca sexta* and *Drosophila melanogaster*. *Insect Biochemistry and Molecular Biology*, **33**, 1327–1338.

Robbins, W. E., Kaplanis, J. N., Thompson, M. J., Shortino, T. J. & Joyner, S. O. (1970). Ecdysones and synthetic analogs: molting hormone activity and inhibitive effects on insect growth, metamorphosis and reproduction. *Steroids*, **16**, 105–125.

Sáenz-de-Cabezón Irigaray, F.-J., Marco, V., Zalom, F. G. & Pérez-Moreno, I. (2005). Effects of methoxyfenozide on *Lobesia botrana* Den & Schiff (Lepidoptera: Tortricidae) egg, larval and adult stages. *Pest Management Science*, **61**, 1133–1137.

Sakai, T., Satake, H., Minakata, H. & Takeda, M. (2004). Characterization of crustacean cardioactive peptide as a novel insect midgut factor: isolation, localization, and stimulation of alpha-amylase activity and gut contraction. *Endocrinology*, **145**, 5671–5678.

Sato, Y., Oguchi, M., Menjo, N. *et al.* (1993). Precursor polyprotein for multiple neuropeptides secreted from the suboesophageal ganglion of the silkworm *Bombyx mori*: characterization of the cDNA encoding the diapause hormone precursor and identification of additional peptides. *Proceedings of the National Academy of Sciences of the USA*, **90**, 3251–3255.

Sauphanor, B. & Bouvier J. C. (1995). Cross-resistance between benzoylureas and benzoylhydrazines in the coddling moth, *Cydia pomonella* L. *Pesticide Science*, **45**, 369–375.

Sen, S. E., Trobaugh, C., Béliveau, C., Richard, T. & Cusson, M. (2007). Cloning, expression and characterization of a dipteran farnesyl diphosphate synthase. *Insect Biochemistry and Molecular Biology*, **37**, 1198–1206.

Shinoda, T. & Itoyama, K. (2003). Juvenile hormone acid methyltransferase: a key regulatory enzyme for insect metamorphosis. *Proceedings of the National Academy of Science of the USA*, **100**, 11 986–11 991.

Sláma K, Romaňuk, M. & Šorm, F. (1974). *Insect Hormones and Bioanalogues*. New York: Springer-Verlag.

Smagghe, G., Carton, B., Decombel, L. & Tirry, L. (2001). Significance of absorption, oxidation, and binding to toxicity of four ecdysone agonists in multi-resistant cotton leafworm. *Archives of Insect Biochemistry and Physiology*, **46**, 127–139.

Stanley, D. (2006). Prostaglandins and other eicosanoids in insects: biological significance. *Annual Review of Entomology*, **51**, 25–44.

Stay, B. & Tobe, S. (2007). The role of allatostatins in juvenile hormone synthesis in insects and crustaceans. *Annual Review of Entomology*, **52**, 277–299.

Sundaram, K. M. S., Nott, R. & Curry, J. (1996). Deposition, persistence and fate of tebufenozide (RH-5992) in some terrestrial and aquatic components of a boreal forest environment after aerial application of mimic. *Journal of Environmental Science, Health B*, **31**, 699–750.

Sundaram, M., Palli, S. R., Ishaaya, I., Krell, P. J. & Retnakaran, A. (1998). Toxicity of ecdysone agonists correlates with the induction of CHR3. *Pesticide Biochemistry and Physiology*, **62**, 201–208.

Truman, J. W. & Copenhaver, P. F. (1989). The larval eclosion hormone neurones in *Manduca sexta*: identification of the brain-proctodeal neurosecretory system. *Journal of Experimental Biology*, **147**, 457–470.

Wearing, C. H. (1998). Cross-resistance between azinophosmethyl and tebufenozide in the greenheaded leaf roller, *Planotortrix octo*. *Pesticide Science*, **54**, 203–211.

Williams, C. M. (1956). The juvenile hormone of insects. *Nature*, **178**, 212–213.

Williams, C. M. (1967). Third-generation pesticides. *Scientific American*, **217**, 13–17.

Žitňan, D. & Adams, M. E. (2005). Neuroendocrine regulation of insect ecdysis. In *Comprehensive Molecular Insect Science*, vol. 3, eds. L. I. Gilbert, K. Iatrou & S. S. Gill, pp. 1–50. Amsterdam, Netherlands: Elsevier.

Žitňan, D., Žitňanová, I., Spalovská, I. *et al.* (2003). Conservation of ecdysis-triggering hormone signalling in insects. *Journal of Experimental Biology*, **206**, 1275–1289.

Chapter 23

Eradication: strategies and tactics

Michelle L. Walters, Ron Sequeira, Robert Staten, Osama El-Lissy and Nathan Moses-Gonzales

There are four main terms with similar yet unique definitions to consider when developing a method of pest management. Eradication is the application of phytosanitary measures to eliminate a pest from an area or geographic region (Food and Agriculture Organization, 2005). Suppression involves maintaining an insect population at or below the economic injury level (Pfadt, 1972; Hendrichs *et al.*, 2002). Containment is the application of phytosanitary measures in and around an infested area to prevent spread of a pest (FAO, 2005). Prevention is the application of phytosanitary measures in and/or around a pest free area to avoid the introduction of a pest. Of the four, only prevention involves a preemptive strategy to keep the pest at bay rather than managing it upon arrival (Hendrichs *et al.*, 2002, 2005; Food and Agriculture Organization, 2005). The other three terms, eradication, suppression and containment, are designed to manage a pest after its initial infestation. These terms are not mutually exclusive; rather, they provide different strategies and tactics designed to custom fit an areawide IPM (AW-IPM) program. The section of this chapter on pink bollworm (*Pectinophora gossypiella*) provides a brief review of San Joaquin Valley, California in regards to prevention, as well as the ongoing efforts to eradicate this invasive lepidopteran insect from the cotton growing regions of the southwest USA and northern Mexico. The section on Mediterranean fruit fly (*Ceratitis capitata*) of this chapter discusses in greater detail the economic viability and opportunity cost of eradication versus suppression in the international market. The final section of this chapter discusses the history, efficiency and effect of the boll weevil (*Anthonomus grandis*) eradication program.

"We have," according to Klassen and Curtis (2005), "entered an era of an unprecedented level of travel by exotic invasive organisms." As a result, we have also entered an era of unprecedented research and development of means and methods designed to reduce and eradicate these exotic invasive organisms. This chapter explores the principles of insect eradication, as well as the means and methods necessary to achieve this daunting task. For our purposes, the term "eradication" is defined as "A type of regulatory-control program in which a target pest is eliminated from a geographical region (Gordh & Headrick, 2001)" versus the one traditionally used for infectious disease, "extinction of the pathogen in humans and/or the environment" (Arita *et al.*, 2004). The reader is directed to Klassen (1989) *Eradication of Introduced Arthropod Pests: Theory and Historical Practice* for an excellent review of the practice in the USA. There is no standard outline or guide to running an eradication program. Instead, many programs rely on organic support structures rather than rigid systems.

Integrated Pest Management, ed. Edward B. Radcliffe, William D. Hutchison and Rafael E. Cancelado. Published by Cambridge University Press.

Motivations for eradication campaigns vary with the situation and pest. Klassen (1989) states that the presence of new pests may bother people directly (bite or sting), vector diseases to humans, livestock or crops, cause extensive direct damage to crops or the environment, cause trade embargos or significantly increase the use of pesticides and consequently damage the environment. To date, exotic pests (insects, mammals, mollusks, weeds and diseases) are responsible for an estimated one third of agricultural losses in the USA. Natural ecosystems may also be adversely affected by invasive exotics. Dutch elm disease (*Ophiostoma novo-ulmi*), chestnut blight (*Cryphonectria parasitica*), pink bollworm, red imported fire ant (*Solenopsis invicta*) and Kudzu vine (*Pueraria lobata*) are but a few examples. People are left with the desire to eradicate the pest, whether introduced or not, as in the case of expanded ranges of boll weevil and screwworm (*Cochliomyia hominivorax*).

The practice of eradication is controversial, as well it should be, due to the uncompromising nature of the practice, the uncertainty of success and the extreme measures needed to achieve it. Moral issues around the eradication of a native species and unintended consequences from eradication efforts including damage to the environment, loss of non-target species including pollinators and rare, native insects (Knipling, 1978; Rabb, 1978) are just a few of the possible negative impacts of eradication programs.

23.1 Is eradication an option?

Klassen concluded that the justification for eradication must be based on the "anticipated economic, ecological, and sociological consequences." In analyzing program efficiency, one must heed the warnings communicated by Myers *et al.* (1998) and avoid known biases and shortcomings commonly associated with economic analyses. Myers *et al.* warn that the benefits of eradication programs are "almost always overestimated" because of lack of scientific data to separate impacts from an eradication program from other, unrelated factors, potentially biased decision processes due to strong stakeholders in industry, or potentially biased evaluations due to primary focus on producers. At the same time, they warned that "in contrast to benefits, biases in procedures often underestimate the costs of proposed eradication programs." Examples of such costs include the escalating expense of eradicating the last individuals, unanticipated impacts on other aspects of society, the need for continuous monitoring of populations, risks of potential reintroductions, public relations, potential lawsuits, costs of human error and risks to human health. With the above caveats, economic analyses may provide information about efficiencies, not about program efficacy. That is an important distinction.

23.1.1 Smallpox prototype

While the focus of this chapter will be on the principles of eradicating an insect from a geographical region, smallpox (*Variola vera*, an infectious disease) provides a prototype of successful eradication. In 1980, smallpox became the first, and to date, only successfully eradicated human disease. The effort took 96 years and was aided by numerous characteristics of the disease. Humans were the sole reservoir, thus immunization of humans stopped transmission of the disease. Patients did not shed the virus after recovery. There were no subclinical infections, that is, infected people displayed symptoms, and disease symptoms were easily and accurately identified, making surveillance effective. The treatment, a vaccine, was nearly 100% effective. In addition, the smallpox eradication program was strongly led by the United Nations World Health Organization and carried out by cooperative member states with funding and logistics, and in a time of relative political stability (Arita *et al.*, 2004). The smallpox prototype serves as a launching pad for the main factors that ensure successful insect eradication programs.

The prerequisites for developing an eradication strategy requires an understanding of life cycle, climatic needs, geographic distribution, mating habits, preadaptation to its new environment, degree of genetic plasticity, number of generations per year, reproductive capacity, ability to compete for niches, natural enemies and host species.

The implementation of programs typically requires the examination, evaluation and/or development of a wide variety of tactics, including agronomic practices, and the availability of powerful methods to suppress the pest, including resistant hosts (through selective breeding or genetic modification), sterile insect technique (SIT), mating disruption, pesticide treatments, insect growth regulators, biological control, host elimination, and cultural controls including host-free periods and crop diversity. Early detection of introduced species, indicating an effective means of survey immediately followed by a decisive plan of action (methods) is beneficial to control losses and minimize costs. In addition to those listed by Klassen (1989), successful eradication benefits from a well-defined and geographically delimited host range. This requires the implementation of new technologies (GPS, GIS, etc.) in tandem with conventional survey techniques (field surveying, direct observation, etc.). Geospatially referenced data, stored in well-designed databases, facilitates documentation and tracking supports decision making surrounding the eradication effort. Also required are survey techniques that gather reliable estimation of pest presence and population size.

The support of large and robust infrastructures is necessary to fund and carry out eradication programs. Broad economic and political support, i.e. areawide effort, along with good communication, record keeping and evaluation are necessary to ensure that program needs are addressed quickly, as in the case of boll weevil in the cotton belt. Klassen (1989) attributes the establishment of eradication as a strategy to C. H. Fernald of University of Massachusetts who was the architect behind the widespread use of the arsenical pesticide Paris Green and other tactics to successfully contain and eradicate the gypsy moth (*Lymantria dispar*) from 1890 to 1901. Local containment was achieved; however, eradication failed due to poor public support and lack of foresight from the federal government. To date, gypsy moth has not been eradicated.

Public perception and education is essential to ensure continuing public support of an eradication program. Education includes outreach programs, publications and the use of mass media (e.g. the internet, television, radio, newspapers, etc.). The public should be informed about the economic costs/benefits of the program, as well as the health and quality of life they may enjoy post-eradication. If eradication is merited and public education is actively pursued, then cooperation by the public should follow suit.

"Adequacy of legal authority" is vital not only to conduct the eradication program, but also for effective regulation, inspection and quarantine programs that are geared to prevent reinfestation. In order to initiate an eradication program successfully, a stakeholder-supported, areawide organization is required. The organization must not be inhibited by geopolitical borders and benefits by boundaries with strong means of enforcement, such as geographic properties that isolate it from rapid reintroduction of the pest. Furthermore, "*Cohesiveness of stakeholders in the private sector and effective stakeholder leadership*," as well as "*strong support of political leaders*" is vital to a successful eradication program (Klassen, 1989).

Finally, realizing that eradication is not always a permanent venture is paramount for an eradication program. A successful program requires long-term maintenance and monitoring systems. Without these systems, the fruits of eradication may be wiped out. Achieving eradication, according to Klassen (1989), requires a zero count of insects for at least ten generations. In order to ensure and maintain a zero insect count, continual surveying, as well as a contingency plan that would successfully eliminate any possible reintroduction of the insect is required.

23.2 Case studies

23.2.1 Pink bollworm

The worldwide spread of pink bollworm from its presumed origin in India is summarized by Ingram (1994). The history of pink bollworm in the USA is summarized by Noble (1969). In brief, pink bollworm arrived in the western hemisphere in infested seed shipped from Egypt to Mexico in 1911 and is currently the key pest of cotton across the Southwest of the USA and northern Mexico. Lepidoptera are among the most destructive insect pests in the world. According to Klassen & Curtis (2005) "lepidopteran larvae cause immense damage to food and forage crops, forests, and stored products." Pink bollworm feeds almost exclusively on cotton (*Gossypium* spp.) and can cause devastating economic loss by dramatically reducing the yield and quality of cotton lint (Pfadt, 1978). The larvae bore into the developing cotton fruit, where they feed on the cotton lint and seeds. The pink bollworm is difficult to control with insecticides because the egg is typically protected under the calyx of the boll and it spends the destructive larval phase inside the cotton boll where it is also well protected. Cultural controls, such as a short growing season, enforced plowdown and a host-free period, have successfully decreased populations (Chu *et al.*, 1992) and are widely used but are insufficient to prohibit economic loss. In years past, pink bollworm has cost growers in counties of Texas as much as $US 52 million in a single season (Pfadt, 1972). Currently, infestations by pink bollworm costs USA cotton producers over $US 32 million each year in control costs and yield losses (National Cotton Council, 2007). In addition to production losses, there are quarantine implications that limit cotton export markets.

Efforts to control damage by pink bollworm began early on. According to James Rudig (Program Manager, California Department of Food and Agriculture), "the Pink Bollworm Program in the San Joaquin Valley is probably the most successful and longest running yet least known areawide Integrated Pest Control (IPC) program in the world." The program has been in continual operation since 1967. Currently the program uses mapping of cotton fields, trapping, population modeling, cultural practices, sterile insect release and occasional use of mating disruption to prevent the establishment of pink bollworm. The program is supported by bale assessments with some federal government support (California Department of Food and Agriculture, 2007).

A six-year (1989–95) areawide program to control pink bollworm in Parker, Arizona demonstrated how careful mapping, trapping and prompt application of pheromone mating disruption could drive a heavy pink bollworm infestation to near zero in that valley (Antilla *et al.*, 1996). Similarly, from 1994 to 2000, a research trial was conducted to determine if pink bollworm could be controlled using the Sterile Insect Technique (SIT), in the heavily infested Imperial Valley of California. Early successes (and a few failures) showed promise and the program was expanded to include the also heavily infested Palo Verde Valley on the California–Arizona border. SIT is advantageous to pink bollworm eradication because the insect is released during the benign, adult phase of its life cycle when it feeds innocuously on nectar. Naturally, mating occurs during the adult portion of the pink bollworm's life cycle. Thus, releasing sterilized adult pink bollworm does not cause further damage to the cotton and yet reduces the amount of progeny. During the course of the SIT trial, *Bacillus thuringiensis* (*Bt*) cotton came into widespread use and was incorporated in the experiment. The trial had proven its point by 1999 – SIT alone or SIT with *Bt* cotton could suppress pink bollworm populations; however, the eradication plan did not pass in Arizona and the trial was ended (Pierce *et al.*, 1995; Staten *et al.*, 1999; Walters *et al.*, 2000). As in Parker, in the year following the end of each trial, pink bollworm populations rebounded (Walters *et al.*, 2000).

After its faltering start in the West, by 2002 pink bollworm eradication was active in Texas and New Mexico, following quickly after the Boll Weevil Eradication Program. Currently, pink bollworm is under eradication in the USA, including the states of Texas, New Mexico, Arizona and California as well as in the northern Mexico state of Chihuahua. Although pink bollworm eradication was achieved at various times and in specific locations, the development of new technologies, such as mating disruption, transgenic crops and SIT, has once again made pink bollworm eradication economically feasible.

The Pink Bollworm Eradication Program requires multi-year support at the grower and federal level, as well as willingness to remain committed to the program (Fig. 23.1). The Pink Bollworm Eradication Program and others like it are funded by a congressional line item. While the US Federal Government provides $US 4–6 million a year for sterile insect rearing and release in the eradication program, the governmental support of the program costs is relatively small compared to the growers' 80% share in the funding (National Cotton Council, 2007). The government also provides scientific support and a framework for the program. Frisvold (2006) contends that, given current economic trends, the current price of seed, insecticide and technological costs, the Pink Bollworm Eradication Program will become

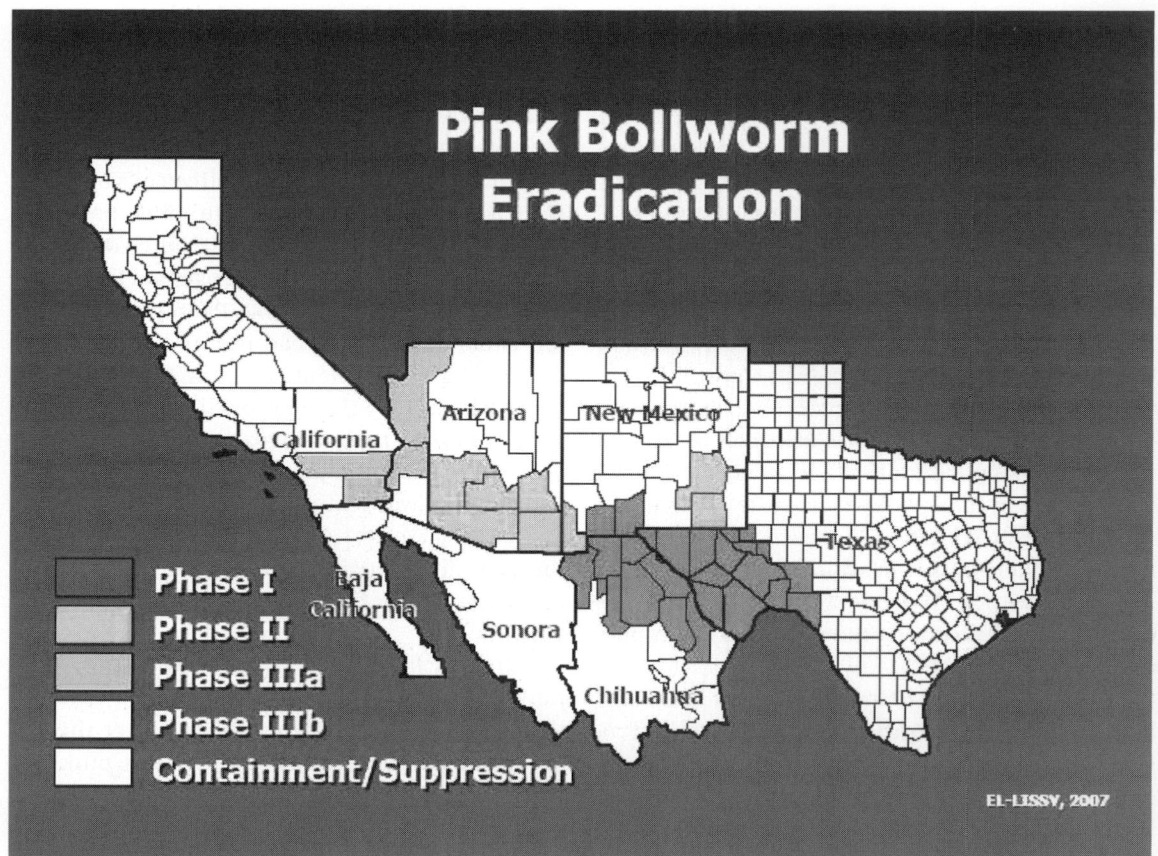

Fig. 23.1 Incremental phases of the APHIS Pink Bollworm Eradication Program.

cost efficient in five years, if non-*Bt* cotton is purchased, and no technology fees are assessed, or in six years if *Bt* cotton is actively purchased and technology fees are assessed.

Early infestations of pink bollworm in the USA underwent cycles of eradication and reintroduction as various counties came under and went out of quarantine regulations. Past government initiatives, such as the Pink Bollworm Act of 1918 did not have the legal backing to enforce cultural and sanitary controls. As a result, cotton growers refused to follow the government programs and law enforcement was unable, and sometimes unwilling, to enforce laws. While many cotton growers wanted pink bollworm out of their fields, infestations continued to plague growers because they were unwilling to follow the Pink Bollworm Act of 1918 (Geong, 2000). Grower ambivalence was a result of a two-prong failure, namely the lack of adequate educational programs established to explain why eradication was vital to the cotton industry and the lack of governmental foresight that ensured adequate legal authority to successfully carry out eradication. Thus, government foresight and adequate policies to ensure perennially funded and implemented programs are paramount to the success of the present pink bollworm eradication effort or any other eradication program. Finally, the crux of any eradication program undertaken on cultivated crops is the predominant role of the grower over governmental authority. This ensures that the grower takes responsibility and provides the critical agricultural input that takes the program out of the laboratory and into the field.

23.2.2 Mediterranean fruit fly

The Mediterranean fruit fly, commonly referred to as "medfly," has been described as the world's most threatening agricultural pest, attacking over 200 different fruits, vegetables and nuts (Thomas *et al.*, 2005). Medfly females deposit their eggs in the epi- or mesocarp region of ripening host fruits. The eggs are laid in clutches of one to ten eggs (up to 800 during its life), larvae pass through three instars feeding on the fruit; they leave the fruit to pupate in the soil. After emergence and before becoming sexually active, adults feed on carbohydrates and water to survive and on protein sources to allow for gonad maturation (Christenson & Foote, 1960; Weems, 1981). The time required to complete a life cycle under summer weather conditions (e.g. in Florida) is 21–30 days. Females die soon after ceasing to oviposit. Development of pupae is variable and an induced quiescence may help this pest survive through unfavorable conditions. Adults can fly short distances, but winds may carry them 2 km or more. According to reports, 50% of the flies that emerge die during the first two months of life; however, adults may survive up to a year or more under favorable conditions. When host fruit is continuously available and weather conditions favorable for many months, successive generations can be large and of mixed age distribution. Lack of fruit for three or four months reduces the population dramatically (Back & Pemberton, 1915, 1918; Weems, 1981). The largely internal life cycle of the feeding immature stages is key to understanding the challenges associated with medfly control.

From their likely origin in sub-Saharan Africa, medflies spread rapidly around the world (Davies *et al.*, 1999; Thomas *et al.*, 2005) largely by human trade in fresh fruits and vegetables. Its frequent incursions worldwide have demonstrated the medfly's destructive capacity and extended host range. In addition to its direct impact in reducing yields, medfly is a quarantine pest for the USA and for many important trading partners. The presence of medfly in a given area implies that fruit products moving out of that area (exports and even domestic movement if the fly population is contained and restricted) need to be treated with expensive phytosanitary treatments such as hot water immersion, fumigation, irradiation cold treatment. Often, the only cost-effective measure is to cease exports to countries that have quarantines against medfly.

Medflies reached the Americas by 1900. Medflies were first reported from Hawaii in 1907 (Harris, 1989), from Florida in 1929, from California in 1975 (Metcalf, 1995; Thomas *et al.*, 2005) and from Texas in 1966. Whereas considered established in Hawaii, medfly has been officially declared eradicated from the continental USA despite periodic outbreaks that have occurred in Florida and California since 1975. Scientists (e.g. Carey, 1991) suggested that California may have an undetectable resident population. Whether eradicated, or established but undetectable, California and Florida have adopted permanent SIT programs to ferret out remaining flies and prevent measurable reinfestation.

Medfly and economic viability: international benefits of eradication

Each new country that is invaded by a spreading disease or pest represents a new node for the invasive organism and adds strength to the invasion process, complicating global management efforts. An analogy is made to the internet where each node (server) adds redundancy and stability to a highly interconnected system. The fact that trade and human traffic have become so widespread strengthens the analogy between the internet and epidemiology. An infestation in a given country has direct implications to the likelihood of spread to all other countries – increasing probability of spread.

In the following equation, Global Welfare (GW) is the value to all nations engaged in open trading systems that accrue from the reduced probability of pest introduction that result from eradication of a pest in a given nation.

$$GW_i = \frac{\sum_j \text{Export value for host commodities}}{\sum_{i,j} \text{Global export value}}$$

where

i = country index
j = commodity index.

The fruit fly GW_i value is the proportion of total commerce in fruit fly susceptible commodities that is consumed by a given country (i). The numerator is the theoretical maximum for investment in global eradication efforts. In practice the maximum investment in eradication is approximated by the actual losses due to fruit fly. A constant global value is used here to minimize the variation across commodities and conditions, such that:

$$\text{Maximum eradication investment}_l = \frac{\sum_j \text{Export value for host commodities}}{\sum_{i,j} \text{Global export value}} * \text{Global losses}_l$$

where

$l =$ year index.

Despite the generalizations (e.g. we do not include domestic pest management or eradication costs), those expressions can be used to explore the value that eradication in a third country has for one's own country. It can also be used to determine (in a global economy) what share of the total eradication costs in a country with limited sanitary and phytosanitary infrastructure should be borne by countries that benefit from exports in that commodity and are susceptible to that pest (fruit flies in this example). This is important because when national plant protection organizations conduct cost–benefit studies, the benefits considered are only those that accrue to an individual country or even an individual state, such as Hawaii (Mumford, 2005). In the GW context, however, the benefits of the absence of fruit flies in Hawaii are enjoyed mostly by mainland America. The concept of GW is applied to fruit flies here but is of general applicability.

International benefits of regional suppression

The statutory discretion of US Department of Agriculture Animal Plant Health Inspection Service (APHIS) was established in part with a view to increase and sustain exports as well as to ensure that when trade occurs, phytosanitary risks are minimized. The establishment of officially recognized, managed "low prevalence areas" could be beneficial in that there is clearly great potential to integrate with systems approaches in support of expanded export markets. The use of areawide suppression approaches is compatible with both of those guidelines (Food and Agriculture Organization, 2006).

The analysis of effectiveness and efficiency indicators confirms technological viability of eradication and areawide suppression (Dyck *et al.*, 2005), albeit with eradication having greater uncertainties and technological limitations than areawide suppression. The analyses of efficiency show the areawide suppression options routinely have significantly lower initial costs, but higher long-term maintenance costs than the eradication option. However, the assumption that eradication is "forever" runs counter to recent history with more than 15 outbreaks of tephritid fruit flies in the past two decades in the USA. Eradication always shows the highest benefits over the long term (compared to areawide management or the status quo), if one assumes that no new introductions occur. If eradication is not maintained over the long term, then costs can overrun benefits depending on the frequency of new introductions. Consequently, suggestions have been forwarded that the fruit fly control program in Hawaii should shift from eradication and toward areawide management (Mitchell & Saul, 1990). There is ample evidence that areawide programs for fruit fly management are successful in reducing populations, allowing increased yields, and in some cases, permitting the export of fruit under specific quality controls, monitoring and implementation of appropriate international standards such as cited above.

23.2.3 Boll weevil

The boll weevil, a native of Mexico and Central America, was first introduced into the USA near Brownsville, Texas, in about 1892 (Hunter, 1905; Hunter & Hinds, 1905). By 1922, the weevil had spread into cotton-growing areas of the USA from the eastern two-thirds of Texas and Oklahoma to the Atlantic Ocean. Northern and western portions of Texas were colonized by the boll weevil between 1953 and 1966 (Newsom & Brazzel, 1968). The history of boll weevil and its effects on the cotton industry as well as efforts to manage or eradicate the boll weevil have been reviewed by Klassen (1989), El-Lissy *et al.* (1996), Haney *et al.* (1996) and Dickerson *et al.* (2001).

The Boll Weevil Eradication Program in the USA began in 1893, or one year after the boll weevil was first observed in Brownsville. The boll weevil was immediately recognized as an invasive, though not exotic pest species, which was expanding its range northward from Mexico into Texas and the southern states. Fast action was taken and various methods were implemented; however, these early efforts failed to stop the spread of boll weevil. While the boll weevil was an extremely devastating pest economically, it contributed to the westward spread of the cotton industry and to the twentieth century shift in the southeast from a cotton monoculture towards a wider variety of crops. The presence of boll weevil also contributed greatly to the development of entomology in the southern USA (Hardee & Harris, 2003). A monument was created in Enterprise, Alabama to commemorate the role of the boll weevil in forcing crop diversity in the region.

The boll weevil's main host is cotton. Boll weevils survive winter as diapausing adults (Hardee & Harris, 2003). Nearly 50% of female boll weevils can store sperm, thus eliminating the necessity to find a mate before laying eggs after re-emergence (Beckham, 1962) so that even harsh winters, where populations may be severely depleted tend not to eliminate the population. These characteristics contribute to the invasive nature of the boll weevil.

Early suppression methods included the aerial application of arsenic dusts and chlorinated hydrocarbons; however, by the 1950s boll weevil developed resistance to these tactics (Hardee & Harris, 2003). In view of the economic and environmental problems posed by the boll weevil and its control, and in recognition of the technical advances developed during 80-plus years of research, most notably by E. F. Knipling, J. R. Brazzel and T. B. Davich, a cooperative boll weevil eradication experiment was initiated in 1971 in southern Mississippi and parts of Louisiana and Alabama (Parencia, 1978; Perkins, 1980). "The ultimate goal" of this project "was to discover or develop a tactic or combination of tactics that would provide a long-term solution to the boll weevil problem in cotton production" (Hardee & Harris, 2003). This experiment used an area-wide IPM approach including chemical control, release of sterile male weevils, mass trapping and cultural controls. Based on this experiment, the National Cotton Council of America concluded that it was technically and operationally feasible to eliminate the boll weevil from the USA. By 2007, the boll weevil had been eradicated from nearly 5.3 million hectares of cotton in: Virginia, North Carolina, South Carolina, Georgia, Florida, Alabama, Kansas, California and Arizona, and portions of Tennessee, Mississippi, Missouri, Arkansas, Louisiana, Oklahoma, Texas and New Mexico, as well as from the neighboring regions of the Mexicali Valley, Sonoita and Caborca in Mexico (Fig. 23.2). The program is currently operating in the remaining 1.42 million hectares of cotton in Tennessee, Mississippi, Missouri, Arkansas, Louisiana, Oklahoma, Texas and New Mexico. As of 2007, 100% of the USA Cotton Belt is involved in boll weevil eradication, with nearly 80% having completed eradication and the remaining 20% nearing eradication. Nationwide eradication in the USA is expected in 2008.

The operational success of the current eradication program hinges on three interdependent components: mapping, detection and decisions resulting in well-timed application of control methods. Mapping is one of the first phases of operation. Mapping identifies the exact location of each cotton field and defines the surrounding environment. Additionally, each field is identified with a unique number to provide for accurate data management. Chemical control consists of a single aerial application of malathion (ultra-low-volume, ULV) beginning at the pinhead-square growth stage to fields that had reached the treatment criteria (action threshold). By preemptively targeting pinhead-squares before sustainability is achieved, adult emergence numbers are reduced (Hardee & Harris, 2003). The 2004 season-long action threshold for treatment was a trap catch of 1–2 adult boll weevils per field (16 ha or less) in all active zones. Chemical treatment is the principal method of suppressing the boll weevil population.

All eradication zones use the boll weevil pheromone trap as the primary tool of detection (Cross, 1973; El-Lissy *et al.*, 1996). Although the primary function of the trap is detection, an indirect benefit of trapping, especially in low weevil populations, is that it removes a percentage of the population (Lloyd *et al.*, 1972). The Boll Weevil Eradication Program provides an amalgamation of various techniques including automated trap data collection and a database designed to provide decision support.

Time-frames for uniform cotton planting and harvesting, as organized by growers, local agricultural extension services, and in some cases state regulatory agencies, are key components of

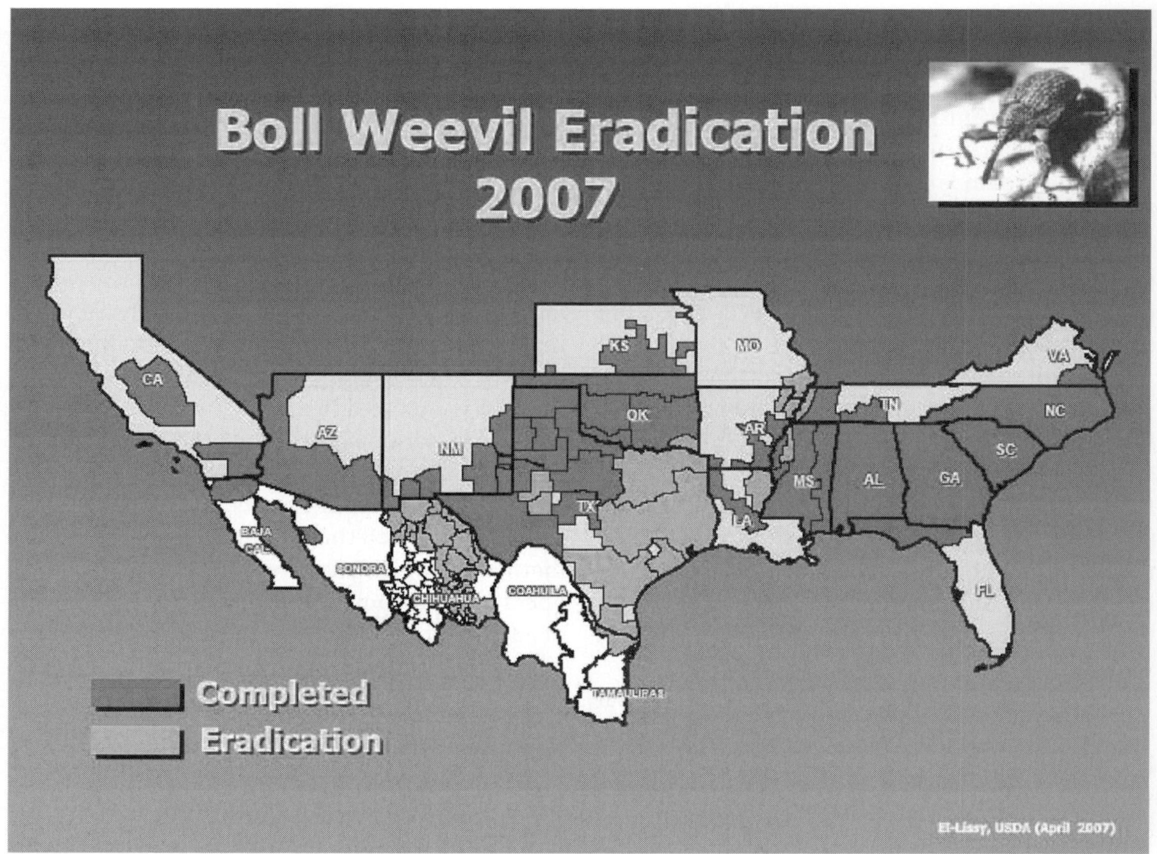

Fig. 23.2 Progress of the APHIS Boll Weevil Eradication Program.

cultural control in providing the necessary host-free period. In some states such as Arkansas and Texas, growers were offered a rebate to destroy crop residues as soon as possible after harvest in an effort to reduce overwintering populations and insecticide treatments. During the 2005 diapause phase, the Rio Grande Valley and the Northern Blacklands of Texas underwent weekly aerial applications with malathion ULV. These treatments began on 15 June and 15 July, respectively, and continued until cotton fields were defoliated and harvested.

Trapping in post-eradication zones provides early warning of boll weevil reintroduction, from natural migration or artificial movement. Early detection allows an immediate response in containing and eradicating the reintroduced population before it reestablishes. Post-eradication trapping will continue until nationwide eradication is complete, at which time a reduced trapping density will be put in place.

23.3 Conclusions

Success of an eradication program depends on an overwhelming desire to eradicate an invasive insect pest. Effective programs such as the Boll Weevil Eradication Program experience perennial public and governmental support and funding and were promoted by educational outreach and stakeholder involvement. Well-documented research and dedicated professionals in both the scientific and farming communities help lead to a clear, well-defined program with a solid foundation of knowledge regarding to an insect pest's biology, life cycle, climatic needs, mating behaviors, geographic distribution, etc. Finally, the confluence of desire, availability of suitable methods, funding, identification of weak links in the armor of the pest, tremendous hard work and an ounce of luck are the keys to a successful eradication program.

References

Antilla L., Whitlow, M., Staten, R. T., El-Lissy, O. & Myers, F. (1996). An integrated approach to areawide pink bollworm management in Arizona. In *Proceedings of the Beltwide Cotton Conference*, **2**, pp. 1083–1085. Memphis, TN: National Cotton Council.

Arita, I., Wickett J. & Nakane, M. (2004). Eradication of infectious diseases: its concept, then and now. *Japanese Journal of Infectious Diseases*, **57**, 1–6.

Back, E. A. & Pemberton, C. E. (1915). Life history of the Mediterranean fruit fly from the standpoint of parasite introduction. *Journal of Agricultural Research*, **3**: 363–374.

Back, E. A. & Pemberton, C. E. (1918). *The Mediterranean Fruit Fly*, US Department of Agriculture Bulletin No. 640. Washington, DC: US Government Printing Office.

Beckham, C. M. (1962). *Seasonal Studies of Diapause in the Boll Weevil in Georgia*, Mimeograph Series N.S. No. 161. Athens, GA: University of Georgia, Georgia Agricultural Experiment Station.

California Department of Food and Agriculture (2007). *Pink Bollworm: Program Details*. Sacramento, CA: California Government. Available at www.cdfa.ca.gov/phpps/ipc/pinkbollworm/pbw_hp.htm.

Carey, J. R. (1991). Establishment of the Mediterranean fruit fly in California. *Science*, **253**, 1369–1373.

Christenson, L. D. & Foote, R. H. (1960). Biology of fruit flies. *Annual Review of Entomology*, **5**, 171–192.

Chu, C. C., Weddle, R. C., Staten, R. T. *et al.* (1992). Pink bollworm: populations two years following initiation of a short-season cotton system in the Imperial Valley, CA. In *Proceedings: Beltwide Cotton Production and Research Conference*, pp. 804–806. Memphis, TN: National Cotton Council.

Cross, W. H. (1973). Biology, control and eradication of the boll weevil. *Annual Review of Entomology*, **18**, 17–46.

Davies, N., Villablanca, F. X. & Roderick, G. K. (1999). Bioinvasions of the Medfly *Ceratitis capitata*: source estimation using DNA sequences at multiple intron loci. *Genetics*, **153**, 351–360.

Dickerson, W. A., Brashear, A. L. Brumley, J. T. *et al.* (2001). *Boll Weevil Eradication in the United States through 1999*. Memphis, TN: Cotton Foundation.

Dyck, V., Hendrichs, J. & Robinson, J. (eds.) (2005). *Sterile Insect Technique*. New York: Springer-Verlag.

El-Lissy, O., Myers, F., Frisbie, R. *et al.* (1996). Boll weevil eradication status in Texas. In *Proceedings of the Beltwide Cotton Production and Research Conference*, pp. 831–839. Memphis, TN: National Cotton Council.

Food and Agriculture Organization (2005). *International Standards for Phytosanitary Measures: Glossary of Phytosanitary Terms (updated 2007)*, ISPM No. 5. Produced by the Secretariat of the International Plant Protection Convention. Rome, Italy: Food and Agriculture Organization of the United Nations.

Food and Agriculture Organization (2006). *International Standards for Phytosanitary Measures (ISPM)*, 2005 edn. Produced by the Secretariat of the International Plant Protection Convention. Rome, Italy: Food and Agriculture organization of the United Nations. Available at www.fao.org/docrep/009/a0450e/a0450e00.htm.

Frisvold, G. (2006). *Economics of Pink Bollworm Eradication*. Memphis, TN: Cotton Incorporated.

Geong, H.-G. (2000). The pink bollworm campaign in the South: agricultural quarantines and the role of the public in insect control, 1915–1930. *Agricultural History*, **74**, 309–321.

Gordh, G. & Headrick, D. H. (eds.) (2001). *A Dictionary of Entomology*. Wallingford, UK: CABI Publishing.

Haney, P. B., Lewis, W. J. & Lambert, W. R. (1996). *Cotton Production and the Boll Weevil in Georgia: History, Cost of Control, and Benefits of Eradication*, Research Bulletin No. 428. Athens, GA: University of Georgia, Georgia Agricultural Experiment Station.

Hardee, D. D. & Harris, F. A. (2003). Eradicating the boll weevil. *American Entomologist*, **49**, 82–97.

Harris, E. J. (1989). Pest status in Hawaiian islands and North Africa. In *World Crop Pests: Fruit Flies, Their Biology, Natural Enemies and Control*, vol. 3A, eds. A. Robinson & G. Hooper, pp. 73–80. Amsterdam, Netherlands: Elsevier.

Hendrichs, J., Robinson, A. S., Cayol, J. P. & Enkerlin, W. (2002). Medfly areawide sterile insect technique programmes for prevention, suppression or eradication: the importance of mating behavior studies. *Florida Entomologist*, **85**, 1–13.

Hendrichs, J., Vreysen, M. J. B., Enkerlin, W. R. & Cayol, J. P. (2005). Strategic options in the use of the sterile insects for area-wide integrated pest management. In *Sterile Insect Technique*, eds. V. Dyck, J. Hendrichs & J. Robinson, pp. New York: Springer-Verlag.

Hunter, W. D. (1905). *The Control of the Boll Weevil, Including Results of Recent Investigations*, US Department of Agriculture Farmers' Bulletin No. 216. Washington, DC: US Government Printing Office.

Hunter, W. D. & Hinds, W. E. (1905). *The Mexican Cotton Boll Weevil*, USDA Agricultural Handbook

No. 512. Washington, DC: US Government Printing Office.

Ingram, W. R. (1994). *Pectinophora* (Lepidoptera: Gelechiidae). In *Insect Pests of Cotton*, eds. G. A. Mathews & J. P. Tunstall, pp. 107–148. Wallingford, UK: CAB International.

Klassen, W. (1989). *Eradication of Introduced Arthropod Pests: Theory and Historical Practice*, Miscellaneous Publication No. 73. Lanham, MD: Entomological Society of America.

Klassen, W. (2005). Area-wide integrated pest management and the sterile insect technique. In *Sterile Insect Technique*, eds. V. A. Dyck, A. J. Hendrichs & A. S. Robinson, pp. 39–68. New York: Springer-Verlag.

Klassen, W. & Curtis, C. F. (2005). History of the Sterile Insect Technique. In *Sterile Insect Technique*, eds. V. Dyck, J. Hendrichs & J. Robinson, pp. 3–36. New York: Springer-Verlag.

Knipling, E. F. (1978). Strategic and tactical use of movement information in pest management. In *Conference on Radar, Insect Population Ecology and Pest Management*, NASA Conference Publication No. 2070, pp. 41–57.

Lloyd, E. P., Merkl, M. E., Tingle, F. C. *et al.* (1972). Evaluation of male-baited traps for control for boll weevils following a reproduction-diapause program in Monroe County, Mississippi. *Journal of Economic Entomology*, **65**, 552–555.

Metcalf, R. L. (1995). Biography of the Medfly. In *The Medfly in California: Defining Critical Research*, eds. J. G. Morse, R. L. Metcalf, J. R. Carey & R. V. Dowell, pp. 43–48. Riverside, CA: University of California, Center for Exotic Pest Research.

Mitchell, W. C. & Saul, S. H. (1990). Current control methods for Mediterranean fruit fly, *Ceratitis capitata*, and their application in the USA. *Review of Agricultural Entomology*, **78**, 923–940.

Mumford, J. D. (2005). Application of benefit/cost analysis to insect pest control using the sterile insect technique. In *Sterile Insect Technique*, eds. V. Dyck, J. Hendrichs & J. Robinson, pp. 481–498. New York: Springer-Verlag.

Myers, J. H., Savoie, A. & van Randen, E. (1998). Eradication and pest management. *Annual Review of Entomology*, **43**, 471—491.

National Cotton Council (2007). *Pink Bollworm Eradication: Proposal and Current Status*. Available at www.cotton.org/tech/pest/bollworm/index.cfm.

Newsom, L. D. & Brazzel, J. R. (1968). Pests and their control. In *Advances in Production and Utilization of Quality Cotton: Principles and Practices*, eds. F. C. Eliot,M. Hoover & W. K. Porter Jr., pp. 365–405. Ames, IA: Iowa State University Press.

Noble, L. W. (1969). *Fifty Years of Research on the Pink Bollworm in the United States*, US Department of Agriculture Agricultural Handbook No. 357. Washington, DC: US Government Printing Office.

Parencia, C. R. Jr. (1978). *One Hundred Twenty Years of Research on Cotton Insects in the United States*, US Department of Agriculture Agricultural Handbook No. 515. Washington, DC: US Government Printing Office.

Perkins, J. H. (1980). Boll weevil eradication. *Science*, **207**, 1044–1050.

Pierce, D. L., Walters, M. L. Patel, A. J. & Swanson, S. P. (1995). Flight path analysis of sterile pink bollworm release using GPS and GIS. In *Proceedings of the Beltwide Cotton Conference*, **2**, pp. 1059–1060. Memphis, TN: National Cotton Council.

Pfadt, R. (1972). *Fundamentals of Applied Entomology*, 2nd edn. New York: Macmillan.

Pfadt, R. E. (1978). Insect pests of cotton. In *Fundamentals of Applied Entomology*, 3rd edn, ed. R. E. Pfadt, pp. 369–403. New York: Macmillan.

Rabb, R. L. (1978). Eradication of plant pests: con. *Bulletin of the Entomological Society of America*, **24**, 40–44.

Staten, R. T., Walters, M., Roberson, R. & Birdsall, S. (1999). Area-wide management/maximum suppression of pink bollworm in Southern California. In *Proceedings of the Beltwide Cotton Conference*, pp. 985–988. Memphis, TN: National Cotton Council.

Thomas, M. C., Heppner, J. B., Woodruff, R. E. *et al.* (2005). *Mediterranean Fruit Fly*. Originally published as DPI Entomology Circulars 4, 230 and 273, 2001; revised December 2005. Available at http://creatures.ifas.ufl.edu/fruit/Mediterranean_fruit_fly.htm.

Walters, M. L., Staten, R.T., Roberson, R.C. & Tan, K. H. (2000). Pink bollworm integrated management using sterile insects under field trial conditions, Imperial Valley, California. In *Area Wide Control of Fruit Flies and Other Insect Pests*, Joint Proceedings of the International Conference on Area Wide Control of Insect Pests, May 28–June 2, 1998 and the Fifth International Symposium on Fruit Flies of Economic Importance, Penang, Malaysia, June 1–5, 1998., pp. 201–206. Pulau Pinang, Malaysia: Penerbit Universiti Sains Malaysia.

Weems, H. V. (1981). *Mediterranean Fruit Fly*, Ceratitis capitata *(Wiedemann) (Diptera: Tephritidae)*, Entomology Circular No. 230. Tallahassee, FL: Florida Department of Agriculture and Consumer Services, Division of Plant Industry.

Chapter 24

Insect management with physical methods in pre- and post-harvest situations

Charles Vincent, Phyllis G. Weintraub, Guy J. Hallman and Francis Fleurat-Lessard

In theory, IPM programs should be an optimal blend of science (knowledge) and technologies – used concomitantly or sequentially – to manage pests below an economic injury level. There are five main approaches available to achieve that goal: chemical control (synthetic and naturally derived), biological control (predators, parasitoids and pathogens), cultural control (including cover crops and genetically resistant plants), physical control and human factors (legal restrictions on commodities, quarantines, etc.) (Vincent *et al.*, 2003). In practice, few technologies are blended into most pest management programs. For both pre- and post-harvest pest control, the primary approach worldwide is chemical/fumigation (Fields & White, 2002). Like any technology, chemical control has its merits and limits; the development of resistance to pesticides by some arthropod populations, environmental contamination and tightening of regulations in registration and restrictions of use are among factors that limit the use of chemical control measures. However, human factors are playing an effective role in movement towards truly integrated control programs. For example, there have been new regulations enacted in North America (US Food and Drug Administration, 2004) and the European Economic Community (European Union, 2002) for hygienic food quality and safety (Table 24.1). According to these new legislative measures, every food or feed product destined for trade must be free of arthropod pests. This requirement is the standard for food sanitary quality and hygiene of general application in international exchanges as established by the World Trade Organization (WTO).

Physical control methods have been used for millennia. However, in the last two decades a revival of interest in alternative and sustainable pest control methods has prompted research projects in various agricultural contexts for both pre- and post-harvest pest control. Physical control encompasses: (1) an array of techniques (including applied engineering) and (2) knowledge from various scientific domains, notably physiology and population ecology.

24.1 Physical control techniques

According to their mode of action, physical control methods can be active or passive (Table 24.2). The level of efficacy of active methods is proportional to both intensity of the energy and duration of its application to the target. Examples of active methods include thermal shock (heat, cold), electromagnetic radiations (microwaves, radio frequencies, infrared, ionizing radiations, UV and

Integrated Pest Management, ed. Edward B. Radcliffe, William D. Hutchison and Rafael E. Cancelado. Published by Cambridge University Press.

Table 24.1 Recent European Union (EU) regulation dealing with the Food Quality and Safety Act

Scope of EU regulation	Official Regulations publication references
Hygiene "package"	Regulation 852/2004 on the hygiene of foodstuffs
Good processing and manufacturing hygiene practices	Regulation 178/2002: legislation about the respect of minimum hygiene requirements in food processing industries all along the food chain
Effective Hazard Analysis and Critical Control Point (HACCP) procedures application for good hygiene and wholesomeness of food	Regulation 178/2002: food business operators should establish and operate food safety programs and procedures based on HACCP and supported by the Guidelines for Good Hygiene Practices (GGHP) to enforce the regulation in all food chains
Traceability of identity preserved products along the food chain	Regulation 882/2004: implementation of official controls in every EU country to check food business operators compliance with respect to the minimum requirements to be provided for food safety

visible light), mechanical shock and pneumatic (blowing or vacuum). Most active methods have a short persistence as the stressor effect is limited to the period of application. This could be an advantage, as in post-harvest situations or situations where the environment should be minimally disturbed, or a disadvantage, in situations where the treatment must be repeated several times to achieve prolonged control. Passive methods do not require further energy to achieve desired effect. Examples are traps, airtight or hermetic storage, barriers of various kinds and trenches.

24.2 Pest physiology and population ecology

Empirical use of physical control is acceptable but sustainable usage requires knowledge and this can only be accomplished through a thorough understanding of the biology and population dynamics of the pest. A case in point is the management of insects by using thermal sensitivity; one must determine the sensitivity of the pest and its substrate (either living or dead) to various temperatures to know whether thermal management is even feasible (Hallman & Denlinger, 1998). If an active physical control measure is to be used, knowledge of the population dynamics of the pest and the economic threshold of the crop are required to know when to apply the measures.

24.3 Practical considerations

In general, physical control methods are relatively more labor intensive and often time consuming when applied pre-harvest. The implementation of physical methods compares favorably with, and can augment, biological methods. However, most successes occur in post-harvest situations. Effective use of physical control measures, in both pre- and post-harvest control, relies on three main components: (1) good sanitation practices – which should be applied with all control methods, (2) monitoring to determine extent and stage of pest populations and (3) application of other direct or indirect physical control methods. An implementation scheme of these control measures for post-harvest protection is shown in Table 24.3.

The effects of physical control methods (with the exception of electromagnetic radiation) are limited spatially. This attribute starkly contrasts with chemicals which may drift several hundred meters and may be bioaccumulated in food chains, or with biological control agents which may disperse actively or passively over long distances. Post-harvest IPM is greatly facilitated by the fact that it takes place inside a well-defined

Table 24.2 Classes, subclasses and key examples of physical control methods

Class (subclass)	Key examples of target insects[a]	Context and comments[b]
Passive		
Airtight or hermetic storage	Stored product insects (A, L, P, E)	PH
Aluminum foil	Aphids (A)	Pr; repels aphids
Fences	Anthomyiid flies (A), carrot rust fly (*Psila rosae*) (A)	Pr
Flooding	Cranberry insects, vegetable pests	Pr; cranberry
Inorganic (plastic) mulch	Tarnished plant bug (*Lygus lineolaris*) (L, A), anthomyiid flies	In strawberry
Organic mulch	Melonworm (*Diaphania hyalinata*), Colorado potato beetle (*Leptinotarsa decemlineata*) (E, L)	Pr; indirect effect on natural enemies, positive interaction with *Bt*
Packaging	Insects of transformed products (A, L, P, E)	PH
Screening	Various insect species (A)	Greenhouse, orchards
Slippery surfaces	Ants, cockroaches, Colorado potato beetle (A)	Pr; made of fluon, teflon or dust
Sticky barriers	Codling moth (*Cydia pomonella*) (L), tent caterpillars (L)	Pr
Trapping	Various dipteran pests – apple maggot fly (*Rhagoletis pomonella*) (A), Mediterranean fruit fly (*Ceratitis capitata*) (A), Colorado potato beetle (A)	Pr; fruit orchards
Trench	Chinch bug (*Blissus leucopterus*) (A), Colorado potato beetle (A)	Pr
Tripple-bagging	Cowpea insects (A, L, P, E)	PH
Windbreak	Aphid vectors (A)	Pr
Active		
Mechanical		
Dislodging	Plum curculio (*Conotrachelus nenuphar*) (A)	Pr; in apple orchards
Disturbing	Stored product insects (A, L, P, E)	Pr; fluidized bed
Forced air	(A, L)	PH; increase insecticide effect
Inert dusts	Stored product insects (L, A)	PH; including diatomaceous earth, silica dust
Infrasound	Various insect species	Pr
Mechanical impacts	Stored product insects (A, L, P, E)	PH, Entoleter
Mechanical polishing	Rice weevil (*Sitophilus oryzae*)	In rice
Mineral and vegetable oils	Phytophagous mites, soft bodied insects	Pr, tree fruits
Overhead irrigation	Diamondback moth (*Plutella xylostella*)	Pr
Particle films (kaolin)	Codling moth, leafrollers, mites (L)	Pr
Physical removal of hosts	Apple insects (A)	Removal of host plants
Plant residue removal	Millet stem borer (*Coniesta ignefusalis*)	
Pneumatic	(L, P) Colorado potato beetle (A), *Lygus* spp. (L, A),	Pr; vacuuming, blowing
Pruning	Obliquebanded leafroller (*Choristoneura rosaceana*) (L)	Pr

(cont.)

Table 24.2 (*cont.*)

Class (subclass)	Key examples of target insects[a]	Context and comments[b]
Pulsed ultrasound	Various insect species	Pr
Sieving	Insects of peanuts	PH
Surface tension agents and sufactrants	Phytophagous mites (A, L)	Pr; soaps
Ultrasound	Various insect species	Pr
Vacuumized packaging	Weevils	Polyester film bags
Thermal		
Burning	Maize stalk borer (*Busseola fusca*)	PH; partial stalk burning
Flaming	Colorado potato beetle (A)	Pr
High temperatures	Stored product insects	PH
Hot water–steam	Fruit (E, L)	PH
Infrared heating		PH
Low temperatures	Stored product insects	PH
Post-harvest chilling	Stored product insects	PH
Rapid freezing	Stored product insects	PH
Solar heating	Cowpea weevil (*Callosobruchus maculatus*)	PH; stored grains
Steam	Colorado potato beetle (A)	PH; potato
Electromagnetic		
Ionizing radiation	Stored product insects	PH
Microwave	Stored product insects	Pr, PH
Radio frequencies	Stored product insects	PH
UV and visible light	Stored product insects	
Modified atmospheres		
CO_2		PH
Inert gases	Stored product insects	PH
N		PH
Plastic sheeting	Stored product insects	PH

[a] A, adult; E, eggs; L, larvae; P, pupae.
[b] PH, post-harvest; Pr, pre-harvest.
Source: Adapted from Supplemental material of Vincent *et al.* (2003). Reprinted, with permission, from the Annual Review of Entomology, 48 © 2003 by Annual Reviews.

area, delimited by the structures of industrial buildings (food processing facility, grain storage elevator, food factory, feed mill, etc.). As a consequence of all of these aspects, physical control methods occupy a smaller market share than pesticides worldwide. For further information on physical control methods in agricultural plant protection, we refer the reader to the reviews of Banks (1976), Oseto (2000), Vincent *et al.* (2001, 2003) and Weintraub & Berlinger (2004).

24.4 Pre-harvest control measures

24.4.1 Exclusion barriers

Insect exclusion screening is probably the single most important physical control method developed in the last century. Following the invasion of virus-bearing whiteflies, tomato crops in the entire Mediterranean region could not be

Table 24.3 Development of an IPM system as a part of the HACCP procedures preferably using physical preventative and control means: effective IPM based on biophysical methods applicable in post-harvest situations

Inspection and identification of infestation risks on plant structure and material	Detection, surveillance and monitoring systems	Prevision of infestation risk by pest population dynamics models	Implementation of preventive measures and control vs pre-established limits	Application of corrective treatments for the limitation of damage	Replanning the prevention plan from new data or experience
Visual	Visual	Species identification	Cleaning and sanitation of surroundings and inside plant	Heat disinfestation	Use of a decision support system based on expertise
	UV light traps	Simulation of population kinetics	Manipulation of physical conditions	Temporary freezing	Replanning of optimized strategies
	Pheromone traps	Specific monitoring measures	Mass-trapping	Microwave or radio-frequency heating	Training the personnel in charge of IPM strategy application
			Complementary measures for pest exclusion from food plant by physical barriers	Modified atmosphere packaging	Information of employees about their implication into the IPM system
				Controlled atmosphere storage	
				Cold storage	
				Combination of physical treatments	

grown in open fields from late spring through fall. The development of screens allowed tomatoes to be produced year-round, and the use of screens has become a standard pest management practice worldwide. This form of physical control has proven cost-effective both for consumers and growers (Weintraub & Berlinger, 2004).

Usage of exclusion screening is not limited to the traditional greenhouse crops; orchards are increasingly being covered to prevent pests from gaining access to trees. Netting which excludes fruit flies has proven to be an economically effective method of protecting peaches and nectarines, increasing yield quantity and quality (Nissen *et al.*, 2005a, b). Certain diseases, such as papaya dieback, can only be controlled by covering plantations with coarse white nets to limit vector movement (Franck & Bar-Joseph, 1992). Bananas are grown under screening in many parts of the world.

Another example of an exclusion barrier in northeastern North America involves the simultaneous management of weeds in apple orchards and two insect pests, plum curculio (*Conotrachelus nenuphar*) and apple sawfly (*Hoplocampa testudinea*) (Benoit *et al.*, 2006). Both insects lay their eggs in fruitlets where the larvae develop. In late June infested fruitlets fall to the ground and mature larvae enter the soil to pupate. In field experiments over a four-year period, cellulose sheeting prevented most weeds from emerging (Benoit *et al.*, 2006). Total weed density was significantly lower in plots covered with cellulose sheeting as compared to control (no sheeting) plots. However accumulation of soil on the sheeting allowed a few weeds to grow. Likewise, emergence of adults from fallen apples covered with a cage was significantly reduced for both plum curculio and apple sawfly. However, some individuals completed their development and successfully overwintered on the sheeting. For example, in spite of cellulose sheeting, the percentage of plum curculio emergence varied from 0.7% to 18.9% in one orchard compared to 9.3% to 63.5% in the control. Over the years, the integrity of cellulose sheeting has been challenged by accumulation of plant debris and water, causing biodegradation of the sheeting, allowing weeds and insects to penetrate through damaged surfaces.

The potential effect of exclusion barriers on weeds depends on the size of the seed bank at the beginning and the viability of the seeds. If one uses a resistant material, these results imply that over medium to long term (five to ten years), populations of both insects would decline in absence of nearby (<200 m) host plants. Such conditions may be difficult to achieve in commercial orchards in an agricultural context. However, the scheme should be useful in suburban areas where strict pesticide use is often enforced and hosts are rare.

With the worldwide phasing out of methyl bromide as a pre-plant fumigant, a tremendous amount of research has gone into finding alternatives. It is doubtful that any chemical will emerge as a solitary replacement, but physical methods, such as solarization, are effective and gaining usage. Soil, covered with transparent polyethylene for a month or more and often watered to increase solarization, effectively kills a variety of pathogens (Katan *et al.*, 1976) and insect pests (Katan, 1981). Once thought to be effective only in tropical and subtropical environments, solarization has been demonstrated to be effective in northern Italy (Tamietti & Valentino, 2001) and as far north as Oregon (Pinkerton *et al.*, 2000).

24.4.2 Pneumatic control

In pneumatic control, insects can be dislodged from plants with negative (aspiration) or positive (blowing) air pressure, then killed by a system of turbines or collected and killed upstream in a dedicated system of the blower. With respect to the plant to be protected, distinction must be made between flexible and rigid plants. In the latter, only blowing can be used efficiently, otherwise the plant can be damaged by the machinery.

Reviews on pneumatic control from an entomological perspective have been provided by Vincent & Boiteau (2001), Weintraub & Horowitz (2001) and Vincent (2002). Reviews discussing engineering aspects were published by Khelifi *et al.* (2001) and Lacasse *et al.* (2001). Most papers thus far published on pneumatic control have focused on the tarnished plant bug (*Lygus lineolaris*) on strawberry (a flexible plant), Colorado potato beetle (*Leptinotarsa decemlineata*) on potato (a semi-rigid plant), pea leafminer (*Liriomyza huidobrensis*) in celery and whiteflies (*Bemisia tabaci*) on potatoes and melon.

At the present time, pneumatic control cannot be used as the sole control method in agricultural systems and further research and design improvements are necessary before implementation. Major drawbacks include soil compaction resulting from numerous passes and increasing cost of fossil fuel in open-field situations. In a few European greenhouses, we have observed vacuum machines that utilize tracts along side-mobile watering systems. In these cases there is no concern for soil compaction and no tractor operator is needed, as the system moves automatically from one end of the greenhouse to the other. Among the research avenues that should be explored further to make pneumatic control amenable to commercial operations are field behavior of the insects, design of efficient machines and fine-tuning of operational parameters.

24.4.3 Shredding of apple leaves

Physical control methods typically affect a broad number of pest species; however, few studies have examined the effects of physical control methods on pests in two taxonomic classes. A notable exception is shredding fallen apple leaves to manage fungal and lepidopteran pests (Vincent *et al.*, 2004).

Apple scab (*Venturia inaequalis*) is a fungus that overwinters on fallen apple leaves; it is a major threat to apple orchards worldwide. Typically, it is managed with 6–12 fungicide treatments per season. Spotted tentiform leafminer (*Phyllonorycter blancardella*), a secondary pest in apple orchards of Europe and North America, has several generations per year. In some regions, populations have become resistant to an array of insecticides, rendering management difficult. The insect overwinters as a pupa in mined leaves. Several parasitoids attack and overwinter with the leafminer (Bishop *et al.*, 2001).

Shredding of fallen leaves in autumn, combined with 5% urea treatments (which accelerates decomposition of the leaves), reduced ascospore production by 92% (Vincent *et al.*, 2004). Leaf shredding can also be combined with antagonist fungi, e.g. *Athelia bombacina* or *Microsphaeropsis ochracea*, which compete with apple scab spores. All leaf-shredding treatments caused significant decrease of apple scab inoculum as compared to the control. As pseudothecia are small (90–150 μm), direct effect of shredding must be ruled out as a mechanism to kill the fungus. Leaf shredding also significantly decreased overwintering populations of spotted tentiform leafminer by mechanical killing of the pupae. However, its parasitoids were also significantly affected. A major challenge in implementation is timing. Efficient leaf shredding can only be achieved when leaves are completely and naturally fallen and before snow cover impedes shredding operations.

24.4.4 Dusts

Dusts and powders have been known, for decades, to desiccate arthropods; as such, they were used directly on pests and often included in pesticide formulations pre-1945. What is novel is the use of kaolin (aluminum silicate hydroxide) in a particle-film technique (Glenn *et al.*, 1999). With this technique, plant surfaces are coated with a fine layer of kaolin particles which disrupt arthropod recognition of plant surfaces (Puterka *et al.*, 2000), resulting in reduced oviposition and feeding. A wide range of insects have been shown to be affected, including psyllids and rust mites (Puterka *et al.*, 2000), aphids, spider mites and leafhoppers (Glenn *et al.*, 1999); various Lepidoptera (Knight *et al.*, 2000; Unruh *et al.*, 2000) and beetles (Lapointe, 2000; Thomas *et al.*, 2004). In addition to the disruption of arthropod pests, some fungal pathogens are also killed and heat stress is reduced in trees (Thomas *et al.*, 2004). One drawback of this physical protection is its water solubility; it must be reapplied after rains. Another drawback concerns the whitish residues left on fruit; some growers are reluctant to have such residues on their fruit at harvest. These residues may be washed, but that increases production cost.

24.5 Impact of post-harvest pest infestations

Insect pests are a perennial problem in grain storage facilities, food processing plants, feed mills and warehouses where they mainly infest dry plant products such as cereal foods. An insect infestation in raw agricultural products is considered a biological hazard because of the high rate of increase of insect populations. For instance, insect infestations are one of the major causes of product downgrading and rejection of grain deliveries to grain stores, terminal elevators or processors. However, pests may contaminate either the intermediate or finished products at any point, and they may affect sanitation of facilities and premises and products in those facilities. Presence of stored product insects in processed food products is unacceptable and may cause a general negative impression about the skills of the manufacturer in the application of good hygienic practices, and thereby harm the trademark image.

Post-harvest phytosanitary treatments usually are not thought of in the context of IPM because they have been viewed as stand-alone or "corrective" measures. In reality they do operate as part of a pest management system preventing the spread of invasive species, and current thought is to concentrate more on phytosanitary treatments as part of a system ensuring quarantine security. A major point that has hampered a more integrated development of phytosanitary treatments is that control must be 100% to prevent the introduction of invasive species. A fundamental precept of integrated approaches is the concept of a threshold level that allows for a certain number of pests to survive.

Sanitation must be considered in the wider ecosystem management scheme as many potential invasive species are not traditional pests of cultivated, economic plants but are severe threats to the overall ecology. Nevertheless, recent failures in sanitary treatments (Hallman, 2007) have highlighted this fact. Natural ecosystems are valued for their more intangible benefits (such as air and water detoxification, regeneration of soil fertility, and production and maintenance of biodiversity) arising from our innate affinity with nature (Wilson, 1984) and also provide very tangible economic benefits of absorption and recycling of waste products, maintenance of conditions sustainable to human development, and evolution and storage of products useful to humanity (Perrings *et al.*, 2000).

24.6 Post-harvest control measures

24.6.1 Mechanical injury

Insects in stored grain or cereal flour are often killed mechanically during transportation by pneumatic conveyors as the result of violent repeated shocks of kernels against metal ducts. In flour mills, both grain and flour can be disinfested by passing through "entoleters" that use centrifugal force to throw the grain against a steel surface (Fields *et al.*, 2001). When this equipment is activated on the grain stream prior to milling, infested kernels break apart and are separated from intact kernels. This equipment is most often used to kill insects or eggs infesting fresh flour before its packaging or bulk storage (Stratil *et al.*, 1987).

24.6.2 Wash and wax

Fruit coatings and waxes are known to function as modified atmosphere treatments against tephritid fruit fly immatures in fruit by reducing oxygen and raising carbon dioxide levels inside fruit (Hallman, 1997). However, when coatings are applied to fruit infested with small surface pests the mode of action is primarily physical; the organism becomes adhered and unable to move. Coatings form a major component of an integrated treatment against the mite *Brevipalpus chilensis* on cherimoya, lime and passion fruit, from Chile to the USA (Animal and Plant Health Inspection Service, 2007). Another component of this treatment is a physical wash with soap and water to remove many of the mites. Washing and other physical forms of removal may also serve as phytosanitary treatments for external feeders on durable fruits. The cleaned fruit must then pass phytosanitary inspection.

24.6.3 Exclusion screens

Pest-proofing food plants is of paramount importance before undertaking a sanitation program. For instance, insect-proof screens should be installed on all the open windows and appropriate insect exclusion systems should be placed at plant gates (e.g. pulse-air curtains or automatic closure doors). The interior design of buildings and disposition of equipment in stored product warehouses and food processing plants should facilitate surveillance of insect presence and limit room-to-room exchanges (Russo *et al.*, 2002). Insect movement could be further limited by zoning or separating buildings and workrooms that have different activities or process different products.

At the end of the food-plant chain, the product is generally packaged for temporary or long-term insect-proof protection. The maximum period spent in the distribution channel may reach 18 months or more and insect-proof packaging must protect food up to the official expiration date for consumption. Consequently, packaged foods are often submitted to insect resistance bioassays to test effects of aging on the insect-resistance properties (Davey & Amos, 2006). Low temperature below the development threshold (e.g. 15 °C) drastically reduces the potential of increase of a majority of post-harvest pest species or may even stop their development. Thus, the air-conditioning in the different workrooms of a food factory at temperatures below 18 °C will appreciably reduce the risks of pest levels increasing.

24.6.4 Mass trapping

Intense trapping systems should be installed before there are high infestations. In temperate climates this means that trapping systems should be in place by the end of the cold season. Either UV light or pheromone traps may be used for mass trapping targeted pest species. This technique is recommended to delay the use of conventional pest control procedures against moth pests.

24.6.5 Ionizing irradiation

Exposure of commodities to ionizing radiation at absorbed doses of 50–400 Gy causes physical breaks in molecules, especially larger ones such as DNA, resulting in their inability to function correctly and prevention of continued development (Hallman, 2001). Although most quarantine pests are not killed rapidly by doses in this range they will not successfully reproduce.

24.6.6 Cooling aeration in grain storage

Most of the insect species living in post-harvest storage areas are of tropical or subtropical origin and have fairly high temperatures for optimal development (Fleurat-Lessard, 2004). Low temperatures reduce the rate of increase which can go down to zero when the temperature declines below the development threshold (Sinha & Watters, 1985). The population growth of all grain insect species is inhibited when temperature falls below 10 °C. Times required to achieve virtually 100% kill for treatments near 0 °C are at least 10 days and may be as long as 42 days, which is tolerated by commodities such as apples that are stored for months at temperatures within this range.

The cooling process is often slow due to insulating properties of grains, and thus reducing grain temperature to the target level of 10 °C requires several aeration periods (Lasseran *et al.*, 1994; Fields *et al.*, 2001). In temperate climate situations, each cooling step generally requires one to two weeks of aeration, mainly achieved during the night. In Mediterranean or subtropical climatic situations, the grain elevators may be equipped with refrigeration units to cool the grain even though ambient air temperature is above the grain bulk temperature (Fields *et al.*, 2001).

Grain cooling aeration became very popular in Europe mainly because it is a means of preservation of stored grain during a year without the use of persistent insecticide treatment. Thus, cooling aeration is extensively used for the storage of organic wheat or rye, for which any post-harvest insecticidal treatment on grain is forbidden. The cost of cooling aeration is about $US 0.22/mT, i.e. slightly higher than the cost of an insecticidal treatment ($US 0.165/mT), and it is much lower than phosphine fumigation ($US 1.32/mT in Europe).

24.6.7 Ultra low temperatures

For the disinfestation of high-value food commodities such as "organic" dry beans, temporary freezing can also be used before marketing. The disinfestation of dry beans from the bruchid *Acanthoscelides obtectus* is obtained after 50 h exposure to −23 °C in a rapid-cooling chamber (Dupuis *et al.*, 2006). The winter temperatures prevailing in Canada are sufficiently low to kill insects in grain silos equipped with cooling aeration systems. Grain insects cannot survive temperature levels below −15 °C more than one month (Fields *et al.*, 2001). Cooling grain to −15 °C or below in Canada during winter is easily achievable and is a cheap physical control method for grain pests.

24.6.8 Heat disinfestation

At the high end of the thermobiological scale, insect pests cannot survive temperatures higher than 60–63 °C for more than a couple of minutes (Fields, 1992; Fleurat-Lessard, 2005). Pests also have difficulties surviving at permanent temperature regimes higher than 40 °C (Banks and Fields, 1995). They can survive only less than 2 h at 50 °C, and this temperature level is generally targeted in heat disinfestation of cereals in primary processing plants. In practice quarantine pests are often shielded from direct heat and heating a large commodity load requires time to bring up to lethal temperatures, requiring treatment times up to 16 h for some heated air treatments (Hallman, 2007).

After the ban on use of methyl bromide for killing pests in bins, silos and hard-to-access storage and food processing facilities, dry heat treatment became the most popular technique. Heat

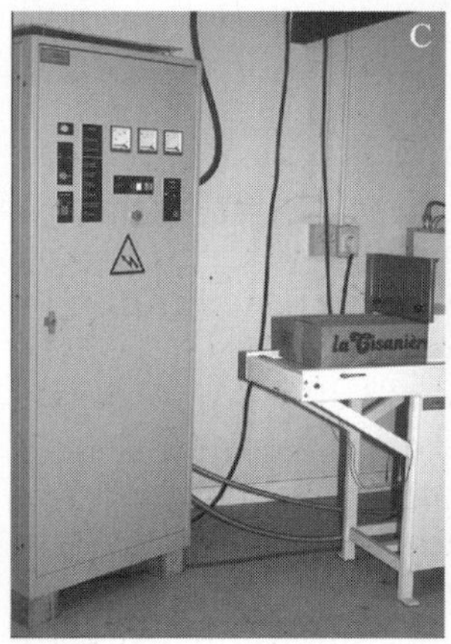

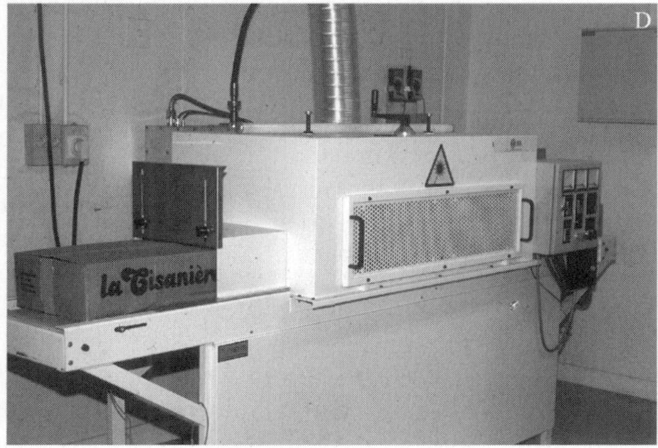

Fig. 24.1 (A, B) Autoclave of 80 m^3 capacity for the disinfestation of packaged palletized pet food by exposure to CO_2 under high pressure (2 MPa). (C, D) Heat treatment of packaged aromatic plants in a radio-frequency oven (27.12 MHz, 6kW radio-frequency power) as an insect-free assurance before product distribution.

treatment consists in raising the temperature of the food processing facilities to 50–55 °C and maintaining these elevated temperatures for at least 36 h to kill stored product insects. Two different types of heating equipment were developed: electric heaters and gas burners.

Disinfestation of a food commodity may be obtained by thermal treatment if the commodity is heat-tolerant. Dry foods are a dielectric material with low thermal conductivity. Consequently, the rapid heating of this material up to the lethal temperature for food pests needs specific technological solutions. Two techniques are currently used commercially for rapid heating of foodstuffs: (1) heated-air fluidized bed (or heated air pneumatic fluid lift) and (2) dielectric heating by using microwave or radio-frequency electric fields. For high-temperature fluidized bed treatments, effectiveness of disinfestation depends on a number of factors: product moisture content, inlet air temperature, air-flow/product-flow ratio, and the pest to be controlled (Fleurat-Lessard & Le Torc'h, 2001). Pilot and industrial-scale heated-air fluidized beds were developed in Australia for the disinfestation of export grain (Dermott & Evans, 1978). To control insect eggs in semolina, the product is heated during pneumatic conveyance with air at 156–200 °C for 6–7 seconds (Fleurat-Lessard & Le Torc'h, 2001). The semolina heats to a maximum of 70 °C, before being cooled with ambient air before storage.

Processes have been developed to control food insect pests using microwaves (2.45 GHz) or radio-frequency electric radiations (13.56, 27.12 or 40.68 MHz). Although application to disinfestation of food commodities remains a marginal application of dielectric heating, pilot-scale equipment has been developed for disinfestation of spices, vegetal material, dried fruits, nut fruits, etc. (Nelson, 1996; Wang *et al.*, 2003) (Fig. 24.1A). At industrial scales, the energy cost of treating more than 4.5 mT of in-shell nuts to kill larvae of codling moth (*Cydia pomonella*) or Indianmeal moth (*Plodia interpunctella*) is estimated at $US 0.23/kg (Wang *et al.*, 2003). This technique offers many advantages as compared to heated-air treatments in terms of the design of the treatment units and for the possibility of online automated treatment directly on the flow of the food commodity. However, installation of large-scale radio-frequency heat treatment facilities is capital intensive, independent of the cost of energy input needed to produce heat, and it should be reserved for high-added-value foodstuffs (semolina, rice, dried fruits, nut fruits, spices, herbs, aromatic plants).

24.6.9 Hermetic enclosure and controlled atmospheres

Storing grain in hermetically sealed structures causes a progressive depletion of oxygen by natural grain respiration and can also kill any pests and prevent reinfestation. Inert gas may be introduced in storage structures to accelerate the oxygen depletion rate and increase the speed of pest mortality. There are two different ways to control stored grain insects with inert atmospheres: the injection of controlled atmosphere obtained from an exothermic inert gas generator or the injection of carbon dioxide gas at a concentration higher than 40% (v/v) (Fleurat-Lessard, 1990).

24.6.10 Combination of different physical methods

The effectiveness of combining different physical treatments to obtain a synergistic effect was demonstrated in two major protocols: combination of heat and modified atmosphere (MA), and high pressure and CO_2. With the first combination of heat and MA, there is an increase in insect mortality rate due to enhanced respiratory demand (Fleurat-Lessard, 1990). For packaged food, the elevation of temperature during exposure to MA may be obtained by the heating of plastic-film-packaged food by radio-frequency heating. It induces a significant increase in the rate of insect mortality (Fleurat-Lessard & Le Torc'h, 2001). With the second combination of high pressure (2 MPa) and high concentration of CO_2, the complete kill of stored product insects at all stages can be observed after less than 2 h exposure time (Fleurat-Lessard & Le Torc'h, 2001). Industrial equipment has been built on this principle and are in use in European Union countries for quick disinfestation of spices or pet food (Fig. 24.1).

24.7 Implementation of IPM procedures in a decision support system

Ability to predict an organism's potential abundance and distribution is an invaluable tool, and provides an opportunity to be 'one step ahead' of the pest. These models may be used in computer-assisted decision support systems (DSS); one such model is CLIMEX (Sutherst & Maywald, 1985). Using predictive models is also part of a broader concept of precision agriculture. Precision agriculture allows site-specific management of pests and diseases. If the abundance of a pest can be predicted, control measures can be put in place to prevent populations from reaching economic injury levels (see Chapter 3).

From the knowledge acquired of biological characteristics of the most damaging stored product pest species, in a very large range of development conditions, simulation models of temperature-dependent growth rates have been developed (Beckett *et al.*, 1994) in order to determine the safe storage period, or to predict the increase of the population density with time (Flinn & Hagstrum, 1990; Williams *et al.*, 2006). Some of these DSSs have been built especially for the implementation of IPM programs in post-harvest systems (Wilkin & Mumford, 1994). The more recent developments in DSSs for IPM in post-harvest situations have dealt with stored grain preservation in silos, including: in Canada, the software CanStore® (Mani *et al.*, 2001); in the USA, Stored Grain Adviser Pro® (Flinn *et al.*, 2006); in the UK, GrainPlan® software (Williams *et al.*, 2006); in Australia, PestMan® expert system (Longstaff & Cornish, 1994); and in France, QualiGrain© (Ndiaye, 2001). Each DSS was developed for grain quality management, including pest contamination risk.

24.8 Conclusions

Today, there is a common will worldwide to limit the pest problems at all steps of the pre- and post-harvest food chain by expanding IPM applications. If managers fully considered the "value-added" aspect of IPM implementation instead of the simpler conventional/chemical pest control tactics, we believe that physical control measures would be more broadly used. However, legislation, such as the phasing out of the fumigant methyl bromide, is pushing the search for safer alternative control measures.

The implementation of physical control measures in food storage or processing facilities needs a certain amount of preparedness and investment. This is especially true for the monitoring and the assessment of the indicators of pest presence and population dynamics trends with time. Application of physical control measures requires the recruitment of trained personnel with high levels of knowledge about bionomics and behavior of the major pest species. More often, starting an IPM program requires an improvement of the design of the structure and the modification of material layout to facilitate the implementation of different components. There are generally moderate costs associated with optimization of the facilities. After this first stage, the practical application of an IPM program should be continuously updated by the use of modern tools facilitating the sanitation procedures as well as the interpretation of monitoring data. The other obstacle to application is to fit the recommendations included in the codes for good hygiene and sanitation practices to the specific constraints and needs of a particular food processing or manufacturing facility (see Chapter 32). This difficulty may be overcome with the increasing availability of computer software dedicated to the training of IPM practitioners or giving valuable advice for correct practical implementation and customization of IPM plans. These decision support systems may also contain expert knowledge accessible to the questions and inquiries from the users.

Although it is easy to demonstrate that the balance in investment/efficiency is favorable in most cases, managers are often reluctant to invest a lot of money in IPM program development. However, since the publication of a new regulation dealing with food quality and safety assurance in all food production chains, the position of managers about the IPM system implementation will have to change for the benefit of all.

References

Animal and Plant Health Inspection Service (2007). *Treatment Manual*. Washington, DC and Riverdale, MD: US Department of Agriculture. Available at www.aphis.usda.gov/import_export/plants/manuals/ports/downloads/ treatment.pdf

Banks, H. J. (1976). Physical control of insects: recent developments. *Journal of the Australian Entomological Society*, **15**, 89–100.

Banks, H. J. & Fields, P. G. (1995). Physical methods for insect control in stored grain ecosystems. In *Stored-Grain Ecosystems*, eds. D. S. Jayas, N. D. G. White & W. E. Muir, pp. 353–409. New York: Marcel Dekker.

Beckett, S. J., Longstaff, B. C. & Evans, D. E. (1994). A comparison of the demography of four major stored grain coleopteran pest species and its implications for pest management. In *Proceedings of the 6th International Working Conference on Stored Product Protection*, vol. 1, eds. E. Highley, E. J. Wright, H. J. Banks & B. R. Champ, pp. 491–497. Wallingford, UK: CABI Publishing.

Benoit, D. L., Vincent, C. & Chouinard, G. (2006). Management of weeds, apple sawfly (*Hoplocampa testudinea* Klug) and plum curculio (*Conotrachelus nenuphar* Herbst) with cellulose sheets. *Crop Protection*, **25**, 331–337.

Bishop, S. D., Smith, R. F., Vincent, C. *et al.* (2001). Hymenopterous parasites associated with *Phyllonorycter blancardella* (Lepidoptera: Gracillariidae) in Nova Scotia and Quebec. *Phytoprotection*, **82**, 65–71.

Davey, P. M. & Amos, T. G. (2006). Testing of paper and other sack materials for penetration by insects which infest stored products. *Journal of the Science of Food and Agriculture*, **12**, 177–187.

Dermott, T. & Evans, D. E. (1978). An evaluation of fluidized-bed heating as a means of disinfesting wheat. *Journal of Stored Product Research*, **14**, 1–12.

Dupuis, A. S., Fuzeau, B. & Fleurat-Lessard, F. (2006). Feasibility of French beans disinfestation based on freezing intolerance of post-embryonic stages of *Acanthoscelides obtectus* (Say) (Col.: Bruchidae). In *Proceedings of the 9th International Working Conference on Stored Product Protection*, eds. I. Lorini, B. Bacaltchuk, H. Beckel *et al.*, pp. 956–965. Passo Fundo, RS, Brazil: Brazilian Post-Harvest Association (ABRAPOS).

European Union (2002). Regulation (EC) 178/2002 of the European Parliament and of the Council of 20/01/02 laying down the general principles and requirements of food law, establishing the European Food Authority, and laying down procedures in matters of food safety. *Official Journal of the European Communities L 31* 01/02/02.

Food and Drug Administration (2004). *Federal Food, Drug, and Cosmetic Act*. Washington, DC: US Government Printing Office. Available at www.fda.gov/opacom/laws/fdcact/fdctoc/htm.

Fields, P. G. (1992). The control of stored-product insects and mites with extreme temperatures. *Journal of Stored Product Research*, **28**, 89–118.

Fields, P. G. & White, N. D. G. (2002). Alternatives to methyl bromide treatments for stored-product and quarantine insects. *Annual Review of Entomology*, **47**, 331–359.

Fields, P. G., Korunic, Z. & Fleurat-Lessard, F. (2001). Control of insects in post-harvest: inert dusts and mechanical means. In *Physical Control Methods in Plant Protection*, eds. C. Vincent, B. Panneton & F. Fleurat-Lessard, pp. 248–257. Berlin, Germany: Springer-Verlag.

Fleurat-Lessard, F. (1990). Effect of modified atmospheres on insects and mites infesting stored products. In *Food Preservation by Modified Atmospheres*, eds. M. Calderon & R. Barkai-Golan, pp. 21–38. Boca Raton, FL: CRC Press.

Fleurat-Lessard, F. (2004). Stored grain: pest management. In *Encylopedia of Grain Science*, eds. C. Wrigley, H. Corke & C. Walker, pp. 244–254. Amsterdam, Netherlands: Elsevier.

Fleurat-Lessard, F. (2005). Ecophysiological basis of insect physical control methods. In *Phytosanitary Challenge for Agriculture and Environment*, ed. C. Regnault-Roger, pp. 787–804. Paris, France: Paris Lavoisier Technique et Documentation (in French).

Fleurat-Lessard, F. & Le Torc'h, J.-M. (2001). Control of insects in postharvest: high temperature and inert atmospheres. In *Physical Control Methods in Plant Protection*, eds. C. Vincent, B. Panneton & F. Fleurat-Lessard, pp. 74–94. Berlin, Germany: Springer-Verlag.

Flinn, P. W. & Hagstrum, D. W. (1990). Stored grain advisor: a knowledge-based system for management of insect pests of stored grain. *AI Applications in Natural Resource Management*, **4**, 44–52.

Flinn, P., Opit, G. P. & Throne, J. E. (2006). Integrating the stored grain advisor Pro expert system with an automated electronic grain probe trapping system. In *Proceedings of the 9th International Working Conference on Stored Product Protection*, eds. I. Lorini, B. Bacaltchuck, H. Beckel *et al.*, pp. 408–413. Passo Fundo, RS, Brazil: Brazilian Post-Harvest Association (ABRAPOS).

Franck, A. & Bar-Joseph, M. (1992). Use of netting and whitewash spray to protect papaya plants against Nivun–Haamir (NH)-die back disease. *Crop Protection*, **11**, 525–528.

Glenn, D. M., Puterka, G. J., Vanderzwet, T., Byers, R. E. & Feldhake, C. (1999). Hydrophobic particle films: a new paradigm for suppression of arthropod pests and plant diseases. *Journal of Economic Entomology*, **92**, 759–771.

Hallman, G. J. (1997). Mortality of Mexican fruit fly (Diptera: Tephritidae) immatures in coated grapefruits. *Florida Entomologist*, **80**, 324–328.

Hallman, G. J. (2001). Irradiation as a quarantine treatment. In *Food Irradiation: Principles and Applications*, ed. R. A. Molins, pp. 113–130. New York: John Wiley.

Hallman, G. J. (2007). Phytosanitary measures to prevent the introduction of invasive species. In *Biological Invasions*, ed. W. Nentwig, pp. 367–384. Heidelberg, Germany: Springer-Verlag.

Hallman, G. J. & Denlinger, D. L. (eds.) (1998). *Temperature Sensitivity in Insects and Application in Integrated Pest Management*. Boulder, CO: Westview Press.

Katan, J. (1981). Solar heating (solarization) of soil for control of soilborne pests. *Annual Review of Phytopathology*, **19**, 211–236.

Katan, J., Greenberger, A., Alon, H. & Grinstein, A. (1976). Solar heating by polyethylene mulching for the control of diseases caused by soil-borne pathogens. *Phytopathology*, **66**, 683–688.

Khelifi, M., Laguë, C. & Lacasse, B. (2001). Pneumatic control of insects in plant protection. In *Physical Control Methods in Plant Protection*, eds. C. Vincent, B. Panneton & F. Fleurat-Lessard, pp. 261–269. Berlin, Germany: Springer-Verlag.

Knight, A. L., Unruh, T. R., Christianson, B. A., Puterka, G. J. & Glenn, D. M. (2000). Effects of a kaolin-based particle film on obliquebanded leafroller (Lepidoptera: Tortricidae). *Journal of Economic Entomology*, **93**, 744–749.

Lacasse, B., Laguë, C., Roy, P.-M. *et al.* (2001). Pneumatic control of agricultural pests. In *Physical Control Methods in Plant Protection*, eds. C. Vincent, B. Panneton & F. Fleurat-Lessard, pp. 282–293. Berlin, Germany: Springer-Verlag.

Lapointe, S. L. (2000). Particle film deters oviposition by *Diaprepes abbreviatus* (Coleoptera: Curculionidae). *Journal of Economic Entomology*, **93**, 1459–1463.

Lasseran, J. C., Niquet, G. & Fleurat-Lessard, F. (1994). Quality enhancement of stored grain by improved design and management of aeration. In *Proceedings of the 6th International Working Conference on Stored Product Protection*, vol. 1, eds. E. Highley, E. J. Wright, H. J. Banks, pp. 296–299. Wallingford, UK: CABI Publishing.

Longstaff, B. C. & Cornish, P. (1994). PestMan: a decision support system for pest management in the Australian grain-handling system. *AI Applications in Natural Resource Management*, **8**, 13–23.

Mani, S., White, N. D. G., Jayas, D. S. *et al.* (2001). *Canadian Storage Guidelines for Cereals and Oilseeds (CanStore©): An Expert System for Agricultural Producers and Elevator Managers* (modified March 3, 2003). Available at http://sci.agr.ca/winnipeg/canstoronweb/cotw_e.htm.

Ndiaye, A. (2001). *QualiS: An Expert System Shell for Maintenance of Stored Grain Initial Quality*, Copyright 001.290023.00. Paris, France: Agence pour la Protection des Programmes.

Nelson, S. O. (1996). Review and assessment of radio frequency and microwave energy for stored-grain insect control. *Transactions of American Society of Agricultural Engineers*, **39**, 1475–1484.

Nissen, R. J., George, A. P., Waite, G., Lloyd, A. & Hamacek, E. (2005a). Innovative new production systems for low-chill stonefruit in Australia and South-East Asia: a review. *Acta Horticulturae*, **694**, 247–251.

Nissen, R. J., George, A. P. & Topp, B. L. (2005b). Producing super sweet and firm peaches and nectarines. *Acta Horticulturae*, **694**, 311–314.

Oseto, C. Y. (2000). Physical control of insects. In *Insect Pest Management*, eds. J. E. Rechcigl & N. A. Rechcigl, pp. 25–100. Boca Raton, FL: Lewis Publishers.

Perrings, C., Williamson, M. & Dalmazzone, S. (eds.) (2000). *The Economics of Biological Invasions*. Cheltenham, UK: Edward Elgar.

Pinkerton, J. N., Ivors, K. L., Miller, M. L. & Moore, L. W. (2000). Effect of soil solarization and cover crops on populations of selected soilborne plant pathogens in Western Oregon. *Plant Disease*, **84**, 952–960.

Puterka, G. J., Glenn, D. M., Sekutowski, D. G., Unruh, T. R. & Jones, S. K. (2000). Progress toward liquid formulations of particle films for insect and disease control in pear. *Environmental Entomology*, **29**, 329–339.

Russo, A., Candida Vasta, M., Verdone, A. & Eros Coccuzza, G. (2002). The use of light traps for monitoring flies in a cheese industry in Sicily. *IOBC/WPRS Bulletin*, **25**, 99–104.

Sinha, R. N. & Watters, F. L. (1985). *Insect Pests of Flour Mills, Grain Elevators, and Feed Mills and their Control*, Publication No. 1776. Ottawa, Canada: Agriculture and Agri-Food Canada General Directorate.

Stratil, H., Wohlgemuth, R., Bolling, H. & Zwingelberg, H. (1987). Optimization of the impact machine method of killing and removing insect pests from foods, with particular reference to quality of flour products. *Getreide, Mehl und Brot*, **41**, 294–302.

Sutherst, R. W. & Maywald, G. F. (1985). A computerized system for matching climates in ecology. *Agriculture, Ecosystems and Environment*, **13**, 281–299.

Tamietti, G. & Valentino, D. (2001). Soil solarization: a useful tool for control of *Verticillium* wilt and weeds in eggplant crops under plastic in the Po valley. *Journal of Plant Pathology*, **83**, 173–180.

Thomas, A. L, Muller, M. E., Dodson, B. R., Ellersieck, M. R. & Kaps, M. (2004). A kaolin-based particle film suppresses certain insect and fungal pests while reducing heat stress in apples. *Journal of the American Pomology Society*, **58**, 42–51.

Unruh, T. R., Knight, A. L., Upton, J., Glenn, D. M. & Puterka, G. J. (2000). Particle films for suppression of the codling moth (Lepidoptera: Tortricidae) in apple and pear orchards. *Journal of Economic Entomology*, **93**, 737–743.

Vincent, C. (2002). Pneumatic control of agricultural insect pests. In *Encyclopedia of Pest Management*, ed. D. Pimentel, pp. 639–641. New York: Marcel Dekker.

Vincent, C. & Boiteau, G. (2001). Pneumatic control of agricultural insect pests. In *Physical Control Methods in Plant Protection*, eds. C. Vincent, B. Panneton & F. Fleurat-Lessard, pp. 270–281. Berlin, Germany: Springer-Verlag.

Vincent, C., Panneton, B. & Fleurat-Lessard, F. (eds.) (2001). *Physical Control in Plant Protection*. Berlin, Germany: Springer-Verlag.

Vincent, C., Hallman, G., Panneton, B. & Fleurat-Lessard, F. (2003). Management of agricultural insects with physical control methods. *Annual Review of Entomology*, **48**, 261–281.

Vincent, C., Rancourt, B. & Carisse, O. (2004). Apple leaf shredding as a non-chemical tool to manage apple scab and spotted tentiform leafminer. *Agriculture, Ecosystems and Environment*, **104**, 595–604.

Wang, S.-J., Tang, J.-M., Cavalieri, R. P. & Davis, D. (2003). Differential heating of insects in dried nuts and fruits associated with radiofrequency and microwave treatments. *Transactions of the American Society of Agricultural Engineers*, **46**, 1175–1182.

Weintraub, P. G. & Berlinger, M. (2004). Physical control in greenhouses and field crops. In *Novel Approaches to Insect Pest Management*, eds. A. R. Horowitz & I. Ishaaya, pp. 301–318. Heidelberg, Germany: Springer-Verlag.

Weintraub, P. G. & Horowitz, A. R. (2001). Vacuuming insects pests: the Israeli experience. In *Physical Control Methods in Plant Protection*, eds. C. Vincent, B. Panneton & F. Fleurat-Lessard, pp. 294–302. Berlin, Germany: Springer-Verlag.

Wilkin, D. R. & Mumford, J. D. (1994). Decision support system for integrated management of stored commodities. In *Proceedings of the 6th International Working Conference on Stored Product Protection*, vol. 2, eds. E. Highley, E. J. Wright, H. J. Banks & B. R. Champ, pp. 879–883. Canberra, Australia.

Williams, R. H., Hook, S. C. W., Parker, C. G. *et al.* (2006). GrainPlan: development of a practical tool to improve grain storage on UK farms – knowledge transfer in action. In *Proceedings of the 9th International Working Conference on Stored Product Protection*, eds. I. Lorini, B. Bacaltchuck, H. Beckel *et al.*, pp. 1206–1211. Passo Fundo, RS, Brazil: Brazilian Post-Harvest Association (ABRAPOS).

Wilson, E. O. (1984). *Biophilia*. Cambridge, MA: Harvard University Press.

Chapter 25

Cotton arthropod IPM

Steven E. Naranjo and Randall G. Luttrell

Cotton is the world's most important natural source of fiber, accounting for almost 40% of total worldwide production. The rich history of cotton and cotton production is closely linked to expanding human civilization (Kohel & Lewis, 1984; Frisbie *et al.*, 1989). Cotton belongs to the genus *Gossypium* and four species are cultivated worldwide. Levant cotton (*G. herbaceum*) and tree cotton (*G. arboreum*) are primarily grown in Asia, while the long staple sea island (American Pima, Creole, Egyptian) cotton (*G. barbadense*) is cultivated in Egypt, India, the West Indies and parts of the western USA and South America. Upland cotton (*G. hirsutum*) is the most common species cultivated throughout the world. Cotton is a perennial plant, but is grown as an annual through manipulation of irrigation, defoliants and cultivation. The harvestable portions of the plant are found in the cotton fruit. The primary product, fiber, arise from the growth of single cells on the seed surface, while the seeds are further used as animal feed or in the production of oil found in many food products.

Cotton is grown in more than 75 countries with a total production in 2006 of 116.7 million bales (~25 400 million kg: National Cotton Council, 2007a). The current top five producing countries, in order, are China, India, the USA, Pakistan and Brazil. In the USA, cotton is grown in 17 states grouped into four major production regions (Fig. 25.1) with a total production of 21.7 million bales in 2006. Total USA harvested cotton hectares and total production of lint has increased about 10% and 22%, respectively, from the ten-year period 1986–95 to the period 1996–2005 (Table 25.1). Crop loss to insects and mites has generally declined in the past ten years which represents a marked improvement in crop protection technologies and IPM practices.

Cotton farmers in the USA have a long history of organized support from public and private research and education efforts. The National Cotton Council coordinates a wide range of initiatives and policies for the cotton industry with a defined mission to ensure the ability of all USA segments of the cotton industry to compete in national and international markets. They also sponsor an annual industry-wide meeting (Beltwide Cotton Conferences). Cotton Incorporated is an organization funded through per-bale assessments on producers and importers. It is a major source of funding for cotton IPM and it has a mission to increase the demand for and profitability of cotton through research and promotion. Public support through various extramural and intramural programs of the US Department of Agriculture (USDA) continues to be a critical resource for research and extension efforts.

Integrated Pest Management, ed. Edward B. Radcliffe, William D. Hutchison and Rafael E. Cancelado. Published by Cambridge University Press.

Table 25.1 Summary of cotton production (bales/ha) and insect control costs ($US) in the four major USA production regions over the past 20 years (1986–2005)

	West		Southwest		Midsouth		Southeast	
	1986–1995	1996–2005	1986–1995	1996–2005	1986–1995	1996–2005	1986–1995	1996–2005
Harvested ha (× 1000)	604	451	2143	2110	1444	1453	607	1261
Production (bales × 1000)[a]	3538	2862	4747	5583	5052	5802	1918	4376
Percent production (lint)	23.2	15.4	31.1	30.0	33.1	31.2	12.6	23.5
Percent crop loss to insects and mites	6.3	4.4	6.1	8.0	9.1	5.9	9.0	5.4
Control cost per ha[b]								
Insecticides and application	131	168	49	47	36	141	124	77
(Insecticides, application, scouting, eradication and technology fees)		(205)		(74)		(220)		(133)

[a] Cotton bale = 218 kg of lint.
[b] The data from 1996 onward is more complete, including costs ($US) associated with scouting, eradication of boll weevil and technology fees for growing transgenic cotton, in addition to insecticide and application costs. This more inclusive figure is presented within parentheses. None of the cost figures are adjusted for inflation.

Source: Summarized from *Cotton Insect Losses*, a database compiled by the National Cotton Council (2007b).

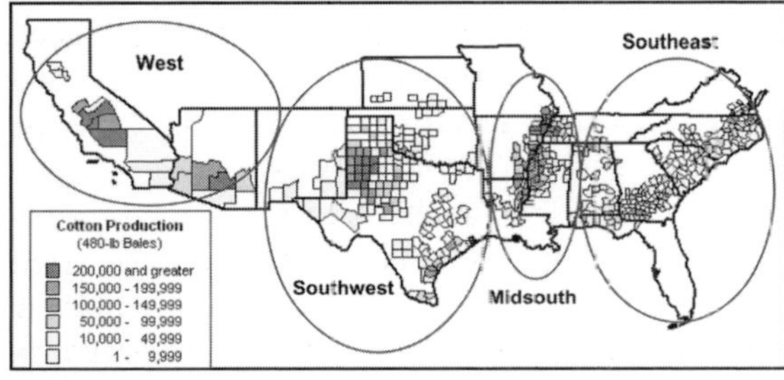

Fig. 25.1 Major cotton producing regions of the USA (modified from National Cotton Council, 2007a).

In many ways, cotton arthropod IPM both exemplifies and has shaped many of the general principles, control tactics and integrated strategies covered with great detail throughout this textbook. Numerous reviews have previously summarized cotton insect pest management in the USA (Gaines, 1957; Bottrell & Adkisson, 1977; Ridgway *et al.*, 1984; Frisbie *et al.*, 1989; El-Zik & Frisbie, 1991; Luttrell, 1994; King *et al.*, 1996b). Here we summarize current and recent past efforts in cotton IPM that continue to build upon more than a century of scientific research and innovation based on ecological principles and understanding.

25.1 Arthropod fauna

25.1.1 Pest species and damage

Rainwater (1952) stated "cotton is a plant that nature seems to have designed specifically to attract insects." Over 1300 herbivorous insects are known from cotton systems worldwide (Hargreaves, 1948) but many fewer are common inhabitants and still fewer are of economic importance. Roughly 100 species of insects and spider mites are pests of cotton in the USA but only 20% of these are common and likely to cause damage if left uncontrolled (Leigh *et al.*, 1996) (Fig. 25.2). The remaining 80% are sporadic or secondary pests that become problematic in some years due to changing environmental factors or the misuse of insecticides or other disruptions of natural controls. Pest species vary from one production area to the next. Bollworm (*Helicoverpa zea*) and tobacco budworm (*Heliothis virescens*) are major pests of cotton in the USA from Texas eastward while the pink bollworm (*Pectinophora gossypiella*) is the dominant bollworm in the western USA. Various species of *Lygus* and other mirid plant bugs affect cotton throughout

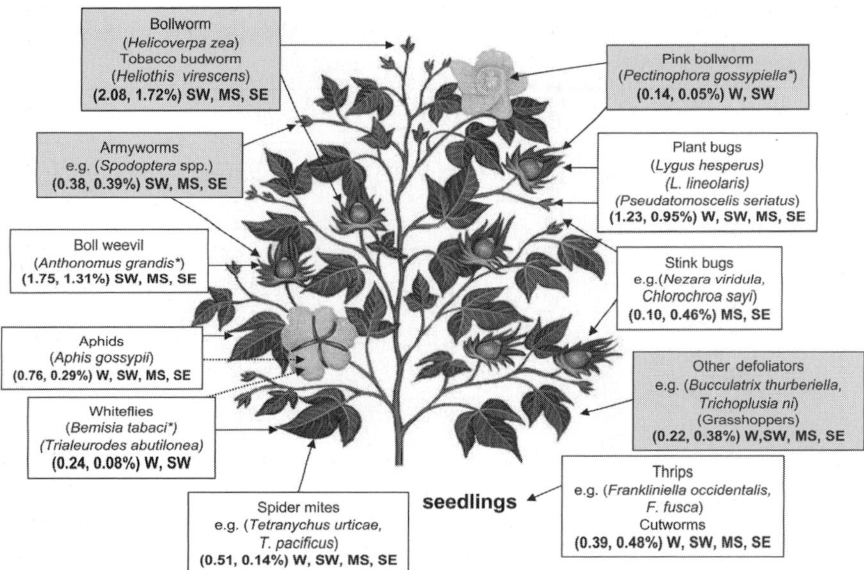

Fig. 25.2 Summary of major insect and mite pests of cotton in the USA, Numbers in parentheses indicate the average percent yield loss for the periods 1986–95 and 1996–2005, respectively, for each pest group across the entire USA. Bold letters denote the major production regions where the pest group primarily impacts cotton production; W, West; SW, Southwest; MS, Midsouth; SE, Southeast. Insects attacking cotton bolls (fruit) generally attack squares (flower buds) as well. Dotted lines pointing to lint indicate that whiteflies and aphids can affect yield quality through the deposition of honeydew. Asterisks next to species names denote exotic pests and shaded boxes indicate pests which are affected significantly by *Bt* cotton. Loss figures summarized from *Cotton Insect Losses*, a database compiled by the National Cotton Council (2007b).

the world. The pink bollworm, the cotton aphid (*Aphis gossypii*) and sweetpotato whitefly (*Bemisia tabaci*) are significant pests of cotton throughout the world while the boll weevil (*Anthonomus grandis*) is only found in the Americas. Some pests such as the bollworm, tobacco budworm and plant bugs are native insects that have expanded their ranges as cotton production grew, while the sweetpotato whitefly and pink bollworm are exotic invaders. The exotic boll weevil's movement into the USA was associated with the expansion of cotton production.

Although the levels of crop loss may appear small, the economic impact can be enormous. For example, in 2005 total yield reduction from arthropod pests in USA cotton was 4.47%, which represented a loss of >1.5 million bales of cotton valued at $US 1250 million in yield reduction and control costs (Williams, 2006). As will be discussed below, many factors have contributed to reductions in pest losses over the past 20 years including boll weevil eradication, transgenic cottons for control of caterpillar pests and improved overall IPM programs for various pests.

25.1.2 Beneficial arthropods

Many beneficial arthropod species are associated with cotton. In a seminal study, Whitcomb & Bell (1964) cataloged about 600 species of arthropod predators including ~160 species of spiders in Arkansas cotton fields. Van den Bosch & Hagan (1966) suggested that there may be nearly 300 species of parasitoids and arthropod predators in western cotton systems. Like cotton herbivores, only a fraction of these are common, with some of the more abundant groups including big-eyed bugs (*Geocoris* spp.), anthocorid bugs (*Orius* spp.), damsel bugs (*Nabis* spp.), assassin bugs (*Zelus* and *Sinea* spp.), green lacewings (*Chrysopa* and *Chrysoperla* spp.), lady beetles (e.g. *Hippodamia* and *Scymnus* spp.), ants (especially *Solenopsis* spp.), a wide variety of web-building and wandering spiders and both parasitic wasps (e.g. *Bracon* spp., *Cotesia* spp., *Microplitis* spp., *Hyposoter* spp., *Trichogramma* spp.) and flies (*Archytas* spp., *Eucelatoria* spp.). We continue to learn about the important role that these natural enemies play in cotton pest control but the most dramatic evidence of their impact comes from studies in which the destruction or disturbance of natural enemy communities by indiscriminant insecticide use is associated with pest outbreaks (e.g. Leigh *et al.*, 1966; Eveleens *et al.*, 1973; Stoltz & Stern, 1978; Trichilo & Wilson, 1993). Overall, arthropod communities in cotton are dynamic and largely driven by the wide diversity of management options discussed below.

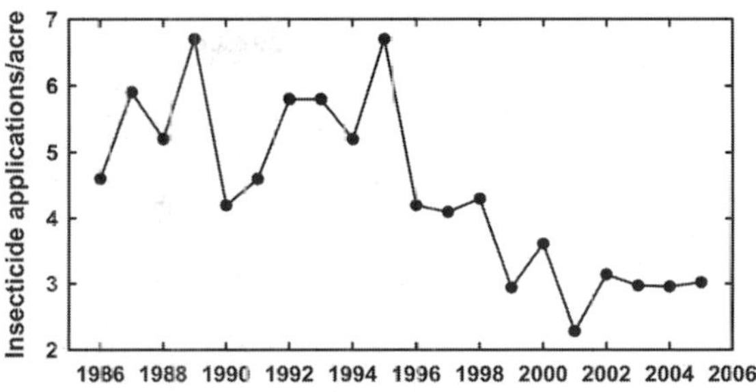

Fig. 25.3 Overall insecticide use patterns in USA cotton 1986–2005 summarized from *Cotton Insect Losses*, a database compiled by the National Cotton Council (2007b).

25.2 IPM tactics

25.2.1 Chemical control

Insecticide use has a long history in USA cotton pest control beginning with arsenicals for control of boll weevil in the early twentieth century and followed by a progression of synthetic insecticides (e.g. organochlorines, organophosphates, carbamates and pyrethroids) in the subsequent decades following World War II (Herzog *et al.*, 1996; Sparks, 1996). With each new introduction, periods of excellent pest control were generally followed by control failures due to the evolution of insecticide resistance. Many past and current insecticides have broad activity against pests and their associated natural enemies and pose hazards to the environment and human health. On a worldwide basis cotton accounts for about 22.5% of all insecticide use (Anonymous, 1995) and historically USA cotton has been the heaviest user of insecticides. That pattern has begun to shift (Fig. 25.3) with the introduction of transgenic *Bt* cotton in 1996 and the availability of a wide range of effective and safer insecticides registered in part through the US Environmental Protection Agency (EPA)'s Reduced-Risk Initiative over the past decade (Environmental Protection Agency, 2006) (Table 25.2). While a variety of many classes of insecticides continue to be used for cotton pest control throughout the USA (National Agricultural Statistics Service, 2006), adoption and use of reduced-risk insecticides has grown in recent years (e.g. Goodell *et al.*, 2006).

25.2.2 Resistance management

The development of resistance in pest populations to insecticides is a continual threat to successful implementation of chemical control and careful vigilance and proactive strategies are needed to preserve this important tactic (Castle *et al.*, 1999) (see Chapter 15). Around 550 arthropod pests have developed resistance to one or more insecticides, and currently a total of 34 cotton pests (19 in the USA) have developed resistance to as many as three insecticide classes (Whalon *et al.*, 2007). The mitigation of resistance is based on management of insecticide type and use that either attempts to reduce the fitness of resistant individuals or minimizes selection pressure on a pest population (Roush & Daly, 1990). Simply put, this means limiting insecticide use through adherence to economic thresholds, diversifying modes of action through rotations, mixtures and use of synergist, and partitioning of insecticide use in space and time by adoption of seasonal stages or crop-specific usage. Examples include the Australian IRM strategy for managing resistance in the Old World bollworm (*Heliocoverpa armigera*) through a three-stage plan which rotates pyrethroids with non-pyrethroids over the season (see Castle *et al.*, 1999), the Texas and Midsouth pyrethroid use window strategy for resistance in tobacco budworm (Plapp *et al.*, 1990) and the multi-crop resistance management plan for whitefly in the western USA in which various classes of insecticides are rotated depending on predominant crop mixtures within a region (Palumbo *et al.*, 2001, 2003).

25.2.3 Cultural control

The indeterminate nature of cotton plant growth and the influence of production practices such as cultivation, irrigation, fertilization, cultivar

Table 25.2 Insecticides registered for use on cotton through the EPA Reduced-Risk/Organophosphate Alternatives Program

Compound	Mode of action (IRAC MoA)[a]	Cotton target	Year registered
Spinosad	Acetylcholine receptor modulator (5)	Caterpillars	1997
Pyriproxyfen	Juvenile hormone mimic (7)	Whiteflies, aphids	1998
Tebufenozide	Molting hormone agonist (18)	Caterpillars	1999
Methoxyfenozide	Molting hormone agonist (18)	Caterpillars	2000
Indoxacarb	Sodium channel agonist (22)	Caterpillars	2000
Thiamethoxam	Acetylcholine receptor agonist (4A)	Whiteflies, aphids, thrips	2000/2001
Buprofezin	Chitin synthesis inhibitor (16)	Whiteflies, aphids	2001
Pymetrozine	Feeding blocker (9)	Whiteflies, aphids	2001
Bifenazate	Neural inhibitor (25)	Mites	2002
Acetamiprid	Acetylcholine receptor agonist (4A)	Whiteflies, aphids, plant bugs	2002
Etoxazole	Growth inhibitor (10B)	Mites	2003
Novaluron	Chitin synthesis inhibitor (15)	Whiteflies, thrips, caterpillars, plant bugs	2004
Fenpyroximate	Electron transport inhibitors (21)	Mites	2004
Dinotefuran	Acetylcholine receptor agonist (4A)	Whiteflies, thrips, plant bugs	2005
Flonicamid	Feeding blocker (9)	Aphids, plant bugs	2005
Spiromesifen	Lipid synthesis inhibitor (23)	Whiteflies, mites	2006

[a] IRAC MoA = *Insecticide Resistance Action Committee Insecticide Mode of Action Classification* (IRAC, 2007).

selection, weed control and planting date on the crop's susceptibility to damage and suitability for insect infestations remain an extremely important aspect of effective pest management (Ridgway *et al.*, 1984; El-Zik *et al.*, 1989; Matthews, 1994; Walker & Smith, 1996). For example, stalk destruction, field sanitation, efficient harvest, tillage and winter irrigation can effectively control or reduce populations of boll weevil and pink bollworm (Walker & Smith, 1996). Early planting and early crop termination are long-standing principles of cultural control and pest avoidance that are still relevant for many pest species. From eastern Texas to the Atlantic coast, timely planting and early harvest helps to avoid fall and winter rains and resulting in important economic advantages (Parvin & Smith, 1996). Delayed planting also may have benefits. For example, planting later so that no fruiting forms are present when pink bollworm adults emerge from the soil maximizes "suicidal emergence" and reduces pest populations throughout the season (Brown *et al.*, 1992). Cultivation can effectively reduce overwintering survival of bollworm and tobacco budworm, and Schneider (2003) suggested that recent trends for reduced tillage could accelerate resistance of tobacco budworm to insecticides and *Bt* toxins in transgenic cotton (see below). Deep tillage induces high overwintering mortality in pink bollworm (Watson, 1980).

There has been renewed interest in manipulating dispersal and crop colonization though trap crops, especially in the management of polyphagous plant bugs. The use of strip crops of alfalfa within cotton is the classic example of cultural control through the practical deployment of trap crops (Stern *et al.*, 1964). Most examples of trap cropping, including this classic example of strip harvesting alfalfa, have been only sporadically accepted and utilized in production agriculture because of the logistic impact on farming operations and the wide availability of effective chemical controls options (Shelton & Badenes-Perez, 2006). For example, only 4% of California growers reported using manipulation of alfalfa to control cotton pest insects (Brodt *et al.*, 2007).

Future management systems may need to examine incentives for grower adoption and expansion of cultural management tactics that may reduce pest populations across broad geographic regions.

25.2.4 Behavioral control

A suite of tactics are available that alter or manipulate the behavior of pest arthropods leading to population suppression or even elimination. Two examples, (1) pheromones and (2) sterile insect release, are discussed here.

Pheromones

A pheromone is a chemical that mediates behavioral interactions between members of the same species. The most common examples are sex pheromones which are involved in mating, but aggregation and alarm pheromones are also known from cotton pest species. As of 1994 sex pheromones have been identified in 15 major cotton pest species (seven species in the USA) including 14 moths and one beetle (Campion, 1994). The three main applications of pheromones are monitoring, mating disruption and mass trapping (see Chapter 21).

Traps baited with sex pheromones are routinely used for selective monitoring of pink bollworm, boll weevils, bollworm and tobacco budworm. Trapping information is useful in pest detection at low densities and tracking seasonal events such as adult emergence and the number and timing of generations. For pink bollworm, pheromone traps have even been used to monitor density for pest control decision making (Toscano *et al.*, 1974) and traps are a major component of the ongoing pink bollworm eradication program (see below). Pheromone traps continue to play an important role in ongoing boll weevil eradication efforts and as long-term monitoring tools in post-eradication areas of the USA.

Mating disruption is achieved by applying pheromone to a field, thereby making it difficult for potential mates to find one another and resulting in reduced mating and subsequent reproduction. This technology was first applied in 1978 for the pink bollworm and the method is still used today. Mating disruption is a major component of the ongoing eradication and exclusion programs for this pest and has been used in several past areawide programs in California and Arizona and in other countries (Campion, 1994). Mating disruption has been evaluated for other pests such as bollworm and tobacco budworm, but their polyphagous nature is problematic and results have been unsuccessful or ambiguous (Campion, 1994).

Mass trapping showed some promise for pink bollworm control in a three-year grower funded trial in Arizona (Huber *et al.*, 1979) and the method was used in some Arizona production areas into the mid-1990s. The approach is not currently in use for pest control in cotton.

Sterile insect release

The notion that mass release of sterile insects could be used to manage or eradicate a pest was first proposed by E. F. Knipling during the 1930s (Knipling, 1955) and was first successfully used to eradicate screwworm (*Cochliomyia hominivorax*) from the island of Curaçao during the 1950s (Baumhover *et al.*, 1955). The concept, known as the sterile insect release method (SIRM) or the sterile insect technique (SIT) has been attempted with several major cotton pests in the USA including the bollworm and tobacco budworm, the boll weevil, and most successfully with the pink bollworm. Various biological and operational factors precluded the successful application of SIRM to the two former species/groups (see Villavaso *et al.*, 1996) but the method has been used annually since 1968 to help mitigate the establishment of pink bollworm on cotton in the Central Valley of California (Miller *et al.*, 2000), and is a component of the current pink bollworm eradication program.

25.2.5 Host plant resistance

Host plant resistance is a fundamental management tactic (El-Zik & Thaxton, 1989; Gannaway, 1994; Jenkins & Wilson, 1996). Host plant resistance can be broadly categorized as antibiosis (reduced fitness or pest status), antixenosis (avoidance or behavioral factors) and tolerance (ability of plant to compensate for damage) (see Chapter 18). Plant resistance traits may include manipulation of the plant's genome or the resistance may be associated with indirect selections for traits like yield and fiber quality. Genetically controlled traits useful in cotton resistance to insects include: crop earliness, a range of plant morphological traits (nectariless, glaborous or pilose leaf surface, okra-shaped leaf, frego bract, red plant color, yellow or orange pollen) and varying concentrations of plant secondary compounds (high gossypol and tannin content). Relatively few of these traits have been incorporated into commercial cultivars.

The tools of biotechnology have provided new opportunities to enhance traditional approaches to host plant resistance (see Chapters 18 and 21). The impact of transgenic cottons producing insecticidal toxins from *Bacillus thuringiensis* (*Bt*), primarily for control of pink bollworm, tobacco budworm and bollworm has been enormous. The reduction in insecticide use in cotton over the past decade (Fig. 25.3) can be partly attributed to increased adoption of *Bt* cottons. Use of *Bt* varieties has expanded each year with 57% of all USA upland cotton hectares planted to *Bt* cotton in 2006 (Williams, 2007). In most areas of the Midsouth and Southeast where bollworm and tobacco budworm are historically important pests, adoption of *Bt* cotton approaches 80–90%. In 2006, Texas planted about 35% of its total cotton hectares to *Bt* varieties while California planted less than 20% (a large portion of California's hectares are planted to long staple Pima varieties that have not been transformed). In Arizona, *Bt* cottons are widely adopted because of their dramatic impact on pink bollworm.

Commercial transgenic cottons with insecticidal activity are currently limited to the transgenes from *B. thuringiensis*. Bollgard cotton (Monsanto Company, St. Louis, MO) was the first commercially available cotton in 1996. It expresses the Cry1Ac insecticidal protein. Bollgard II cotton (Monsanto) expressing Cry1Ac and Cry2Ab2 protein was commercially introduced in 2003, and Widestrike cotton (Dow AgroSciences, Indianapolis, IN) expressing Cry1Ac and Cry1F protein was launched in 2005. VipCot cotton (Syngenta Biotech, Jealott Hill, Berkshire, UK) will express the Vip3A vegetative protein from *B. thuringiensis*, probably along with a Cry protein, and is expected to be commercialized in the USA shortly.

One of the most hotly debated issues facing cotton insect management is the sustainability of transgenic *Bt* cottons. However, despite the high use of *Bt* crops there has been little or no increase in insect resistance over the ten years of commercial deployment (Tabashnik *et al.*, 2003). This success can be partly attributed to a EPA-mandated resistance management program that requires growers using *Bt* cotton to also plant non-*Bt* cotton refuges. The principle behind this mandated strategy is that non-*Bt* cottons produce susceptible target pests that can readily interbreed with any resistant pests that may arise from *Bt* fields, thereby diluting incipient resistant populations.

Bt cottons offer real environmental and economic advantages to conventional cottons sprayed more with insecticides. Frisvold *et al.* (2006) estimated global economic benefits of $US 836 million for *Bt* cotton in USA. The potential role of *Bt* cotton in reducing human exposure to toxic chemicals, especially in developing countries where insecticides are often applied manually, is large. Still, environmental risk issues such as effects on non-target organisms and ecosystem function and gene flow associated with transgenic crops in general continue to be debated and researched in the scientific community (e.g. Andow *et al.*, 2006; Romeis *et al.*, 2006).

There has been limited progress in the development of transgenic cottons for pests other than caterpillars. For example, Monsanto is in the very early stages of development of transgenic cottons targeting *Lygus* spp. based on *Bt* and non-*Bt* approaches. Focus on non-caterpillar pests remains a major goal of various biotech firms and basic researchers around the world, and it highlights the importance of a continued investment in traditional host plant resistance.

25.2.6 Biological control

The three major approaches to biological control include classical biological control, where exotic agents are introduced for permanent establishment against exotic and native pests, augmentation biological control, which involves the rearing and periodic release of natural enemies, and conservation biological control, which attempts to protect, manipulate and enhance existing natural enemies for improved control (see Chapters 9 and 12). Classical biological control programs have been carried out in the past for several pest groups including bollworm, tobacco budworm, boll weevil, pink bollworm and to a lesser extent for lygus bugs. These efforts have been largely unsuccessful in the cotton system as natural enemies have either failed to become established and/or their impacts have been minimal (King *et al.*, 1996a). The whitefly *B. tabaci* has been the most recent target of classical biological control, with numerous species of parasitoids released for establishment (Gould *et al.*, 2008); however, as with other classical efforts, the impact of these established agents have so far been minimal in the cotton system (Naranjo, 2007).

Likewise, augmentative biological control with predators and parasitoids has been researched and evaluated for several major cotton insect pests, but factors such as lack of efficacy, technical difficulties with natural enemy mass-production and cost relative to insecticides have combined to limit this approach from becoming a viable option in cotton pest control in the USA (King *et al.*, 1996a). Augmentation with microbial agents (viruses, fungi, bacteria) for control of bollworm, tobacco budworm, boll weevil, pink bollworm, whitefly and plant bugs (King *et al.*, 1996a; Faria & Wraight, 2001; McGuire *et al.*, 2006) has been examined; however, commercialized microbial products continue to have very small shares of the cotton pest control market.

In contrast to classical and augmentative biological control, conservation biological control continues to be a major focal area of cotton IPM that has been further stimulated by the many recent changes to cotton pest management systems. As noted, the cotton system in the USA harbors a diverse complex of native natural enemies, many of which are generalist feeders that opportunistically attack many insect and mite pests. Naturally occurring epizootics of some microbes also may significantly suppress pest species (Steinkraus *et al.*, 1995). The potential value of these natural enemies in pest suppression has been repeatedly demonstrated over many decades in the cotton system when broad-spectrum insecticides applied for one pest lead to resurgence of the target pest and/or outbreaks of secondary pests through the destruction of natural enemies (see Bottrell & Adkisson, 1977). This potential is also widely recognized in state recommendations for cotton IPM. Most guidelines call for sampling of natural enemies and emphasize their preservation through inaction or judicious use of insecticides, particularly those with selective action. Our understanding of the role and interaction of natural enemy species and complexes and how to manipulate them for improved pest control in cotton has a rich history that continues to grow (e.g. Sterling *et al.*, 1989; Naranjo & Hagler, 1998; Prasifka *et al.*, 2004).

25.2.7 Sampling and economic thresholds

A hallmark of all cotton IPM programs in the USA is monitoring of pest density or incidence combined with action or economic threshold to determine the need for control measures. In 2006, about 50% of USA cotton hectarage was scouted an average of 1.3 times per week at an average cost of $US 18.97/ha across the cotton belt (Williams, 2007). The intensity of scouting varies greatly by state. In Virginia and Kansas less than 5% of cotton hectarage was scouted but more than 90% was scouted in Arizona, Louisiana and South Carolina. In Texas, the largest cotton producing state, only 19% of the hectarage was reported as scouted.

The cooperative extension programs of each of the 17 cotton growing states produce recommendations for scouting, treatment thresholds and insecticides to help growers and consultants implement IPM.[1] The basic tools of sampling include sweep nets, beat cloths and beat boxes, traps and visual inspection of various plant parts, all of which require human labor and the associated cost. Sampling plans, which specify the general protocols for how samples should be collected and how many sample units should be taken, are

[1] Web links to sampling and threshold recommendations and individual state recommendations are available at http://ipmworld.umn.edu/textbook/Naranjo3.htm.

typically based on research to understand the distribution and variability of pest populations (see Chapter 7). Sequential sampling plans, which minimize the number of sample units that need to be taken, are often developed for cotton pest management application, but in practice it is more typical for a set sample size to be recommended and implemented. Many cotton pest sampling plans also use a presence/absence approach (e.g. percent infested) for monitoring rather than a complete count which allows for quicker sampling and decision making.

Thresholds tend to vary depending on production region and also may be dynamic, with critical pest densities being a function of plant development and prior management activity. For example, thresholds for bollworm and tobacco budworm are lower following an initial insecticide application during post flower bloom, while thresholds for plant bugs generally increase as the cycle of flower bud (square) production progresses. Thresholds in *Bt* cotton also may differ from non-*Bt* cotton for the bollworm which is not controlled completely in some transgenic cottons. Some of these thresholds are based on experimental study (see Chapter 3), but many are nominal thresholds developed on the basis of trial-and-error experience by researchers, extension agents, consultants and growers. Many state guidelines encourage scouting of natural enemy populations, but only a few have provided explicit information on how to use these counts to modify treatment decisions (e.g. Fillman & Sterling, 1985; Wilson *et al.*, 1985; Conway *et al.*, 2006). Nonetheless, the preservation of natural enemies through judicious use of insecticides is implicitly recognized as a key component of most management systems.

25.3 IPM programs and implementation

Despite the challenge of researching and compiling the necessary component tactics into a workable IPM strategy, the greatest difficulty may be in implementing and evaluating IPM programs because such tasks depend on logistical, socio-logical and economic factors (Mumford & Norton, 1994; Kogan, 1998) (see Chapter 38). Education is a key element in IPM implementation regardless of the crop and Cooperative Extension Services associated with land-grant universities generally take the lead in developing educational materials (circulars, bulletins, websites) as well as organizing training and even on-farm demonstrations and adaptive research. Private industry may also contribute educational and consulting activities, and depending on the scope of implementation, grower organizations and/or federal agencies such as USDA may be involved (Harris *et al.*, 1996).

25.3.1 Areawide programs

Cotton entomologists have long recognized the importance of spatial scale of management activities for mobile pest species. Ewing & Parencia (1950) demonstrated effective control of boll weevil when coordinated early-season treatments were applied on a community basis. With the dramatic success of screwworm eradication, Knipling (1979) extended the areawide concepts of pest population suppression to cotton insects, especially the boll weevil and the tobacco budworm. Henneberry & Phillips (1996) provide an overview of the elaborate experiments and theoretical debates that followed.

Boll weevil eradication (see below) is perhaps the largest-scale example of areawide programs. Other examples include numerous attempts to manage bollworm and tobacco budworm through biological control and community management systems, a 40-year program to exclude pink bollworm from central California, successful control of whitefly in the desert valleys of the West, and emerging management systems for plant bugs and stink bugs in the Midsouth and Southeast.

25.3.2 Models and decision aids

A wide diversity of simulation models and decision support systems have been developed for USA cotton beginning in the 1970s with the NSF/EPA Integrated Pest Management Project (commonly known as the "Huffaker project"), continuing with the USDA/EPA Consortium for IPM project during the 1980s and many other cooperative state and federal projects thereafter (Mumford & Norton,

1994; Wagner *et al.*, 1996). Modeling efforts have focused on individual pests and groups of pests and some have included simple or detailed cotton plant models. In general, these models have been useful in structuring knowledge of the plant–pest system, studying and predicting population dynamics, examining alternative management scenarios and identifying areas needing further research. However, with few exceptions such models have found very limited application in guiding day-to-day pest and crop management activities.

In the 1990s, management tools based on expert systems and information management began to be developed for cotton insect management (Wagner *et al.*, 1996). Some of these decision aids were coupled with the more complex simulation models and others were based on broad generalizations of expert opinion and historical data. These systems included the expert systems rbWHIMS, COTFLEX, CALEX and CIC-EM for a range of cotton production systems. In general, none of these systems has seen wide-scale adoption. COTMAN, a more recent decision aid, emphasizes the synthesis of field samples rather than projection of information and has remained a component of practical cotton production in limited areas of the Midsouth, Southeast and Texas. The strength of COTMAN and its continued use may be due to the simplicity of the system and its close conceptual linkage to crop development.

25.3.3 Case studies

There are many examples of operational IPM programs for cotton pests throughout the USA that involve many of the component tactics discussed. Here we highlight two representative programs, (1) whiteflies in Arizona and (2) plant bug and stink bug management in the Midsouth and Southeast.

Whitefly IPM strategy in Arizona

Since the early 1990s the polyphagous sweetpotato whitefly (*B. tabaci*), Biotype B, has had major impacts on most agricultural production in the West (Oliveira *et al.*, 2001). In response, a multi-component research and educational plan was launched that resulted in a successful IPM program which continues to be expanded and refined today (Ellsworth & Martinez-Carrillo, 2001; Naranjo, 2001; Ellsworth *et al.*, 2006). The overall program can be envisioned, and is taught to growers and consultants, as a pyramid with multiple, overlapping layers and components (Fig. 25.4). The broad base of the pyramid, founded on research, emphasizes tactics and strategies that can be implemented to reduce overall pest populations including various crop management practices and selection of well-adapted, smooth-leaf varieties which are generally less attractive to whiteflies. The foundation also emphasizes natural enemy conservation through the use of selective control methods for whitefly and other pests and an array of areawide tactics like crop placement and arrangement to reduce pest movement, destruction of crop residue and weeds and coordinated use of insecticides among all affected crops to manage resistance.

The two upper layers of the pyramid outline pest monitoring through an efficient binomial sampling scheme (Ellsworth *et al.*, 1996; Naranjo *et al.*, 1996), and the timing of effective control methods based on economic threshold and a three-stage insecticide use system which emphasizes selectivity (i.e. safety to beneficial arthropods) in the initial stages (Naranjo *et al.*, 2004; Ellsworth *et al.*, 2006). Follow-up treatments are rarely needed if these selective options are used first because the conserved natural enemies and other natural forces are then able to suppress whitefly populations long term (Naranjo, 2001). The three-stage system also implicitly encourages the rotation of insecticides with differing modes of action in order to mitigate resistance. Operationally, the IPM plan has significantly reduced insecticide use for all cotton pests in Arizona from a decades-long high of over 12 applications in 1995 at a cost of $US 536/ha to a decades low application rate of 1.4 at a cost of $US 77/ha in 2006.

Plant bug and stink bug management in the Midsouth and Southeast

A plant bug complex, including tarnished plant bugs (*Lygus lineolaris*), cotton fleahoppers (*Pseudatomoscelis seriatus*) and clouded plant bugs

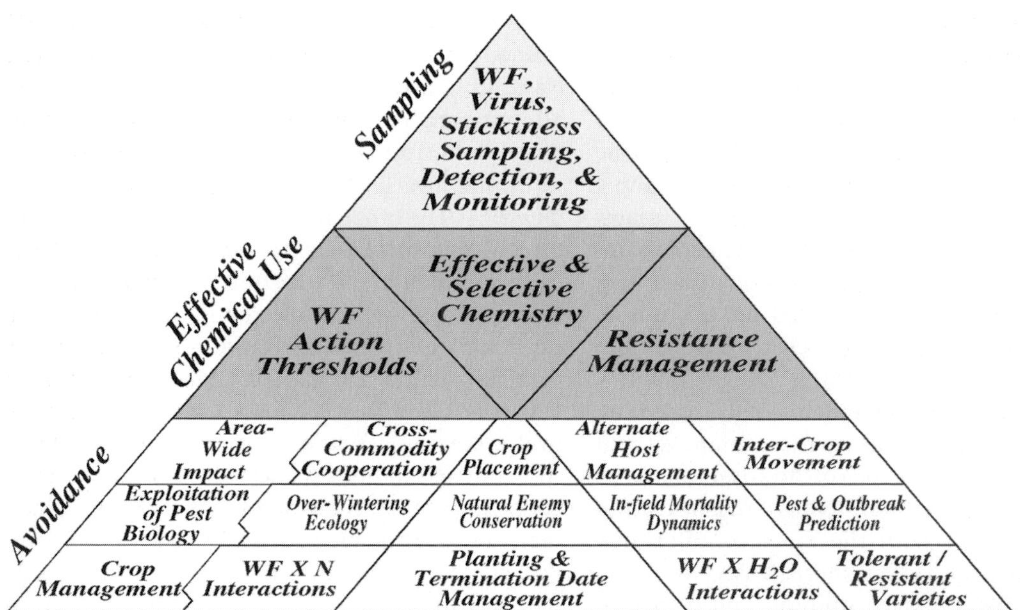

Fig. 25.4 Conceptual diagram of the Arizona Whitefly (WF) IPM program showing the three main components of effective IPM: avoidance, effective use of insecticides and sampling. (Reprinted from Ellsworth & Martinez-Carrillo [2001] with permission from Elsevier.)

(*Neurocolpus nubilis*), has been a long-standing pest problem in Midsouth and Southeast cotton. Plant bugs typically attack cotton at early squaring but can persist as a pest problem through boll development. A complex of seed-feeding stink bugs (brown stink bug [*Euschistus servus*], green stink bug [*Acrosternum hilare*] and southern green stink bug [*Nezara viridula*]) also attack maturing bolls later in the growing season. Both plant bugs and stink bugs are increasing in status but these tend to be more important in the Midsouth and Southeast, respectively. The elevated importance of these polyphagous and mobile insects reflects success in eliminating boll weevil (through eradication) and tobacco budworm (by *Bt* cotton) as major pests. In 2006, crop loss and insecticide use for these bug pests were twice to three-fold those of other pests across the Midsouth and Southeast (Williams, 2007).

Designing effective control measures for the bug complex has been a challenging and difficult task. Plant bugs are now resistant to several insecticides (Snodgrass, 1996). The USDA in Stoneville, Mississippi has developed an areawide approach to the removal of early-season broadleaf hosts of tarnished plant bug (Snodgrass *et al.*, 2005). This research approach has been evaluated by extension entomologists across the Midsouth and is being adopted on limited hectares by growers in some regions. Additional testing is needed to confirm the broader impacts on other pest and beneficial species in the system. Extension entomologists in the Southeast are developing treatment thresholds and monitoring procedures for the stink bugs (Greene *et al.*, 2001), and those in the Midsouth are studying sampling and management options for plant bugs.

25.4 Pest eradication

Pest eradication involves the complete elimination of a pest from its current range and generally focuses on invasive pest species (Chapter 10). Significant debate continues over the value of pest eradication as a substitute for or as a complement to pest management (Myers *et al.*, 1998). Nonetheless, eradication programs are currently under way for two key cotton pests in the USA and they will be briefly discussed here.

25.4.1 Boll weevil

Effective removal of boll weevil as a key pest of USA cotton is an important biological and social achievement covering a half-century of scientific and strategic effort (Cross, 1973; Smith & Harris, 1994; Dickerson *et al.*, 2001; Hardee & Harris, 2003). Interest in a coordinated effort to eliminate this invasive cotton specialist began in the late 1950s and the eradication program was initiated in earnest by the late 1970s. Incipient populations were eradicated in Arizona, California and northwest Mexico in the 1980s, and the nationwide effort began in North Carolina and successfully expanded through South Carolina, Georgia, Florida and south Alabama. In the early to mid-1990s boll weevil was eradicated from central and north Alabama, middle Tennessee and the Texas Southern Rolling Plains. During the late 1990s, boll weevil eradication expanded to the Midsouth and Texas, with only a few isolated regions still infested. It is anticipated that the USA will be weevil free by 2009. Reductions in cotton insect losses (Table 25.1) can be directly and indirectly related to removal of this key USA cotton pest. In the past, early-season treatments were necessary to keep weevil populations from expanding, but they also triggered many additional pest outbreaks. The basic components of the program include monitoring with pheromone (grandlure)-baited traps to time early-season applications of insecticides that reduce establishment in cotton, late-season "diapause" treatments to reduce overwintering weevils and early crop maturity and crop destruction to enable a host-free period coordinated across large geographic zones.

25.4.2 Pink bollworm

Like the boll weevil, the pink bollworm is an exotic cotton specialist that successfully invaded in USA in the early 1900s and became firmly established in the West by the mid 1960s following various attempts to contain and suppress populations throughout the first half of the twentieth century (Henneberry & Naranjo, 1998). The cooperative eradication program involves growers, and state and federal agencies. The program is being implemented in phases beginning with west Texas, New Mexico and northern Chihuahua, Mexico in 2001 and continuing through Arizona to southern California and northern Sonora, Mexico in 2007 with the total program completed by 2010. The basic elements of the program include mapping and monitoring of all cotton fields within each region and the use of a combination of *Bt* transgenic cotton, mating disruption with pheromones, sterile insect release, and follow-up insecticides as needed. The sterile insect release in this case serves both to augment population control and as a substitute for the required non-*Bt* refuge for resistance management which was relaxed in Arizona and southern California to allow for 100% production of *Bt* cotton. Pink bollworm populations in the Phase I regions have been reduced by >99% from 2001 to 2005 (El-Lissy & Grefenstette, 2006), but it is too early to gauge the overall success of the eradication effort.

25.5 Conclusions

IPM is based on an ever-changing foundation of improved scientific knowledge, economic circumstances, and societal issues and demands. Several significant technological advances (e.g. transgenic crops) have occurred in the past decade that have dramatically lowered pest losses and significantly lowered insecticide use in a system that has historically been associated with insecticide over-reliance and misuse. Undoubtedly, future advancements will continue to improve the sustainability and environmental quality of cotton production in the USA and worldwide. Environmental issues will continue to grow, especially as urban areas expand and become more closely integrated with crop production areas.

Current IPM programs in cotton like many other crop systems are largely focused on what Kogan (1998) characterizes as "Level 1 IPM," or IPM of single pest species in individual fields. This is contrasted to "Level 2 IPM" which focuses on interactive effects of multiple pest species within whole farms or "Level 3 IPM" which involves management of multiple pests on perhaps multiple crops within entire agroecosystems. Some of the areawide programs summarized above have begun to view and manage cotton pests within a broader landscape perspective, but much additional research will be needed to understand the simultaneous and multiple impact of all pests (insects, weeds, pathogens) on plant health and to develop efficient decision aids and control methods for managing multiple stressors at multiple spatial scales. This task will be an even greater challenge for polyphagous and mobile pests such as aphids, mites, whiteflies, plant bugs, bollworm and tobacco budworms.

Meeting these challenges will likely call upon the increased use of models, risk assessment tools and information technology at both the grower and regulatory level to better understand, predict and manage systems behavior. This is going to require information managers and more user-friendly systems for storing, mining, analyzing and applying this information to farm-level decisions. Transgenic cotton conferring either insect or herbicide resistance or both has been widely adopted by growers to manage risk from caterpillar and weed pests and that trend is likely to continue. Current commercial cultivars of insecticidal transgenic cotton are based on one or more of three Cry toxins and one vegetative insecticidal protein but other proteins like snowdrop lectin (GNA) and protease inhibitors are being examined in other crop species (Christou *et al.*, 2006), and over 170 distinct δ-endotoxins as well as many other toxins are known from *B. thuringiensis* (Glare & O'Callaghan, 2000) providing much to be mined for future transgenic plant development targeting multiple pests. The technologies associated with precision agriculture (GIS, remote sensing and GPS) are likely to expand (Shaw & Willers, 2006). Such technologies may reduce overall inputs like fertilizers, herbicides and insecticides by selectively allowing growers to apply these only as needed within specific areas of a single field.

Overall, the long tradition of pest management research and practice in the cotton system will continue to expand, leading to reduced risk and greater predictability for producers, and greater sustainability and environmental stewardship benefiting society as a whole.

Acknowledgements

We thank Drs. Pete Goodell (University of California), Peter Ellsworth (University of Arizona) and John Ruberson (University of Georgia) for their insightful comments on an earlier draft of this manuscript.

References

Andow, D. A., Lovei, G. L. & Arpaia, S. (2006). Ecological risk assessment for *Bt* crops (response to Romeis). *Nature Biotechnology*, **24**, 749–751.

Anonymous (1995). Cotton: the crop and its pesticide market. *Pesticide News*, **30**, 1.

Baumhover, A. H., Graham, A. J., Bitter, B. A. *et al.* (1955). Screw-worm control through release of sterilized flies. *Journal of Economic Entomology*, **48**, 462–466.

Bottrell, D. G. & Adkisson, P. L. (1977). Cotton insect pest management. *Annual Review of Entomology*, **22**, 451–481.

Brodt, S. J., Goodell, P. B., Krebill-Prather, R. L. & Vargas, R. N. (2007). California cotton growers utilize integrated pest management. *California Agriculture*, **61**, 24–30.

Brown, P., Silvertooth, J., Moore, L. & Watson, T. (1992). Revised planting window for full season cotton varieties. In *Cotton: A College of Agriculture Report*, pp. 240–251. Tucson, AZ: University of Arizona Press.

Campion, D. G. (1994). Pheromones for the control of cotton pests. In *Insect Pests of Cotton*, eds. G. A. Matthews & J. P. Tunstall, pp. 505–534. Wallingford, UK: CABI Publishing.

Castle, S. J., Prabhaker, N. & Henneberry, T. J. (1999). Insecticide resistance and its management in cotton insects. *ICAC Review Article on Cotton Production Research, International Cotton Advisory Committee*, **5**, 1–55.

Christou, P., Capell, T., Kohli, A., Gatehouse, J. A. & Gatehouse, A. M. R. (2006). Recent developments and future prospects in insect pest control in transgenic crops. *Trends in Plant Science*, **11**, 302–308.

Conway, H. E., Steinkraus, D. C., Ruberson, J. R. & Kring, T. J. (2006). Experimental treatment threshold for the cotton aphid (Homoptera: Aphididae) using natural enemies in Arkansas cotton. *Journal of Entomological Science*, **41**, 361–373.

Cross, W. H. (1973). Biology, control, and eradication of the boll weevil. *Annual Review of Entomology*, **18**, 14–46.

Dickerson, W. A., Brashear, A. L., Brumley, J. T. *et al.* (2001). *Boll Weevil Eradication in the United States through 1999*, Cotton Foundation Reference Book Series No. 6. Memphis, TN: Cotton Foundation.

El-Lissy, O. A. & Grefenstette, W. J. (2006). Progress of pink bollworm eradication in the U.S. and Mexico, 2005, In *Proceedings of the Beltwide Cotton Conferences*, pp. 1313–1319. Memphis, TN: National Cotton Council.

El-Zik, K. M. & Frisbie, R. E. (1991). Integrated crop management systems for pest control. In *Handbook of Pest*

Management in Agriculture, vol. 3, ed. D. Pimentel, pp. 3–104. Boca Raton, FL: CRC Press.

El-Zik, K. M., Grimes, D. W. & Thaxton, P. M. (1989). Cultural management and pest suppression. In *Integrated Pest Management Systems and Cotton Production*, eds. R. E. Frisbie, K. M. El-Zik & L. T. Wilson, pp. 11–36. New York: John Wiley.

El-Zik, K. M. & Thaxton, P. M. (1989). Genetic improvement for resistance to pests and stresses in cotton. In *Integrated Pest Management Systems and Cotton Production*, eds. R. E. Frisbie, K. M. El-Zik & L. T. Wilson, pp. 191–224. New York: John Wiley.

Ellsworth, P. C. & Martinez-Carrillo, J. L. (2001). IPM for *Bemisia tabaci*: a case study from North America. *Crop Protection*, **20**, 853–869.

Ellsworth, P. C., Diehl, J. W. & Naranjo, S. E. (1996). *Sampling Sweetpotato Whitefly Nymphs in Cotton*, IPM Series No. 6, Publication 196006. Tucson, AZ: University of Arizona, Cooperative Extension. Available at http://cals.arizona.edu/crops/cotton/insects/wfsampl.html.

Ellsworth, P. C., Palumbo, J. L., Naranjo, S. E., Dennehy, T. J. & Nichols, R. L. (2006). *Whitefly Management in Arizona Cotton 2006*, IPM Series No. 18, Publication AZ1404. Tucson, AZ: University of Arizona, Cooperative Extension. Available at http://ag.arizona.edu/crops/cotton/insects/wf/ipm6.html.

Environmental Protection Agency (2006). *Reducing Pesticide Risk*. Washington, DC: US Environmental Protection Agency. Available at www.epa.gov/pesticides/health/reducing.htm.

Eveleens, K. G., van den Bosch, R. & Ehler, L. E. (1973). Secondary outbreak induction of beet armyworm by experimental insecticide application in cotton in California. *Environmental Entomology*, **2**, 497–503.

Ewing, K. P. & Parencia, C. R. Jr. (1950). *Early-Season Applications of Insecticides on a Community-Wide Basis for Cotton Insect Control in 1950*, Bureau of Entomology and Plant Quarantine Publication No. E810. Washington, DC: US Department of Agriculture.

Faria, M. & Wraight, S. P. (2001). Biological control of *Bemisia tabaci* with fungi. *Crop Protection*, **20**, 767–778.

Fillman, D. A. & Sterling, W. L. (1985). Inaction levels for the red imported fire ant, *Solenopsis invicta*, a predator of the boll weevil, *Anthonomus grandis*. *Agriculture, Ecosystems and Environment*, **13**, 93–102.

Frisbie, R. E., El-Zik, K. M. & Wilson, L. T. (eds.) (1989). *Integrated Pest Management Systems and Cotton Production*. New York: John Wiley.

Frisvold, G. B., Reeves, J. M. & Tronstad, R. (2006). *Bt* cotton adoption in the United States and China: international trade and welfare effects. *AgBioForum*, **9**, 69–78.

Gaines, J. C. (1957). Cotton insects and their control in the United States. *Annual Review of Entomology*, **2**, 319–338.

Gannaway, J. R. (1994). Breeding for insect resistance. In *Insect Pests of Cotton*, eds. G. A. Matthews & J. P. Tunstall, pp. 431–453. Wallingford, UK: CABI Publishing.

Glare, T. R. & O'Callaghan, M. (2000). Bacillus thuringiensis: *Biology, Ecology and Safety*. New York: John Wiley.

Goodell, P. B., Montez, G. & Wilhoit, L. (2006). Shifting patterns in insecticide use in California, 1993 to 2004. In *Proceedings of the Beltwide Cotton Conference*, pp. 1367–1373. Memphis, TN: National Cotton Council.

Gould, J., Hoelmer, K. & Goolsby, J. (eds.) (2008). *Classical Biological Control of* Bemisia tabaci *in the United States: A Review of Interagency Research and Implementation*. New York: Springer-Verlag.

Greene, J. K., Turnipseed, S. G., Sullivan, M. J. & May, O. L. (2001). Treatment thresholds for stink bugs (Hemiptera: Pentatomidae) in cotton. *Journal of Economic Entomology*, **94**, 403–409.

Hardee, D. D. & Harris, F. A. (2003). Eradicating the boll weevil (Coleoptera: Curculionidae): a clash between a highly successful insect, good scientific achievement, and differing agricultural philosophies. *American Entomologist*, **49**, 82–111.

Hargreaves, H. (1948). *List of the Recorded Cotton Insects of the World*. London: Commonwealth Institute of Entomology.

Harris, F. A., Canerday, T. D., Henry, L. G. & Palmquist, D. L. (1996). Working together: roles of private consultants, industry, researchers, extension, and growers. In *Cotton Insects and Mites: Characterization and Management*, eds. E. G. King, J. R. Phillips & R. J. Coleman, pp. 843–851. Memphis, TN: Cotton Foundation.

Henneberry, T. J. & Naranjo, S. E. (1998). Integrated management approaches for pink bollworm in the southwestern United States. *Integrated Pest Management Review*, **3**, 31–52.

Henneberry, T. J. & Phillips, J. R. (1996). Suppression and management of cotton insect populations on an areawide basis. In *Cotton Insects and Mites: Characterization and Management*, eds. E. G. King, J. R. Phillips & R. J. Coleman, pp. 601–624. Memphis, TN: Cotton Foundation.

Herzog, G. A., Graves, J. B., Reed, J. T., Scott, W. P. & Watson, T. F. (1996). Chemical control. In *Cotton Insects and Mites: Characterization and Management*, eds. E. G. King, J. R. Phillips & R. J. Coleman, pp. 447–469. Memphis, TN: Cotton Foundation.

Huber, R. T., Moore, L. & Hoffman, M. P. (1979). Feasibility study of areawide pheromone trapping of male pink bollworm moths in a cotton insect pest management program. *Journal of Economic Entomology*, **72**, 222–227.

Insecticide Resistance Action Committee (2007). *MoA Resources*. Brussels, Belgium: IRAC. Available at www.irac-online.org/.

Jenkins, J. N. & Wilson, F. D. (1996). Host plant resistance. In *Cotton Insects and Mites: Characterization and Management*, eds. E. G. King, J. R. Phillips & R. J. Coleman, pp. 563–600. Memphis, TN: Cotton Foundation.

King, E. G., Coleman, R. J., Morales-Ramos, J. A. *et al.* (1996a). Biological control. In *Cotton Insects and Mites: Characterization and Management*, eds. E. G. King, J. R. Phillips & R. J. Coleman, pp. 511–538. Memphis, TN: Cotton Foundation.

King, E. G., Phillips, J. R. & Coleman, R. J. (eds.) (1996b). *Cotton Insects and Mites: Characterization and Management*. Memphis, TN: Cotton Foundation.

Knipling, E. F. (1955). Possibilities of insect control or eradication through the use of sexually sterile males. *Journal of Economic Entomology*, **48**, 459–462.

Knipling, E. F. (1979). *The Basic Principles of Insect Population Suppression and Management*, US Department of Agriculture Agricultural Handbook No. 512. Washington, DC: Government Printing Office.

Kogan, M. (1998). Integrated pest management: historical perspectives and contemporary developments. *Annual Review of Entomology*, **43**, 243–270.

Kohel, R. J. & Lewis, C. F. (eds.) (1984). *Cotton*, Monograph No. 24, Agronomy Series. Madison, WI: American Society of Agronomy.

Leigh, T. F., Black, H., Jackson, C. E. & Burton, V. E. (1966). Insecticides and beneficial insects in cotton fields. *California Agriculture*, **20**(7), 4–6.

Leigh, T. F., Roach, S. H. & Watson, T. F. (1996). Biology and ecology of important insect and mites pests of cotton. In *Cotton Insects and Mites: Characterization and Management*, eds. E. G. King, J. R. Phillips & R. J. Coleman, pp. 17–85. Memphis, TN: Cotton Foundation.

Luttrell, R. G. (1994). Cotton pest management. II. A US perspective. *Annual Review of Entomology*, **39**, 527–542.

Matthews, G. A. (1994). Cultural control. In *Insect Pests of Cotton*, eds. G. A. Matthews & J. P. Tunstall, pp. 455–461. Wallingford, UK: CABI Publishing.

McGuire, M. R., Leland, J. E., Dara, S., Park, Y. H. & Ulloa, M. (2006). Effect of different isolates of *Beauveria bassiana* on field populations of *Lygus hesperus*. *Biological Control*, **38**, 390–396.

Miller, E., Lowe, A. & Archuleta, S. (2000). Evaluation of different release strategies for use in pink bollworm sterile release programs. In *Proceedings of the Beltwide Cotton Conference*, pp. 1368–1370. Memphis, TN: National Cotton Council.

Mumford, J. D. & Norton, G. A. (1994). Pest management systems. In *Insect Pests of Cotton*, eds. G. A. Matthews & J. P. Tunstall, pp. 559–576. Wallingford, UK: CABI Publishing.

Myers, J. H., Savoie, A. & van Randen, E. (1998). Eradication and pest management. *Annual Review of Entomology*, **43**, 471–491.

Naranjo, S. E. (2001). Conservation and evaluation of natural enemies in IPM systems for *Bemisia tabaci*. *Crop Protection*, **20**, 835–852.

Naranjo, S. E. (2007). Establishment and impact of exotic aphelinid parasitoids in Arizona: a life table approach. *Journal of Insect Science*, **7**, 63 (abstr.).

Naranjo, S. E. & Hagler, J. R. (1998). Characterizing and estimating the impact of heteropteran predation. In *Predatory Heteroptera: Their Ecology and Use in Biological Control*, eds. M. Coll & J. Ruberson, pp. 170–197. Lanham, MD: Entomological Society of America.

Naranjo, S. E., Flint, H. M. & Henneberry, T. J. (1996). Binomial sampling plans for estimating and classifying population density of adult *Bemisia tabaci* on cotton. *Entomologia Experimentalis et Applicata*, **80**, 343–353.

Naranjo, S. E., Ellsworth, P. C. & Hagler, J. R. (2004). Conservation of natural enemies in cotton: role of insect growth regulators in management of *Bemisia tabaci*. *Biological Control*, **30**, 52–72.

National Agricultural Statistics Service (2006). *Agricultural Chemical Usage: 2005 Field Crops Summary*. Washington, DC: US Department of Agriculture. Available at http://usda.mannlib.cornell.edu/usda/nass/AgriChemUsFC/2000s/2006/AgriChemUsFC-05-17-2006.pdf.

National Cotton Council (2007a). *Crop Information: Data bases*. Memphis, TN: National Cotton Council. Available at www.cotton.org/tech/pest/index.cfm.

National Cotton Council (2007b). *Pest Management. Cotton Pest Loss Data*. Memphis, TN: National Cotton Council. Available at www.cotton.org/tech/pest/index.cfm.

Oliveira, M. R. V., Henneberry, T. J. & Anderson, P. (2001). History, current status, and collaborative research projects for *Bemisia tabaci*. *Crop Protection*, **20**, 709–723.

Palumbo, J. C., Horowitz, A. R. & Prabhaker, N. (2001). Insecticidal control and resistance management for *Bemisia tabaci*. *Crop Protection*, **20**, 739–765.

Palumbo, J. C., Ellsworth, P. C., Dennehy, T. D. & Nichols, R. L. (2003). *Cross-Commodity Guidelines for Neonicotinoid Insecticides in Arizona*. IPM Series No. 17, Publication AZ1319. Tucson, AZ: University of Arizona, Cooperative Extension. Available at http://cals.arizona.edu/pubs/insects/az1319.pdf.

Parvin, D. W. Jr. & Smith, J. W. (1996). Crop phenology and insect management. In *Cotton Insects and Mites:*

Characterization and Managements, eds. E. G. King, J. R. Phillips & R. J. Coleman, pp. 815–829. Memphis, TN: Cotton Foundation.

Plapp, F. Jr., Jackman, J., Campanhola, C. *et al.* (1990). Monitoring and management of pyrethroid resistance in the budworm (Lepidoptera: Noctuidae) in Texas, Mississippi, Louisiana, Arkansas and Oklahoma. *Journal of Economic Entomology*, **83**, 335–341.

Prasifka, J. R., Heinz, K. M. & Minzenmayer, R. R. (2004). Relationships of landscape, prey and agronomic variables to the abundance of generalist predators in cotton (*Gossypium hirsutum*) fields. *Landscape Ecology*, **19**, 709–717.

Rainwater, C. F. (1952). Progress in research on cotton insects. In *Insects, The Yearbook of Agriculture, 1952*, pp. 497–500. Washington, DC: US Department of Agriculture, US Government Printing Office.

Ridgway, R. L., Bell, A. A., Vetch, J. A. & Chandler, J. M. (1984). Cotton protection practices in the USA and world. In *Cotton*, Monograph No. 24, Agronomy Series, eds. R. J. Kohel & C. F. Lewis, pp. 266–361. Madison, WI: American Society of Agronomy.

Romeis, J., Meissle, M. & Bigler, F. (2006). Transgenic crops expressing *Bacillus thuringiensis* toxins and biological control. *Nature Biotechnology*, **24**, 63–71.

Roush, R. T. & Daly, J. C. (1990). The role of population genetics in resistance research and management. In *Pesticide Resistance in Arthropods*, eds. R. T. Roush & B. E. Tabashnik, pp. 97–125. New York: Chapman & Hall.

Schneider, J. C. (2003). Overwintering of *Heliothis virescens* (F.) and *Helicoverpa zea* (Boddie) (Lepidoptera: Noctuidae) in cotton fields of Northeast Mississippi. *Journal of Economic Entomology*, **96**, 1433–1447.

Shaw, D. R. & Willers, J. L. (2006). Improving pest management with remote sensing. *Outlooks on Pest Management*, **17**, 197–201.

Shelton, A. M. & Badenes-Perez, E. (2006). Concepts and applications of trap cropping in pest management. *Annual Review of Entomology*, **51**, 285–308.

Smith, J. W. & Harris, F. A. (1994). Boll weevil: cotton pest of the century. In *Insect Pests of Cotton*, eds. G. A. Matthews & J. P. Tunstall, pp. 223–258. Wallingford, UK: CABI Publishing.

Snodgrass, G. L. (1996). Insecticide resistance in field populations of the tarnished plant bug (Heteroptera: Miridae) in cotton in the Mississippi Delta. *Journal of Economic Entomology*, **89**, 783–790.

Snodgrass, G. L., Scott, W. P., Abel, C. A. *et al.* (2005). Tarnished plant bug (Heteroptera: Miridae) populations near fields after early season herbicide treatment. *Environmental Entomology*, **34**, 705–711.

Sparks, T. C. (1996). Toxicology of insecticides and acaricides. In *Cotton Insects and Mites: Characterization and Management*, eds. E. G. King, J. R. Phillips & R. J. Coleman, pp. 283–322. Memphis, TN: Cotton Foundation.

Steinkraus, D. C., Hollingsworth, R. G. & Slaymaker, P. H. (1995). Prevalence of *Neozygites fresenii* (Entomophthorales: Neozygitaceae) on cotton aphids (Homoptera: Aphididae) in Arkansas cotton. *Environmental Entomology*, **24**, 465–474.

Sterling, W. L., El-Zik, K. M. & Wilson, L. T. (1989). Biological control of pest populations. In *Integrated Pest Management Systems and Cotton Production*, eds. R. E. Frisbie, K. M. El-Zik & L. T. Wilson, pp. 155–189. New York: John Wiley.

Stern, V., van den Bosch, R. & Leigh, T. F. (1964). Strip cutting alfalfa for lygus bug control. *California Agriculture*, **18**, 5–6.

Stoltz, R. L. & Stern, V. M. (1978). Cotton arthropod food chain disruption by pesticides in the San Joaquin Valley, California. *Environmental Entomology*, **7**, 703–707.

Tabashnik, B. E., Carrière, Y., Dennehy, T. J. *et al.* (2003). Insect resistance to transgenic *Bt* crops: lessons from the laboratory and field. *Journal of Economic Entomology*, **96**, 1031–1038.

Toscano, N. C., Mueller, A. J., Sevacherian, V. & Sharma, R. K. (1974). Insecticide applications based on hexalure trap catches versus automatic schedule treatments for pink bollworm moth control. *Journal of Economic Entomology*, **67**, 522–524.

Trichilo, P. J. & Wilson, L. T. (1993). An ecosystem analysis of spider mite outbreaks: physiological stimulation or natural enemy suppression. *Experimental and Applied Acarology*, **17**, 291–314.

van den Bosch, R. & Hagen, K. S. (1966). *Predaceous and Parasitic Arthropods in California Cotton Fields*, California Agricultural Experiment Station Bulletin No. 820. Berkeley, CA: University of California.

Villavaso, E. J., Bartlett, A. C. & Laster, M. L. (1996). Genetic control. In *Cotton Insects and Mites: Characterization and Management*, eds. E. G. King, J. R. Phillips & R. J. Coleman, pp. 539–562. Memphis, TN: Cotton Foundation.

Wagner, T. L., Olson, R. L., Willers, J. L. & Williams, M. R. (1996). Modeling and computerized decision aids. In *Cotton Insects and Mites: Characterization and Management*, eds. E. G. King, J. R. Phillips & R. J. Coleman, pp. 205–249. Memphis, TN: Cotton Foundation.

Walker, J. K. & Smith, C. W. (1996). Cultural control. In *Cotton Insects and Mites: Characterization and Management*, eds. E. G. King, J. R. Phillips & R. J. Coleman, pp. 471–510. Memphis, TN: Cotton Foundation.

Watson, T. F. (1980). Methods for reducing winter survival of the pink bollworm. In *Pink Bollworm Control in the Western United States*, ed. H. S. Graham, pp. 24–34. Oakland, CA: US Department of Agriculture, Science and Education Administration, Western Region.

Whalon, M. E., Mota-Sanchez, D., Hollingsworth, R. M. & Duynslayer, L. (2007). *IRAC Arthropod Pesticide Resistance Database*. East Lansing, MI: Michigan State University. Available at www.pesticideresistance.org/.

Whitcomb, W. H. & Bell, K. (1964). *Predaceous Insects, Spiders and Mites of Arkansas Cotton Fields*, Bulletin No. 690. Fayetteville, AR: Arkansas Agricultural Experiment Station.

Williams, M. R. (2006). Cotton insect losses 2005. In *Proceedings of the Beltwide Cotton Conference*, pp. 1151–1204. Memphis, TN: National Cotton Council.

Williams, M. R. (2007). Cotton insect losses 2006. In *Proceedings of the Beltwide Cotton Conference*, pp. 974–1026. Memphis, TN: National Cotton Council.

Wilson, L. T., Gonzales, D. & Plant, R. E. (1985). Predicting sampling frequency and economic status of spider mites on cotton. In *Proceedings of the Beltwide Cotton Conference*, p. 168. Memphis, TN: National Cotton Council.

Chapter 26

Citrus IPM

Richard F. Lee

IPM programs are designed to keep plants healthy and economically productive while minimizing environmental impact. Citrus (*Citrus* spp.) as clonally propagated perennial crops are subject to many graft-transmissible diseases caused by viruses, viroids and systemic prokaryotes. Some of these graft-transmissible diseases can be very destructive and even threaten the continued production of citrus in a production area, whereas other diseases cause minor losses. The starting point for an IPM program for citrus is to begin with healthy plants. The concept of planting with healthy plants in citrus began almost simultaneously with the discovery that some diseases of citrus were caused by graft-transmissible pathogens (GTPs). Indicator plants grafted with parts of diseased trees subsequently show characteristic symptoms for the disease (Fawcett, 1938). The concept of a clean stock program evolved from the finding that the disease could be prevented by using graft propagations from source trees that were free of the disease. Applications of this concept led to the development of regional and national clean stock programs and certification programs. GTPs of citrus which do not have insect vectors or other natural means of spread are easily controlled by use of clean, or "pathogen-tested," budwood, but diseases which have a natural means of spread, such as insects, are more difficult to control. However, beginning with healthy plants is even more important for the management and control of these naturally spread graft-transmissible diseases. In this chapter, the essential components of a citrus certification program will be reviewed, the graft-transmissible diseases of citrus and methods available for therapy or cleaning of germplasm will be summarized and the application of mild strain cross protection to lessen losses due to *Citrus tristeza virus* will be described as an example of the management of a devastating insect vectored graft-transmissible disease.

26.1 Components of a citrus certification program

The term "certification program" is often applied in a vague sense. The term has been used to describe a regulatory program to prevent the spread of nematodes affecting citrus, or to recognize the fact that a nursery site has been "certified" to meet certain predefined standards, or to define horticultural standards such as size of propagated plants before sale to growers and height of bud-unions. Here the term, certification program, is used to describe an IPM program designed to produce "pathogen-tested" citrus plants for planting into the field. A typical certification program has three critical elements which are interrelated:

Integrated Pest Management, ed. Edward B. Radcliffe, William D. Hutchison and Rafael E. Cancelado. Published by Cambridge University Press.

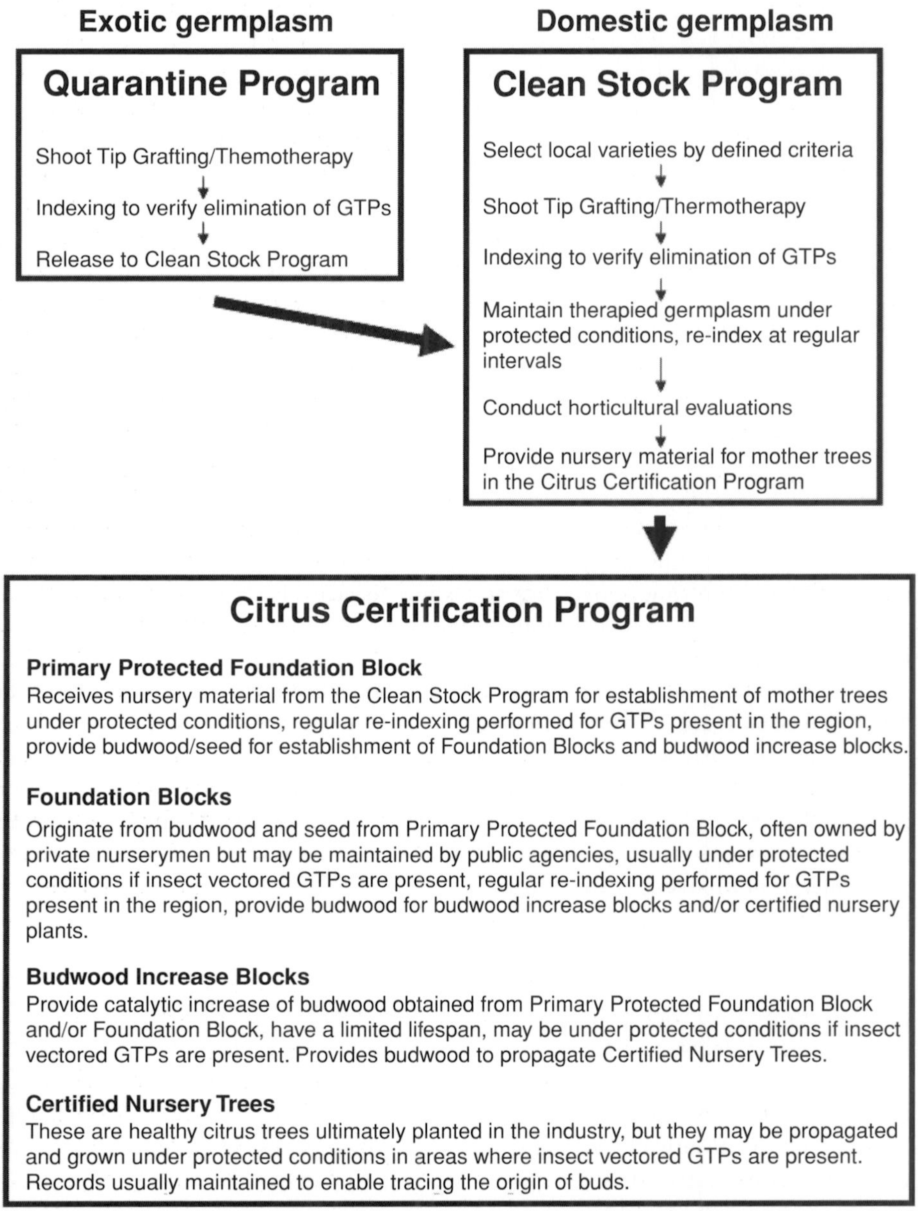

Exotic germplasm

Quarantine Program

Shoot Tip Grafting/Themotherapy
↓
Indexing to verify elimination of GTPs
↓
Release to Clean Stock Program

Domestic germplasm

Clean Stock Program

Select local varieties by defined criteria
↓
Shoot Tip Grafting/Thermotherapy
↓
Indexing to verify elimination of GTPs
↓
Maintain therapied germplasm under protected conditions, re-index at regular intervals
↓
Conduct horticultural evaluations
↓
Provide nursery material for mother trees in the Citrus Certification Program

Citrus Certification Program

Primary Protected Foundation Block
Receives nursery material from the Clean Stock Program for establishment of mother trees under protected conditions, regular re-indexing performed for GTPs present in the region, provide budwood/seed for establishment of Foundation Blocks and budwood increase blocks.

Foundation Blocks
Originate from budwood and seed from Primary Protected Foundation Block, often owned by private nurserymen but may be maintained by public agencies, usually under protected conditions if insect vectored GTPs are present, regular re-indexing performed for GTPs present in the region, provide budwood for budwood increase blocks and/or certified nursery plants.

Budwood Increase Blocks
Provide catalytic increase of budwood obtained from Primary Protected Foundation Block and/or Foundation Block, have a limited lifespan, may be under protected conditions if insect vectored GTPs are present. Provides budwood to propagate Certified Nursery Trees.

Certified Nursery Trees
These are healthy citrus trees ultimately planted in the industry, but they may be propagated and grown under protected conditions in areas where insect vectored GTPs are present. Records usually maintained to enable tracing the origin of buds.

Fig. 26.1 Diagram of the critical components of a citrus certification program. GTP, graft-transmissible pathogens.

(1) a quarantine program, (2) a clean stock program and (3) a certification program (Lee *et al.*, 1999; Navarro, 1993) (Fig. 26.1).

26.1.1 Quarantine programs

Adherence to proper quarantine procedures is an essential safeguard when bringing in new, exotic germplasm into the local certification program. Citrus growers, by their nature, are always looking for something new and unique, such as low-seeded varieties, varieties higher in color than those presently available, or varieties which may extend the marketing window by maturing earlier and later than varieties presently available. Such germplasm selections will be imported into any citrus industry; it is essential that they come in through an authorized quarantine program so that additional pests and/or graft-transmissible diseases are not imported with illegally and untested germplasm. Quarantine programs operate under the jurisdiction of the ministry of agriculture of a country or commissioner of agriculture in a state or province. Usually the quarantine program is operated by the plant protection services of a government regulatory agency.

The traditional approach to quarantine is to use isolation to prevent accidental entry of new pests into the local industry. The isolation may be geographic with large distances maintained from the industry, or by the use of screened, vector-proof greenhouses. Under quarantine isolation, the new germplasm is tested for presence of pathogens and, if present, then therapy measures are applied to remove the pest(s). Following therapy, retesting is conducted to assure the pest(s) have been eliminated. The best approach for testing is the use of biological indicator plants as their use would enable the visualization of symptoms which may be caused by unsuspected pests. Biological indexing provides a higher level of confidence that all pests have been eliminated. Laboratory testing for pests will reveal freedom or presence only of the pest for which the diagnostic procedure is designed. As an example, a biological index for *Citrus psorosis virus* on Dweet tangor produces symptoms visually indistinguishable from those caused by several other virus-like pathogens of the psorosis group (e.g. Citrus concave gum, impietratura, cristacortis, and *Citrus leaf blotch virus*) while the reverse transcription polymerase chain reaction (RT-PCR) assay for psorosis would only detect psorosis.

An alternative approach to quarantine by geographic isolation is to use the in vitro method where imported budwood is kept under quarantine in glass test tubes. This procedure was introduced by Navarro in Spain for the safe introduction of new citrus varieties (Navarro *et al.*, 1984; Lee *et al.*, 1999). The budwood, upon arrival, is surface sterilized, then placed in test tubes containing culture media and then maintained in a growth chamber. When the shoots emerge, shoot tip grafting (STG) is done, with the thin section of the meristematic buds being grafted to healthy receptor seedling plants used as a rootstock. This approach offers an advantage in that it requires less space than a quarantine greenhouse, and entry of new germplasm is usually expedited as everything is shoot tip grafted upon receipt. However, the incoming budwood is not always in good shape following shipment, and sometimes fungal contaminates prevent recovery of the germplasm regardless of efforts to sterilize the budwood stick.

Often citrus seed is freely imported without concern to what pests may be present. Several important citrus diseases have been shown to be seed-transmitted including: citrus variegated chlorosis, caused by the xylem-inhabiting bacterium *Xylella fastidiosa* (Li *et al.*, 2003), witches' broom disease of lime, caused by *Candidatus* Phytoplasma aurantifolii (El-Kharbotly *et al.*, 2000; Khan *et al.*, 2002), psorosis and a psorosis-like pathogen in hardy orange (*Poncirus trifoliate*) and in hybrids having *P. trifoliata* as one of the parents (e.g. citranges [sweet orange (*C. aurantium*) × hardy orange] and citrumelos [hardy orange × grapefruit (*C. paradisi*)] (Roistacher, 1991; Powell *et al.*, 1998), and *Citrus leaf blotch virus* (Guerri *et al.*, 2004). Care should be taken that seed source trees have been indexed for freedom from these diseases. Additionally imported seed should undergo hot water treatment (10 min at 52 °C), rinse in ambient temperature water, then treatment for 2 min in 1% solution of 8-hydroxquinoline sulfate in water and then air drying to eliminate fungal contamination (Roistacher, 1991).

26.1.2 Clean stock programs

Development of the clean stock program has enabled recovery of healthy plants from locally grown (domestic) varieties and cultivars which may be infected with GTPs and the continued maintenance of this therapied germplasm for use in the certification program. Additionally, the clean stock program maintains exotic germplasm which has been previously freed from pathogens during quarantine as "pathogen tested" material for use in the local certification program. Clean stock programs are often maintained by universities, research institutions or non-government organizations as a service for the citrus industry, but they also may be maintained by regulatory agencies.

There are several steps involved in the operation of a clean stock program: (1) selection of mother trees from the local cultivars, (2) indexing of the selected mother trees, (3) therapy to eliminate GTPs which may be present, (4) indexing of the recovered plants to ensure the GTPs have been eliminated, (5) horticultural evaluation of the recovered, healthy plants and (6) maintenance of recovered, healthy plants. Recovery of healthy breeding stock is expensive and usually takes three to four years. Because of this, care should be given to the selection of potential mother trees; documentable traits such as superior productivity, better fruit color, etc. should be used for selection and concern of the disease status should not be considered as the pathogen(s) will be eliminated by therapy. The selected mother trees should be propagated in the greenhouse or screenhouse on vigorous rootstock to provide a source of material for virus indexing and therapy procedures. Indexing should be performed on the selected trees to identify GTPs which are present (Roistacher, 1991). Biological indexing may be augmented by the use of laboratory diagnostic assays. Selected trees are then subjected to therapy, either thermotherapy or STG which will be described in greater detail later. The resultant "presumably clean" recovered plants following therapy are then re-indexed to verify the freedom from targeted GTPs. Horticultural evaluation of the recovered plants is essential, and often this material is propagated on a conditional basis until fruit has set on the recovered plants to verify horticultural trueness-of-type. There are no reports of problems with trueness-of-type following thermotherapy or STG. However, some varieties, such as navel sweet orange (*C. aurantium*), have more of a tendency to produce spontaneous mutations, so it is good horticultural practice to confirm trueness-of-type. After the germplasm has been selected, subjected to therapy, indexed to verify freedom from target pathogens, and horticulturally evaluated, the pathogen-tested germplasm should be maintained so that it is protected from further infections and especially from insect vectors. This may be in a screenhouse having a suitably small-mesh screen to exclude insects or in a greenhouse. The level of exclusion depends on the pathogens/vectors that are present in the area. For example, citrus greening disease, also known as huanglongbing (causal agent *Candidatus* Liberibacter) was discovered to be in Florida in 2005. State nursery regulations have since been updated (Florida Legislature, 2007) to encourage the location of protected propagation sources to areas isolated from commercial citrus. Practices for the exclusion from vectors have also been updated by the requirement for double-entry doorways equipped with air doors, insect exclusion of ventilation fans either by use of screening and/or use of better sealing vents which always keep a positive air pressure. Additional information on recommended guidelines for exclusion of citrus greening may be found in the Citrus Health Response Program (CHRP) – Florida State (US Department of Agriculture, 2007). Because of the time and effort required to produce pathogen-tested germplasm from the selected varieties, consideration should be given to maintaining a back-up location at a different site. Natural disasters, such as tornados and hurricanes, can be destructive and beyond human control, and encroachment of vectored diseases represent distinct risks.

26.1.3 Certification programs

The purpose of a certification program is to make sure nursery material being planted in the industry is pathogen-tested and of the highest possible horticultural quality. Certification programs are usually operated under the jurisdiction of a governmental agency having the legal authority to impose restrictions for the good of the industry in general and having the right to inspect nurseries (Fegan *et al.*, 2004). The requirements for periodical retesting of propagation materials, inspection of nursery materials, and detailed reports which may be required are described in regulations which govern the certification program, and are put into law by the governmental agency operating the program. In many instances though, the regulations have been developed by the affected nurseries and growers who constitute the local industry who have the foresight to see the advantages of a certification program; once the regulations have been formulated by the citrus industry, the industry leaders then work with the regulatory agency to implement the drafted regulations. The certification programs in California, Florida, Spain, South Africa, Taiwan and Australia have been in operation the longest. The Florida program was voluntary until 1997 when it became mandatory because of concern over spread of *Citrus tristeza virus* by the brown citrus aphid (*Toxoptera citricida*) and additional modifications in the program regulations occurred in 2007 because of the discovery of citrus greening in 2005 (Halbert, 2005) and the presence of its vector, Asian citrus psyllid (*Diaphorina citri*).

A certification program receives pathogen-tested germplasm through the quarantine and clean stock programs. New varieties are released into the industry through the certification program, and the certification program provides yield and horticultural information, and distributes extension type materials on varieties and graft-transmissible diseases. Record keeping is an important part of a certification program, and the documentation created allows for tracing problems back to a specific budwood mother tree.

A certification program is structured so that the indexing and horticultural evaluations are limited to the primary, protected foundation trees realizing that the superior quality in these elite trees will be available in all resultant propagated plants if the guidelines are followed. The general structure of all certification programs are similar, although the terminology used may vary. There are four main blocks of trees in a certification program: (1) primary protected foundation block, (2) foundation blocks, (3) budwood increase blocks and (4) certified nursery tree blocks which are ultimately distributed and planted into the industry (Roistacher, 1993).

Primary protected foundation blocks are comprised of pathogen-tested plants recovered from either the quarantine or clean stock programs. They are usually maintained by the governmental agency operating the certification program. They are grown in protected conditions, either in screenhouses or greenhouses to prevent entry of insect vectors. The plants will have been evaluated for horticultural quality and subjected to recurring pathogen testing to continually verify their pathogen-tested status. Depending on the program, the plants may be maintained in containers or planted in the ground. The plants in the primary protected foundation block provide the primary source of budwood for the establishment of foundation blocks and sometimes budwood increase blocks.

Foundation blocks may be maintained by either the governmental agency operating the certification program or by private nurseries. In areas where insect-vectored pathogens are not present, they are often maintained in the field while if insect-vectored pathogens are present, they must be maintained either in a greenhouse or screenhouse. The trees are subjected to pathogen-testing on a recurring basis and inspected at least annually for abnormalities. The trees in a foundation block are propagated only using budwood originating from the primary protected foundation block. In addition to providing high-quality budwood for the nursery which owns the block, these blocks are also useful for recovering valuable budwood in the event of widespread natural disasters, such as hurricanes.

Budwood increase blocks are used to provide the volume of high-quality budwood needed to propagate the trees needed by the industry. Budwood used to establish a budwood increase block originates either from the primary protected foundation block or the foundation block, and collection of budwood from these trees is allowed for a limited period of time, usually two to four years. Recurring indexing for insect-vectored pathogens, such as *Citrus tristeza virus*, is usually required. The budwood increase blocks may be maintained under protected conditions or in the field, depending on the risks posed by insect-vectored pathogens.

Certified nursery trees ultimately are planted in the industry. Usually they are grown under field conditions unless *Citrus tristeza virus* and/or citrus greening or other insect-vectored pathogens are present. Certified nursery trees are often visually inspected, but not usually subjected to indexing for pathogens. Because the budwood used for certified nursery trees originated from elite budwood from the primary protected foundation block or the foundation block, the trees benefit from the horticultural evaluations and pathogen-testing previously performed on the elite material.

Seed source trees should be included in the certification program. Horticultural evaluation of seed source trees is important; these trees will provide the multitude of rootstock liners used by the local citrus industry. There are some pathogens reported to be seed transmitted: psorosis-like pathogens in hardy orange or hybrids having hardy orange as one of the parent, witches' broom disease of lime (WBDL) caused by a Phytoplasma and citrus variegated chlorosis (Table 26.1). Indexing of seed source trees to verify freedom from the pathogens present in the area which are known to be seed transmitted is required. Seed source trees should also be inspected annually for freedom from abnormalities and other diseases.

26.2 Summary of graft-transmissible diseases of citrus

Citrus has more GTPs than most clonally propagated crops (Roistacher, 1991, 1993). This probably reflects the fact that citrus was transported as vegetative material around the world before the nature of virus and virus-like diseases were known. Most of the important citrus GTPs are reviewed in Roistacher (1991), and some new diseases have been recognized. Citrus variegated chlorosis was first recognized in 1987 and this disease limits citrus production in the northern citrus growing areas of São Paulo State, Brazil (Hartung *et al.*, 1994). WBDL caused by a Phytoplasma occurs in the Persian Gulf area and has almost totally eliminated the production of acid limes (*C. aurantifolia*) in Oman and surrounding areas. WBDL is reported to be seed transmitted (El-Khorbotly *et al.*, 2000; Khan *et al.*, 2002) and now occurs in Iran on seed and/or budwood from the Omani lime in Oman. Citrus chlorotic dwarf, spread by bayberry whitefly (*Parabemisia myricae*) in Turkey is a relatively new virus-like disease of citrus (Kersting *et al.*, 1996).

Most GTPs that do not have a natural means of spread, such as by vectors, soil fungi or seed transmission, are easy to eliminate from a citrus growing area over a period of time by starting a mandatory citrus certification program having the clean stock and quarantine components. The mandatory certification program assures that all citrus plants produced are free of GTPs, and that the GTPs not having a method of natural spread are gradually eliminated in the industry over a period of several years as the existing trees are removed as existing GTPs present in them render them non-productive and more susceptible to stresses such as drought and cold. GTPs having a means of natural spread (Table 26.1) are harder to control and some naturally spread GTPs, such as *Citrus tristeza virus*, citrus variegated chlorosis and citrus greening will shorten the economic productive lifespan of the trees being planted. However, if the naturally spread GTPs are propagated in the nursery, the trees being planted in the field will already be infected and in many cases, there will not even be an economic return to the grower (Roistacher, 1996).

Citrus greening provides an excellent case study. The disease is caused by a fastidious phloem-inhabiting bacterium; there are three strains presently recognized: the Asian form, C. L. asiaticus, which expresses the most severe symptoms under warm temperatures, the African form, C. L. africanus which expresses the most severe symptoms under cooler temperatures, and the Brazilian form, C. L. americanus which was first reported in 2004 (Teixeira *et al.*, 2005) and expresses symptoms similar to the Asian form. Phloem-feeding psyllids are the vectors of citrus greening; Asian citrus psyllid and the African citrus psyllid (*Trioza erytreae*) are the vectors associated with the Asian and African forms, respectively. Either vector can transmit either form of citrus greening. Asian citrus psyllid transmits both the Brazilian and Asian forms in Brazil (Bove, 2006). Asian citrus psyllid has become widely established in citrus areas in the western hemisphere with perhaps California being the only major citrus area where this psyllid has not been established. Citrus greening has been reported from 29 countries in Asia and Africa and has caused major crop losses in many areas (Bove, 2006). In areas where citrus greening occurs, it is considered to be a limiting factor of citrus production.

The economic losses caused by citrus greening have been reported. Grenzebach (1994) reported on an economic study of mandarin production in Thailand. He used cost data provided by the Thailand Department of Agricultural Extension and a national average production of 12.5 t/ha. An average rate of spread of citrus greening, based on

Table 26.1 Graft-transmissible pathogens reported to have a vector or other means of natural spread or are seed transmitted

Graft-transmissible disease	Causal agent	Vector or means of spread	Seed transmitted
Citrus blight (Derrick & Timmer, 2000)	Unknown graft-transmissible agent	Unknown	Not reported
Citrus chlorotic dwarf (Kersting *et al.*, 1996)	Unknown graft-transmissible agent	Whitefly: *Parabemisia myricae*	Not reported
Citrus variegated chlorosis (Hartung *et al.*, 1994)	Strain of *Xylella fastidiosa*	Sharpshooter species: Hemiptera: Cicadellidae	Yes (Li *et al.*, 2003)
Huanglongbing (Citrus greening) (Bove, 2006)	*Candidatus* Liberobacter asiaticum, L. africanum, L. americanus for Asian, African and American (Brazilian) forms, respectively	Psyllids: *Diaphorina citri* and *Trioza erytreae*	Not reported
Indian citrus mosaic (Ahlawat *et al.*, 1996)	Badnavirus	Citrus mealybug: *Planococcus citri*	Not reported
Citrus leprosis (Rodrigues *et al.*, 2003; Guerra-Moreno *et al.*, 2005)	Cytoplasmic and nuclear citrus leprosis viruses	Mites: *Brevipalpus* spp. (Acari: Tenuipalpidae)	Not reported
Naturally spread citrus psorosis (Derrick & Timmer, 2000)	*Citrus psorosis virus*	Unknown	Maybe (Roistacher, 1991)
Satsuma dwarf	*Satsuma dwarf nepovirus*	Unidentified soilborne agent	Not reported
Stubborn (Bove, 1995)	*Spiroplasma citri*	Leafhoppers: *Scaphytopius nitrides, Neoaliturus tenellus, N. haemoceps*	Not reported
Tristeza (Bar-Joseph *et al.*, 1989)	*Citrus tristeza virus*	Aphis: *Toxoptera citricida*, *Aphis gossypi* and *A. citricola* are the most common	Not reported
Witches' broom disease of lime (El-Kharbotly *et al.*, 2000)	*Candidatus* Phytoplasma aurantifolia	Leafhopper: *Hishimonus phycitis* is suspected but not confirmed	Yes (El-Khorbotly *et al.*, 2000; Khan *et al.*, 2002)
Woody gall (Whiteside *et al.*, 1988)	Citrus vein enation virus	Aphids: *T. citricida* and *A. gossypii*	Yes (Whiteside *et al.*, 1988)

visual symptoms, was estimated to be 10% per year and non-productive trees were removed and new plants used to replace them on a continual basis. An average mandarin (*C. reticulate*) grove would produce 22.75 t/ha in years five to eight, then production would decline to 6.5 t/ha by year 12, giving an average lifespan for the average mandarin grove of 12 years. Roistacher (1996) further examined the economic impact of citrus greening in Thailand. He considered the rate of spread of citrus greening in different regions of Thailand and the healthy status of trees being used to establish groves and for replanting. Roistacher's model indicated that if healthy trees were used to establish the grove but there were no efforts to manage the psyllid vectors of citrus greening or remove infected trees, a grove could be expected to last for about ten years and have a cumulative profit of $US 3383/ha. If management of the psyllid vectors was done and infected trees removed on a regular basis, the estimated lifespan of a block would be 20 years for a cumulative profit of $US 125 000/ha.

26.3 Therapy of graft-transmissible disease from citrus germplasm

There are two commonly used methods for therapy of germplasm of GTPs: thermotherapy or heat treatment (Calavan *et al.*, 1972; Roistacher, 1991) and shoot tip micrografting, commonly called shoot tip grafting or STG (Navarro *et al.*, 1975). Both methods are useful, but have limitations.

Thermotherapy is performed by taking buds from the source plant which needs to be freed from infection and grafting them onto citrange seedlings which have originated from pathogen-tested sources to preclude seed-transmitted GTPs being present (Table 26.1). The buds are allowed to develop callus, but the wrap is not removed. The budded seedlings are then placed in a precondition greenhouse with 28–40 °C daytime and 25°C night temperatures for 30 days. The budded seedlings are then placed in a controlled temperature chamber set for 16 h 40 °C days and 8 h 30 °C nights for 16 weeks. At the end of 16 weeks, the plants are removed from the chamber and placed in a greenhouse having 28–35 °C and 25 °C night temperature (but the temperature is not critical at this point). The wraps are removed, the rootstock seedling is bent over so that the grafted bud will become the terminal bud. After thermotherapy and when the bud has pushed and is growing, the seedling is pruned to allow only this bud to grow. After a period of time, 12–18 weeks, the plant is indexed for freedom of viruses. Thermotherapy is effective at eliminating most GTPs except for viroids and citrus stubborn, caused by *Spiroplasma citri*.

STG is perhaps the most universally utilized method for therapy of citrus from GTPs. This method is effective at removing viroids and stubborn, which persists after thermotherapy, as well as all other GTPs (Navarro *et al.*, 1975; Roistacher *et al.*, 1977). STG is performed by forcing new growth tips on the source plant. Under a microscope, about 0.1 mm of the shoot tip is removed and grafted onto a seedling which has been grown under sterile conditions in a test tube. The theory behind STG is that the meristem tip grows faster than the GTPs can move into the new tissue, and if a thin enough tip is cut, it is free of GTPs. Following the grafting of the shoot tip onto the seedling, the seedling is returned to the test tube and placed under light in a culture chamber. When the micrografted propagation reaches 1–2 cm in length, the scion is regrafted to a healthy rough lemon (*C. jambhiri*) or citrange seedling and moved to the greenhouse where it is allowed to grow. While all GTPs have been demonstrated to be eliminated by STG (Roistacher, 1991), viroids and *Citrus tatterleaf virus* are the most difficult to eliminate, based on personal experience.

Perhaps most important, following therapy, it cannot be assumed that the germplasm has been freed of GTPs. Indexing absolutely is required. It is probably preferable to use both methods of therapy to give a high confidence level that all known and any unknown or unrecognized GTPs have been eliminated. Biological indexing is preferable to laboratory diagnostic methods as biological indexing is more likely to reveal unknown or unrecognized GTPs, whereas laboratory diagnostic assays will only find pathogens that the reagents are specific for.

26.4 Application of mild strain cross-protection to protect against *Citrus tristeza virus*

Vectored GTPs are more difficult to control than pathogens not having a means of natural spread. While propagation of healthy plants is a beginning of an IPM program, other measures may be needed to extend the economic life in the threat of devastating, vectored diseases. The use of mild strain cross-protection against severe strains of *Citrus tristeza virus* is a good example for study. This is not intended as a comprehensive review of mild strain cross-protection of *Citrus tristeza virus* in citrus, but rather as a brief summary of recent research.

Mild strain cross-protection (MSCP) is a management strategy which has been applied in citrus for continuing production in the threat of severe strains of *Citrus tristeza virus* (Lee *et al.*, 1987). MSCP is a phenomenon which occurs when a mild strain of a virus is introduced into a plant and prevents or delays the expression of the symptoms of a severe strain of the same virus that is later introduced into the same plant (Lee *et al.*, 1987). This definition is important because MSCP is often confused as being the same as virus resistance. MSCP is a management strategy which may be useful in certain instances when severe strains of *Citrus tristeza virus* are present to extend the economic lifespan of a grove, but it is not a permanent resistance to the virus. Repeated and continual challenge of plants protected by mild strains by severe strains will eventually result in the breakdown of cross-protection and result in the decline of the tree. Much of the research on MSCP has been empiric because molecular methods only have been available recently.

Citrus tristeza virus has many strains and displays a diversity of biological symptoms depending on the strain of the virus and host which is being affected (Rocha-Pena *et al.*, 1995). In areas where sour orange is the predominate rootstock, decline strains of *Citrus tristeza virus* which cause decline and death of trees on sour orange are a concern, but the more severe strains which cause stem pitting of either sweet orange and/or grapefruit scions are usually not common in areas where sour orange rootstock is predominate (Rocha-Pena *et al.*, 1995). Mild strains are a relative term depending on what isolates are present and how severe they are. In Florida the mild isolates are very mild, often requiring a susceptible host such as Mexican lime (*C. aurantifolia*) to be grown under ideal conditions to even see mild vein flecking or vein clearing, and often serological assays are needed to confirm the presence of the mild strain because the symptoms are so slight. The stem pitting strains of *Citrus tristeza virus* are considered to be the most severe form of the virus and control of stem pitting strains is not controlled simply by planting on a *Citrus tristeza virus* tolerant rootstock such as can be done to avoid decline of trees on sour orange when decline strains of *Citrus tristeza virus* are present.

Most of the research on MSCP to protect against *Citrus tristeza virus* has been done for the protection against stem pitting strains. MSCP is used on a large scale in Australia and South Africa to protect against grapefruit stem pitting strains of *Citrus tristeza virus*, and in Brazil to protect against stem pitting of sweet orange, with Pera sweet orange being the most common variety affected by stem pitting *Citrus tristeza virus* (Lee & Niblett, 2000). Some common traits have been observed and associated with mild strains of *Citrus tristeza virus* which are useful for cross-protection (Lee *et al.*, 1987); the mild virus usually has high titer or concentration in infected plants although the symptoms being expressed are mild. This is in contrast to most instances where severe strains of *Citrus tristeza virus* tend to have high concentrations while mild strains tend to have lower concentrations in infected plants. The mild strains used for MSCP must have the ability to be easily graft transmitted and must be able to move into new growth flushes quickly. It is probably useful if the mild strains are easily aphid transmitted as they may be more likely to be spread than the severe strains (Rocha-Pena *et al.*, 1991). The mild *Citrus tristeza virus* isolate must be mild in all citrus hosts grown in a region where MSCP might be employed under field conditions. It is important to realize that the protective ability of the mild strains is lessened with repeated challenge with severe strains, and eventually the protection may be overwhelmed causing cross-protection to break down.

The use of MSCP is a last, desperate measure to be used to help extend the economic lifespan of a block of trees (Lee & Niblett, 2000). MSCP should be used only when severe strains of *Citrus tristeza virus* are endemic and have an obvious detrimental effect on production and the economic life of trees. In areas where severe stem pitting strains of *Citrus tristeza virus* are endemic, MSCP is usually implemented in the nurseries. The MSCP measure works best if the trees are propagated free of GTPs. The selected and desired mild strain of *Citrus tristeza virus* is then added back to the sources used for the propagation of plants to be planted in the field.

While most research on MSCP has involved selecting mild strains of *Citrus tristeza virus* which offer protection against stem pitting strains, research in Florida was done to determine if mild strains could be introduced into trees on sour orange rootstock established in the field and already infected with *Citrus tristeza virus* (Lee & Niblett, 2000). A monoclonal antibody was developed which enabled selective detection of *Citrus tristeza virus* isolates causing decline on sour orange rootstock, while mild, non-decline strains of *Citrus tristeza virus* did not react (Permar *et al.*, 1990).

For years, sour orange was the most popular rootstock in Florida, because of its production of high-quality fruit and tolerance to citrus blight, soil fungi and calcareous soils. In the mid-1980s it became obvious that decline strains of *Citrus tristeza virus* were widespread in Florida and being spread to new locations because budwood source trees were not screened for *Citrus tristeza virus* under the voluntary Florida budwood program (Brlansky *et al.*, 1986). Many of the trees on sour orange rootstock were severely dwarfed and did not reach production. A biological index of budwood source trees in 1984 revealed that nearly half of the trees contained isolates of *Citrus tristeza virus* which reduced growth of the scions by 50% when compared to virus-free or mild isolate infected scion of the same cultivar when propagated on sour orange liners (Yokomi *et al.*, 1992). Several Florida mild strains had been identified which appeared to be useful for MSCP against stem pitting strains of *Citrus tristeza virus* (Yokomi *et al.*, 1987). A system was developed to inoculate trees on sour orange in the field with the selected mild strains for cross-protection (Rocha-Pena *et al.*, 1992). This was done by using leaf midribs from plants infected with the mild strains; the leaf midribs were grafted onto four branches of the field tree (Lee *et al.*, 1992). One isolate, T30, has a unique predominate double-stranded RNA (dsRNA), which we now know is from a defective RNA, which enables confirmation of the infection by T30 by extraction of dsRNA and electrophoresis on polyacrylamide gels. Using this approach, it was determined that the mild isolate T30 was distributed within a mature field tree (5–12 years old) within six months. Inoculation of the mild strains, T30, T49 and T50a, into field trees did not prevent some trees in the block from declining from decline strains of *Citrus tristeza virus*, but the rate of decline was generally more constant allowing for gradual replacement with trees propagated on a *Citrus tristeza virus* tolerant rootstock, generally Swingle citrumelo (Duncan grapefruit × hardy orange). The production remained relatively constant in the blocks inoculated with mild strains, while the control blocks often declined rapidly with all the trees dying within two to three years, and then the production did not exceed costs of grove maintenance for newly planted blocks for another three to four years. In one area in the flatwoods near Avon Park, Florida, at the end of ten years after inoculation of mild strains into the existing navel trees on blocks of 750 trees (seven years old when inoculated and 10% of the trees declining from *Citrus tristeza virus* when the inoculation with the mild isolate was done), 89% of the original trees on sour orange rootstock remained. In the control block about 90 meters away and under the same grower management, only 21% of the original trees on the same rootstock remained. In northern Lake and Orange County and in Marion County, about 13 000 ha of existing trees on sour orange rootstock were inoculated with mild strains from 1999 to 2003 (R. Lee, unpublished data).

The trees used as the sources of the mild strains were monitored by ELISA using the monoclonal antibody which selectively detects decline strains of *Citrus tristeza virus* every three months, and the use of strain group specific probes (Ochoa *et al.*, 2000) was used to monitor for freedom of potential decline and/or stem pitting strains of *Citrus tristeza virus*. Other research on the population of strains present in *Citrus tristeza virus* isolates has revealed that severe strains may be hidden in common or mild isolates of *Citrus tristeza virus*, but these severe strains may be transmitted by aphids and then expressed as severe strains (Albiach-Martí *et al.* 2000; Tsai *et al.*, 2000; Brlansky *et al.*, 2003).

As better molecular tools become available more will be learned about the mechanism of MSCP. In areas where severe stem pitting strains of *Citrus tristeza virus* cause loss of production, MSCP has been useful to enable an economic return to growers. In Florida, the use of mild strains introduced into existing trees on sour orange rootstock in the field in areas where decline on sour strains of *Citrus tristeza virus* were killing trees enabled the continued economic return to growers and allowed time to replace missing trees on *Citrus tristeza virus*-tolerant rootstocks instead of having to remove all trees at the same time and then waiting three to four years for the yields to return to a profitable situation.

References

Ahlawat, Y. S., Pant, R. P., Lockhart, B. E. L. *et al.* (1996). Association of a badnavirus with citrus mosaic disease in India. *Plant Disease*, **80**, 590–592.

Albiach-Martí, M. R., Guerri, J., Hemoso De Mendoza, A. *et al.* (2000). Aphid transmission alters the genomic and defective RNA populations of *Citrus tristeza virus* isolates. *Phytopathology*, **90**, 134–138.

Bar-Joseph, M., Marcus, R. & Lee, R. F. (1989). The continuous challenge of citrus tristeza virus control. *Annual Review of Phytopathology*, **27**, 291–316.

Bove, J. M. (1995). *Virus and Virus-Like Diseases of Citrus in the Near East Region*. Rome, Italy: Food and Agricultural Organization of the United Nations.

Bove, J. M. (2006). Huanglongbing: a destructive, newly emerging, century-old disease of citrus. *Journal of Plant Pathology*, **88**, 7–37.

Brlansky, R. H., Pelosi, R. R., Garnsey, S. M. *et al.* (1986). Tristeza quick decline epidemic in South Florida. *Proceedings of the Florida State Horticultural Society*, **99**, 66–69.

Brlansky, R. H., Damsteegt, V. D., Howd, D. S. & Roy, A. (2003). Molecular analysis of *Citrus tristeza virus* subisolates separated by aphid transmission. *Plant Disease*, **87**, 397–401.

Calavan, E. C., Roistacher, C. N. & Nauer, E. M. (1972). Thermotherapy of citrus for inactivation of certain viruses. *Plant Disease Reporter*, **56**, 976–980.

Derrick, K. S. & Timmer, L. W. (2000). Citrus blight and other diseases of recalcitrant etiology. *Annual Review of Phytopathology*, **38**, 181–205.

El-Kharbotly, A., Al-Shanfari, A. & Al-Subhi, A. (2000). Molecular evidence for the presence of the *Phytoplasma aurantifolia* in lime seeds and transmission to seedlings. In *2000 Proceedings of the International Society of Citriculture*, pp. 97–98. Orlando, FL: International Society of Citriculture.

Fawcett, H. S. (1938). Transmission of psorosis of citrus. *Phytopathology*, **28**, 669.

Fegan, R. M., Olexa, M. T. & McGovern, R. J. (2004). Protecting agriculture: the legal basis of regulatory action in Florida. *Plant Disease*, **88**, 1040–1043.

Florida Legislature (2007). *Citrus Nursery Stock Propagation and Production and the Establishment of Regulated Areas around Citrus Nurseries*, Florida State Statute 581.1843 (2007). Tallahassee, FL: Florida Legislature. Available at www.leg.state.fl.us/Statutes/index.cfm?App_mode=Display_Statute&Search_String=&URL=Ch0581/SEC1843.HTM&Title=-%3E2006-%3ECh0581-%3ESection%20184.

Guerra-Moreno, A. S., Manjunath, K. L., Brlansky, R. H & Lee, R. F. (2005). Citrus leprosis symptoms can be associated with the presence of two different viruses: cytoplasmic and nuclear, the former having a multipartite RNA genome. In *Proceedings of the 16th Conference of the International Organization of Citrus Virologists*, pp. 230–239. Riverside, CA: IOCV.

Guerri, J., Pina, J. A., Vives, M. C., Navarro, L. & Moreno, P. (2004). Seed transmission of *Citrus leaf blotch virus*: implications in quarantine and certification programs. *Plant Disease*, **88**, 906.

Grenzebach, E. (1994). *Integrated Pest Management in Selected Fruit Trees*, Report No. 42 on a Short-Term Consultancy Mission, Feb 27–March 12, 1994. Eschborn, Germany: Deutsche Gesellschaft für Technische Zusammenarbeit (GTZ).

Halbert, S. E. (2005). The discovery of huanglongbing in Florida. In *Proceedings of the 2nd International Citrus Canker and Huanglongbing Research Workshop*, pp. 1–3. Orlando, FL: Florida Citrus Mutual.

Hartung, J. S., Beretta, J., Brlansky, R. H., Spisso, J. & Lee. R. F. (1994). Citrus variegated chlorosis bacterium: axenic culture, pathogenicity, and serological relationships with other strains of *Xylella fastidiosa*. *Phytopathology*, **84**, 591–597.

Kersting, U., Korkmaz, S., Cinar, B. *et al.* (1996). Citrus chlorotic dwarf: a new whitefly transmitted disease in the east Mediterranean region of Turkey. In *Proceedings of the 13th Conference of the International Organization of Citrus Virologists*, pp. 220–225. Riverside, CA: IOCV.

Khan, I. A., Lee, R. F. & Hartung, J. (2002). Confirming seed transmission of witches' broom disease of lime. In *Proceedings of the 16th International Plant Protection Congress*, p. 281. Christchurch, New Zealand: IOCV.

Lee, R. F. & Niblett, C. L. (2000). Citrus tristeza: strains, mild strain cross protection and other management strategies. *Revista Horticultura Mexicana*, **8**, 25–35.

Lee, R. F. & Rocha-Pena, M. A. (1992). Citrus tristeza virus. In *Plant Diseases of International Importance*, vol. 3, eds. A. N. Mukhopadhyay, H. S. Chaube, J. Kumar & U. S. Singh, pp. 226–249. Upper Saddle River, NJ: Prentice-Hall.

Lee, R. F., Brlansky, R. H., Garnsey, S. M. & Yokomi, R. K. (1987). Traits of citrus tristeza virus important for mild strain cross protection of citrus: the Florida approach. *Phytophylactica*, **19**, 215–218.

Lee, R. F., Niblett, C. L. & Derrick, K. S. (1992). Mild strain cross protection against severe strains of citrus tristeza virus in Florida. In *Proceedings of the 1st International Seminar on Citriculture in Pakistan*, Dec. 1992, Faisalabad, Pakistan, eds. I. A. Khan, pp. 400–405.

Lee, R. F., Lehman, P. S. & Navarro, L. (1999). Nursery practices and certification programs for budwood and rootstocks. In *Citrus Health Management*, eds. L. W. Timmer & L. W. Duncan, pp. 35–46. St. Paul, MN: American Phytopathological Society Press.

Li, W.-B., Pria, W. D. Jr., Lacava, P. M., Qin, X. & Hartung, J. S. (2003). Presence of *Xylella fastidiosa* in sweet orange fruit and seeds and its transmission to seedlings. *Phytopathology*, **93**, 953–958.

Navarro, L. (1993). Citrus sanitation, quarantine and certification programs. In *Proceedings of the 12th Conference of the International Organization of Citrus Virologists*, pp. 383–391. Riverside, CA: IOCV.

Navarro, L., Roistacher, C. N. & Murashige, T. (1975). Improvement of shoot tip grafting in vitro for virus-free citrus. *Journal of the American Society of Horticultural Science*, **100**, 471–479.

Navarro, L., Juarez, J., Pina, J. A. & Ballester, J. F. (1984). The citrus quarantine station in Spain. In *Proceedings of the 9th Conference of the International Organization of Citrus Virologists*, pp. 365–370. Riverside, CA: IOCV.

Ochoa, F. M., Cevik, B., Febres, V., Niblett, C. L. & Lee, R. F. (2000). Molecular characterization of Florida citrus tristeza virus isolates with potential use in mild strain cross protection. In *Proceedings of the 14th Conference of the International Organization of Citrus Virologists*, pp. 94–102. Riverside, CA: IOCV.

Permar, T. A., Garnsey, S. M., Gumpf, D. J. & Lee, R. F. (1990). A monoclonal antibody that discriminates strains of citrus tristeza virus. *Phytopathology*, **80**, 224–228.

Powell, C. A., Pelosi, R. R., Sonoda, R. M. & Lee, R. F. (1998). A psorosis-like agent prevalent in Florida's grapefruit groves and budwood sources. *Plant Disease*, **82**, 208–209.

Rocha-Pena, M. A., Permar, T. A., Lee, R. F., Yokomi, R. K. & Garnsey, S. M. (1991). Comparative infection rates from aphid or graft challenge by a severe CTV isolate on plants preinoculated with mild isolates. In *Proceedings of the 11th Conference of the International Organization of Citrus Virologists*, pp. 93–102. Riverside, CA: IOCV.

Rocha-Pena, M. A., Lee, R. F. & Niblett, C. L. (1992). Effects of mild isolates of citrus tristeza virus on the development of tristeza decline. *Subtropical Plant Science*, **45**, 11–17.

Rocha-Pena, M. A., Lee, R. F., Lastra, R. *et al.* (1995). *Citrus tristeza virus* and its aphid vector *Toxoptera citricida*: threats to citrus production in the Caribbean and Central and North America. *Plant Disease*, **79**, 437–445.

Rodrigues, J. C. V., Kitajima, E. W., Childers, C. C. & Chagas, C. M. (2003). Citrus leprosis virus vectored by *Brevipalpus phoenicis* (Acari: Tenuipalpidae) on citrus in Brazil. *Experimental and Applied Acarology*, **30**, 161–179.

Roistacher, C. N. (1991). *Graft-Transmissible Diseases of Citrus: Handbook for Detection and Diagnosis*. Riverside, CA: IOCV and Rome, Italy: Food and Agriculture Organization of the United Nations.

Roistacher, C. N. (1993). Arguments for establishing a mandatory certification program for citrus. *Citrus Industry*, **74** (Nov.), 8.

Roistacher, C. N. (1996). The economics of living with citrus diseases: huanglongbing (greening) in Thailand. In *Proceedings of the 13th Conference of the International Organization of Citrus Virologists*, pp. 279–285. Riverside, CA: IOCV.

Roistacher, C. N., Calavan, E. C. & Navarro, L. (1977). Concepts and procedures for importation of citrus budwood. In *Proceedings of the 2nd International Citrus Congress*, vol. 1, pp. 133–136.

Teixeira, D. D. C., Danet, J. L., Eveillard, S. *et al.* (2005). Citrus huanglongbing in Sao Paulo State, Brazil: PCR detection of the '*Candidatus*' Liberibacter species associated with the disease. *Molecular and Cellular Probes*, **19**, 173.

Tsai, J. H., Liu, Y. H., Wang, J. J. & Lee, R. F. (2000). Recovery of orange stem pitting strains of citrus tristeza virus (CTV) following single aphid transmissions with *Toxoptera citricida* from a Florida decline isolate of CTV. *Proceedings of the Florida State Horticultural Society*, **113**, 75–78.

US Department of Agriculture (2007). *Citrus Health Response Program (CHRP): State of Florida*. Washington, DC: US Department of Agriculture, Animal Plant Health Inspection Service. Available at www.aphis.usda.gov/plant_health/plant_pest_info/citrus/downloads/chrp.pdf.

Whiteside, J. O., Garnsey, S. M. & Timmer, L. W. (eds.) (1988). *Compendium of Citrus Disease*. St. Paul, MN: APS Press.

Yokomi, R. K., Garnsey, S. M., Lee, R. F. & Cohen, M. (1987). Use of insect vectors to screen for protecting effects of mild citrus tristeza virus isolates in Florida. *Phytophylactica*, **19**, 183–185.

Yokomi, R. K., Garnsey, S. M., Lee, R. F. & Youtsey, C. O. (1992). Spread of decline-inducing isolates of citrus tristeza virus in Florida. *Proceedings of the International Society of Citriculture*, **1992**, 778–780.

Chapter 27

IPM in greenhouse vegetables and ornamentals

Joop C. van Lenteren

Before the large-scale application of chemical pesticides, biological control was one of the pest management methods embedded in a system's approach of pest prevention and reduction. (The word *pest* comprises animal pests, diseases and weeds: Food and Agriculture Organization, 1999.) Farmers, but also growers of greenhouse vegetables, needed to think about pest prevention before they designed their next season's planting scheme and choice of crops. They generally made use of three pest management methods: cultural control, host plant resistance and biological control. Cultural methods like crop rotation, cover crops, and sowing and harvesting dates were used to prevent excessive development of pests (Delucchi, 1987). Plants that had a high degree of resistance or tolerance to pests were another cornerstone of pest prevention. The third cornerstone was formed by natural, classical, inundative and conservation biological control (Bale *et al.*, 2008).

After 1945, these methods seemed to have become redundant as almost all pests could easily be managed by pesticides. As a result, pest control research became highly reductionistic, and changed from a decisive factor in farming design to prevent pests, to a mind-numbing but initially successful fire-brigade activity. Another effect was that plants were no longer selected for resistance to pests, but only for the highest production of biomass (food) or nicest cosmetic aspects (flowers) and under a blanket of pesticide application. This, in turn, resulted worldwide in crops that can be considered "incubator plants" that were unable to survive without frequent pesticide applications and in agroecosystems with strongly reduced or exterminated populations of natural enemies.

Now that chemical pesticides are no longer seen as the major solution for lasting pest control, we cannot simply return in a year or so to pre-pesticide pest management methods. The crops that we currently grow are too weak to survive without pesticides, the natural enemies are no longer present and farmers are often pesticide addicted. So, first we need to invest heavily in development of new cultivars with resistance to pests and diseases. This is actually happening for greenhouse crops (van Lenteren, 2000). At the same time, we can restore previously used forms of natural and biological control. Further, several other alternatives for conventional chemical pest control methods can also be implemented, such as mechanical, physical, genetic, pheromonal and semiochemical control (Table 27.1). In the development of IPM programs for greenhouses, we have based our work on the following IPM philosophy: IPM is a durable, environmentally and economically justifiable system in which damage caused by pests, diseases and weeds is prevented through the use of natural factors which limit the

Integrated Pest Management, ed. Edward B. Radcliffe, William D. Hutchison and Rafael E. Cancelado. Published by Cambridge University Press.

Table 27.1 Methods to prevent or reduce development of pests

Prevention
Prevent introduction of new pests (inspection and quarantine)
Start with clean seed and plant material (thermal disinfection)
Start with pest free soil (steam sterilization and solarization)
Prevent introduction from neighboring crops

Reduction
Apply cultural control (crop rotation)
Use plants which are (partly) resistant to pests
Apply one of the following control methods:
- Mechanical control (mechanical destruction of pest organisms)
- Physical control (heating)
- Control with attractants, repellants and antifeedants
- Control with pheromones
- Control with hormones
- Genetic control
- Biological control (natural enemies and antagonists)
- Selective chemical control

Guided or supervised pest management
Control based on sampling and spray thresholds

IPM
Control based on the integration of methods which cause the least disruption of ecosystems

Source: After van Lenteren (1993).

population growth of these organisms, if necessary supplemented with appropriate control measures (van Lenteren, 1993). A control program will only be considered truly IPM if it involves a number of natural enemy species (van Lenteren, 2000).

An often underestimated aspect of IPM is manipulation of the environment to make it more advantageous to natural enemies. This strategy involves both manipulation of biotic and abiotic elements of the environment and can imply tactics from changing the climate (e.g. greenhouses and wind shields) to applying chemicals that stimulate the activity of natural enemies of pests. If natural enemies fail to become established (either due to agricultural practices or shortcomings of natural enemy adaptability) or, if established, fail to control the host, manipulation of the natural enemy or its environment may lead to better control. The insect habitat may lack only certain key requisites and addition of these may make the action of natural enemies possible or more effective. Manipulation of the environment is applied on a limited scale, though there are many opportunities for implementation, even in greenhouses (see van Lenteren, 1987 and Landis *et al.*, 2000 for reviews).

Successful IPM programs for greenhouse crops have a number of characteristics in common, such as: (1) their use was promoted only after a complete IPM program had been developed covering all aspects of pest and disease control for a crop, (2) an intensive support of the IPM program by the advisory/extension service was necessary during the first years, (3) the total costs of crop protection in the IPM program were not higher than in the chemical control program and (4) non-chemical control agents (like natural enemies, resistant plant material) had to be as easily available, as reliable, as constant in quality and as well guided as chemical agents (van Lenteren, 1993).

27.1 Steps towards successful implementation of IPM in greenhouses

When looking back on 35 years of experience with IPM in greenhouse crops, I have learned a number of essential lessons in order to obtain success. I summarize these lessons below and refer to van Lenteren (2007) for more information.

Lesson 1: Discuss the development of an IPM program with all stakeholders

When a new pest is perceived, we organize a meeting with all stakeholders (e.g. growers, pest control specialists using all kinds of control methods, extension service personnel, researchers [e.g. plant breeders, entomologists, etc.]) and discuss potential short- and long-term solutions. The initiative for such a meeting can be from any group of stakeholders. The conclusion of such a meeting might be that IPM is the best solution. At that same meeting we then discuss what other conditions need to be fulfilled for IPM to be able to function in the greenhouse setting. A major point is always that a complete pest management program should be available covering all aspects of pest management. If, for example, one of the chemical pesticides used for arthropod, disease or weed control is having a strong negative side effect on a new natural enemy, biological control is not realistic until an alternative for this pesticide has been found (van Lenteren, 1995).

What follows from these initial meetings is a pragmatic design of a draft IPM program for the new pest, including an overall IPM program for the other pests and diseases. This is then discussed in following meetings with the stakeholders until agreement has been reached about the applicability of the program. Next, the IPM program is continuously adapted based on growers' experience and new research results. Often, development of these IPM programs was only possible because of intensive cooperation within and provision of essential information by the European and North American Working Groups of the International Organization for Biological Control of Noxious Animals and Plants (IOBC/WPRS, 2008).

Lesson 2: Work with the best and most progressive growers when developing IPM

Very early in the development of the first IPM programs we learned that it is crucial to cooperate with the most progressive growers. To our initial surprise, they were keenly interested, took up the knowledge quickly, suggested many improvements concerning release of natural enemies and sampling methods for pests, saw possibilities to advertise crops produced under IPM, and were able to convince other growers how useful IPM was. It was these growers who allowed us to do experiments in their commercial greenhouses, and who invited other growers and the extension service to demonstrate how well biological control and IPM worked. We could not have found better advocates for implementation of IPM!

Lesson 3: Develop good teaching material about IPM to retrain extension service personnel and growers

When we found that modern growers in the 1970s had never heard about IPM, our conclusion was that we needed to develop teaching material for vocational schools, high schools and universities. A fruitful coincidence was that this happened during a period when many people were concerned about environmental problems. Teachers of science and biology were happy that they could link the development of an applied ecological method that was beneficial for the environment to general biological issues. The result was that teaching of biological control took off quickly and had a clear impact on changes in thinking about crop protection: children and students taught their parents how biological control worked within an IPM approach.

We also realized that it was necessary to retrain the personnel of the extension services. Next, and often together with the extension service, we organized free courses on IPM and biological control during the winter to train the farmers in recognizing the natural enemies and pests, and in sampling and release methods. In addition to training, we started to publish in journals that the growers use primarily for obtaining the newest information on production and crop protection techniques.

Lesson 4: Provide/sell IPM pest management guidance and not just biological control agents

The danger of just providing natural enemies is that if they do not work, the grower is disappointed and will speak negatively of biological control. Therefore, producers of natural enemies should provide a guidance information system, which is sold to the growers for a certain price and includes provision of the natural enemies and other crop protection materials. Advisory personnel of these natural enemy producers also advise which chemical pesticides might be integrated with the natural enemies that have been released. Information of negative side effects of all kinds of pesticides is essential in applying IPM. This was realized early in the development of IPM programs

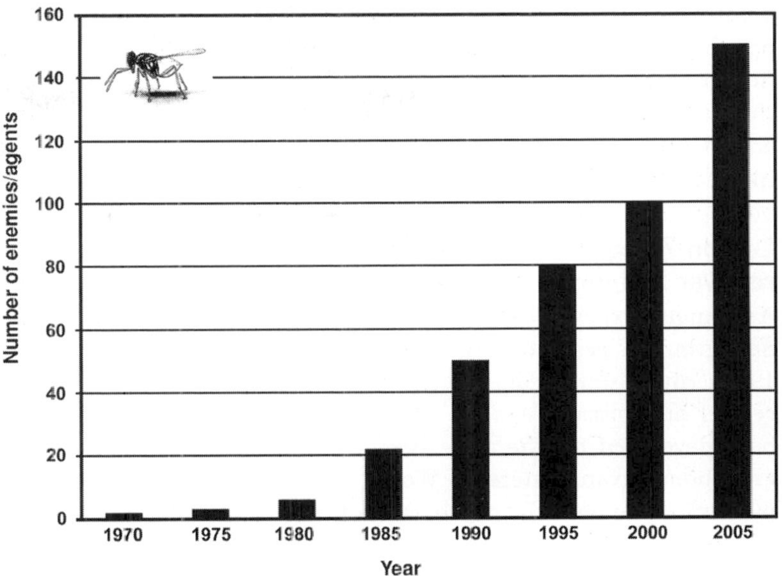

Fig. 27.1 Numbers of commercially available natural enemies and microbial control agents for pest control in greenhouses.

and IOBC formed a special Working Group on Pesticides and Beneficial Organisms, which evaluates pesticides for their side effects. This information is available on the OIBC website (OIBC-Global, 2007a) and at the websites of various natural enemy producers.

Lesson 5: Develop effective, economic and high-quality mass production of natural enemies

Today Europe has more than 30 commercial natural enemy producers including the world's three largest. These three largest companies serve more than 75% of the greenhouse biological control market worldwide. Of the more than 150 biological control agents marketed today for pest control in greenhouses, about 30 account for 90% of the total sales (van Lenteren, 2003a). It appears that many more species of biological control agents are available in Europe than elsewhere. This is caused by the much larger greenhouse industry and a longer history of research in greenhouse biological control in Europe (van Lenteren *et al.*, 1997).

Although on-farm production of natural enemies is possible, most growers purchase them from commercial suppliers. Mass production of natural enemies has seen a very rapid development during the past three decades. The numbers produced have greatly increased (up to 50 million individuals per week), the spectrum of species available has widened dramatically (from two in 1970 to now more than 150) (Fig. 27.1), and mass production methods clearly have evolved (van Lenteren & Tommasini, 2003). The larger arthropod mass production companies employ scientists who develop and apply quality control tests (van Lenteren, 2003b). Research and application of quality control is coordinated by a working group of IOBC, which has published quality control guidelines on the internet (see IOBC–Global, 2007b).

Lesson 6: Be realistic – not all pests can be managed with biological control

Pushing for biological control as the only solution to control pests is utopian. Biological control workers need perseverance, green fingers and creativity, but sometimes biological control is not the best solution or will not work. An example is pest control in short-term crops, like lettuce. Lettuce is produced in six-week cycles and one of the main pests is aphids, a notoriously quick-developing pest which is almost always difficult to control, even in long-term crops. We were able to keep pests under biological control with frequent releases of great numbers of a whole array of natural enemies, but it was way too expensive to be of practical use. Here we had to conclude that development of host plant resistance to aphids was the first important step in the IPM program, and when this was realized and became a success, we could advise application of biocontrol for other pests, like leafminers (de Ponti & Mollema, 1992).

Lesson 7: Expect the unexpected – recover and win

A dramatic experience was the application of a new group of pesticides, the pyrethroids, in the 1980s. After 10–15 years of very hard work, successful and broadly applied IPM programs had been developed for the main vegetable crops in greenhouses (van Lenteren & Woets, 1988). Then, pyrethroids appeared on the market and were effective against a number of important pests, making biological control redundant. The result of their use was that the biocontrol agent producers almost went bankrupt. "Luckily" enough, some pest species developed resistance against these pesticides rather quickly and the help of the old natural enemies was needed again. We learned an important lesson: it is crucial to know the effects of a new pest control agent on natural enemies in an already perfectly working IPM program. Initially, side effects of pesticides on natural enemies were tested within an IOBC working group. The work of this group resulted in a very important development in pesticide legislation within the European Union: the applicant for registration of a new pesticide now has to provide data about safety of the pesticide for natural enemies. Such data can then be used to advise for or against the use of this pesticide within an IPM program. Several side effect lists have been published so that growers can quickly check whether a pesticide can safely be used in an IPM program (see information under lesson 4).

Lesson 8: With good support and guidance, growers prefer IPM as first option for pest management

Growers of greenhouse products are now using IPM on a large scale. When we asked them if and why they prefer IPM, we received the following responses (paraphrased):

- When using biological control agents there are no phytotoxic effects on young plants, and premature abortion of flowers and fruit does not occur; also yield increases have been obtained with biological control.
- Release of biological control agents takes less time and is more pleasant than applying chemicals in humid and warm greenhouses.
- Release of biological control agents usually occurs shortly after the planting period when the grower has sufficient time to check for successful development of natural enemies; thereafter the system is reliable for months with only occasional checks, chemical control requires continuous attention.
- Chemical control of some of the key pests is difficult or impossible because of pesticide resistance.
- When using IPM there is no safety period between application of a biological control agent and harvesting fruit; with chemical control one has to wait several days before harvesting is allowed again.
- Biological control is permanent: once a good natural enemy – always a good natural enemy.
- IPM is appreciated by the general public, they have more respect for our work and sometimes we receive a better price for the product.

These advantages of IPM are so important for growers that they will not easily return to chemical control.

Lesson 9: With most arthropods under IPM, it is possible to concentrate on the biological and integrated control of diseases

The use of IPM in greenhouses was limited mainly to the control of arthropods until a few years ago (van Lenteren & Woets, 1988). Disease problems can be considerable, particularly in tomatoes, cucumbers and cut flowers. Some fungicides can be integrated with the use of natural enemies, but as problems of fungicide resistance are strongly increasing, fewer "relatively safe" fungicides remain available. Thus, serious negative effects of fungicides on natural enemies of insects and widespread resistance of foliar pathogens to fungicides demand alternatives. Although use of fungicides remains substantial for foliar pathogens, disease management is now evolving towards strategies relying on the use

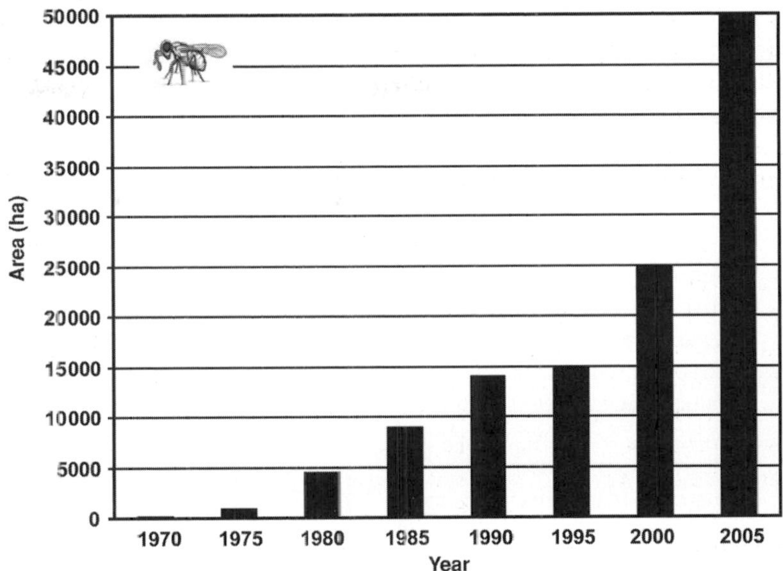

Fig. 27.2 Increase of worldwide area of greenhouse crops under IPM since 1970.

of resistant cultivars and manipulation of the environment. During the past decade several initiatives have led to research in non-chemical control, such as the effect of soil solarization on nematodes and fungi, and the potential use of antagonistic leaf fungi (Albajes *et al.*, 1999). For an overview of recent successes and practical applications with disease suppressive soils, biological control of soil-borne pathogens and root, stem or foliar diseases see van Lenteren (2000). Several microbial products now are registered and used for disease control in greenhouse vegetables and ornamentals in Europe, and other bacterial and fungal products for control of fungi are in the last phase of the registration procedure.

Lesson 10: Try to get IPM accepted as official plant protection philosophy at the national and international level

In several countries, IPM has been accepted as the official plant protection philosophy. Sometimes this was only for a specific crop, like for rice in Indonesia (Röling & van de Fliert, 1994). In other cases it included all crops, as in the Netherlands (Dutch Ministry of Agriculture, 2005). Currently, at an international level, the European Union is strongly supporting implementation of IPM both by providing grants to develop IPM programs (see www.endure-network.eu), as well as by supporting policies that lead to a quicker registration of alternative pest control methods needed in IPM programs (see e.g. www.rebeca-net.de). A large European Commission funded Network of Excellence in IPM is aiming at accelerated implementation of IPM by providing practical knowledge to countries with limited use of IPM. In this project, northwestern European countries will assist the development and application of greenhouse IPM programs in other European areas.

27.2 Current situation in greenhouse IPM

IPM can be used in all main vegetable crops. In the Netherlands for example, more than 90% of all tomatoes, cucumbers, sweet peppers and eggplants are produced under IPM (van Lenteren, 2000). Worldwide 5% of the greenhouse area is under IPM, and there is potential for this to increase to about 20% of this area in the coming ten years. The development of the area under IPM is presented in Fig. 27.2. Until the year 2000 most IPM took place in European greenhouses, but currently, application of IPM is growing very quickly in Asia (e.g. Zheng *et al.*, 2005).

An example of an often-used IPM program is the one for tomato in Europe. It involves ten or

Table 27.2 IPM as applied in tomato in Europe

Pests and diseases	Method used to prevent or control pest/disease
Pests	
Whiteflies (*Bemisia tabaci, Trialeurodes vaporariorum*)	Parasitoids: *Encarsia, Eretmocerus* Predators: *Macrolophus, Amblyseius* Pathogens: *Verticillium, Paecilomyces, Aschersonia*
Spider mite (*Tetranychus urticae*)	Predator: *Phytoseiulus*
Leafminers (*Liriomyza bryoniae, L. trifolii* and *L. huidobrensis*)	Parasitoids: *Dacnusa, Diglyphus* and *Opius*; natural control[a]
Lepidoptera (e.g. *Chrysodeixis chalcites, Lacanobia oleracea, Spodoptera littoralis*)	Parasitoids: *Trichogramma* Pathogens: *Bacillus thuringiensis*
Aphids (e.g. *Myzus persicae, Aphis gossypii, Macrosiphum euphorbiae*)	Parasitoids: *Aphidius, Aphelinus* Predators: *Aphidoletes* and natural control[a]
Nematodes (e.g. *Meloidogyne* spp.)	Resistant and tolerant cultivars, soil-less culture
Diseases[b]	
Gray mold (*Botrytis cinerea*)	Climate management, mechanical control and selective fungicides
Leaf mold (*Fulvia* = *Cladosporium*)	Resistant cultivars, climate management
Mildew (*Oidium lycopersicon*)	Selective fungicides
Fusarium wilt (*Fusarium oxysporum lycopersici*)	Resistant cultivars, soil-less culture
Fusarium root rot (*Fusarium oxysporum radicis-lycopersici*)	Resistant cultivars, soil-less culture, hygiene
Verticillium wilt (*Verticillium dahliae*)	Pathogen-free seed, tolerant cultivars, climate control, soil-less culture
Bacterial canker (*Clavibacter michiganensis*)	Pathogen-free seed, soil-less culture
Several viral diseases	Resistant cultivars, soil-less culture, hygiene, weed management, vector control

[a] Natural control: natural enemies spontaneously immigrating into the greenhouse and controlling a pest.
[b] Several antagonists for biological control of diseases about to be used commercially, in registration process.
Source: After van Lenteren (2000).

more natural enemies and various other control methods like host plant resistance, climate control and cultural control (Table 27.2). At a first glance, such an IPM program may look complicated, but after a year of experience and support from the provider of biological control agents, growers are able to carry it out. A recent development which gave a strong stimulus to the application of IPM is the use of bumblebees for pollination, because chemical control can no longer be used as it kills the pollinators (van Lenteren, 1995). Detailed examples of IPM programs for vegetables used in different parts of the world are presented in Albajes *et al.* (1999).

Development of IPM for ornamentals is more complicated than for vegetables. The first problem is that many different species and cultivars of ornamentals are grown. In western Europe, for example, more than 100 species of cut flowers and 300 species of potted plants are cultivated, and for several ornamentals more than 100 cultivars are produced. Each of these species/cultivars may need specifically designed IPM programs, and these are applied on much smaller areas then those for vegetables, which results in higher costs. Other problems for implementation of IPM in ornamentals are that: (1) more pesticides are available than for vegetables and higher residue levels are accepted, and (2) the whole plant is marketed, instead of only the fruits, so no leaf damage is allowed. But, since the 1990s use of IPM is steadily growing in cut flowers (e.g. gerbera, orchids, rose and chrysanthemum) and pot plants (e.g. anthurium, poinsettia) (Parrella *et al.*, 1999). IPM was applied on more than 10% (600 ha) of the total greenhouse area planted with flowers and ornamentals in 1998 in the Netherlands. Commercially used IPM programs for ornamental crops are presented in Parrella *et al.* (1999) for chrysanthemum, in van Lenteren (1995) for gerbera, and for various other ornamentals in Gullino and Wardlow (1999). Worldwide, I estimate that about 1000 ha of ornamentals are under IPM.

27.3 How has implementation of IPM in greenhouses been realized?

From behind a desk it is rather easy to develop a set of guidelines for implementation of IPM. Each practical situation dictates, however, a number of special aspects for consideration. We have experienced during the past decades that implementation of IPM in greenhouses in some crops and regions (e.g. vegetables in temperate climates) is much easier than in others (e.g. vegetables in semi-tropical climates and ornamentals in all climates) because of differences in attitudes of growers, in climate, in greenhouse design, in cultural methods and in composition of the pest complex. Therefore, only one example of a specific IPM program was given above.

Technically, implementation of IPM is not different from that of chemical control. At the introduction of the first IPM program for a new crop, special attention should be paid to extension. The degree of knowledge makes acceptance of more complicated IPM programs initially difficult for the grower. IPM methods are rather new and demand a different attitude based on the principle to introduce a natural enemy or pesticide only when the pest insect is present and expected to lead to economic loss. A misconception is that a practice is adopted readily if it is superior to current ones. In reality, only when the IPM method is perceived to be better than conventional methods will it be adopted by growers. The phase of introducing IPM into practice is often neglected. Experience in the Netherlands has shown that the amount of application of IPM is strongly related to the activity and attitude of extension personnel. If governmental extension services are weak, implementation of IPM will be very difficult due to aggressive approaches from pesticide salesmen. Also, all participants in an IPM program must be receptive to new developments and willing to implement them. When growers, extension workers and researchers agree that use of IPM is as cheap as chemical control and that production and delivery of alternative control methods is reliable, IPM can be applied in a similar way as chemical control and becomes a normal commercial affair.

27.4 What factors hamper introduction of IPM?

It seems difficult to transfer the successful greenhouse IPM programs applied in some countries to other countries, though the production systems do not differ so much that this would explain this problem. The following factors may explain difficulties encountered in having IPM implemented.

27.4.1 Funding of research in IPM

The results obtained in non-chemical pest control are, of course, in the first instance dependent on the amount of research and development work. Funding of this work is limited, especially if one realizes the complications of this type of research. Therefore, getting IPM accepted as the official pest management strategy (see lesson 10 above) is so important, as it makes it possible to direct research money to IPM studies. Very often only limited funding is mentioned as a main limitation for implementation. Although it explains part of the story, implementation is most hindered by other constraints, which are discussed below.

27.4.2 Growers' attitudes

Until very recently, only few growers or their organizations asked for, or stimulated, development of non-chemical control methods. The adoption of insecticides by greenhouse growers was rapid because they allowed the grower to decide when and where they should be used. Initially it was a straightforward technology. In contrast, integrated control is more complicated because of the requirement for the monitoring of various pests, the integration of different control methods and situation-specific prescriptions. The latter systems require a degree of knowledge and sophistication much greater than demanded by pesticide technology. However, as soon as growers realized that chemical control was no longer sufficient to control all greenhouse pests, they became interested in IPM systems. Growers of greenhouse vegetables and ornamentals in a number of countries have now experienced the positive aspects of IPM and they seriously worry about the increasing public concern about pesticide usage. Therefore, they generally prefer to use IPM methods (see e.g. van Lenteren, 2000).

27.4.3 The viewpoint of the chemical industries

In general, we can state that any complication in a simple chemical pest control program is appreciated as a negative development by the chemical pesticide industry. Alternatives like biological control agents and several other IPM methods not only complicate chemical control programs, but they seem to be unattractive commercially as well because of a combination of: (1) the impossibility of patenting natural enemies, (2) complicated mass production, (3) short shelf-life, (4) specificity (which limits market potential) and (5) different and more complicated guidance for growers (van Lenteren, 1986). Chemical industries will not start the production of other than broad-spectrum pesticides on their own initiative, unless the use of those pesticides is prohibited or when pest organisms substantially develop resistance – but time is on our side! An important step in helping growers to select pesticides that are relatively safe for use in IPM programs is that in Europe the applicant for registration of a new pesticide now has to provide data about the safety of a pesticide for natural enemies (see lesson 7 above).

27.4.4 The roles of governments, food processors and retailers

It is governmental agencies that should be the leaders in research and implementation of IPM and moreover they are able to change the pest control picture through measures that make some kinds of chemical control less attractive or impossible (by measures concerning registration, taxation, side-effect labeling, etc.), and by stimulating other control methods (by funding research, by teaching environmentally safe pest control methods, and by improvement of the extension service). It is a rather bizarre situation that public money is used for the development of alternatives for chemical control when, at the same time, their application is often not encouraged by other governmental entities, and by the overall presence of (too) cheap broad-spectrum pesticides. Food processors and retailers can also strongly contribute positively to a change in pest control measures as has been experienced in North America and Europe: they may demand the application of IPM methods and prohibit the use of (certain) chemical pesticides.

27.4.5 Registration and environmental risk assessment

Recently, the issue of the need to evaluate the (environmental) risks of releasing alien biological control agents as part of a regulation process has resulted in a debate concerning the difficulty to import and release new natural enemies. However, it seems that application of risk evaluations have only led to somewhat later release and did not decrease the number of releases (van Lenteren *et al.*, 2003, 2006). Many countries are now demanding risk analyses before permitting import and release of new agents. The challenge in developing risk assessment methodologies is to develop protocols and guidelines that will prevent serious mistakes through import and release of harmful exotics, while at the same time still allowing safe forms of biological control to proceed.

27.4.6 Misunderstandings about IPM

In the pest control literature one may often find erroneous statements about IPM. I have listed these extensively in the IOBC Internet Book of Biological Control (IOBC–Global, 2008), and summarized the most important ones here.

IPM systems create new pests

Use of IPM systems does not necessarily result in the creation of new pests when broad-spectrum pesticides are no longer used in greenhouses. The new pests that did occur during the past three decades were the result of unintentional introductions and created problems in greenhouses whether under IPM or under conventional chemical control.

IPM is unreliable

The professional attitude of most IPM workers is to advocate the use of only those IPM methods which have proven to be effective under practical conditions and within the total pest and disease program for a certain crop. IPM programs developed by these workers are as reliable as and more sustainable than conventional chemical control programs (van Lenteren, 2000).

IPM research is expensive

The contrary is true: research in and registration of most IPM methods is much cheaper than that of conventional pesticides. Developmental and registration costs of one biological control method in an IPM program range between $US 2 and 50 million, whereas the average developmental and registration costs of a synthetic pesticide are in the order of $US 180 million (Bale *et al.*, 2008).

IPM research is slow

Because of the need to find a management solution for all pests, diseases and weeds once an IPM program is chosen, research can be time-consuming. However, it takes generally a similar amount of time (ten years on average) to develop an IPM program as it takes to develop a synthetic pesticide and place a product on the market.

Augmentative biological control does not work effectively in IPM programs

In the pest control literature one may find analyses pointing at ineffectiveness of augmentative biological control. An example is a recent paper by Collier & van Steenwyk (2004) entitled "A critical evaluation of augmentative biological control." Contrary to what the title promises, the article does not present an evaluation of augmentative releases; instead the authors evaluated several research articles of augmentative biological control. The title is also wrong in that the article is not a critical evaluation of augmentative biological control in general, but is limited mainly to experimental situations in the USA. There are, however, plenty of examples of successful practical augmentative programs that have found a place in greenhouse IPM (see e.g. van Lenteren, 2006). Augmentative biological control is in many – not all – cases: (1) as effective or more so than chemical pesticide applications, (2) able to achieve target densities often even lower than chemical pesticides can and (3) has costs lower than or similar to those of chemical pesticides. In a number of greenhouse crops, augmentative biological control within greenhouse IPM programs has completely or in large part replaced broad-spectrum pesticides.

27.5 Conclusions

The success achieved with IPM in greenhouses has set a very high standard that is difficult for other segments in agriculture to match (Parrella *et al.*, 1999). This success has occurred primarily as a result of outstanding cooperation between research, extension, growers and producers of natural enemies, often within the framework of IOBC (see e.g. Enkegaard, 2005 and Castane & Sanchez, 2006). Several current trends will lead to a strong increase in the application of IPM in greenhouses. First, fewer new insecticides are becoming available because of skyrocketing costs for development and registration, particularly for the relatively small greenhouse market. Second, pests continue to develop resistance to any type of pesticides, a problem particularly prevalent in greenhouses, where intensive management and repeated pesticide applications exert strong selective pressure on pest organisms. Third, there is a strong demand from the general public (and in an increasing number of countries also from governments) to reduce the use of pesticides. Finally, in order to escape from the "pesticide treadmill," more sustainable forms of pest and disease control will have to be developed (Lewis *et al.*, 1997).

Because of the desire to reduce pesticide use, the future role of IPM is expected to increase strongly. This is aided by the extensive demonstration of its positive role and because many new natural enemy species still await discovery. Cost–benefit analyses show that biological control is the most cost-effective control method (Bellows & Fisher, 1999) and that greenhouse IPM methods which include biological control have similar costs as conventional chemical control (e.g. Ramakers, 1993). With improved methods for evaluation of beneficial insects, an increased insight into the functioning of natural enemies, and more efficient mass production methods, the cost effectiveness of biological control, and thus IPM, will even be increased. Together with other control methods such as mechanical and physical control, control by use of kairomones and host plant resistance, new IPM programs will be developed (for examples, see van Lenteren, 2000; Enkegaard, 2005; Castane & Sanchez, 2006). During the first decade of this century crop production in greenhouses without conventional chemical pesticides could become a fact!

References

Albajes, R., Gullino, M. L., van Lenteren, J. C. & Elad, Y. (eds.) (1999). *Integrated Pest and Disease Management in Greenhouse Crops*. Dordrecht, Netherlands: Kluwer.

Bale, J., van Lenteren, J. C. & Bigler, F. (2008). Biological control and sustainable food production. *Philosophical Transactions of the Royal Society of London B*, **363**, 761–776.

Bellows, T. S. Jr. & Fisher, T. W. (eds.) (1999). *Handbook of Biological Control*. New York: Academic Press.

Castane, C. & Sanchez, J. A. (eds.) (2006). Proceedings of the IOBC/WPRS working group on Integrated Control in Protected Crops: Mediterranean Climate. *IOBC/WPRS Bulletin*, **29**(4).

Collier, T. & van Steenwyk, R. (2004). A critical evaluation of augmentative biological control. *Biological Control*, **31**, 245–256.

Delucchi, V. (ed.) (1987). *Integrated Pest Management: Quo Vadis? An International Perspective*. Geneva, Switzerland: Parasitis 86.

de Ponti, O. M. B. & Mollema, C. (1992). Emerging breeding strategies for insect resistance. In *Plant Breeding in the 1990s*, eds. H. T. Stalker & J. P. Murphy, pp. 323–347. Wallingford, UK: CABI Publishing.

Dutch Ministry of Agriculture (2005). *Durable Crop Protection: Policy for Crop Protection towards 2010*. The Hague, Netherlands: Dutch Ministry of Agriculture, Nature and Food Quality. Available at www.minlnv.nl.

Enkegaard, A. (ed.) (2005). Proceedings of the IOBC/WPRS working group on Integrated Control in Protected Crops: Temperate Climate. *IOBC/WPRS Bulletin*, **28**(1).

Food and Agriculture Organization (1999). *Glossary of Phytosanitary Terms*. Rome, Italy: Food and Agriculture Organization of the United Nations. Available at http://permanent.access.gpo.gov/lps3025/gloss99_7.pdf.

Gullino, M. L. & Wardlow, L. R. (1999). Ornamentals. In *Integrated Pest and Disease Management in Greenhouse Crops*, eds. R. Albajes, M. L. Gullino, J. C. van Lenteren & Y. Elad, pp. 486–506. Dordrecht, Netherlands: Kluwer.

IOBC-Global (2007a). *International Organization for Biological Control of Noxious Animals and Plants (IOBC)*. Available at www.iobc-global.org.

IOBC-Global (2007b). *IOBC Quality Control Guidelines*. IOBC Global Working Group on Arthropod Mass Rearing and Quality Control. International Organization for Biological Control of Noxious Animals and Plants. Available at http://users.ugent.be/~padclerc/AMRQC/guidelines.htm.

IOBC-Global (2008). *IOBC Internet Book of Biological Control*. International Organization for Biological Control of Noxious Animals and Plants, ed. J. van Lenteren. Available at www.iobc-global.org.

IOBC/WPRS (2005). *Classification of Side Effects of Pesticides to Beneficial Organisms*. International Organization for Biological Control of Noxious Animals and Plants, IOBC/WRPS Working Group on Pesticides and Beneficial Organisms and IOBC/WPRS Commission on IP Guidelines and Endorsement. Available at www.iobc.ch/2005/IOBC_Pesticide%20Database_Toolbox.pdf.

IOBC/WPRS (2008). International Organization for Biological Control, Working Group on Greenhouses. Available at www.iobc-wprs.org.

Landis, D. A., Wratten, S. D. & Gurr, G. M. (2000). Habitat management to conserve natural enemies of arthropod pests in agriculture. *Annual Review of Entomology*, **45**, 175–201.

Lewis, W. J., van Lenteren, J. C., Phatak, S. C. & Tumlinson, J. H. (1997). A total systems approach to sustainable pest management. *Proceedings of the National Academy of Sciences of the USA*, **94**, 12 243–12 248

Parrella, M. P., Stengard Hansen, L. & van Lenteren, J. C. (1999). Glasshouse environments. In *Handbook of Biological Control*, eds. T. S. Bellows Jr. & T. W. Fisher, pp. 819–839. San Diego, CA: Academic Press.

Röling, N. G. & van de Fliert, E. (1994). Transforming extension for sustainable agriculture: the case of integrated pest management in rice in Indonesia. *Agriculture and Human Values*, **11**, 96–108.

Ramakers, P. M. J. (1993). More life under glass. In *Modern Crop Protection: Developments and Perspectives*, ed. J. C. Zadoks, pp. 91–94. Wageningen, Netherlands: Wageningen Pers.

van Lenteren, J. C. (1986). Evaluation, mass production, quality control and release of entomophagous insects. In *Biological Plant and Health Protection*, ed. J. M. Franz, pp. 31–56. Stuttgart, Germany: Fischer.

van Lenteren, J. C. (1987). Environmental manipulation advantageous to natural enemies of pests. In *Integrated Pest Management: Quo Vadis? An International Perspective*, ed. V. Delucchi, pp. 123–166. Geneva, Switzerland: Parasitis 86.

van Lenteren, J. C. (1993). Integrated pest management: the inescapable future. In *Modern Crop Protection: Developments and Perspectives*, ed. J. C. Zadoks, pp. 217–225. Wageningen, Netherlands: Wageningen Pers.

van Lenteren, J. C. (1995). Integrated pest management in protected crops. In *Integrated Pest Management*, ed. D. Dent, pp. 311–343. London: Chapman & Hall.

van Lenteren, J. C. (2000). A greenhouse without pesticides: fact or fantasy? *Crop Protection*, **19**, 375–384.

van Lenteren, J. C. (2003a). Commercial availability of biological control agents. In *Quality Control and Production of Biological Control Agents: Theory and Testing Procedures*, ed. J. C. van Lenteren, pp. 167–179. Wallingford, UK: CABI Publishing.

van Lenteren, J. C. (ed.) (2003b). *Quality Control and Production of Biological Control Agents: Theory and Testing Procedures*. Wallingford, UK: CABI Publishing.

van Lenteren, J. C. (2006). How not to evaluate augmentative biological control. *Biological Control*, **39**, 115–118.

van Lenteren, J. C. (2007). Biological control of pests and diseases in greenhouses: an unexpected success. In *Biological Control: A Global Perspective*, eds. C. Vincent, M. S. Goettel & G. Lazarovits, pp. 105–117. Wallingford, UK: CABI Publishing.

van Lenteren, J. C. & Tommasini, M. G. (2003). Mass production, storage, shipment and release of natural enemies. In *Quality Control and Production of Biological Control Agents: Theory and Testing Procedures*, ed. J. C. van Lenteren, pp. 181–189. Wallingford, UK: CABI Publishing.

van Lenteren, J. C. & Woets, J. (1988). Biological and integrated pest control in greenhouses. *Annual Review of Entomology*, **33**, 239–269.

van Lenteren, J. C., Roskam, M. M. & Timmer, R. (1997). Commercial mass production and pricing of organisms for biological control of pests in Europe. *Biological Control*, **10**, 143–149.

van Lenteren, J. C., Babendreier, D., Bigler, F. *et al.* (2003). Environmental risk assessment of exotic natural enemies used in inundative biological control. *Biocontrol*, **48**, 3–38.

van Lenteren, J. C., Bale, J., Bigler, F., Hokkanen, H. M. T. & Loomans, A. J. M. (2006). Assessing risks of releasing exotic biological control agents of arthropod pests. *Annual Review of Entomology*, **51**, 609–634.

Zheng, L., Zhou, Y. & Song, K. (2005). Augmentative biological control in greenhouses; experiences from China. In *Proceedings of the International Symposium on Biological Control of Arthropods*, September 12–16, 2005, Davos, Switzerland, pp. 538–545. Riverside, CA: US Department of Agriculture, Forest Service.

Chapter 28

Vector and virus IPM for seed potato production

Jeffrey A. Davis, Edward B. Radcliffe, Willem Schrage and David W. Ragsdale

Potato is South America's greatest gift to world agriculture and human nutrition (Graves, 2001). A dietary staple of indigenous Andean peoples for eight millennia, potato was unknown to the rest of the world before the mid sixteenth century. Today, potato is the world's fourth most important food crop, after maize, wheat and rice, and is grown on a significant scale in more than 130 countries on six continents with annual tuber production exceeding 320 million tonnes (Food and Agriculture Organization, 2007). In recognition of potato's potential to provide food security and eradicate poverty, 2008 has been proclaimed International Year of the Potato.

In most production systems, potato is clonally propagated from "seed" tubers. Clonal propagation offers agronomic and genetic advantages, e.g. vigorous early growth, higher yields and consistent expression of desirable traits. More than 10% of world potato production is used to provide "seed tubers" for planting the next production season. Seed potato tubers can be infected with a wide range of pests and pathogens which may affect growth of the crop and health of progeny tubers. Thus, access to high-quality, disease-free seed potatoes has been described as "the single most important integrated pest management practice available to potato growers" (Gutbrod & Mosley, 2001). Seed potato lots can be downgraded or rejected for recertification for myriad causes from "varietal mix" to herbicide injury. However, aphid-transmitted, tuber-borne potato viruses far exceed all others.

Management of insect-vectored pathogens represents a level of complexity beyond that of the classic disease triad of susceptible host, viable inoculum and favorable environment (Ragsdale *et al.*, 2001). The additional complication is that the vector must acquire and transmit the pathogen to enable progression of the disease. Removing either the virus or the vector will eliminate disease, increase yields and reduce losses due to quality issues.

28.1 Viral pathogens

Of the more than 30 known potato viruses 13 are transmitted by aphids; the most important of these are *Potato leafroll virus* (PLRV) and *Potato virus Y* (PVY). Aphid-transmitted viruses differ in how they are acquired, whether they circulate within the body of the vector, how they are transmitted to healthy plants and how long a vector remains infective following virus acquisition. These differences in transmission dictate which control tactics will be effective in preventing or reducing virus spread.

Integrated Pest Management, ed. Edward B. Radcliffe, William D. Hutchison and Rafael E. Cancelado. Published by Cambridge University Press.

PLRV is a phloem-localized virus which cannot be transmitted by sap (mechanical) inoculation. In general, only potato colonizing aphids are consequential PLRV vectors. Green peach aphid (*Myzus persicae*) is the most efficient, cosmopolitan and commonly abundant vector of PLRV (Robert & Bourdin, 2001) and is the species around which most IPM programs for aphid control in potato are designed. Other potential vectors of PLRV include cotton aphid (*Aphis gossypii*), buckthorn aphid (*Aphis nasturtii*), crescent-marked lily aphid (*Aulacorthum circumflexum*), foxglove aphid (*Aulacorthum solani*), potato aphid (*Macrosiphum euphorbiae*), shallot aphid (*Myzus ascalonicus*), damson-hop aphid (*Phorodon humuli*), bulb and potato aphid (*Rhopalosiphoninus latysiphon*) and mangold aphid (*R. staphyleae tulipaellus*) (Ragsdale *et al.*, 2001; Radcliffe & Ragsdale, 2002; Radcliffe & Lagnaoui, 2007).

Symptoms of PLRV infection include stunted plants with upward rolling leaves. Leaves can become chlorotic or red in color with a leathery texture. Beyond simply reducing yield, PLRV can, in some varieties, reduce tuber quality by causing a symptom termed "net necrosis," a darkening of the vascular bundles which becomes more pronounced in storage. Maximum acquisition of PLRV occurs within 12 h of feeding, but can occur in as little as 10 min. Once PLRV is acquired by a vector the virus penetrates the gut wall, enters the hemolymph, circulates within the body cavity and penetrates the lining of the accessory salivary gland. Virus eventually passes out through the saliva. Viruses requiring this intimate association with the vector are termed persistently transmitted circulative viruses (Nault, 1997). When a naive vector begins feeding on a PLRV-infected plant, circulation takes ~24–36 h before the aphid can transmit. Once this latent period has passed the aphid can often transmit PLRV for the rest of its life, even after molts. PLRV only replicates in plant tissues, not in the vector. There are no recognized strains of PLRV but many isolates have been described (Taliansky *et al.*, 2003). These isolates differ in severity and transmissibility (Tamada *et al.*, 1984).

PVY, and all aphid-transmitted potato viruses other than PLRV, are "stylet-borne" and transmitted non-persistently (Nault, 1997). Viruses like PVY are relatively stable in the plant host and reach high titers in epidermal and sub-epidermal cells. More than 50 species of aphids are capable of transmitting PVY. Green peach aphid is the most efficient vector of PVY, but the greater abundance or propensity to develop winged adults (alatae) of less efficient vector species can make them more important in PVY epidemiology, especially if their dispersal occurs when potato plants are young. Other potentially important PVY vectors include pea aphid (*Acyrthosiphon pisum*), black bean aphid (*Aphis fabae*), the cotton/melon aphid complex (including *A. frangulae*), soybean aphid (*Aphis glycines*), buckthorn aphid, leaf-curling plum aphid (*Brachycaudus helichrysi*), currant aphid (*Cryptomyzus ribis*), sowthistle aphid (*Hyperomyzus lactucae*), potato aphid, lesser rose aphid (*Myzaphis rosarum*), *Myzus certus* (no accepted common name), damson-hop aphid; various cereal (small grain) aphids especially bird cherry-oat aphid (*Rhopalosiphum padi*); bulb and potato aphid and mangold aphid (Harrington *et al.*, 1986; Piron, 1986; Harrington & Gibson, 1989; Sigvald, 1990; Ragsdale *et al.*, 2001; Davis *et al.*, 2005).

Acquisition and inoculation of PVY and other non-persistently transmitted stylet-borne viruses can occur in intracellular feeding probes of a few seconds duration in epidermal tissues of infected plants. Such "sap sampling" feeding is associated with host selection behavior by winged aphids. Thus, non-persistent viruses can be transmitted by aphid species having only transitory association with potato when searching for a suitable host. Non-persistent viruses are quickly lost from aphid stylets; thus a vector's ability to transmit persists for at most a few probes immediately following acquisition. Once an aphid is no longer able to transmit, it must reacquire the virus by feeding on another infected plant. Non-persistently transmitted stylet-borne viruses are never transmitted following a molt by the aphid and the virus does not enter the hemocoel. Rather, virus particles bind to receptor sites located in the foregut and healthy plants are inoculated when aphids egest fluid from their stylets.

Most viruses occur in nature as variants, which if they differ sufficiently from type, are designated as strains. PVY exists as several strains that have been placed into groups based on symptomology in potato and tobacco (Blanco-Urgoiti *et al.*, 1998; Kerlan *et al.*, 1999). The common strain, PVY^O, causes mosaic symptoms in leaves and leaf drop in potato and tobacco. A strain designated as the tobacco veinal necrotic strain, PVY^N, causes mild to no symptoms in potato, yet severe leaf necrosis occurs in tobacco. In potato, PVY^N infection is visually asymptomatic in many potato cultivars making it difficult to detect except by serological bioassay or reverse transcriptase-polymerase chain reaction (RT-PCR). Similarly, a recombinant strain, $PVY^{N:O}$, which possesses the coat protein of PVY^O and causes the veinal necrosis typical of PVY^N, is often asymptomatic in potato (Singh *et al.*, 2003). Another variant of PVY^N designated as PVY^{NTN} causes typical mosaic symptoms on potato leaves and sunken necrotic rings on tubers (Le Romancer *et al.*, 1994).

28.2 Aphid vectors

Knowledge of aphid biology and ecology is an essential but largely neglected aspect of virus epidemiology and management (Radcliffe & Ragsdale, 2002). Aphids possess unique biological properties that make them proficient virus vectors; high fecundity, short generation time and propensity to form alatae which select host plants using visual and gustatory stimuli (Robert & Bourdin, 2001).

The probability of infecting a plant with a virus using a vector under specific conditions is termed vector efficiency. Vector efficiency is influenced by virus titer available to the aphid, viral aggressiveness, temperature and developmental stage of the plant. Differences in transmission can also be influenced by vector reproduction and mortality rates. However, the most important aspect of vector efficiency is the vector itself (Gibson *et al.*, 1988). There appears to be a genetic component inherent in the aphid which allows different aphids to bind viruses to mouthparts more efficiently. The more a vector wanders the more plants it will visit, increasing chances for PVY acquisition and inoculation. Non-colonizers tend to be more restless and probe more frequently than colonizers (de Bokx & Piron, 1990), making them more effective vectors of PVY (Sigvald, 1990).

Approaches to management of virus spread in potato can be categorized as preventive or therapeutic (Ragsdale *et al.*, 2001). Prevention tends to focus on reducing virus inoculum, a primary objective of all seed potato certification programs, while therapeutic action tends to focuses on reduction of either vector numbers or virus transmission.

28.3 Preventive measures

28.3.1 Seed certification

Most seed potato certification programs employ a limited generation production system (Franc, 2001; Gutbrod & Mosley, 2001). Typical production systems permit field increase for four to eight generations. In modern practice, seed potato increase is initiated with tissue-culture-derived seedlings enzyme-linked immunosorbent assay (ELISA) tested to assure freedom from viruses. Tissue-culture-derived plantlets are increased one generation in a protected environment to produce mini tubers which are subsequently increased in the field. Seed potato increase fields are inspected periodically during the growing season by seed certification personnel and a representative sample of harvested tubers is indexed for virus or other defects. Virus tolerances for recertification are stringent for all generations (usually in the range of 0.0% to 1.0% total virus), but typically are relaxed incrementally with successive generations of increase. Seed potato lots that pass summer inspections, but are not winter tested or fail to meet winter test recertification standards, generally can still be sold for ware production. When this is permitted, massive amounts of virus inoculum can be dispersed through the production system.

Virus management in seed potato production is becoming increasingly challenging worldwide, with PVY presenting the greatest problem. The situation in Minnesota is illustrative. Here, PVY has been epidemic since 1987 (Radcliffe *et al.*, 2008). On average, 28.5% of seed lots entered into the Minnesota Seed Potato Certification Program winter grow-out over the last 19 years exceeded tolerance for PVY. Worse, that failure rate has averaged 44.5% since 1997.

Reasons for persistence of this PVY epidemic are poorly understood, but factors suggested as contributing to PVY epidemics are: (1) unreliability of visual virus indexing (Rowberry & Johnson, 1975), (2) statistical inadequacy of the samples indexed (in terms of sample size and in how representative it is of the population), (3) the popularity of certain essentially asymptomatic cultivars (Slack, 1992; Russo *et al.*, 1999; Mollov & Thill, 2004), (4) changing cropping systems, e.g. expanded production of canola (a favored host of green peach aphid and turnip aphid [*Lipaphis pseudobrassicae*]) and soybean (host of soybean aphid), crops that have changed the dynamics of PVY epidemiology in Minnesota by contributing massive aphid flights in early summer when potatoes are most susceptible to infection (Davis *et al.*, 2005), (5) changing pesticide use patterns, e.g. Colorado potato beetle (*Leptinotarsa decemlineata*) developed resistance to all classes of insecticides in common use in the Minnesota in the 1980s and early 1990s leading to intensive spray schedules that tended to flare green peach aphid populations and in the mid-1990s emergence of more virulent strains of the potato late blight pathogen (*Phytophthora infestans*), necessitated greatly increased use of protective fungicides which interfere with aphid entomopathogens thus favoring green peach aphid survival (Lagnaoui & Radcliffe, 1998) and (6) even global warming (Davis *et al.*, 2006).

What presently appears to be the greatest obstacle to managing PVY is the increasing prevalence of viruses that are cryptic in visual symptomology. PVY^N was first detected in North America in 1990 (Singh, 1992). In response, Canada and the USA implemented a PVY^N Management Plan and declared PVY^N to be a quarantine-regulated disease. However, by 2004, it was evident that quarantine had failed to prevent spread and that PVY^N and $PVY^{N:O}$ recombinants (i.e. variants of PVY sharing the pathotype of PVY^N and serotype of PVY^O) were distributed widely across North America (Crosslin *et al.*, 2002; Piche *et al.*, 2004; Davis *et al.*, 2005). In 2005, PVY^N was reclassified as a non-quarantine pest to be regulated in accord with a new Canada/US Management Plan for Potato Viruses that Cause Tuber Necrosis (APHIS, 2007).

The prevalence in North America of PVY^N and $PVY^{N:O}$, alone or as mixed infections with PVA or PVS, has called into question the effectiveness of visual virus indexing for purposes of seed certification (Sturz *et al.*, 1997; Singh & Singh, 1995; Singh *et al.*, 1999). Singh *et al.* (1999) compared the effectiveness of visual indexing, ELISA, nucleic acid spot hybridization (NASH) and RT-PCR across various cultivars. Visual indexing identified PVY^O infections correctly in 64% of the plants. By contrast, ELISA, RT-PCR and NASH were 94%, 95% and 96% accurate, respectively.

In North America, as in Europe, $PVY^{N:O}$ recombinants appear to have a high rate of spread (Chrzanowska, 2001; Glais *et al.*, 2001; Singh *et al.*, 2003). In Eastern Europe, $PVY^{N:O}$ has been reported predominant over PVY^N by 9 : 1 (Kaczmarek & Mosakowska, 2001). In Manitoba, incidence of $PVY^{N:O}$ recombinants increased from 0.7% in 1996 to 64% in 2002 (Singh *et al.*, 2003). Of particular concern is the suggestion that high levels of PVY^N may contribute to the evolution of PVY^{NTN} strains (Kus, 1995). PVY^N tends to be translocated more quickly than PVY^O (Weidemann, 1988) and in potato can reach higher concentrations sooner and with milder symptoms (Chrzanowska, 1991). When non-viruliferous green peach aphid were allowed to probe PVY^N-infected plants before or after probing PVY^O-infected plants, PVY^N had a negative effect on PVY^O transmission whereas PVY^O had no effect on PVY^N (Katis *et al.*, 1986).

In 2002–03, PVY^N was detected in the winter grow-out of tubers harvested from cultivars entered in a regional "virus trial" at the University of Minnesota's UMore Park, Rosemount (Davis, 2006). Purposes of this trial were to evaluate clones for resistance to PVY and PLRV and serve as a screen against release of advanced breeding lines that poorly express visual symptoms of PVY infection. To determine validity of our screening procedures we tested leaflets from plants scored as visually positive for PVY and from plants visually scored as negative for PVY using ELISA monoclonal antibodies specific for both PVY^O and PVY^N. The discovery of PVY^N was unexpected and a first in Minnesota, although we were to learn that PVY^N had been known in Manitoba since 1996 (Singh *et al.*, 2003). Of the plants that were both visually and ELISA positive for PVY in the 2002–03 trial,

Table 28.1 Comparison of visual indexing and ELISA for detection of total virus

	2004–2005		2006–2007	
	Visual positive	Visual negative	Visual positive	Visual negative
ELISA +ve	789	292	118	309
ELISA −ve	340	679	40	136
Totals	1129	971	158	445

65% tested positive for PVY^O and 35% tested positive for PVY^N (Davis, 2006). In the 2003–04 iteration of this trial, PVY^N (72%) had displaced PVY^O (28%) as the predominant strain of PVY. Presence of PVY^N variants in the sampled leaflets was confirmed both years by RT-PCR.

In 2005 and 2007, we sampled potato seed lots entered by Minnesota seed potato growers into the Minnesota Department of Agriculture (MDA) Seed Potato Certification Program winter grow-out on Oahu, Hawaii for PVY and other viruses. Purposes of this research were to: (1) ascertain prevalence of PVY strains in Minnesota potato seed lots and (2) determine how well results of visual virus indexing correlated with results of serological testing. In Hawaii, plants were visually indexed by MDA seed potato inspectors for expression of virus symptoms shortly after full stand emergence. Each year, leaflets were collected from two subsets of plants: (1) plants scored as visually positive for PVY and (2) plants scored as visually negative for PVY. In 2005, we collected leaflets from 2100 plants (1129 visually positive for PVY, 971 visually negative). In 2007, we collected leaflets from 663 plants (158 visually positive for PVY and 445 visually negative). Leaflets were transported on ice to Minnesota and stored at 4 °C until tested serologically. Each leaflet was first tested for PVY using polyclonal antibodies (PVY^{all}). Polyclonal antibodies for PVY are not PVY-strain specific. Each leaflet was then tested using monoclonal antibodies specific for: (1) PVY^O and (2) PVY^N. Finally, representative PVY-positive leaflets were tested using RT-PCR to confirm PVY strain and isolate (PVY^O, PVY^N, $PVY^{N:O}$ and PVY^{NTN}). Both years, all leaflets were also individually tested using polyclonal double antibody sandwich (DAS)-ELISA for PVS, PVA, PVM, PLRV and PVX.

PVY was the only virus that MDA inspectors visually scored as exceeding tolerance for recertification in any of the Minnesota seed lots in the 2005 and 2007 winter grow-outs. However, PVY was rampant in Minnesota seed lots each year causing seed lot rejections of 62% in 2005 and 35% in 2007 (Radcliffe *et al.*, 2008). Results obtained using PVY^{all} showed reasonable agreement with visual positives both in 2005 (71%) and 2007 (74%) (Table 28.1). However, visual indexing failed to detect many of the PVY positives indicated by DAS-ELISA. Of the plants scored as visually negative for PVY, 23% were ELISA positive in 2005 and 57% were ELISA positive in 2007. PVY strain identification by RT-PCR indicated a concurrent shift from PVY^O to $PVY^{N/NTN}$ variants. PVY^O accounted for 32% of total PVY in 2005, but only 18% in 2007. $PVY^{N:O}$ increased from 56% to 75% over the same time.

Another unexpected result of the DAS-ELISA testing of Minnesota seed lots entered in the winter grow-out was the high frequency of other viruses, particularly in 2007 (Table 28.2). Most surprising was the 67% level of PLRV positives found in 2007, compared to just 1% PLRV-positives in 2005. The significance of such high levels of PLRV is unclear since only a handful of plants in the 2007 MDA winter grow-out expressed visual symptoms of PLRV, and no seed lot was rejected for recertification because of PLRV. However, it has to be assumed that considerable PLRV infected seed was replanted in 2007. The lack of visual expression of PLRV in the 2007 grow-out could reflect low-titer PLRV infections in the tubers sent to Hawaii suggesting late season transmission. Seemingly less probable is the possibility of current season spread in Hawaii. Other viruses showing greatly increased occurrence in the 2007 ELISA tests were

Table 28.2 Virus types detected by ELISA in leaves visually scored as virus positive or virus negative

	2004–2005		2006–2007	
	Visual +ve ELISA +ve	Visual −ve ELISA +ve	Visual +ve ELISA +ve	Visual −ve ELISA +ve
PVY^{all}	565 (71%)	68 (23%)	87 (74%)	175 (57%)
PVS	185 (23%)	194 (66%)	25 (21%)	69 (22%)
PVA	27 (3%)	11 (4%)	5 (4%)	11 (6%)
PVM	12 (2%)	18 (6%)	41 (35%)	135 (44%)
PLRV	4 (1%)	5 (2%)	71 (60%)	216 (70%)
PVX	4 (1%)	5 (2%)	46 (39%)	84 (27%)

Table 28.3 PVY strain and isolate as determined by RT-PCR

	2005	2007
PVY^{O}	32%	19%
PVY^{N}	0%	1%
$PVY^{N:O}$	56%	75%
PVY^{NTN}	12%	5%

PVM and PVX. PVA and PVS were common both years.

Among plants PVY-positive by DAS-ELISA in the 2005 winter grow-out, RT-PCR showed 32% had PVY^{O}, 56% had $PVY^{N:O}$ and 12% had PVY^{NTN} (Table 28.3). Among PVY-positive plants in the 2007 winter grow-out, 19% had PVY^{O}, 75% had $PVY^{N:O}$ and 5% had PVY^{NTN}. Some seed lots in the 2007 winter grow-out had many missing plants suggesting the possibility that PVY^{NTN} incidence in the tubers may have been higher than leaf testing indicated.

For purposes of seed potato certification PVY and PLRV are the viruses of primary concern. In Minnesota, seed lots are rarely failed due to any other virus. Accuracy of identification of these viruses in the winter grow-out is critical in two respects. False positives can cause rejection of a seed lot that should be recertified. False negatives can permit seed lots that are actually over tolerance to be recertified. The latter defeats the purpose of recertification because it fails to remove virus inoculums from the seed production system and can serve to perpetuate virus epidemics.

The rapidly shifting abundance of different virus types, strains and isolates that has occurred in Minnesota since 2002 presents a difficult situation for seed potato inspection. Of plants visually scored as PVY-negative in the MDA winter grow-out, DAS-ELISA showed 38% to be PVY-positive in 2002, 62% in 2003, 36% in 2004 and 7% in 2005. In 2003, inspectors overlooked many infected plants likely because they were not accustomed to the generally mild symptomology induced by infection with PVY^{N} or $PVY^{N:O}$ recombinants. They did much better in 2005 and 2007. The dramatic improvement in 2007 may have been aided by the upsurge in PVX which in combination with PVY tends to induce "severe mosaic." However, as the proportion of false negatives has decreased, the proportion of false positives has increased. In 2003, 87% of plants scored as visually positive for PVY tested were ELISA positive, but in 2004 agreement dropped to 62% and in 2005 was only 50%. False visual positives may occur because of the difficulty in discriminating among PVY^{N} and other mild mosaics, particularly the very prevalent PVS.

With PVY^{N} and $PVY^{N:O}$ recombinants now prevalent in North America, visual virus indexing may be unable to achieve the level of diagnostic accuracy required for a seed potato certification to be effective. Molecular testing (RT-PCR) is the only method that can provide absolute assurance that a potato seedling is free of virus.

Serological testing (DAS-ELISA) is a more realistic alternative, but even that is time consuming and expensive compared to visual virus indexing. Serological testing of foliage from all the plants grown in a state seed potato certification program winter grow-out would be a formidable task and is perhaps impractical because of the handling time required to collect and process such a high volume of samples. Moreover, no single serological test is certain to detect all strains of PVY. Most US and international seed potato certification regulations specify visual indexing as the standard to be used. Obviously, there could be an economic disadvantage to being the first certification agency to use more sensitive testing methods. However, change is occurring. In October of 2007, the Idaho Crop Improvement Association announced significant changes to their winter testing procedures. All seed lots in the post-harvest test will be required to be tested for PVY by ELISA. In addition, seed lots imported to Idaho must also be ELISA tested for PVY.

28.3.2 Roguing

Physical removal (roguing) of plants grown from virus-infected tubers can be a useful management tactic. Roguing is practical only when the initial incidence of virus infection is low (<1–2%) and if the field is small enough that every plant can be inspected several times during the growing season. Roguing should begin as soon as symptoms of secondary infection can be seen, typically when plants are 15–20 cm tall. Care should be taken in roguing since stand gaps as small as 0.6 m^2 can significantly increase PVY spread (Davis, 2006).

28.4 Therapeutic measures

28.4.1 Spatial and temporal isolation

Virus management strategies based on spatial or temporal isolation of the protected crop can be considered therapeutic approaches since they achieve their effect by minimizing vector pressure. Ideally, seed potato increase should occur in areas well isolated from ware production since in most situations potato is the principal or only source of potato virus inoculums. Isolation from vector flights can also be achieved by modifying planting or harvesting dates (Hille Ris Lambers, 1972; Hanafi *et al.*, 1995). To achieve spatial isolation, many US states and Canadian provinces have established seed farms or designated geographic areas where potato production is limited to seed. However, in most situations, there is little isolation of late-generation seed and commercial production.

Based on vector flight behavior as evidenced by aphid captures in suction traps in eastern Idaho, it was suggested 400 m to 5 km could provide effective isolation from known PVY sources, but that 30 km or more might be required for isolation from PLRV sources (Halbert *et al.*, 1990). In England, minimum separation of 800 m is recommended from potential sources of PVY (Harrington *et al.*, 1986), but in Denmark, a distance of just 40 m was shown to reduce spread of PVY (Hiddema, 1972). Most seed potato growers have only limited flexibility in locating their fields, thus other cultural control methods and vector management assume greater importance. Isolation from vector flights can also be achieved by modifying planting or harvesting dates. However, most seed potatoes are grown in short-season locations where growers have limited flexibility for manipulation of planting and harvest dates.

28.4.2 Chemical control

Among reported successes of controlling virus spread by use of insecticides (all crops and insect vectors), 94 of 119 cases involved persistent and semi-persistent viruses (Perring *et al.*, 1999). Most of the failures, 32 of 48 cases, involved non-persistent viruses. Insecticide use can reduce spread of PLRV (National Agricultural Statistics Service, 2007), but is seldom of benefit in preventing spread of non-persistently transmitted stylet-borne viruses such as PVY (Ragsdale *et al.*, 1994). PLRV spread from within field foci are often attributed to movement of apterae and thus infection of neighboring plants can be suppressed by insecticides because of the extended post-acquisition latent period required before an aphid is able to transmit (Hanafi *et al.*, 1989). Conversely, insecticides are much less effective in controlling PLRV if the virus is being moved from outside sources by winged, viruliferous aphids that have already passed through the latent period and may only require a few minutes of phloem feeding to transmit.

Development of insecticide resistance can severely limit insecticide options. Green peach aphid is notorious for its capacity to develop insecticide resistance, with resistance reported to 68 insecticide chemistries (Whalon *et al.*, 2004). Another important vector of potato viruses, cotton aphid, is reported to have resistance to 39 insecticide chemistries including methamidophos, the product of choice for aphid control in many seed potato production systems. Consolidation of agrichemical companies worldwide, a slowing rate of development of new insecticides and the alarming rate at which insecticide resistance develops are persuasive arguments for active development and promotion of sound Insecticide Resistance Management (IRM) programs.

To slow the development of insecticide resistance, it is desirable to avoid sequential use of products with the same mode of action that exposes more than one generation of the target pest to selection. To assist crop professionals practicing IRM, the Insecticide Resistance Action Committee (IRAC), an inter-agrochemical company committee whose mission is to "promote the development of resistance management strategies in crop protection" (Insecticide Resistance Action Committee, 2007), has developed a classification system and tentatively assigned all insecticides to one of 28 unique groups based on their mode of action. This numbering system when included on the label for insecticides (fungicide and herbicide labels already indicate mode of action codes) will make it easier to implement IRM programs since practitioners will not have to be conversant in insecticide toxicology to make appropriate decisions on product rotation.

28.4.3 Population monitoring/forecasts

Association of virus spread in potatoes with aphid flight activity is well documented. Early aphid-trapping networks focused on green peach aphid because this species was considered the most efficient vector of potato viruses. As the importance of PVY spread by less efficient but abundant vectors was recognized, aphid-trapping networks began routinely identifying additional aphid species. Trapping networks are intended to monitor aphid flight on a regional basis. At any particular location, e.g. an individual farm, the first spring migrants may not be detected because their occurrence is rare and the sample unit small. Another limitation is that the traps may not be monitored daily and expertise is required to identify the captured aphids.

A regional aphid trapping network, Aphid Alert, was operated on seed potato farms throughout Minnesota and North Dakota from 1992 to 1994, and again from 1998 to 2003 (Radcliffe *et al.*, 2008). We used two trap types, pan traps baited with green or yellow ceramic tiles submerged in propylene glycol and 2.3 m tall, low-volume (2.4 m^3/min) suction traps. The traps were emptied weekly, the aphids sorted to species and tallied, and the capture data reported to seed potato growers by surface mail, email and the World Wide Web. Growers used the information to make management decisions especially with regard to timing foliar sprays and vine kill.

Although the economic benefit of Aphid Alert to regional seed potato producers was estimated to be considerably greater than the cost of maintaining it, the program terminated when public funding ended. For this reason, we explored the use of models based on meteorological data, as a possible cost-effective surrogate for aphid-trapping data (Zhu *et al.*, 2006). Analysis of historical data sets showed that cumulative duration of early season low-level jet streams was strongly correlated with summer abundance of green peach aphid and spread of both PLRV and PVY. Statistical models were developed relating frequency and duration of spring wind events to subsequent green peach aphid abundance and severity of PLRV and PVY spread in the northern Great Plains.

28.4.4 Site-specific IPM

In 2003, experiments were done in 23 Minnesota and North Dakota seed potato fields to evaluate effectiveness of targeted applications for aphid control (Carroll *et al.*, 2008). When there was a spike in aphid flight activity in traps of the Aphid Alert network, an 18-m wide spray swath of methamidophos was applied by airplane to border areas (i.e. at the ends of rows) abutting fallowed headlands. Pre-treatment green peach aphid densities were ~10 times higher in border areas than in field interiors. The targeted spray applications provided excellent control of early colonizing aphids. Aphid densities mid-field, pre-application and 3 days after, did not differ from densities in spray swaths 3 days after. Overall, only 38.5 of 730 ha were treated, saving an estimated 93% of costs compared to treating the entire field.

28.5 Cultural control

Cultural control methods are often among the most effective and inexpensive of virus/vector control measures that growers can implement (Radcliffe, 2006). These practices are usually not complicated, but require application of knowledge of vector biology and ecology. What follows is a description of several non-chemical or reduced chemical use approaches that have been tried in Minnesota to limit the spread of aphid-transmitted viruses in seed potato.

28.5.1 Crop oils

Mineral oils applied to plants can substantially reduce PVY transmission. Limitations to crop oils include weathering of oil deposits, new plant growth between applications and incomplete coverage. The mechanism by which oils inhibit aphid transmission of viruses is still not known. It has been suggested that the oil interfered with adherence of the virus to receptors on the mouthparts. Crop oils can be mildly repellent to aphids and some cause direct mortality, but the latter effect is unlikely to occur quickly enough to prevent transmission of a nonpersistently transmitted virus.

28.5.2 Protective barriers

Polymer webs can provide a high degree of protection against aphid-transmitted viruses, but the cost and inconvenience of row covers limit their application to seed potato fields of very high value and small size. Barrier (border) crops are more widely adaptable than mulches or floating row covers; they are easier to install and keep in place, and do not lose effectiveness due to weathering or when the canopy closes. Barrier crops should have a fallow outside border with no gap between the barrier crop and the potato field, since winged aphids tend to alight at the interface of fallow ground and green crop (DiFonzo *et al.*, 1996). If immigrating alatae carrying PVY feed first on the border crop, they will probably lose their virus inoculum before moving into the potatoes. Barrier crops need be only a few meters wide to be effective. The use of crop borders to protect seed potatoes, especially high-value, early-generation seed, from PVY spread has been widely adopted by Minnesota and North Dakota growers. A 2004 survey of growers found that 97% ($n = 32$) of seed lots protected by crop borders passed seed certification, whereas only 54% ($n = 57$) without crop borders passed (Olson *et al.*, 2005).

28.5.3 Host plant resistance

For many years, our laboratory has focused a major research effort on identifying sources of host plant resistance in wild potato species (Flanders *et al.*, 1992). Accessions highly resistant to green peach aphid were identified in 36 of the 86 potato species screened and to potato aphid in 24 of 85. This research has resulted in the development of advanced potato breeding lines that have high levels of resistance to green peach aphid and potato aphid and apparent immunity to both PLRV and PVY. Tuber yield and type in the best of these selections is near agronomic acceptability. Research is under way to identify and map these sources of aphid and virus resistance (Davis, 2006).

It is often suggested that extant potato cultivars offer little promise as sources of useful aphid resistance. However, few prior studies have critically measured effects of host cultivar on aphid age-dependent life table statistics or related these measures to field performance. Therefore, we recently undertook a comprehensive field and greenhouse study to assess 49 commercial potato cultivars, primarily of North American origin, for resistance to green peach aphid and potato aphid (Davis *et al.*, 2007). Cultivars were found to show considerable differences in resistance to each aphid species, but these resistances were not associated ($R^2 = 0.032$). Aphid/predator population models using a "Kill Factor" (K)-value of 15.2 indicated that following colonization green peach aphid populations would remain stable for 20 days on Russet Norkotah (resistant) whereas on Red La Soda (susceptible) populations would reach over 54 000 per plant (Davis *et al*, 2007). Population models indicated that with non-persistent foliar insecticides as the only control, three applications would be necessary to maintain green peach aphid below the Minnesota recommended action threshold on Red La Soda for 21 days while just one application would be needed for green peach aphid on Russet Norkotah.

Transgenic potato lines have been developed that are highly resistant, but not immune, to infection by PLRV, PVY and PVX. While aphids can still acquire virus from low titer plants, efficiency of transmission is greatly reduced. Transgenic cultivars were released in the USA that expressed the Colorado potato beetle specific toxin (*Bacillus thuringiensis tenebrionis* [*Bt*]) combined with PLRV replicase. Cultivars were also developed that expressed *Bt* and PVY coat protein (Li *et. al.*, 1999). This technology was far more effective than any presently used tactic, but these cultivars have been withdrawn because of concerns over a public backlash against genetically modified food.

28.6 Conclusions

Insecticides are valuable tools for preventing spread of PLRV, but seldom effective in limiting spread of non-persistently transmitted stylet-borne viruses. Cultural practices and other non-insecticidal or reduced use approaches are often among the most effective and inexpensive control measures that growers can implement. Such practices are often not complicated, but like all pest management tactics require application of knowledge of vector biology and ecology to be effective. But cultural control tactics that take advantage of vector behavior must work in concert with an effective seed certification program that reduces inoculum. Seed certification programs, however, continue to fail in removing virus inoculum because of their reliance on visual inspections which fail to identify asymptomatic PVY strains, the most abundant strains in North America. Though inexpensive in comparison to serological and molecular methods, visual virus indexing is too inefficient to reverse a virus epidemic. Without disease-free seed, all other preventative and therapeutic methods are only marginally effective.

References

APHIS (2007). *Plant Health, Potato Diseases, Potato Virus Y Strains*. Washington, DC: US Department of Agriculture, Animal and Plant Health Inspection Service. Available at www.aphis.usda.gov/plant_health/plant_pest_info/potato/pvy_main.shtml.

Blanco-Urgoiti, B., Tribodet, M., Leclere, S. *et al.* (1998). Characterization of potato potyvirus Y (PVY) isolates from seed potato batches: situation of the NTN, Wilga and Z isolates. *European Journal of Plant Pathology*, **104**, 811–819.

Carroll, M. W., Radcliffe, E. B., MacRae, I. V., *et al.* (2008). Border treatment to reduce insecticide use in seed potato production: biological, economic, and managerial analysis. *American Journal of Potato Research*, **84** (in press).

Chrzanowska, M. (1991). New isolates of the necrotic strain of *potato virus Y* (PVY^N) found recently in Poland. *Potato Research*, **34**, 179–182.

Chrzanowska, M. (2001). Importance of different strains of PVY in potato production and breeding program in Poland. In *Proceedings of the 11th Meeting of the European Association of Potato Research, Virology Section*. Havlíčkův Brod, Czech Republic: Potato Research Institute.

Crosslin, J. M., Hamm, P. B., Eastwell, K. C. *et al.* (2002). First report of the necrotic strain of potato virus Y (PVY^N) Potyvirus on potatoes in the northwestern United States. *Plant Disease*, **86**, 1177.

Davis, J. A. (2006). Identifying and mapping novel mechanisms of host plant resistance to aphid and viruses in diverse potato populations. Ph.D. dissertation, University of Minnesota, St. Paul.

Davis, J. A., Radcliffe, E. B. & Ragsdale, D. W. (2005). Soybean aphid, *Aphis glycines* Matsumura, a new vector of *Potato virus Y* in potato. *American Journal of Potato Research*, **82**, 197–201.

Davis, J. A., Radcliffe, E. B. & Ragsdale, D. W. (2006). Effects of high and fluctuating temperatures on green peach aphid, *Myzus persicae* (Hemiptera: Aphididae). *Environmental Entomology*, **35**, 1461–1468.

Davis, J. A., Radcliffe, E. B. & Ragsdale, D. W. (2007). Resistance to green peach aphid, *Myzus persicae* (Sulzer), and potato aphid, *Macrosiphum euphorbiae* (Thomas), in potato cultivars. *American Journal of Potato Research*, **84**, 259–269.

de Bokx, J. A. & Piron, P. G. M. (1990). Relative efficiency of a number of aphid species in the transmission of potato virus Y^N in the Netherlands. *Netherlands Journal of Plant Pathology*, **96**, 237–246.

DiFonzo, C. D., Ragsdale, D. W., Radcliffe, E. B., Gudmestad, N. C. & Secor, G. A. (1996). Crop borders reduce potato virus Y incidence in seed potato. *Annals of Applied Biology*, **129**, 289–302.

Flanders, K. L., Hawkes, J. G., Radcliffe, E. B. & Lauer, F. I. (1992). Insect resistance in potatoes: sources, evolutionary relationships, morphological and chemical defenses, and ecogeographic associations. *Euphytica*, **61**, 83–111.

Food and Agriculture Organization (2007). *International Year of the Potato 2008*. Rome, Italy: United Nations, Food and Agriculture Organization. Available at www.potato2008.org/en/index.html.

Franc, G. D. (2001). Seed certification as a virus management tool. In *Virus and Virus-Like Diseases of Potatoes and Production of Seed-Potatoes*, eds. G. Loebenstein, P. H. Berger, A. A. Brunt & R. H. Lawson, pp. 407–420. Dordrecht, Netherlands: Kluwer.

Gibson, R. W., Payne, R. W. & Katis, N. (1988). The transmission of *potato virus Y* by aphids of different vectoring abilities. *Annals of Applied Biology*, **113**, 35–43.

Glais, L., Tribodet, M. & Kerlan, C. (2001). Molecular detection of particular PVY isolates: PVY^{NTN} and $PVY^{N}W$. *Proceedings of the 11th Meeting of the European Association of Potato Research, Virology Section*, pp. 70–71. Havlíčkův Brod, Czech Republic: Potato Research Institute.

Graves, C. (ed.) (2001). *The Potato, Treasure of the Andes, from Agriculture to Culture*. Lima, Peru: International Potato Center (CIP).

Gutbrod, O. A. & Mosley, A. R. (2001). Common seed potato certification schemes. In *Virus and Virus-Like Diseases of Potatoes and Production of Seed-Potatoes*, eds. G. Loebenstein, P. H. Berger, A. A. Brunt & R. H. Lawson, pp. 421–443. Dordrecht, Netherlands: Kluwer.

Halbert, S. E., Connelly, J. & Sandvol, L. E. (1990). Suction trapping of aphids in western North America (emphasis on Idaho). *Acta Phytopathologica Entomologica Hungarica*, **25**, 411–422.

Hanafi, A., Radcliffe, E. B. & Ragsdale, D. W. (1989). Spread and control of potato leafroll virus in Minnesota. *Journal of Economic Entomology*, **82**, 1201–1206.

Hanafi, A., Radcliffe, E. B. & Ragsdale, D. W. (1995). Spread and control of potato leafroll virus in the Souss Valley of Morocco. *Crop Protection*, **14**, 145–153.

Harrington, R. & Gibson, R. W. (1989). Transmission of potato virus Y by aphids trapped in potato crops in southern England. *Potato Research*, **32**, 167–174.

Harrington, R., Katis, N. & Gibson, R. W. (1986). Field assessment of the relative importance of different aphid species in the transmission of potato virus Y. *Potato Research*, **29**, 67–76.

Hiddema, J. (1972). Inspection and quality grading of seed potatoes. In *Viruses of Potatoes and Seed-Potato Production*, eds., J. A. de Bokx, pp. 206–215. Wageningen, Netherlands: Pudoc.

Hille Ris Lambers, D. (1972). Aphids: their life cycles and their role as virus vectors. In *Viruses of Potatoes and Seed-Potato Production*, ed., J. A. de Bokx, pp. 36–56. Wageningen, Netherlands: Pudoc.

Insecticide Resistance Action Committee (2007). *IRAC Mode of Action Classification, revised and reissued, July 2007. Version: 5.3.* Available at www.irac-online.org/documents/IRAC%20MoA%20Classification%20v5_3.pdf.

Kaczmarek, U. & Mosakowska, E. (2001). Interaction between strains of the *Potato virus Y* (PVY^{O}, PVY^{N}-type Wilga, PVY^{NTN}) and the potato plants of two cultivars. *Proceedings of the 11th Meeting of the European Association of Potato Research, Virology Section*, pp. 18–20. Havlíčkův Brod, Czech Republic: Potato Research Institute.

Katis, N. Carpenter, J. M. & Gibson, R. W. (1986). Interference between potyviruses during aphid transmission. *Plant Pathology*, **35**, 152–157.

Kerlan, C., Tribodet, M., Glais, L. & Guillet, M. (1999). Variability of Potato virus Y in potato crops in France. *Journal of Phytopathology*, **147**, 643–651.

Kus, M. (1995). The epidemic of the tuber necrotic ringspot strain of potato virus Y (PVY^{NTN}) and its effect on potato crops in Slovenia. *Proceedings of the 9th Meeting of the European Association of Potato Research, Virology Section*, pp. 159–160. Bled, Slovenia.

Lagnaoui, A. & Radcliffe, E. B. (1998). Potato fungicides interfere with entomopathogenic fungi impacting population dynamics of green peach aphid. *American Journal of Potato Research*, **75**, 19–25.

Le Romancer, M., Kerlan, C. & Nedellec, M. (1994). Biological characterization of various geographical isolates of *Potato virus Y* inducing superficial necrosis on potato tubers. *Plant Pathology*, **43**, 138–144.

Li, W., Zarka, K. A., Douches, D. S. *et al.* (1999). Coexpression of potato PVY^{O} coat protein and *cry* V-Bt genes in potato. *Journal of the American Society for Horticultural Science*, **124**, 218–223.

Mollov, D. S. & Thill, C. A. (2004). Evidence of *Potato Virus Y* asymptomatic clones in diploid and tetraploid potato-breeding populations. *American Journal of Potato Research*, **81**, 317–326.

National Agricultural Statistics Service (2007). *Quick Stats (Agricultural Statistics Data Base)*. Washington, DC: US Department of Agriculture. Available at www.nass.usda.gov/Data_and_Statistics/Quick_Stats/index.asp.

Nault, L. R. (1997). Arthropod transmission of plant viruses: a new synthesis. *Annals of the Entomological Society of America*, **90**, 521–541.

Olson, K., Badibanga, T., Radcliffe, E. B. & Ragsdale, D. W. (2005). *Producers' Use of Crop Borders for Management of Potato Virus Y (PVY) in Seed Potatoes*, Staff Paper No. P05-14. St. Paul, MN: University of Minnesota, Department of Applied Economics.

Perring, T. M., Gruenhagen, N. M. & Farrar, C. A. (1999). Management of plant viral diseases through chemical control of insect vectors. *Annual Review of Entomology*, **44**, 457–481.

Piche, L. M., Singh, R. P., Nie, X. & Gudmestad, N. C. (2004). Diversity among *Potato virus Y* isolates obtained from potatoes grown in the United States. *Phytopathology*, **94**, 1368–1375.

Piron, P. G. M. (1986). New aphid vectors of potato virus Y^N. *Netherlands Journal of Plant Pathology*, **92**, 223–229.

Radcliffe, E. B. (2006). Use of non-chemical alternatives to synthethic pesticides in maintaining plant health in a clonally propagated crop: potato. *Arab Journal of Plant Pathology*, **42**, 170–173.

Radcliffe, E. B. & Lagnaoui, A. (2007). Pests and Diseases, Part D: Insect pests in potato. In *Potato Biology and Biotechnology: Advances and Perspectives*, eds. D. Vreughenhil, J. Bradshaw, C. Gebhardt *et al.*, pp. 541–567. Amsterdam, Netherlands: Elsevier.

Radcliffe, E. B. & Ragsdale, D. W. (2002). Invited Review: Aphid transmitted potato viruses: the importance of understanding vector biology. *American Journal of Potato Research*, **79**, 353–386.

Radcliffe, E. B., Ragsdale, D. W., Suranyi, R. A., DiFonzo, C. D. & Hladilek, E. E. (2008). *Aphid Alert*: how it came to be, what it achieved, and why it proved unsustainable. In *Areawide IPM: Theory to Implementation*, eds. O. Koul, G. W. Cuperus & N. C. Elliott, pp. 227–242. Wallingford, UK: CABI Publishing.

Ragsdale, D. W., Radcliffe, E. B., DiFonzo, C. D. & Connelly, M. S. (1994). Action thresholds for an aphid vector of potato leafroll virus. In *Advances in Potato Pest Biology and Management*, eds. G. W. Zehnder, M. L. Powelson, R. K. Jansson & K. V. Raman, pp. 99–110. St. Paul, MN: American Phytopathological Society.

Ragsdale, D. W., Radcliffe, E. B. & DiFonzo, C. D. (2001). Epidemiology and field control of PVY and PLRV. In *Virus and Virus-Like Diseases of Potatoes and Production of Seed-Potatoes*, eds. G. Loebenstein, P. H. Berger, A. A. Brunt & R. H. Lawson, pp. 237–270. Dordrecht, Netherlands: Kluwer.

Robert, Y. & Bourdin, D. (2001). Aphid transmission of potato viruses. In *Virus and Virus-Like Diseases of Potatoes and Production of Seed-Potatoes*, eds. G. Loebenstein, P. H. Berger, A. A. Brunt & R. H. Lawson, pp. 195–225. Dordrecht, Netherlands: Kluwer.

Rowberry, R. G. & Johnston, G. R. (1975). Virus infection of potato seed stocks in Ontario under commercial insect-control practices. *Canadian Plant Disease Survey*, **55**, 15–18.

Russo, P., Miller, L., Singh, R. P. & Slack, S. A. (1999). Comparison of PLRV and PVY detection in potato seed samples tested by Florida winter field inspection and RT-PCR. *American Journal of Potato Research*, **76**, 313–316.

Sigvald, R. (1990). Aphids on potato foliage in Sweden and their importance as vectors of Potato virus Y^O. *Acta Agriculturæ Scandinavica*, **40**, 53–58.

Singh, M. & Singh, R. P. (1995). Digoxin-labeled cDNA probes for the detection of potato virus Y in dormant potato tubers. *Journal of Virological Methods*, **52**, 133–143.

Singh, M., Singh, R. P. & Moore, L. (1999). Evaluation of NASH and RT-PCR for the detection of PVY in dormant tubers and its comparison with visual symptoms and ELISA in plants. *American Journal of Potato Research*, **76**, 61–66.

Singh, R. P. (1992). Incidence of the tobacco veinal necrotic strain of potato virus Y (PVY^N) in Canada in 1990 and 1991 and scientific basis for eradication of the disease. *Canadian Plant Disease Survey*, **72**, 113

Singh, R. P., McLaren, D. L., Nie, X. & Singh, M. (2003). Possible escape of a recombinant isolate of *Potato virus Y* by serological indexing and methods of its detection. *Plant Disease*, **87**, 679–685.

Slack, S. A. (1992). A look at potato leafroll and PVY: past, present, future. *Valley Potato Grower*, **57**(95), 35–39.

Sturz, A. V., Diamond, J. F. & Stewart, J. G. (1997). Evaluation of mosaic symptom expression as an indirect measure of the incidence of PVY^O in potato cv. Shepody. *Canadian Journal of Plant Pathology*, **19**, 145–148.

Taliansky, M., Mayo, M. A. & Barker, H. (2003). Potato leafroll virus: a classic pathogen shows some new tricks. *Molecular Plant Pathology*, **4**, 81–89.

Tamada, T, Harrison, B. D. & Roberts, I. M. (1984). Variation among British isolates of potato leafroll virus. *Annals of Applied Biology*, **104**, 107–116.

Weidemann, H. L. (1988). Importance and control of potato virus Y^N (PVY^N) in seed potato production. *Potato Research*, **31**, 85–94.

Whalon, M. E., Sanchez, D. & Duynslager, L. (2004). *The Database of Arthropods Resistant to Pesticides*. Available at www.pesticideresistance.org/DB/index.html.

Zhu, M., Radcliffe, E. B., Ragsdale, D. W., MacRae, I. V. & Seeley, M. W. (2006). Low level jet streams associated with spring aphid migration and current season spread of potato viruses in the U.S. northern Great Plains. *Agricultural and Forest Meteorology*, **138**, 192–202.

Chapter 29

IPM in structural habitats

Stephen A. Kells

Over the past 30+ years, there has developed a sizeable complement of knowledge about structural pests, as well as the devices and methods for detecting and controlling these pests. Despite existing knowledge, there is still much to learn about implementing this knowledge into IPM programs that are robust in ability to prevent or reduce pest infestations, provide opportunities for decision making and result in continued successes or directed improvement. IPM programs for residential, institutional, commercial and industrial structures are largely undeveloped and fragile in operation; there are considerable challenges involved in adopting the IPM philosophy and practices for these different structures. Some urban structures have established programs typically mandated by regulations (e.g. public schools and government buildings), while others employ programs using industry-developed standards (e.g. food processing). Overall, structural IPM practices typically resemble modified conventional practices compared to how practices should operate under a true IPM program.

In the current state of developing IPM programs for structures, insects can be monitored, excluded, cleaned up, or controlled with least-toxic products, but how these elements are delivered in a program is subject to the type of structure, past experiences and opinion of the practitioner, and forces of competition by companies that provide pest management services. Structural IPM programs are in a fragile state, readily discarded when things go wrong, resulting in a "knee-jerk reaction" return to recognized and convenient "conventional" methods of control. In many cases, IPM programs become "conventionalized" because of control failure resulting from incomplete or inconsistent IPM processes (e.g. inaccurate monitoring, technical decision not to spray in an area, lack of training), and seldom do IPM programs become re-established after such failure.

Structural IPM involves the prevention and control of a variety of pests in a variety of buildings and structures. It is this variety that perhaps is the single major reason why development of structural IPM programs has not followed the course of development historically favored by agriculturally based practices. Buildings differ in many aspects, e.g. size, shape, construction materials, contents, purpose of the structure, building occupants, surrounding landscape and properties, maintenance and upkeep procedures, all contributing to the uniqueness of the situation. Compounding the problem is the knowledge base of pest control service providers and relationships between pest control personnel and facilities management. The resulting issues are often significantly more complex than those encountered by the agricultural community. Yet, solutions have to

Integrated Pest Management, ed. Edward B. Radcliffe, William D. Hutchison and Rafael E. Cancelado. Published by Cambridge University Press.

be developed because of the proximity of people to pests, the value of the commodities threatened by pests, and costs and associated risks of repeated control procedures. There are many challenges to producing workable IPM programs with the main issue involving the development of a set of robust programs that can be applied to different situations, can be economical while still providing a set of principles as the driving force for making necessary changes to support the success of structural IPM programs.

29.1 Defining IPM for urban structures

In terms of structural IPM, current definitions provide a tremendous amount of variables and a lack of essential details, making it difficult to decide which components of IPM would be applicable or useful to a particular structural situation. The US Environmental Protection Agency (2005) provides a relatively complete definition of an IPM program, but Kramer (2004) notes most definitions of IPM retain an agricultural bias. Supporting documentation for this definition provides a plethora of items and topics in no particular order leaving the affected person or industry the challenge of attempting to formulate a cohesive program with the endpoint goals of pest suppression or elimination, maximizing safety and performing economically. For example, one of the steps under existing definitions is to determine the threshold for pest activity, but what happens when a threshold is unavailable? Or if the pest threshold is at detection, then is this merely conventional control where the insect is seen and sprayed? Failure to provide answers for this fundamental step in construction of an IPM program leads people to apply loosely the steps of IPM. This can result in uncoordinated practices and failure to develop a cohesive program with purpose and measurable endpoints.

During work in the pest control industry, the author employed the following operational paradigm for IPM in structures:

Structures, structural contents, and human activities provide opportunities for a number of pests to harbor and locate resources. In a program of regular assessment, inspection and monitoring, a structural IPM program identifies pest-conducive conditions and through a variety of control practices the program prevents pests from locating or continuing to exploit these opportunities.

Along with important terms and considerations from other definitions, e.g. "economical means, reduced or less hazardous pesticides," this statement provides direction for IPM programs to coordinate different practices to deny opportunities to pests. Following a situational assessment program goals can be mapped out. The goals may be to: (1) reduce pesticides, (2) completely eliminate a pest from a particular structural habitat and (3) determine what practices can be afforded in terms of budget, or a combination of all three.

29.2 Challenges in developing and operating a structural IPM program

In structural IPM programs, economic injury levels and economic thresholds are much less definable, and for the most part lacking (Robinson, 1996). In addition to the difficulties in conducting the typical research required for generating these estimates there are other obstructions. In most cases, the economic injury level and economic threshold are arbitrarily set at the lowest level of detection. There are numerous reasons for arbitrarily setting such a low level without considering the pest, facility, or the actual damage caused by a pest. Some illustrative reasons are:

Variability of what people consider a threshold level Wood *et al.* (1981) demonstrated that an "aesthetic threshold" for German cockroaches (*Blattella germanica*) was unlikely to be accepted by public housing residents. If such a threshold were used, most residents would expect progress to the point of elimination (Zungoli & Robinson, 1984). Similar cases have been found in commercial and industrial situations. Even with decreases in pest load, people may find the new level unacceptable leading to frustrated occupants using conventional measures of control for relief (Williams *et al.*, 2005).

Lawsuits ignoring established pest levels Rodents in food processing plants and complaints by consumers concerning pest invasion in finished goods often result in lawsuits, even if the offending parts found are below the US Food and Drug Administration (FDA) standard for pieces/parts contamination, or if insect activity is adjudged as below the economic threshold. Lawsuits will usually challenge that product is defective if the consumer finds any pest contamination, regardless of FDA standards or threshold levels set for purposes of IPM.

Point sources of infestation leading to repeated infestation elsewhere Structures and affiliated society exist as a series of paths and systems that pests use to move from one place to another. A bedbug (*Cimex lectularius*) or German cockroach infestation in one locale provides a source for these pests to other locations (Kells, 2006). For instance controlling cockroach infestations in the locker rooms of large manufacturing plants often involves finding the source populations at employees' residences and addressing infestation at these sources. Because the presence of cockroaches is often difficult to detect at low densities, use of treatment thresholds in manufacturing plants can lead to there being a source of cockroachs for other locations.

Infestations and the issue of time In agricultural habitats the production of crops is a finite event, following which there is usually a depletion of resource or inclement weather and then the cycle starts anew. In structural pest management, there are no interruptions in the cycle during which a pest control program might be reset and renewed. Although there may be seasonal fluctuations, pest activity in a structure is ongoing and the damage caused cumulative.

The variety of damage caused by pests in structural habitats There are several types of economic injury levels (and thresholds) to be considered in pest management programs, beyond pest presence or the direct damage they cause. It is important to realize that many of the following categories do not have numbers that allow them to be associated with the presence of pests. Often they may be associated with "potential risk," or a risk that is quantifiable only when damage has occurred in that structure. Sometimes damages caused by a pest will shift across different categories or may encompass many categories at the same time.

(a) *Brand security* Pest infestations can cause consumers to reject buying a product, avoid using the services of a structure (i.e. hotel) and instead switch to another brand. For processors of food and other consumable items, and commercial establishments, brand security represents a loss that greatly exceeds the immediate value of a product, the cost of control practices or the cost of disposal of infested product. Internet and media exposés increase the swift and anecdotal dissemination of information linking brands to infestation problems.

(b) *Health* Pest presence sometimes involves personal costs due to infestation, including hospitalization, medication and other associated costs.

(c) *Aesthetic damage* There can be a loss of quality and appreciation perceived by an individual because of the presence of pests.

(d) *Product liability* Pest presence in a finished product or place can result in civil action and associated costs of defending against such action.

(e) *Trade liability* Pest presence or associated products or parts can result in rejection of an entire food shipment to another country and preclude opportunity for future shipments.

(f) *Food safety* Pest presence represents a potential for transmission of agents responsible for food borne illness. The ability for flies and cockroaches to mechanically vector (spread) pathogenic bacteria increases the requirements for pest control and prevention. This might be considered a subset of human health, but as the damage is potential and considered a risk to human health, it is often considered separately.

Assessing the many types and levels of economic injury is complicated and the practice of using thresholds is diminished because of the different categories of damages that might occur, the perceptions of when damage was actually caused and the difficulty of linking pest presence to damage across a diversity of habitats and even within habitats that appear the same. Economic injury levels and economic thresholds may be set at the lowest possible levels or assumed to be at an "acceptable" possible level, namely less than the level of detection. Therefore the goal of the structural IPM program is to use the following steps to push the insect activity to the lowest possible level, while at the same time not formally accepting the presence of insects.

29.3 Key steps in structural IPM programs

There are two methods of presenting the key steps for a structural IPM program. The first most common method is to provide the components as a menu and permit the practitioners to select which components are used in the program (e.g. Environmental Protection Agency, 2006). The second method is to present these components within the process of a program so there is an implied cohesiveness with the components and other factors such as time can be considered. Emphasis on different IPM options is very important because, as mentioned, there will be structural and operational variability. For example, one facility manager will accept (and budget for) enhanced sanitation practices which will reduce the need for pesticide applications. Other facility managers may not budget for appropriate sanitation, or the situation (the structure) may not be suitable for proper sanitation; hence the program emphasis will be shifted to more inspections, and/or pesticides or other reactive measures.

The following key steps are presented in order of how IPM should be managed. The key to a structural IPM program is to develop a concise and unified program, making the procedures part of a collective operation rather than attempting to define each key step in isolation. Programs must be considered as operating continuously and cyclical over time.

29.3.1 Assessment

The structural IPM program must have a plan or strategy of operation. Prior to initiation of any structural IPM program and when the program cyclically restarts, an assessment is necessary to provide an overview of current practices, areas where improvements can be made and the goals of the program for the coming year. For simple residential habitats, the assessment for a simple control procedure may be brief and can be accomplished on site. It would include the pest identity, those opportunities available to the pest, control measures to be considered, confounding factors that may restrict application of a control procedure, and finally, the potential for reoccurrence. An example of this type of simple assessment might be the presence of occasional invading pests around a residence, stored-products pests in a pantry or termite control via barrier treatments.

Assessments may become quite involved depending on the pest, the habitat and the situation. As an example, food processing plants will require annual (at least) assessments to review occurrence of pest activity over the past year, the response of the program to preventing (or controlling) pests and projections of what should be expected for the next year. In the case of institutional structures where pests can cause damage directly to health, or historic structures containing valued items such as natural fibers and furs, a more in-depth and regular assessment will be required compared to the residential program. Assessments are an opportunity to review the program determining if and where improvements can be made, and what prevention measures can be afforded in the next budget cycle. Certainly in locations where there are risks to health or brand security, assessments are essential to ensuring the IPM program does not become stagnant, thereby allowing complacency and resulting in pests locating habitat opportunities.

A typical assessment involving more complex habitats will include a number of steps. First, what are the policies and current goals of the program? As demonstrated by the Environmental Protection Agency (2006, Step 1), stating the intention to control or prevent pests, what pests could be present and why such a program is required sets a course of action for the following elements. This fundamental first step is not often used (Platt *et al.*, 1998; personal observation). Second, what is the current status with respect to pests presently in the structure? Pest presence and damage in a structure is an indication that the current pest management program does not have components in place to address the underlying sources of the problem. Third, what are the characteristics of the past infestation and is there a chance that an infestation will repeat itself, either through resurgence of the pest, importation of pests or situational changes? This step relies on review of data generated from the structure during inspections. Fourth, can the program be made better by: improving the protection of increased numbers or new/different items in the habitat, decreasing the baseline insect population, anticipating pest activity with respect to planned changes in the structure, or decreasing the number of large-scale control procedures currently being used?. These steps are completed by a combination of documentation review, planned inspections related to an assessment and review of other building functions and departments, i.e. maintenance, janitorial, etc.

29.3.2 Training

Training is a critical step in IPM programs, but one which is often underestimated, underutilized or completely ignored. Training also incurs a significant cost that must be realized as essential for the program but unfortunately is often developed arbitrarily and/or subject to budgetary constraints. Seldom does it become part of a formal curriculum, with a budget adequate to ensuring active learning. In recent publications (Miller & Meek, 2004; Williams *et al.*, 2005; Wang & Bennett, 2006) on IPM of cockroaches and the measure of cost effectiveness in comparison with conventional systems, training costs were not included in the budget models and were assumed to come from elsewhere in the system. The "elsewhere" was never specified. Often, there is a tendency for non-budgeted training to be ignored or the training event might be used once but seldom repeated.

Training serves many goals for an IPM program, including making sure the program follows the plan constructed during the assessment phase. Training ensures that details will not be initially overlooked, or become overlooked, forgotten or ignored. IPM programs often require detailed work from different groups such as pest control personnel, custodial staff and maintenance staff. Without everyone understanding established directives, parts of the program will falter, or worse, overall goals of the program will fail. Other characteristics of successful pest management training include:

- *Inspire people* to professionalism and enthusiasm for their career. Make sure that interest and vigilance is maintained so that boredom does not begin to guide people into set, unthinking patterns.
- In some companies, *frequent* (monthly) *training* of the sanitation staff is required to ensure a consistent level of knowledge concerning control and prevention of pests through sanitation. This is particularly true where sanitation detail is an entry-level position, with frequent changes in employees through promotion, layoffs, etc. For instance, at a food plant, cleaning and sanitation may be dealt with successfully by a junior-level employee and because that person does well, e.g. actively cleaning to prevent insects and other pests, their proficiency may result in reassignment to other duties. Often their knowledge will not be transferred to their replacement and absent pertinent training in pest prevention and control, that person must independently learn the job.
- Details can be lost or distorted over the years and *refresher courses* are necessary to ensure action when pest activity is found, for example, cockroaches found by plumbers or warehouse beetle (*Trogoderma variabile*) discovered by electricians.
- During training, *feedback from employees* helps to ensure that the IPM programs are still fully

protecting the structure. It offers also a second chance to report pests and vulnerable systems that may have gone unreported previously.

- Training provides a low-cost method of *showing interest in employees* through their development. Particularly with training about insects and other pests, opening the floor for questions from employees about pests in and around their homes helps to build interest in the topic outside of being just a "work-related" issue.

29.3.3 Inspections

Inspections are typically repeated visitations to various sites in a structure to look for opportunities that could be used by pests. Monitoring devices may include intercept traps, insect light traps and pheromone traps for insect pests. Rodent traps are typically used to monitor for rodents in the structure and will often form the main commercial service in an IPM program. Inspections must focus on areas beyond the traps, and include signs of damage, evidence of infestation or conducive conditions that could provide opportunities for pests.

Inspections must be documented so data on pests can be assembled and lists of conducive conditions can be addressed before or after control measures are done. Data such as the type of pests encountered, their number, the risk of damage occurring and the presence of existing conducive conditions will be important for the next scheduled assessment, or where a pest outbreak is detected and longer-term controls are required.

29.3.4 Monitoring

Inspections of areas represent a single point in time and there may be pest activity that is not detected during the limited period that a person is actively looking for infestations. The structural habitat is very complex and the areas available for inspection are only a small fraction of the total area available to pests. Monitoring devices provide continuous surveillance of pest activity. Devices can be as simple as cardboard with a soft adhesive applied to the surface to intercept insects traveling in the structure. On the other hand, devices may include food or pheromones to lure insects into the trap. Use of these lures will depend on the insect of interest. In warehouses and food processing plants, pheromone traps are typically used against Indianmeal moth (*Plodia interpunctella*), warehouse beetle and cigarette beetle (*Lasioderma serricorne*), while in places where natural fibers are stored, traps may be used against clothes moths (e.g. *Tineola pellionella* and *T. bisselliella*). Multiple-catch mouse traps are also available for monitoring the presence of mice within the structure.

There are four important considerations when using traps for IPM programs in structures. First, there are no defined thresholds for treatment of most pests inside structures, unlike when traps are used in the agricultural habitat. Traps are generally used for detection of pest activity at low levels, though there is a lack of knowledge and understanding relating captures to level of infestation. Second, an abundance of food or poor sanitation conditions interfere with trap yields and adversely affects the level at which an infestation becomes detectable. Third, traps are generally used as an indication of the source population, particularly in areas where food resources may be more discrete. Fourth, there is often a misunderstanding that pheromone traps can be used as control devices because many insects are caught. In reality, most pheromone traps utilize a sex pheromone to lure males, leaving the females and an unknown number of males to continue the infestation. There are some recent exceptions to this trend, with some products being available to capture females (e.g. Moth Suppression® traps for Indianmeal moths and Serrico® traps for cigarette beetles). These limitations must be considered during the assessment and goals must be formulated with these limitations in mind.

Monitoring also may be accomplished on a complaint, or pest reporting basis. This is a method used by the US Government Services Administration (Greene & Breisch, 2002), but may be used by facilities with little or no formal inspection procedures, or restricted monitoring capabilities (e.g. military installations and detention centers). Caution must be taken with monitoring of complaints because of the variability in perceiving what constitutes a pest situation. Often extensive training and reminders must be done to maintain standards.

29.4 Habitat alteration techniques

29.4.1 Sanitation

One of the major reasons for pests being present in a structure is the availability of resources, namely food, water and harborage. Often one or all three are present in and around a structure that has an infestation. Where possible, extraneous material should be removed, habitat complexity reduced (fewer places to hide) and food sources eliminated. Sanitation is often required for reducing physical and nutritional interference of insecticides (Arthur, 2000). Sanitation can be as simple as vacuuming shelves in a residential pantry to removing spilled and infested food, but sanitation programs can also be considerably more involved, including the removal of food debris from the inside of electrical conduits inside a food processing facility. Sometimes the sanitation procedures are elaborately complex. For example, a sewer main break under a concrete slab will encourage species of phorid flies to find and exploit the raw sewage leaking under a concrete floor. In this case, pesticide applications will provide only temporary (2 to 24 hours) relief from this pest and the only way to gain an effective fix to this situation is to remove a portion of the concrete, clean up the spill and repair the pipe. Another example where sanitation is absolutely required occurs with instances of mothflies (Psycodidae) emerging in large numbers from toilets. This would indicate an accumulation of sewage within a portion of pipe that has a dead space or a region of pipe that has accumulated material not flushed through.

29.4.2 Exclusion

Preventing pests from entering the structure provides major support for avoiding damage and disruption from corrective measures. Barriers to entry can include physical barriers such as walls, doors and windows. With physical barriers, semi-annual to annual inspections must be performed to ensure that holes and cracks have not appeared or reopened because of the normal aging process a building undergoes. Exclusion may refer to operational barriers preventing pests from arriving on incoming goods, through inspections of incoming goods, sifting, insect destruction via grinding or centrifugation (termed entoletion), and preemptive control of infestations in artifacts (such as routine freezing or carbon dioxide fumigation).

29.5 Control techniques within the context of a structural IPM program

There are several methods for the application of pesticides and other control agents in the structural habitat and, similar to the presentation of IPM tactics, the control techniques tend to be categorized based on the formulation and mode of application. Within a program and for planning purposes, another method of categorizing control techniques is to consider the extent that the technique affects the facility. This is a preferred method when trying to determine costs of control and comparing different control tactics necessary when planning for more aggressive IPM measures that reduce disruption to the facility, or insecticide use in the facility (personal observations). Use of the most common control methods is summarized in the following sections.

29.5.1 Mass control techniques

Mass control techniques (or methods) are used to "biologically zero" a structure or area, killing pests in large numbers, attacking those with widespread distribution in multiple locations, or reaching those hidden in refuges that cannot be reached by typical spraying procedures. Specific techniques involve the use of ultra-low-volume (fogging) applications and fumigations, but major non-chemical techniques such as heat treatments can also be considered. Fumigation and structural heat treatments are designed to penetrate into all areas of a structure thereby affecting difficult-to-reach refuges. General surface and fogging applications control only exposed insects, missing those in refuges. In cases where there is an outbreak of pest activity, mass control techniques permit a large and sudden reduction in the number of pests, permitting follow-up precision methods to be used to address any new activity or tend to those insects that successfully found refuges during the major control event. Mass control techniques require a major disruption in operations of the structure in that processes must be shut down and people must be removed from the building or area. During fumigations and heat treatments, facilities may be closed for 24 hours or more. General sprays, fogging or ultra-low-volume applications may require facility closure for 2 to 12 hours, until the residues are dry or the fog has settled.

29.5.2 Precision methods

Precision methods are used when pests are present and the extent of the infestation is in a known and limited area. In addition, knowledge of pest biology or behavioral traits is used to limit applications only to places where the pests will come into contact with the control product. The application methods used for precise applications include: baits, crack and crevice, spot, and other limited area applications of mass control techniques. Using precision methods enables minimum disruption to the structure. During application to the facility, small areas may be shut down, e.g. a hotel room or hospital room may be temporarily closed or a single processing line may be shut down for focused treatments. The time it takes for the residues to dry usually provides a rough estimate for reentry period. This criterion appears on some pesticide labels in the USA. For some insects such as cockroaches and ants, the use of baits further helps in avoiding any disruption to the facility operation, except in facilities such as schools, detention centers and the primate section of zoos.

29.5.3 Commodity disinfestation

Disinfestation methods are used to prevent further insect activity within a commodity or artifacts. Often pests will enter a mass of commodity (e.g. grain, spices, other food), or artifacts (e.g. animal mounts, skins or natural fibers). Commodity disinfestation may include fumigation and heat treatments, but can also include atmospheric modification such as oxygen displacement, low pressure, low temperature and ozone treatments. Disinfestation procedures can be provided without moving the infested materials, or the materials can be moved to areas suitable for such procedures. For example, to avoid disruption to the entire facility, items may be moved out of the structure and placed in a trailer or shipping container or placed in a different area to receive the treatment. Some commercial facilities may have fumigation (or oxygen displacement) vaults on sites requiring repeated product disinfestation measures.

29.6 Post-application inspections

When control techniques are followed up by some form of inspection, the presence of dead or live insects post-application provides very useful information about the presence of conducive conditions, opportunities and source populations of these pests. This is a relatively new procedure in practice. Post-application inspections aid the program beyond simply determining the number of insects controlled by providing information as to the location of pest sites and the number and extent of these locations. Collating this information into maps and lists of infested areas provides documentation required for further control procedures, indicates if the program is continuing to be effective and helps to prioritize future control events. This type of documentation can also help in determining what habitat-modifying measures could be used and where they would have the best effect. Larger-scale structural changes can be justified with the use of post-application inspections and examples of such have resulted in removal of secondary non-structural "veneer walls," removal of false ceilings and exclusion work on roof and wall junctions (personal observations).

Post-application measures are conducted as follows: after a mass control or precision application, dead insects found close to treated areas indicate specific sources or refuges of infesting insects. Upon finding the refuges, precision applications can be applied to these newly found areas. For mass control applications, post-application inspections may follow in the same procedure (particularly with heat treatments), but there can be challenges because of the large areas involved and the increase in time that has to be scheduled to collect this vital data. Post-application inspections following mass control techniques can also use information collected from live insects trapped in pheromone traps or other trapping devices such as insect light traps or food baited traps. When trapping insects, post-applications inspections during the next 24 to 48 hours help pinpoint sources for the larger infestation. Once an infestation is located with these measures, further chemical and non-chemical control activities can be scheduled to progressively reduce insect presence in the facility.

29.7 Legal or system standards

Normally in an IPM program, the legal measures are thought of as a control tool for preventing infestations. For instance, company policies may require doors to be shut, thus supporting the exclusion tactics to prevent pest entry. Property standards in many jurisdictions enforced by municipal agencies or public health prevent accumulation of rodents, cockroaches and flies. The same standards were used by public health officials in controlling stagnant water during the West Nile virus outbreak between 2000 and 2006. Quarantine regulations help to minimize highly invasive pests and control resistant stored products pests such as Khapra beetle (*Trogoderma granarium*). For food production in the USA, federal laws include 21 US CFR Part 110 and the Food Drug and Cosmetic Act Sections 402(a) (3, 4), which support the use of IPM techniques (Heaps, 2006) against pests of food processing plants. Use of practices such as legal regulations and system standards as control and prevention tools have been effective and are considered a necessary part of an IPM program.

29.8 Documentation and communication

A popular anecdote among pest control companies is that technicians are no longer "baseboard spray jockeys," but "documentation management specialists." There have been cases where technicians have become so proficient in managing documents and recording data that when chemical methods are justified, they must be reacquainted with operating equipment such as ultra-low-volume applicators and how to determine when an application should be used. This problem has occurred in a number of companies. In a structural IPM program, collection and consolidation of data from routine inspections and post-application inspections is important for demonstrating that the program is operational, or where changes should be considered. During the annual assessment phase, this data is important for making decisions to improve the program, but the assessment time should not be consumed by data entry and formatting.

Table 29.1 An example of a subjective scale for quantifying rodent activity at bait stations

Scale	Description
0	No feeding activity
1	Feeding activity on one corner of a bait block
2	Feeding activity on two corners or along one edge
3	Feeding activity on more than one edge
4	Half of the placed bait consumed
5	All of the bait consumed

There are a number of methods used to collect information and they depend on the type of pest, methods of monitoring and inspection for that pest. Typically pest numbers can be compared over months or years and quantifying the activity permits trend reporting through graphical analysis. Pest numbers can be quantified and reported based on the number of traps, e.g. five cockroaches per trap, per month, but as programs improve there is a tendency to use the actual counts, e.g. five cockroaches caught per month, especially if the trapping program is successful and the number of traps used remains stable. With rodents the actual number is used because of the damage potential associated with even small numbers of these pests. With other devices such as bait stations for rodents, it is difficult to obtain a precise number for rodents affected and therefore an activity scale is used (Table 29.1). This scale may be modified for use with insects that do not respond well to traps, for instance red and confused flour beetles (*Tribolium* spp.). Typically the modified scale is based on a 0 to 5 rating that reflects an exponential increase in activity. Similarly, a scale may be used to account for large numbers of insects, particularly in devices such as insect light traps, where a scale is used to provide rapid accounting of pests considered incidental, or those classified as occasional invading pests.

29.9 Challenges for structural IPM

The potential for IPM in structural habitats remains undeveloped. Formal research regarding structural pests has been limited to efficacy testing, investigating pest behavior and biology and IPM in simple situations. Comparisons of different programs are discontinuous, limited to a single season or problem. While these are important components, they do not address proper program development and management in a variety of situations. Further, comparisons of different programs are discontinuous, and limited to a single season and problem. Robinson (1995) observed that the adoption of IPM measures had resulted in a simple substitution, e.g. replacing sprays with baits and adding a few traps for control of cockroaches. That is still the industry norm for development of structural IPM.

Unfortunately, current development of IPM programs for structures still depends in large part on a trial-and-error approach to addressing problems. The practitioner is required to take a plethora of information developed from research, such as pest biology, pest behavior, pesticide efficacy studies, and apply this information to a specific site. Sometimes the control strategies work, sometimes not. In addition, there may be a requirement for repeated applications because of continued control failures or limitations to being able to use the best control and prevention measures in the particular situation. It is less an issue of the specific research supported practice used and more about how the variety of practices can be used in concert with each other. In development of structural IPM programs, a scarcity of available scientific knowledge often means that considerable trust must exist between the service provider and personnel responsible for the facility. The "trials" often go well, until an "error" occurs and the program becomes conventional in substance, albeit with an IPM label.

Another major difficulty in successful delivery of IPM measures is the requirement for some intangible practices that cannot be easily accounted for under current business models. Most current structural IPM programs focus on tangible practices that involve the ability to record pest numbers or incidence and see how control changes affect these numbers. For example, a rodent caught in a trap is a number and the trap itself represents a number and a cost. There are a number of traps placed around a facility that monitor rodent activity. During servicing of these traps, the time to locate that trap and clean and reset the trap is a tangible service/item. Managing tangible items matches with the current (yet historic) business practices for pest management where the practice includes: entering the premises, applying an insecticide to areas most frequented by pests and quickly leaving to repeat the procedure at the next place of business in the case of contracted pest control, or somewhere else in the facility in the case of in-house personnel. Much of the business today still clings to a model of providing a service by managing the simple measures of personnel time, materials used, and travel. Trap monitoring has fallen into this pattern where insects are counted and the time charged relates to the technician's ability to handle and reset the traps. Little work if any is apportioned to inspections of facilities or the movement of devices to establish the source of an infestation.

There are many intangible items in an IPM program. The time a pest management technician may spend in inspecting and cleaning areas to prevent pest invasion is one example. Should there be a half hour or one hour, or only 15 minutes allotted to this practice, and also how frequently should these practices be done? Training, documentation and habitat alteration represent other intangibles. Adding to the complexity is that the problems that arise from cutting back on inspections and maintenance measures will not be realized until the problem reoccurs and there can be considerable delay between cutbacks and reoccurrence. With reduced measures, once the problem occurs again, conventional measures are required to re-establish the program. In many cases, a failure in the IPM program causes a "conventionalized" or reduced IPM program to be reinstigated because of the reluctance to invest in a program that is perceived to have failed in the past.

29.10 Conclusions

Structural IPM provides many options for prevention and control of pests in a diversity of structures and facilities, but requires much improvement with respect to its operation as a cohesive program that sets directions for progressive, long-term pest reduction. Many practices used in agricultural IPM, particularly action thresholds and economic injury levels, simply do not work with urban structural habitats because (1) they do not take into consideration the complexity associated with structures, (2) there are no "seasons" – problems are continuous and pest pressures constant and (3) as planned pest thresholds are met and maintained, there is a desire/requirement for even lower thresholds. Starting with a plan to reduce pest numbers and being prepared to plan improvements to the program on an annual (or more frequent) basis will predispose people to evaluate the pest management program and consider all options that could be used to prevent future pest activities. Important elements for structural IPM, including frequent assessments, post-application inspections and formal training, need to be implemented in order to attain successful and continuous pest suppression. Once a resident pest population has been eliminated or suppressed to their lowest point of activity, the program can then be changed to a goal of maintained vigilance against new introductions in the structure, or resurgence as a result of refuges that still remain within the structure or exist in off-property structures.

References

Arthur, F. H. (2000). Impact of accumulated food on survival of *Tribolium castaneum* on concrete treated with cyfluthrin wettable powder. *Journal of Stored Products Research*, **36**, 15–23.

Environmental Protection Agency (2005). *Integrated Pest Management (IPM) and Food Production*. Washington, DC: US Environmental Protection Agency. Available at www.epa.gov/pesticides/factsheets/ipm.htm.

Environmental Protection Agency (2006). *Integrated Pest Management (IPM) in Schools: Protecting Children in Schools from Pests and Pesticides*. Washington, DC: US Environmental Protection Agency. Available at www.epa.gov/pesticides/ipm/.

Greene, A. & Breisch, N. L. (2002). Measuring pest management programs for public buildings. *Journal of Economic Entomology*, **95**, 1–13.

Heaps, J. W. (2006). *Insect Management for Food Storage and Processing*. St. Paul, MN: American Association of Cereal Chemists International.

Kells, S. A. (2006). Bedbugs: a systemic pest within society. *American Entomologist*, **52**, 109–111.

Kramer, R. D. (2004). Integrated pest management. In *Mallis' Handbook of Pest Control*, 9th edn, eds. S. A. Hedges & D. Moreland, pp. 1311–1337. Cleveland, OH: GIE Media.

Miller, D. M. & Meek, F. (2004). Cost and efficacy comparison of integrated pest management strategies with monthly spray insecticide applications for German cockroach (Dictyoptera: Blattellidae) control in public housing. *Journal of Economic Entomology*, **97**, 559–569.

Platt, R. R., Cuperus, G. W., Payton, M. E., Bonjour, E. L. & Pinkston, K. N. (1998). Integrated pest management perceptions and practices and insect populations in grocery stores in south-central United States. *Journal of Stored Product Research*, **34**, 1–10.

Robinson, W. H. (1995). Perspective on IPM a critical review. *Pest Control Technology*, **23**, 34, 39–42.

Robinson, W. H. (1996). Integrated pest management in the urban environment. *American Entomologist*, **42**, 76–77.

Smith, L. M., Appel, A. G., Mack, T. P., Keever, G. J. & Benson, E. P. (1995). Comparative effectiveness of an integrated pest management system and an insecticide perimeter spray for control of smokybrown cockroaches (Dictyoptera: Blattidae). *Journal of Economic Entomology*, **88**, 907–917.

Tripp, J. M., Suiter, D. R., Bennett, G. W., Klotz, J. H. & Reid, B. L. (2000). Evaluation of control measures for black carpenter ant (Hymenoptera: Formicidae). *Journal of Economic Entomology*, **93**, 1493–1497.

Wang, C. & Bennett, G. W. (2006). Comparative study of integrated pest management and baiting for German cockroach management in public housing. *Journal of Economic Entomology*, **99**, 879–885.

Williams, G. M., Linker, H. M., Waldvogel, M. G., Leidy, R. B. & Schal, C. (2005). Comparison of conventional and integrated pest management programs in public schools. *Journal of Economic Entomology*, **98**, 1275–1283.

Wood, F. E., Robinson, W. H., Kraft, S. K. & Zungoli, P. A. (1981). Survey of attitudes and knowledge of public housing residents toward cockroaches. *Bulletin of the Entomological Society of America*, **27**, 9–13.

Zungoli, P. A. & Robinson, W. H. (1984). Feasibility of establishing an aesthetic injury level for German cockroach pest management programs. *Environmental Entomology*, **13**, 1453–1458.

Chapter 30

Fire ant IPM

David H. Oi and Bastiaan (Bart) M. Drees

Imported fire ants (*Solenopsis invicta*, *S. richteri* and their hybrid) are notorious invasive ants from South America that continue to plague the southern USA since their inadvertent introductions prior to the mid-1930s. They now infest over 129.5 million hectares in the USA. The red imported fire ant (*S. invicta*), commonly referred to as "fire ant," has continued to spread and is now a worldwide concern with infestations confirmed in Australia, Southeast Asia and Mexico. The painful, burning sensation that is inflicted by the sting of a fire ant is easily the most recognizable hazard to humans. While one sting is painful, it is not uncommon for a person to receive numerous stings simultaneously when ants swarm out of their nest to attack an intruder. This greatly intensifies the pain and can cause panic; thus fear or apprehension of these ants can be present in heavily infested or newly infested areas. In addition, it is conservatively estimated that 1% of stung individuals in the USA are allergic to the venom and at risk for anaphylaxis. Deaths from fire ant stings have been reported and lawsuits have resulted in awards of over $US 1 million.

Besides the costs associated with litigation, the annual economic impact of fire ants in the USA is estimated to be over $US 6500 million across both urban and agricultural sectors. In addition, their dominance in natural ecosystems has reduced biodiversity and harmed wildlife (Wojcik *et al.*, 2001). Given the tremendous impact that fire ants have had in the USA, incursions into previously non-infested areas have promulgated very expensive eradication programs. The cost of a planned, but aborted, ten-year eradication program in California was valued at $US 65.4 million (Jetter *et al.*, 2002). The current eradication program in Australia will cost over $US 144 million over 7 years (2001–2007, McNicol, 2006).

In the southern USA, eradication is no longer considered possible and instead, IPM for fire ants is evolving. The evolution of fire ant control strategies is an interesting mix of politics and science. The nasty sting and ubiquitous presence of fire ants makes its control politically expedient, and hence, the availability of funding makes fire ants the most intensely studied ant. In this chapter we discuss the historical transition of fire ant control strategies and illustrate how knowledge of the biology and ecology of fire ants advanced the development of IPM for this invasive pest.

30.1 Evolution of fire ant control: eradication/control attempts in the southern USA

The accidental introductions of fire ants into the USA through the port at Mobile, Alabama is

Integrated Pest Management, ed. Edward B. Radcliffe, William D. Hutchison and Rafael E. Cancelado. Published by Cambridge University Press.

thought to have occurred in 1918 for the black imported fire ant (S. *richteri*) and 1933 for the red imported fire ant (Tschinkel, 2006). By 1937, the fire ants had spread west into the state of Mississippi and east across Mobile Bay in Alabama. The seriousness of the problem resulted in a combined county, state and federal government program to control the ants. An insecticide containing 48% calcium cyanide dust was injected into individual fire ant nests on over 800 ha of vegetable cropland, with 80% of colonies reported eliminated. During World War II (1941–45), control efforts and research on the fire ant problem apparently were suspended.

During the late 1940s more intensive research on fire ants by university and US Department of Agriculture (USDA) laboratories commenced and have continued to the present. In 1948, the Mississippi legislature appropriated $US 15 000 to begin a control and eradication program which entailed the application of chlordane dust to nests. Surveys undertaken in 1949 through 1953 documented the rapid spread of fire ants and indicated that transport of infested commodities, such as nursery plants, facilitated the dispersal to 102 counties in 10 southern states. However it was not until 1958 that a quarantine was instituted, prohibiting the movement of fire ant harborages such as sand, gravel, soil on timbers, grass sod and nursery plants, unless treated with insecticide to kill any ants.

Prior to the quarantine, emphasis was still on eradication using contact insecticides. In 1957, aerial applications of heptachlor were made in an eradication project in Arkansas and the USA. The US Congress appropriated $US 2.4 million for USDA to immediately conduct a cooperative federal–state control and eradication program. Within three months, quarantine (effective in May 1958) was proposed to limit the spread of fire ants and an eradication plan was conceived that called for aerial and ground applications of heptachlor or dieldrin. Early applications of heptachlor at 2.24 kg/ha occurred in the winter of 1957–58 and soon after wildlife and cattle deaths were reported. From 1959 to 1960 application rates of heptachlor were reduced drastically in response to environmental concerns. In early 1960, residues of heptachlor on harvested crops were no longer permitted because heptachlor degraded into a more toxic compound, heptachlor epoxide, thus preventing treatment of cropland. The exclusion of cropland made eradication impractical because all infested areas could no longer be treated. The failed eradication attempt with heptachlor and dieldrin and the severe consequences to the environment was one of the events described in Rachel Carson's 1962 book *Silent Spring* which brought to the forefront public awareness of the environmental consequences of pesticides.

Concurrent with the heptachlor eradication program, research was being conducted at university and USDA laboratories to improve control methods. Delivery of insecticide via a bait formulation was a way to reduce the amount of insecticide applied to the environment. Ant baits are comprised of a slow-acting insecticide mixed into a food fed upon by ants. The slow action of the insecticide allows time for the natural ant behaviors of foraging and transferring food among ants within a colony to distribute the insecticide before causing death. The active ingredient mirex incorporated into a food attractive to fire ants and compatible with conventional insecticide application equipment was developed in 1961 and by 1963 refined into a standardized ant bait formulation. The bait had an application rate that resulted in 8.4 g of active ingredient per ha (which two years later was reduced by half).

Because of the low amount of mirex needed for effective fire ant control and application of such a small amount of insecticide that had less acute toxicity than DDT, a very widely used insecticide, the mirex bait was not considered to be hazardous to humans and animals at that time (Lofgren *et al.*, 1963; Lofgren, 1986). The cooperative federal–state fire ant control program continued to use mirex bait from 1962 to 1978, with an estimated 18.6 million ha treated three times, which encompassed almost all the infested states (Williams *et al.*, 2001). Unlike heptachlor and dieldrin, mirex bait did not have residual toxicity to fire ants and thus serial applications were needed when treated areas became reinfested. Lofgren & Weidhaas (1972) calculated three to nine applications with 90% to 99.99% control would be needed to eradicate fire ants from an 809.7 thousand ha area.

Despite the early eradication attempt and ongoing control programs, fire ants continued to spread and political pressure was still sufficient for another eradication attempt to be proposed. Congress appropriated funds in 1967 to the USDA to conduct large-scale tests to determine the feasibility of eradicating fire ants with mirex bait. Studies were conducted from 1967 to 1970 on three sites that ranged in size from 103 600 to 862 750 ha. It was concluded "that technical problems we did encounter are surmountable and, therefore, total elimination of IFA [imported fire ants] from large isolated areas may be technically feasible" (Banks *et al.*, 1973). However, the proposed eradication program was never implemented because in 1970 the use of mirex was challenged by environmental groups and banned from use on land managed by the US Department of the Interior.

The newly created US Environmental Protection Agency (EPA) held public hearings from 1973 through 1975 to review the use of mirex. Studies in the late 1960s and into the 1970s indicated that mirex residues persisted in the environment and accumulated in the tissues of non-target organisms, and that residue levels were magnified in predators that ate contaminated prey and were toxic to estuarine organisms. Mirex was also found to be carcinogenic and its use was forbidden after 30 June 1978. During the legal challenges and public hearings, the application of mirex bait continued and even after mirex was banned there was an effort to register a biodegradable formulation of mirex bait called ferriamicide. However, this product was never commercialized as it also persisted in the environment.

With the environmental concerns and legal challenges against the use of mirex, the USDA accelerated an effort to find alternative fire ant bait toxicants. From 1976 to 1981 over 3000 chemicals were evaluated resulting in the registration in 1980 of the fire ant bait Amdro® (which contained the active ingredient hydramethylnon). This product is currently available. Several other fire ant baits were developed since the registration of Amdro®, including insect growth regulating baits and fast-acting baits that are effective within three days in contrast to the two to four weeks for traditional baits (Oi & Oi, 2006). The acute toxicity of active ingredients used in currently available fire ant baits is generally much lower than the insecticides used for fire ant control before hydramethylnon (Table 30.1). No further attempts at the eradication of fire ants from the southern USA have occurred since the cancellation of mirex.

In hindsight, fire ants, with their painful sting and easily recognized nests, made their spread alarming. Public outcry and political pressure to eliminate the venomous invader, and the confidence during that era of the recently discovered synthetic insecticides to eliminate pests led to the regrettable widespread applications of heptachlor and dieldrin. With the development of the mirex bait there was a very significant reduction in the amount of insecticide applied and an assumption that it was not hazardous to humans and wildlife. Certainly acute mammalian toxicity was not apparent, but its environmental persistence and biomagnification were not anticipated. In addition, its effect on non-target arthropods, especially other ants that would feed on the bait, may have facilitated fire ant reinfestations and their spread into habitats not dominated by fire ants (Markin *et al.*, 1974; Buren in Canter, 1981, p. F-29).

During the period of the large-scale eradication and control programs, the area under the fire ant quarantine still continued to expand. However, these programs most likely operated under the logistical and technical problems identified in the eradication feasibility trials. Thus, while the eradication of fire ants from the southern USA was a goal, technologies, such as environmentally compatible, species-specific control methods were not available to eliminate fire ants and return and/or retain the native ant ecosystem over millions of hectares. Nevertheless, the development of insecticidal fire ant baits for the eradication/control programs provided the basis of a control tactic (i.e. broadcast application of bait-formulated insecticide) that continues to be the most efficient and environmentally compatible method of reducing fire ant and other pest ant populations (Fig. 30.1).

30.2 Fire ant biology and IPM

With the early emphasis on fire ant eradication, research focused on identifying effective

Table 30.1 Acute toxicity (LD_{50}[a]) of active ingredients used in formulations or products to control fire ants before and after large-scale eradication and control programs ended in the southern USA in 1978

1937–1978		1980–present	
Active ingredient	LD_{50} (mg/kg)	Active ingredient	LD_{50} (mg/kg)[b]
Calcium cyanide	39[c]	Hydramethylnon	1146
Chlordane	137–590[d]	Fenoxycarb	16 800
Dieldrin	51–64[e]	Abamectin	10[f]
Heptachlor	70–230[d]	Methoprene	> 34 600
Mirex	365–600[d]	Pyriproxyfen	>5000
		Spinosad	>5000
		Fipronil	95[f]
		Indoxacarb	1730

[a] LD_{50}, amount (mg) of technical active ingredient (AI) per kg of body weight that kills 50% of rats given the AI orally. LD_{50} of formulated products may be much higher (less toxic) than AI alone.
[b] LD_{50}s from Barr *et al.* (2005).
[c] LD_{50} from Pesticide Action Network (www.pesticideinfo.org/Index.html).
[d] LD_{50}s from Agency of Toxic Substances and Disease Registry (ATSDR) (www.atsdr.cdc.gov/toxpro2.html#bookmark05).
[e] LD_{50} from St. Omer (1970).
[f] LD_{50} from http://extoxnet.orst.edu/pips/abamecti.htm; LD_{50} of formulated fire ant products >5000 mg/kg

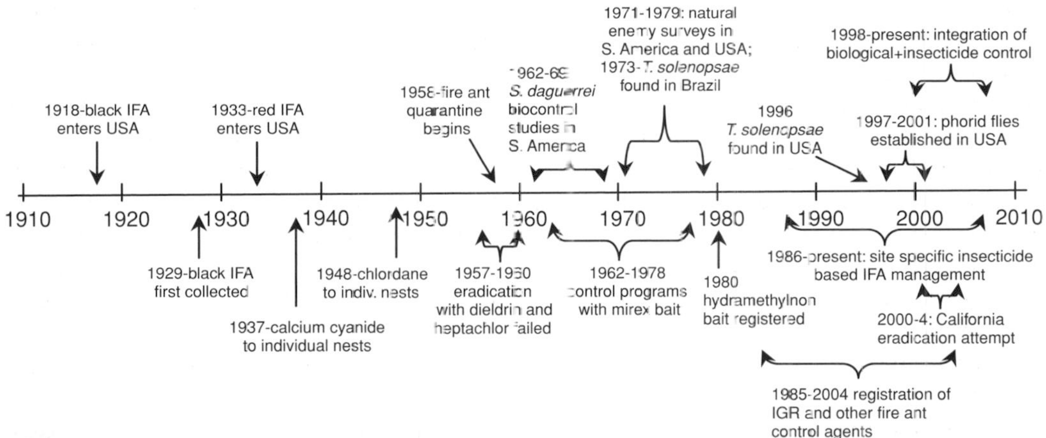

Fig. 30.1 Transition of chemical and biological fire ant control measures in the USA. Year active ingredients registered for fire ant control: 1985 fenoxycarb (IGR), 1986 abamectin, 1998 s-methoprene (IGR) and pyriproxyfen (IGR), 2000 spinosad and fipronil, 2004 indoxacarb.

insecticides and formulations, application rates and application technology. In addition, aspects of the biology and ecology of fire ants were also documented, and over the years the red imported fire ant has become the most studied ant species. Within the framework of IPM certain aspects of the biology and ecology of the pest must be known if environmentally compatible control tactics are to be developed and utilized effectively. Required components of IPM include basic biological information on a pest's life cycle, seasonal phenology and population dynamics. Techniques are also needed to feasibly gather the aforementioned information that are relevant and timely to manage the pest.

Identification of the pest species is one of the most basic and critical components of IPM. In the case of fire ants, species identification and nomenclature can be problematic. The name "fire ant" is most commonly used in reference to the red imported fire ant (*S. invicta*). However, "fire ant" also applies to other species in the genus *Solenopsis* (Taber, 2000) and "imported fire ant" is also used in reference to the black imported fire ant (*S. richteri*). Both of these species were accidentally introduced separately from South America into the USA before 1935 and at the time were considered to be two forms of the same species, *Solenopsis saevissima richteri*. It was not until 1972 that the two forms were formally named as separate species. Of the two species, red imported fire ant is more widespread, with black imported fire ant restricted to northern pockets in the states of Mississippi, Alabama and portions of Tennessee. In areas where the species overlap, they have hybridized. The majority of research and control efforts are for red imported fire ant, and most control recommendations are applied to both species and the hybrid. Distinguishing fire ants among each other and from other ants generally requires training. Thus, with an incursion into a new geographic area, initial detection of red imported fire ant often occurs after an infestation had established and spread. Such has been the case with the discoveries of red imported fire ant in California in 1997, New Mexico in 1998, Australia in 2001, Taiwan in 2003 and in 2005 Hong Kong, mainland China and Mexico.

The life cycle, behavior, seasonal development and other biological aspects of red imported fire ant have been examined intensively and provide a foundation for fire ant IPM and an understanding of the characteristics of a dominant invasive species. A brief overview of the biology of red imported fire ant is provided below.

Like all ants, fire ants are social insects that live in colonies and display the following characteristics: (1) cooperative brood care, where immature ants are tended by groups of adult worker ants that are not their parents; (2) overlapping generations, where at least two different generations of adults occur simultaneously in the same colony; and (3) reproductive and non-reproductive castes, where only the reproductives are capable of producing fertile offspring. The non-reproductives, or workers, perform tasks necessary for colony survival, such as foraging for food, caring for immature ants and reproductives, colony defense and nest building.

Communication needed to coordinate the activities within a colony is mediated by chemical signals called pheromones. Some of the behaviors regulated by pheromones include queen recognition and trail-following, or recruitment to a food source. Chemical cues also are used in the recognition of colony nest mates and play a role in aggression and establishing territorial boundaries between colonies. Understanding the chemical communication among fire ants can lead to novel methods of control if communication can be disrupted or to improvement of insecticidal baits by making them attractive more specifically to red imported fire ant and less available or acceptable to non-target ants. Research on such pheromones and their applications is currently in progress.

Fire ants develop from eggs, through four stages of larvae, to pupae, and finally to adults. Depending on temperature, red imported fire ant development time from egg to adult ranges from 20 to 45 days. Adult workers can live as long as 97 weeks, but depending on size and temperature, lifespans range from 10 to 70 weeks. Queens can live as long as five to seven years with maximum egg-laying rates of over 2000 eggs per day. Thus, red imported fire ant colonies can survive for several years. With the large reproductive capacity of the queen, control methods that kill workers only and do not affect the queen generally result in colony recovery.

Mature red imported fire ant colonies also contain a reproductive caste of non-stinging, winged (alate) males and females that will initiate new colonies. Alates will fly from the nest and mate in midair usually in late spring and early summer. These mated females, or newly mated queens, have been reported to fly as far as 19.3 km from the nest, or even farther when aided by wind, but most land within 1.6 km of their nest. After landing, newly mated queens move to a protected, moist harborage (e.g. in soil or crevices, under debris) that can serve as an initial nesting site. Males die after mating. A queen sheds her wings after landing, and lays a clutch of eggs which she will tend until they develop into adult workers. Worker ants will then tend the queen and the additional eggs she lays and eventually a colony can grow exponentially. After six weeks a new nest may be barely noticeable, after six months nests are typically 5 to 13 cm in diameter and detected more easily. Very large, mature colonies, that are a few years old, can construct nests over a 0.9 m in basal diameter and 0.9 m high. Over 4500 alates can be produced annually in large colonies, but it is speculated that fewer than 0.1% of the newly-mated queens will successfully found a colony (Taber, 2000). Nevertheless, enough queens will survive and can quickly reinfest treated or disturbed areas cleared of ants.

Colonies of red imported fire ant consist of two types: (1) colonies with only a single, fertile queen called monogyne colonies and (2) colonies with multiple fertile queens known as polygyne colonies. Monogyne colonies are territorial and their workers fight with other red imported fire ant colonies. As a result of this antagonistic behavior, nests are farther apart with densities of 99 to 370 nests per hectare, with 100 000–240 000 ants per colony. In contrast, polygyne colonies are not antagonistic to other polygyne colonies and thus queens, workers and immature ants (brood) can move between nests. The visible mound structure of polygyne nests are usually smaller in size and closer together than monogyne mounds with densities of 494 to 1976 per hectare, and 100 000–500 000 ants per mature colony. Polygyne populations contain nearly twice the number of worker ants per unit area than monogyne populations (35 million versus 18 million per ha). The behavioral and population size differences between monogyne and polygyne red imported fire ants can affect the type of control approach. For example, the spread of insect growth regulator baits among colonies may be greater in polygyne populations since they are not territorial. Distinguishing between the monogyne and polygyne colonies without locating fertile queens can now be accomplished through molecular markers. This has allowed for the elucidation of differential responses to control tactics between the two forms. One example is that disease may spread more rapidly in polygyne populations since ants move among colonies.

Given adequate warmth and moisture, the tremendous reproductive capacity, mobility and stinging capability, red imported fire ant has become a dominant arthropod in the areas it invaded. Red imported fire ant colonies can survive and reproduce over a temperature range of 20 to 35 °C, with optimal temperatures of 27 to 32 °C. Moisture is critical for red imported fire ant survival, where a minimum 510 mm of annual precipitation has been estimated to be a reasonable threshold for limiting colony establishment. Because colonies are very mobile and capable of relocating within a day, red imported fire ant can occupy seemingly inhospitable habitats by moving to more favorable niches as environmental conditions change.

While the knowledgebase on red imported fire ant biology and ecology is sizeable, population monitoring methods that are convenient, accurate and timely are not well developed. Traditional population assessments were designed for research studies and often entailed intensive surveys for nests and estimates of colony size based on visual rating scales, nest volume or ant activity. A simplified method of assessing red imported fire ant populations uses food lures to survey for the presence or absence of fire ants (Porter & Tschinkel, 1987; Drees, 1994; Vander Meer *et al.*, 2007). Long-range, pheromone-based surveillance traps used for other insect pests are currently not available for fire ants.

30.3 Control strategies for fire ants

30.3.1 Biological control of fire ants

Classical biological control, that is the release, establishment and spread of effective natural enemies of a pest, is one approach that offers the possibility for permanent regional suppression of fire ants. The introduction of exotic species into new continents often occurs without natural enemies. For the red imported fire ants, over 35 natural enemies have been identified in South America (Williams *et al.*, 2003) compared to about seven in the USA (Collins & Markin, 1971; Valles *et al.*, 2004). The absence of natural enemies can allow exotic species to attain much higher population densities in newly invaded regions than in their native homelands. Accordingly, fire ant populations in the USA generally are five to ten times higher than in South America (Porter *et al.*, 1997).

Interest in natural enemies of both red imported fire ant and black imported fire ant extended back into the era of the mirex control programs. In the 1960s Silveira-Guido *et al.* (1973) conducted extensive studies on the parasitic ant *Solenopsis* (*Labauchena*) *daguerrei* which was found on several species of *Solenopsis* fire ants in South America. Other parasites, such as *Pseudacteon* phorid flies and *Orasema* eucharitid wasps, have been reported from fire ants. Surveys for pathogens of fire ants in South America and the southeastern USA have been conducted since the 1970s with several microorganisms being evaluated as biological control agents (Williams *et al.*, 2003).

To date, two microsporidian pathogens (*Thelohania solenopsae* and *Vairimorpha invictae*) have proven to be most detrimental to fire ant colonies in the field. *Thelohania solenopsae* (recently placed in genus *Kneallhazia*: Sokolova & Fuxa, 2008) caused reductions of up to 83% in field populations of black imported fire ants in Argentina. This organism was discovered in the USA in 1996 and has since been found or introduced in several states. Infections of red imported fire ant have resulted in population reductions of 63% and smaller nest sizes in the USA. In addition, *T. solenopsae* infected fire ant colonies were more susceptible to fire ant bait than uninfected colonies. *Vairimorpha invictae* infections alone and in combination with *T. solenopsae* have resulted in dramatic declines (53–100%) in red imported fire ant populations in Argentina (Briano, 2005). *Vairimorpha invictae* is currently being evaluated for release in the USA.

In 1997, the first successful releases were made of a phorid fly (*Pseudacteon tricuspis*) which parasitizes and eventually decapitates fire ants. Another phorid fly (*Pseudacteon curvatus*) has since been released and established in the USA on both black imported fire ant and red imported fire ant. Direct mortality of individual fire ants through parasitism by *P. tricuspis* was found to be <1%, but the main phorid fly impact is thought to be indirect. For example, it is well documented that the presence of phorid flies disrupts normal foraging behavior (Porter, 1998). In laboratory studies deleterious impacts on colonies by foraging disruption have been inconsistent and influenced by the experimental designs. An extensive field study in Florida could not detect reductions in fire ant populations by *P. tricuspis* relative to natural fluctuations in fire ant populations (Morrison & Porter, 2005). All the above studies caution that detection of fire ant population reductions by phorids may require more species and time.

30.3.2 IPM of fire ants

IPM for imported fire ants includes utilizing cultural, biological and chemical control methods or their combination. In contrast to the concept of treating entire counties or states for fire ant eradication as discussed earlier, fire ant IPM entails site-specific, goal-oriented management programs for commonly infested sites. Current approaches to manage imported fire ants were described in state extension service bulletins (Flanders & Drees, 2004; Drees *et al.*, 2006). The goal of these programs was to prevent or eliminate problems caused by fire ants for a specific land-use pattern rather than elimination of all ants from the ecosystem. Certainly, doing nothing is recognized as one option, especially where the potential for being stung is minimal due to the lack of human or fire ant activity. Predominately shaded and/or dry habitats are unfavorable to fire ant colonization and may not require treatment. Where fire ant suppression is desired, justified use of insecticides has remained the primary tactic for control. Because fire ant baits generally need to be fed upon within a day before they degrade, timing bait applications relative to active foraging can improve control. Development of additional insecticide products that are more cost-effective, target-specific and safer to the user and the environment will continue to improve fire ant IPM.

Control in urban areas

Because of the fire ant's aggressive stinging behavior, fire ant control programs have been developed for specific urban areas such as lawns and athletics fields, buildings, electrical equipment, vegetable gardens, flower beds and shorelines. Each of these areas is often frequented by people and pets and represents different situations that may require different treatment regimes. In lawns and athletics fields where there is a low tolerance for fire ants (a primary habitat of fire ants that is frequented by people), three main treatment approaches, or programs, have been described:

PROGRAM 1: "TWO-STEP METHOD"

This program used the broadcast application of a bait formulated product (step 1), followed by treating nuisance ant nests with an individual nest treatment (step 2). This has been the least-toxic, most environmentally sound approach for treating heavily infested medium-sized to large areas (Riggs *et al.*, 2002; Drees, 2003). However, it was not suggested for use in previously untreated areas with few fire ant nests (≤8 nests/ha), or where competitor ant species were to be preserved. The goal of this program was to reduce fire ant problems while minimizing the search and treatment of individual nests. Because most bait products take at least two weeks to kill colonies, only nuisance nests were individually treated with fast-acting insecticides that kill ants on contact. With the development of fire ant baits that can eliminate colonies within a few days, treatment of individual nests (step 2) may be unnecessary if fast-acting baits are broadcast in step 1.

PROGRAM 2: INDIVIDUAL NEST TREATMENTS

This approach was best used in small areas (usually 0.4 ha or less) with fewer than 8 to 12 nests per ha or where preservation of native ants was desired. This program selectively treated fire ants by targeting only their nests, but rapid reinvasion was anticipated from undetected or untreated colonies in surrounding areas (Barr *et al.*, 1999). Individual nest treatments included insecticides applied as dusts, granules, granules drenched with water after application, liquid drenches, baits, or aerosol injections. Non-chemical treatment methods such as drenching nests with hot water also could be used.

PROGRAM 3: LONG-RESIDUAL CONTACT INSECTICIDE TREATMENT

This program used a long-residual contact insecticide applied to the soil surface (Drees *et al.*, 2006). These liquid or granular treatments could kill a greater number of other ants besides fire ants. Its effects could be more rapid than the other programs and reinvasion of treated areas by migrating colonies and newly mated queen ants was minimized as long as the insecticide remained active.

PROGRAM COMBINATIONS

Any of the three programs above could be used on specific sites within a managed area where different levels of fire ant control were desired. On golf courses, for instance, Program 3 could be suitable for high use areas such as putting greens and tee boxes. In fairways and rough areas, Program 1 could be sufficient.

Control in agricultural production systems

In commercial agriculture, justified treatment costs would be less or equal to the losses sustained by fire ants. In cattle operations the estimated cost for treatments using broadcast application of conventional fire ant baits in pastures is approximately $US 25/ha. For small operations this treatment cost may be negligible, while in larger operations, this treatment cost may be unacceptably high. With the exception of a few, most fire ant baits have not been approved for all use sites that comprise a cattle operation. In these situations, a better approach to treating the entire ranch with one bait product would be the implementation of site-specific IPM techniques that include chemical as well as cultural methods. For example, bait could be applied only to a designated calving pasture and calving programs could be scheduled to avoid summer months when fire ant activity is high. Few fire ant baits have been registered for use in food crops, thus limiting control options in crop production systems. It should be noted that since fire ants are major predators of arthropods, they are beneficial in some systems (e.g. sugarcane) by significantly reducing pest populations.

Community-wide and areawide fire ant management and abatement programs

Although site-specific IPM approaches have been widely promoted for use on individual properties, reinvasion by the ants from neighboring untreated areas has been a recurring problem. By treating larger outdoor areas (i.e. entire neighborhoods), reinfestation has been shown to be reduced. Beginning in 1993, demonstrations of community treatments were reported from Alabama and Arkansas. Benefits from community-wide fire ant management in Texas included reductions in fire ant populations, less pesticide expenditures by homeowners and the maintenance or increases in other ant species (Riggs *et al.*, 2002). All of these programs have relied on aerial or ground broadcast application of bait-formulated insecticides as well as sustained community leadership and involvement.

Integration of biological control agents with chemical control programs

Efforts have been made to demonstrate the impact of integrating classical biological control agents with insecticide applications for fire ant control. The objective of releasing natural enemies of fire ants is to establish self-sustaining suppression of imported fire ant populations. However, the reduction in fire ant populations after establishment and spread of the phorid fly *P. tricuspis* or the fire ant pathogen *T. solenopsae* has not been sufficient to alleviate the threat of fire ant stings to tolerable levels. However, when biological controls were established in unmanaged landscapes surrounding areas chemically treated for fire ants, control was extended in the treated areas by over a year (Oi *et al.*, 2008). It is hypothesized that the establishment of several parasites and pathogens of fire ants will, over time, help reduce sources of fire ant reinfestations into managed landscapes or sensitive areas such as playgrounds. Thus, insecticide use for fire ants may be reduced.

30.4 Eradication programs in southern California and Brisbane, Australia

As discussed previously, attempts to eradicate fire ants from 1957 through 1977 failed to eliminate these species from the southeastern USA. By the mid-1980s, the concept of eradicating the fire ants from the southeast had been abandoned. Nevertheless, many small isolated infestations have routinely been eradicated by state department of agriculture personnel throughout the USA and the USDA-Animal Plant Health Inspection Service has maintained an imported fire ant quarantine. Few, if any, of these successful efforts have been documented in the scientific literature. The development of fire ant baits that contain insect growth regulating (IGR) active ingredients, e.g. fenoxycarb, methoprene and pyriproxyfen, have provided more tools for treating large land areas. IGR baits have fewer environmental impact concerns, and some are registered for use in a broader range of land-use patterns, including cropland. In addition, the availability of global positioning and geographic information systems have facilitated delineation and treatment of infested areas.

In 2000 and 2001, eradication programs were implemented in southern California, USA and in Brisbane, Australia, respectively. Both programs were designed, in part, on considerations for spot eradication, suggesting that infestations be surveyed so that, where feasible and justifiable, insecticide treatments be made to the entire population including a buffer area where fire ants were not detected. Suggested treatments included a minimum of two broadcast bait applications annually for several years with IGR baits to prevent further spread (Drees *et al.*, 2000). Winged female reproductive caste ants produced in IGR treated colonies do not have functional ovaries, inhibiting spread by mating flights. Aerial treatment was encouraged to attain thorough coverage in a timely manner, but only ground application was used in urbanized areas in both programs. Additional treatments using bait or individual nest treatments with faster modes of action could hasten population reduction. Finally, program success could be documented through several years of intensive ant surveys in and around previously treated areas.

In southern California, after the initial discovery of fire ants in October 1998, the California Department of Food and Agriculture delineated the infestation and issued proclamations of eradication in three counties encompassing over 204 000 ha. Local county agencies were then contracted to survey for ants and conduct large-scale broadcast application of fire ant baits primarily containing pyriproxyfen or hydramethylnon. The program made applications from February 2000 through February 2004, treating over 29 000 sites in one county alone. The treatments were made to most properties four times per year at three- to four-month intervals. Post-treatment monitoring began three months after the final treatment and showed fewer than 5% of the sites remained infested. However, because of budget reductions, the eradication effort was terminated in February 2004. Since then residents from two counties voted to approve a new program initiated in October 2004 to complete treatments and surveys in parks, school greenbelts, public medians and commercial businesses. Nevertheless, eradication has not been achieved because not all infested areas were treated or received the full treatment regime.

In Brisbane, Australia, the eradication program was initiated by the Queensland Department of Primary Industries and the Fire Ant Control Centre (FACC) in 2001, within a year of confirming the infestation of over 37 000 ha. The eradication program involved three years of multiple annual treatments (broadcasting pyriproxyfen, hydramethylnon or methoprene baits; and nest injection of chlorpyrifos or fipronil), followed by two years of surveillance. From 2001 through 2006 additional detections expanded the program significantly and additional treatment areas and surveillance was required through 2007. In 2006, the program included 62 250 ha under surveillance following treatment and 6200 ha still under treatment. Treatment using aerial application was made to non-urbanized lands where use of insecticides was permitted, including land bordering bodies of water. Scheduling ground treatments using vehicle-mounted or hand-held application equipment and surveillance using visual inspections and ground traps for such a large infested area was obviously challenging, but most likely was facilitated by having a well-funded, single, centralized authority to direct and implement operations. Results from this large program have been very promising. In 2006, only about 150 red imported fire ant colonies were found in 49 properties of 160 000 in the treated area, an infestation rate of about 0.03%. The impressive eradication effort in Australia benefited from high public support and secure funding.

The major problem with documenting successful eradication is to find and eliminate the last fire ant colony. Surveillance for two (or more) years following termination of prescribed treatments may still fail to detect low numbers of surviving colonies. Utilizing computer modeling with satellite imagery to locate favorable fire ant habitats has made surveillance more efficient, yet improvements in detecting fire ants at very low population densities are a critical need. New incursions could be occurring unless vigilance at ports and other points of commerce is maintained. Systematic surveillance and rapid responses have limited incursions in New Zealand. Only time will confirm whether these eradication efforts have truly been successful.

30.5 Conclusions

In summary, the rapid and extensive spread of the easily recognized, stinging fire ant into the southern USA provoked drastic attempts to eliminate or control the problem quickly. Unfortunately, fire ants, with their tremendous reproductive capacity, mobility and adaptation to a wide range of habitats made eradication, especially without a thorough understanding of their biology and ecology, highly unlikely. With eradication in the southern USA no longer an option, greater emphasis was placed on basic biological research which ultimately improved control measures including the utilization of biological control. With the expansion of global commerce, extensive fire ant infestations have occurred in Australia and Asia and new incursions will certainly continue. The fire ant invasion of the USA has provided valuable lessons, technology and knowledge on fire ant control and biology. This has provided a basis for the ongoing development of fire ant IPM as well as recent eradication programs. Furthermore the fire ant experience can serve as a blueprint from which informed decisions and responses can be formulated for other invasive ant species.

References

Banks, W. A., Glancey, B. M., Stringer, C. E. *et al.* (1973). Imported fire ants: eradication trials with mirex bait. *Journal of Economic Entomology*, **66**, 785–789.

Barr, C. L., Summerlin, W. & Drees, B. M. (1999). A cost/efficacy comparison of individual mound treatments and broadcast baits. In *Proceedings of the 1999 Imported Fire Ant Conference*, March 3–5, Charleston, SC, pp. 31–36. Clemson, SC: Clemson University Cooperative Extension Service.

Barr, C. L., Davis, T., Flanders, K. *et al.* (2005). *Broadcast Baits for Fire Ant Control, B-6099*. College Station, TX: Texas Cooperative Extension, Texas A&M University System.

Briano, J. A. (2005). Long-term studies of the red imported fire ant, *Solenopsis invicta*, infected with the microsporidia *Vairimorpha invictae* and *Thelohania solenopsae* in Argentina. *Environmental Entomology*, **34**, 124–132.

Canter, L. W. (1981). *Final Programmatic Environmental Impact Statement for the Cooperative Imported Fire Ant Program*. ADM-81-01-F. Washington, DC: US Department of Agriculture, Animal Plant Health Inspection Service.

Carson, R. (1962). *Silent Spring*. Boston, MA: Houghton Mifflin.

Collins, H. L. & Markin, G. P. (1971). Inquilines and other arthropods collected from nests of the imported fire ant, *Solenopsis saevissima richteri*. *Annals of the Entomological Society of America*, **64**, 1376–1380.

Drees, B. M. (1994). Red imported fire ant predation on nestlings of colonial waterbirds. *Southwestern Entomologist*, **19**, 355–360.

Drees, B. M. (2003). *Estimated Amounts of Insecticide Ingredients Used for Imported Fire Ant Control Using Various Treatment Approaches*, Fire Ant Plan Fact Sheet No. FAPFS042. College Station, TX: Texas Imported Fire Ant Research & Management Project, Texas A&M University System.

Drees, B. M., Collins, H., Williams, D. F. & Bhatkar, A. (2000). *Considerations for Planning, Implementing and Evaluating a Spot-Eradication Program for Imported Fire Ants*, Fire Ant Plan Fact Sheet No. FAPFS030. College Station, TX: Texas Imported Fire Ant Research & Management Project, Texas A&M University System.

Drees, B. M., Vinson, S. B., Gold, R. E. *et al.* (2006). *Managing Imported Fire Ants in Urban Areas, a Regional Publication Developed for: Alabama, Arkansas, California, Florida, Georgia, Louisiana, Mississippi, New Mexico, Oklahoma, South Carolina, Tennessee and Texas, B-6043*. College Station, TX: Texas Cooperative Extension, Texas A&M University.

Flanders, K. L. & Drees, B. M. (2004). *Management of Imported Fire Ants in Cattle Production Systems*, ANR-1248. Auburn, AL: Alabama Cooperative Extension System.

Jetter, K. M., Hamilton, J. & Klotz, J. H. (2002). Red imported fire ants threaten agriculture, wildlife, and homes. *California Agriculture*, **56**, 26–34.

Lofgren, C. S. (1986). History of the imported fire ants in the United States. In *Fire Ants and Leaf-Cutting Ants: Biology and Management*, eds. C. S. Lofgren & R. K. Vander Meer, pp. 36–47. Boulder, CO: Westview Press.

Lofgren, C. S. & Weidhaas, D. E. (1972). On the eradication of imported fire ants: a theoretical appraisal. *Bulletin of the Entomological Society of America*, **18**, 17–20.

Lofgren, C. S., Bartlett, F. J. & Stringer, C. E. (1963). Imported fire ant toxic bait studies: evaluation of carriers for oil baits. *Journal of Economic Entomology*, **56**, 62–66.

Markin, G. P., O'Neal, J. & Collins, H. L. (1974). Effects of mirex on the general ant fauna of a treated area in Louisiana. *Environmental Entomology*, **3**, 895–898.

McNicol, C. (2006). Surveillance methodologies used within Australia: various methods including visual surveillance and extraordinary detections – above ground and in ground lures. In *Proceedings of the Red Imported Fire Ant Conference*, March 28–30, 2006, Mobile, AL, ed. L. C. Graham, pp. 69–73. Auburn, AL: Department of Entomology and Plant Pathology, Auburn University.

Morrison, L. W. & Porter, S. D. (2005). Testing for population-level impacts of introduced *Pseudacteon tricuspis* flies, phorid parasitoids of *Solenopsis invicta* fire ants. *Biological Control*, **33**, 9–19.

Oi, D. H. & Oi, F. M. (2006). Speed of efficacy and delayed toxicity characteristics of fast-acting fire ant (Hymenoptera: Formicidae) baits. *Journal of Economic Entomology*, **99**, 1739–1748.

Oi, D. H., Williams, D. F., Pereira, R. M., *et al.* (2008). Combining biological and chemical controls for the management of red imported fire ants (Hymenoptera: Formicidae). *American Entomologist*, **54**, 46–55.

Porter, S. D. (1998). Biology and behavior of *Pseudacteon* decapitating flies (Diptera: Phoridae) that parasitize *Solenopsis* fire ants (Hymenoptera: Formicidae). *Florida Entomologist*, **81**, 292–309.

Porter, S. D. & Tschinkel, W. R. (1987). Foraging in *Solenopsis invicta* (Hymenoptera: Formicidae): effects of weather and season. *Environmental Entomology*, **16**, 802–808.

Porter, S. D., Williams, D. F., Patterson, R. S. & Fowler, H. G. (1997). Intercontinental differences in the abundance of *Solenopsis* fire ants (Hymenoptera: Formicidae): escape from natural enemies? *Environmental Entomology*, **26**, 373–384.

Riggs, N. L., Lennon, L., Barr, C. L. *et al.* (2002). Community-wide red imported fire ant programs in Texas. *Southwestern Entomologist* (Suppl.) **25**, 31–42.

Silveira-Guido, A., Carbonell, J. & Crisci, C. (1973). Animals associated with the *Solenopsis* (fire ants) complex, with special reference to *Labauchena daguerrei*. *Proceedings of the Tall Timbers Conference on Ecological Animal Control by Habitat Management*, **4**, 41–52.

Sokolova, Y. Y. & Fuxa, J. R. (2008). Biology and life-cycle of the microsporidium *Kneallhazia solenopsae* Knell Allan Hazard 1977 gen. n., comb. n., from the fire ant *Solenopsis invicta*. *Parasitology*, **135**, 903–929.

St. Omer, V. V. (1970). Chronic and acute toxicity of the chlorinated hydrocarbon insecticides in mammals and birds. *Canadian Veterinary Journal*, **11**, 215–226.

Taber, S. W. (2000). *Fire Ants*. College Station, TX: Texas A&M University Press.

Tschinkel, W. R. (2006). *The Fire Ants*. Cambridge, MA: Belknap Press of Harvard University Press.

Valles, S. M., Strong, C. A., Dang, P. M. *et al.* (2004). A picorna-like virus from the red imported fire ant, *Solenopsis invicta*: initial discovery, genome sequence, and characterization. *Virology*, **328**, 151–157.

Vander Meer, R. K., Pereira, R. M., Porter, S. D., Valles, S. M., & Oi, D. H., (2007). Areawide suppression of invasive fire ant populations. In *Area-Wide Control of Insect Pests: From Research to Field Implementation*, eds. M. J. B. Vreysen, A. S. Robinson & J. Hendrichs, pp. 487–496. Dordrecht, Netherlands: Springer-Verlag.

Williams, D. F., Collins, H. L. & Oi, D. H. (2001). The red imported fire ant (Hymenoptera: Formicidae): a historical perspective of treatment programs and the development of chemical baits for control. *American Entomologist*, **47**, 146–159.

Williams, D. F., Oi, D. H., Porter, S. D., Pereira, R. M. & Briano, J. A. (2003). Biological control of imported fire ants (Hymenoptera: Formicidae). *American Entomologist*, **49**, 150–163.

Wojcik, D. P., Allen, C. R., Brenner, R. J. *et al.* (2001). Red imported fire ants: impact on biodiversity. *American Entomologist*, **47**, 16–23.

Chapter 31

Integrated vector management for malaria

Chris F. Curtis

Malaria is caused by species of the protozoan genus *Plasmodium* which infect red blood corpuscles as well as other human tissues. There are estimated to be 300–500 million clinical cases of malaria each year, about 60% in tropical Africa (WHO & UNICEF, 2005), with almost all the remainder in tropical and subtropical Asia, Latin America and the Southwest Pacific. At the present time there is no malaria transmission in developed countries in the temperate zone, but a few thousand cases are "imported" each year in travelers who have been in tropical countries.

The only *Plasmodium* species which causes appreciable numbers of human deaths is *P. falciparum* which is the cause, or a contributory cause, of death of 1–2 million people per year (about 10 000 times the number of deaths per year from mosquito-borne West Nile fever in the recent much publicized outbreak in the USA). More than 80% of malaria deaths are among rural, lowland African infants and children, for whom malaria is one of the major causes of death. If children in the highly endemic parts of Africa survive the malaria attacks which they suffer early in life, they develop a degree of immunity which protects them from the very severe anemia, blockage of cerebral blood vessels and respiratory distress which are the main causes of malaria deaths (Berkeley *et al.*, 1999). Much research effort is being devoted to development of various types of vaccine which are primarily aimed at raising immunity in infants to the levels found in frequently exposed adults (Schofield, 2007).

Plasmodium vivax is the second most important *Plasmodium* species; it causes severe fevers but is seldom lethal. It is the most common form of malaria in Asia and Latin America and also occurs in Ethiopia. However, *P. vivax* is rare in the rest of tropical Africa because this parasite requires the Duffy blood group substance to enter red blood corpuscles (Miller *et al.*, 1976). Few Africans, apart from Ethiopians, have this blood group. *Plasmodium malariae* and *P. ovale* also contribute to the malaria problem and in East Malaysia it has recently been shown that *P. knowlesi*, a monkey parasite, causes a significant number of human cases (Singh *et al.*, 2004).

For centuries it has been known that quinine (from the cinchona tree) is an effective anti-malaria drug and it is still used, especially to treat severe cases. Inexpensive synthetic drugs, especially chloroquine and sulfadoxine-pyrimethamine (SP), have been of great value to treat cases and as prophylactics for travelers into malarious areas and for infants and pregnant women resident in endemic areas. (Schellenberg *et al.*, 2005). However, resistance has now evolved in *P. falciparum* to both of these drugs in many areas. The WHO recommended alternative is a derivative of the plant *Artemisia annua*

Integrated Pest Management, ed. Edward B. Radcliffe, William D. Hutchison and Rafael E. Cancelado. Published by Cambridge University Press.

as artemisinin-based combination therapy (ACT) with a synthetic drug. This is more expensive than chloroquine or SP, which is a serious consideration in the very low income countries most affected by malaria. This calls for diagnosis to be more accurate than hitherto to avoid wastage of expensive ACTs on non-malaria fevers and also to ensure that those fevers are given appropriate treatment.

Transmission of *Plasmodium* from human to human is only (apart from human errors such as reuse of needles) via female mosquitoes of the genus *Anopheles*. Female mosquitoes require a blood meal to provide the protein for development of each of their egg batches, which they produce at intervals as short as two to three days at temperatures of the tropical lowlands. *Plasmodium* is not passed from an *Anopheles* female to its offspring but is acquired by a female if it imbibes blood from a person infected with the sexual stages (gametocytes) of *Plasmodium* (Gillies, 1993). The gametocytes become gametes in the *Anopheles* stomach, fertilization occurs there and the zygote penetrates the stomach wall. After a development process, which takes about 12 days at tropical temperatures, infective sporozoites are formed and these move to the mosquito's salivary glands from which they can be carried to a new victim with the flow of saliva during a blood feed. The sporozoites pass through a stage in the human liver before entering the blood and starting to cause disease symptoms.

Anopheles eggs are laid and larvae develop in relatively clean water, e.g. in marshes, wells, pools and irrigation water; in the case of *An. stephensi* in India breeding also occurs in cisterns and other man-made sites, thus causing a malaria problem in Indian cities. In African urban areas *Anopheles* occur in certain sites, such as suburban irrigated vegetable plots, but much stagnant water in African towns (cess pits, blocked drains, etc.) has high organic pollution and thus is unsuitable for the breeding of malaria-transmitting *Anopheles*. However, these sites allow the breeding of large numbers of *Culex* mosquitoes which create a severe biting nuisance but are not able to nurture human-infecting *Plasmodium*. A recent extensive review (Hay *et al.*, 2005) emphasized that malaria transmission is far more intense in African rural than urban areas and that subsidies to assist malaria control programs should be concentrated in African rural areas. The problem is also limited by altitude; above about 2000 m, temperatures are too low to allow much breeding of *Anopheles* and development of *Plasmodium* in them. Just below this altitude, dangerous epidemics occur in unusually warm and wet seasons among people who have not developed immunity as a result of frequent malaria attacks in infancy and childhood (Cox *et al.*, 2007). Similarly, epidemics may occur in years with high rainfall in desert fringes which are normally too dry to support a significant *Anopheles* population.

With very few exceptions *Anopheles* bite in the hours of darkness. The most dangerous vector species such as *An. gambiae s.s.* and *An. funestus*, which favor humans rather than animals as blood sources, mostly bite in houses in the middle of the night when people are in bed. Thus there is a strong argument for indoor application of residual insecticides of low toxicity to humans. If a large proportion of the houses in a village are treated, *Anopheles* are at great risk every time they enter a bedroom to feed and few will survive the 12 days required for infective sporozoites to develop within them. This is the argument underlying the concentration on indoor residual spraying of walls and ceilings (Macdonald, 1957) which eradicated malaria from southern Europe (Snowden, 2006), the USA, most Caribbean islands and Taiwan. This method also produced major reductions in south and east Asia, the then USSR, the Middle East, North Africa, South and Central America and southern Africa (Mabaso *et al.*, 2004). The method was also shown to be effective in tropical Africa, for example the near elimination of malaria achieved in the 1950s–1960s in the islands of Zanzibar (Curtis & Mnzava, 2000) and the highlands of Madagascar (Joncour, 1956). These programs included indoor spraying of DDT and lindane (gamma-hexachlorocyclohexane) integrated with treatment of the remaining human cases using chloroquine to which there was then no resistance. Recently the WHO urged revival of DDT indoor spraying, which is authorized by the Stockholm Convention on Persistent Organic Pollutants provided that rigorous measures are taken to prevent illegal sales of DDT to farmers for use outdoors where DDT may have harmful effects on wildlife (Cooper, 1991). The highland epidemics, mentioned above, may be best dealt with by a "fire brigade" of trained and equipped spraymen who could be quickly moved to a highland area where meteorological and health facility data indicate an incipient epidemic.

Insecticide-treated nets (ITNs) integrate the use of a targeted dose of residual insecticide with the physical barrier of the net. Thus, this method may be considered as a form of integrated vector management (IVM) as described in Section 31.4 below.

Most readers will consider that IVM for malaria primarily concerns control of *Anopheles* breeding by engineering, chemical or biological means and following sections will discuss these. For these three approaches it must be borne in mind that very high percentage coverage is essential of all *Anopheles* breeding sites within mosquito flight range (~1–2 km) of the community to be protected. If a breeding site is missed by treatment operations, there is nothing to prevent it yielding a dangerously long-lived crop of mosquitoes.

31.1 Environmental management and drainage

In the past, drainage of *Anopheles* breeding sites was of major importance in reducing malaria transmission, but in present times is seldom used for this purpose. However, drainage against malaria is of historical interest and should be renewed in appropriate circumstances to integrate with chemical methods of control.

The cause of decline of *P. vivax* malaria in southern and eastern England in the nineteenth century is thought to have been extensive drainage of marshy land (Kühn *et al.*, 2003). This was done before the role of *Anopheles* in malaria transmission was understood, though the association of malaria with marshes had been noticed long before. The drainage in England was intended to create more agricultural land, with little or no thought about its beneficial side effect on malaria.

In early twentieth-century Italy, malaria was a major health burden and was thought to be the main reason for the very poor health and short life expectancy of peasant farmers (Snowden, 2006). Immediately after the discovery by Ross in 1897 that feeding *Anopheles* on a malaria patient led to development of *Plasmodium* in the mosquito, Italian researchers such as Grassi had a major role in elucidating the details of transmission of malaria from human to human. In the first decade of the twentieth century the Liberal government of Italy made a major effort to integrate mass use of quinine, for treatment or prophylaxis of humans, with drainage of marshes where *Anopheles* bred. After Mussolini came to power in the 1920s his regime adopted drainage of the Pontine Marshes near Rome, spurred on by the doctrine that anything the Soviet Five-Year Plans could do the Fascist regime could do better. By the time of Mussolini's decision to enter World War II on the Nazi side, the drainage of the Pontine Marshes had indeed made this area habitable, with near complete elimination of malaria there. In the chaotic conditions of wartime there was resurgence of malaria. Furthermore, Snowden (2006) states that during the Nazi retreat of 1943–44 the re-flooding of the Pontine Marshes involved reversal of the pumping stations so as to pump in sea water. This was not merely aimed at hindering the allied military advance by creating marshy ground, but was a deliberate act of biological warfare aimed at enhancing breeding of the vector *An. labranchiae*, which favors brackish water, to produce a malaria epidemic to punish the Italian population for changing sides in the war. However, before the end of the war DDT house spraying was started. It was highly successful and led to WHO-certified malaria eradication from Italy by 1962.

Despite many malaria cases imported into Italy with returning travelers and high densities of *Anopheles* especially in the rice-growing areas, there have been only rare cases of local transmission since the 1950s (Romi *et al.*, 2001). This can be attributed to the Italian health system ensuring that the imported malaria cases are effectively treated, generally before gametocytes, which could infect local mosquitoes, have developed. This should be reassuring for those who fear that global warming will lead to return of malaria to northern Europe. Italian summers are hotter than even the most extreme predictions of global

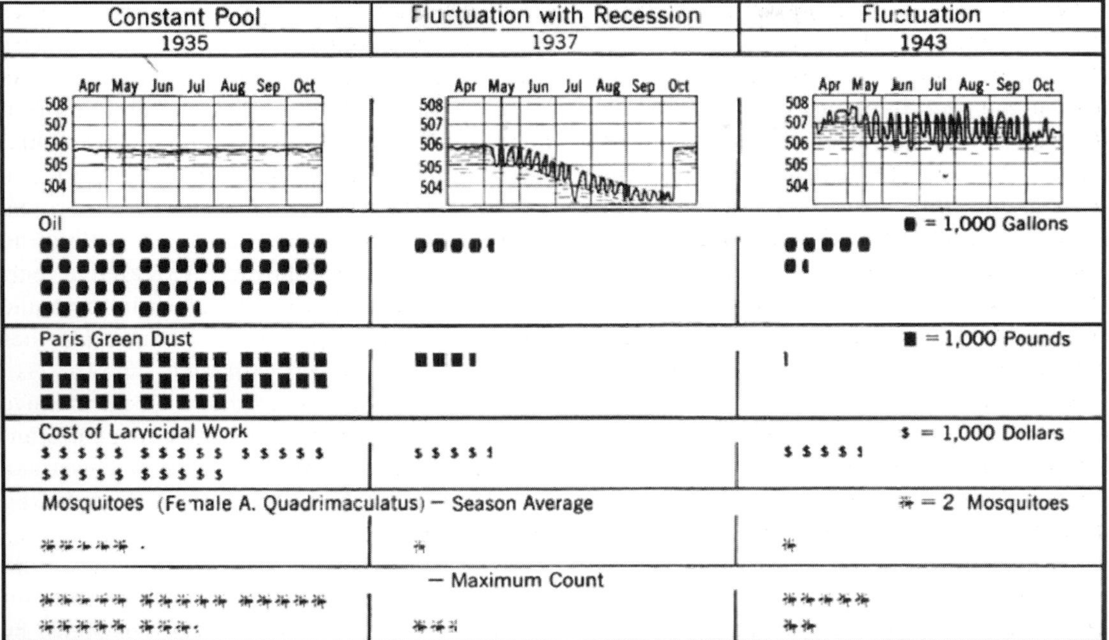

Fig. 31.1 The effect of manipulating the water level in a reservoir in the Tennessee Valley, USA, on the need for alternative means of malaria vector control and the success of such control in the 1930s and 1940s (diagram from US Public Health Service & Tennessee Valley Authority, 1947).

warming in northern Europe. It would appear that the universally available medical systems of Western Europe should be able to limit cases of infection of local mosquitoes and malaria transmission to isolated rare events (e.g. Krüger *et al.*, 2001). This contrasts with the situation in former Soviet Central Asia where the near elimination of malaria has been replaced by recurrent malaria epidemics. This may be because many malaria cases remain untreated (e.g. in refugees from Afghanistan) and become sources of gametocytes which can infect local mosquitoes and hence nearby humans.

In the southeastern USA malaria was a serious problem until the 1940s, but was brought under control largely by a remarkable program of water management to prevent the breeding of the vector *An. quadrimaculatus*. This program was described in detail in a book published by the US Public Health Service & Tennessee Valley Authority (1947). This emphasized that the water impoundments by dam construction needed to be carefully controlled to prevent a severe malaria risk within mosquito flight range. *Anopheles quadrimaculatus* larvae flourish where there is floating plant material or emergent vegetation; thus it was necessary to keep the shoreline cleared and to strand floating wood carried into the reservoir by inflowing streams. This was done by alternately raising and lowering the water level by manipulating dam sluices. Raising the water level stranded floating wood and lowering the level drew water out of the marginal vegetation exposing *Anopheles* larvae to stranding, natural predation and wave action. It was recognized that judicious use of larvicides (e.g. Paris Green, which is an arsenic compound, or oil) might be needed. Figure 31.1 shows that, for the Wilson Reservoir in the Tennessee Valley, much less larvicide was needed, less expense was incurred and better mosquito control was achieved when the water level in the reservoir was fluctuated, with or without a general seasonal recession. Data are also presented in this 1947 book on the remarkable impact of supplementary house screening on malaria. Other examples of such data from 1899 to 1931 are tabulated by Keiser *et al.* (2005).

Overall these and related techniques produced a remarkable reduction in malaria incidence and mortality in the southeastern USA between the 1920s and mid-1940s. This same work records the introduction of DDT indoor residual spraying. Malaria eradication was completed in the USA only a few years later.

In Sichuan, China, the residual malaria problem is reported to have been almost eliminated because, with the improved reliability of the supply of irrigation water, farmers no longer feel the need to retain irrigation water in fallow rice fields in the winter (Liu *et al.*, 2004). Previously, such retained water was a major breeding site for the malaria vectors *An. sinensis* and *An. anthropophagus*.

Yasuoka *et al.* (2006) reported a program to educate Sri Lankan farmers in environmental management by leveling rice fields to eliminate puddles and clearing irrigation channels to speed water flow. These activities by the farmers were found to have significantly reduced overall *Anopheles* densities. However, the species shown to be reduced were *An. nigerrimus* and *An. peditaeniatus*, not the major Sri Lankan malaria vectors *An. culicifacies* and *An. subpictus*, which tend to breed in pits and pools, not rice fields (Yapabandara & Curtis, 2004a).

There are two kinds of environmental modification which many still believe are effective against malaria and which unfortunately are frequently included in health education posters as anti-malaria measures. These are:

- Cutting grass and bush clearance – this was shown to be ineffective by Ribbands (1946);
- Clearing away garbage which accumulates rainwater – such garbage provides breeding sites for *Aedes* mosquitoes which need to be controlled where they transmit dengue, but *Aedes* mosquitoes do not transmit malaria.

31.2 Larvivorous fish and other biological control agents

Of the several types of biological control which have been tried for control of *Anopheles*, fish which eat mosquito larvae have been the most successful. The two main species used are *Gambusia affinis* and *Poecilia reticulata*, both of which are viviparous and originated in Central and South America, but have been introduced into other continents.

Self-maintaining populations of these species have been set up in ponds and wells in villages in Karnataka State, India (Ghosh *et al.*, 2005). This work was initiated because of the reluctance of many Indian householders to allow insecticidal spraying of their houses; in villages where silk is produced there was adamant refusal to allow house spraying because of the risk of killing the silkworms (*Bombyx mori*) which are reared in their houses. The impact of fish on malaria in these silk-producing villages has been so remarkable that the use of fish is now being spread to many parts of Karnataka State. Table 31.1 shows the progressive decline in malaria cases in the years after the intervention with fish in three sectors of Karnataka and the remarkably low costs of introducing fish.

The primary malaria vector in this part of India is *An. culicifacies* sibling species A. This was found to breed mainly in village ponds and wells. By contrast, its close relative *An. culicifacies* species B breeds predominantly in streams and is not a malaria vector because *Plasmodium* gametocytes do not develop in this species (Subbarao *et al.*, 1988). With this knowledge it was possible to target only wells and ponds for introduction of fish and to omit the more difficult task of trying to ensure presence of larvivorous fish in streams (Ghosh *et al.*, 2005).

Fish from a hatchery are used for initial stocking of wells and ponds to which villagers can readily guide vector control teams. Thereafter, checks are made every six months for presence of a flourishing fish population in every site. If one is found to have lost its fish population this can be re-established by collecting some fish from another site in the same village. A geographical positioning system was found useful to prepare a list of the locations of all the wells and ponds in a village and to guide the rechecking team back to these sites (Boswell *et al.*, 2005).

There has been concern that introducing fish from another continent might be ecologically harmful, especially as *Gambusia* is known to eat the eggs of other fish (Gerberich & Laird, 1985). However, it was reassuring that in one of the villages where larvivorous fish had been introduced several years earlier, the present author was informed by a fish merchant that he was running a

Table 31.1 Decline of number of detected malaria cases in districts in Karnataka State, India, after introduction of larvivorous fish into all village wells and ponds and on costing of the program

	Kolar District	Hassan District	Four districts
No. of villages	93	160	1766
Population	36484	85867	1.2 million
Year	Malaria cases (year of initiating fish introductions)		
Pre-intervention	1 446 (1993)	10 274 (1995)	73 270 (2001)
1 (post-intervention)	517	4 526	20 660
2	381	322	3 327
3	428	177	973
4	141	417	497
5	44	829	
6	4	1250	
7	2	128	
8	4	52	
9	2	12	
10	7	9	

Note: The annual per capita cost of this program was Indian Rupees 1.00 ($US 0.022) in the initial period and about half that amount thereafter.
Source: Data of Ghosh *et al*. 2006.

profitable business rearing edible fish in a pond containing large numbers of *Gambusia* and *Poecilia*.

In Goa, India, as part of a program of integrated control of malaria, *An. stephensi* larvae in wells have been controlled with fish of the indigenous Asian species *Aplocheilus blocki* (Kumar *et al*., 1998).

One may consider whether these encouraging results could be transferred to Africa where the malaria problem is far worse than in India. It would seem that the generally small and shifting breeding sites of *An. gambiae* could not feasibly be supplied with fish. However, in arid areas in Somalia *An. arabiensis* breeds in water storage tanks which can be readily stocked with fish (Alio *et al*., 1987). In urban areas locating and treating the relatively few water bodies suitable for *Anopheles* breeding would seem to be feasible and would benefit the large number of people within mosquito flight range of each water body.

Hitherto all consideration of biological control of *Anopheles* has focused on larvae. However, recently it has been found possible to infect adult *Anopheles* by making them walk on surfaces treated with spores of the fungi *Metarhizium* or *Beauveria*. The infection markedly reduces the subsequent mosquito survival (Blanford *et al*., 2005; Scholte *et al*., 2005). Such a reduction can be expected to have a major impact on probability of malaria transmission which, as already emphasized, depends on mosquitoes surviving long enough for sporozoite maturation to occur. In addition, at least for rodent malaria, sporozoite maturation is directly impacted by the fungal infection, even in the minority of surviving mosquitoes (Blanford *et al*., 2005).

31.3 Biopesticides: insect growth regulators and bacterial products

Several insect growth regulators are available which interfere with mosquito larval maturation and metamorphosis. These include juvenile hormone mimics such as pyriproxyfen, which prevent larval-pupal-adult metamorphosis.

Table 31.2 Impact of treating Sri Lanka gem pits with pyriproxyfen

	Four treated villages	Four untreated villages
Change from pre-intervention year in human biting catch of *An. culicifacies* vectors	70% reduction ($p < 0.001$)	60% increase (not statistically significant)
Incidence of malaria cases recorded at clinics per thousand person years		
P. vivax	53 ($p < 0.001$)	204
P. falciparum	2.9	24.4
Prevalence of malaria infection in mass blood surveys		
Either *Plasmodium* sp.	0.30%	

Source: Yapabandara *et al.* (2001).

A series of trials were conducted by Yapabandara and colleagues (Yapabandara *et al.*, 2001; Yapabandara & Curtis, 2002, 2004b) in Sri Lanka in major breeding sites of the main malaria vectors *An. culicifacies* and *An. subpictus* in pits, hand-dug by gem miners, which fill with water in the monsoon season, and in pools left in rivers in the dry season. In both breeding environments pyriproxyfen was shown to be effective in preventing emergence of adults. This was proved using floating cages with netting bottoms placed in the pits or pools and stocked at monthly intervals with *An. culicifacies* larvae. Full effectiveness persisted for several months at treatment dosages measured in parts per 1000 million. The floating cage method gave useful warning of when the residual pyriproxyfen began to fail to prevent adult emergence and retreatment was needed. Comparison of the cost per year of preventing any adult mosquito emergence from pits showed that pyriproxyfen was more cost-effective than treatment with oil or the organophosphate temephos, which had to be reapplied every two weeks.

Multi-village controlled trials were conducted of pyriproxyfen treatment of all the pits in some villages and leaving all pits untreated in comparison villages. Because the gem pits are so important in the lives of the villagers it was possible, with their guidance, to be sure that every one of the very large number of pits in the intervention villages had been found and treated. Table 31.2 summarizes data on adult *Anopheles* densities, incidence of cases of fever associated with *P. vivax* or *P. falciparum* infection and prevalence of *Plasmodium* infection found in mass blood surveys (regardless of perceived symptoms). It was clear that the decline of all these parameters was highly significant in the pyriproxyfen-treated villages.

The highly specific anti-mosquito toxins of the bacterium *Bacillus thuringiensis israelensis* (*Bti*) have been used in stagnant water in construction sites in Goa, India, to control the Indian urban malaria vector *An. stephensi* (Kumar *et al.*, 1995). A recent concerted effort to use *Bti* and *Bacillus sphaericus* to control an *An. gambiae* population was reported by Fillinger & Lindsay (2006). In a Kenyan community covering 4.5 km^2 419 potential *Anopheles* breeding sites were identified, checked weekly and, where necessary, treated with one of the bacterial agents. The proportion of sites containing late-stage larvae was reduced from about 36% without the intervention to 0.8% with it. The mean number of blood-fed *An. gambiae* per person found in bedrooms was reduced from 0.8 to 0.06. The cost of materials and labor was about $US 0.90 per community member per year. This effort required for frequent retreatments might be reduced by using the insect growth regulator diflubenzuron or pyriproxyfen which persist much longer than *Bti* (Begum & Curtis, 2007).

31.4 Insecticide-treated nets

Not long after his discovery of the infection of *Anopheles* with *Plasmodium* from a human malaria patient, Ross (1910) advocated use of mosquito nets as an anti-malaria measure because he perceived that for the most important vector species of *Anopheles*, bites on humans occur indoors, late at night when most people are in bed. However, mosquitoes can bite through an untreated net if part of the body is touching the net during the night and Lines *et al.* (1987) and Curtis *et al.* (1996) showed that, if torn, untreated nets fail to prevent mosquito entry. These authors showed that if a torn net was treated with a pyrethroid insecticide, with low human toxicity, it gave substantial protection. This is partly due to killing of mosquitoes and partly to the excitorepellent effect of pyrethroids driving mosquitoes away if they land on the net. The number of blood-fed mosquitoes found in a room in the morning and in exit traps on its windows is much less if an insecticide-treated net (ITN) has been used by the sleepers during the preceding night (Maxwell *et al.*, 2003). However, a few blood-fed mosquitoes are still found even where an ITN has been used. The question arose whether these had succeeded in feeding on the ITN occupants or whether they had fed on humans elsewhere and entered the room already fed. Soremekun *et al.* (2004) investigated this by observing whether the genetically variable DNA microsatellites in blood meals in mosquitoes collected in rooms matched with the microsatellites in blood samples from the people who had slept under ITNs in these rooms. It was found that, of the few blood-fed mosquitoes caught in the rooms, 85% had fed on these sleepers, presumably either by biting through the nets or by waiting for people to get up in the night. Thus, one can conclude that an ITN gives good, but not perfect, personal protection of a sleeper under it.

The fact that personal protection of sleepers under ITNs is not perfect makes it all the more important that ITNs are provided to whole communities, so that as many as possible of the local *Anopheles* are killed after being attracted by the body odor of sleepers to make contact with ITNs. In fact one can consider ITNs to be mosquito traps baited with the odor of sleepers inside them. This mass mosquito killing reduces the mean age of the mosquito population (which can be assessed by dissection of the ovaries of large samples of the mosquitoes) and hence reduces the number of mosquitoes old enough for infective *Plasmodium* sporozoites to have developed inside them. Data collected by Tony Wilkes (see Magesa *et al.* 1991; Curtis *et al.*, 2006) before and after providing ITNs or DDT spraying in Tanzanian villages demonstrated the life-shortening effect and hence reduction of infective mosquitoes due to either method of using residual insecticides. This "bonus" effect of widespread use of ITNs is about equal to the personal protection effect of having one's own ITN (Maxwell *et al.*, 2003).

There has been a tendency to emphasize the targeting of ITNs to the members of the community who are most vulnerable to malaria, i.e. young children and pregnant women (Desai *et al.*, 2007). However, the data on the benefit of community-wide use of ITNs suggest that the best protection for the malaria vulnerable would be not only to ensure that they received ITNs, but also community-wide use so that the ITNs of the less vulnerable would kill many mosquitoes before they could bite the very vulnerable (Teklahaimanot *et al.*, 2007).

There is abundant evidence that ITNs reduce malaria morbidity (Lengeler, 2004), including infant anemia (Maxwell *et al.*, 2003) which is a major cause of malaria-related deaths. A series of trials in Africa were organized by the World Health Organization of the United Nations (WHO) in which whole communities were provided with nets which were retreated regularly. These trials led to the conclusion that, on average, 5.5 child deaths were prevented per year for every 1000 children provided with ITNs (Lengeler, 2004). However, where nets are provided only for children, there will be little of the above-mentioned "bonus" from reduction of the infective vector population and one must expect the reductions in child mortality to be somewhat less.

It is often stated that there is a major problem in retreatment of nets with sufficient frequency to replace insecticide lost by washing and thus to maintain their insecticidal effectiveness. However, it is entirely possible to organize community-wide retreatment days each year: in Vietnam since 1998 this has been provided for 10 million people in the most malarious parts of the country (see data of Tran Duc Hinh in Curtis *et al.*, 2004). In recent years long-lasting insecticidal nets (LLINs) have been manufactured, either with the pyrethroid insecticide incorporated into the polyethylene before it is made into fiber, or with the insecticide applied to the completed net with a resin which increases the adherence to the netting so that the net remains insecticidal after more than 20 washes. Thus, nets can be expected to remain insecticidal for their full physical "life" until they are badly torn and need replacement. Sachets of insecticide and of the resin "adhesive" are now becoming available which would allow any net to be made long-lastingly insecticidal by home treatment (Yates *et al.*, 2005).

So far only pyrethroids have been used for operational treatment of nets. There is therefore the threat of evolution of forms of resistance strong enough to interfere with effective use of ITNs. The existing urgent effort must continue to make ready alternative insecticides to replace pyrethroids if such forms of resistance evolve, as appears now to be the case in southern Benin (N'Guessan *et al.*, 2007).

Bites of some Asian and Latin American malaria vectors mostly occur before people go to bed (Pates & Curtis, 2005; Harris *et al.*, 2006). In such cases there is now evidence that repellents may give valuable protection against malaria (Rowland, 2004; Hill *et al.*, 2007). However, there must be concern that repellents which are not insecticidal will tend to divert biting from users to non-users (Waka *et al.*, 2006; Moore *et al.*, 2007).

A direct comparison of the impact of ITNs and residual house spraying with the same pyrethroid compound showed very similar impact of both methods on mosquito populations and incidence of malaria infection (Curtis *et al.*, 1998). However, house spraying required about six times as much insecticide per family protected as did ITNs with annual retreatment. None of the recent ITN projects has yet shown as good an impact on malaria as the prolonged use of DDT house spraying in Zanzibar integrated with mass use of chloroquine (Curtis & Mnzava, 2000). In many areas with poor infrastructure and administrative systems, organizing the logistics of comprehensive house spraying programs may well be more difficult than mass provision of LLINs.

In countries such as Tanzania there has been much emphasis on using the private sector net retailers as the route to deliver insecticidal nets to the people. Subsidies have been provided for social marketing, i.e. use of modern advertising to encourage purchase of nets and insecticide to treat them. More recently vouchers have been provided at maternal and child health clinics which entitle a pregnant woman to purchase one net with a reduction on the normal retail price. However, Maxwell *et al.* (2006) and the Tanzanian National Bureau of Statistics (2006) found that, though social marketing has achieved good, or relatively good, coverage in towns (where there is the major nuisance of biting by urban *Culex* mosquitoes), in 2005 there was an average of less than 10% usage of treated nets by children and pregnant women in rural areas, where the main malaria problem exists.

Table 31.3 Long-lasting insecticidal nets (LLINs) delivered, or planned to be delivered, free of charge between December 2004 and mid-2007 in programs linked to measles vaccination campaigns and/or other public health measures

Countries	Number of nets
Ethiopia	~18 210 000
Nine West and Central African countries	8 069 000
Six East and southern African countries	10 070 000
Indonesia	2 050 000
Total	**~38 399 000**

Source: Data from Mark Grabowsky, Measles and Malaria Partnership and Yemane Ye-ebiyo, Centre for National Health Development, Ethiopia (personal communication).

Recently it has been found that delivery of LLINs free of charge can be efficiently and economically linked to measles vaccination campaigns (Grabowsky *et al.*, 2005) which are well attended by all income groups in many African countries (e.g. almost 80% coverage in Tanzania, as reported by the Tanzanian National Bureau of Statistics, 2006). Such campaigns, which purchase LLINs at wholesale prices and link their free distribution to vaccination and other public health measures, have recently taken off on a huge scale, as indicated in Table 31.3, and are leaving marketing programs far behind (Noor *et al.*, 2007; Teklahaimanot *et al.*, 2007). It is very important that definitive data are collected on large enough samples to assess the extent of the benefits for child health and survival which these massive programs have achieved. However, it already seems likely that the numbers of lives saved will bear comparison with what was achieved in the first few years of large-scale use of the greatest of medical advances.

References

Alio, A. Y., Isaq, A. & Delfini, L. F. (1987). *Field Trial on the Impact of* Oreochromis spilurus spilurus *on Malaria Transmission in Northern Somalia*, WHO Document No. WHO/MAL/85.1017. Geneva, Switzerland: World Health Organization of the United Nations.

Begum, M. & Curtis, C. F. (2007). Comparisons of the persistence of the biological larvicide *Bacillus thuringiensis israelensis* and the insect growth regulators Dimilin and pyriproxyfen against larvae of *Anopheles stephensi*. *Bangladesh Journal of Zoology*, **35**, 19–25.

Berkeley, J., Mwarumba, S., Bramham, K., Lowe, B. & Marsh, K. (1999). Bacteraemia complicating severe malaria in children. *Transactions of the Royal Society of Tropical Medicine and Hygiene*, **93**, 283–286.

Blanford, S., Chan, B. H. K., Jenkins, N. *et al.* (2005). Fungal pathogen reduces potential for malaria transmission. *Science*, **308**, 1638–1641.

Boswell, E., Tiwari, S. N. & Ghosh, S. K (2005). Feasibility of global positioning systems in mapping of mosquito breeding sites for the control of malaria vectors using larvivorous fish in Karnataka State, India. *Transactions of the Royal Society of Tropical Medicine and Hygiene*, **99**, 944.

Cooper, K. (1991). Effects of pesticides on wild life. In *Handbook of Pesticide Toxicology*, eds. W. J. Hayes & E. R. Laws, vol. 2, pp. 463–496. New York: Academic Press.

Cox, J., Hay, S. I., Abeku, T. A., Checchi, F. & Snow, R. W. (2007). The uncertain burden of *Plasmodium falciparum* epidemics in Africa. *Trends in Parasitology*, **23**, 142–148.

Curtis, C. F. & Mnzava, A. E. P. (2000). Comparison of house spraying and insecticide treated nets for malaria control. *Bulletin of the World Health Organization*, **78**, 1389–1400.

Curtis, C., Myamba, J. & Wilkes, T. J. (1996). Comparison of different insecticides and fabrics for anti-mosquito bednets and curtains. *Medical and Veterinary Entomology*, **10**, 1–14.

Curtis, C. F., Maxwell, C. A., Finch, R. J. & Njunwa, K. J. (1998). A comparison of use of a pyrethroid either for house spraying or for bednet treatment against malaria vectors. *Tropical Medicine and International Health*, **3**, 619–631.

Curtis, C. F., Jana-Kara, B. & Maxwell, C. A. (2004). Insecticide treated nets: impact on vector populations and relevance of initial intensity of transmission and pyrethroid resistance. *Journal of Vector Borne Diseases*, **40**, 1–8.

Curtis, C. F., Maxwell, C. A., Magesa, S. M., Rwegoshora, R. T. & Wilkes, T. J. (2006). Insecticide treated bednets against malaria mosquitoes. *Journal of the American Mosquito Control Association*, **22**, 501–506,

Desai, M., ter Kuile, F. O., Nosten, F. *et al.* (2007). Epidemiology and burden of malaria in pregnancy. *Lancet Infectious Diseases*, **7**, 93–104.

Fillinger, U. & Lindsay, S. W. (2006). Suppression of exposure to malaria vectors by an order of magnitude by using microbial larvicides in rural Kenya. *Tropical Medicine and International Health*, **11**, 1–13.

Gerberich, J. B. & Laird, M. (1985). Larvivorous fish in the biocontrol of mosquitoes, with a selected bibliography of recent literature. In *Integrated Mosquito Control Methodologies*, eds. M. Laird & J. W. Miles, vol. 2, pp. 47–78. London: Academic Press.

Ghosh, S. K., Tiwari, S. N., Sathyanarayan, T. S. *et al.* (2005). Larvivorous fish in wells target the malaria vector sibling species of the *Anopheles culicifacies* complex in villages in Karnataka, India. *Transactions of the Royal Society of Tropical Medicine and Hygiene*, **99**, 101–105.

Ghosh, S. K., Tiwari, S. N., Sathyanarayan, T. S., Dash, A. P. & Magurran, A. E. (2006). Experience of larvivorous fish in malaria control over a decade in India and need for study on biodiversity implications. *Proceedings of the 11th International Congress of Parasitology*, August 2006, Glasgow, UK, Abstract no. a528.

Gillies, H. M. (1993). The epidemiology of malaria. In *Bruce-Chwatt's Essential Malariology*, 3rd edn, eds. H. M. Gillies & D. A. Warrell, pp. 132–136. London: Edward Arnold.

Grabowsky, M., Nobiya, T., Ahun, M. *et al.* (2005). Distributing insecticide-treated bednets during measles vaccination: low cost means of achieving high and equitable coverage. *Bulletin of the World Health Organization*, **83**, 195–201.

Harris, A. F., Marias-Arnez, A. & Hill, N. (2006). Biting time of *Anopheles darlingi* in the Bolivian Amazon and implications for controlling malaria. *Transactions of the Royal Society of Tropical Medicine and Hygiene*, **100**, 46–47.

Hay, S. I., Guerra, C. A., Tatem, A. J., Atkinson, P. M. & Snow, R. W. (2005). Urbanization, malaria transmission and disease burden in Africa. *Nature Reviews Microbiology*, **3**, 81–90.

Hill, N., Lenglet, A., Arnez, A. M. & Carneiro, I. (2007). Plant based insect repellent used in combination with insecticide treated bed nets give greater protection against malaria than treated nets alone in areas of early evening biting: a double blind, placebo controlled, clinical trial in the Bolivian Amazon. *British Medical Journal*, **335**, 1023–1026.

Joncour, C. (1956). La lutte contre le paludisme à Madagascar. *Bulletin of the World Health Organization*, **15**, 711–723.

Keiser, J., Singer, B. H. & Utzinger, J. (2005). Reducing burden of malaria in different eco-epidemiological settings with environmental management: a systematic review. *Lancet Infectious Diseases*, **5**, 695–708.

Krüger, A., Rech, A., Si, X.-Z. & Tannich E. (2001). Two cases of autochthonous *Plasmodium falciparum* malaria in Germany with evidence of local transmission by indigenous *Anopheles plumbeus*. *Tropical Medicine and International Health*, **6**, 983–985.

Kühn, K. G., Campbell-Lendrum, D. H., Armstrong, B. & Davies, C. R. (2003). Malaria in Britain: past, present and future. *Proceedings of the National Academy of Sciences of the USA*, **100**, 9997–10 001.

Kumar, A., Sharma, V. P., Thavaselvam, D. & Sumodan, P. K. (1995). Control of *Anopheles stephensi* breeding in construction sites and abandoned overhead tanks with *Bacillus thuringiensis* var. *israelensis*, strain 164, serotype H-14. *Journal of the American Mosquito Control Association*, **11**, 86–89.

Kumar, A., Sharma, V. P., Sumodan, P. K. & Thavaselvam, D. (1998). Field trials of bio-larvicide *Bacillus thuringiensis* var. *israelensis*, strain 164 and larvivorous fishes *Aplocheilus blocki* against *Anopheles stephensi* for malaria control in Goa, India. *Journal of the American Mosquito Control Association*, **14**, 457–462.

Lengeler, C. (2004). *Insecticide Treated Bednets and Curtains for Malaria Control: A Cochrane Review*. The Cochrane Library, Issue 3. Oxford, UK: Oxford Update Software Ltd.

Lines, J. D., Myamba, J. & Curtis, C. F. (1987). Experimental hut trials of permethrin-impregnated mosquito nets and eave curtains against malaria vectors in Tanzania. *Medical and Veterinary Entomology*, **1**, 37–51.

Liu, Q., Kang, X., Chao, C. *et al.* (2004). New irrigation methods sustain malaria control in Sichuan Province, China. *Acta Tropica*, **89**, 241–247.

Mabaso, M. L. H., Sharp, B. & Lengeler, C. (2004). Historical review of malaria control in southern Africa with emphasis on the use of indoor residual housespraying. *Tropical Medicine and International Health*, **9**, 846–856.

Macdonald, G. (1957). *Epidemiology and Control of Malaria*. London: Oxford University Press.

Magesa, S. M., Wilkes, T. J., Mnzava, A. E. P. *et al.* (1991). Trial of pyrethroid impregnated bednets in an area of Tanzania holoendemic for malaria. II. Effects on the malaria vector population. *Acta Tropica*, **49**, 97–108.

Maxwell, C. A., Chambo, W., Mwaimu, M. *et al.* (2003). Variation in malaria transmission and morbidity with altitude in Tanzania and with introduction of alphacypermethrin treated nets. *Malaria Journal*, **2**, 28. Available at www.malariajournal.com.

Maxwell, C. A., Rwegoshora, R. T., Magesa, S. M. & Curtis, C. F. (2006). Comparison of coverage with insecticide-treated nets in a Tanzanian town and villages where nets and insecticide are either bought or provided free of charge. *Malaria Journal*, **5**, 44. Available at www.malariajournal.com.

Miller, L., Mason, S. J., Clyde, D. F. & McGinnis, M. H. (1976). The resistance factor to *P. vivax* in blacks: the Duffy blood group genotype. *New England Journal of Medicine*, **295**, 302–304.

Moore, S. J., Davies, C. R., Hill, N. & Cameron, M. M. (2007). Are mosquitoes diverted from repellent-using to non-using individuals? Results from a field trial in Bolivia. *Tropical Medicine and International Health*, **12**, 1–8.

N'Guessan, R., Corbel, V., Akogbeto, M. & Rowland, M. (2007). Reduced efficacy of insecticide treated nets and indoor residual spraying for malaria control in a pyrethroid resistance area, Benin. *Emerging Infectious Diseases*, **13**, 199–206.

Noor, A. M., Amin, A. A., Akhwale, W. S. & Snow, R. W. (2007). Increasing access and decreasing inequity of insecticide-treated bed net use among rural Kenyan children. *Public Library of Science, Medicine* 2007:4(8)e255, doi:10.1371/journal.pmed.0040255.

Pates, H. & Curtis, C. F. (2005). Mosquito behavior and vector control. *Annual Review of Entomology*, **50**, 53–70.

Ribbands, C. R. (1946). Effects of bush clearance on fighting West African anophelines. *Bulletin of Entomological Research*, **37**, 33–40.

Romi, R., Sabatinelli, G. & Majori, G. (2001). Could malaria reappear in Italy? *Emerging Infectious Diseases*, **7**, 915–919.

Ross, R. (1910). *The Prevention of Malaria*. London: John Murray.

Rowland, M. (2004). DEET mosquito repellent provides personal protection against malaria: a household randomized trial in an Afghan refugee camp in Pakistan. *Tropical Medicine and International Health*, **9**, 335–342.

Schellenberg, D., Menendez, C., Aponte, J. J. *et al.* (2005). Intermittent preventive antimalarial treatment for

Tanzanian infants: follow-up to age 2 years of a randomized, placebo-controlled trial. *Lancet*, **365**, 1481–1483.

Schofield, L. (2007). Rational approach to an anti-disease vaccine against malaria. *Microbial Infection*, **9**, 784–791.

Scholte, E. J., Ng'habi, K., Kihonda, J. *et al.* (2005) An entomopathogenic fungus for control of adult African malaria mosquitoes. *Science*, **308**, 1641–1642.

Singh, B., Lee, K. S., Matusop, A. *et al.* (2004). A large focus of naturally acquired *Plasmodium knowlesi* infections in human beings. *Lancet*, **363**, 1017–1024.

Snowden, F. M. (2006). *The Conquest of Malaria: Italy 1900–1962*. New Haven, CT: Yale University Press.

Soremekun, S., Maxwell, C. A., Zuwakuu, M. *et al.* (2004). Measuring the efficacy of treated bednets: the use of DNA fingerprinting to increase the accuracy of personal protection estimates in Tanzania. *Tropical Medicine and International Health*, **9**, 664–672.

Subbarao, S. K., Adak, T., Vasantha, K. *et al.* (1988). Susceptibility of *Anopheles culicifacies* species A and B to *Plasmodium vivax* and *Plasmodium falciparum* as determined by immunodiagnostic assay. *Transactions of the Royal Society of Tropical Medicine and Hygiene*, **82**, 394–397.

Tanzanian National Bureau of Statistics (2006). *Demographic and Health Survey (2004–5)*. Dar es Salaam, Tanzania.

Teklehaimanot, A., Sachs, J. D. & Curtis C. F. (2007). Malaria control needs mass distribution of insecticidal bednets. *Lancet*, **369**, 2143–2146.

US Health Service & Tennessee Valley Authority (1947). *Malaria Control in Impounded Water*. Washington, DC: US Government Printing Office.

Waka, M., Hopkins, R. J., Glinwood, R. & Curtis, C. F. (2006). The effects of repellents *Ocimum forskolei* and deet on the response of *Anopheles stephensi* to host odours. *Medical and Veterinary Entomology*, **20**, 373–376.

WHO & UNICEF (2005). *World Malaria Report*. Geneva, Switzerland: World Health Organization and The United Nations Children's Fund of the United Nations.

Yapabandara, A. M. G. M. & Curtis, C. F. (2002). Laboratory and field comparisons of pyriproxyfen, polystyrene beads and other larvicidal methods against malaria vectors in Sri Lanka. *Acta Tropica*, **81**, 211–223.

Yapabandara, A. M. G. M. & Curtis, C. F. (2004a). Vectors and malaria transmission in a gem mining area in Sri Lanka. *Journal of Vector Ecology*, **29**, 264–276.

Yapabandara, A. M. G. M. & Curtis C. F. (2004b). Control of malaria vectors in an irrigated settlement scheme in Sri Lanka using the insect growth regulator pyriproxyfen. *Journal of the American Mosquito Control Association*, **20**, 395–400.

Yapabandara, A. M. G. M., Curtis, C. F., Wickramasinghe, M. B. & Fernando, W. P. (2001). Control of malaria vectors with the insect growth regulator pyriproxyfen in a gem mining area in Sri Lanka. *Acta Tropica*, **80**, 265–276.

Yasuoka, J., Levine, R. T., Mangione, T. M. & Spielman, A. (2006). Community-based rice ecosystem management for suppressing vector anophelines in Sri Lanka. *Transactions of the Royal Society of Tropical Medicine and Hygiene*, **100**, 995–1006.

Yates, A., N'Guessan, R. N., Kaur, H., Akogbeto, M. & Rowland, M. (2005). Evaluation of KO-Tab 1-2-3: a wash-resistant "dip-it-yourself" insecticide formulation for long lasting treatment of mosquito nets. *Malaria Journal*, **4**, 52. Available at www.malariajournal.com.

Chapter 32

Gypsy moth IPM

Michael L. McManus and Andrew M. Liebhold

Over the last 50 years, North American forests have been inundated by a multitude of alien pest invasions. Among these, noteworthy invaders include the hemlock woolly adelgid (*Adelges tsugae*), emerald ash borer (*Agrilus planipennis*), chestnut blight and Dutch elm disease. These species have greatly altered both the ecological and economic values associated with forests and their management, representing perhaps the most demanding challenge facing state and federal forest pest management personnel. In this chapter, we provide an overview of the gypsy moth (*Lymantria dispar*) problem and describe the various approaches to managing this species, which serve as a model system for understanding the management of non-indigenous forest pests.

32.1 Gypsy moth in North America

Gypsy moth was accidentally introduced to North America over 130 years ago, but it has spread relatively slowly, currently occupying less than one-third of its potential habitat in the eastern USA. It is possible to observe at any time areas where the species has not yet established, where introduced populations are occasionally eradicated, areas at the leading edge of the expanding range where concerted efforts are directed to slow its spread, and areas where the species has been established for many years and where considerable resources are expended to suppress defoliating populations. Because the gypsy moth has a dramatic and continuing impact on the public, this species has been the target for many intense research and management programs. Consequently, the technology being applied to manage this species is particularly advanced and thus elucidates the possibilities for managing other non-indigenous species.

The gypsy moth is native to virtually all temperate forest regions of Europe, Asia and North Africa. The northern limit of its range proceeds from southern Sweden and Finland through Europe and across Russia, and the southern limit begins in northern Morocco, Algeria and Tunisia and proceeds east to include all of the Mediterranean islands on a line through Israel, Iran, Central Asia and finally into China and Japan. Though many non-indigenous pests are considered to be innocuous species in their native ranges, the gypsy moth periodically causes outbreaks over much of Europe and Asia where it is regarded as a pest. The gypsy moth was first introduced to North America in 1868 or 1869 by Étienne Léopold Trouvelot, a French artist who was living in Medford, Massachusetts. Trouvelot was engaged in the amateur study of native silkworms and other silk-producing insects on his property when the gypsy

Integrated Pest Management, ed. Edward B. Radcliffe, William D. Hutchison and Rafael E. Cancelado. Published by Cambridge University Press.

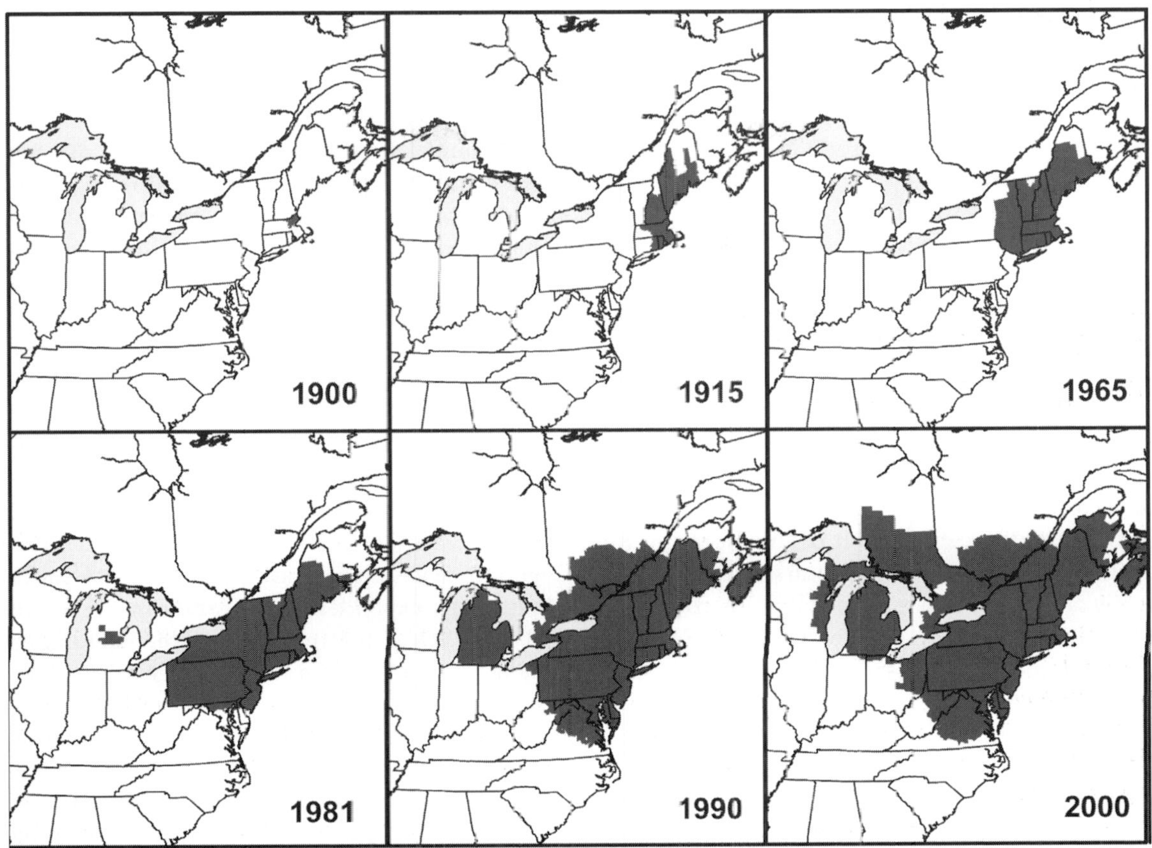

Fig. 32.1 Historical gypsy moth spread in North America (from Liebhold *et al.*, 2007).

moth larvae that he was cultivating escaped from containment (Forbush & Fernald, 1896; Liebhold *et al.*, 1989). About 20 years after the accidental introduction, larvae of the insect became so abundant and destructive on fruit and shade trees that it attracted public attention; the extensive defoliation and nuisance created by hordes of larvae prompted the state of Massachusetts to embark on an intensive program to eradicate the gypsy moth, an effort that ultimately failed. By 1912, infestations were detected in the neighboring states of Vermont, New Hampshire, Rhode Island and Connecticut. The current distribution of the species includes most of the northeastern USA and southeastern Canada (Fig. 32.1). Populations in the Great Lakes region originated from a secondary population that was accidentally introduced into Michigan in the 1960s.

The relatively slow spread of the gypsy moth in North America can be attributed to the fact that the female moths are incapable of sustained flight. Through most of Asia and portions of Eastern Europe, gypsy moth females are capable of extended flight. But in Western Europe, though female moths have fully developed wings, they are incapable of directed flight. Apparently Trouvelot's population was collected from France where females are flightless. Adults emerge, mate and oviposit in mid to late summer. Egg masses, which may contain 50–1000 eggs, are laid on tree trunks, branches, as well as objects on the forest floor such as logs, stumps and rock outcroppings. Populations remain in the egg stage throughout winter and are well protected from extreme conditions by an obligate diapause. Eggs hatch in the spring and young larvae engage in wind-borne dispersal on silken threads, which facilitate the redistribution of local populations. Upon locating suitable hosts, larvae feed and complete five (male) or six (female) instars before pupating. In low- to moderate-density populations, late-instar larvae exhibit diel behavior whereby they feed at night and descend from the tree canopy at dawn, resting in cryptic sites on the tree or on the forest floor.

32.1.1 Host range

Gypsy moth larvae are polyphagous and are known to feed on hundreds of tree species. Late instars are more polyphagous than early stage larvae. Completion of early larval development is largely restricted to oak (*Quercus*), poplar (*Populus*), willow (*Salix*) or larch (*Larix*) in North America. Liebhold *et al.* (1995) provide a comprehensive list of preferred species collated from various sources in the literature. As a result of the more specific host requirements by early instars, outbreaks generally do not develop in stands that are composed of less than 20% of these preferred genera. In North America, most such susceptible forests are dominated by oak. Outbreaks are most frequent in dry sites such as on ridge top stands characterized by poor, shallow soils, rock outcroppings, and preferred species such as chestnut oak.

32.1.2 Population dynamics

During outbreaks, populations reach extremely high densities (e.g. egg mass populations exceeding 1000/ha) such that they may consume most or all of the foliage on trees prior to reaching the last instar. Gypsy moth outbreaks are common events both in the species' native and alien ranges, and the temporal patterns of outbreaks are similar. Most populations exhibit periodicity with outbreaks recurring every eight to 12 years (Johnson *et al.*, 2005). It has also been noted that in the highly susceptible sites, however, the very dry sites described previously, populations exhibit a "doubled" frequency such that outbreaks recur every four to six years (Johnson *et al.*, 2006). Despite considerable research devoted to gypsy moth population dynamics (Elkinton & Liebhold, 1989), there remains considerable uncertainty about the causes of gypsy moth population oscillations. Perhaps the most plausible explanation advanced to date is that by Dwyer *et al.* (2004) who hypothesized that periodic oscillations are the combined result of strongly density-dependent mortality caused by a nucleopolyhedrosis virus at high densities coupled with predation by generalist small mammal predators at low densities.

Of course these periodic oscillations are not precise and at any one location; outbreaks rarely recur with regularity. However, there is considerable synchrony in the development of outbreaks over large areas and regional defoliation levels exhibit statistical periodicity (Peltonen *et al.*, 2002; Johnson *et al.*, 2005). Since 1924, over 35 million ha of forest land in North America have been defoliated. The extent of outbreaks, measured by the total forested area defoliated, has worsened dramatically as the area infested has increased. Annual defoliation exceeding 500 000 ha occurred in 20 years between 1970 and 1995, a period when the distribution of gypsy moth populations expanded significantly in the South and West. Over 5.2 million ha were defoliated in 1981, 3 million ha in 1990. Over the last decade (1997–2007), the area of annual defoliation in North America has declined slightly, coinciding with the appearance in 1989 of a previously unknown fungal pathogen, *Entomophaga maimaiga*, which has caused extensive larval mortality in gypsy moth populations throughout their current range. Unfortunately, there is little certainty that this pathogen has significantly impacted regional defoliation levels; nor is it known how it may affect gypsy moth populations in the future because fungal epizootics are closely associated with specific environmental conditions (i.e. rainfall, high humidity).

32.1.3 Losses due to defoliation

The effects of defoliation on trees is highly variable and depends on both the frequency of defoliation, the condition of the stand prior to defoliation, and the presence of other factors that influence tree growth (Twery, 1991). For example, considerable tree mortality can occur following only one year of heavy defoliation in trees in poor crown condition, particularly if the defoliation occurs in a drought year. In other stands, trees in good condition may be able to survive after one or more years of total defoliation. Maximum tree mortality usually occurs three to five years after an episode

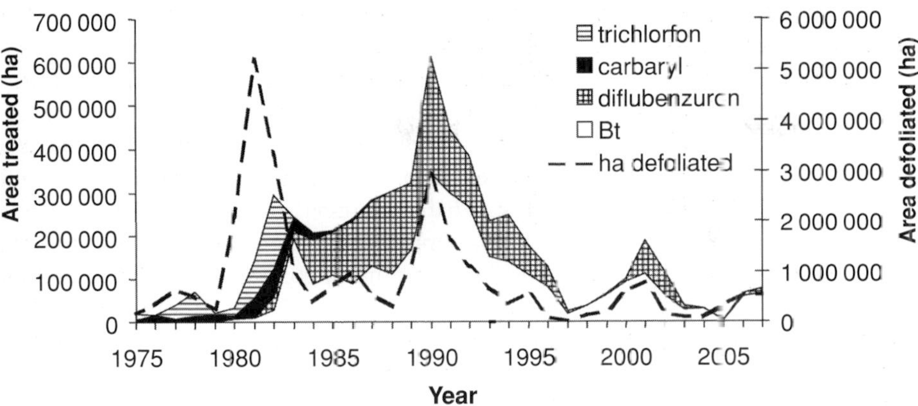

Fig. 32.2 Historical suppression of gypsy moth populations via aerial application of pesticides, and defoliation (from US Department of Agriculture, 2007).

of defoliation and is usually caused by secondary agents such as the pathogenic fungus *Armillaria mellea* and the wood-boring beetle *Agrilus bilineatus*, which readily attack severely weakened trees.

Probably the most comprehensive analysis of socioeconomic impacts of gypsy moth outbreaks was provided by Leuschner *et al.* (1996) who concluded that impacts of tree mortality on timber values were dwarfed by the much greater residential costs associated with gypsy moth defoliation and nuisance. The area currently invaded by the gypsy moth encompasses about 864 475 km^2, an area that coincides with the most heavily populated region in the USA. Therefore, the interaction between the public and the gypsy moth has been frequent and, at times, intense. During the outbreak phase, when the density of gypsy moth populations can increase 100-fold in successive years, larvae can pose a hazard to human health and disrupt the public's enjoyment of outdoor activities. In extreme situations such as the severe outbreak of 1981, there were hundreds of documented reports where individuals suffered severe allergenic reactions to airborne hairs and scales that originated from gypsy moth life stages. Further, defoliation of trees in residential areas and the possibility of their loss detract from aesthetic and property values such that homeowners are willing to go to great expense to protect their trees and combat nuisance populations of larvae.

32.2 Suppression

Given the propensity for gypsy moth populations to periodically reach high levels and the general adversity of the public towards these outbreaks, there has been a long history of direct control using ground and aerial applications of pesticides to suppress high-density populations. In 1956, 222 000 ha were aerially sprayed with DDT followed by 1.2 million ha in 1957. Although this spraying was highly effective in eliminating gypsy moth populations and preventing defoliation, the use of DDT was phased out in 1958 because of public concern about residues on food and feed crops and adverse effects of DDT on species of beneficial organisms, fish and wildlife. It was replaced by other broad-spectrum synthetic pesticides such as carbaryl and trichlorfon. However, the use of these products eventually declined because of their negative impacts on species of parasitoids and bees (Fig. 32.2). Diflubenzuron (Dimilin®), an insect growth regulator, was registered by the US Environmental Protection Agency (EPA) in 1976 for use against the gypsy moth and was extremely effective against all larval stages at very low dosages. Despite its efficacy, its use has been somewhat limited because it is toxic to aquatic invertebrates and crustaceans and thus cannot be applied near bodies of water or in areas where surface water is present.

Beginning in the 1970s, research and methods improvement was accelerated towards developing microbial pesticides for use against the gypsy moth, specifically *Bacillus thuringiensis* (*Bt*), to address the public's concerns about the environmental effects of aerially applied chemical pesticides. Although the pathogenicity of *Bt* against gypsy moth was well known, its efficacy in the field was erratic until a more potent strain called *Btk* (*Bacillus thuringiensis* variety *kurstaki*) was isolated in 1970. Since then there has been a dramatic improvement in commercial *Btk* formulations so that it is currently one of the most widely used materials in gypsy moth suppression programs. Gypchek, the naturally occurring gypsy moth nucleopolyhedrosis virus, was one of the first viral pesticides registered in 1978 by the EPA. Because it is produced in vivo, the process is labor intensive and more costly than conventional pesticides, thus only 5000–8000 ha equivalents are produced annually. Because it is specific only to gypsy moth, Gypchek is highly sought after for use in environmentally sensitive habitats where application of broad spectrum products is not acceptable.

Most efforts to suppress outbreak gypsy moth populations are carried out as part of the Cooperative US Department of Agriculture Forest Service–State Gypsy Moth Suppression Program which facilitates suppression of gypsy moth populations on federal, state and privately owned land. The cost of suppression is typically shared by the Forest Service, state/local governments and the landholder, though the cost share varies considerably among states. Participation is voluntary and proposals for funding must meet established criteria and include treatments approved by the Federal Environmental Impact Statement for gypsy moth. Monitoring plays an important role in any pest management program and is especially critical in gypsy moth suppression. Traditionally, treatment decisions are driven by counts of overwintering egg mass populations (Ravlin *et al.*, 1987; Liebhold *et al.*, 1994). Several statistical models are available which predict defoliation based upon egg mass densities (Gansner *et al.*, 1985; Liebhold *et al.*, 1993) and most states use egg mass density thresholds (e.g. 100 egg masses/ha) as part of their decision-making criteria for suppression. Unfortunately, the relationship between pre-treatment egg mass density and subsequent defoliation is not precise and this can lead to substantial error in predicting defoliation. Part of the uncertainty in these predictions is due to the high sampling error encountered in estimating egg mass densities (Liebhold *et al.*, 1991); additionally, the collapse of high-density populations is difficult to predict due to the complexity of the trophic interactions between gypsy moths and their natural enemies, such as the nucleopolyhedrosis virus and *E. maimaiga*. The uncertainty in these predictions detracts from the efficiency of large-scale suppression programs (Liebhold *et al.*, 1996; Weseloh, 1996).

32.2.1 Eradication

The gypsy moth currently occupies less than one-third of the forested region in the USA that is capable of supporting a gypsy moth outbreak (Morin *et al.*, 2004). As the gypsy moth slowly expands its range, life stages are occasionally transported accidentally well beyond the expanding population front and these are capable of founding new isolated populations. It has long been recognized that the tendency for larvae to pupate and for emerging female gypsy moths to oviposit in cryptic locations often results in accidental movement of egg masses over long distances via commodities, e.g. nursery stock, household goods or recreational vehicles. A federal domestic quarantine was enacted in 1912 to minimize the rapid expansion of the insect to the remainder of eastern USA and Canada, and still in effect today, is credited with reducing the accidental long-range transport of gypsy moth life stages on regulated commodities.

Unfortunately, domestic quarantines are rarely perfect and thousands of gypsy moth life stages are probably transported into uninfested regions every year. For example, California recorded more than 2000 interceptions of gypsy moth life stages (mainly egg masses) from recreational vehicles and shipments of household goods that originated from 14 states and Canada between 1980 and 1990, a period of high population density in the eastern USA (McFadden & McManus, 1991). While most of these individuals do not successfully colonize new areas, populations frequently arise in favorable habitats outside of the generally infested region. As part of the Cooperative Agricultural Pest Survey (CAPS), thousands of gypsy moth pheromone-baited traps are placed annually in uninfested states in order to detect incipient populations (US Department of Agriculture, 2006). These traps are highly efficient, and consequently, each year, there are many locations in the uninfested area where males are detected. The typical response to new detections is to deploy more traps in the following year in and around the site of initial detection at a rate of 40/km^2 in order to (1) determine that the population has persisted and (2) spatially delimit the population. Only about one-tenth of populations that are initially detected survive the following year because of Allee and random effects that naturally cause extinction (Liebhold & Bascompte, 2003); therefore it is important to confirm the persistence of populations prior to any attempt to eradicate them. Once a population has been adequately delimited, multiple aerial applications of pesticides (typically *Bt*) are applied in order to achieve eradication. Following these treatments, pheromone traps are once more deployed to confirm the presence/absence of gypsy moth populations. Any evidence of residual populations is followed by cycles of delimitation/treatment over subsequent years until no male moths are trapped. While most current eradication programs utilize aerial applications of *Bt*, a mass trapping is also carried out in some locations when the area infested is small and well delineated. In these cases, pheromone traps are deployed at rates of 9–25 traps/ha.

The above description of gypsy moth eradication efforts is based upon the assumption that the population detected is of typical European genetic origin. In the USA, trapped males detected from new locations far removed from the infested area, particularly from the west coast, are routinely subjected to a DNA analysis in order to evaluate the genetic origin of the population. In recent years, there have been several incidents in which gypsy moth life stages of Asian origin have been accidentally transported to ports on the east coast of the USA and western USA and Canada. The current US Department of Agriculture policy for response to the detection of Asian strains of the gypsy moth demands a more rapid and aggressive response than that employed for detection of the European strain. Instead of waiting a year for delimiting populations, any detection of Asian individuals is typically followed by multiple aerial applications of *Bt* in the following year.

Though eradication is sometimes a controversial subject because of some well-publicized failures to eradicate certain alien species (Myers *et al.*, 2000), the gypsy moth is a good example of a species for which eradication is almost always successful when isolated populations are discovered and delimited. Much of the success in gypsy moth eradication can be attributed to the low cost yet high efficiency of utilizing pheromone traps as a first line of defense to detect and delimit incipient populations.

32.2.2 Containment

Given the immense public concern about gypsy moth over the last century, the concept of containing its spread has arisen repeatedly. In 1923, federal and state officials began a program to halt gypsy moth spread by establishing a barrier zone that encompassed more than 27 300 km^2 and extended from Canada to Long Island along the Champlain and Hudson River valleys. The territory east of this zone was treated by the individual states while infestations within the zone were eliminated by joint state and federal actions using mainly chemical and mechanical methods. The barrier zone became generally infested by 1939 and the effort was terminated in 1941. Although this program failed to stop the spread of gypsy moth, it has been credited with slowing the rate of spread (Liebhold *et al.*, 1992) despite the fact that only labor-intensive methods for control were available during that period.

Gypsy moth populations in North America are slowly expanding their range and while it is probably impossible to stop this spread, slowing the rate of spread is a more realistic goal. Leuschner *et al.* (1996) performed an economic analysis that indicated that expenditure of funds to slow gypsy moth spread would result in net savings, primarily because of the reduction of funds and resources that would be required to suppress outbreak populations in subsequent years should the insect become established in new areas. As a result, the gypsy moth Slow the Spread (STS) program was initiated in 1999 as a cooperative

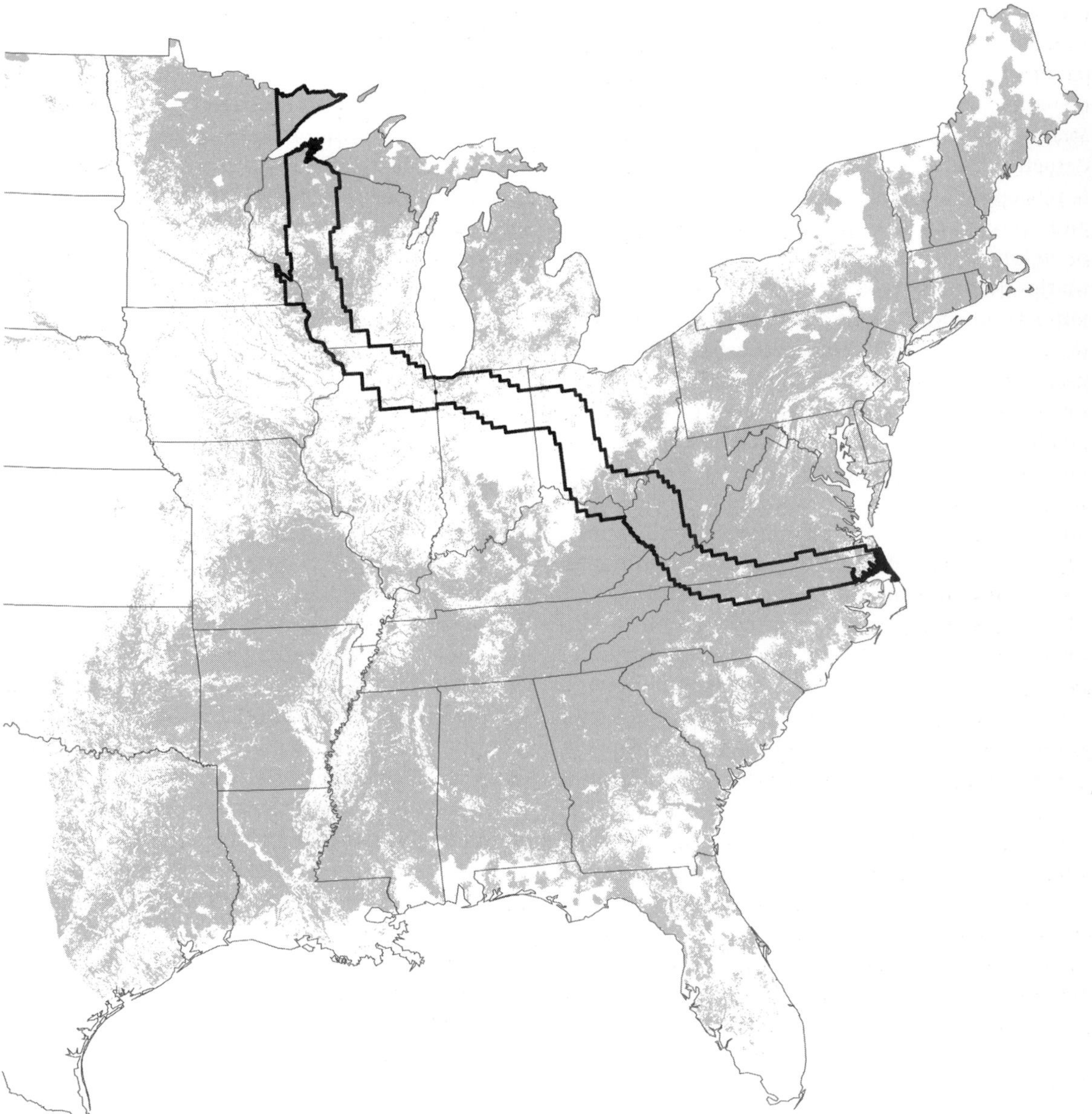

Fig. 32.3 Location of the gypsy moth "Slow the Spread" action area, a ~100-km band along the expanding gypsy moth population front. Areas shaded in gray represent forested areas that are susceptible to defoliation as predicted by the presence of >2 m^2/ha basal area of preferred gypsy moth hosts estimated from forest inventory data (Morin *et al.*, 2004).

federal/state program located along the expanding population front (Sharov *et al.*, 2002) (Fig. 32.3). The program is based upon the scientific finding that gypsy moth spread is exacerbated because isolated colonies form ahead of the advancing population front usually as a result of accidental movement of life stages; over time, these colonies grow, coalesce and thereby promote their expansion (Sharov & Liebhold, 1998). The strategy in STS is to locate and eradicate these isolated colonies before they expand and coalesce. This is accomplished by annually deploying over 100 000 pheromone-baited traps in a 2-km grid along a 100-km band ranging from northern Wisconsin to coastal North Carolina. When colonies are detected within this grid, a more intensive 1-km grid is deployed in the subsequent year to better delimit the population; finally in the third year, the colony is treated (Fig. 32.4). Most colonies are treated using mating disruption which is accomplished by aerially applying a formulation of pheromone flakes. Extensive research has demonstrated that mating disruption is not effective when trap captures exceed 100 males/trap; therefore colonies that exceed this population level are usually treated with aerial applications of *Bt*. Results to date indicate that the program has been successful at reducing rates of spread by well over 50% (Tobin & Blackburn, 2007).

The STS program relies heavily on new technologies and serves as a model system for managing alien invasions. Trap locations are recorded via GPS technology and all data are assembled in a GIS that is used to process trap count data, which is the basis for decision making, for planning treatment areas and for web delivery of summary information (Tobin *et al.*, 2004). Actual costs associated with applying treatments comprise less than 50% of the ~$US 12 million that the program costs annually. To put the cost of this program in perspective, it's important to note that in the outbreak of 1990, aerial spraying of biological and chemical pesticides through the cooperative federal–state suppression program was conducted on 0.65 million ha at an estimated cost of $US 22.5 million. In addition to the environmental concerns associated with spraying, expenditures by the public for spraying pesticides on private forested land and in urban residential areas were astronomical.

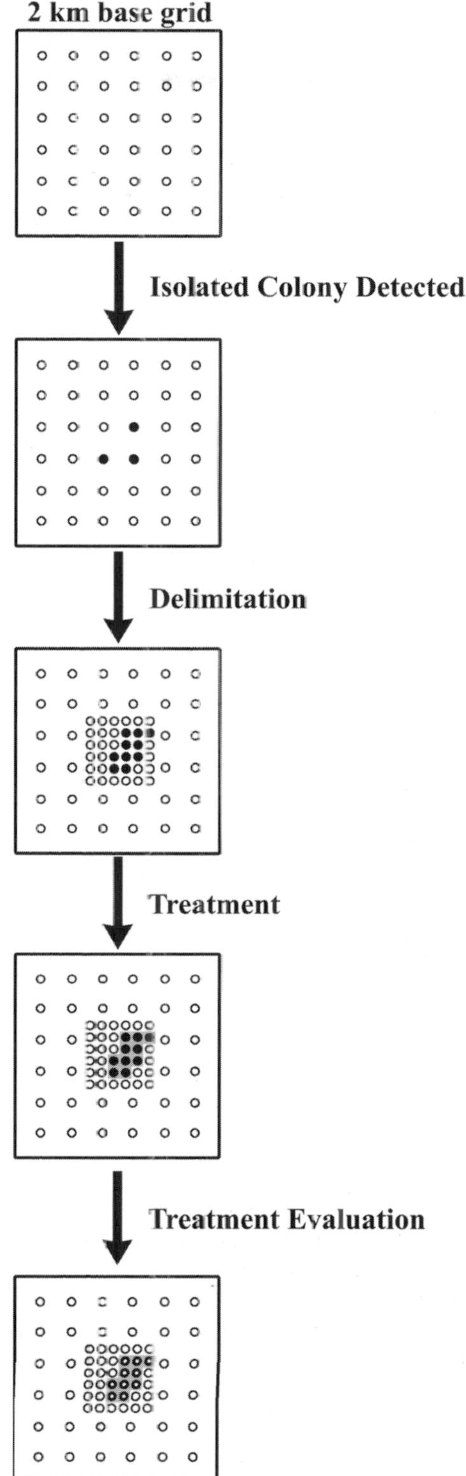

Fig. 32.4 Diagrammatic representation of the gypsy moth "Slow the Spread" strategy (modified from Tobin & Blackburn, 2007).

32.3 Conclusions

The prognosis for the gypsy moth and its associated impacts in the USA is not encouraging. Based on an analysis conducted by Liebhold *et al.*, (1997), there are 19 states currently not infested by gypsy moth that contain more than 1 million ha of forests that are classified as susceptible to gypsy moth defoliation and damage. This suggests that costs associated with managing this pest will continue to escalate which is a strong justification for slowing the spread of the pest. In addition to the benefits that will accrue from delaying impacts and costs associated with management programs, STS, like most IPM programs, is based on a strong foundation of intensive monitoring and deploys only environmentally acceptable treatments when such actions are deemed necessary. Additional information on the strategies discussed in this chapter can be found in the programmatic Environmental Impact Statement *Gypsy Moth Management in the USA: A Cooperative Approach* (US Department of Agriculture, 1995).

Acknowledgements

We thank Laura Blackburn for assistance in preparing figures.

References

Dwyer, G., Dushoff, J. & Yee, S. H. (2004). The combined effects of pathogens and predators on insect outbreaks. *Nature*, **430**, 341–345.

Elkinton, J. S. & Liebhold, A. M. (1989). Population dynamics of gypsy moth in North America. *Annual Review of Entomology*, **35**, 571–596.

Forbush, E. H. & Fernald, C.H. (1896). *The Gypsy Moth*. Boston, MA: Wright & Potter.

Gansner, D. A., Herrick, O. W. & Ticehurst, M. (1985). A method for predicting gypsy moth defoliation from egg mass counts. *Northern Journal of Applied Forestry*, **2**, 78–79.

Johnson, D. M., Liebhold, A. M., Bjornstad, O. N. & McManus, M. L. (2005). Circumpolar variation in periodicity and synchrony among gypsy moth populations. *Journal of Animal Ecology*, **74**, 882–892.

Johnson, D. M., Liebhold, A. M. & Bjornstad, O. N. (2006). Geographical variation in the periodicity of gypsy moth outbreaks. *Ecography*, **29**, 367–374.

Leuschner, W. A., Young, J. A., Walden, S. A. & Ravlin, F. W. (1996). Potential benefits of slowing the gypsy moth's spread. *Southern Journal of Applied Forestry*, **20**, 65–73.

Liebhold, A. M. & Bascompte, J. (2003). The allee effect, stochastic dynamics and the eradication of alien species. *Ecology Letters*, **6**, 133–140.

Liebhold, A., Mastro, V. & Schaefer, P. W. (1989). Learning from the legacy of Leopold Trouvelot. *Bulletin of the Entomological Society America*, **35**, 20–21.

Liebhold, A. M., Twardus, D. & Buonaccorsi, J. (1991). Evaluation of the timed-walk method of estimating gypsy moth, *Lymantria dispar* (Lepidoptera: Lymantriidae), egg mass densities. *Journal of Economic Entomology*, **84**, 1774–1781.

Liebhold, A. M., Halverson, J. & Elmes, G. (1992). Quantitative analysis of the invasion of gypsy moth in North America. *Journal of Biogeography*, **19**, 513–520.

Liebhold, A. M., Simons, E. E., Sior, A. A. & Unger, J. D. (1993). Forecasting defoliation caused by the gypsy moth, *Lymantria dispar*, from field measurements. *Environmental Entomology*, **22**, 26–32.

Liebhold, A., Thorpe, K., Ghent, J. & Lyons, D. B. (1994). *Gypsy Moth Egg Mass Sampling for Decision-Making: A Users' Guide*, USDA Forest Service, Forest Health Management, NA-TP-94-04. Washington, DC: US Government Printing Office.

Liebhold, A. M., Gottschalk, K. W., Muzika, R. *et al.* (1995). *Suitability of North American Tree Species to the Gypsy Moth: A Summary of Field and Laboratory Tests*, General Technical Report No. NE-211. Radnor, PA: US Department of Agriculture, Forest Service.

Liebhold, A., Luzader, E., Reardon, R. C., *et al.* (1996). Use of a geographical information system to evaluate regional treatment effects in a gypsy moth (Lepidoptera: Lymantriidae) management program. *Journal of Economic Entomology*, **17**, 560–566.

Liebhold, A. M., Gottschalk, K. W., Mason, R. R. & Bush, R. R. (1997). Evaluation of forest susceptibility to the gypsy moth across the conterminous USA. *Journal of Forestry*, **95**, 20–24.

Liebhold, A. M., Sharov, A. A. & Tobin, P. C. (2007). Population biology of gypsy moth spread. In *Slow the Spread: A National Program to Manage the Gypsy Moth*, General Technical Report NRS-06, eds. P. C. Tobin & L. M. Blackburn, pp. 15–32. Newtown Square, PA: US Department of Agriculture Forest Service.

McFadden, M. W. & McManus, M. E. (1991). An insect out of control? The potential for spread and establishment of the gypsy moth in new forest areas in the United States. In *Forest Insect Guilds: Patterns of Interaction with Host Trees*, General Technical Report No. NE 153, eds. Y. N. Baranchikov, W. J. Mattson, F. P. Hain, & T. L. Payne, pp. 172–186. Radnor, PA: US Department of Agriculture, Forest Service.

Morin, R. S., Liebhold, A. M., Luzader, E. R. *et al.* (2004). *Mapping Host-Species Abundance of Three Major Exotic Forest Pests*, Research Paper No. NE-726. Radnor, PA: US Department of Agriculture, Forest Service, Northeastern Research Station.

Myers, J. H., Simberloff, D., Kuris, A. M. & Carey, J. R. (2000). Eradication revisited: dealing with exotics. *Trends in Ecology and Evolution*, **15**, 316–321.

Peltonen, M., Liebhold, A. M., Bjornstad, O. N. & Williams, D. W. (2002). Spatial synchrony in forest insect outbreaks: roles of regional stochasticity and dispersal. *Ecology*, **83**, 3120–3129.

Ravlin, F. W., Bellinger, R. G. & Roberts, E. A. (1987). Gypsy moth management programs in the United States: status, evaluation, and recommendations. *Bulletin of the Entomological Society of America*, **33**, 90–98.

Sharov, A. A. & Liebhold, A. M. (1998). Model of slowing the spread of the gypsy moth (Lepidoptera: Lymantriidae) with a barrier zone. *Ecological Applications*, **8**, 1170–1179.

Sharov, A. A., Leonard, D., Liebhold, A. M., Roberts E. A. & Dickerson, W. (2002). "Slow the Spread": a national program to contain the gypsy moth. *Journal of Forestry*, **100**, 30–36.

Tobin, P. C. & Blackburn, L. M. (eds.) (2007). *Slow the Spread: A National Program to Manage the Gypsy Moth*. Forest Service, General Technical Report No. NRS-06. Washington, DC: US Department of Agriculture.

Tobin, P. C., Sharov, A. A., Liebhold, A. M. *et al.* (2004). Management of the gypsy moth through a decision algorithm under the STS Project. *American Entomologist*, **50**, 200–209.

Twery, M. J. (1991). Effects of defoliation by gypsy moth. In *Proceedings of the US Department of Agriculture Interagency Gypsy Moth Research Review 1990*, Technical Report No. NE-146, eds. K. W. Gottschalk. M. J. Twery & S. I. Smith, pp. 27–39. Radnor, PA: US Department of Agriculture, Forest Service.

US Department of Agriculture (1995). *Gypsy Moth Management in the United States: A Cooperative Approach – Final Environmental Impact Statement 1995*, 5 vols. Radnor, PA: US Department of Agriculture, Forest Service, Northeastern Area State and Private Forestry and Animal and Plant Health Inspection Service.

US Department of Agriculture (2006). *Gypsy Moth Program Manual 05/2001-01*. Washington, DC: US Department of Agriculture, Marketing and Regulatory Programs, Animal and Plant Health Inspection Service, Plant Protection and Quarantine. Available at www.aphis.usda.gov/import_export/plants/manuals/domestic/downloads/gypsy_moth.pdf.

US Department of Agriculture (2007). *Gypsy Moth Digest*. Radnor, PA: US Department of Agriculture, Forest Service, Northeastern Area State and Private Forestry. Available at www.na.fs.fed.us/fhp/gm.

Weseloh, R. M. (1996). Developing and validating a model for predicting gypsy moth (Lepidoptera: Lymantriidae) defoliation in Connecticut. *Journal of Economic Entomology*, **89**, 1546–1555.

Chapter 33

IPM for invasive species

Robert C. Venette and Robert L. Koch

Invasive species present major challenges to the sustainability of biologically based economic systems, including apiculture, forestry, agriculture, horticulture and aquaculture, and the preservation of valued plants and animals (Liebhold *et al.*, 1995; Williamson, 1996; Mack *et al.*, 2000, 2002; Simberloff *et al.*, 2005). In the USA, Executive Order 13112 defines an invasive species as any organism that is not native to a particular ecosystem and is causing or is likely to cause economic or environmental harm, or harm to human health (The White House, 1999). Invasive species cause these harms by directly consuming or infecting resident species, out competing resident species for resources, hybridizing with resident species, altering disturbance regimes, vectoring native or exotic pathogens and/or creating refuges for other pests (Mack *et al.*, 2000). In production-oriented systems, infection, predation and competition are often the primary concerns (Pimentel *et al.*, 2000).

The White House's definition of an invasive species is useful but assumes that the historic composition of ecosystems is well known and that harm can be objectively described. Plants, insects, vertebrates and certain fungi that had not been reported in the USA can be classified as non-native with some confidence. However, the nativity of microflora and microfauna can be more difficult to determine (Kurdyla *et al.*, 1995). Additionally, a species may be exotic to an ecosystem but still native to that country. Why does the designation of a species as a native or exotic matter? The designation can affect who is responsible for management, options that are available and consequences of failed management.

All invasions progress through the phases of arrival, establishment, integration and spread (e.g. Liebhold *et al.*, 1995; Williamson, 1996; Venette & Carey, 1998; Poland & Haack, 2003). Arrival occurs when a species overcomes natural barriers to spread, often as an intentional or accidental result of human activity, and moves into an area outside of its historical range. Opportunities for arrival depend on the pathways with which a species might be associated. For example, cargo and international travelers are major pathways for the movement of species (Liebhold *et al.*, 2006; McCullough *et al.*, 2006). Establishment occurs when the newly arrived species maintains a population through local reproduction. Establishment depends on the availability of a suitable climate, nutrient source, reproductive partner (if needed) and appropriate developmental cues. At the same time, the species must avoid generalist predators and pathogens. Consequently, the odds of successful establishment improve as the number of individuals in the founding population increase. Integration (a.k.a. naturalization) occurs as the invading population adapts to its new environment and resident taxa adapt to the

Integrated Pest Management, ed. Edward B. Radcliffe, William D. Hutchison and Rafael E. Cancelado. Published by Cambridge University Press.

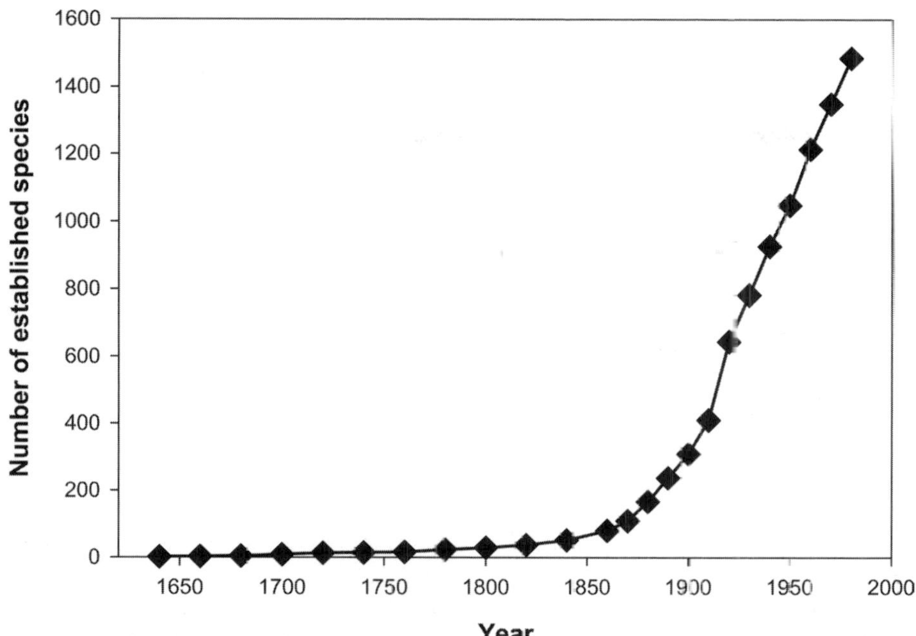

Fig. 33.1 Cumulative number of exotic insect and mite species that were established in the conterminous USA over time (Sailer, 1983).

invader. Finally, spread describes the redistribution of a species within its new geologic boundaries. Spread depends on a species' reproduction and movement of individuals by active or passive means (Liebhold *et al.*, 1995; Williamson, 1996). Williamson (1996) suggests that 10% of intentionally introduced species will escape; of these, 10% will establish breeding populations in the wild, and of these, 10% will achieve pest status.

Historically, management of invasive species has been pursued on a case-by-case basis (e.g. Liebhold *et al.*, 1995; Poland & Haack, 2003). These efforts have been worthwhile but have proven remarkably expensive. Given the rate at which new species appear to be arriving (Fig. 33.1), the case-by-case approach may not be sustainable over the long run. The fundamental philosophy of IPM is to strategically use physical, chemical, cultural and biological methods to minimize pest damage and meet long-term economic, environmental and social goals. Consistent with this philosophy, several scientists have suggested the importance of an ecosystem approach to invasive species management with component strategies working as an integrated whole, not as successive, fallback alternatives when one strategy fails (Baker *et al.*, 2005; Hulme, 2006). The four component strategies are prevention, eradication, suppression, and restoration, which generally correspond to the primary phases of the invasion process (Brockerhoff *et al.*, 2006; Hulme, 2006) (Table 33.1).

A basic premise of this chapter is that management of invasive species is not fundamentally different from IPM for well-established pests. What is different is the number of pests that must be considered, the range of tactics that are available, the number of managers/organizations that are involved and the scale at which management must operate. In this chapter, we begin by exploring the current and future magnitude of the problem. We then clarify our definition of "management" as it relates to invasive species and discuss the challenges of selecting a management objective. Finally, we describe the goals of each component strategy in more detail and outline major tools to achieve those goals. Because many dimensions of suppression of invasive species and restoration of affected ecosystems are nearly identical to those for established pests (discussed elsewhere in this volume), we focus most heavily on prevention and eradication. Sampling, a cornerstone of IPM receives special attention in each section.

Table 33.1 Relationship between the phases of a biological invasion and management strategies for invasive species

Phase	Management strategy	Goal for future[a]	Purpose of sampling	Responsible for management decisions
Arrival	Prevention	$N_t = 0; N_{t+1} = 0$	Detection	National government Industry
Establishment	Eradication	$N_t > 0; N_{t+1} = 0$	Detection Delimitation	National, state and tribal government Universities NGOs[c]
Integration and spread	Suppression and/or restoration	$N_t > 0; N_{t+1} < \text{EIL}^b$	Enumeration	National, state, tribal and local government Universities NGOs Landowners Industry

[a] The "future" is represented by time, $t + 1$. Thus, N_t is the "current" size of the invading population and N_{t+1} is the desired size of the population in the future.
[b] Economic injury level.
[c] Non-governmental organizations.

33.1 The magnitude of the problem

From 1882 to 1912, about eight non-native insect species became established in the USA each year on average (Mathys & Baker, 1980). From 1912 to 1972, the rate increased to about ten insect species per year, but given the rate of trade expansion, even more species should have been established. Recent estimates of the number of non-indigenous species in the USA range from ~6500 (Williams & Meffe, 1998) to more than 50 000 (Pimentel *et al.*, 2000). Of these, approximately 500 plant and 1000 arthropod species are pests in agriculture (Pimentel *et al.*, 2000) and >360 insects and 30 pathogens are harmful in forestry (Liebhold *et al.*, 1995). Historical invasions in North America by chestnut blight (causal agent *Cryphonectria parasitica*), European corn borer (*Ostrinia nubilalis*), gypsy moth (*Lymantria dispar*), pink bollworm (*Pectinophora gossypiella*) and spotted knapweed (*Centaurea maculosa*) and more recent invasions by emerald ash borer (*Agrilus planipennis*), garlic mustard (*Alliaria petiolata*), multicolored Asian lady beetle (*Harmonia axyridis*), soybean aphid (*Aphis glycines*) and soybean rust (*Phakopsora pachyrhizi*) provide important reminders of the potential economic and environmental consequences of pest invasion (Marlatt, 1921; Liebhold *et al.*, 1995; Poland & Haack, 2003; Koch *et al.*, 2006b). Most major agricultural commodities and common tree species now contend with one or more exotic pests. Furthermore, many exotic species, such as the multicolored Asian lady beetle, show little site fidelity and are active in fields, vineyards, orchards and forests (Koch *et al.*, 2006b).

The number of invasive species will continue to increase. International trade is a primary culprit for the accidental introduction of species to new areas. Pathway analyses repeatedly reveal numerous species that are, or could be, associated with particular commodities or conveyances. For example, a positive correlation exists between the volume of imports a country receives and the number of invasive species within that country (Perrings *et al.*, 2005). The frequency of introduction of individual species is correlated with the volume and frequency of trade in particular commodities. In the USA, 8–273 species could have become established per year between 1997 and 2001 from four kinds of imported cargo, if we can assume 2–65% of all arriving species establish (Work *et al.*, 2005). The consequence is that a manager must remain mindful of resident pests *and* species that are not yet harmful in his/her area when developing long-term pest management strategies.

33.2 An overview of management for invasive species

"Management" refers to any action that intentionally affects the presence, density, distribution or impact of a species. The overall objective of invasive species management is to preserve or reintroduce resident species that have been or may be affected by exotic species in order to obtain the maximum economic, environmental and social benefit(s) possible from an ecosystem (Hulme, 2006). In the simplest case, benefits refer to the monetary value of goods or services provided by an (agro-)ecosystem (Kremen & Ostfeld, 2005). For example, goods may be the volume of grain harvested from a particular area of land and services may save energy costs from the shade provided by a tree. Damage relates to the monetary losses that ensue from lost production, lost value (marketability) or management costs (Pimentel *et al.*, 2000). Benefits and costs are relatively easy to define because established markets determine monetary worth.

Ecosystems also provide benefits through services that are not traded in markets. Some of these services are tangible such as providing clean air or water (Kremen & Ostfeld, 2005), but others are more intangible, such as the emotional benefit provided by a cherished animal or plant. For example, trees may be planted to commemorate significant personal events, and when agents threaten theses species, loss is felt (Schroeder, 1996). Although the value of a species or service is not necessarily reflected by markets, methods exist to express non-market worth in economic terms.

Costs and benefits can be a matter of perspective, especially for species like ornamental plants that were intentionally introduced but escaped (Gobster, 2005). For example, invasive species have formed dense thickets ("green screens"), provided habitat or food for valued species, inhibited erosion and lowered insect densities. Each of these outcomes might be perceived as a benefit or detriment (Foster & Sandberg, 2004). Invasive species management has been criticized for using language that only portrays exotic species as detrimental (Gobster, 2005) and for failing to consider perceived benefits from an invasion (Leung *et al.*, 2005). As a result, both the benefits and costs posed by an invasive species for market and non-market goods and services should be incorporated into any analysis of the socially optimal investment in invasive species management (Leung *et al.*, 2005).

33.3 Prevention

33.3.1 Goals

Stopping the arrival of new species is considered the least expensive and most effective method for managing invasive species (Wittenberg & Cock, 2005; Hulme, 2006). Effective prevention rests on the legal authority to block the arrival of a species but equally depends on how society perceives the relative costs and benefits of activities that might lead to new arrivals (Kahn, 1991). Conceptually, the most effective strategy would be to prohibit *all* activities (trade, travel and transit) that could lead to the arrival of new pests. This scenario should quickly be seen as ridiculous because most societies recognize some benefit from trade, immigration and freedom of movement. Immigration and trade policy are often highly controversial for reasons that are unrelated to exotic invasive species, and we do not intend to enter the debate explicitly or implicitly. We simply recognize that the social challenge is to permit movement of goods and people as much as possible but to stop specific cases when potential harm is recognized (Kahn, 1991).

Potential harms from invasive species have been recognized for at least 130 years, and debates over costs and benefits of efforts to manage invasive species have been common. Through the late nineteenth century, the USA gave little thought to harm that might be associated with the intentional introduction of new species and widely encouraged the importation of new plants for the betterment of agriculture (Henstridge, 2002). The invasion of Germany by Colorado potato beetle (*Leptinotarsa decemlineata*) in 1873 led that country to adopt some of the world's first, noteworthy quarantine regulations (Mathys & Baker, 1980) and stimulated awareness of potential impacts to agriculture from pest invasion. Invasions by grape phylloxera (*Daktulosphaira vitifoliae*) and San Jose scale (*Quadraspidiotus perniciosus*) further encouraged Europe to adopt stricter quarantine regulations against plant materials from the USA (Marlatt, 1921). By 1905, the USA recognized risks associated with insects that might attack fruits, vegetables, field crops and trees of all kinds and passed the Insect Pest Act (Remington, 1949) but still maintained liberal plant importation policies (Henstridge, 2002). As Marlatt (1921) described for the period until 1912, "Not only could plants be imported by nursery and florist establishments without regard to their freedom from pests, but in the absence of any protective legislation, America became a dumping ground for the plant refuse of other countries." With the passage of the Plant Quarantine Act of 1912 authority was granted to regulate the movement of plants or plant parts that might carry pests. Soon thereafter claims arose that the Act was merely a tariff, especially against bulbs from Europe, and that plants should only be prohibited when proof of infestation was available (Anonymous, 1925). Smith *et al.* (1933) summarize common perspectives about plant quarantine through the early twentieth century:

> The attitude of individuals, groups, or classes of people toward plant quarantines depends primarily on the direct or immediate manner in which such quarantines seem to affect them. The tourist, stopped for inspection of his baggage, feels immediately that he is inconvenienced and delayed. If this happens to be his only contact with quarantines and if he knows nothing about the purpose and other probable effects, he may conclude that plant quarantines are some form of graft and ought to be abolished. ... Some taxpayers may look only at the governmental expenditures for enforcing and maintaining plant-quarantine regulation. To them taxes seem high, and they may feel that plant quarantines constitute an item of public expenditure that they are helping to support without receiving in return any tangible benefits. Growers as a class usually favor plant quarantines, because they have had unfavorable experiences with plant pests and diseases. They know that their crops are in constant danger of damage and destruction. ... However, the benefits would be smaller if the costs of the quarantines were assessed against the crops instead of being paid from general tax funds. ... A few growers may object to plant quarantines because they think such regulations are futile and expensive. ... People in other states may object to California plant quarantines because it appears to them that such regulations exclude their products from California markets. To them, the California quarantines may appear to be only trade barriers which ought to be removed. On this account, they may foster retaliatory measures of any kind to exclude California products from their states.

More than 70 years later, attitudes about domestic and international quarantines have not changed substantially (e.g. Margolis, 2004; Perrings *et al.*, 2005).

The World Trade Organization's Agreement on the Application of Sanitary and Phytosanitary Measures (SPS Agreement) recognizes costs and benefits from international trade and attempts to establish consistent standards under which signatories can justifiably apply phytosanitary measures to prevent pest invasions. The SPS Agreement emphasizes the importance of scientific evidence and pest risk analysis, or more precisely pest risk assessment, as the appropriate tool to justify plant or animal quarantines. Under international guidelines, pest risk assessments are only applicable to potential quarantine pests, which are "pest[s] of potential economic importance to the area endangered thereby and not yet present there, or present but not widely distributed and being officially controlled" (Integrated Plant Protection Center, 2006). For example, in the USA, most quarantine pests would be species that do not yet occur in this country. Although European gypsy moth is well established in the USA, it is a quarantine pest because it is under official control. Effective prevention relies on identifying species that are likely to establish and cause harm before those species arrive.

33.3.2 Tools

Pest risk assessments are used to forecast the probability and consequences of successful invasion by a species that is considered a potential hazard. Standard frameworks exist to guide this process (reviewed by Baker *et al.*, 2005). These frameworks consistently recognize climate suitability, host breadth and spread potential as major factors that affect both probability and consequence of invasion. Potential environmental impacts and the value of affected commodities influence the predicted severity of consequences. The probability of invasion depends on details about the pathways with which a species might be associated (Orr *et al.*, 1993; Venette & Gould, 2006).

A pathway describes the means by which an organism might move from location A to location B. Critical details include the conveyance (e.g. airplane, ship or passenger baggage), associated plant material and conditions during transport (e.g. modified atmosphere). For example, asparagus exported from Peru to the USA by ship represents a different pathway for the introduction of *Copitarsia* spp. than asparagus sent by plane (Venette & Gould, 2006). The number of pathways, the volume of material moved by each pathway, in-field production practices and post-harvest processing all affect the number of viable individuals that might arrive and thus the probability of successful invasion.

As typically conducted, pathway risk assessments are not much different from pest risk assessments. In brief, for a specified pathway, a complete list of species that might be associated with the pathway (e.g. commodity or ballast) is prepared. Potential quarantine pests (i.e. not present in the country endangered thereby or under official control) that may be transported are identified. A tandem series of pest risk assessments is then prepared, one assessment for each pest (Orr *et al.*, 1993; US Department of Agriculture, 2000; Orr, 2003).

Pest risk assessments are not a panacea for the prevention of new arrivals. Simberloff (2005) questions the value of pest risk assessments given the high degree of uncertainty about the biology of a pest organism in a novel environment and vagaries of qualitative pest risk assessments. However, recent software advances are simplifying elements of the process for quantitative assessments. For example, Koch *et al.* (2006a) used a quantitative risk assessment to demonstrate that the multicolored Asian lady beetle could have a strong negative impact on monarch butterflies. Hulme (2006) expressed concern that quantitative or qualitative risk assessments may not be feasible given the large number of species to be evaluated and the expense associated with each assessment. Indeed, it is difficult to determine whether pest risk assessments and associated quarantines were effective in preventing invasions (Mathys & Baker, 1980). Smith *et al.* (1933) state that the evaluation of plant quarantines "does not lend itself to study by the usual experimental method, and therefore conclusions in regard to it must be largely, if not entirely, of a theoretical nature." If pest risk assessments provide useful information to managers and help prevent species from being shipped to a country, the number of interceptions of targeted species should decline. Information about the number and volume of shipments arriving at each port and the level of inspection effort is needed to confirm that fewer interceptions are not simply because officials quit looking. Mathys & Baker (1980) also propose the utility of comparing the number of established species over a specified time interval to the expected number of introductions after accounting for the volume of trade.

The role of sampling in prevention efforts is to detect at least one specimen of a quarantine pest when it is present in a shipment, technically before the species arrives in the new environment. However, the vast volume of cargo and number of airline passengers that arrive in a country often preclude careful inspection (Mathys & Baker, 1980). For example, only about 2% of the material in a cargo shipment can be inspected, and with such a limited sample, the likelihood of detecting pests is low, unless much of the shipment is infested (Venette *et al.*, 2002). As a result, port of entry inspections do not directly prevent invasions from occurring but help to confirm that quarantine regulations are being honored (Nyrop, 1995). When coupled with stiff penalties, inspections may do more good by deterring others from intentionally or accidentally bringing prohibited species or articles. Inspections at ports of entry will benefit from new, molecular methods to assist with species diagnosis (Brockerhoff *et al.*, 2006).

A number of tactics complement port of entry inspections and work to prevent pest arrival. Inspections in the country of origin, so-called preclearance programs, exist for a number of commodities and have been highly effective at preventing pest invasion (Mathys & Baker, 1980). Alternatively, commodity treatments at the port of entry can disinfest goods (Mathys & Baker, 1980), but treatments for insects may not work against plant pathogens (Kahn, 1991). Post-entry quarantine is routinely used for propagative material (Mathys & Baker, 1980). In this case, propagative material is disinfested of insects using cold, heat or fumigants, for example. Plants are then grown in isolation and tested for multiple pathogens over time. If pathogens are found, the material is destroyed or, if practical, cleaned using aseptic tissue culture.

33.4 Eradication

33.4.1 Goals

The primary goal of eradication is to eliminate every individual from an invading population, or at least lower pest densities to a point where population extinction from natural causes is highly likely (Myers *et al.* 2000). The success of this strategy depends, in part, on early detection of the newly arrived population and the ability to kill those individuals when they are found (Simberloff, 2003). Success rates for eradication programs are difficult to estimate but may be between 45% and 50% (Myers *et al.* 1998). Eradication of screwworm (*Cochliomyia hominivorax*) from the southeastern USA and the malaria mosquito *Anopheles gambiae* from northeastern Brazil are examples of successful insect eradication (Myers *et al.*, 1998).

Eradication is a controversial management goal because the short-term costs can be high and certainty of success is moderate at best. For true eradication to be successful, Myers *et al.* (2000) suggest that six criteria must be met: (1) adequate funding must be available to support an eradication program to the end, (2) a management agency must have clear lines of authority and be able to take decisive action, (3) adequate techniques must exist to detect the targeted species at low densities, (4) methods must be in place to prevent reinvasion, (5) the biology of the targeted species must be well understood and lend itself to control with available methods and (6) accompanying restoration efforts may be needed to prevent other exotic species from replacing the target pest. Combinations of available methods (e.g. sterile insect technique, broad-spectrum pesticides, biocides, burning or pulling) may be needed to remove every invading individual. Native resident species may also be eliminated locally, but the hope is that the environmental costs will be isolated, short-term, and more than offset by the benefits from preventing the spread and impact from the invader (Simberloff, 2002). One interpretation of the precautionary principle supports this philosophy (Hulme, 2006). However, to be effective, the invading population must be small, confined to a manageable area and recognized as a threat when it is found (Hulme, 2006). Moreover, public support, achievable through discourse between program managers and local citizens, is essential for success (Myers *et al.*, 1998).

"Early detection and rapid response" is occasionally presented as a distinct, intermediate activity between prevention and eradication (Wittenberg & Cock, 2005). In fact, early detection is a prerequisite to eradication but, by itself, is not a form of management. Management comes from the rapid response. For an early-detection, rapid-response program to be effective, careful thought must be given to what species or suite of species are being sought, how likely the program is to detect one or more of these species and what the rapid response might be if a targeted species is found. Without carefully considering which species are being targeted and how likely they are to be detected, early detection becomes difficult to distinguish from routine monitoring (Wittenberg & Cock, 2005).

33.4.2 Tools

Sampling for early detection focuses on determining whether a species is present and likely to become established within an area of interest, often a nation or state. Ironically, a detection survey cannot prove that a species is absent unless field personnel can (1) inspect every possible location where a target organism might be found, (2) reliably locate the target and (3) accurately identify the target when it is present (Venette *et al.*, 2002). Such surveys based on a complete census are only feasible in relatively small areas that attempt to locate distinctive, accessible species. More commonly, detection surveys rely on sampling a subset of sites where the species might occur (Venette *et al.*, 2002). Companion tools for detection surveys (e.g. baited traps, molecular diagnostic aids, taxonomic guides) are essential, but these tools will not be discussed here.

Once a target pest has been selected, the engineer of a detection survey should carefully describe the sample unit and the sampling universe. The sample unit is a biologically relevant "thing" that is likely to contain the species of interest. For example, a sample unit might be an area or volume of land, a whole plant or animal (as when looking for parasitic insects), a natural part of a plant or animal (e.g. a leaf or horn), or a mass of tissue. Alternatively, if a trapping device will be used, the sample unit would be the effective trapping area of the trap. The sample universe represents the complete set of sample units that have some chance (i.e. >0) of being selected for inspection. More importantly, the sample universe sets boundaries for the population that is being characterized.

The chance that an invading pest will be found during a survey depends on the size of the sample universe, the number of "infested" sample units, the number of sample units that will be inspected (also known as the sample size) and the sensitivity of the sampling method (reviewed in Venette *et al.*, 2002). Sensitivity refers to the probability that a diagnostic tool (e.g. visual inspection, baited trap or ELISA test) will accurately reveal a pest when it is present in/on a sample unit. If the sample size is large relative to the size of the sampling universe, hypergeometric statistics may be used to calculate the sample size needed to detect a pest with a desired degree of statistical confidence during random sampling (reviewed in Venette *et al.*, 2002). More often the sample size is small relative to the sample universe, and in this case binomial statistics can be used to calculate an appropriate sample size. Briefly, the probability of detecting a pest, $P[X > 0]$, is

$$P[X > 0] = 1 - [1 - (f S_e)]^n$$

where f is the relative frequency of infestation in a sample universe, n is the sample size and S_e is the sensitivity of the diagnostic tool. In most cases, f is not known but is set based on the goals of the survey. The probability of detection increases as more sample units become infested, as the sample size increases or as the sensitivity of the sampling method improves. Early detection surveys have been criticized because the probability of detection is quite low, given the expected proportion of infested sample units, the limited sample size and the limited sensitivity of many diagnostic tools (Hulme, 2006). Sample size is typically dictated by available resources.

As an alternative to purely random sampling, the sampling universe may be stratified into zones where the invader is more or less likely to occur. Zones may be established based on regional differences in climate, host plants and trade statistics among other factors. Studies to support pest risk assessment often provide explicit predictions of areas that are likely to provide suitable climate and/or hosts (e.g. Venette & Hutchison, 1999; Koch *et al.*, 2006b; Venette & Cohen, 2006). Such information can be extremely helpful for early detection efforts. Samples are still collected at random within each zone, but the sample size within each zone is adjusted based on the likelihood that the pest will be present. For this approach to be effective, reliable estimates of the anticipated frequency of infestation in each zone are needed. If reliable, the stratification can substantially increase the probability of detecting an invading population early, even if the sample size is relatively small.

Once an invading population is detected, any number of techniques may be used, but the goal for each is the same: kill all the individuals in the population or prevent them from reproducing. The first step is to delimit the extent of the infestation by conducting more intensive, detection-based surveys in and around areas where initial finds were made. To extirpate the population, chopping, pulling, spraying, burning, fumigating and other techniques have all been used with varying degrees of success (Hulme, 2006). Potential for success is largely influenced by the efficacy of the treatment (i.e. the number of invading individuals that can be removed or the extent to which reproduction can be reduced). Factors such as the size of the area to be treated, the complexity of the terrain, the method of treatment application, the number of applications and the toxicity of a pesticide affect the efficacy of a treatment option. Population models play a useful role to identify critical life stages to target and the extent of mortality that is needed to drive the population to extinction (Hulme, 2006).

33.5 Suppression

33.5.1 Goals

Suppression is typically the management strategy of choice for species that have integrated into an ecosystem, at least partially, and started to spread. At this point in the invasion process, the invading species is capable of withstanding at least some of the local variation in weather, can recognize suitable hosts and is synchronized temporally with the availability of those hosts (Brown, 1993). Strong selection pressure may drive genetic changes that underlie adaptations to local conditions (Wares *et al.*, 2005). Integrated pests can outbreak and cause severe damage, though later in the integration process, resident natural enemies may respond and more tightly regulate population cycles (Brown, 1993).

The goal of suppression is to keep the size of the pest population at or below acceptable levels using physical, chemical, cultural or biological methods (Mack *et al.*, 2000). In agriculture, acceptable levels typically relate to an economic injury level (EIL). As traditionally calculated, the EIL depends on the expense and efficacy of treatment, the market value of the commodity that is being protected, and the relationship between the injury caused by a pest and the resulting damage said injury causes, measured in lost production. As described early in this chapter, costs and benefits for the control of invasive species can be difficult to estimate, especially in natural settings. Ill-defined relationships between pest injury and lost production in forestry have limited the utility of the EIL concept in this setting.

We consider containment a special case of suppression. With containment the objective is to keep a pest population within a certain geographic range. Island populations outside the prescribed area are treated intensively with the goal of eradicating those localized infestations (Sharov *et al.*, 2002). The core population may be treated, but theoretical work indicates that when resources are limited, treatment of island populations is more effective at containing the population than treating the core (Hulme, 2006). However, in practice, perfect containment of most insects and pathogens has proven difficult, if not impossible, to achieve. Better success has been attained by slowing the spread of invading populations with the intent of keeping as many areas below damaging levels for as long as possible (Sharov *et al.*, 2002).

33.5.2 Tools

Sampling to support suppression decisions for invasive species does not differ from similar sampling for resident pests. In general, some form of binomial or enumerative sampling is used to estimate the extent of infestation or the density of the pest population. Such sampling is used to determine if local suppression activity is needed and, following such action, to verify the effectiveness of the treatment. (See Chapter 7 for a more complete discussion of relevant sampling procedures.)

Physical and chemical controls are widely used in production agriculture to lower densities of pests but these methods can become impractical in natural settings (Mack *et al.*, 2000). For insects, mating disruption by dispensing artificially produced sex pheromones over a landscape will lower the chances that a female will be mated and able to reproduce. Mating disruption has proven effective in natural and agricultural ecosystems (El-Sayed *et al.*, 2006). Biological control is also an attractive alternative. Although some biological control projects have led to unanticipated, negative effects on non-target species, more recent releases have a better safety record which is likely the result of stricter pre-release testing procedures (Messing & Wright, 2006). Stricter procedures for testing candidate biological controls help to ensure that the likely benefits outweigh potential adverse risks (Heimpel *et al.*, 2004). Cultural controls also play a role. In forestry, removing vulnerable trees, planting stock with resistance to pests and adjusting the stand age distribution are important silvicultural practices that can minimize the impacts from invasive pests (Waring & O'Hara, 2005).

33.6 Restoration

33.6.1 Goals

The goal of restoration is to return ecosystem structure and function to a state that is roughly equivalent to the condition prior to invasion. Typically, restoration focuses on plants or vertebrates. Somewhat like IPM, where the goal is to use a suite of complementary tactics to prevent pest populations from causing unacceptable damage, restoration addresses the case where unacceptable damage has already occurred and now a suite of tactics will be used to recover from that damage. Part of the management plan will deal with suppressing pest populations, so called "negative" biodiversity, and the other part of the plan will deal with the valued species or function, so called "positive" biodiversity (Shine *et al.*, 2005). In extreme cases, restoration may require the reintroduction of species that were eliminated by an invader.

Restoration can be substantially more complex if an invasive species is not the proximate cause of change in an ecosystem. Other agents such as global warming, nutrient deposition or acid rain may be the real cause, and these agents may directly benefit an invader (Hulme, 2006). Under these circumstances, although the density of the invading population may be increasing and a valued species may be declining, removal of the invasive species alone may not lead to ecosystem recovery. The removal of one invasive species may simply allow another species to invade (Zavaleta *et al.*, 2001). Consequently, restoration efforts must take a comprehensive ecosystem approach (Masters & Sheley, 2001).

33.6.2 Tools

Sampling for restoration does not differ substantially from aspects of sampling for IPM or ecological applications. Generally, the goal is to measure the density, distribution and condition of valued species and/or to measure the diversity of native and exotic species within an area.

Appropriate plant materials are needed for restoration to succeed and include species that were present prior to an invasion (ideally with the same genetic background), selections of pre-existing species with resistance to the invader, or alternate plant species that provide the same function as a former resident. For example, efforts to restore American elm have relied heavily on searches for resistance to the causal agents of Dutch elm disease (*Ophiostoma ulmi* and *O. novo-ulmi*) (Smalley and Guries, 1993).

33.7 Conclusions

Management of invasive species must begin with a clear definition of goals for a system. Most prevention and eradication programs are designed and operated by national or regional governments. As a result, complex decisions about which species to exclude or which species to eradicate must reflect broader societal values. Decisions to suppress pest populations or restore degraded habitats typically fall to local governments, environmental groups or individual landowners, so setting management goals can be somewhat simpler. Biologists are well trained to make the tactical decisions to reach a goal once it has been set, but are generally less experienced in soliciting and incorporating public comment to formulate a goal.

Economists and social scientists have useful but largely untapped expertise to bear on the formulation of management goals and the allocation of resources to achieve those goals. Olson (2006) and Mehta *et al.* (2007) provide thorough reviews of relevant economic theory and specific models. Both reviews highlight the need for greater communication among terrestrial biologists and economists to ensure that model assumptions are reasonable and requisite data, though perhaps not immediately available, are ultimately attainable through additional research.

IPM for invasive species is not fundamentally different from IPM for established, resident pests. As a result, general advances in IPM will improve our ability to detect and suppress invading species. Likewise, advances in assessment, exclusion, detection and eradication for invading pests will provide tools to manage resident pests in new ways. An ultimate goal for IPM is to incorporate landscape-level information (such as interactions among agricultural and forested habitats) about multiple pests into habitat management decisions. Such a vision also applies to the integrated management of invasive pests. However, for invasive species this vision expands to incorporate global information about international trade, production practices, species composition and species dynamics. The challenge is daunting and will only be met through collaborative partnerships between landowners, universities, industry, non-governmental organizations and government agencies.

References

Anonymous (1925). Narcissus bulbs face an embargo. *New York Times*, May 10, p. W22.

Baker, R., Cannon, R., Bartlett, P. & Barker, I. (2005). Novel strategies for assessing and managing the risks posed by invasive alien species to global crop production and biodiversity. *Annals of Applied Biology*, **146**, 177–191.

Brockerhoff, E. G., Liebhold, A. M. & Jactel, H. (2006). The ecology of forest insect invasions and advances in their management. *Canadian Journal of Forest Research – Revue Canadienne de Recherche Forestière*, **36**, 263–268.

Brown, W. M. (1993). Population dynamics of invading pests: factors governing success. In *Evolution of Insect Pests*, eds. K. C. Kim & B. A. McPheron, pp. 203–218. New York: John Wiley.

El-Sayed, A. M., Suckling, D. M., Wearing, C. H. & Byers, J. A. (2006). Potential of mass trapping for long-term pest management and eradication of invasive species. *Journal of Economic Entomology*, **99**, 1550–1564.

Foster, J. & Sandberg, L. A. (2004). Friends or foe? Invasive species and public green space in Toronto. *Geographical Review*, **94**, 178–198.

Gobster, P. H. (2005). Invasive species as ecological threat: is restoration an alternative to fear-based resource management? *Ecological Restoration*, **23**, 261–270.

Heimpel, G. E., Ragsdale, D. W., Venette, R. C. *et al.* (2004). Prospects for importation biological control of the soybean aphid: anticipating potential costs and benefits. *Annals of the Entomological Society of America*, **97**, 249–258.

Henstridge, P. (2002). History of regulatory plant health in the United States. In *Invasive Arthropods in Agriculture*, eds. G. J. Hallman & C. P. Schwalbe, pp. 21–49. Enfield, NH: Science Publishers.

Hulme, P. E. (2006). Beyond control: wider implications for the management of biological invasions. *Journal of Applied Ecology*, **43**, 835–847.

Integrated Plant Protection Center (2006). *International Standards for Phytosanitary Measures No. 5: Glossary of Phytosanitary Terms.* Rome, Italy: Food and Agriculture Organization of the United Nations. Available at www.ippc.int/servlet/BinaryDownloaderServlet/184195_ISPM05_2007_E.pdf?filename=1179928883185_ISPM05_2007.pdf&refID=184195.

Kahn, R. P. (1991). Exclusion as a plant-disease control strategy. *Annual Review of Phytopathology*, **29**, 219–246.

Koch, R. L., Venette, R. C. & Hutchison, W. D. (2006a). Predicted impact of an exotic generalist predator on monarch butterfly (Lepidoptera: Nymphalidae) populations: a quantitative risk assessment. *Biological Invasions*, **8**, 1179–1193.

Koch, R. L., Venette, R. C. & Hutchison, W. D. (2006b). Invasions by *Harmonia axyridis* (Pallas) (Coleoptera: Coccinellidae) in the Western Hemisphere: implications for South America. *Neotropical Entomology*, **35**, 421–434.

Kremen, C. & Ostfeld, R. S. (2005). A call to ecologists: measuring, analyzing, and managing ecosystem services. *Frontiers in Ecology and the Environment*, **3**, 540–548.

Kurdyla, T. M., Guthrie, P. A. I., McDonald, B. A. & Appel, D. N. (1995). RFLPs in mitochondrial and nuclear DNA indicate low levels of genetic diversity in the oak wilt pathogen *Ceratocystis fagacearum*. *Current Genetics*, **27**, 373–378.

Leung, B., Finnoff, D., Shogren, J. F. & Lodge, D. (2005). Managing invasive species: rules of thumb for rapid assessment. *Ecological Economics*, **55**, 24–36.

Liebhold, A. M., MacDonald, W. L., Bergdahl, D. & Mastro, V. C. (1995). Invasions by exotic forest pests: a threat to forest ecosystems. *Forest Science Monographs*, **30**, 1–49.

Liebhold, A. M., Work, T. T., McCullough, D. G. & Cavey, J. F. (2006). Airline baggage as a pathway for alien insect species invading the United States. *American Entomologist*, **52**, 48–54.

Mack, R. N., Simberloff, D., Lonsdale, W. M. *et al.* (2000). Biotic invasions: causes, epidemiology, global consequences, and control. *Ecological Applications*, **10**, 689–710.

Mack, R. N., Barrett, S. C. H., deFur, P. L. *et al.* (2002). *Predicting Invasions of Nonindigenous Plants and Plant Pests*. Washington, DC: National Academy of Sciences.

Margolis, M. (2004). Fending off invasive species: can we draw the line without turning to trade tarrifs? *Resources*, Spring, 18–22.

Marlatt, C. L. (1921). Protecting the United States from plant pests. *National Geographic Magazine*, **40**, 205–218.

Masters, R. A. & Sheley, R. L. (2001). Principles and practices for managing rangeland invasive plants. *Journal of Range Management*, **54**, 502–517.

Mathys, G. & Baker, E. A. (1980). An appraisal of the effectiveness of quarantines. *Annual Review of Phytopathology*, **18**, 85–101.

McCullough, D. G., Work, T. T., Cavey, J. F., Liebhold, A. M. & Marshall, D. (2006). Interceptions of nonindigenous plant pests at US ports of entry and border crossings over a 17-year period. *Biological Invasions*, **8**, 611–630.

Mehta, S. V., Haight, R. G. & Homans, F. R. (2007). Decision-making under risk in invasive species management: risk management theory and applications. In *Encyclopedia of Forest Environmental Threat*. Available at www.threats.forestencyclopedia.net/ (in press).

Messing, R. H. & Wright, M. G. (2006). Biological control of invasive species: solution or pollution? *Frontiers in Ecology and the Environment*, **4**, 132–140.

Myers, J. H., Savoie, E. & van Randen, E. (1998). Eradication and pest management. *Annual Review of Entomology*, **43**, 471–491.

Myers, J. H., Simberloff, D., Kuris, A. M. & Carey, J. R. (2000). Eradication revisited: dealing with exotic species. *Trends in Ecology and Evolution*, **15**, 316–320.

Nyrop, J. P. (1995). A critique of the risk management analysis for importation of avocados from Mexico. In *Risks of Exotic Pest Introductions from Importation of Fresh Mexican Hass Avocados into the United States*, eds. J. G. Morse, R. L. Metcalf, M. L. Arpaia & R. E. Rice, pp. 89–97. Riverside, CA: University of California.

Olson, L. J. (2006). The economics of terrestrial invasive species: a review of the literature. *Agricultural and Resource Economics Review*, **35**, 178–194.

Orr, R. (2003). Generic nonindigenous aquatic organisms risk analysis review process. In *Invasive Species: Vectors and Management Strategies*, eds. G. M. Ruiz & J. M. Carlton, pp. 415–438. London: Island Press.

Orr, R. L., Cohen, S. D. & Griffin, R. L. (1993). *Generic Nonindigenous Pest Risk Assessment Process (for Estimating Pest Risk Associated with the Introduction of Nonindigenous Organisms)*. Riverdale, MD: US Department of Agriculture, Animal and Plant Health Inspection Service.

Perrings, C., Dehnen-Schmutz, K., Touza, J. & Williamson, M. (2005). How to manage biological invasions under globalization. *Trends in Ecology and Evolution*, **20**, 212–215.

Pimentel, D., Lach, L., Zuniga, R. & Morrison, D. (2000). Environmental and economic costs of nonindigenous species in the United States. *BioScience*, **50**, 53–65.

Poland, T. M. & Haack, R. A. (2003). Exotic forest insect pests and their impact on forest management. In *Proceedings of the Society of American Foresters 2002 National Convention*, pp. 132–141. Bethesda, MD: Society of American Foresters.

Remington, C. L. (1949). Official regulations for shipping live insects. *Lepidopterists' News*, **3**, 13.

Sailer, R. I. (1983). History of insect introductions. In *Exotic Plant Pests and North American Agriculture*, eds. C. L. Graham & C. L. Wilson, pp. 15–38. New York: Academic Press.

Schroeder, H. W. (1996). Ecology of the heart: understanding how people experience natural environments. In *Natural Resource Management: The Human Dimension*, ed. A. W. Ewert, pp. 13–27. Boulder, CO: Westview Press.

Sharov, A. A., Leonard, D., Liebhold, A. M., Roberts, E. A. & Dickerson, W. (2002). "Slow the Spread": a national program to contain the gypsy moth. *Journal of Forestry*, **100**, 30–35.

Shine, C., Williams, N. & Burhenne-Guilmin, F. (2005). Legal and institutional frameworks for invasive alien species. In *Invasive Alien Species: A New Synthesis*, eds. H. A. Mooney, R. N. Mack, J. A. McNeely, L. E. Neville, P. Johan Schei & J. K. Waage, pp. 233–284. London: Island Press.

Simberloff, D. (2002). Today Tiritiri Matangi, tomorrow the world! Are we aiming too low in invasives control? In *Turning the Tide: The Eradication of Invasive Species*, eds. C. R. Veitch & M. N. Clout, pp. 4–12. Cambridge, UK: IUCN Publications Services Unit.

Simberloff, D. (2003). Eradication: preventing invasions at the outset. *Weed Science*, **51**, 247–253.

Simberloff, D. (2005). The politics of assessing risk for biological invasions: the USA as a case study. *Trends in Ecology and Evolution*, **20**, 216–222.

Simberloff, D., Parker, I. M. & Windle, P. N. (2005). Introduced species policy, management, and future research needs. *Frontiers in Ecology and the Environment*, **3**, 12–20.

Smalley, E. B. & Guries, R. P. (1993). Breeding elms for resistance to Dutch elm disease. *Annual Review of Phytopathology*, **31**, 325–352.

Smith, H. S., Essig, E. O., Fawcett, H. S. *et al.* (1933). *The Efficacy and Economic Effects of Plant Quarantines in California*, Bulletin No. 553. Berkeley, CA: University of California, Agricultural Experiment Station.

The White House (1999). Executive Order 13112 of February 3, 1999: Invasive species. *Federal Register*, **64**, 6183–6186.

US Department of Agriculture (2000). *Guidelines for Pathway-Initiated Pest Risk Assessments*, ver 5.02. Washington, DC: US Department of Agriculture, Animal and Plant Health Inspection Service, Plant Protection and Quarantine. Available at www.aphis.usda.gov/ppq/pra/.

Venette, R. C. & Carey, J. R. (1998). Invasion biology: rethinking our response to alien species. *California Agriculture*, **52**, 13–17.

Venette, R. C. & Cohen, S. D. (2006). Potential climatic suitability for establishment of *Phytophthora ramorum* within the contiguous United States. *Forest Ecology and Management*, **231**, 18–26.

Venette, R. C. & Gould, J. R. (2006). A pest risk assessment for *Copitarsia* spp., insects associated with importation of commodities into the United States. *Euphytica*, **148**, 165–183.

Venette, R. C. & Hutchison, W. D. (1999). Assessing the risk of establishment by pink bollworm (Lepidoptera: Gelechiidae) in the southeastern United States. *Environmental Entomology*, **28**, 445–455.

Venette, R. C., Moon, R. D. & Hutchison, W. D. (2002). Strategies and statistics of sampling for rare individuals. *Annual Review of Entomology*, **47**, 143–174.

Wares, J. P., Hughes, A. R. & Grosberg, R. K. (2005). Mechanisms that drive evolutionary change: insights from species introductions and invasions. In *Species Invasions: Insights into Ecology, Evolution, and Biogeography*, eds. D. F. Sax, J. J. Stachowicz & S. D. Gaines, pp. 229–257. Sunderland, MA: Sinauer Associates.

Waring, K. M. & O'Hara, K. L. (2005). Silvicultural strategies in forest ecosystems affected by introduced pests. *Forest Ecology and Management*, **209**, 27–41.

Williams, J. D. & Meffe, G. K. (1998). Nonindigenous species. In *Status and Trends of the Nation's Biological Resources*, eds. M. J. Mac, P. A. Opler, C. E. Pucket-Haeker & P. D. Doran. Reston, VA: US Department of the Interior & US Geological Survey. Available at http://biology.usgs.gov/s+t/SNT/noframe/ns112.htm.

Williamson, M. (1996). *Biological Invasions*. New York: Chapman & Hall.

Wittenberg, R. & Cock, M. J. W. (2005). Best practices for the prevention and management of invasive alien species. In *Invasive Alien Species: A New Synthesis*, eds. H. A. Mooney, R. N. Mack, J. A. McNeely, L. E. Neville, P. Johan Schei & J. K. Waage, pp. 209–232 London: Island Press.

Work, T. T., McCullough, D. G., Cavey, J. F. & Komsa, R. (2005). Arrival rate of nonindigenous insect species into the United States through foreign trade. *Biological Invasions*, **7**, 323–332.

Zavaleta, E. S., Hobbs, R. J. & Mooney, H. A. (2001). Viewing invasive species removal in a whole-ecosystem context. *Trends in Ecology and Evolution*, **16**, 454–459.

Chapter 34

IPM information technology

John K. VanDyk

Today as in no time in history, the management in IPM is supported by information technology. From automated temperature sensors in the field to the desktop computer to the internet website, technology assists us in making pest management decisions. It is therefore necessary to have a framework for thinking about the information flowing through the system and the ways the information is processed. In this chapter we'll focus on a concrete example, namely, that of pest data making its way from the scout or researcher to the decision maker. Along the way we'll encounter common concepts of information flow that are equally useful in other scenarios.

34.1 Information life cycle

Information is all around us, yet we must select which information we want to record. Information has a definite life cycle. It can be recorded, analyzed, summarized, interpreted, shared and finally archived or discarded (Fig. 34.1).

34.1.1 Observations in field or laboratory

The number of insects in a trap, the percent leaf area removed, the damage done to a root system – these are all measurements or observations that represent the foundation of IPM decision making. The grower, researcher or knowledge worker uses the observations or measurements taken to make a statement or recommendation based on analysis. In doing so, the worker draws on past experience, the work of others, and the analysis of many contextual variables such as weather, local landscape, parallel observations at other locations, etc.

34.1.2 Quality control

Peer review or an editor serves as a quality control filter for the information that comes from researchers or models. Only those with the appropriate knowledge and experience are qualified to judge whether the analysis is helpful or not. In practice, when looking at information that makes it into the public eye, this quality control step may be skipped. That is, there is nothing to prevent erroneous information from being published on the internet or sent by email. Consumers of information compensate for this by placing more trust in information sources that have a good history. A good way to think about the review process is that it adds value to information. All other things being equal, reviewed information will be preferred by the information consumer.

34.1.3 Distribution system

After passing quality control, information is presented to a wider audience through print or electronically. We will focus here on electronic

Integrated Pest Management, ed. Edward B. Radcliffe, William D. Hutchison and Rafael E. Cancelado. Published by Cambridge University Press.

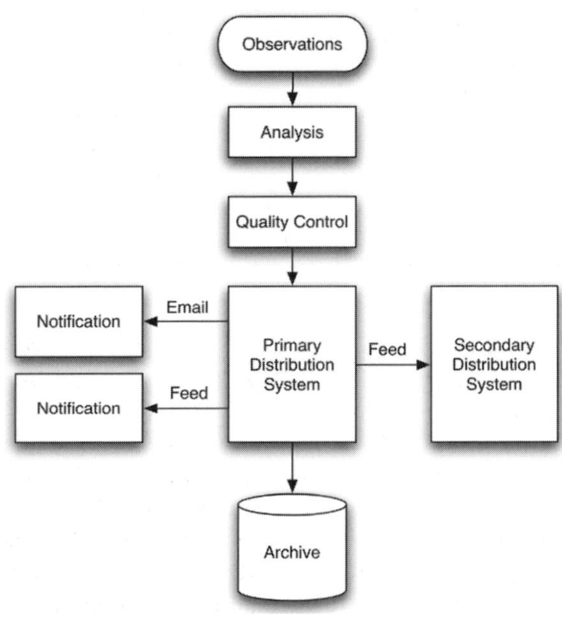

Fig. 34.1 Simplified IPM information life cycle.

information. Information is distributed outward from a trusted source to others. The primary ways to do this electronically are by electronic mail, by creating a page on the World Wide Web, or a combination of these two.

Electronic mail has the advantage of being ubiquitous, but has several disadvantages. First, because electronic mail may be forwarded from person to person, information may be lost along the way. For example, a recipient could truncate a message and remove information about the source of the message in the process. Second, because the authentic original copy of the message is not preserved (except possibly in the sender's outbox), a public reference to the original is not preserved.

Publication on the World Wide Web is fundamentally different than sending information by electronic mail. First, there is one copy, preferably at a permanent uniform resource locator (URL). Second, only the entity that controls the website on which the resource is published has access to change the information.

Primary source

The electronic version of the information being shared should have one permanent online location with a URL. It is helpful if the URL is semantically meaningful; for example the URL *http://example.org/2007/05/23/applemaggot.html* seems to denote some information on apple maggot that was published in May of 2007.

If the information is part of a continuing information series, such as a pest management newsletter, an alias may be created that points the reader to the permanent URL. Suppose a newsletter is published electronically each week. The easily bookmarkable URL *http://newsletter.example.org/latest* might be used for the most recent issue, in which summaries of the five most recent pest recommendations are available. However, a permanent and semantically meaningful URL, such as *http://newsletter.example.org/2007/05* would also be established as each issue is published, and clicking on the title of a recommendation summary would take you to that recommendation's permanent URL, such as *http://newsletter.example.org/2007/05/gypsymoth.html*.

Secondary sources

As information is posted, interested parties need to be notified. Electronic mail is one way to achieve notification, but syndicated web feeds such as RSS (Really Simple Syndication) and Atom (Atom Syndication Format) are increasingly popular. Feeds allow a computer program to check a website for updates periodically. One can think of a feed as broadcasting a channel for other websites or programs to pick up. For example, a second website may present recent headlines from the original website in a sidebar entitled *Recent News*. As new information is posted to the original website, a given headline in the *Recent News* sidebar will move down in the list of headlines and ultimately fall off the bottom of the list. Thus, secondary sources such as the sidebar used in our example point to the primary source but should be considered transient while the original source should be permanent.

34.1.4 Archiving

What happens to old information? Although information that is published at a given URL should remain there permanently so that links to the information do not break, the reality is that websites often disappear or change the URL at which the information appears. The causes of this phenomenon are numerous: political issues such as funding changes or the renaming or dissolution of the organization that originally published the information; technological issues such as moving to a different platform for serving web pages without preserving the original URLs, and simply bad planning, where pages were not given a permanent URL in the first place.

A good system for presenting information should have organized archives. Failing that, information may be picked up and stored by third-party entities, e.g. *Internet Archive* (2007) and thus remain available after a website has gone offline.

34.2 Adding value

Although information is valuable, it can be made more valuable by adding metadata, sometimes called metainformation. Metadata is data about data. Metadata makes working with information much easier because it allows data to be classified, sorted, indexed and filtered. From an information technology perspective, the usefulness of metadata rests on marriage of the relational model of data management (Codd, 1990) with semantic web concepts such as machine readability and ontology languages (Antoniou & van Harmelen, 2004). For more information about metadata, see *Understanding Metadata* (National Information Standards Organization, 2004).

34.2.1 Taxonomy

Taxonomy is the science of classification. You may be familiar with the classification of biological organisms into groups such as families, genera and species. In the same way, information can be classified, making it easier to discover and retrieve. Information is classified by assigning semantic terms from vocabularies (collections of terms within a semantic group) in a process called tagging.

34.2.2 Tagging

Tagging involves associating a certain word or phrase with a piece of information. For example, an article about the biological control of fire ants might be tagged with such terms as fire ants, *Solenopsis invicta*, decapitating fly, and biological control. Tagging is an expensive proposition, because the person creating the tag must be familiar enough with the problem domain to create correct and appropriate tags. The time of such experts is always in high demand. Typically this cost has prevented the tagging of information. Recently, however, a new approach is being tried. By opening up the tagging of information to the universe of information consumers, the expense of tagging is borne not by the agent creating the information or making it publicly available, but by those who are reading it. This requires a level of trust between information provider and consumer, and risk is involved: the risk that the information may be tagged incorrectly or inappropriately. Still, it is less expensive to review tags that others have created and correct the bad ones than to tag everything correctly in the first place.

Table 34.1 Example of vocabularies and terms

Vocabulary	Terms
Organism	fire ant, decapitating fly
Scientific name	*Solenopsis invicta, Pseudacteon tricuspis*
IPM subdiscipline	biological control

34.2.3 Vocabularies

A vocabulary is a selection of related terms. Clearly defined vocabularies are more powerful than simple tagging. The process of tagging described above involves assigning terms to an implicit vocabulary, usually a vocabulary called "Topic" or "Category." Looking more closely at the example of an article on fire ants, we can see that the terms we used to tag the article can be organized into several clearly different vocabularies, as shown in Table 34.1.

Vocabularies can be shared among separate websites if (1) there is agreement on the semantic meaning of the vocabulary and (2) terms in the vocabulary are consistent between websites.

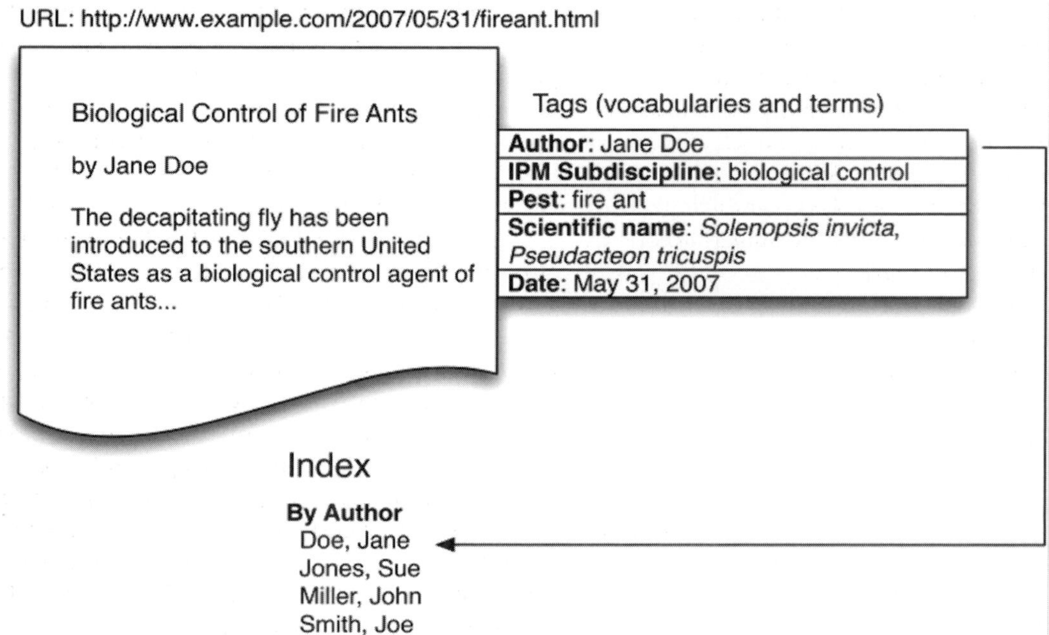

Fig. 34.2 A web page and its associated pieces of metadata, including vocabularies and terms.

34.2.4 Indexing and archiving

The major payoff of establishing terms and vocabularies is indexing; that is being able to browse through information by a vocabulary's terms instead of being limited to performing simple keyword searches. When thinking about IPM information, the vocabularies needed can be identified by asking yourself, how would I like to be able to navigate this information later? Vocabulary names will occur after the word *by*: by pest, by author, by date, by scientific name, by geographic region, etc. (Fig. 34.2).

An even larger benefit would be for providers of IPM information to adopt semantically standardized vocabularies so that information from separate providers could be merged together and browsed as one body of information. However, at the present time there is no standardized set of vocabularies for IPM information. There are several reasons for this. First, it's difficult to agree on standards because information can be classified in multiple ways. This is true even in small, clear-cut sets of information. Second, integrated pest information is by definition quite diverse. It encompasses multiple disciplines from agricultural engineering to zoology. Establishing vocabularies across disciplines is very difficult. That leaves vocabularies that are already very standardized, such as geographic information and scientific nomenclature.

34.2.5 Term relationships

Another benefit of providing metadata with information is the ability to find contextual information quickly. The algorithm for doing so is remarkably simple. If all information is tagged, we simply search through the information for other information that matches the term with which the current information is tagged. For example, we might have an article about Asian soybean rust (*Phakopsora pachyrhizi*) in an online publication dealing with plant diseases. We can then search for other articles with that same tag. If named vocabularies have been used instead of the implicit *Topic* vocabulary used by general tagging, the specificity of the related articles can be improved; that is, searching for information inside a *Plant Diseases* vocabulary for the term Asian soybean rust will give more specific information than a general search for that term.

34.2.6 Filtering

Filtering is another benefit of tagging. Filtering allows information to be discarded based on a vocabulary–term pair. For example, one might want to see all of the information on the effects of *Pythium* infection on maize. Because *Pythium* affects both maize and soybeans, a simple search for information tagged with the term *Pythium* will not be especially helpful. What would be helpful would be two vocabularies, a *Crop* vocabulary and a *Scientific Name* vocabulary. Then a compound search could be performed resulting in only information where the *Crop* vocabulary contains the term maize and the *Scientific Name* vocabulary contains the term *Pythium*.

34.3 Aggregation

Aggregation is a simple but powerful concept. Aggregation means taking data from disparate sources and merging them into one data stream. In order to join the data together, all sources of data must adhere to a common schema; i.e. a number of data fields that have the same semantics. For example, if scouting data are being aggregated from a number of sources, the same units of measurement (e.g. insects per plant versus insects per leaf) and geographic information should be used by all participants.

34.3.1 Pest monitoring

A common theme in IPM is the monitoring of insect populations across a geographic area. In the upper midwestern USA, black cutworm (*Agrotis ipsilon*) is a pest of maize that does not overwinter there but invades each year from overwintering sites in the South. The invasion is tracked by observing pheromone traps and aggregating the trap counts to get a picture of what is happening in the field; the aggregated data can then be visualized and used for predicting a timeline for damage (Rice & Pope, 2007). Thus, aggregation is a way to get an overview or summary of many separate events.

34.3.2 Feeds

In the same way that aggregation of pest monitoring data gives a summary of what is happening in the field, aggregation of IPM information itself can give a summary of what is happening across informational boundaries. This is accomplished through the use of open standards for sharing information and the software programs called *news aggregators*. Such programs are also referred to as *feed readers* or simply *aggregators*. They work by polling special files known as *feeds* provided by websites. The presence of a feed on a website is usually indicated by an icon as well as the presence of some additional code in the website's hypertext markup language. A website can provide multiple feeds. For example, the website may provide a main feed that contains updates of any page on the site; as new pages are posted they appear on this feed. For a website that provided IPM news, a new article about the presence of black cutworms in pheromone traps would appear on the feed, and those people using aggregators to track news across multiple websites would receive an indication that a new page had been posted. The notification is usually in the form of the title and a brief synopsis of the article.

If the information on the website has been properly categorized into vocabularies and terms, the website may expose feeds for individual terms. The new article on black cutworms would then appear not only in the main feed but also in the feed for the Black Cutworm term of the *Pest* vocabulary, or the *Agrotis ipsilon* (Hufnagel) term of the *Scientific Name* vocabulary.

34.4 Feedback

The information system described above offers information in itself. One can make observations about what information is being viewed most often, what information is seasonally accessed and what terms in a vocabulary garner the most attention. These observations are obtainable from access records in web server logs.

Also, the presence of terms in vocabularies can be used to assess the breadth and depth of coverage a website has. If all information has been tagged, it becomes easier to identify areas of strength and weakness in coverage. The logs generated from the use of a search feature on a website are very helpful, as they generally indicate what is being searched for. A low level of correlation between keywords being searched for in a site's search engine and the terms in the site's vocabularies may indicate that the information presented on the site is not what the site's users are expecting to find. Such observations and following analysis can be used to reorient the site to be more useful.

34.5 Conclusions

The use of information technology to obtain and manage IPM information will continue to grow. By applying the basic principles of information taxonomies such as tagging information with terms from vocabularies, filtering and aggregation, knowledge workers will have the necessary tools to become increasingly informed about the realm of IPM.

References

Antoniou, G. & van Harmelen, F. (2004). *A Semantic Web Primer*. Cambridge, MA: MIT Press.

Codd, E. F. (1990). *The Relational Model for Database Management*, v. 2. Reading, MA: Addison Wesley.

Internet Archive (2007). *Internet Archive*. San Francisco, CA: Internet Archive. Available at www.archive.org/.

National Information Standards Organization (2004). *Understanding Metadata*. Bethesda, MD: National Information Standards Organization. Available at www.niso.org/standards/resources/UnderstandingMetadata.pdf.

Rice, M. E. & Pope, R. E. (2007) *Integrated Crop Management Newsletter*. Ames, IA: Iowa State University. Available at www.ipm.iastate.edu/ipm/icm/2007/5-7/blackcutworm.html.

Chapter 35

Private-sector roles in advancing IPM adoption

Thomas A. Green

Public-sector roles in IPM-related regulation, research, education, outreach and incentives are well documented and often emphasized in the history and ongoing development of IPM. The private sector, however, is playing an increasingly key and leading role in advancing progress along the IPM continuum. This continuum begins with basic monitoring and action thresholds, and progresses towards effective systems built on biologically based, preventive approaches to avoiding pest problems (Balling, 1994). This advanced end of the continuum includes biointensive IPM, or "a systems approach to pest management based on an understanding of pest ecology. It begins with steps to accurately diagnose the nature and source of pest problems, and then relies on a range of preventive tactics and biological controls to keep pest populations within acceptable limits. Reduced-risk pesticides are used if other tactics have not been adequately effective, as a last resort and with care to minimize risk" (Benbrook *et al.*, 1996).

Benefits to advancing along the continuum and reducing the impacts of pests and pest management include improved human health and biodiversity, and conservation of air, soil and water resources. Public- and private-sector efforts and greater collaboration between sectors are needed to continue progress, reduce impacts and protect the economic viability of agriculture and communities.

Much progress has been made over the past 40 years. A prime example is the recovery of the bald eagle from fewer than 500 nesting pairs in the continental USA, due in part to the thinning of egg shells caused by ingestion of DDT, to now more than 5000. A second example is the decline in the amounts of diazinon and chlorpyrifos found in umbilical cord blood of infants, which has been correlated with impaired fetal growth (Whyatt *et al.*, 2004). Declines have been attributed to regulatory restrictions on the use of these products in structural pest management.

Recent reports documenting the need for continued improvements include a US Geological Survey review (Gilliom *et al.*, 2006) of 51 studies over ten years reporting that 96% of fish, 100% of surface water and 33% of major aquifers sampled from 1992 to 2001 contained one or more pesticides. Nearly 10% of stream sites and 1.2% of groundwater sites in agricultural areas, and 6.7% of stream and 4.8% of groundwater sites in urban areas contained pesticides at concentrations exceeding benchmarks for human health.

The adult human body is similarly contaminated with pesticides, pesticide-related compounds and other synthetic chemicals. A study led by researchers at Mount Sinai School of Medicine

Integrated Pest Management, ed. Edward B. Radcliffe, William D. Hutchison and Rafael E. Cancelado. Published by Cambridge University Press.

found an average of 91 industrial compounds, pollutants and other chemicals in the blood and urine of nine volunteers (Thornton *et al.*, 2002; Houlihan *et al.*, 2003). A total of 167 chemicals were found in these individuals, none of whom worked with chemicals occupationally or lived near industrial facilities. Of the 167 chemicals found, 17 were pesticides or pesticide breakdown products. Seventy-six were carcinogens, 94 neurotoxins and 79 developmental or reproductive toxins.

Improvements in pest management are also sorely needed. According to the World Health Organization (2005), malaria continues to kill more than 1 million annually, recently spurring the organization to recommend DDT applications to interior surfaces of living spaces as a management strategy (WHO, 2006). This recommendation was made despite studies showing DDT in breast-milk above acceptable intake limits for infants (Bouwman *et al.*, 2006) and the designation of DDT as a probable human carcinogen by the US Environmental Protection Agency (EPA).

Asthma incidence and asthma-associated morbidity is increasing in inner city children in the USA. Asthma is associated with cockroach allergen sensitivity and exposure (Gruchalla *et al.*, 2005) as well as exposure to pesticides (Salam *et al.*, 2004). Other persistent and emerging pest problems challenging pest managers in the USA and elsewhere include vectored human and animal diseases such as West Nile virus, Eastern equine encephalitis and Lyme disease; plant pests and diseases such as emerald ash borer (*Agrilus planipennis*) and soybean rust (*Phakopsora pachyrhizi*); and more than 170 noxious aquatic, terrestrial or parasitic weeds.

To address problems presented by both pesticide use and pests, pesticide users in agricultural, community and natural resource settings have joined together with public agencies, non-governmental organizations and others to develop incentives and marketplace recognition for pesticide users who transition to less hazardous practices and products. Wholesale and commercial buyers of products and services are encouraging or requiring advanced IPM from suppliers. The pesticide industry has responded to regulatory incentives with many new reduced-risk active ingredients and formulations. An agchem retailer has pioneered an innovative program in several crops that increases profits while selling less pesticide and fertilizer (see Section 35.6).

These and other private-sector initiatives and public–private partnerships stand to increase implementation, improve documentation of performance and impacts, and build consumer and taxpayer support for IPM and other conservation practices. Further, leadership roles assumed by individuals, industry and non-governmental organizations in improving economic, health and environmental impacts are much broader than pest management. New and expanding private-sector efforts involve a wide array of products and services including IPM-related nutrient management, cleaning, building construction and maintenance products and practices, as well as unrelated conservation initiatives addressing greenhouse gas emissions reduction and energy and water conservation. This chapter will survey selected IPM-related and other recent and ongoing private-sector initiatives and public–private partnerships impacting IPM adoption, principally in the USA.

35.1 Public-sector support for private-sector IPM adoption

To date, regulation has played arguably the most effective role in reducing pesticide impacts on non-targets by eliminating and restricting more toxic pesticide products and uses. However, voluntary, market-based approaches with great potential for advancing IPM adoption are both spurring and receiving greater public agency support. Specific market-based approaches have been highlighted in the EPA's annual budget requests to the US Congress since at least 2004, and EPA's Pesticide Environmental Stewardship Program has engaged, supported and reported on private-sector IPM initiatives since 1994 (Environmental Protection Agency, 2007). The *National Road Map for Integrated Pest Management* (US Department of Agriculture, 2004; Chapter 37) cites private-sector influences including consumer demand and public opinion as evidence of the need for a national IPM strategy, and public–private partnerships as key to success. The US Department of Agriculture (USDA), which has operated the National Organic Program since 2000, has formed a Market-based Environmental Stewardship Coordination Council (Johanns, 2005) and codified the Council's roles and membership in a departmental regulation (US Department of Agriculture, 2006b). The USDA Natural Resources Conservation Service (NRCS) included market-based approaches in its latest strategic plan as one of three overarching strategies (US Department of Agriculture, 2005) along with cooperative conservation and watershed approaches. Requests for funding applications from federal agencies and others are providing incentives or requiring private-sector involvement such as stakeholder statements of need, active participation in proposal development or project implementation and/or commitment of matching funds or in-kind contributions.

A growing number of public agency procurement processes for pest management services from the private sector require or state a preference for IPM providers. Since 1996, federal legislation has required that "Federal agencies shall use Integrated Pest Management techniques in carrying out pest management activities and shall promote Integrated Pest Management through procurement and regulatory policies, and other activities" under Title III, Section 303, Food Quality Protection Act, 1996 (US Congress, 1996).

The potential exists for this mechanism to deliver tremendous improvements, evidenced by a meticulously documented 89% reduction in pest complaints and 93% reduction in pesticide use in public buildings (Greene & Breisch, 2002). These reductions were accomplished by implementing IPM-based bid specifications and service contracts, expert oversight of contractors and maintenance and sanitation practices that reduce pest-conducive conditions.

A number of state and local agencies also operate under policies or regulations mandating private-sector IPM services for public facilities including Santa Clara County, California; Carrboro, North Carolina; Massachusetts; New York State; and others (Table 35.1).

In many cases, however, contractual specifications fall short of potential due to lack of resources provided for training, oversight and enforcement. Many federal contracts which are mandated to include IPM also include conflicting language specifying regular pesticide applications. For example, a 2007 request for bids for structural pest management from the University of Arizona, a leader in IPM in both agricultural and community arenas, specified regular spray applications of diazinon and chlorpyrifos – no longer legal uses for these products (P. Cardosi, government sales representative, Ecolab Pest Elimination, personal communication, 2007). Thirteen of 29 school systems evaluated were found to be non-compliant with state IPM mandates or their own internal IPM policies, including contracting for services for calendar-scheduled pesticide applications (Green *et al.*, 2007). Contracting or supervisory staff are often unaware of the mandate and/or lack the expertise to draft contracts or effective requests for bids or qualifications, or to provide adequate contractor oversight to comply with these policies and regulations.

A recent innovation in public agency IPM bid specifications, with applications to the private sector, granted a preference for structural pest management service providers who are certified by an independent third party for IPM practices (e.g. Green Shield Certified or Ecowise Certified: J. Weiss, environmental specialist, City of Palo Alto, CA, personal communication, 2007; C. Geiger, municipal toxics reduction coordinator, City and County of San Francisco, CA, personal communication, 2007; M. Siciliano, director of pest control, New York City Department of Education, personal communication, 2007).

35.2 Efforts by private-sector pest managers in agriculture and communities

Curtis (1998) documented efforts by 22 individual farm operations to implement both basic IPM practices such as monitoring and thresholds and bio-intensive IPM elements including use of cover crops, crop rotation, pheromone-mediated mating disruption, beneficial organisms, biopesticides, insect traps, cultivation and mowing.

Table 35.1 Examples of federal, state, local and institutional regulations and policies requiring or encouraging IPM by private-sector service providers

Agency or institution	Mechanism	Reference
Federal		
USA	Regulation requiring federal agencies to promote IPM through procurement policies	US Congress (1996). *Food Quality Protection Act*. Available at www.epa.gov/pesticides/regulating/laws/fqpa/gpogate.pdf.
State		
Massachusetts	Regulation requires IPM in schools and public buildings by in-house staff or contractors	Commonwealth of Massachusetts (2000). *Chapter 65 of the Acts of 2000*. Available at www.mass.gov/legis/laws/seslaw00/sl000085.htm.
New York City	Administrative code requires city employees and contractors providing services to property leased or owned by the city to eliminate highly toxic pesticide uses, post notices and keep records of applications	City of New York (2005). *Local Law 37*. Available at www.nyccouncil.info/pdf_files/bills/law05037.pdf.
Town of Carrboro, NC	Policy requires compliance with IPM policy by contractors working for the town	Town of Carrboro, North Carolina (1999). *Least Toxic Integrated Pest Management Policy*. Available at http://townofcarrboro.org/PW/ipm.htm#fourteen.
Santa Clara County, CA	Ordinance requires contractors to eliminate highly toxic pesticide uses, post notices and keep records of applications	County of Santa Clara (2002). *Division B28. Integrated Pest Management and Pesticide Use*. Available at www.sccgov.org/portal/site/ipm.
Santa Cruz City, CA	Resolution requires compliance with city IPM policy by contractors who apply pesticides to city property	Santa Cruz City (1998). *Resolution No. NS-24,067*. Available at www.ci.santa-cruz.ca.us/pw/ep/ipmpolicy.html.

Crop management systems profiled included fresh market fruit and vegetables from 11 states, plus cotton, wheat, dairy, rice, soybeans, pork and sunflowers from five additional states. Curtis's producers reported synthetic pesticide use reductions of from 0% to 100%, including several producers who transitioned to organic systems. Producers were motivated to reduce reliance on pesticides, and to reduce the toxicity of pesticide products used by economic, environmental and health concerns.

A comparable compendium of IPM approaches in the urban/community arena recently described IPM programs at 27 public school systems in 19 states (School Pesticide Reform Coalition and Beyond Pesticides, 2003). Although technically in the public sector, the collection details largely individual or small group efforts to identify and implement IPM strategies within predominantly small organizations. Many of these programs were developed prior to broader public and public agency awareness of the need for improvements in school pest management, and before how-to publications and expertise became widely available. Motivations included parent concerns about pesticides and a desire for more effective pest control. Reductions in pesticide use and pest complaints of up to 90% and 95%, respectively, were achieved by stopping calendar-based spray applications, improving sanitation, closing entry points, eliminating water sources and transitioning to insecticide bait formulations.

A four-stage program implemented by the Lodi-Woodbridge Winegrape Commission, based in Lodi, California, illustrates a producer/pest manager group approach to IPM implementation. The Commission was formed in 1991 to establish the region as a premium supplier, fund research to solve local problems and develop an areawide IPM program (Ohmart, 2006). Membership now includes 750 growers with 36 423 ha of winegrapes yielding 20% of all wine from California and a farm-gate value exceeding $US 230 million per year. The Commission employs a full-time research/IPM program director.

Ongoing outreach and education activities initiated in 1992 addressed the entire membership and included meetings, field days, research seminars and newsletters. Stage two demonstrations involved a subset of 45 growers and 63 vineyards, totaling 1052 ha. Growers and pest control advisors were trained in the use of weekly pest monitoring and action thresholds for pesticide applications and other interventions, record keeping and implementation of specific sustainable agricultural practices addressing biodiversity and air, soil and water quality. In the third stage, 255 growers worked through a detailed self-assessment including nearly 100 specific issues in soil, water and pest management, wildlife habitat and human resources (Ohmart & Matthiasson, 2000). Working in groups, often around a grower's kitchen table, participants rated their operations on a one-to-four scale for each issue, with category four on the scale describing the most sustainable condition. Growers then developed and implemented action plans for priority improvements.

The project culminated in a set of standards incorporating a list of point-weighted practices and pesticide hazard ratings. Producers worked towards a required minimum score by implementing sufficient practices and minimizing pesticide use and toxicity. Twelve vineyards and 2195 ha were certified in 2006, up from six vineyards and 589 ha in 2005.

The project has been funded by grower assessments and more than $US 1.4 million in grants from public agencies and private foundations. Performance has been measured by periodic grower surveys (Dlott & Dlott, 2005) and indicator data collection, including pesticide use and toxicity.

Structural pest management service providers have also developed advanced IPM offerings, including two by national companies. Orkin Gold Medal Protection, initially designed for food processing and healthcare facilities with low tolerance for pests or pesticide residues on human-contact surfaces, has been implemented in other environments including zoological parks (Meek *et al.*, 2006). Ecolab Pest Elimination introduced Ecolab Balance™, a high-level IPM offering for school systems, in 2007.

Additional pest-manager-led initiatives in agriculture are profiled by Thrupp (2002) and in both agricultural and community settings on the "member strategy" pages of the EPA Pesticide Environmental Stewardship Program website (Environmental Protection Agency, 2006), and in the Proceedings of the Fifth National IPM Symposium (US Department of Agriculture, 2006a).

35.3 The role of buyer–seller contractual relationships in advancing IPM for agricultural and structural pests

IPM roles for private-sector buyers of products and services have included education, technical assistance, incentives and setting specifications for products and services. Kashmanian (Environmental Protection Agency, 1998) catalogued a variety of efforts by 40 USA-based food companies to influence and reduce the health and environmental impacts of agricultural practices, including IPM and nutrient management. Buyers employed both voluntary measures such as encouragement, information, incentives and technical support as well as mandatory stipulations regarding nutrient and pesticide product selection, timing and application rates. Reported buyer and seller motivations included improved worker safety, reduced liability, less post-harvest pesticide residue on food products, delayed resistance to pesticides, lower production costs and improved crop yield and quality. Sellers are of course also motivated to meet buyer specifications to maintain the relationship.

Bolkan & Reinert (1994) had previously documented one effort referenced by Kashmanian which spurred IPM adoption by celery and carrot suppliers to Campbell Soup Company, achieving pesticide use reductions of from 30% to 60%. Campbell incorporated a unique grower assurance program providing a contractual guarantee to compensate growers should yield or quality decline under IPM practices specified by Campbell's experts. This side-by-side guarantee approach pioneered by Campbell is now widely available to sweet corn producers for nutrient and tillage best management practices (BMP) under a partnership operated by Agflex, an Iowa corporation, and the non-profits American Farmland Trust and the IPM Institute of North America and is supported by USDA and other funders (BMP CHALLENGE^{SM}, 2007).

A "second generation" of food processor, distributor and retailer stewardship has emerged since the Kashmanian review. Driven by current and anticipated growth in consumer demand, companies are incorporating IPM and other health and environmental initiatives into corporate social responsibility (CSR) portfolios that include dedicated leadership at the highest levels of management, a substantial investment in resources, and regular reporting to shareholders, consumers and the general public.

Accelerated demand for goods and services that address environmental and social causes in general is foreshadowed by a 2006 survey of USA consumers between the ages of 13 and 25. A whopping 89% of this 78-million-strong demographic reported that they were likely or very likely to switch to a brand associated with a good cause if quality and price of the product were equal (Cone, Inc., 2006).

Companies launching or expanding CSR efforts include General Mills which appointed Gene Kahn vice president of sustainable development after it acquired Kahn's Small Planet Foods in 1999. Small Planet included Cascadian Farms, the largest producer of frozen organic food in the USA at the time of the purchase. General Mills has long had IPM programs for its field production and processing plants (Environmental Protection Agency, 2006) and began releasing an annual CSR report in 2004. Other companies incorporating IPM in broader sustainability initiatives include Wal-Mart, with annual revenues exceeding $US 340 000 million. The company has invested in research and development in advanced IPM practices for structural pests for its distribution centers and retail stores (R. Corrigan, principal, RMC Consulting, personal communication, 2007) and identified IPM and organic as primary strategies to reduce impacts in the food, ornamental and fiber supply chains.

In 1999, SYSCO Corporation, an international distributor to institutional buyers with more than $US 34 000 million in 2006 sales, challenged a Peru-based asparagus supplier to reduce reliance on pesticides. IQF del Peru has documented a 90% reduction in insecticides and fungicides, and a 75% reduction in herbicide use since 1999 (Fernandini, 2006). Tactics included drawing weed-seed-free irrigation water from wells rather than river water; intensive scouting; light and visual trapping; timing of interventions to address pest immigration spurred by harvest of adjacent crops; and an on-site beneficial insectary producing lacewings, *Trichogramma*, predatory stink bugs, minute pirate bugs and assassin bugs to suppress armyworms (*Spodoptera frugiperda*), whitefly (*Bemisia tabaci*), cotton bollworm (*Helicoverpa virescens*) and thrips. The company credits cost reductions delivered by this transition from chemical to biointensive IPM for maintaining price competitiveness against imports from China.

In 2004, SYSCO initiated a survey of its suppliers to assess levels of IPM performance, convened a meeting of canned and frozen fruit and vegetable suppliers to review existing sustainable agriculture and IPM programs, and contracted with the IPM Institute of North America to produce iterative drafts of a written audit of supplier operations to be performed by independent third parties. In addition to having the largest internal quality assurance field staff in the industry, SYSCO relies heavily on third-party audits for food safety, fair treatment of workers and humane livestock management.

The final audit document included criteria for IPM and also soil quality, water and energy conservation, recycling and worker relations in both field production and processing. IPM criteria included understanding key pest biology, implementing effective monitoring, sampling and thresholds for key pests, using a combination of suppression, prevention and avoidance strategies, and tracking and reducing pesticide toxicity and use per unit of production.

The audit has now entered its third year, supplemented by an annual environmental indicators report submitted electronically by SYSCO suppliers. In 2006, the program yielded a 35% increase to 207 495 ha, 58 third-party audits with passing scores out of 88 performed, more than doubling the percent passing over the previous year, and an estimated 90 720 kg of pesticide use avoided through scouting, thresholds, weather monitoring, biocontrols and trapping (Sustainable Food Laboratory, 2007). More than 997 920 kg of fertilizer use reductions were reported due to soil testing, variable-rate application, split-application timing and other measures. Grower comments reported included the following:

> A very large percentage of our beans were swept for beetles. As a result, only 49 percent were sprayed vs. 78 percent in 2005.
>
> We had all of our planting material tested for *Phytophthora fragariae* (a highly toxic raspberry pathogen) prior to planting. We thus eliminated prophylactic usage of [fungicide, name deleted] for root rot caused by this pathogen. We used to use 907–1134 kg of this product each year; we now use none.

35.4 Eco-certification and eco-labels: informing buyers about IPM and stewardship efforts behind products and services

IPM and other grower, processor, distributor and retailer programs can be marketed to end consumers via an "eco-label" or seal signifying that a product or service has met a set of environmental or social standards. The use of eco-labels to communicate to the marketplace is growing tremendously in number, scope, producers and hectares, including those that require IPM for participation.

US sales of organic food and beverages have grown from $US 1000 million in 1990 to $US 20 000 million in 2007, according to the Organic Trade Association. Organic, one of the earliest eco-labels, is currently produced on 1 618 800 certified ha (approaching 2% of USA cropland). Organic producers may use any of hundreds of approved pesticides, nearly all of which are derived from natural products (Organic Materials Review Institute, 2007). IPM, including understanding key pest biology, monitoring/sampling, action thresholds and selection of least-toxic options, can be applied to reduce health and environmental impacts of organic production. For example, applications of copper fungicides, allowable in organic systems, can have detrimental effects on earthworm and soil microbial populations (reviewed in Bunemann *et al.*, 2006). Understanding environmental conditions that lead to fungus infections and limiting applications to favorable conditions for disease development can help mitigate those effects.

Other, less familiar eco-labels on food and fiber, cleaning, maintenance and building products have also experienced tremendous growth. These successes are evidence that wholesale buyers and consumers will throw their support behind producers and service providers who implement and document stewardship practices.

Four of several programs operating in the USA that incorporate IPM in their standards (Food Alliance, Forest Stewardship Council, Rainforest

Alliance and Protected Harvest) reported certifying a combined 10.5 million ha in the USA and 87.6 million ha worldwide in 2006 (IPM Institute of North America, 2007). These organizations lend their eco-labels to qualifying products to signify that participating producers have undergone an on-site audit to document practices including IPM.

The Forest Stewardship Council certified 8.9 million ha in the USA and 85.8 million ha worldwide. Wholesale buyers include Home Depot, Kraft and Time Warner. Rainforest Alliance's SmartWood program, headquartered in New York City and accredited by the Council, accounted for 40 million of those hectares across 58 countries. The Rainforest Alliance, founded 20 years ago, also certified 227 441 ha of food and ornamental crops outside the USA including bananas, cocoa, coffee, oranges, ferns, flowers, guava, macadamia nuts, passion fruit, pineapple and plantains. More than 15% of bananas sold in international trade now bear its eco-label.

On the USA food-product front, Food Alliance certified over 1.6 million ha of production including beef, lamb, pork, dairy products, wheat, dry beans and mushrooms and nearly 200 varieties of fruits and vegetables. Food Alliance began certifying products in 1999 and now operates out of offices in Oregon, Minnesota and California.

Protected Harvest reported certifying nearly 1620 ha of Wisconsin Healthy Grown potatoes in 2006. Since the program's inception in 2000, pesticide use has been reduced by 32% on certified fields. IPM practice adoption has increased by 26% during that interval. A fifth program, Northeast Eco Apples, promotes apples grown according to an IPM protocol as locally produced, tapping regional sentiments for keeping agriculture in the landscape. Sales will top $US 1 million in the third year of the program. Top buyers include supermarkets Trader Joes and Whole Foods Market.

Eco-labels help buyers – wholesalers or end-consumers – identify and support products addressing stewardship values beyond the primary need satisfied by the product. Producers and service providers can use eco-labels to escape the commodity designation and climb up onto the added-value shelf. Best practices for constructing and operating an eco-label have been described in detail (Dlott & Curtis, 2000). More than 130 labels are now given a thumbs up or down by Consumers Union (Consumer Reports, 2007), just like home appliances and other products.

Behind the label are overarching guiding principles – flowery language describing the ultimate aims of the program such as environmental improvement or fair trade. These are translated into specific criteria, such as pesticide use only when a documented need exists. Criteria are converted into standards which set the bar, e.g. treat only for a specific pest when a land-grant university recommended threshold has been reached. Objective evaluation tools are created for use by independent, third-party auditors who verify that the required elements are in place. Typically these involve a point-based system with a minimum score to earn the label. These program details are often available for public inspection online; see Green (2007c) for examples.

On the non-agricultural side, the US Green Building Council LEED Standards for existing buildings includes IPM as an opportunity to score points towards certification. A recent proposed revision of the standards included more detailed language describing IPM practices including landscape management (US Green Building Council, 2007). Several programs including the New England Pest Management Association's (2007) IPM Registry, the Association of Bay Area Governments (2007), and the IPM Institute of North America's Green Shield Certified (2007) now offer IPM certification for structural pest management providers.

35.5 Pesticide registrant roles in advancing IPM

In 1997, the EPA introduced its Reduced Risk Pesticide Program, which provides accelerated registration of conventional pesticide products that reduce hazards to human health and non-target organisms, and reduces potential for contamination of groundwater, surface water and other environmental resources. In addition, the EPA has a separate expedited process for biological and antimicrobial pesticides. Since 2003, the bulk of new pesticide registrations have met these criteria (Fig. 35.1). A majority of new registrations

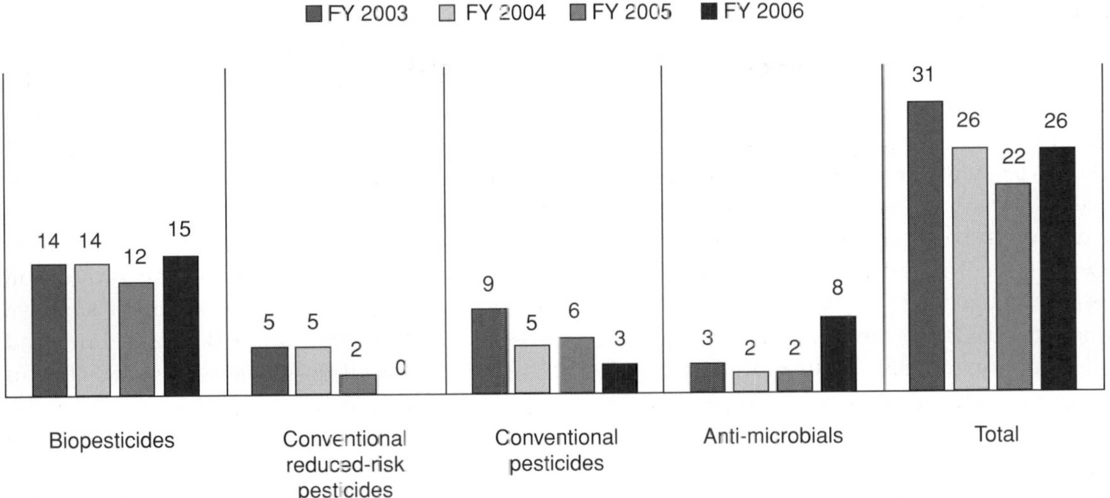

Fig. 35.1 Numbers of new pesticide registrations reported by US EPA by classification. The majority of new registrations have met EPA definitions for biopesticides or reduced risk (from Environmental Protection Agency, 2007).

are biopesticides, including naturally occurring substances, microorganisms and pesticidal substances produced by plants containing genetic material introduced specifically to control pests.

Pesticide registrants investing in research and development and committed to sustainable agriculture have a central role in continuing to improve IPM. For example, Syngenta Crop Protection has developed "AgriEdge®," a grower-customized crop production program combining traditional and reduced risk pesticides; genetically modified crops containing biopesticides; and Farm Management System, a record-keeping tool that allows growers to track production inputs, track product usage precisely using georeferencing, check for label compliance and share information easily with processors, regulators and others. Quality management system (QMS), a new technology tool currently in the pilot stage, provides lot-level tracking that consolidates production, post-harvest and control point data into one record to meet traceability requirements. In some crops, such as citrus, the decision support component of AgriEdge focuses on pest thresholds through scouting services and hotlines. Weather, product and stewardship information is accessed centrally on the Syngenta FarmAssist® website. The comprehensive service component of Syngenta's programs includes a large customer service group providing technical training and support for both the portfolio and tools, as well as Syngenta Learning Centers™ for hands-on training in the field.

In the non-agricultural arena, one of the greatest industry successes has been the advent of insecticidal baits for cockroaches. This innovation has contributed to measurable reductions in pesticide use in government-leased buildings (Greene & Breisch, 2002) and in residues in public schools (Williams *et al.*, 2005). Bait formulations are also readily available and widely used for several ant species including carpenter ants (*Camponotus* spp.), fire ants (*Sclenopsis* spp.) and termites (*Coptotermes, Reticulitermes and Heterotermes* spp.).

Industry sponsorship of meetings including the triennial IPM Symposium series (US Department of Agriculture, 2006a), rekindled in 2003 and scheduled next for 2009, has permitted participation by many, including students who would not otherwise have the resources to attend.

35.6 Consultant roles and an agchem dealer innovation

Western Farm Service, an agchem retailer based in Santa Maria, California, developed an innovative

Complete Crop Care program which guaranteed a clean, marketable crop to celery producers for a flat per-hectare fee. Any incentive to oversell inputs to increase profits is thus reversed. Retailer profits increase with fewer inputs applied and bonuses to sales agronomists can be tied to reduced input use. The program has since expanded to additional crops and geographic regions (J. Dana, product development manager, personal communication, 2007).

The Certified Crop Advisor Program was initiated in 1993 with dual goals of certifying the competency of farm advisors, and promoting and developing member professionalism. Certified membership has plateaued at about 14 000 worldwide. According to a 2005 survey of 13 837 current members, 37% or 5120 were retail agronomists supporting sales of seed, pesticides, fertilizers and other inputs; 3004 were manufacturer representatives; 2076 were self-employed; and 1998 were employed by government agencies, Cooperative Extension, or universities (American Society of Agronomy, 2005).

The National Alliance of Independent Crop Consultants (NAICC) was founded in 1978 as a membership organization for agricultural professionals providing research and consulting services to clients for a fee. Voting memberships, averaging about 360 in number, are limited to professionals receiving no compensation from a client's purchase of products including pesticides, soil amendments, seed or plant materials, equipment, animal feed or medicines. Services include advising farmers in IPM, nutrient management, watershed management, animal waste management, GIS technology, and conducting field and laboratory research for farmers and product manufacturers including efficacy trials, residue studies and environmental fate tests to support pesticide registrations with regulatory agencies.

35.7 Additional independent non-governmental organization roles

In addition to IPM programs developed by commodity-specific organizations and third-party certification, independent scientific research, education and advocacy groups provide essential support for IPM through lobbying, research reports and other publications, conferences and websites.

Examples include Beyond Pesticides which provides an information clearing-house including the *Safety Source for Pest Management*, a directory of structural and landscape pest service providers who document IPM performance by completing a written survey of practices and products used (Beyond Pesticides, 2007). The organization and its grassroots advocacy partner organizations have successfully advocated for IPM-related laws and regulations including those addressing IPM in schools.

The Natural Resources Defense Council (2007) has recently advocated for more IPM in the USA Farm Bill programs (Hamerschlag & Kaplan, 2007) and Green Buildings.

The Bio-Integral Resource Center, publisher of the monthy *IPM Practitioner*, has helped direct professionals and consumers to least toxic products since 1982. Originally a one-page *IPM Products and Services* list of 30 pheromones and a two-page *Suppliers of Beneficial Organisms* with 23 biocontrol agents, the annual *Directory of Least-Toxic Pest Control Products* has grown to more than 50 pages, 2500 products and 600 suppliers. Current print circulation is about 1500; thousands view the resource online each month (Bio-Integral Resource Center, 2007). Products listed include biocontrol agents, microbials, traps, pheromones, physical controls and monitoring devices. Chemical controls include soaps, borates, baits, insect growth regulators, oils and botanicals. Synthetic chemical control products are limited primarily to formulations that have low acute toxicity and reduced potential for non-target exposure, insecticidal baits for example.

Finally, the IPM Institute of North America, co-founded by the author in 1998, has recruited and co-led the steering committee for the IPM Symposium series since 2003, and has provided technical and scientific expertise under contract to the Food Alliance, Northeast Eco Apples, Protected Harvest, the US Army, SYSCO Corporation and others. The Institute also serves as a volunteer jury member to the Orkin Gold Medal Awards Program, recognizing Orkin commercial clients who excel in facility sanitation and maintenance, and in reducing pest problems and pesticide use.

35.8 Measuring private-sector implementation

Increasing private-sector activity is a positive sign and begs the question of impacts and return on investment. The USDA National Agricultural Statistics Service (2001) reported that private-sector adoption of IPM practices in agriculture increased from 51% of crop hectarage in 1997 to 71% in 2000, just shy of USDA's goal of 75% adoption. USDA's methodology for estimating IPM adoption had changed substantially since the time the goal was set in 1993 and been questioned by both public (US General Accountability Office, 2001) and private-sector analysts (Benbrook *et al.*, 1996). These critics asserted that the bar for counting a production unit as IPM was too low and contended that measuring against a standard based on biointensive IPM, with greater potential for reducing reliance on pesticides, would generate much lower adoption rates. In a response, USDA pointed out that although the overall use of pesticides increased during the period from 1993 to 2000, use of those pesticides identified as most hazardous by EPA declined by 14%. USDA continues to periodically survey IPM and other practices in agriculture, including documenting a 34 million kg reduction in active ingredient usage on cropland between 1997 and 2004 (Wiebe & Gollehon, 2006). No attempt has been made to attribute IPM adoption or pesticide reductions to private versus public-sector efforts.

A second thrust in adoption assessment has included crop-specific IPM measurement tools, first proposed for cotton by Boutwell & Smith (1981). This system lists available IPM techniques and assigns point values to each practice (Table 35.2). Users can measure IPM adoption by tallying the points for each practice implemented. These measurement systems, and variations on this approach, are now available for more than 75 crops and regions (compiled in Green, 2007b), and also for structural pest management service providers (Hollingsworth, 2000), schools and child-care facilities (Green, 2006) and residential turf grass and golf courses (Young, 2002; University of Massachusetts, 2007). One use of these tools has been to evaluate agricultural producers for public cost-share and incentives, including the USDA Natural Resource Conservation Service Environmental Quality Incentives Program and Conservation Security Program. A second use has been to assess producer compliance with standards for purchasing and marketing programs, covered in the preceding sections. These systems have also been adapted to measure industry-wide adoption in at least one instance, for potatoes by the National Potato Council (1998) as part of a grower survey conducted periodically since 1998.

The other side of the measurement coin is assessment of improvement in crop yields or quality, reduction in crop losses due to pests, improvements to health, reductions in pest populations in structures and landscapes or reductions in pest complaints by facility users.

The bottom line for many private- as well as public-sector programs is that direct, comprehensive measures of economic, health and environmental impacts are difficult and costly to measure and should not be expected unless sufficient resources are provided to do so. Private-sector efforts can be especially difficult to track due to privacy and proprietary concerns, and overlap with public sector efforts.

35.9 Conclusions

The following priorities for accelerating and improving the efficiency of private-sector initiatives have been identified by stakeholders (Green, 2007a) and others:

- Improved documentation of private-sector programs and impacts, including contribution of performance data to a national database.
- Harmonized, baseline standards for IPM and sustainable agriculture and traceability.
- More crop- and region-specific IPM elements and guidelines.
- User-friendly, on-line pesticide hazard ranking and selection tools.

Table 35.2 Excerpt from Elements of IPM for Fresh Market Sweet Corn in NY State (Cornell University, 2001). This evaluation tool is available for grower self-assessment and has been used to qualify production for purchase and marketing by Wegmans Food Markets, Inc. The goal represents the percentage of hectares the practice must be implemented on to earn the points indicated

	Goal (ha)	*Points*
(A) Site preparation		
(1) Review weed map/list of fields to choose appropriate weed control strategies. See the Weed Assessment List available for use in satisfying this element.	50%	10
(2) Crop rotation	75%	5
(a) Plant sweet corn only in fields where sweet corn or maize were not grown in the previous year to avoid anthracnose, smut, northern corn leaf blight		
(b) Plant sweet corn only in fields where sweet corn or maize were not grown in the previous year to avoid corn rootworms OR if rotation is not possible, scout late season fields (previous year) for presence of corn rootworm adults and only apply soil insecticides if over threshold	75%	5
(3) Soil test at least once every three years. Maintain records. Fertilize according to test results. Consider fertility contribution from cover crop, manure and other inputs	100%	5
(4) Test soil using PSNT[a] for nitrogen sidedress decisions.	1%	5
(5) Apply supplemental nitrogen based on PSNT.[a]	1%	5
(B) Planting		
(1) Use tolerant or resistant varieties whenever possible for controlling maize dwarf mosaic, common rust, smut, barley yellow dwarf, and Stewart's wilt	50%	10
(2) Seed treatments. Use fungicide and insecticide seed treatments for control of root, seed rots, and Stewart's wilt on susceptible varieties	100%	10
(3) Test the use of banded herbicide applications and cultivation in order to reduce herbicide use	1%	3
(C) Pest monitoring and forecasting		
(1) Scout as recommended for European corn borer, fall armyworm, corn earworm, flea beetles, and common rust	100%	10
(2) Update weed map/list of the field when crop is no taller than 6 inches for use in evaluating the current year's weed control and for use in determining if a post-emergent treatment is needed. See the Weed Assessment List available for use in satisfying this element	50%	10
(D) Pest management		
(1) Use recommended action thresholds for making decisions about applying pesticides for insects and diseases of importance	90%	10
(2) Choose labeled pesticides that have the least environmental impact. Choose pesticides that preserve natural enemies	35%	10
(3) Cultivate or use a post-emergent treatment based on information obtained in the weed map/list update. See the Weed Assessment List available for use in satisfying this element	90%	10
(4) Keep records of pest densities, biological control techniques used, cultural procedures and pesticide applications	100%	10

[a] PSNT, pre-sidedress nitrate soil test.

- Income protection for farmers and advisors who adopt IPM and other conservation practices which have potential to result in crop yield or quality loss.
- Effective communication vehicles and common messages to consumers and taxpayers regarding the need for continuing public support for research, extension, technical assistance and incentives.
- Greater coordination of public-sector support for private-sector programs to increase efficiency and reduce conflicting recommendations and regulations.
- Education for CEOs and other top management on environmental and health concerns, IPM and sustainable agriculture and impact measurement and reporting.
- Preferential purchasing of IPM products and services by public- and private-sector buyers.

References

Association of Bay Area Governments. (2007). Oakland, CA: ABAG. Available at www.abag.ca.gov/.

American Society of Agronomy (2005). *Certified Crop Advisor Needs and Satisfaction Survey*. Madison, WI: ASA, Certified Crop Advisor. Available at www.agronomy.org/cca/pdf/minutes/2006/feb/attachments/a5.pdf.

Balling, S. (1994). The IPM continuum. In *Constraints to the Adoption of Integrated Pest Management: Regional Producer Workshops*, ed. A. A. Sorenson, pp. 4–5. Austin, TX: National Foundation for IPM Education.

Benbrook, C., Groth, E., Halloran, J. M., Hansen, M. K. & Marquardt, S. (1996). *Pest Management at the Crossroads*. Yonkers, NY: Consumers' Union.

Beyond Pesticides (2007). *Safety Source for Pest Management*. Washington, DC: Beyond Pesticides. Available at www.beyondpestcides.org/safetysource/.

Bio-Integral Resource Center (2007). *Directory of Least-Toxic Pest Control Products*. Berkeley, CA: BIRC. Available at www.birc.org.

BMP Challenge (2007). *Take the BMP Challenge*. Madison, WI: BPM Challenge. Available at www.bmpchallenge.org/.

Bolkan, H. & Reinert, W. (1994). Developing and implementing IPM strategies to assist farmers: an industry approach. *Plant Disease*, **78**, 545–550.

Boutwell, J. L. & Smith, R. H. (1981). A new concept in evaluating integrated pest management programs *Bulletin of the Entomological Society of America*, **27**, 117–188.

Bouwman, H., Sereda, B. & Meinhardt, H. M. (2006). Simultaneous presence of DDT and pyrethroid residues in human breast milk from a malaria endemic area in South Africa. *Environmental Pollution*, **144**, 902–917.

Bunemann, E. K., Schwenke, G. D. & Van Zwieten, L. (2006). Impact of agricultural inputs on soil organisms: a review. *Australian Journal of Soil Research*, **44**, 379–406.

Cone, Inc. (2006). *Civic-Minded Millennials Prepared to Reward or Punish Companies Based on Commitment to Social Causes: National Survey Finds Millennials Steadfast Pro-Social Attitudes Drive New Rules of Engagement; Offer Business Untapped Opportunities*. Boston, MA: Cone, Inc. Available at www.coneinc.com/content1090.

Consumer Reports (2007). *Greener Choices: Eco-Label Center*. Yonkers, NY: Consumer Reports. Available at www.eco-labels.org

Cornell University (2001). *Elements of IPM for Fresh Market Sweet Corn in NY State*. Ithaca, NY: Cornell University, Cornell Cooperative Extension. Available at www.nysipm.cornell.edu/elements/fmswcorn.asp.

Curtis, J. (1998). *Fields of Change: A New Crop of American Farmers Finds Alternatives to Pesticides*. New York: Natural Resources Defense Council.

Dlott, J. & Curtis J. (2000). Eccountability and eco-labels: what eco-benefits did I buy today? White Paper presented at the *Read the Label: Understanding the Challenges and Opportunities for Eco-Labels and Eco-Brands Conference*, October 19, Portland, OR.

Dlott, J. & Dlott F. (2005). *Lodi-Woodbridge Winegrape Commission 2003 and 1998 IPM Program Grower Questionnaires: 2003 and 1998 Report of Results*. Soquel, CA: SureHarvest Sustainability Solutions. Available at www.lodiwine.com/Grower_Survey_LWWC_Final_report.pdf.

Environmental Protection Agency (1998). *Food Production and Environmental Stewardship: Examples of How Food Companies Work with Growers*. Washington, DC: US Environmental Protection Agency.

Environmental Protection Agency (2006). *Pesticide Environmental Stewardship Program: Strategy Information*. Washington, DC: US Environmental Protection Agency, Office of Pesticide Programs, Environmental Stewardship Branch. Available at www.epa.gov/oppbppd1/pesp/strategies/strategy_intro.htm

Environmental Protection Agency (2007). *Office of Pesticide Programs Annual Reports, 2006*. Washington, DC: US Environmental Protection Agency. Available at www.epa.gov/oppfead1/annual/.

Fernandini, J. (2006). An integrated pest management program for asparagus and artichokes in Peru. In *Proceedings of the 5th National IPM Symposium*,

St. Louis, MO. Washington, DC: US Department of Agriculture. Available at www.ipmcenters.org/ipmsymposiumv/sessions/8_Jorge.pdf.

Gilliom, R. J., Barbash, J. E., Crawford, C. G. *et al.* (2006). *Pesticides in the Nation's Streams and Ground Water, 1992–2001*. Washington, DC: US Department of the Interior & US Geological Survey. Available at http://pubs.usgs.gov/circ/2005/1291/pdf/circ1291.pdf.

Green Shield Certified (2007). *Green Shield Certified: Effective Pest Control – Peace of Mind*. Madison, WI: Green Shield Certified. Available at www.greenshieldcertified.org/contact.

Green, T. A. (2006). *IPM STAR Program Guide and Evaluation Form for Schools and Childcare Facilities*. Madison, WI: IPM Institute of North America. Available at www.ipminstitute.org/IPM_Star/IPM%20Star%20Evaluation%20for%20Schools%20V2%20091404.pdf.

Green, T. A. (ed.) (2007a). *Food Industry Summit/Communications in IPM Workshop*. Madison, WI: IPM Institute of North America. Available at www.ipminstitute.org/Food_Ind_Summit_IPM_Aug07.htm.

Green, T. A. (ed.) (2007b). *IPM Elements and Guidelines*. Madison, WI: IPM Institute of North America. Available at www.ipminstitute.org/Federal_Agency_Resources/IPM_elements_guidelines.htm.

Green, T. A. (ed.) (2007c). *Links to IPM Certification and Recognition Programs*. Madison, WI: IPM Institute of North America. Available at www.ipminstitute.org/links.htm.

Green, T. A., Gouge, D. H., Braband, L. A., Foss, C. R. & Graham, L. C. (2007). IPM STAR Certification for school systems: rewarding pest management excellence in schools and childcare facilities. *American Entomologist*, **53**, 150–157.

Greene, A. & Breisch, N. (2002). Measuring integrated pest management programs for public buildings. *Journal of Economic Entomology*, **95**, 1–13.

Gruchalla, R. S., Pongracic, J., Plaut, M. *et al.* (2005). Inner city asthma study: relationships among sensitivity, allergen exposure, and asthma morbidity. *Journal of Allergy and Clinical Immunology*, **115**, 478–485.

Hamerschlag, K. & Kaplan, J. (2007). *More IPM Please: How USDA Could Deliver Greater Environmental Benefits from Farm Bill Conservation Programs*. New York: Natural Resources Defense Council. Available at www.nrdc.org/health/pesticides/ipm/ipm.pdf.

Hollingsworth, C. (ed.) (2000). *Integrated Pest Management Guidelines for Structural Pests: Model Guidelines for Training and Implementation*, UMass Extension Publication No. IP-STRC. Amherst, MA: University of Massachusetts.

Houlihan, J., Wiles, R., Thayer, K. *et al.* (2003). *Body Burden: The Pollution in People*. Washington, DC: Environmental Working Group. Available at http://www.ewg.org/featured/15.

IPM Institute of North America (2007). *Twenty-Six Million Acres Certified for Sustainable Practices in 2006*. Madison, WI: IPM Institute of North America. Available at www.ipminstitute.org/newsletter/newsletter_v8i1%20.htm.

Johanns, M. (2005). Remarks as prepared by the Hon. Mike Johanns, Secretary US Department of Agriculture. In *Innovations in Land and Resource Governance, White House Conference on Cooperative Conservation*, August 29, 2005. Washington, DC: US Department of Agriculture. Available at www.usda.gov/wps/portal/!ut/p/_s.7_0_A/7_0_1OB/.cmd/ad/.ar/sa.retrievecontent/.c/6_2_1UH/.ce/7_2_5JM/.p/5_2_4TQ/.d/2/_th/J_2_9D/_s.7_0_A/7_0_1OB?PC_7_2_5JM_contentid=2005%2F08%2F0335.xml&PC_7_2_5JM_parentnav=TRANSCRIPTS_SPEECHES&PC_7_2_5JM_ navid=TRAN SCRIPT#7_2_5JM.

Meek, F., Kemper, T. & Doyle, L. (2006). Zoo IPM: unique challenges, creative solutions. In *Proceedings of the 5th National IPM Symposium*, St. Louis, MO. Washington, DC: US Department of Agriculture. Available at www.ipmcenters.org/ipmsymposiumv/sessions/OrkinZooIPM2.pdf.

National Agricultural Statistics Service (2001). *Pest Management Practices: 2000 Summary*. Washington, DC: Agricultural Statistics Board, US Department of Agriculture. Available at http://usda.mannlib.cornell.edu/MannUsda/ViewDocumentInfo.do?documentID=1452.

National Potato Council (1998). *The National IPM Protocol for Potatoes: A Pest Management Assessment Tool and Educational Program Developed for America's Potato Growers*. Englewood, CO: National Potato Council.

Natural Resources Defense Council (2007). *Urge Your Senators to Make Sure the New Farm Bill Protects the Environment*. New York: Natural Resources Defense Council. Available at www.nrdc.org.

New England Pest Management Association (2007). *NEPMA IPM Registry*. Concord, NH: NEPMA. Available at www.nepma.org.

Ohmart, C. P. (2006). Reducing pesticide risk: case study from Lodi, California. In *Proceedings of the 5th National IPM Symposium*, St. Louis, MO. Washington, DC: US Department of Agriculture. Available at www.ipmcenters.org/ipmsymposiumv/sessions/11-6.pdf.

Ohmart, C. P. & Matthiasson, S. K. (2000). *Lodi Winegrower's Workbook: A Self-Assessment of Integrated Farming*

Practices. Lodi, CA: Lodi-Woodbridge Winegrape Commission.

Organic Materials Review Institute (2007). *OMRI Brand Name List*. Eugene, OR: OMRI. Available at www.omri.org/OMRI_brand_name_list.html.

Salam, M. T., Li, Y. F., Langholz, B. & Gilliland, F. D. (2004). Early-life environmental risk factors for asthma: findings from the children's health study. *Environmental Health Perspectives*, **112**, 760–765.

School Pesticide Reform Coalition and Beyond Pesticides (2003). *Safer Schools: Achieving a Healthy Learning Environment through Integrated Pest Management*. Washington, DC: Beyond Pesticides & School Pesticide Reform Coalition. Available at www.beyondpesticides.org/schools/publications/IPMSuccessStories.pdf.

Sustainable Food Laboratory (2007) SYSCO uses IPM to improve environmental performance. Hatland, VT: Sustainable Food Laboratory. Available at www.sustainablefood.org/article/articleview/12991/1/484.

Thornton, J. W., McCally, M. & Houlihan, J. (2002). Biomonitoring of industrial pollutants: health and policy implications of the chemical body burden. *Public Health Reports*, **117**, 315–323.

Thrupp, L. A. (2002). *Fruits of Progress: Growing Sustainable Farming and Food Systems*. Washington, DC: World Resources Institute. Available at www.wri.org/biodiv/pubs_description.cfm?pid=3065.

University of Massachusetts (2007). *The Professional Guide for IPM in Turf in Massachusetts*. Amherst, MA: University of Massachusetts.

US Congress (1996). *Food Quality Protection Act of 1996*. Public Law 104–170, Section 303: Integrated Pest Management. Available at www.epa.gov/pesticides/regulating/laws/fqpa/fqpa_implementation.htm#ipm.

US Department of Agriculture (2004). *National Road Map for Integrated Pest Management*. Washington, DC: US Department of Agriculture. Available at www.ipmcenters.org/IPMRoadMap.pdf.

US Department of Agriculture (2005). *Productive Lands – Healthy Environment: NRCS Strategic Plan 2005–2010*. Washington, DC: US Department of Agriculture, Natural Resources Conservation Service. Available at www.nrcs.usda.gov/ ABOUT/ strategicplan.

US Department of Agriculture (2006a). *Proceedings of the 5th National IPM Symposium*, St. Louis, MO. Washington, DC: US Department of Agriculture. Available at www.ipmcenters.org/ipmsymposiumv/sessions/index.html.

US Department of Agriculture (2006b). *USDA Roles in Market-Based Environmental Stewardship*, Departmental Regulation No. 5600-003. Washington, DC: US Department of Agriculture. Available at www.ocio.usda.gov/directives/doc/DR5600–003.htm.

US General Accountability Office (2001). *Agricultural Pesticides: Management Improvements Needed to Further Promote Integrated Pest Management*. Washington, DC: US General Accountability Office. Available at www.gao.gov/new.items/d01815.pdf.

US Green Building Council (2007). *LEED for High Performance Operations, v. 2008: Draft Credits for 2nd Comment Period*. Washington, DC: US Green Building Council.

World Health Organization (2005). *World Malaria Report*. Geneva, Switzerland: World Health Organization of the United Nations. Available at www.rbm.who.int/wmr2005/pdf/WMReport_lr.pdf.

World Health Organization (2006). WHO gives indoor use of DDT a clean bill of health for controlling malaria: WHO promotes indoor spraying with insecticides as one of three main interventions to fight malaria. From a press release. Geneva, Switzerland: World Health Organization of the United Nations. Available at www.who.int/mediacentre/news/releases/2006/pr50/en/.

Whyatt, R. M., Rauh, V., Barr, D. B. *et al.* (2004). Prenatal insecticide exposures and birth weight and length among an urban minority cohort. *Environmental Health Perspectives*, **112**, 1125–1132.

Wiebe, K. & Gollehon, N. (eds.) (2006). *Agricultural Resources and Environmental Indicators*, Economic Information Bulletin No. EIB-16. Washington, DC: US Department of Agriculture, Economic Research Service. Available at www.ers.usda.gov/publications/arei/eib16/.

Williams, G. M., Linker, H. M., Waldvogel, M. G., Leidy, R. B. & Schal, C. (2005). Comparison of conventional and integrated pest management programs in public schools. *Journal of Economic Entomology*, **98**, 1275–1283.

Young, C. E. (ed.) (2002). *Residential Turfgrass IPM Definitions*. Columbus, OH: Ohio State University Extension. Available at http://ipm.osu.edu/element/resturf.htm.

Chapter 36

IPM: ideals and realities in developing countries

Stephen Morse

It is all too readily forgotten that agriculture still is the most important livelihood base for the majority of people living in developing countries (Abate *et al.*, 2000; Lenné, 2000), and it follows that crop protection is an important dimension given the extent of crop losses to pests seen across the globe (Oerke, 2006). However, as crop protection is but a part of a bigger picture it can be difficult, if not impossible, to make a link between it and poverty alleviation (Lenné, 2000). IPM is the current ideology as to the form that crop protection should take, and is being promoted on a global scale within both developed and developing countries. But while its appeal is understandable, is IPM really the road we should encourage farmers in developing countries to follow? This chapter will address this question but I will put forward a set of arguments to spark a critical analysis in the mind of the reader rather than provide an "answer." But it has to be said that this aim generates a number of conundrums.

First, the chances are that most readers of this chapter, like the author, will not be one of those farmers that form the basis of the discussion. Thus I am asking readers unconnected and remote in just about every sense of the term (spatially, culturally, economically, etc.) from the conditions of those farmers to ruminate on ways in which they can be "helped." There are obviously dangers here, and even the verb "help" can be loaded with arrogance.

Second, the terms "developing" and "developed" are subjective and open to contestation. I will not go into this extensive debate here but instead will seek to minimize the impact of the problem in a number of ways. The ideas upon which IPM rests are global and hence at times I will write in general terms and draw unapologetically upon literature based upon experience in the richer parts of the world as well as the poorer. At other times I will concentrate upon the poorest continent on the planet – Africa. After all, it's here that much aid is currently being targeted and it's here that I believe we see one of the frontiers of IPM.

Third, who are these farmers I will be talking about? "Farmers" are a diverse group in both developed and developing worlds. Some are richer than others and have access to greater resources (land, labor, machinery, etc.), while some are "part-time" in the sense of having significant "off-farm" income. IPM as a philosophy does not discriminate between these diverse groups. It is truly universal in outlook, but the form and extent of practice will inevitably be different. Hence to refer to "farmers" with any implied sense of homogeneity is cavalier to say the least. I will attempt to deal with the issue by drawing upon the experience of research with

Integrated Pest Management, ed. Edward B. Radcliffe, William D. Hutchison and Rafael E. Cancelado. Published by Cambridge University Press.

a diverse group of farmers from many countries, but at times specifically address the relationship between IPM and subsistence farmers in Africa.

The chapter will begin with a summary of the contested meaning of IPM, and much of this surrounds the role that pesticides do or should play. Following this discussion of meaning I will explore whether farmers have adopted IPM given that the philosophy is arguably worthless unless farmers can put it into practice. Building upon all of the above I will place IPM within what is known as the post-development debate. Is IPM nothing more than another manifestation of a dominance of the richer countries over the poorer – a further expression of neocolonialism? What alternatives does a "post-IPM" discourse provide and could it really make a difference to the developing world?

36.1 IPM as a positive force

IPM has its origins within the "experiences, understanding and attitudes" of the latter half of the twentieth century (Haines, 2000, p. 829), especially in terms of a reaction to the problems created by the indiscriminate use of pesticides (Barfield & Swisher, 1994). As Harris (2001, p. 119) succinctly puts it, "IPM is a convenient descriptor for the wide range of still-evolving pest control programs that emerged from the chemical prophylaxis practiced in the 1950s." The rise of the pesticide culture created environmental problems and spawned the notion of integrated control in the USA in the 1940s and 1950s (Ehler, 2006). Initially integration meant the use of insecticides in such a way as to minimize damage to the natural enemies of pests. This can be achieved in various ways, but an obvious first step is to minimize the application of insecticide to an "only when necessary" basis. A key concept here was the notion of an economic injury level (EIL); the lowest number of pests that cause economic damage. Low levels of a pest population can be tolerated and it's only when it begins to reach the EIL that the farmer should step in and reduce numbers. Thus the pest population has to be monitored on a regular basis and once the population reaches a point where it looks like it will progress to the EIL then action is taken. An economic threshold (ET: a point that triggers action) is set below the EIL, so as to ensure that the latter is not reached. The theory is solid, but of course pest population growth and damage per pest can be dependent upon a range of environmental and host plant factors and strictly speaking the ET and EIL need to take account of these. ETs will also vary between different pests (will be higher for some than others) and strictly speaking should vary with the economic value of the crop. While all of this is highly complex, in practice what we use for an ET is a compromise – an average – that may not be technically ideal but which can be easily used. The notion of "integration" from the USA became fused with growing ideas of ecological pest control (pest control without the use of any pesticide) in Australia (Geier & Clark, 1961; Geier, 1966) to generate what we now know under the broad church of IPM.

A more detailed discussion of the history of IPM can be found in Kogan (1998), but for the purposes of this chapter all that needs to be stressed from the foregoing is that IPM has a reactive (rather than proactive) origin in that its "integrated" and "ecological" strands were responses to an environmental problem that occurred in the richer parts of the world; IPM was a developed world answer to a developed world problem. These regions have a strong research base in terms of institutions, staff and diversity. Thus they had both the need and the resources to think through what could be done to address the problem they faced and make it a reality. This is not to say that the underlying principles of managing pests, reducing pesticide use, etc. are irrelevant on a global scale, but it is important to remember where IPM came from and the maelstrom of forces that created it.

Just what is meant by IPM has tended to vary a great deal between different workers in the field and this may in part reflect its origins from "integration" (pesticide use can continue, but be more in tune with other factors that reduce pest pressure) and "ecological" (no pesticides) pest control. Indeed some definitions are so general as to mean almost anything (Jeger, 2000). Take, for example, the following two definitions:

> the farmer's best mix of control tactics in comparison with yields, profits and safety of alternatives
>
> Kenmore (cited in Jones, 1996).

But what is meant by the highly subjective term "best" and how are yields, profits and safety to be balanced within this mix?

> an adaptable range of pest control methods which is cost effective whilst being environmentally acceptable and sustainable
>
> (Perrin, 1997).

"Cost effective" presumably means profitable, but "acceptable" is a loaded term that will vary between interest groups and over time. And all this is before the highly contested term "sustainable" is added to the mix.

Perhaps the one thing that all visions of IPM have in common is an emphasis on reducing pesticide use (Brower, 2002), although even this "common denominator" is disputed and some argue that IPM can actually result in an increase in pesticide use (Kogan & Bajwa, 1999). Nonetheless, for some, IPM must imply the complete removal of pesticides from the "toolbox" of options available to farmers to create an ecologically based management. They argue that there must be a reconstruction of the agri-environment based on sound scientific knowledge so as to negate the need for regular injections of toxin. Here the pest populations are managed in such a way that they rarely, if ever, reach the EIL (Brower, 2002). Others appear to be far more willing to acknowledge an important role for pesticides albeit in a much more controlled way by only spraying once the pest population had reached the ET (Jeger, 2000).

> a zero-pesticides approach to agriculture would not be sustainable . . . Chemical control is an indispensable part of Integrated Crop Management (ICM), an integrated farming system based on a sound combination of all available pest control methods
>
> (Urech, 2000, pp. 832–833).

A "sound combination" sounds very much like a "best mix" and "adaptable range." In this case the ethos of IPM is seen by some as nothing more than pesticide management (Ehler, 2006).

The IPM purists, those who want to see crop protection built upon solid ecological foundations that negates the need for pesticides, have been referred to as "strategists," while those who see IPM as a better management of pesticide use are regarded as "tacticians" (Barfield & Swisher, 1994). This apparent dichotomy in meaning has been the subject of some debate (Table 36.1). Jeger (2000) used the terms "low order" and "high order" IPM to denote the same categories as "tactical" and "strategic" respectively. Prokopy (2003) used the term "top down" for tactical IPM and contrasts this with what he calls a "bottom up" strategic approach. Ehler (2006) saw strategic IPM as "true" IPM (= the "real" IPM of Brader, 1988) and tactical IPM as the "other." Ishii-Eiteman & Ardhianie (2002) suggested that pesticides have to be replaced by IPM, and Ehler (2006, p. 789) deplored the trend towards adopting tactical IPM as "mission creep" which needs to be prevented. However, it can also be argued that the two are not mutually exclusive as tactical IPM can be seen as a step along the road to strategic IPM. Nonetheless, those who believe in "strategic/true/real/high order" IPM can be quite disparaging over tactical IPM as a sort of business as usual and a betrayal of the ecological foundation upon which IPM should be founded.

For all this dissension, what is beyond doubt up to the time of writing this chapter is that IPM has been heavily promoted by government agencies, scientists, environmentalists and international development agencies for at least 40 years (Zadoks, 1993). It is also true that IPM has made a positive contribution to our thinking about crop protection (Kogan & Bajwa, 1999). The intellectual ideas behind IPM also helped influence the notion of sustainability in agriculture and indeed resource management, and these broader ideas have fed into the emergence of the sustainable development paradigm that pervades so much of our lives since the last decade of the twentieth century.

36.2 Are we there yet?

While the intellectual basis of IPM has value it can all be comforting rhetoric unless it is put into practice, and it's here that problems start to arise. Take, for example, the following quotation published in 1999, nearly 30 years since the term IPM first appeared in press:

Table 36.1 Classification of IPM and where each type is predominantly found

Types of IPM				Prime focus	Where predominant
Tactical	*Low order*	*Top–down*	*Other*	Better use of pesticide so as to minimize environmental and health damage	Places where pesticide use has proved to be a problem *Most notably the richer (developed) parts of the world (e.g. North America, Europe)*
	Medium order				
Strategic	*High order*	*Bottom–up*	*True or Real*	No use of pesticide but reliance on ecosystem design	Places where pesticides have not been affordable or obtainable *Most notably the poorer (developing) parts of the world (e.g. Africa)*
Barfield & Swisher (1994)	Jeger (2000)	Prokopy (2003)	Ehler (2006) Brader (1988)		

Practically, however, IPM is a tangible reality in some privileged regions of the world, but still remains a distant dream for many others

(Kogan & Bajwa, 1999, p. 20).

Perhaps even more disturbing is the following:

there is little evidence that IPM (as originally envisioned) has been implemented to any significant extent in American agriculture . . . with a few exceptions, IPM as originally envisaged had not been implemented to any significant extent in Western Europe . . . the World Bank issued a report concluding that IPM adoption remained relatively low in most of the developing world. And that there was no convincing evidence for changes in pesticide use in targeted crops such as rice or cotton in Asia

(Ehler, 2006, p. 788).

But how are the assessments of a "distant dream for many" and "IPM adoption remained relatively low" arrived at?

Defining IPM is obviously a key first step in the process of measuring adoption by farmers, and malleability in what it means can help serve various agendas. Depending upon how IPM is defined you can present any existing snapshot of the same farming community as representing either 0% or 100% adoption (Shennan *et al.*, 2001; Ehler, 2006). If IPM is defined in strategic terms as the elimination of the need for pesticides, then in the UK, for example, adoption is close to zero for field crops despite all of the efforts over the past 40 years. If, on the other hand, IPM is defined in tactical terms as a reduction of pesticide pressure on the environment, then based on an index of environmental toxicity (the environmental impact quotient: EIQ) studies have shown a reduction in hazard between 1992 and 2002 for the most commonly grown arable crops in the UK (Cross & Edwards-Jones, 2006). Expressed in these terms, adoption of IPM is encouraging. Indeed, if a reduction in insecticide use or environmental damage from insecticide is taken as the measure of IPM, then developing countries lead the rest of the world as they have the lowest level of pesticide use

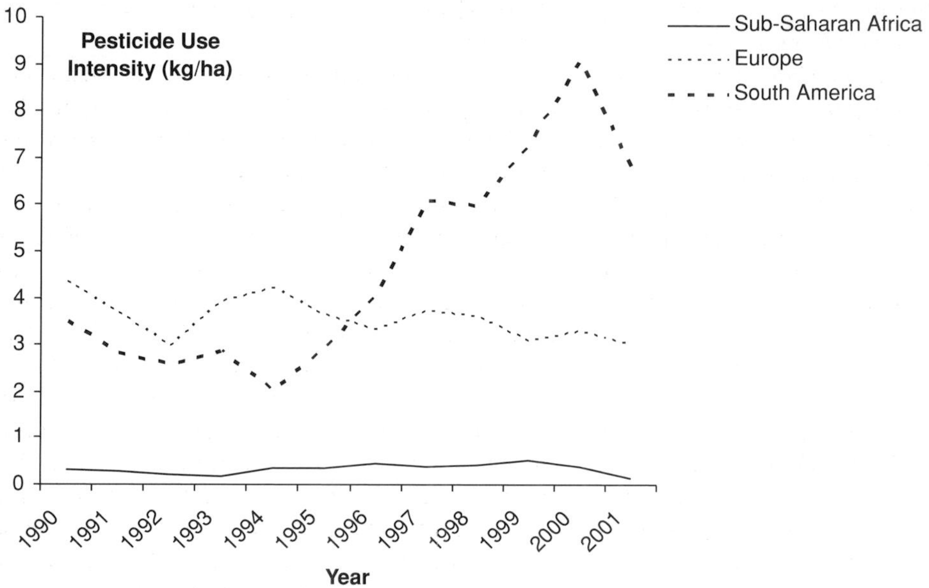

Fig. 36.1 Comparison of pesticide use intensity (kg/ha) for sub-Saharan Africa, Europe and South America (data from Earthtrends, 2007).

(Abate *et al.*, 2000; Lenné, 2000). Making reliable estimates of pesticide use on a country-by-country basis remains problematic. One estimate by Abate *et al.* (2000) was that pesticide use in Africa averaged 1.23 kg/ha of cultivated land compared with 7.17 kg/ha and 3.12 kg/ha for Latin America and Asia respectively, and figures were still higher for developed parts of the world. Another set of figures based on pesticide use intensity (PUI) between 1990 and 2001 can be found on one of the World Resources Institute databases (Earthtrends, 2007); averages for Africa, South America and Europe have been summarized in Fig. 36.1. Over this period, the average PUI for Africa was consistently lower than that of Europe and South America. Indeed, data for the latter suggest a worsening trend with regard to IPM as the PUI increased significantly between 1994 and 2000. The PUI for Europe is more stable, and if anything, is suggestive of a slight decline. Thus, by estimates of both Abate *et al.* (2000) and the World Resources Unit, it is clear that Africa is the global "IPM heartland" and far from it being a "distant dream" it is what African farmers have done for thousands of years and continue to do today on a large scale. With this perspective Africa is one of the "privileged regions" mentioned by Kogan & Bajwa (1999) in the above quotation. In complete contrast, and indeed somewhat oddly, Hillocks (2005) regards adoption of IPM in Africa for cotton as poor because pesticide use on that crop is so low. This is a strange lament that the "cure" hasn't been adopted because there is no disease! But, of course, this is only a part of the story as Africa also has relatively low agricultural productivity, although the picture for the continent as a whole is admittedly a complex one (Fulginiti & Perrin, 2004). With Peter Kenmore's definition of IPM given above as the "best mix" using "yields, profits and safety" as key criteria, then while Africa presumably scores well in terms of safety (i.e. low use of pesticide) it doesn't do so well with regard to yields and profits.

In practice IPM has come to mean something of a moving target in between the "tactical" and "strategic" extremes (embracing the "medium order" IPM of Jeger, 2000). The "strategic" position might be expressed as an ideal, but in practice an IPM program may be focused in the short to medium term on the "tactical." Measuring adoption within these gray areas can be complex. A number of studies have been conducted whereby a farmer is said to have adopted IPM if he/she has indicators that point in the right direction (Shennan *et al.*, 2001). For example, farmers can be questioned as to whether they or an advisor regularly monitor pest populations as part of their management. Studies in the USA have generated mixed results. There are general claims that "on 70 percent of land used for farming in the US, some form of IPM is employed, which is comparable to the situation in Europe" (Brower, 2002, pp. 403–404). Yet there are also interesting claims that when questioned in surveys farmers tend to overstate their adherence to IPM relative to what they actually do in practice, and what is required is an assessment of both what they say and what they do (Shennan *et al.*, 2001). Way & van Emden (2000) and Oerke & Dehne (2004) cite examples of successful IPM in both the developed and developing worlds, and reduction in pesticide use is constantly mentioned as a criterion in assessment of "success." On the other hand, Jeger (2000) makes the claim that the only example of an IPM in widespread use for field crops is for alfalfa and that there are "few examples of high-order *[strategic]* IPM. Most programmes lie somewhere between the low *[tactical]* and medium orders" (p. 791). Epstein & Bassein (2003) in their analysis of pesticide use in California found that pesticide use was "fairly constant during the 1990s" (p. 378) and they suggest that IPM is "unlikely to result in much pesticide reduction in California in the next 10 years."

It would seem from the literature that while there are undoubted stories of success, even the IPM liberalists continue to decry the lack of adoption of IPM in practice, and this includes developing regions such as Africa. Farmers who use pesticides have often proved to be difficult to wean away from the habit, and various explanations have been offered as to why this should be so (Orr & Ritchie, 2004). Part of the problem seems to be that pesticides are popular and farmers can be resistant to "managing" them even within the limited "tactical" mindset. This applies to both the developed world and those parts of the developing world where pesticides are available and farmers can afford them. Perhaps worryingly it can be convincingly argued that those farmers in the poorer parts of the globe who do not use pesticides, and who thus practice IPM by one definition, do so entirely out of a lack of choice; they simply can't afford not to do IPM! For those farmers who can afford and obtain them, pesticides are an attractive option.

While IPM is "knowledge based" (Mullen *et al.*, 2005) and thus somewhat dependent upon the level of education and training of farmers (Chaves & Riley, 2001; Shennan *et al.*, 2001), the ends of the tactical–strategic spectrum differ in terms of research effort and also in the skill required from the farmer. In tactical IPM there is a need to monitor pest populations founded upon knowledge over what the EIL and ET should be. For IPM strategists the skill required from the farmers and indeed the support agencies (extension and research) is even greater and some have questioned such a purist approach, calling instead for more practical controls that equate to "real situations in the field" (Way & van Emden, 2000, p. 96). Thus whatever the shade of IPM employed it does imply a more intensive management regime (Williamson *et al.*, 2005) although the tactical end of the spectrum is arguably less "knowledge demanding" and provides a more practicable approach for farmers already using pesticide, especially when tied to promises of saving costs. Indeed farmers may even feel nervous if they feel that IPM equates to a removal of pesticides (Shennan *et al.*, 2001). What receives less emphasis is that IPM also has ramifications for researchers. The oft-used (and misused) phrases of interdisciplinarity and multidisciplinarity abound whenever IPM is mentioned (Wei *et al.*, 2001), yet while the words are easy to express the "doing" is not (Altieri, 1987). There are also pressures to consider such as the continuous need to generate "high impact" publications in short timescales with ever-limited resources (Brunner, 1994; Wei *et al.*, 2001).

Promotion of IPM to small-scale farmers in Asia, and more recently Africa, has typically been via Farmer Field Schools (FFS). Much success has been claimed for this approach (Prudent *et al.*, 2006), and its application has been extended into related issues such as nutrient management (Abate *et al.*, 2000; de Jager, 2005). However, some have begun to question whether the IPM FFS impact will be long-lasting (Way & van Emden, 2000). It has been said to be expensive on a per-farmer-trained basis (Quizon *et al.*, 2001), and the general conclusion that farmers trained in this way are an effective vehicle for spreading the IPM message has also been critiqued. While farmers who attend such IPM FFS may gain knowledge of IPM, even if "quantitatively small" (Feder *et al.*, 2004, p. 238), that is no guarantee in itself that they would spread that knowledge to other farmers and the few studies that have been conducted on this issue do not provide encouraging results (Feder *et al.*, 2004). This should not be all that surprising given that "trickle down" of knowledge, benefits, etc. has long been shown to be an oversimplistic assumption within modernizing approaches to development.

Beyond adoption there is, of course, impact; has IPM generated an economic or environmental return in the "real world" and not just in controlled experiments? Literature still remains sparse on this topic but Mullen *et al.* (2005) estimate that IPM in California has received a substantial investment of between $US 50 million and $US 90 million per annum between 1970 and 1997 (35% of the funds allocated to agricultural research in general) and this has resulted in a cost : benefit ratio of 1 : 6; for every dollar spent six have been realized. The evidence for environmental and human health benefits resulting from IPM is less easy to distinguish (Mullen *et al.*, 2005).

36.3 Who wants IPM?

Given that IPM requires an increased management input (= investment) from farmers, then it can be theorized (Morse & Buhler, 1997) that the circumstances under which adoption is most likely to occur are with:

- high value crops (in terms of economics or sustenance) where the extra input of time and skill is worth it
- crops where at least one pest represents a significant factor limiting yield or value
- crops grown in highly controlled and monitored environments (greenhouses or perennial plants in orchards for example). These also tend to have high value.

Even so, the malleability in what IPM means in practice has the effect that almost anyone can say that they want it, or indeed are practicing it. The crop protection and biotechnology industries, including pesticide manufacturers and retailers (Eddleston *et al.*, 2002), promote "IPM" and so do the large multi- and bilateral development agencies. Indeed Jeger (2000, p. 789) states that:

> A personal comment perhaps, but it seems to me that many IPM programmes especially those defined against an overtly environmental agenda (as defined by donors not by the recipients) take on many of the residual garments of neo-colonialism, with competing donors following different agendas.

While donors seem to be keen, commitment from governments in developing countries, especially in Africa, towards IPM has been questioned (Abate *et al.*, 2000; Matthews, 2006).

But what do the "IPM recipients," mostly farmers, actually want? Well, put simply, most, whether they are in the developed or developing worlds, probably want what all of us usually do: to survive and prosper and for themselves and their families to do well and be happy. This may sound obvious but it is well to remember that crop protection is merely a means to an end and not an end in itself. Farmers don't necessarily want (or can afford) technical excellence. Additional resources, no matter how small this may seem to an outsider, spent on IPM will have to come at the expense of something else. Understandably, a pest sampling protocol that is technically excellent and highly publishable in a good journal will not necessarily be welcomed by farmers (Cullen *et al.*, 2000). Even claimed savings in pesticide will come with extra costs incurred through sampling (MacHardy, 2000), and while the latter may be marginal to a researcher they will be readily apparent to the farmer. Thus what farmers may want is a trade-off – less technical precision but something they can do and that will provide them with a tangible benefit.

It is also well to remember that crop protection may not be one of the most pressing issues for the farmer. An interesting point here is that attempts have been made to consider "integration" in IPM not only in terms of combining different technologies (biological control with insecticide for example) but in terms of how IPM links to other issues. Prokopy (1993) suggests a classification based on levels of integration from single

Level of IPM	Focus
1st	One type of pest (e.g. insects)
2nd	All types of pest (insects, weeds, nematodes, etc.)
3rd	All agronomic practices (management of soil fertility, water levels, etc.) as well as crop protection
4th	Social, economic, cultural and political realms within which agricultural production take place

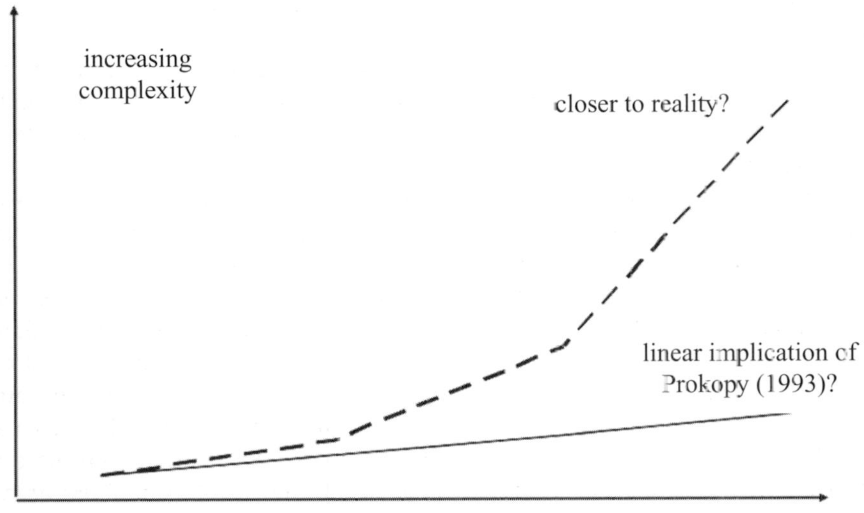

Fig. 36.2 The four levels of IPM as proposed by Prokopy (1993).

pest type (e.g. insects) through to multiple types of pest (e.g. insects with disease), other factors important in crop production up to the larger economic, social and even political domain (Fig. 36.2). While a useful reminder of context, the problem with such a classification is that it implies that IPM remains important as we progress from 1st level to 4th level. After all, these are "levels of IPM." It also implies a smooth "linear" gradation: 1 → 2 → 3 → 4. The progression from level 1 to level 2, what some also refer to as "horizontal" integration (Ehler, 2006), may seem logical and relatively smooth, but it is not easy given that we are dealing with quite different organisms with their own interactions within the biotic and abiotic environment (Chellemi, 2000).

From a farmer's perspective, crop protection, let alone IPM, may cease to be of that much importance between levels 2 and 3. The main concern under level 3 may be "plant health management" of which crop protection is but a part alongside issues such as water and nutrient stress (Cook, 2000). Promotion of IPM where the main constraint is soil fertility is unlikely to result in much adoption. This is not to ignore the fact that there are interactions between pest management and other aspects of plant growth (Altieri, 2002). Agronomic practices such as intercropping provide farmers an element of insurance against environmental and/or economic risk, but the diversity of plant architecture in such systems can also encourage natural enemies of pests and provide obstacles to pest dispersal (Abate *et al.*, 2000; Altieri, 2002). An interesting development from the latter is the "push–pull" system in maize to manage lepidopteran stem borers. Here maize is planted alongside other crops – one of which repels stem borers (the push) while another attracts their natural enemies (the pull). The result is that the maize receives less damage (Kahn *et al.*, 1998).

The implied linearity of IPM levels 1 to 4 hides the fact that the jump from level 3 to 4 is far more complex than provided here (Fig. 36.2). There are issues that have immediate force at the micro-scale such as availability of farm labor and inputs. If labor is limiting, IPM technologies that require more of it may be unlikely to succeed, especially if farmers have access to well-paid "off-farm" employment (Beckmann & Wesseler, 2003; Mahmoud & Shively, 2004). Indeed, given the latter point Orr & Ritchie (2004) argue that the most attractive IPM components for resource-poor farmers are host plant resistance and biological control. Nonetheless, with the jump to level 4 there are other considerations that work in favor of IPM. Agriculture, next to forestry, is the industry that occupies the largest area of the planet's land surface. What farmers do can impact on the rest of us, and "society" is entitled to a voice. Therefore it may be argued that while farmers may not readily adopt an approach that provides no tangible benefits for them, they should be obliged to do so for the sake of society. Thus the key consideration of sustainability has to mean that all four levels are considered.

However, the problem is that there are complex forces at play at larger scales in "society," and these can be malleable and fickle. Take, for example, the ongoing World Trade Organization discussions which may have a major impact on agriculture in the developing world. Trying to predict outcomes let alone impacts of such complex talks is, as physicists like to say, "non-trivial." Public opinion is another force that has proved to be highly significant with regards to agriculture as we can see with antagonism towards genetically modified (GM) crops in Europe. Ironically some proponents of pesticides as an essential component of IPM are now arguing that:

> The values of the crop protection industry and the benefits crop protection products [including pesticides] bring to agriculture must be better communicated to citizens and consumers. Further efforts are needed to build consumer confidence and develop dialogue with key stakeholders
>
> (Urech, 2000, p. 836).

This may well be a tall order! But given all of the above it does seem like the best way to proceed with IPM is from Prokopy's level 1 upwards – the "standard" model of IPM research. But is it?

36.4 Post-IPM?

Most forms of IPM have one characteristic in common in that they seek to minimize, defined in terms of environmental and health impact and not simply in terms of volume or amount, or eliminate (strategic) pesticide use. That is the prime directive of IPM: its soul. What follows is practice. A tactical approach to IPM requires ETs, decision-support tools, etc., while a strategic approach requires a substantial knowledge of system agroecology. Thus there can be a virtual unanimity about wanting the soul of IPM but substantial disagreement with regard to practice. Proponents of IPM would argue that it's based on "good science" and thus must be the "right" thing to do for the sake of society (including the farmers) and the environment. It is sometimes said that the IPM being promoted in developing countries is not the vision of IPM seen in the developed world and it is a mistake to equate the two (Way & van Emden, 2000). After all, agriculture in the developed world has followed a pesticide trajectory and IPM is the cure to the problems that this treadmill has generated. Africa has not had the pesticide problems of the developed world, and the only example of widespread pest problems arising out of overspraying of pesticide on that continent is said to be for cotton in Sudan (Hillocks, 2005). If IPM is nothing more than "use as little pesticide for crop protection as possible" we need not worry about poor adoption in Africa – that continent is leading the world and we have much to learn from its farmers. But arguably this is not necessarily because African farmers want to lead; it has much more to do with them not being able to afford to do anything other than IPM. Defined in these terms the adoption of IPM is inversely related to wealth; the richer a country, then arguably the less the adoption of IPM. Abate *et al.* (2000) make the reasonable point that given the relatively low level of pesticide use in Africa, IPM should be a bottom-up process of building from "traditional pest management approaches that abound in small-scale agriculture in the continent" (p. 646). However, let's assume that given a choice and the necessary resource (cash for purchase and/or subsidies) these farmers would readily adopt pesticides. Let's also assume that an indigenous strategic-based IPM would switch to an indiscriminate use of pesticides which in turn would need to be corrected with the promotion of tactical IPM. It is this causality that I feel is the frontier with IPM in Africa and makes it distinctive to many other parts of the world – not how to encourage farmers to adopt IPM, but how to encourage them not to discard it.

Of course there is an immediate problem with what I have suggested. If farmers can afford pesticides, and have access, then why should they be stopped from using them? There is an uneasy sense of a northern-driven agenda, and the development literature has seen the emergence of a fundamental reassessment of the relationship between North and South generically termed post-development, or even as some would say, anti-development. The thesis is that the post-World War II development project based on modernization theory (transfer of ideas and technology from North to South) has failed, and many in the South, especially in Africa, are poorer now than they were in the 1950s (Rahnema & Bawtree, 1997; Pieterse, 1998; Hart, 2001). The reason for this failure, the post-developmentalists argue, is that development has not been in tune with what people in the South really want. Development has imposed a tautological vision of poverty and what is required to address it. The voice of the locals – those meant to benefit from the process – has not been heard. Included within this criticism is sustainable development which is seen by post-developmentalists as nothing more than soothing words (or even a siren song) and a sticking plaster to address the environmental problems that have largely been created by the North (Roe, 1995). So, if as post-develomentalists suggest, development has to be a bottom–up process of mediated change with the wishes of the society undergoing this change paramount (Escobar, 1992; Mathews, 2004), then what if farmers want pesticides and were able to obtain them? Any Northern-facilitated attempt to prevent or limit that switch, to maintain and build upon an indigenous IPM, could be interpreted within a post-developmental critique as unhealthy especially as farmers in the North continue to use high levels of pesticide. Given that IPM is such a powerful ideology, and has been adopted by a wider range of interest groups, even if not by farmers, then there is a danger that a Northern-driven promotion of IPM could be seen as yet another form of neo-imperialism: IPM as another "siren song" to lure us into a sense that the pesticide problems of the North can be cured with even the poorest farmers, many of who have never even seen a pesticide container, coaxed into playing a role.

So is there a post-IPM? Building from the soul of IPM, the prime necessity has to be the avoidance of the problems that indiscriminate use of some pesticides can generate. This is for the ultimate benefit of farmers and their families as well as wider society at national and global scales, and applies throughout the globe. This isn't just a "siren song" but a worthy and socially just goal. As I have argued before (Morse & Buhler, 1997), what replaces a damaging pesticide could be a wide range of options, including less environmentally and toxic pesticide and genetically modified crops, but whatever is presented to farmers has to take onboard their socioeconomic realities. It has to build from where farmers are and not attempt to impose unrealistic ideals. In effect this is calling for a reversal of the "levels of IPM" proposed by Prokopy (1993). IPM has traditionally built from level 1 upwards and hence looks to slot the results of research into farming systems while expecting that socioeconomic and political constraints will somehow be overcome (a technocentric approach). I would argue that post-IPM needs to start with the higher levels of Prokopy and thus on understanding the farmers' socioeconomic contexts (as well as the inevitable variation) before studies can be instigated into how best to bring about an intervention (a sociocentric approach).

35.5 Conclusions

IPM has proved to be a powerful ideology that has spawned ideas of sustainability in agriculture and beyond, and thus its positive contribution should not be underestimated. But it can also be argued that IPM has lost its way. It has become enmeshed in debates over meaning and arguments and counter-arguments over adoption in both the developed and developing worlds. IPM has, by definition, forced a concentration on crop protection when it may not be warranted, or asked farmers to change their management in ways that they cannot afford or are unwilling to do. Instead we need a refocusing upon what is really important within the realities of a world where many are poor and some are getting poorer. Post-IPM needs to begin with the realities that farmers in the developing world face and help facilitate a sustainable crop protection which minimizes, and ideally eliminates, the negatives that can result from pesticide use. This is an acknowledgement that many farmers in developing countries may want to use pesticide if affordable and denying that right is impossible. Thus the goal is to have something in place which farmers can "do" and which benefits them but which also prevents environmental degradation for the good of society at large. Research is an important part of the strategy, but cognizance has to be taken of what farmers can and will "do" as the centerpiece rather than place them as an afterthought.

References

Abate, T., van Huis, A. & Ampofo, J. K. O. (2000). Pest management strategies in traditional agriculture: an African perspective. *Annual Review of Entomology*, **45**, 631–659.

Altieri, M. A. (1987). *Agroecology: The Scientific Basis of Alternative Agriculture*. London: Intermediate Technology Publications.

Altieri, M. A. (2002). Agroecology: the science of natural resource management for poor farmers in marginal environments. *Agriculture, Ecosystems and Environment*, **93**, 1–24.

Barfield, C. S. & Swisher, M. E. (1994). Integrated pest management: ready for export? Historical context and internationalization of IPM. *Food Reviews International*, **10**, 215–267.

Beckmann, V. & Wesseler, J. (2003). How labour organization may affect technology adoption: an analytical framework analyzing the case of integrated pest management. *Environmental and Development Economics*, **8**, 437–450.

Brader, L. (1988). Needs and directions for plant protection in developing countries: the FAO view. *FAO Plant Protection Bulletin*, **36**, 3–8.

Brower, V. (2002). Growing greener. The impact of integrated pest management. *EMBO Reports*, **3**, 403–406.

Brunner, J. F. (1994). Integrated pest management in tree fruit crops. *Food Reviews International*, **10**, 135–157.

Chaves, B. & Riley, J. (2001). Determination of factors influencing integrated pest management adoption in coffee berry borer in Colombian farms. *Agriculture, Ecosystems and Environment*, **87**, 159–177.

Chellemi, D. O. (2000). Adaptation of approaches to pest control in low-input agriculture. *Crop Protection*, **19**, 855–858.

Cook, R. J. (2000). Advances in plant health management in the twentieth century. *Annual Review of Phytopathology*, **38**, 95–116.

Cross, P. & Edwards-Jones, G. (2006). Variation in pesticide hazard from arable crop production in Great Britain from 1992 to 2002: pesticide risk indices and policy analysis. *Crop Protection*, **25**, 1101–1108.

Cullen, E. M., Zalom, F. G., Flint, M. L. & Zilbert, E. E. (2000). Quantifying trade-offs between pest sampling time and precision in commercial IPM sampling programs. *Agricultural Systems*, **66**, 99–113.

de Jager, A. (2005). Participatory technology, policy and institutional development to address soil fertility degradation in Africa. *Land Use Policy*, **22**, 57–66.

Earthtrends (2007). *Agricultural Inputs: Pesticide Use Intensity*. Washington, DC: World Resources Institute. Available at http://earthtrends.wri.org/searchable_db/index.php?theme=8&variable_ID=204&action=select _countries.

Eddleston, M., Karalliedde, L., Buckley, N. *et al.* (2002). Pesticide poisoning in the developing world: a minimum pesticides list. *Lancet*, **360**, 1163–1167.

Ehler, L. E. (2006). Integrated pest management (IPM): definition, historical development and implementation, and the other IPM. *Pest Management Science*, **62**, 787–789.

Epstein, L. & Bassein, S. (2003). Patterns of pesticide use in California and the implications for strategies for reduction of pesticides. *Annual Review of Phytopathology*, **41**, 351–375.

Escobar, A. (1992). Reflections on 'development': grassroots approaches and alternative politics in the Third World. *Futures*, **24**, 411–436.

Feder, G., Murgai, R. & Quizon, J. B. (2004). The acquisition and diffusion of knowledge: the case of pest management training in Farmer Field Schools, Indonesia. *Journal of Agricultural Economics*, **55**, 221–243.

Fulginiti, L. E. & Perrin, R. K. (2004). Institutions and agricultural productivity in sub-Saharan Africa. *Agricultural Economics*, **31**, 169–180.

Geier, P. W. (1966). Management of insect pests. *Annual Review of Entomology*, **11**, 471–490.

Geier, P. W. & Clark, L. R. (1961). An ecological approach to pest control. *Proceedings of the Technical Meeting of the International Union for Conservation of Nature and Natural Resources*, Warsaw, 1960, pp. 10–18.

Haines, C. P. (2000). IPM for food storage in developing countries: twentieth century aspirations for the twenty-first century. *Crop Protection*, **19**, 825–830.

Harris, M. K. (2001). IPM, what has it delivered? A Texas case history emphasizing cotton, sorghum and pecan. *Plant Disease*, **85**, 112–121.

Hart, G. (2001). Development critiques in the 1990s: culs de sac and promising paths. *Progress in Human Geography*, **25**, 649–658.

Hillocks, R. J. (2005). Is there a role for *Bt* cotton in IPM for smallholders in Africa? *International Journal of Pest Management*, **51**, 131–141.

Ishii-Eiteman, M. J. & Ardhianie, N. (2002). Community monitoring of integrated pest management verses conventional pesticide use in a World Bank project in Indonesia. *International Journal of Occupational and Environmental Health*, **8**, 220–231.

Jeger, M. J. (2000). Bottlenecks in IPM. *Crop Protection*, **19**, 787–792.

Jones K. (1996). IPM in developing countries: the Sri Lankan experience. *Pesticides News*, **31** (March 1996), 4–5.

Kahn, Z. R., Ampong-Nyarko, K., Hassanali, A. & Kimni, S. (1998). Intercropping increases parasitism of pests. *Nature*, **388**, 631–632.

Kogan, M. (1998). Integrated pest management: historical perspectives and contemporary developments. *Annual Review of Entomology*, **43**, 243–270.

Kogan, M. & Bajwa, W. I. (1999). Integrated pest management: a global reality? *Anais da Sociedade Entomológica do Brasil*, **28**, 1–25

Lenné, J. (2000). Pests and poverty: the continuing need for crop protection research. *Outlook on Agriculture*, **29**, 235–250.

MacHardy, W. E. (2000). Current status of IPM in apple orchards. *Crop Protection*, **19**, 801–806.

Mahmoud, C. & Shively, G. (2004). Agricultural diversification and integrated pest management in Bangladesh. *Agricultural Economics*, **30**, 187–194.

Mathews, S. (2004). Post-development theory and the question of alternatives: a view from Africa. *Third World Quarterly*, **25**, 373–384.

Matthews, G. (2006). Can we help Africa? *Outlooks on Pest Management*, April, 50–51.

Morse, S & Buhler, W. (1997). IPM in developing countries: the danger of an ideal. *Integrated Pest Management Reviews*, **2**, 1–11.

Mullen, J. D., Alston, J. M., Sumner, D. A., Kreith, M. T. & Kuminoff, N. V. (2005). The payoff to public investments in pest-management R&D: general issues and a case study emphasizing integrated pest management in California. *Review of Agricultural Economics*, **27**, 558–573.

Oerke, E.-C. (2006). Crop losses to pests. *Journal of Agricultural Science*, **144**, 31–43.

Oerke, E.-C. & Dehne, H.-W. (2004). Safeguarding production: losses in major crops and the role of crop protection. *Crop Protection*, **23**, 275–285.

Orr, A. & Ritchie, J. M. (2004). Learning from failure: smallholder farming systems and IPM in Malawi. *Agricultural Systems*, **79**, 31–54.

Perrin, R. M. (1997). Crop protection: taking stock for the new millennium. *Crop Protection*, **16**, 449–456.

Pieterse, J. N. (1998). My paradigm or yours? Alternative development, post-development, reflexive development. *Development and Change*, **29**, 343–373.

Prokopy, R. J. (1993). Stepwise progress toward IPM and sustainable agriculture. *IPM Practitioner*, **15**, 1–4.

Prokopy, R. J. (2003). Two decades of bottom–up, ecologically based pest management in a small commercial orchard in Massachusetts. *Agriculture, Ecosystems and Environment*, **94**, 299–309.

Prudent, P., Loko, S. & Vaissayre, M. (2006). Farmers Field Schools in Benin: a participatory approach to the transmission of information on integrated cotton pest management. *Cahiers Agricultures*, **15**, 100–101.

Quizon, J., Feder, G. & Murgai, R. (2001). Fiscal sustainability of agricultural extension: the case of the Farmer Field School approach. *Journal of International Agricultural and Extension Education*, **8**, 13–24.

Rahnema, M. & Bawtree, V. (1997). *The Post-Development Reader*. London: Zed Books.

Roe, E. (1995). Critical theory, sustainable development and populism. *Telos*, **103**, 1249–1262.

Shennan, C., Cecchettini, C. L., Goldman, G. B. & Zalom, F. G. (2001). Profiles of Californian farmers by degree of IPM use as indicated by self-descriptions in a phone survey. *Agriculture, Ecosystems and Environment*, **84**, 267–275.

Urech, P. (2000). Sustainable agriculture and chemical control: opponents or components of the same strategy? *Crop Protection*, **19**, 831–836.

Way, M. J. & van Emden, H. F. (2000). Integrated pest management in practice: pathways towards successful application. *Crop Protection*, **19**, 81–103.

Wei, W., Alldredge, J. R., Young, D. L. & Young, F. L. (2001). Downsizing an integrated crop management field study affects economic and biological results. *Agronomy Journal*, **93**, 412–417.

Williamson, S., Ferrigno, S. & Vodouhe, S. D. (2005). Needs-based decision-making for cotton problems in Africa: a response to Hillocks. *International Journal of Pest Management*, **51**, 219–224.

Zadoks, J. C. (1993). Crop protection: why and how. In *Crop Protection and Sustainable Agriculture*, eds. D. J. Chadwick & J. Marsh, pp. 48–55. Chichester, UK: John Wiley.

Chapter 37

The USA National IPM Road Map

Harold D. Coble and Eldon E. Ortman

In 1994, the US Department of Agriculture (USDA) and the US Environmental Protection Agency (EPA) jointly announced an "IPM Initiative" with the goal of achieving adoption of IPM on 75% of planted cropland area in the USA by the end of 2000 (Jacobsen, 1996). Both the USDA and EPA indicated that an anticipated outcome of that level of IPM adoption would be a reduction in pesticide use on the nation's farms. In order to accomplish the goals of the initiative, a modest increase in funding was allocated for research, outreach and education by both the USDA and EPA.

A survey conducted by the USDA National Agricultural Statistics Service (NASS) in 2000 indicated that growers adopting some level of IPM on farms had increased from around 50% at the beginning of the IPM Initiative to about 71% at the end of 2000 (National Agricultural Statistics Service, 2001). However, the anticipated reduction in pesticide use did not occur, and according to NASS surveys, pesticide use actually increased about 4% from 1994 to 2000 measured by quantity of active ingredient applied. In that same period, use of those pesticides considered most risky by EPA decreased by approximately 14%.

During 2000 and 2001, the US General Accounting Office (GAO) conducted a review of the IPM program. The review was sponsored by Senator Patrick J. Leahy of Vermont, Chairman of the Subcommittee on Research, Nutrition, and General Legislation of the Senate Committee on Agriculture, Nutrition, and Forestry. The review was requested in order to examine the status of IPM adoption in US agriculture. The purpose of the review was to determine how widely IPM had been adopted, document the environmental and economic results of IPM adoption and identify any impediments that limit IPM adoption and realization of its potential benefits. The GAO report on the review entitled *Agricultural Pesticides: Management Improvements Needed to Further Promote Integrated Pest Management* was issued August 17, 2001 (US General Accounting Office, 2001).

37.1 The GAO report

The GAO report on the IPM Initiative effectively ignored the fact that the goal of the initiative was a percent cropland adoption goal and not a pesticide reduction goal. Even though the 75% adoption goal was nearly reached (71%), GAO concluded that "the implementation rate is a misleading indicator of the progress made toward an original purpose of IPM – reducing chemical pesticide use." During the review process, GAO personnel were told numerous times that reducing the risk from pesticide use was a more important issue than total pounds of pesticides used. However, since an anticipated outcome of reduced

Integrated Pest Management, ed. Edward B. Radcliffe, William D. Hutchison and Rafael E. Cancelado. Published by Cambridge University Press.

pesticide use was mentioned in the rollout of the IPM Initiative, that outcome seemed to carry the most weight in the conclusions and recommendations in the report.

The GAO report concluded that the IPM initiative was missing several management elements essential for successful implementation of any federal effort. In the report, GAO wrote:

- No one is effectively in charge of federal IPM efforts.
- Coordination of IPM efforts is lacking among federal agencies and with the private sector;
- The intended results of IPM efforts have not been clearly articulated or prioritized; and
- Methods for measuring IPM's environmental and economic results have not been developed.

The report also issued four recommendations, mainly to the Secretary of Agriculture:

(1) Establish effective department-wide leadership, coordination and management for federally funded IPM efforts.
(2) Clearly articulate and prioritize the results the department wants to achieve from its IPM efforts, focus IPM efforts and resources on those results and set measurable goals for achieving those results.
(3) Develop a method for measuring the progress of federally funded IPM activities toward the stated goals of the IPM initiative.
(4) Collaborate with EPA to focus IPM research, outreach and implementation on the pest management strategies that offer the greatest potential to reduce the risks associated with agricultural pesticides.

GAO reports require a response from the departments and agencies involved. The Secretary of Agriculture delegated the responsibility of responding to the report to the USDA Office of Pest Management Policy (OPMP). OPMP organized the input from all USDA agencies reviewed and developed a response to the report. In the response, USDA pledged to set priorities for the results desired from the IPM program, to set measurable goals for the program and devise methods for measurement of progress toward the goals, and to develop a comprehensive, authoritative and focused road map for the IPM program. EPA submitted their response to GAO as well.

37.2 The IPM Road Map development process

In early 2002, a meeting was held in Washington, DC to begin the development process for the IPM Road Map. Approximately 40 individuals attended, representing various stakeholders, including federal and state research programs, cooperative extension, major crop and specialty crop commodity groups, public interest groups, regional IPM Centers, agricultural chemical companies, food product companies and federal agencies involved in supporting or regulating pest management activities. The group was asked to participate in discussions to identify the goals and objectives of a national IPM program and to suggest how priorities should be set for research and implementation efforts funded by the federal government. After the meeting, a first draft of the IPM Road Map was generated from output of the meeting.

Over the next two years, several revisions of the document were developed with input from over 100 individuals and organizations with interest in IPM. During the review process, it became obvious that there was great interest in expanding the IPM Road Map beyond the traditional boundaries of agriculture. The major thrust that emerged from the deliberations was that the program should be targeted toward risk reduction rather than some arbitrary pesticide use reduction goal. The final revision of the IPM Road Map was completed and approved by the Federal IPM Coordinating Committee on May 17, 2004.

37.3 National Road Map for Integrated Pest Management, May 17, 2004

Introduction

Integrated Pest Management, or IPM, is a long-standing, science-based, decision making process that identifies and reduces risks from pests and pest management related strategies. It coordinates the use of pest biology, environmental information and available technology to prevent unacceptable levels of pest damage by the most economical means, while posing the least possible risk to people, property, resources and the environment. IPM provides an effective strategy for managing pests in all arenas from developed agricultural, residential and public areas to wild lands. IPM serves as an umbrella to provide an effective, all encompassing, low-risk approach to protect resources and people from pests.

Keeping a step ahead

Pest management systems are subject to constant change, and must respond to a variety of pressures. For example, pests may become resistant to chemical pesticides, crop rotation or trapping methods. Regulatory agencies may restrict or phase out certain pesticides when their risks outweigh their benefits. Environmental concerns, consumer demands, and public opinion are significant influences in the marketplace related to pest management practices. IPM practitioners must now, more than ever, strive to implement best management practices and tools to incorporate a pest management regime where strategies work in concert with each other to achieve the desired effects while posing the least risks. Current and evolving conditions clearly signal the need for the increased development and adoption of IPM practices. The justification for a National IPM Road Map, which serves to make these transitions as efficient as possible, has never been greater.

The IPM Road Map

The goal of the IPM Road Map is to increase nationwide communication and efficiency through information exchanges among federal and nonfederal IPM practitioners and service providers including land managers, growers, structural pest managers, and public and wildlife health officials. Development of this document began in February 2002. Continuous input from numerous IPM experts, practitioners and stakeholders resulted in the current IPM Road Map.

The Road Map for the National Integrated Pest Management (IPM) Program identifies strategic directions for IPM research, implementation and measurement for all pests, in all settings, throughout the nation. This includes pest management for all areas including agricultural, structural, ornamental, turf, museums, public and wildlife health pests, and encompasses terrestrial and aquatic invasive species.

National IPM Program goals

The goal of the National IPM Program is to improve the economic benefits of adopting IPM practices and to reduce potential risks to human health and the environment caused by the pests themselves or by the use of pest management practices. The components of the goals for IPM are further described below.

IPM originally began in the agricultural area; however, in recent years, federal and state governments have broadened their focus on the interface between pests, pest management and people in the human environment, including residential, recreational, institutional facilities and in natural wild land areas. Through state and federal cooperation, a successful IPM in Schools program exists. The impact of exotic, invasive species in natural environments has received tremendous support with the 1999 Invasive Species Act. Federal and state agencies are developing Exotic Plant Management Teams towards this effort. IPM programs are under development at all levels to mitigate the impact of pest organisms.

The National IPM Program will focus its efforts in three areas – production agriculture, natural resources, and residential and public areas. At the core of each area lies a requirement for building and maintaining research, education, and extension programs that are tuned to the priorities outlined in the National IPM Road Map. Priorities for each of these focus areas are identified below.

IPM focus areas

Production agriculture

IPM experts, practitioners and stakeholders expect that systems will be further developed for food, fiber and ornamental crops that harness the full diversity of cost-effective pest management tactics, and improve their efficiency and effectiveness. By focusing on practices that prevent, avoid or mitigate pest attack, these IPM systems will have reduced negative impacts on the production area and associated environment by minimizing impairments to water quality. An important priority is the development and implementation of economical and effective IPM systems for crops and commodities consumed by humans. IPM systems in fruits, vegetables and other specialty crops will help to maintain high-quality produce, to protect agricultural workers and to keep dietary pesticide exposure within acceptable safety standards. These crops make up a major portion of the human diet and require high labor input for production. The priority in this area is to develop alternative tactics that have major economic benefits as well as protect public health including workers and the environment.

Natural resources and recreational environments

Our nation's natural resources and ecosystems are under constant pressures from encroaching invasive species. Invasive species diminish habitat quality and diversity for wildlife. Additionally, Americans spend large amounts of leisure time in natural and recreational environments such as lakes, streams and parks. Greater efforts are required to develop and quantify the impact of IPM programs in these environments. It is critical to protect public health and ecosystem function and minimize adverse environmental effects on natural areas, while maintaining functional and aesthetic standards. Environmental and health benefits should include reduction of pesticide residues in waters used for human consumption or for recreational purposes, as well as minimizing the effects of pesticides on non-target species.

Residential and public areas

The greatest general population exposure to pests and the tactics used to control them occurs where people live, work, and play. IPM programs for Schools and Public Buildings have already been very successful and are excellent examples of education and implementation programs designed for institutional facilities. Priorities in this area include enhanced collaboration and coordination to expand these programs to other institutions and residential environments. Expanding IPM programs in these areas would reduce human health risks posed by pests and the tactics used to manage them, and also reduce or mitigate the adverse environmental effects of pest management practices.

Future direction

Improve cost–benefit analyses when adopting IPM practices

Improving the overall benefits resulting from the adoption of IPM practices is a critical component of the National IPM Program. Conducting a cost–benefit analysis of proposed IPM strategies is not based solely on the monetary costs. It is based on four main parameters: (1) monetary, (2) environmental/ecological health and function, (3) aesthetic benefits and (4) human health.

While there may be many benefits from adoption of IPM practices, if new IPM programs do not appear to be as economically beneficial as practices already in place, they are not likely to be adopted. Risks and benefits need to be determined. A major factor in the adoption of IPM programs is whether the benefit to humans and the broader natural systems, outweighs the cost in implementing an IPM practice. Evaluation of the short- and long-term risks and benefits is needed.

Reduce potential human health risks from pests and related management strategies

IPM plays a major role in human health. Public health is dependent upon a continual supply of affordable, high-quality food. IPM protects human health through its contribution to food security by reducing potential health risks and enhancing worker safety. Success in reducing the health risks from pest management practices themselves were measured in the past by tracking changes in the annual amount of pesticides used in the USA. While pesticide use information is relatively easy to collect, when used alone it is a poor indicator of human health risk, and more advanced systems of measurement are required.

Minimize adverse environmental effects from pests and related management strategies

IPM programs are designed to protect agricultural, urban and natural resource environments from the encroachment of native and non-native pest species while minimizing unreasonable adverse effects on soil, water, air and beneficial organisms. For example, in agriculture, IPM practices promote a healthy within-crop environment, and conserve organisms that are beneficial to agricultural systems, including pollinators and natural enemies. By reducing off-target impacts, IPM also helps to maximize the positive contributions that agricultural land use can make to watershed health and function. IPM practices are used to suppress invasive species in natural wetlands ecosystems; the non-native invasive purple loosestrife for example is managed using a spot application of low-risk herbicide application for short-term control in conjunction with the release of biological control agents for long-term management.

Research, technical development, education, implementation

In order to continue IPM development and adoption it will be critical to enhance investment in: (1) new options for pest management, (2) public and private education infrastructure and (3) implementation and adoption of IPM.

Research needs

Research needs in IPM range from basic investigations of pest biology to the development of new pest management tactics in specific topics or settings. The following list illustrates some of the research needs for the National IPM Program.

- Clarify pest biology and host–pest–climate interactions to identify vulnerable cropping systems and vulnerable stages in the pest life cycle.
- Develop advanced management tactics for specific settings (e.g. crops, parks, the home, the workplace) that prevent or avoid pest attack.
- Develop economical high-resolution environmental and biological monitoring systems to enhance our capabilities to predict pest incidence, estimate damage and identify valid action thresholds.
- Develop new diagnostic tools, particularly for plant diseases and for detection of pesticide resistance in pest populations, including weeds.
- Develop new generation low-risk suppression tactics including biological control and products of traditional breeding and biotechnology.
- Improve action thresholds for vector-borne diseases; provide mechanisms for local vector-borne disease control agencies to monitor pest populations adequately to predict possible outbreaks and implement low-risk approaches to prevent outbreak levels.
- Improve the efficiency of suppression tactics and demonstrate least-cost options and pest management alternatives.
- Develop new delivery methods designed to expand the options for IPM implementation.

Technical development

While there has been dramatic improvement in pest management practices during the last three decades, there continues to be a critical need to devise new options that provide effective, economical and environmentally sound management of pest populations. A parallel need is to provide science-based information concerning the risks and benefits of IPM to the public. Meeting this need will facilitate support and informed discussion and involvement from stakeholders and consumers who understand the benefits of public investment in IPM programs.

Education

A diverse and evolving pest complex requires a cadre of trained individuals with enhanced management skills that ensure human health and environmental protection. It is important for practitioners to acquire new skills to implement targeted IPM strategies using new technologies, including genetic engineering, reduced risk pesticides, cultural practices and biocontrols.

The Federal Agency Core IPM Certification Training Program should be installed. This program will provide state-of-the-art, highly advanced training to federal IPM Practitioners preparing them with basic IPM principles skills and advanced courses in different technical categories.

Implementation and adoption of IPM

Agricultural producers, natural resource managers and homeowners must willingly adopt IPM practices for these programs to reach their full potential. And the public must have information to fully understand these programs. The following activities will contribute to the adoption of IPM:

- Develop user incentives for IPM adoption reflecting the value of IPM to society and reduced risks to users.
- Work with existing risk management programs including federal crop insurance, and incentive programs such as the NRCS Environmental Quality Incentive Program (EQIP) and other farm program payments to fully incorporate IPM tactics as rewarded practices.
- Provide educational opportunities for IPM specialists to learn new communication skills that enable them to engage new and unique audiences having specific language, location, strategy or other special needs.
- Create public awareness and understanding of IPM programs and their economic, health and environmental impacts, through education programs in schools, colleges and the workplace, and through creative use of mass media.
- Leverage federal resources with state and local public and private efforts to implement collaborative projects.
- Ensure a multidirectional flow of pest management information by expanding existing and developing new collaborative relationships with public and private sector cooperators.
- Spotlight successful IPM Programs.

Measuring performance of the National IPM Program

Governments at the national and state levels through directives, rules and laws are placing high priority on the development and implementation of accountability systems. Such systems are based on performance measurements, including setting goals and objectives and measuring progress toward achieving them. Accordingly, federally funded IPM Program activity performance must be evaluated.

The establishment of measurable IPM goals and the development of methods to measure progress toward achieving the goals should be appropriate to the specific IPM activity undertaken. Performance measures may be conducted on a pilot scale or on a geographic scale and scope that corresponds to an IPM program or activity. Examples of potential performance measures follow.

Outcome: the adoption of IPM practices improves economic benefits to users

Performance measures

- In cooperation with the NASS, design a national IPM practices adoption survey based on IPM protocols designed for specific commodities or sites within program priorities.
- Evaluate IPM programs on their ability to improve economic benefits using pilot studies within specific program priority sites and project these economic results to a regional or national basis to predict large-scale impacts using results of the practices adoption survey.
- Develop measures of public awareness of IPM.

Outcome: potential human health risks from pests and the use of pest management practices are reduced

Performance measures

- Using EPA's reduced risk category of pesticides as the standard, document changes in pesticide use patterns over time and relate the changes to IPM practice adoption.
- Relate dietary exposure to pesticides to IPM practice adoption using USDA Agricultural Marketing Service (AMS) Pesticide Data Program (PDP) and any other available data.
- Relate cases of the negative human health impacts caused by pest incidence (for example, asthma cases related to cockroach infestation, insect vectored diseases, allergic reactions to plants) to IPM practice adoption.

Outcome: unreasonable adverse environmental effects from pests and the use of pest management practices are reduced

Performance measures

- Document and relate pesticide levels in specific ground and surface water bodies, including community water supplies, to IPM practice adoption using data from the US Geological Survey (USGS), the Natural Resource Conservation Service (NRCS) and others.

- Document and relate national indicators of natural resource health such as proportion of ground and surface water bodies with pest management-related contaminants and level of contamination to IPM practice adoption, using data from EPA and others.
- Measure the impact of IPM practice adoption on encroachment of selected invasive species in national park lands and other sites where data are available.

National IPM Program leadership and coordination

The National IPM Program is a broad partnership of governmental institutions working with many stakeholders on diverse pest management issues. Leadership, management and coordination of these IPM efforts will occur at several levels to address more completely the needs of program stakeholders.

At the federal level, the IPM Program is a multi-agency effort that demands coordination and collaboration. The Federal IPM Coordinating Committee will provide oversight of the federally funded programs. This committee will be made up of representatives of the major participating federal agencies and departments. The role of the committee will be to establish overall goals and priorities for the program. To achieve this, the Federal IPM Coordinating Committee will require a dynamic system of information flow and feedback that provides an up-to-date, accurate assessment of the status of IPM and the evolving requirements of numerous IPM programs. Stakeholder input to the Federal IPM Coordinating Committee will occur through the USDA Regional IPM Centers. The USDA IPM Coordinator will be responsible for preparing an annual report documenting the status and performance of the IPM Program nationally and distributing the report to Congress, federal and state IPM partners and the general public.

USDA Regional IPM Centers will play a major role in gathering information concerning the status of IPM, and in the development and implementation of an adaptable and responsive National IPM Road Map. These Centers will have a broad, coordinating role for IPM and they will invest resources to enhance the development and adoption of IPM practices.

37.4 Utilization of the Road Map

The IPM Road Map is meant to serve as a guide in identifying the broad IPM goals that are considered most important for the federal government, and to indicate where government support for IPM research, education and outreach will be focused. Federal granting agencies now cite the IPM Road Map goals in their Requests for Applications for grants, and successful grant applications include those goals in their proposals. The Road Map was never intended to be prescriptive in terms of dictating specific work and activities for IPM researchers, educators and practitioners.

37.5 Other responses to the GAO report

In their report on IPM, GAO recommended that the Secretary of Agriculture establish effective department-wide leadership, coordination and management for federally funded IPM efforts. This recommendation was made after the GAO found that one of several impediments to the realization of IPM's potential benefits was a serious leadership, coordination and management deficiency within the Department.

In an effort to provide an infrastructure for appropriate management of the federal IPM program, the Secretary proposed the formation of the Federal Integrated Pest Management Coordinating Committee (FIPMCC). The FIPMCC was formed in 2003 and is composed of representatives of all federal agencies with IPM research, implementation, or education programs. There are currently 17 federal departments, agencies or services represented on the committee. The function of the FIPMCC is to provide inter-agency guidance on IPM policies, programs and budgets. A key responsibility of the FIPMCC is to provide a forum for communication among federal offices with IPM programs to insure efficiency of operations. The committee has been active in defining, prioritizing and articulating the goals of the federal IPM effort through the IPM Road Map, making sure IPM efforts and resources are focused on the goals, and insuring that appropriate measurements toward progress in attaining the goals are in place. The FIPMCC reports to the Secretary of Agriculture through the USDA Office of Pest Management Policy.

References

Jacobsen, B. J. (1996). USDA Integrated Pest Management Initiative. In *Radcliffe's IPM World Textbook*, eds. E. B. Radcliffe & W. D. Hutchison. St. Paul, MN: University of Minnesota. Available at http://ipmworld.umn.edu.

National Agricultural Statistics Service (2001). *Pest Management Practices: 2000 Summary*. Washington, DC: US Department of Agriculture. Available at http://usda.mannlib.cornell.edu/usda/current/PestMana/PestMana-05-30-2001.txt.

US General Accounting Office (2001). *Agricultural Pesticides: Management Improvements Needed to Further Promote Integrated Pest Management*, Report No. GAO-01-815. Washington, DC: US General Accounting Office. Available at www.gao.gov/new.items/d01815.pdf.

Chapter 38

The role of assessment and evaluation in IPM implementation

Carol L. Pilcher and Edwin G. Rajotte

IPM is, by many accounts, a highly successful program. Some claim that the achievements over the last 40 years have given the program a mark of success. In fact, IPM has been described as "one of the best answers to reducing chemical contamination of the environment and improving the safety of food while maintaining agricultural viability" (Rajotte, 1993, p. 297). Others argue that implementation has been slow and success has been limited. Wearing (1988) stated that IPM has not been successful because some IPM technologies have not been adopted by growers. This limited success has resulted in a weakening of political support and a stagnation of funding for IPM in the USA (Gray, 2001). What defines a successful IPM program? How do we assess the true worth of an IPM program? These questions can be answered by conducting an assessment of IPM programs through program evaluation. However, the starting point of any IPM evaluation is made with three goals in mind; economic, an assessment of the costs and benefits; environmental, impacts on soil, water and non-pest organisms; and social, an assessment of a program's impact on people's health and well-being.

38.1 A historical review of IPM evaluation

Evaluation has been a component of some pest management programs, starting back in the 1940s. During this time (1940s to 1960s), program evaluations were used to assess the needs of clients and determine the future directions (Allen & Rajotte, 1990). Sociologists examined farmers' sources of information and adoption of selected IPM practices (e.g. variety selection and record keeping) (Wilkening, 1950).

As funding and support for IPM grew in the 1970s, so did economic impact assessments. According to Regev *et al.* (1976) "Recently, problems in pest management have attracted considerable attention in economic literature" (p. 186). Some of these economic assessments included cotton (Parvin, 1976) and alfalfa research (Regev *et al.*, 1976).

By 1977, Extension IPM pilot programs were successfully implemented in 33 states in the USA. These extension programs were evaluated for their impact. Conclusions from this evaluation include: (1) pesticide usage can be reduced 30% to 70% by discontinuing unwarranted applications, (2) economic benefits to farmers and society can occur from savings in costs of pesticides and their applications as well as increases in the crop yield and/or quality and (3) fewer pesticides enter the environment and residues of pesticides in food products are minimized (Gray, 2001, p. 34).

At that same time, the US Department of Agriculture–Extension Service (USDA–ES) developed a task force to examine the impacts associated with extension programs. IPM was selected to be reviewed during the 1984–1987 funding cycle.

Integrated Pest Management, ed. Edward B. Radcliffe, William D. Hutchison and Rafael E. Cancelado. Published by Cambridge University Press.

The result was a series of nationwide impact level studies conducted by Virginia Polytechnic Institute and State University in collaboration with the USDA–ES and the state Extension programs. Information was collected on extension IPM personnel and private pest management consultants throughout the nation. From these studies, it was documented that IPM programs covered over 11 million ha and private consultants scouted over 1.2 million ha (Rajotte *et al.*, 1987). In addition, case studies were conducted on 16 different states and nine different commodities. It was documented that IPM users reported a net return (when compared with non-IPM users) of $US 578 million per year (Rajotte *et al.*, 1987) based on a $US 7 million federal investment of earmarked funds (Rajotte, 1993).

In the 1980s and 1990s, research on the socio-economic factors associated with adoption of IPM practices was common. The focus of this research was the adoption and diffusion of innovations (Rogers, 1983, 1995). Models were developed to explain whether or not a grower decides to use IPM, the communication method by which he learns about the system, the source of this information and the timing of the information. Several research programs sought to identify the most prominent and/or useful sources(s) of IPM information (Ford & Babb, 1989; Thomas *et al.*, 1990; Ortmann *et al.*, 1993). Other studies (Ridgley & Brush, 1992; McDonald & Glynn, 1994) advocated selective adoption or bundling of technology based on the individual needs of growers.

During the 1990s, innovative approaches to measuring the environmental impacts associated with the adoption of IPM were advanced. The environmental impact quotient (EIQ) was developed by Kovach *et al.* (1992). The environmental economic injury level (EIL) was introduced by using the process of contingent valuation to assign a monetary value to the environmental costs associated with pesticide use (Higley & Wintersteen, 1992). Several other environmental indicators were suggested. In fact, Levitan *et al.* (1995) lists 14 different environmental measures for IPM published in the 1990s.

In 1993, the Government Performance and Results Act (GPRA) brought accountability to center stage. This act required federal agencies to collect performance measures on short-term and long-term goals and conduct program evaluations. The GPRA of 1993 recognized the relationship between performance measures and evaluation by requiring federal agencies to complete program evaluations with assessments of outcomes, especially those at the impact level.

Unfortunately, the trend towards a technology transfer treadmill dominated the field of IPM. Identifying pest problems and responding with short-term preventative and curative IPM tactics became a common strategy for researchers. IPM researchers developed IPM projects and applied new management technologies so they could stay one step ahead of a particular pest problem (Norton *et al.*, 1999). Perhaps these scientists did not have the time or resources to determine if these new technologies were adopted by the end-users nor did they consider the long-term consequences of adopting such tactics. In some cases, scientists partnered with extension personnel or sociologists so the knowledge and technologies could be disseminated to the end-users (Hamilton *et al.*, 1997). A few select IPM researchers continued to produce impact-level results (Fernandez-Cornejo, 1998); however, no accepted comprehensive nationwide evaluation was conducted during this era.

In 2001, the status of IPM was called into question. In a US General Accounting Office (GAO) Report to Senator Patrick Leahy (US General Accounting Office, 2001), shortcomings were identified and criticisms were raised. In terms of assessments and evaluation, the report criticized the USDA for not developing a method to measure outcomes or the environmental or economic impacts associated with IPM implementation. The report made specific recommendations to the IPM program that it should establish measurable goals and develop a method for measuring the progress of the program (US General Accounting Office, 2001).

Since this report, governments at all levels have continued to increase accountability demands and have embraced performance measurement (i.e. monitoring inputs, outputs and outcomes) and requirements to report results (McDavid & Hawthorn, 2006). In 2003, GAO in cooperation with the Office of Management and

Budget (OMB) developed a new system for performance measurement. This new system of measurement is the Program Assessment Rating Tool (PART). PART is designed to bring federal agency performance and budgeting in closer alignment (Datta, 2007). In times of budget shortfalls and increasing demands for accountability, the "so what" question is being asked for all programs at all levels.

More recently, a notable exception to the lack of impact evaluation in IPM programs was the economic assessment of one of the newest IPM tools, web-based pest warning systems (Roberts *et al.*, 2006). The first comprehensive system was created in response to invasion by Asian soybean rust (*Phakopsora pachyrhizi*) in November 2004. This report documented that the new tool provided up to a $US 299 million benefit to US agriculture while costing less than $US 5 million in development and operations. More national-level, impact-level evaluations such as this are needed.

38.2 Objectives of IPM program evaluation

This chapter will mainly address IPM implementation by the public sector, primarily land-grant university research/extension programs. However, these principles can apply to other public and private organizations. In addition, these principles can apply to local, regional or national level programs.

Program evaluation is not a "chore" that must be completed at the end of project, but an ongoing process that helps guide the implementer through the entire project. Perhaps program evaluation is a different way of thinking about implementation, but program evaluation should be a primary consideration at the onset of any IPM program. Program evaluation should be integrated into: (1) the planning phase of the IPM project from the inception of the idea to development and testing of the new technology (formative evaluation–needs assessment); (2) the implementation phase of the IPM project when the technology is disseminated to the end-user (formative evaluation–implementation assessment); and (3) the evaluation phase of the IPM project when the IPM implementer determines the level of adoption and measures impacts associated with the adoption of IPM technology (summative evaluation–outcomes assessment).

38.2.1 Program planning

Program evaluation should be an integral part of the planning phase of an IPM program. By initiating the program evaluation from the start, the IPM implementer can use this tool to guide him/her through the entire project. The following are important considerations during the planning phases of an IPM project.

What is the problem or situation?

The IPM implementer must understand more than just the basic pest problem and new management technology. The implementer must conduct a "needs assessment" by identifying and gathering input from all stakeholders (especially the end-users or IPM adopters) involved in the selected IPM arena. Often times, implementers overlook important stakeholders in the political arenas. After all, most public-sector IPM implementation is publicly funded, and public funds are subject to political processes. Other internal and external stakeholders are equally important and all should be considered. These stakeholders include consumers, non-governmental organizations, farm organizations, government agencies and others. Once the needs assessment is complete, then the implementer can develop a problem statement and define the overall strategy of the program (including goals and objectives).

What resources (inputs) are needed?

Typically, the implementer defines the inputs or resources needed to produce the outputs and outcomes needed to satisfy program goals and objectives. Additional inputs include: equipment, facilities, staff, time and technologies needed to complete the activities. This input information is particularly helpful if efficiency measures (i.e. cost–benefit or cost effectiveness) measures are being sought. These measures determine if the resources are being used to the fullest extent possible. Although efficiency measures are difficult to measure, they are becoming more prevalent as decision makers must chose how to allocate scarce resources and put these resources to optimal use (Rossi *et al.*, 2004).

Who are the recipients of the IPM technology?
The IPM implementer must also understand the characteristics and important attributes of the end users (i.e. IPM adopters or non-IPM adopters) of the IPM technologies, as these end-users are the focus of the output and outcome measurements. By understanding the characteristics correlated with adoptive behavior, the implementer can tailor the program to the target population and aid in the adoption process. Factors such as age, level of education, family income, farm income and farm size are just a few characteristics that influence the adoption of IPM practices (Alston & Reding, 1998; Conley *et al.*, 2007).

What activities will be conducted with the IPM technology?
IPM is an information-intensive technology and one of the main elements is the transfer of this information. Thus, IPM implementers need to design an information delivery system. Understanding factors such as primary source of information and type of delivery method are paramount for success. Research on primary sources of information indicates that different IPM end-users utilize different sources of information. Patrick *et al.* (1993) found that large-scale farmers (average operation 737 ha) use soil fertility consultants, farm worker force, university specialists and field days/conferences as their primary sources of information. Alston & Reding (1998) found that fruit growers relied on extension agents, other growers or trained employees for IPM information, while small grain growers relied on extension agents, agricultural chemical dealers and other growers for IPM information. Types of delivery methods also have been thoroughly researched (Patrick *et al.*, 1993; Garber & Bondari, 1996; Malone *et al.*, 2004; Conley *et al.*, 2007). Once again, the individual audiences varied in terms of their preference for the type of delivery method.

More recently, a growing field of research has divided IPM delivery into three basic approaches: (1) technology transfer, (2) farmer first and (3) participatory approach. The participatory approach has been given the most attention as a successful model for IPM adoption. "Participatory IPM research, through its involvement of farmers, marketing agents, and public agencies, is designed to facilitate diffusion of IPM strategies" (Rajotte *et al.*, 2005, p. 143). Most importantly, this approach allows farmers and scientists to learn together and work together to solve the pest problem with an IPM approach. The program is multidisciplinary in nature where sociologists and economists are assisting biological researchers with the IPM project. This is essential for proper diffusion of the IPM technology, as well as assessment of the economic impacts associated with the adoption of the technology. Furthermore, evaluation is a component of entire IPM transfer of information process. "While crop/pest monitoring is underway, farmers and scientists work together to design, test and evaluate IPM tactics and management strategies" (Norton *et al.*, 2005, p. 20). This participatory approach validates the need for a tailored delivery system for successful adoption of IPM.

Regardless of the activities selected for IPM transfer of information, the implementer must clearly define the information delivery system and procedures for delivering services to the target population. These activities must provide a rationale approach to achieving the goals and objectives of the program and must be tailored to fit the needs of the end users.

38.2.2 What will be measured?

IPM implementers need to link the goals and objectives with appropriate measures to determine what information needs to be collected throughout the program. Implementers must consider the following elements.

Defining IPM
IPM definitions vary widely and may involve several practices or a variety of strategies. In fact, Bajwa & Kogan (2002) identified 67 different definitions of IPM from 1959 to 2000. More recently, the National IPM Road Map was developed to serve as a guide and provide a strategic direction for IPM research, implementation and measurement throughout the nation (see Chapter 37).

IPM Impact Area		IPM Focus Area: Production Agriculture	Residential and Public Areas	Natural Resources and Recreational Environments
	Environmental Impacts			
	Health Impacts			
	Economic Impacts			

Fig. 38.1 IPM Matrix, after Hoffman & Ortmann (presented at annual meeting of the Entomological Society of America, 2004).

This National IPM Road Map and the IPM Matrix (Fig. 38.1) should be used to as a starting point to define IPM and its parameters.

Defining level of measurement

Once the basic parameters of IPM have been defined, then it is necessary to determine what level of measurement is appropriate for the research project. These measurements are a descriptive approach to providing a comprehensive view of the program performance (Poister, 2004). It should be noted that evaluation projects may have more than one activity, reaction, short-term outcome, medium-term outcome and long-term outcome. Table 38.1 shows the various levels of measurement and select examples of program performance measures with a focus on IPM. As one moves down the chart in terms of level of measurement, the level of accountability increases. Unfortunately, the costs associated with collecting these data also increases. The implementer needs to realize the appropriate level of measurement for the project and the budget allocated towards evaluation.

Outcome level measures are strongly emphasized in today's realm of accountability for most programs. These measures represent the extent to which a program is effective in producing outcomes and achieving desired results (Poister, 2004). The highest-level accountability or long-term outcome measures are referred to as program impacts. In the realm of IPM programs, environmental, economic and health impacts are the targets for accountability. The National IPM Road Map outlines each of these long-term outcomes or impacts (see Chapter 37).

Regardless of the level of outcomes (short-term, intermediate, or long-term), when implementers define the outcomes and performance measures for IPM programs, each should contain the following (SMART) elements.

Specific – identify the population and the expected results
Measurable – identify an outcome that can be measured
Attainable – identify a level that is reasonable and
Results-oriented – identify an indicator that is meaningful
Time – identify the specific time-frame in which expected results will occur.

It is important for the implementer to prioritize and focus on the most important level of measurement that will provide an accurate and comprehensive view of the IPM program. Ideally,

Table 38.1 Levels of measurement and IPM examples of reporting at each level of measurement

Level of measurement	Description	Examples for IPM
Inputs	Staff, time and resources	During the last training year, it was estimated that 200 staff in county extension offices, 12 field crop specialists, and 10 on-campus staff devoted a total of 14 000 hours to the program
Outputs	Participants	Last year, it was estimated that 100 farmers attended the special field day on soybean aphid management options
	Activities	Within three years of the program start-up, the School Integrated Pest Management Program trained more than 30 districts statewide
	Reactions	During the 2006 training season, 99% of the respondents indicated that the program was excellent or good
Outcomes Short-term	Change in knowledge	Of the individuals that participated in the 2003 Scout School post survey, 78% reported moderate to significantly improved weed management knowledge
Medium-term	Change in behavior or practices	In a follow-up survey of participants from the 2005 field workshop, 45% of crop scouts reported that they adopted the speed scouting methods for soybean aphid
Long-term	Impacts	In 2003, over 200 individuals attended the Master Gardening short course. A follow-up survey showed that more than 30% reduced pesticide applications in their gardens. It was estimated that these participants saved a total of $US 2390 during the 2004 gardening season

an effective IPM program evaluation must summarize program accomplishments, assess environmental, health and economic impacts the program has made on internal and external stakeholders and ultimately determine the worth of the program.

Defining data collection

Once the performance measures have been identified and prioritized, then it is important to define the process of data collection. The IPM implementer should collect relevant data throughout the program evaluation (i.e. both formative and summative stages of evaluation). If a critical opportunity is lost, there is no way to return in time to collect the information. Thus, it is important to consider data collection from the onset of the program. Four basic areas of data should be examined and the following questions should be addressed:

(1) *Data collection* What type of data collection methods is appropriate? Data can be collected through qualitative or quantitative approaches. True experimental, quasi-experimental, non-experimental and historical research designs are all options for program assessment.
(2) *Baseline data* What existing data are available and are they useful for this study? Data on previous measures of outputs or outcomes (especially impacts) can be used as baseline data.
(3) *Data collection strategy* If baseline data are available, will the data collection methods be appropriate for this study? More specifically, are existing survey instruments available? If not, how will the new instrument be developed? How will the new instrument be tested for reliability and validity? Will all data be gathered on all program participants or only on a sample? How will the samples size be selected? Who will collect the data? How often

will the data be collected? The entire data collection process must be clearly defined from the onset of the program.[1]

(4) *Data analyses strategy* Once the data are collected, how will the results be analyzed and interpreted? What level of measurement is used for each variable (e.g. nominal, ordinal, interval or ratio)? Given the level of measurement, what statistical tests are appropriate?

A well-planned evaluation in terms of the data collection will help the implementer gather the appropriate information, minimize unanticipated events in the process and keep the program on track. Appropriate information can be gathered only after the implementer defines IPM, the appropriate level of measurement and process of data collection. Only then can the information gathered provide answers to the most important questions in a program evaluation.

38.3 Program implementation

Assessment during the implementation phase of the IPM project helps the implementer understand the program in practice and how the program functions. More specifically, evaluation of the implementation process helps explain what is working, what is not working, why some IPM technologies are adopted or not adopted and how to improve future programs. This is an opportunity to conduct an evaluation that serves as more than an assessment, it can also assist with program improvement.

A theory based evaluation approach has been used to explain the implementation of IPM programs. Several implementers have focused on the adoption and diffusion of innovations (Rogers, 1983, 1995) to explain program implementation of IPM. As stated earlier, the model can be used to explain whether or not a grower decides to use IPM, the communication method by which he learns about the system, the source of this information, and the timing of the information.

38.3.1 Barriers to adoption

Just as it is important to understand the theory of IPM adoption, it is also important to identify some of the barriers to adoption of IPM practices. By understanding the barriers to adoption, IPM implementers can avoid the pitfalls from the beginning. These barriers have been well documented in the IPM realm.

Herbert (1995) identified the following obstacles to IPM adoption.

- *Technical, financial, social marketing, educational* IPM programs are interdisciplinary in nature. It takes interdisciplinary knowledge at all phases of the program to develop a successful program (i.e. agronomists, economists, entomologists, sociologists and others). There simply may not be the team of experts or resources available to pull a successful program together.
- *Complexity* Most IPM programs are complex in terms of the information the practitioner has to assimilate. It takes time to process this complex information. The end-user may not be willing or have the time to process this information.
- *Perceived risk and lack of trust* IPM programs often are a "new" way of managing the pest and the practitioner may not trust that the IPM tactic will work. The end-user simply does not want to risk a yield loss and thus may not trust the new approach.
- *Lack of convincing information* IPM programs are often a "new way of thinking" and often have little convincing information to prove that they will work. Most IPM practitioners must be shown and convinced that the program will work before they will adopt a new approach to managing the pest.
- *Lack of incentives* The new IPM program may be more expensive that the traditional method of control (e.g. pesticide application). If end-users have a proven, effective management tool, they

[1] Before any data collection occurs, it is important for the implementer to contact their Human Subjects Office or Institutional Review Board. At each institution, this office upholds and enforces guidelines established by the Declaration of Helsinki and the Nuremberg Code. Each institution is different, but they all review research involving human subjects and make certain implementers follow all established guidelines and procedures.

may not adopt a new IPM program unless there is an incentive to try the new tactic.

In addition to these barriers to initial adoption, IPM programs often fail because of a lack of attention to "access conditions" (Audirac & Beaulieu, 1986; Rajotte & Bowser, 1991). Access conditions are determined, in part, by the development of the technology and by private and public diffusion infrastructures. Often times, IPM technologies are developed with the prime emphasis on the technology itself. They are first tried in an ideal setting such as a university experiment farm, and they are implemented by knowledgeable people with up-to-date equipment and training. However, the average IPM practitioner may be discouraged from adopting a technology if there is a lack of accompanying services/opportunities (the access conditions). These conditions may include a training program, the ability to purchase and use special equipment, the availability of a local expert to consult if there are problems, and even the knowledge that IPM technology support will continue even after the grant funding ends.

Understanding these barriers can help the implementer (1) avoid obvious implementation problems, (2) ask appropriate questions during the implementation phase and (3) generate useful information for program adaptation and improvement (Patton, 1997). If the implementation phase of the program is a failure, then the outcomes assessment cannot be conducted and the program evaluation will be a failure.

38.4 Collecting, analyzing and interpreting data

Because the data collection methods have been clearly defined in the program planning phase, the implementer can more easily conduct data collection and analyses according to the plan. Implementation assessments are conducted to determine how the program is performing and outcomes assessments are conducted to measure the changes or intended effects on the target population. The data collected for each assessment is analyzed according to the program plan.

Once data have been analyzed, it is important for the IPM implementer to interpret the results. The implementer must consider the research design, the data collected and the results. One important consideration for the implementer is determining the difference between statistical significance and practical significance. Statistical significance may be the first assessment of the magnitude of a measured program effect, but the implementer must judge whether the results have practical significance (Rossi, 2004).

38.5 Reporting results

Providing a clear and concise summary of the results is key to a good evaluation and strategic communication plan. Complicated statistical methods and long reports will seldom be used by any audience. Furthermore, many times the only section that is read is the "Recommendations" (Morris *et al.*, 1987). Thus, the implementer should provide only major findings and suggested course of action in this section. Technical reports, executive summaries, professional papers (i.e. journal manuscripts), popular articles, news releases, brochures, posters, memos and meetings are all possible forms for reporting the results. Each type of report has its own criteria and audience and should be examined for use in the communication plan.

38.6 Communicating findings

One of the most important components of the program evaluation is the communication plan. The communication plan uses the results from the program evaluation to provide valuable feedback allowing an IPM program to be constantly improved and its value demonstrated to internal and external stakeholders. It is important to target key decision makers with "hard-hitting" information (Taylor-Powell *et al.*, 1996). Unfortunately, communication plans have been relatively nonexistent in the IPM realm. According to Rajotte (1993) "IPM has been practiced for many years in most states. However, very little effort has been made to advertise this fact to the public or describe the economic and other benefits of these programs" (p. 299).

This failure on the part of IPM has resulted in the loss of support and the stagnation of funding. IPM projects must strive to develop communication plans. They must identify the stakeholder audiences for dissemination of their successes. They must identify the information needs of these stakeholder audiences. For instance, legislators and legislative aids may only have 15 minutes to review a document, thus a 100-page narrative on impact-level IPM research is not an appropriate report for this audience. Instead, a short two-page document with graphs and pictures and short text boxes would be more appropriate. IPM program evaluation must develop an effective communication plan and we must become stronger advocates for our IPM programs. Only then can we show the true worth of these important programs.

References

Allen, W. A. & Rajotte, E. G. (1990). The changing role of extension entomology in the IPM era. *Annual Review of Entomology*, **35**, 379–397.

Alston, D. G. & Reding, E. M. (1998). Factors influencing adoption and education outreach of Integrated Pest Management. *Journal of Extension*, **36**, 1–8.

Audirac, I. & Beaulieu, L. J. (1986). Microcomputers in agriculture: a proposed model to study their diffusion/adoption. *Rural Sociology*, **51**, 60–77.

Bajwa, W. I. & Kogan, M. (2002). *Compendium of IPM Definitions (CID): What Is IPM and How Is It Defined in the Worldwide Literature?* IPPC Publication No. 998. Corvallis, OR: Oregon State University, Integrated Plant Protection Center (IPPC) .

Conley, S. P., Krupke, C., Santini, J. & Shaner, G. (2007) Pest management in Indiana soybean production systems. *Journal of Extension*, **45**, 1–9.

Datta, L. (2007). Looking at the evidence: what variations in practice might indicate. *New Directions for Evaluation*, **113**, 35–54.

Fernandez-Cornejo, J. (1998). Environmental and economic consequences of technology adoption: IPM in viticulture. *Agricultural Economics*, **18**, 145–155.

Ford, S. A. & Babb, E. M. (1989). Farmer sources and uses of information. *Agribusiness*, **5**, 465–477.

Garber, M. P. & Bondari, K. (1996). Landscape maintenance firms. II. Pest management practices. *Journal of Environmental Horticulture*, **14**, 58–61.

Gray, M. E. (2001). The role of extension in promoting IPM programs. *American Entomologist*, **47**, 134–137.

Hamilton, G. C., Robson, M. G., Ghidiu, G. M., Samulis, R. & Prostko, E. (1997). 75% adoption of integrated pest management by 2000? A case study from New Jersey. *American Entomologist*, **43**, 74–78.

Herbert, D. A. Jr. (1995). Integrated pest management systems: back to basics to overcome adoption obstacles. *Journal of Agricultural Entomology*, **12**, 203–210.

Higley, L. G. & Wintersteen, W. K. (1992). A novel approach to environmental risk assessment of pesticides as a basis for incorporating environmental costs into economic injury levels. *American Entomologist*, **38**, 34–39.

Kovach, J., Petzoldt, C., Degni, J. & Tette, J. (1992). *A Method to Measure the Environmental Impact of Pesticides*, New York Food and Life Sciences Bulletin No. 139. Geneva, NY: Cornell University New York State Agriculture Experiment Station. Available at http://ecommons.library.cornell.edu/handle/1813/5203.

Levitan, L., Merwin, I. & Kovach, J. (1995). Assessing the relative environmental impacts of agricultural pesticides: the quest for a holistic method. *Agriculture, Ecosystems and Environment*, **55**, 153–168.

Malone, S., Herbert, D. A. Jr., & Pheasant, S. (2004). Determining adoption of integrated pest management practices by grain farmers in Virginia. *Journal of Extension*, **42**, 1–7.

McDavid, J. C. & Hawthorn, L. R. L. (2006). *Program Evaluation and Performance Measurement: An Introduction to Practice*. Thousand Oaks, CA: Sage Publications.

McDonald, D. G. & Glynn, C. J. (1994). Difficulties in measuring adoption of apple IPM: a case study. *Agriculture, Ecosystems and Environment*, **48**, 219–230.

Morris, L. L., Taylor Fitz-Gibbon, C. & Freeman, M. E. (1987). *How to Communicate Evaluation Findings*. Newbury Park, CA: Sage Publications.

Norton, G. A., Adamson, D., Aitken, L. G. *et al.* (1999). Facilitating IPM: the role of participatory workshops. *International Journal of Pest Management*, **45**, 85–90.

Norton, G. A., Rajotte, E. G. & Luther, G. C. (2005). Participatory Integrated Pest Management (PIPM) Process. In *Globalizing Integrated Pest Management*, eds. G. W. Norton, E. A. Heinrichs, G. C. Luther & M. E. Irwin, pp. 13–26. Ames, IA: Blackwell Publishing.

Ortmann, G. F., Patrick, G. F., Musser, W. N. & Doster, D. H. (1993). Use of private consultants and other sources of information by large Cornbelt farmers. *Agribusiness*, **9**, 391–402.

Parvin, D. W. Jr. (1976). Farm management implications of reducing agricultural pollution related to cotton production in Mississippi. *American Journal of Agricultural Economics*, **58**, 978–983.

Patrick, G. F., Ortmann, G. F., Musser, W. G. & Doster, D. H. (1993). Information sources of large scale farmers. *Choices*, 3rd Quarter, 40–41.

Patton, M. Q. (1997). *Utilization-Focused Evaluation*, 3rd edn. Thousand Oaks, CA: Sage Publications.

Poister, T. H. (2004). Performance monitoring. In *Handbook of Practical Program Evaluation*, eds. J. S. Wholey, H. P. Hatry & K. E. Newcomer, pp. 98–125. San Francisco, CA: Jossey-Bass.

Rajotte, E. G. (1993). From profitability to food safety and the environment: shifting the objectives of IPM. *Plant Disease*, **77**, 296–299.

Rajotte, E. G. & Bowser, T. (1991). Expert systems: an aid to the adoption of sustainable agriculture. In *Sustainable Agriculture and Extension in the Field*, pp. 406–427. Washington, DC: National Academy Press.

Rajotte, E. G., Kazmierczak, R. Jr., Norton, G. W., Lambur, M. T. & Allen, W. A. (1987). *The National Evaluation of Extension Integration Pest Management IPM Programs*, VCES Publication No. 491–010. Virginia Cooperative Extension Service.

Rajotte, E. G., Norton, G. W., Luther, G. C., Barrera, V. & Heong, K. L. (2005). IPM transfer and adoption. In *Globalizing Integrated Pest Management: A Participatory Research Process*, eds. G. W. Norton, E. A. Heinrichs, G. C. Luther & M. E. Irwin, pp. 143–158. Oxford, UK: Blackwell Publishing.

Regev, U., Gutierrez, A. P. & Feder, G. (1976). Pests as a common property resource: a case study of alfalfa weevil control. *American Journal of Agricultural Economics*, **58**, 186–197.

Ridgley, A. M. & Brush, S. B. (1992). Social factors and selective technology adoption: the case of integrated pest management. *Human Organization*, **51**, 367–378.

Roberts, M. J., Schimmelpfennig, D., Ashley, E. *et al.* (2006). *The Value of Plant Disease Early-Warning Systems: A Case Study of USDA's Soybean Rust Coordinated Framework*, Economic Research Report No. (ERR-18). Washington, DC: US Department of Agriculture, Economic Research Service.

Rogers, E. M. (1983). *The Diffusion of Innovations*. New York: The Free Press.

Rogers, E. M. (1995). *The Diffusion of Innovations*, 4th edn. New York: The Free Press.

Rossi, P. H., Lipsey, M. W. & Freeman, H. E. (2004). *Evaluation: A Systematic Approach*, 7th edn. Thousand Oaks, CA: Sage Publications.

Taylor-Powell, E., Steele, S. & Douglah, M. (1996). *Planning a Program Evaluation*, Publication No. G-3658–1. Madison, WI: University of Wisconsin, Cooperative Extension.

Thomas, J. K., Ladewig, H. & McIntosh, W. A. (1990). The adoption of integrated pest management practices among Texas cotton growers. *Rural Sociology*, **55**, 395–411.

US General Accounting Office (2001). *Agricultural Pesticides: Management Improvements Needed To Further Promote Integrated Pest Management*, Report N. GAO 801-815. Washington, DC: US General Accounting Office.

Wearing, C. H. (1988). Evaluating the IPM implementation process. *Annual Review of Entomology*, **33**, 17–38.

Wilkening, E. A. (1950). Sources of information for improved farm practices. *Rural Sociology*, **15**, 19–30.

Chapter 39

From IPM to organic and sustainable agriculture

John Aselage and Donn T. Johnson

Agricultural production is moving from less to more sustainable practices. This is a response to changing cultural values that promote environmental stewardship and ensure a healthy planet for future generations. Producers supply and consumers demand a product that is economical and uncomplicated. Consequently, most agriculture is a monoculture, an oversimplified system where crops grow in neat rows with little genetic diversity. This structure invites pest and disease problems and urges over-reliance on synthetic pesticides. IPM, then, is an essential tool in reducing dependence on pesticides because IPM balances economic and environmental interests through biological and chemical controls (see Chapter 1). Pesticide-treated products and industrially driven systems are now less appealing because of greater attention paid to food quality and agricultural practices, a fact underscored by the 20% annual increase in organic food consumption in comparison to a 2% to 3% increase for industrially produced foods (US Department of Agriculture, 2005, 2007; National Agricultural Statistics Service, 2006). This rising public awareness makes organic and sustainable agriculture increasingly attractive, challenging existing IPM methods to bridge our desire to grow crops outside their natural habitats and our want of a healthier environment.

39.1 Definitions

Comprehensible definitions for the terms organic and sustainable were in order given the host of organizations actively concentrating agricultural, academic and marketing efforts in these areas. In 1990, the US Farm Bill and the US Department of Agriculture (USDA) through the National Organic Program (NOP) defined organic production as a "system that is managed in accordance with the Act and regulations in this part to respond to site-specific conditions by integrating cultural, biological and mechanical practices that foster cycling of resources, promote ecological balance and conserve biodiversity" (US Congress, 1990; Pollack & Lynch, 1991). After 200 000 consumer complaints to the USDA, however, the definition of organic agriculture was extended to the "ecological production management system that promotes and enhances biodiversity, biological cycles, and soil biological activity." It emphasizes the use of on-farm management practices in preference to the use of off-farm inputs, taking into account that regional conditions require locally adapted systems. These goals are met, where possible, through the use of cultural, biological and mechanical methods, as opposed to using synthetic materials to

Integrated Pest Management, ed. Edward B. Radcliffe, William D. Hutchison and Rafael E. Cancelado. Published by Cambridge University Press.

fulfill specific functions within the system (National Organic Standards Board, 2007). In 1990, the Organic Foods Production Act (OFPA) also required that the USDA appoint a National Organic Standards Board (NOSB) to develop standards for the production and handling of organically produced products. It included a national list of substances approved for and prohibited from use in organic production and handling as well as rules establishing organic labeling requirements. The OFPA, in fact, established the National Organic Program in order to facilitate domestic and international marketing of fresh and processed organic foods and to assure consumers that such products meet consistent, uniform standards. Similarly, the International Federation of Organic Agriculture Movements (IFOAM) united 750 member organizations in 108 countries, establishing standards for organic production agriculture (International Federation of Organic Agriculture Movements, 2005).

Growing interest also led to competing definitions for the term sustainable, so Congress addressed sustainable agriculture in the Food, Agriculture, Conservation, and Trade Act of 1990 (FACTA) (US Congress, 1990; Pollack & Lynch, 1991). This act defines sustainable agriculture as:

> An integrated system of plant and animal production practices having a site-specific application that will, over the long-term, satisfy human food and fiber needs; enhance environmental quality and natural resources; make the most efficient use of nonrenewable resources and on farm resources and integrate natural biological cycles and controls; sustain the economic viability of farm operations; and enhance the quality of life.

Generally, sustainable agriculture promotes stable farm families and communities, supports profitable farm incomes and encourages production and consumption of locally grown foods. Nevertheless, sustainability places environmental preservation above economic concerns. Soil protection and improvement and reduced dependence on non-renewable resources, such as fuel, synthetic fertilizers and pesticides, minimizes adverse impact on environmental assets. In turn, executing these practices improves the diversity necessary for long-term conservation management.

39.2 History

Until the middle of the nineteenth century, civilizations relied on essentially organic agricultural production systems, utilizing animal waste for soil humus enrichment and natural materials for pest control. In 1840, synthetic fertilizers first replaced manures (Liebig, 1872), marking a transition from ecologically based, mixed farming practices to industrialized agriculture (Francis *et al.*, 2006). Traditional agriculture was a model of biodiversity: a mix of crops, many of which were native plants, grown without the use of chemicals. Many regions worldwide continue to use traditional practices and remain "organic" because they do not have the access or the capital to purchase synthetic materials. Lack of adaptation to an industrial agricultural system ensures greater biodiversity, stability and ecological and economic sustainability (Altieri & Nicholls, 1995) despite the control synthetic materials potentially offer. In contrast, the move toward synthetics and away from traditional farming practices gave better economic assurance to farmers, who, having spent considerable time and resources on their crops, did not want them destroyed by pests. Synthetics appeared to solve the problem. The use of synthetic materials, then, became the new agricultural paradigm, replacing centuries of traditional farming and leading to much denser crop cultivation well beyond any natural controls.

Introducing synthetic chemicals into an environment, however, requires the environment to adapt rapidly to these unnatural inputs. Evidently the environment could not adjust so immediately; many sectors noted ecologic injury, ushering in public demand for the modern organic agricultural era. In 1924, Rudolf Steiner presented eight lectures on *The Spiritual Foundations for a Renewal of Agriculture* in response to concerns about the depletion of soils and the general deterioration in the health of crops and livestock (Steiner, 1993, 2004). The lecture series evolved into Steiner's biodynamics production system: a healing, nurturing, holistic (emphasizes integration of animals to create a closed nutrient cycle), ecological, organic and spiritual approach to sustainable care of the Earth (Steiner, 2005). Similarly, Jerome Rodale hoped to reverse the declining health of soil and the human population by founding Rodale, Inc. in 1930. He began publishing *Organic Farming and Gardening Magazine* in 1942, informing readers about organic and natural farming methods.

Efforts of early organic agriculture advocates were largely disregarded. Throughout the 1950s and 1960s, users touted the extraordinary powers of synthetic pesticides, particularly DDT. Swept up by the potential for pest elimination, many weighed health and economic benefits over environmental consequences. Eventually, the ecological repercussions caused by over-reliance on pesticides could not be ignored or marginalized. "Insecticide resistance, concentration of chlorinated hydrocarbon insecticides in the food chain, significant declines in densities of natural enemy (predators and parasitoids) populations, secondary outbreaks of pests, resurgence of primary pests and unwanted insecticide residues on fruits and vegetables" underscored the argument for IPM and against indiscriminate use of synthetic products (see Chapter 1). At the same time, public opinion swayed following the publication of Rachel Carson's (1962) *Silent Spring*. Carson's observations triggered adverse public reaction and fueled legislation that ultimately led to the formation of the US Environmental Protection Agency (EPA) in 1972 (see Chapter 1). The EPA's mission is to oversee the nation's environmental programs, and, from its launch, the EPA was charged with protecting the environment and the human population from risks associated with pesticide misuse.

The 1989 Alar crisis confirmed the EPA's purpose and strengthened the movement toward organic agriculture. Accumulated evidence about the potential carcinogenicity of Alar and its breakdown product caused widespread concern about the consumption of Alar-treated apples by children (Herrmann *et al.*, 1997). In 1988, the US Congress requested that the National Academy of Science/National Research Council (NAS/NRC) assess the vulnerability of infants and children to dietary pesticides. The committee found that "quantitative differences in toxicity between children and adults are usually less than a factor of approximately 10-fold" (National Research Council, 1993). One year later, the Natural Resources Defense Council (NRDC) and the CBS news program *Sixty Minutes* reviewed pesticide regulatory issues and examined use of the reported carcinogen Alar. They stated that there was a higher risk for children than for adults to consume damaging levels of a toxin or carcinogen. In 1989, EPA began the process to cancel food uses of Alar (Lecos, 1989). This also bolstered the rationale behind IPM practices that stress a more complex pest management system and require consideration of ecological, social and economic components (see Chapter 1).

Supported by the Clinton Administration and a broad coalition of environmental, public health, agricultural and industry groups, the Food Quality Protection Act (FQPA) was passed by Congress in 1996 (US Congress, 1996). Among other responsibilities, the FQPA requires that the USDA promote and fund IPM projects to develop and implement IPM practices, tactics and systems for specific pest problems while reducing human and environmental risks.

Current development rejects synthetic materials in favor of more sustainable strategies, such as organic practices or IPM. Recently, Mader *et al.* (2002) stated that the key to determining effective farming systems is an understanding of agroecosystems. Their results from a 21-year study of biodynamic, bioorganic and conventional farming systems in central Europe found crop yields to be 20% lower in the organic systems; however, input of fertilizer was reduced by 34%, energy was reduced by 53% and use of synthetic pesticides was reduced by 97%. This study suggests that enhanced soil fertility and higher biodiversity found in organic plots may render these systems less dependent on external inputs. The National Research Council (1996) also confirms that ecologically based pest management (EBPM) requires a broad knowledge of an agroecosystem to develop pest management (PM) strategies that emulate more natural ecosystems. The goal of EBPM is to manage rather than eliminate pests in ways that are safe, stable and profitable, thereby supporting sustainable agriculture.

39.3 Recent developments

Early practitioners of organic farming embraced its sustainable component, standing against the abrupt introduction of synthetic materials into an otherwise unprepared environment. Overreliance on synthetic materials disrupts the complex dynamics of agroecosystems, causing organic practitioners to reject the use of such materials. However, consumers still insist on the availability of a range of fruits and vegetables in all regions and in all seasons. A better understanding of agricultural methods engages the population in the debate about farm-to-table practices, and this has increased pressure for organic and sustainable products. In order to respond to consumer demand, organic farmers no longer have to remain especially small or abstain from modern technology in favor of strictly traditional farming techniques. Essentially, modern organic agriculture can and does benefit from past missteps, not only because better environmental awareness results but also because more tools are available to the organic farmer.

In response to greater organic product requirements, nine USDA agencies expanded research, regulatory and other programs for organic agriculture (Dimitri & Greene, 2002), resulting in swifter development and integration of organic and sustainable agricultural production systems. The European Union has funded, from 2004 to 2007, the project Coordination of European Transnational Research in Organic Food and Farming (CORE Organic). The objective of CORE Organic is to enhance quality, relevance and utilization of resources in European research for transnational study in organic food and farming.

The rapid expansion of the organic industry is not without concerns. There is growing anxiety the modern organic farming industry is becoming industrialized, emphasizing rules for organic production that maximize sales and efficiency to meet increasing world demand (Francis *et al.*, 2006). This results in less emphasis on sustainable farming practices, places more value on the organic label, threatens to expand monocultures and locates production away from major population centers. Consumer demand drives growers to produce crops organically in areas less suitable for organic production, particularly in the more humid eastern and southeastern regions of the USA. Then again, this challenges IPM programs to test and develop more complex sustainable approaches to achieving organic production.

There is also debate regarding the sustainability and environmental impact of some synthetic pesticides in comparison to organically approved pesticides. Newer synthetic fungicides like the strobilurins (e.g. Abound™) achieve disease-free fruit using only 282 g active ingredient per hectare in contrast to the organically approved sulfur that requires more frequent sprays of 22.7 kg/ha (Earles *et al.*, 1999). Strobilurins, then, reduce fuel and labor costs, are less environmentally persistent, are more easily broken down in sunlight and by microorganisms, and offer better control of some diseases, especially fruit rots such as bitter rot caused by *Colletotrichum* spp. and *Glomerella cingulata*, and white rot (*Botryosphaeria dothidea*), so problematic in southern fruit-growing areas. Moreover, growing fruit organically doesn't necessarily mean reduced spray applications (Lehnert, 2005). Michigan findings are similar to Arkansas (Tables 39.1, 39.2 and 39.3); costs for organically approved insecticides, such as kaolin clay, pheromones, *Bacillus thuringiensis* (*Bt*), codling moth granulosis virus (CpGV), azadirachtin (neem), pyrethrum, spinosad, sulfur, oil and lime sulfur for the Michigan organic orchard in 2003 were roughly twice the cost of conventional methods. Nevertheless, although not approved as organic, some materials may be considered sustainable because of less environmental concern, energy savings, superior disease or pest management and profitability. The challenge persists. Organic and sustainable agriculture must still supply consumer demand and adhere to an environmentally conscientious philosophy.

39.4 Organic and sustainable farms

Organic annual crops and tree fruits are most frequently grown in areas with low pressure from diseases and pests because of isolation and less rainfall. West of a line from Texas to North Dakota, the drier climate results in reduced disease pressure (Earles *et al.*, 1999). The Rocky Mountains serve as a

Table 39.1 Current organic pest management tactics available to suppress apple insect and spider mite pest populations

Pests	Type of pest	Tactics
Codling moth	Internal	Mating disruption, Codling moth granulosis virus (Cyd-X), spinosad (Entrust), trunk banding
Oriental fruit moth	Internal	Mating disruption, *Bt*, spinosad
Plum curculio	Internal	Surround, pyrethrin (PyGanic), baited trap trees
Leafroller complex	External	Spinosad, *Bt*
Apple maggot	Internal	Spinosad (GF-120 NF Naturalyte Fruit Fly Bait), mass trapping
Aphids	Indirect	Natural enemies, oil, lime-sulfur, soaps
Stink bugs	External	Kaolin clay (Surround), azadirachtin (Aza-Direct), pyrethrin
Spider mites	Indirect	Oil, lime-sulfur, natural enemies
San Jose scale	Indirect	Oil, lime-sulfur, natural enemies
Japanese beetle	Indirect	Kaolin clay, azadirachtin, pyrethrin, mass trapping
Green June beetle	Damage ripe fruit	Mass trapping

Source: Modified from Krawczyk (2006).

Table 39.2 Insecticides (active ingredient) and mating disruption (MD) sprayable pheromone applied to the conventional and alternative management apple blocks in Berryville, AR, 2004

Date	Degree-days (pest)[a]	Conventional	Alternative
21 May		10 000 *N. fallacis* / ha (mite predator)	10 000 *N. fallacis* mites / ha
24 May	83 (CM)	Intrepid (methoxyfenozide)	Intrepid
31 May	139 (CM)	Guthion (azinphos methyl)	Cyd-X (virus) + Entrust (spinosad)
3–7 June	206–250 (CM)	Intrepid + Guthion	Cyd-X + Entrust
10 June	296 (CM)		Cyd-X + Entrust
17 June	400 (CM)	Guthion	Cyd-X + Entrust
24 June	481 (CM) 1056 (OFM)	3M OFM (Oriental fruit moth MD)	Cyd-X + Xentari (*Bt*)
2 July			Cyd-X + Entrust
6 July	639 (CM)	Intrepid	
12 July	722 (CM) 1333 (OFM)	3M OFM	Cyd-X
17 July	806 (CM)	Intrepid	Cyd-X
26 July	944 (CM) 1611 (OFM)	3M OFM	
9 Aug.	1111 (CM) 1822 (OFM)		Cyd-X
17 Aug.	1194 (CM)	Diamond (novaluron)	Cyd-X + Xentari
21 Aug	1250 (CM)	Cyd-X	Cyd-X + Xentari
8 Sept.		Cardboard on trunks	Cardboard on trunks
2–10 Dec.		Removed cardboard	Removed cardboard
EIQ field use rating[b]		139.3	49.2

[a] DD (OFM or CM), cumulative degree-days after first significant trap catch on 25 March of Oriental fruit moth (OFM) (base 7.2 °C) and catch on 18 May of codling moth (CM) (base 10 °C).
[b] EIQ field use rating, environmental impact quotient (EIQ) of pesticides = EIQ × percent active ingredient × amount used per season (Kovach *et al.*, 1992).

Table 39.3 Insecticides (active ingredient) and mating disruption (MD) pheromone ties applied to the conventional and alternative management apple blocks in Berryville, AR, 2005

Date	Conventional	Alternative
3–6 May	Sevin (carbaryl) + 200 Isomate-C (codling moth MD)	250 Isomate-C
10 May	Esteem (pyriproxyfen)	
16 May	Cyd-X (Codling moth granulosis virus) + Calypso (thiacloprid)	Cyd-X
23 May	Cyd-X + Intrepid (methoxyfenozide) + 10 000 *N. fallacis* / ha	Cyd-X + Entrust (spinosad)
1 June	Pyramite (pyridaben)	10 000 *N. fallacis* / ha
13 June		Acramite (bifenazate) on $\frac{1}{2}$-side of Fuji and Red Delicious
23 June	Calypso	
28 June	Cyd-X + Dipel (*Bt*)	
14 July	Guthion (azinphosmethyl)[a]	Cyd-X + Dipel
28 July	Intrepid	Cyd-X + Dipel
9 Aug.	Diamond (novaluron) + Cyd-X + Surround (kaolin clay)[b]	Cyd-X + Dipel
18 Aug.	Guthion[a] + Cyd-X + Surround	Cyd-X + Javelin (*Bt*)
25–27 Aug.	Diamond + Cyd-X	Cyd-X + Javelin
28 July	Cardboard on trunks	Cardboard on trunks
17 Nov.	Removed cardboard	Removed cardboard
EIQ field use rating[c]	137.2	40.7

[a] Applied only as a rescue spray to an isolated three-row-wide block 91 m south of the conventional block where mating disruption (Isomate-C TT) did not maintain <1% CM damaged fruit.
[b] Surround WP applied to prevent sunscald on one cultivar, so not included in the EIQ field use rating.
[c] EIQ field use rating, environmental impact quotient (EIQ) of pesticides = EIQ × percent active ingredient × amount used per season (Kovack *et al.* 1992).

barrier, isolating western states from several pests that cause significant damage in the eastern USA. The mostly arid areas of California, Colorado, Oregon, Idaho and Washington (Table 39.4) are key states for organic agriculture in the USA.

Organic and sustainable farms range from many thousands of hectares to small farms comprising less than a hectare of vegetables and/or fruit. Large farms typically operate as monocultures that are economically efficient but are also attractive targets for destructive insects or diseases. Most very large organic farms, located in areas of low pest pressure in arid climates, are dependent on dwindling water resources for irrigation, but they almost always benefit from better growing conditions and improved yields compared to more humid areas of the USA. They are successful largely because of a combination of host plant resistance to certain diseases and the lack of pest pressure. Organic and sustainable agriculture production in less suitable regions generally rely on native or augmented biological control, pest avoidance and host plant resistance as the foundation for farming practices (Zehnder *et al.*, 2007). In the eastern and southeastern regions of the USA, plum curculio (*Conotrachelus nenuphar*), apple maggot (*Rhagoletis pomonella*) and Oriental fruit moth (*Grapholita molesta*) cause significant damage, so pesticides still must be applied at times to supplement natural controls. Crop rotations, soil building practices and plant nutrition as mandated by organic regulations are also considered important IPM practices and must be observed as applicable in all areas.

Table 39.4 Comparison of climate, pests and diseases for several tree fruit orchard production areas in the USA and India

State	Latitude/ longitude	Annual rainfall (cm)	Insect pests[a]	Diseases[b]
Washington	N47°/W120°	<25	1, 4, 5, 8, 9, 10, 12, 15, 16	F, M
Colorado	N38°/W107°	<25	1, 5, 10, 12	F, M
Arkansas	N36°/W93°	100	1, 2, 3, 4, 5, 7, 10, 12, 14, 15, 16, 17	F, M, R, S
Uttaranchal, Himachal Pradesh India	N32°/E77°	152	4, 12, 15, 16	

[a] 1, codling moth (*Cydia pomonella*); 2, Oriental fruit moth (*Grapholita molesta*); 3, rosy apple aphid (*Dysaphis plantaginea*); 4, San Jose scale (*Quadraspidiotus perniciosus*); 5, apple aphid (*Aphis pomi*); 6, white apple leafhopper (*Typhlocyba pomaria*); 7, plum curculio (*Conotrachelus nenuphar*); 8, Pandemis leafroller (*Pandemis pyrusana*); 9, obliquebanded leafroller (*Choristoneura rosaceana*); 10, red-banded leafroller (*Argyrotaenia velutinana*); 11, spotted tentiform leafminer (*Phyllonorycter blancardella*); 12, wooly apple aphid (*Eriosoma lanigerum*); 13, apple maggot (*Rhagoletis pomonella*); 14, Japanese beetle (*Popillia japonica*); 15, European red mite (*Panonychus ulmi*); 16, twospotted spider mite (*Tetranychus urticae*); 17, green June beetle (*Cotinis nitida*).

[b] F, fire blight (*Erwinia amylovora*); M, mildew (*Podosphaera leucotricha*); R, summer rots (*Botryosphaeria obtusa, Botryosphaeria dothidea, Colletotrichum* spp., *Glomerella cingulata*); S, apple scab (*Venturia inequalis*).

Most large fruit or vegetable farms employ crop consultants who monitor soil fertility and who recommend crop nutrition and pest and disease management practices. These farms are frequently major agricultural producers, supplying the majority of produce to fresh or processing markets. Allowances for pest damage are often very small given that fresh and processing markets have strict tolerances for defects, and this often leads to an over-reliance on biopesticides. Smaller growers, on the other hand, may or may not have pest management expertise, and they usually supply local or regional markets that have higher tolerances for damage. Small farmers who sell their products locally tend to have more biodiversity in their farming system and greater opportunities for biological control and pest avoidance. They grow a wide variety of crops on smaller plots of ground often with native vegetation in close proximity to cultivated areas.

39.5 Case studies

Domesticated apples originated in central Asia where wild apple forests grow on mountain slopes. There, native pests of apple interact with their natural enemies in an ecological balance. The apple forests are excellent examples of biodiversity; each tree is genetically different and lives in a community with other native forest plants, animals and insects. On the other hand, apples in the USA are grown in monoculture from roots through scion so that there is genetic uniformity; the goal is perfect fruit.

Apples grown in monoculture have no natural predation for non-native apple pests, such as codling moth (*Cydia pomonella*); therefore, augmentative controls are necessary. Mills (2005) notes that codling moth caused from 0% to 5% fruit damage in the region around Almaty, Kazakhstan, indicating a possible role of natural predation. Surveys of codling moth in North America found less than 5% predation by native parasites (Jaynes & Marucci, 1947), and abandoned orchards can experience as much as 100% damage. Overall, the potential for biological control of codling moth in North America is not great because of low economic thresholds and low rates of success in past programs against other olethreutid moths (Mills, 2005). Native apple pests, such as plum curculio, some leafrollers, aphids and the non-native European red mite (*Panonychus ulmi*), have more potential for natural or augmented biocontrol. Organic materials are a practical fit because organic materials are generally more pest specific and less toxic to predators of secondary pests.

Where biological controls are lacking, a combination of other strategies can substantially reduce fruit damage (Table 39.1). These tactics include pheromone-based mating disruption against codling moth (Smith, 2001) and Oriental fruit moth (Johnson *et al.*, 2002a); granulosis virus against codling moth; *Bt* formulations against Oriental fruit moth (Rashid *et al.*, 2001); spinosad against lepidopterous insects; kaolin clay against many insects; and azadirachtin (Azatin XL Plus) against aphids, codling moth, leafrollers, white apple leafhopper, San Jose scale and plum curculio.

Spinosad is a new, environmentally friendly insecticide that is an aerobic fermentation product of the soil bacterium *Saccharopolyspora spinosa*. It is effective against Diptera, Lepidoptera and some flea beetles, and it is highly useful against codling moth and tortricid leafrollers (Olszak & Pluciennik, 1999). Spinosad has little effect on mites and sucking insects (Bret *et al.*, 1997; Dow, 1997; Thompson *et al.*, 2000), but is highly toxic to bees. Its toxicity to non-target pests makes it undesirable to some organic growers.

Surround is labeled for control of leafhoppers and overwintering obliquebanded leafroller (*Choristoneura rosaceana*) and suppression of codling moth, plum curculio, apple maggot (*Rhagoletis pomonella*), green fruitworm (*Lithophane antennata*) and a number of other insects (Garcia *et al.*, 2003). Treatments of Surround are inconsistent against the same pests in separate studies (Wright *et al.*, 2000; Robinson *et al.*, 2002), but treatments are effective against some emerging pests, including the Japanese beetle (*Popillia japonica*) (D. T. Johnson, unpublished data).

CpGV is another extremely effective material that targets codling moth, and it is now widely used by most fruit growers in the USA. Field tests conducted in 1978–79 achieved reductions in the numbers of fully developed larvae and damaged fruit similar to that obtained by two applications of azinphosmethyl (Glen & Payne, 1984). CpGV is successful against codling moth when applied at weekly intervals starting with hatch of the first codling moth larvae (Simon *et al.*, 1999; Polesny *et al.*, 2000). CpGV-treated orchards have damage below 2% after seven to eight applications and are similar to the 2.7% control obtained with the combination of mating disruption and CpGV (by Minarro & Dapena, 2000; http://entomology.tfrec.wsu.edu/New_Insecticides/Surround.html). Eberle & Jehle (2006) reported the appearance of field resistance to CpGV in codling moth which is an impediment to continuous application of CpGV. Thus, sprays of CpGV should be integrated with mating disruption or another tactic for season-long management of codling moth.

There is no apple growing area in the USA equal in biodiversity to the apple forests of central Asia, but a thoughtful mixture of site-specific IPM practices shapes reasonable models of organic and sustainable agriculture. A combination of IPM measures helps growers meet consumer demand for organic products and to adhere to sound ecological practices. Each region adapts models of organic and sustainable agriculture to meet its needs.

39.5.1 Colorado

Hotchkiss, Colorado, is located on the western slope of the Rockies at an altitude of about 1800 m and with an average of 25 cm of rain per year. Fruit production in western Colorado is concentrated on irrigated mesas. Each mesa may have from several hundred to over a thousand hectares of orchard. At one time, the Hotchkiss area had many well-managed orchards, but the number declined over the last 30 years, and abandoned hectares increased, creating a habitat where codling moth flourishes. Additionally, the dry, warm climate is ideal for codling moth flight and reproduction, supporting two plus generations of codling moth annually and three other pests (Table 39.4). Conventional growers have effective pesticides (organophospates, synthetic pyrethroids, neonicotinoids, etc.) to manage codling moth at low levels; however, options for organic growers are limited. Most apple diseases are absent, but powdery mildew (*Podosphaera leucotricha*) and fire blight (*Erwinia amylovora*) may be problems as are weeds and nutrient management. Unquestionably, all Western Slope organic apple growers agree that codling moth causes the most economic damage in organic apple production. In some locations, individual traps frequently catch over 100 moths per night.

Colorado growers have used codling moth traps and degree-day (DD) models for many years to effectively time insecticide sprays (Beers *et al.*, 1993; Alston *et al.*, 2006). In 1994, the first major breakthrough in organic codling moth control came with the introduction of pheromone mating disruption, now a major management tool against lepidopteran pests in fruit (Quarles, 2000). However, this method proved ineffective in orchards with high overwintering codling moth populations (Mansour & Fater, 2001).

Prior to the availability of mating disruption, Colorado organic growers relied on one or more approaches: weekly summer oil applications, trunk banding with cardboard (replace bands weekly and destroy larvae infested bands), DD-timed applications of the biopesticide rotenone, and removal and disposal of infested fruit. All of these approaches had drawbacks. Repeated oil applications were phytotoxic, trunk banding was time consuming as well as labor intensive, and weekly applications of Rotenone were as much an environmental problem as the materials they replaced, with potential harm to non-target organisms like fish and pollinating insects. Mating disruption proved to be a major advance, but it was not a perfect solution areawide. In 2005, CpGV and Spinosad became available and proved to be valuable aids in codling moth management. Arthurs *et al.* (2007) reported that CpGV was less effective at protecting fruit in the first larval generation compared with Spinosad but did suppress the population early in the season. Spinosad also caused no disruptions of beneficial species or secondary pest outbreaks. Miles (2003) reported that "spinosad, when used according to the approved product label recommendations, would be safe to foraging worker bees, queen and brood and may be safely used in flowering crops if applications are made during periods of low bee activity." Even so, Colorado organic growers are reluctant to use Spinosad because it may harm honeybees. For example, despite the availability of more approved materials, an organic grower reported 70% codling moth damage (external feeding with few entries) in his 2006 apple crop despite using pheromone mating disruption and CpGV. The abundance of abandoned orchards near organic orchards makes it extremely difficult to effectively manage codling moth with organic methods. Nevertheless, after the introduction of organic management practices, secondary pests like European red mites, leafrollers and aphids are now biologically controlled.

Organic apple farming in Colorado, then, is problematic because of a climate favorable for codling moth reproduction. To eliminate damage, organic growers in Colorado reject the use of harsher materials, opting for softer organic controls and choosing to sell fruit to a public that accepts some cosmetic damage. Consequently, Colorado orchards are a strong model of sustainable organic agriculture.

39.5.2 Washington

Washington produces 77 699 ha of apples, about 53% of the apples grown in the USA. The warm, sunny summer days, cool nights, abundant irrigation water and arid climate (less than 25 cm rainfall per year) are ideal for apple production. Although not pest free, Washington, like Colorado, does benefit from the absence of many insect pests prevalent in the eastern USA (Table 39.4). However, secondary pests including Pandemis leafroller (*Pandemis pyrusana*) and the obliquebanded leafroller developed resistance to sprays targeted at codling moth and increased their range and importance. In addition, a number of bugs, tarnished plant bug (*Lygus lineolaris*), consperse stink bug (*Euschistus conspersus*), green stink bug (*Acrosternum hilare*) and western boxelder bug (*Boisea rubrolineata*), are increasingly problematic in Washington orchards located near forest or brushy non-crop land. Overall, however, relatively low pest pressure allows some growers to successfully become organic producers (Smith, 2001). Certified organic apple orchards increased from 50 ha in 1988 to 3204 ha in 2004 (Granatstein *et al.*, 2005). Approximately 2700 ha are now in transition in Washington. Stemilt Growers, Inc., Washington's leading organic fruit packer, reports that 12% of its apple production is currently certified organic, and Stemilt intends to pack 25% of its fruit organically by 2010 (Pepperl, 2007).

Around 1993, Campbell Orchards of Tieton, Washington, began experimenting with organic practices in their orchards located on the upper perimeter of the Tieton fruit producing area. Disease and codling moth pressure in this location were low. Available pest management tactics like mating disruption, widely used in Washington, provide adequate codling moth control in this area and conserve natural enemies for biological control of other orchard pests. Nutrition and weed management remain problems, but insect management with organic materials is practical here and in many other Washington locations.

Some sites have far greater pest pressure requiring additional pest management inputs. Because growers are paid on the amount of apples packed and sold fresh, they focus on having the minimum number of defects. Production economics result in lower action thresholds and an increased emphasis on preventing damage through a variety of pest management tactics including increased use of biopesticides. Critics see this as a problem and contend that such growers are merely replacing conventional tactics with organic approved materials. Also, local and sustainable food advocates find fault with the high packaging and transportation costs associated with marketing apples in distant markets. Studies conflict as to which leaves a greater carbon footprint: locally grown German apples or importation of New Zealand grown apples to Germany (Warner, 2007).

Washington can grow organic apples successfully because of favorable climatic conditions. Their sustainability is questionable, however, because of fruit transport over long distances. Accordingly, Washington is not a perfect model of sustainable agriculture but is an effective example of organic agriculture in the USA.

39.5.3 Arkansas

Warm and humid areas are less suited for organic fruit production because they have greater pest and disease pressure (100 cm rain per year), require more frequent pest monitoring and apply more materials for pest control than Washington or Colorado (Table 39.4). For example, south of a line from Missouri to North Carolina, fruit rots are common, two or more generations of plum curculio are typical, three or more generations of codling moth are likely and five or more generations of Oriental fruit moth are common (Rashid *et al.*, 2001).

In 1981, a grower in Osage, Arkansas established a non-irrigated organic fruit planting of apples, peaches, cherries and grapes. Insect pests and diseases were numerous (Table 39.1) and greatly affected the quality of the fruit. At that time, organic apple growers in areas of high pest and disease pressure were fortunate to achieve 60–70% of undamaged fresh market fruit (J. Aselage, unpublished data).

In 2001, John Aselage began implementing organic practices in his orchard management program. In 2004 and 2005, he split his apple planting into two small 0.5-ha blocks: a low-risk conventional block and a nearly organic alternative block (Tables 39.2 and 39.3). Each block was scouted weekly, and pest activity was monitored with pheromone traps for Oriental fruit moth and for codling moth. Plum curculio was monitored by pyramid traps tethered to perimeter apple tree trunks (Johnson *et al.*, 2002b). Every week 300 apples were assessed for species-specific fruit damage and 100 "Red Delicious" leaves were scanned for presence of European red mites. When the economic threshold of 1 mite per apple leaf was surpassed, it was followed by augmentative release of 10 000 predatory mites, *Neoseiulus fallacis*, per hectare. Codling moth, Oriental fruit moth and red-banded leafrollers were maintained below economic thresholds using pheromones for mating disruption and DD-timed sprays of CpGV and *Bt*. Corrugated cardboard strips, 15.2 cm width, were stapled to each tree in August and removed in late November to assess and reduce levels of overwintering codling moth larvae. The Environmental Impact Quotient (EIQ) field use ratings for pesticide applications was <50 for the alternative block compared to 137 to 140 for the conventional block in both years (Tables 39.2 and 39.3).

Fruit damage by internal lepidopterans at harvest in 2005 was similar with 0% damage in the low-risk block and 0.1% in the alternative block. The alternative block in 2004 had 12% damage, which consisted primarily of surface damage with

Table 39.5 Plum curculio fruit damage in baited and unbaited trees in edge row by woods in intermediate-sized trees in Springdale, MA 2004, and dwarf trees in Berryville, AR, 2005

Year	Tree size	Treatment	Percent fruit damage[a]
2004	Intermediate-sized trees	Bait + pyramid trap	17.0 a
		No bait + pyramid trap	5.2 b
		No bait, no pyramid trap	2.5 b
2005	Small-sized trees	Bait	11.6 a
		No bait	3.6 b

[a] Mean percent damage values by year followed by similar letters are not significantly different at $p > 0.05$, Waller–Duncan K-ratio t-test.

no larval tunneling. The low-risk conventional program had higher numbers of codling moth larvae wintering in cardboard strips in 2004 and 2005.

In the humid eastern USA, plum curculio is the most difficult insect to control in an organic management program. Significant plum curculio catch in pyramid traps could be used to time perimeter sprays of Pyganic, an organic insecticide containing pyrethrins (Whalon *et al.*, 2005). First-generation plum curculio adults were attracted to apple trees baited with eight packets of the kairomone benzaldehyde (BA) and with two packets of the aggregation pheromone grandisoic acid (GA) in Arkansas (Table 39.5).

Four BA packets and one GA packet per tree were sufficient to attract plum curculio in Massachusetts (Prokopy *et al.*, 2004). Fruit damage in baited trees exceeded 5% in Massachusetts (Prokopy *et al.*, 2004) and exceeded 10% in Arkansas compared to less than 1% in adjacent unbaited trees in Massachusetts (only one generation) and less than 4% in unbaited trees (two generations) in Arkansas (Table 39.6). Therefore, baited trees may provide an effective solution to managing plum curculio. Entomopathogenic nematodes may play a role in further reducing plum curculio populations. Soil applications of *Steinernema riobrave* achieved 97% control of plum curculio larvae in three out of the four field tests in peach orchards (Shapiro-Ilan *et al.*, 2004). A combination of organic pest management tactics is effective against most of the major Arkansas fruit pests. The primary limiting factor to producing organic fruit in Arkansas is the availability of effective organic control tactics against apple scab (*Venturia inequalis*), fire blight (*Erwinia amylovora*), mildew (*Podosphaera leucotricha*) and the summer fruit rot diseases (*Botryosphaeria obtusa*, *Botryosphaeria dothidea*, *Colletotrichum* spp., *Glomerella cingulata*).

While locally grown Arkansas apples may be fresher for Arkansas consumers, the amount of inputs used to produce Arkansas apples is more than the inputs used in Washington or Colorado. Yields in Arkansas are also frequently less than half that of Washington. Arkansas, then, is the least suitable area for organic apple farming; however, Arkansas is a fair model of sustainability given the use of IPM and the support of a local customer base.

39.5.4 Indian Himalaya

USA scientists working in conjunction with the Himalayan Consortium for Himalayan Conservation (HIMCON) and Winrock International are evaluating fruit crops for adaptation to the various temperature zones of the Himalayan region. A key component of the project is to help determine suitable low-input tree crops. Tree fruits are generally considered a desirable crop because of widespread deforestation in the region. Trees are frequently planted in the same small plots where vegetables, pulses and grains are cultivated

Table 39.6 Changes in percent codling moth (CM) fruit damage and number of live overwintering CM larvae in cardboard strips on trunks in the conventional and alternative management apple blocks (see Tables 39.2 and 39.3) in Berryville, AR, 2004 and 2005

	Percent damage			
Date	2004 conventional	2005 conventional	2004 alternative	2005 alternative
1 July		0.02		0.01
9 July	0		0	
16 July	2.5		0	
21–23 July	0.2	0.03	0.2	0.02
2 Aug.	1.5		0.8	
10–12 Aug.	1.3	0.03	1.7	0.005
24–25 Aug.	1.8	0.04	3.0	0.0
1 Sept.	1.5	0.0	1.0	0.1
8 Sept.	1.5	Harvest	6.7[c]	Harvest
15 Sept.	Harvest		12.0[c]	
22 Sept.			Harvest	
# CM larvae per cardboard strip	1.8[a]	0.75[b]	0.2[a]	0.063[b]

[a] Cardboard strips (15.2 cm width) stapled on apple trunks in conventional block (106 trees) and alternative block (114 trees) in 2004.
[b] Cardboard strips (15.2 cm width) stapled on apple trunks in conventional block (118 trees) and alternative block (80 strips) in 2005.
[c] Frass from codling moth larvae occurred on exterior of fruit but tunneling was <0.6 cm deep and no live larvae inside.

for home consumption. Major fruit pests, such as codling moth and Oriental fruit moth, are absent from the Himalayan region, making this a promising candidate for organic fruit production (Table 39.4). However, the crops interspersed with fruit trees provide habitat for sucking insects that damage developing apples. Attempts to control sucking insects with pyrethroids led to outbreaks of secondary pests like San Jose scale and mites as well as a reduction in local pollenizing insects, thereby reducing yields. Misguided pesticide use, then, caused more damage than benefit to Himalayan apples.

More modern selective pesticides are not available or are too expensive (Aselage, 2002). The traditional pesticides lindane (BHC), endosulfan, methyl parathion and metasystox are still available, but their use is problematic. Growers are poorly trained in pesticide safety and use, which frequently results in applicator exposure and environmental contamination. Furthermore, the food crops growing on the orchard floor are subjected to pesticide drift from applications to the developing fruits. These problems are of more concern than secondary pest outbreaks following pyrethroid use. Until organic tactics are developed or safer and more selective materials are available, the sound alternative is to refrain from using dangerous and environmentally insensitive pesticides.

Not unlike the USA, consumer demand in India drives agricultural choices, and India is susceptible to making the same choices as has the USA when introduced to the potential of synthetic materials. Unlike fruit-growing regions in the USA, however, India is relatively free of pests and has a consumer base willing to accept higher damage thresholds. The most sensible and sustainable

Fig. 39.1 Making the transition towards a more sustainable or organic apple production program.

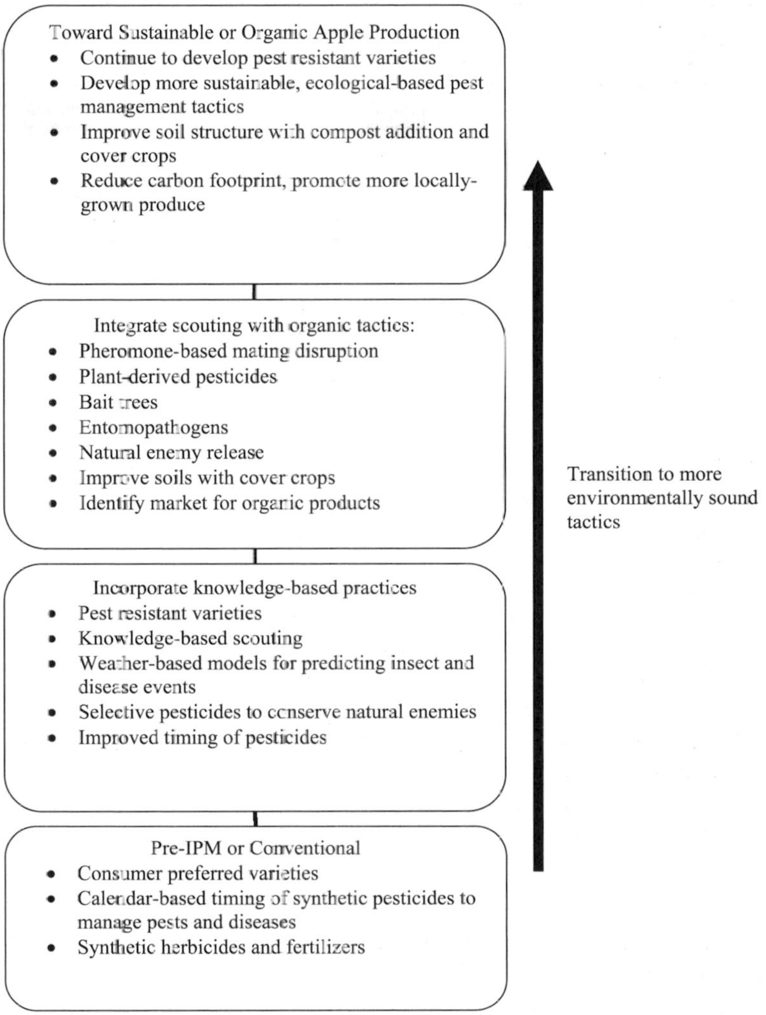

Toward Sustainable or Organic Apple Production
- Continue to develop pest resistant varieties
- Develop more sustainable, ecological-based pest management tactics
- Improve soil structure with compost addition and cover crops
- Reduce carbon footprint, promote more locally-grown produce

Integrate scouting with organic tactics:
- Pheromone-based mating disruption
- Plant-derived pesticides
- Bait trees
- Entomopathogens
- Natural enemy release
- Improve soils with cover crops
- Identify market for organic products

Incorporate knowledge-based practices
- Pest resistant varieties
- Knowledge-based scouting
- Weather-based models for predicting insect and disease events
- Selective pesticides to conserve natural enemies
- Improved timing of pesticides

Pre-IPM or Conventional
- Consumer preferred varieties
- Calendar-based timing of synthetic pesticides to manage pests and diseases
- Synthetic herbicides and fertilizers

Transition to more environmentally sound tactics

method for the Indian Himalaya is to continue to use traditional agricultural practices, making the country a model of organic sustainable farming.

39.6 Conclusions

Agriculture production practices have changed to meet the demand of an exponentially growing human population and to accommodate demographic shifts from a rural to an urban society. The end result is industrial agriculture. The focus is cheap and efficient production that is dependent upon inputs of synthetic pesticides and fertilizers, practices developed with little concern or awareness of their adverse impact on the environment. Many consumers, agricultural scientists and agricultural producers are now questioning the roles of synthetic inputs in the degradation of our environment and in the depletion of soil health and energy resources, recognizing the need to rapidly develop ecologically-based solutions and alternatives.

The terms "environmental risk quotient" assessments for given pesticides (Peterson, 2005), "food miles" for local versus imported foods (Pirag & Van Pelt, 2002) and "carbon footprint"

(Wiedmann & Minx, 2007) are now commonplace and are key considerations in the development of pest management programs for the future. Organically and sustainably grown foods will have a pronounced part in feeding the world and will provide solutions to at least some of our environmental problems. Entomologists and other agricultural scientists will have significant responsibility in developing more environmentally sound systems that correspond to societies' evolving needs.

Figure 39.1 uses apples to illustrate a transition from the traditional calendar spray program with some use of host plant resistance toward a more organic or sustainable program. Along the continuum, knowledge about pest biology, disease epidemiology, crop physiology, soil health and agroecology increases. Based on this understanding, growers and scientists can implement scouting programs to assist in making crop management decisions founded not only on economics but also on social and environmental consequences. The last step toward organic or sustainable production integrates social concerns into more local organic and sustainable food production. Such an agricultural system will improve soil health, food safety and environmental biodiversity in and around crop production areas by employing host plant resistance, conservation of natural enemies and selective biopesticides. Changes are unlikely unless society places more value on the environment than on the economics of crop production. There is a sector of agriculture that rejects the current industrial standard. This is a natural community with which to test new and innovative concepts in pest management. Excluded from conventional sales outlets, organic and sustainable growers profit in local venues by selling products to consumers who share not only environmental values but also a new vision of sustainability. In cooperation with this community of growers, entomologists must design and implement new agricultural production systems where cultural and biological controls are enhanced because of the system itself. The crop ecosystem, then, will act as an area of refuge for predators and pests to exist in ecological balance. The goal of ecologically based pest management systems, finally, is not to eliminate pests but to maintain them below economic levels. The integration of more organic or sustainable tactics into agriculture production systems will be a major part of the solutions toward sustainable agricultural production.

Acknowledgements

A special thanks to Elizabeth Aselage for editing this chapter and thanks to Elena Garcia, Heather Friedrich and Barbara Lewis for reviews.

References

Alston, D., Murray, M. & Reding, M. (2006). *Codling Moth* (Cydia pomonella), Utah Pest Fact Sheet No. 200ENT-13-06 6. Logan, UT: Utah State University Extension and Utah Plant Pest Diagnostic Laboratory. Available at http://extension.usu.edu/files/publications/factsheet/ENT-13-06.pdf.

Altieri, M. A. & Nicholls, C. I. (1995). *Biodiversity and Pest Management in Agroecosystems*. New York: Haworth Press.

Aselage, J. M. (2002). *Orchard Management Practices for Apple*, Technical Consultancy Report No. IND078. Little Rock, AR: Winrock International.

Arthurs, S. P., Lacey, L. A. & Miliczkya, E. R. (2007). Evaluation of the codling moth granulovirus and spinosad for codling moth control and impact on non-target species in pear orchards. *Biological Control*, **41**, 99–109.

Beers, E. H., Brunner, J. F., Willett, M. J. & Warner, G. M. (1993). *Orchard Pest Management*. Yakima, WA: Good Fruit Grower.

Bret, B. L., Larson, L. L., Schoonover, J. R., Sparks, T. C. & Thompson, G. D. (1997). Biological properties of spinosad. *Down to Earth*, **52**, 6–13.

Carson, R. (1962). *Silent Spring*. Boston, MA: Houghton Mifflin.

Dimitri, C. & Greene, C. (2002). *Recent Growth Patterns in the U.S. Organic Foods Market*, Agriculture Information Bulletin No. 777. Washington, DC: US Department of Agriculture, Economic Research Service, Market and Trade Economics Division and Resource Economics Division.

Dow (1997). *Spinosad Technical Bulletin*. Indianapolis, IN: Dow AgroSciences.

Earles, R., Ames, G., Balasubrahmanyam, R. & Born, H. (1999). *Organic and Low-Spray Apple Production Horticulture Production Guide*, NCAT ATTRA Publication No. IP020. Layetteville, AR: National Center for Alternative Agriculture, Appropriate Technology Transfer for Rural Areas.

Eberle, K. E. & Jehle, J. A. (2006). Field resistance of codling moth against *Cydia pomonella* granulovirus

(CpGV) is autosomal and incompletely dominant inherited. *Journal of Invertebrate Pathology*, **93**, 201–206.

Francis, C. A., Poincelot, R. P. & Bird, G. W. (2006). *Developing and Extending Sustainable Agriculture: A New Social Contract*. New York: Haworth Press.

Garcia, M. E., Berkett, L. P. & Bradshaw. T. (2003). Does Surround have non-target impacts on New England Orchards? *Proceedings of New England Fruit Meetings 2002–2003*. Available at www.massfruitgrowers.org/nefrtmtg/proc-2002-03/a03.pdf.

Glen, D. M. & Payne, C. C. (1984). Production and field-evaluation of codling moth granulosis virus for control of *Cydia pomonella* in the United Kingdom. *Annals of Applied Biology*, **104**, 87–98.

Granatstein, D., Kirby, E. & Feise, C. (2005). Trends of organic tree fruit production in Washington State. *Proceedings of the 3rd National Organic Tree Fruit Research Symposium* June 6–8, Chelan, WA. Available at http://organic.tfrec.wsu.edu/OrganicIFP/OrganicFruitProduction/PROCEED.FINAL.pdf.

Herrmann, R. O., Warland, R. H. & Sterngold, A. (1997). Who reacts to food safety scares? Examining the Alar crisis. *Agribusiness*, **13**, 511–520.

International Federation of Organic Agriculture Movements (2005). *Norms for Organic Production and Processing: IFOAM Basic Standards*. Bonn, Germany: IFOAM. Available at hwww.ifoam.org/.

Jaynes, H. A. & Marucci, P. F. (1947). Effect of artificial control practices on the parasites and predators of the codling moth. *Journal of Economic Entomology*, **40**, 9–25.

Johnson, D. T., Lewis, B. A., Striegler, R. K. *et al.* (2002a). Development and implementation of a peach integrated pest management program in the southern USA. *ActaHort*, **592**, 681–688.

Johnson, D. T., Mulder, P. G. Jr., McCraw, D. *et al.* (2002b). Trapping plum curculio, *Conotrachelus nenuphar* (Herbst) (Coleoptera: Curculionidae), in the Southern USA. *Environmental Entomology*, **31**, 1259–1267.

Kovach, J., Petzoldt, C., Degni, J. & Tette, J. (1992). *A Method to Measure the Environmental Impact of Pesticides*, New York Food and Life Sciences Bulletin No., 139. Geneva, NY: Cornell University New York State Agricultural Experiment Station. Available at http://ecommons.library.cornell.edu/handle/1813/5203/.

Krawczyk, G. (2006). Insect management in organic apple orchard. Presentation at *Pennsylvania Association for Sustainable Agriculture, Science-Based Organic Apple Production*, July 12. Biglerville, PA: Pennsylvania State Fruit Center. Available at www.cas.psu.edu/docs/CASDEPT/plant/EXTENSION/FRUITPATH/FIELD_DAY/field%20day%202006/397,1,Slide 1.

Lecos, C. (1989). Alar use on apples. Food and Drug Administration P89–12 News Release March 16, 1989. Available at www.fda.gov/bbs/topics/NEWS/NEWC0128.html.

Lehnert, D. (2005). Growing apples organically in Michigan: researchers are figuring out how it can be done. *Good Fruit Grower* **56**(5). Available at www.goodfruit.com/issues.php?article=763&issue=26.

Liebig, J. von (1872). *Chemistry in the Application to Agriculture and Physiology*. New York: John Wiley.

Mader, P., Fliebbach, A., Dubois, D. *et al.* (2002). Soil fertility and biodiversity in organic farming. *Science*, **296**, 1694–1697.

Mansour, M. & Fater, M. (2001). Mating disruption for codling moth, *Cydia pomonella* (L.) (Lepidoptera: Tortricidae), control in Syrian apple orchards. *Polskie Pismo Entomologiczne*, **70**, 151–163.

Miles, M. (2003). The effects of spinosad, a naturally derived insect control agent to the honeybee. *Bulletin of Insectology*, **56**, 119–124.

Mills, N. (2005). Selecting effective parasitoids for biological control introductions: codling moth as a case study. *Biological Control*, **34**, 274–282.

Minarro, M. & Dapena, E. (2000). Control de *Cydia pomonella* (L.) (Lepidoptera: Tortricidae) con granulovirus y confusión sexual en plantaciones de manzano de Asturias. *Boletín de Sanidad Vegetal, Plagas*, **26**, 305–316.

National Agricultural Statistics Service (2006). *Agricultural Statistics Report*. Washington, DC: US Department of Agriculture. Available at www.ers.usda.gov/Data/Organic/Data/Certified %20and%20total%20US%20acreage%20selected%20crops%20livestock%2095-05.xls.

National Organic Standards Board (2007). *Principles of Organic Production and Handling: National Organic Standards Board Policy and Procedures Manual Section VII*. Washington, DC: US Department of Agriculture, Available at www.ams.usda.gov/nosb/BoardPolicyManual/BoardPolicyManual8-23-05.pdf.

National Research Council (1993). *Pesticides in the Diets of Infants and Children*. National Academy of Sciences, National Research Council, Committee on Pesticide, Board on Agriculture and Board on Environmental Studies and Toxicology Commission on Life Sciences. Washington, DC: National Academy Press. Available at http://books.nap.edu/openbook.php?isbn=0309048753.

National Research Council (1996). *Ecologically-Based Pest Management: New Solutions for a New Century*. National Academy of Sciences, National Research Council, Board on Agriculture. Washington, DC: National Academy Press.

Olszak, R. W. & Pluciennik, Z. (1999). Leafroller (Tortricidae) and fruit moth (*Laspeyresia funebrana* and *Cydia pomonella*) control with modern insecticides. *Proceedings of the 5th International Conference on Pests in Agriculture*, Montpellier, France, December 7–9, Part 2, 311–318.

Pepperl, R. (2007). Stemilt transitions peaches, nectarines and other summer fruits to an all-organic crop. *FreshPlaza.com Newsletter 13*, April 2007. Available at www.freshplaza.com/news_detail.asp?id=124.

Peterson, R. K. D. (2005). Comparing ecological risks of pesticides: the utility of a Risk Quotient ranking approach across refinements of exposure. *Pest Management Science*, **62**, 46–56.

Pirog, R. & Van Pelt, T. (2002). How far do your fruit and vegetables travel? *Leopold Letter*, 14 (1), Spring. Available at www.leopold.iastate.edu/pubs/other/files/food_chart.pdf.

Polesny, F., Muller, W., Polesny, F., Verheyden, C. & Webster, A. D. (2000). Integrated control of codling moth (*Cydia pomonella*) in Austria. *Proceedings of the International Conference on Integrated Fruit Production*, Leuven, Belgium, July 27–August 1, 1998. *ActaHort*, **525**, 285–290.

Pollack, S. L. & Lynch, L. (1991). *Provisions of the Food, Agriculture, Conservation, and Trade Act of 1990*, Agriculture Information Bulletin No. 624. Washington, DC: US Department of Agriculture, Economic Research Service Agriculture and Trade Analysis Division.

Prokopy, R. J., Jácome, I., Gray, E. *et al.* (2004). Establishing characteristics of odor-baited trap trees for monitoring plum curculio. *University of Massachusetts Fruit Notes Fruit Notes*, **69** (Winter), 9–13.

Quarles, W. (2000). Mating disruption success in codling moth IPM. *IPM Practitioner*, **22**(5/6), 1–12.

Rashid, T., Johnson, D. T., Steinkraus, D. C. & Rom, C. R. (2001). Effects of microbial, botanical and synthetic insecticides on "Red Delicious" apple arthropods in Arkansas. *HortTech*, **11**, 615–621.

Robinson, T., Schupp, J., Merwin, I. *et al.* (2002). *A Commercial Organic Apple Production System for New York. Progress Report*. Ithaca, NY: Toward Sustainability Foundation, Cornell University.

Shapiro-Ilan, D. I., Mizell, R. F., Cottrell, T. E. & Horton, D. L. (2004). Measuring field efficacy of *Steinernema feltiae* and *Steinernema riobrave* for suppression of plum curculio, *Conotrachelus nenuphar*, larvae. *Biological Control*, **30**, 496–503.

Simon, S., Corroyer, N., Getti, F. X. *et al.* (1999). Organic farming: optimization of techniques. [Agriculture biologique: optimisation des techniques.] *Arboriculture Fruitière*, **533**, 27–32.

Smith, T. J. (2001). *Crop Profile for Apples in Washington*, Publication No. MISC0368E. Seattle, WA: Washington State University Cooperative Extension and US Department of Agriculture. Available at www.tricity.wsu.edu/~cdaniels/profiles/apple.pdf.

Steiner, R. (1993). *The Spiritual Foundations for the Renewal of Agriculture: A Course of Lectures*, ed. M. Gardner. Kimberton, PA: Bio Dynamic Farming and Gardening Association.

Steiner, R. (2004). *Agricultural Course, The Birth of the Biodynamic Method*, transl. George Adams. Glasgow, UK: Rudolf Steiner Press.

Steiner, R. (2005). *What Is Biodynamics? A Way to Heal and Revitalize the Earth*. Great Barrington, MA: Steiner Books.

Thompson, G. D., Dutton, R. & Sparks, T. C. (2000). Spinosad: a case study – an example from a natural products discovery programme. *Pest Management Science*, **56**, 696–702.

US Congress (1990). *Food, Agriculture, Conservation, and Trade Act. Public Law 101–624*. Washington, DC: US Government Printing Office. Available at www.ers.usda.gov/publications/aib624/aib624.pdf.

US Congress (1996). *Food Quality Protection Act of 1996 (Public Law 104–170)*. Washington, DC: US Government Printing Office. Available at www.epa.gov/pesticides/regulating/laws/fqpa/gpogate.pdf.

US Department of Agriculture (2005). *US Market Profile for Organic Food Products*. Washington, DC: US Department of Agriculture, Foreign Agricultural Service, Commodity and Marketing Programs, Processed Products Division, International Strategic Marketing Group. Available at www.fas.usda.gov/agx/organics/USMarketProfileOrganicFoodFeb2005.pdf.

US Department of Agriculture (2007). *Organic Agriculture: Consumer Demand Continues to Expand*. Washington, DC: US Department of Agriculture. Available at www.ers.usda.gov/Briefing/Organic/Demand.htm.

Warner, G. (2007). Food miles gain traction. *The Good Fruit Grower*, **58** (13). Available at www.goodfruit.com/issues.php?article=1700&issue=63.

Whalon, M. E., Coombs, A. B. & Nortman, D. (2005). An attract and kill strategy for plum curculio in Michigan tree fruit. *Proceedings of the Entomological Society of America Annual Meeting*, Fort Lauderdale, FL, December 18, 2005 (abstract). Available at http://esa.confex.com/esa/2005/techprogram/paper_22927.htm.

Wiedmann, T. & Minx, J. (2007). *A Definition of "Carbon Footprint,"* Integrated Sustainability Analysis UK (ISA^{UK}) Report No. 07–01. Available at www.isa-research.co.uk/docs/ISA-UK_Report_07-01_carbon_footprint.pdf.

Wright, S., Fleury, R. Mittenthal, R. & Prokopy, R. (2000). Small-plot trials of Surround™ and Actara™ for control of common insect pests of apples. *University of Massachusetts Fruit Notes*, **65**, 22–27. Available at www.umass.edu/fruitadvisor/fruitnotes/FNarticle65-06.pdf.

Zehnder, G., Gurr, G. M., Kühne, S. *et al.* (2007). Arthropod pest management in organic crops. *Annual Review of Entomology*, **52**, 57–80.

Chapter 40

Future of IPM: a worldwide perspective

E. A. (Short) Heinrichs, Karim M. Maredia and Subbarayalu Mohankumar

In a small village in western Bangladesh under the shade of a bamboo-framed thatch roof, two women sit and work with a razor blade and eggplant seedlings (Fig. 40.1). With a deft movement of hand on plant, Shovarani Kar and Trishna Rani Biswas are able to graft a high-yielding variety of eggplant onto the rootstock of another variety that is resistant to a devastating soil-borne scourge: bacterial wilt.

These women have been trained to perform this task and are paid to do so, thus raising their income while improving the yield for eggplant farmers. Word has traveled that people in this village are now earning more because of improved agricultural practices, and villagers from surrounding towns and even distant villages travel regularly to this community to learn how to achieve the same results.

Because people in Gaidghat in the district of Jessore are earning more, it has raised their social status. They used to be addressed using the more familiar form of address in Bengali, "tui," which is used to speak to children or someone of lower rank, but are now addressed with the term "apni," reserved for someone of a higher status (Miriam Rich, personal communication, 2007).

The Bangladeshi women's eggplant grafting effort is part of a larger project under the Integrated Pest Management Collaborative Research Support Program (IPM CRSP), supported by funds from the US Agency for International Development (USAID). The IPM CRSP has been addressing problems in developing countries around the world since 1991 and working in Bangladesh since 1998.

The eggplant grafting story is just one of thousands of success stories recorded in the implementation of IPM globally. As Marcos Kogan (1998) stated in his review "Integrated pest management: historical perspectives and contemporary developments," "twenty five years after its first enunciation, IPM is recognized as one of the most robust constructs to rise in the agricultural sciences during the second half of the twentieth century."

This chapter discusses our experiences in implementation of IPM in developing countries globally and the lessons learned that have implications for the implementation of future IPM programs. Although our chapter has an obvious IPM CRSP bias this is because all of the authors are currently involved in IPM CRSP projects globally.

To feed the growing human population, there is a desperate need to develop sustainable agricultural systems (Box 40.1). IPM technology development and transfer is a major component in sustainable agricultural systems. In spite of the progress made in the development and transfer of IPM technology, such as the eggplant grafting success story above, there continues to be a global need. Pests (insects, diseases, weeds, vertebrates,

Integrated Pest Management, ed. Edward B. Radcliffe, William D. Hutchison and Rafael E. Cancelado. Published by Cambridge University Press.

Fig. 40.1 Bangladeshi women grafting eggplants.

etc.) respect no borders and spread through plant and animal migration, wind, water and by human activity including trade in plant and animal products. Concerns over biosecurity and invasive species are global issues that require IPM attention in both developed and developing countries. The last 15 years has witnessed an increase in IPM research and capacity building around the world, supported by the US Agency for International Development (USAID) and other bilateral donors including the United Nations Food and Agriculture Organization (FAO), the World Bank, national governments, non-governmental organizations (NGOs), international agricultural research centers, universities and other organizations (Maredia *et al.* 2003; Norton *et al.*, 2005; Rajalathi *et al.*, 2005; PhilRice, 2007).

40.1 International IPM activities

During the last three decades, there have been many scientific, policy and technological developments that have tremendous potential for

Box 40.1 Non-pesticide alternatives

A convergence of technical, environmental and social forces is moving agriculture towards non-pesticide pest management alternatives like biological control, host plant resistance and cultural *management.*

Michael Fitzner, National IPM Program Leader, USDA Extension Service implementing IPM throughout the world. Many national, regional and international initiatives have contributed towards generating information, building local and regional capacity and creating a favorable environment for IPM implementation. In many countries, IPM has become an integral part of the national agricultural policy. Today, IPM is the prevailing paradigm for crop protection worldwide. A few examples of some of these key initiatives are as follows.

FAO Plant Protection Service

The United Nation's Food and Agriculture Organization (FAO) Inter-Country IPM Programme in Rice (www.fao.org/ag/AGP/AGPP/IPM/activit.htm), as implemented by the national governments of Indonesia and Philippines, provides an excellent example of FAO's contributions in IPM. The latest development in the support of IPM at FAO has been the establishment of the Global IPM Facility (www.fao.org/ag/AGP/AGPP/IPM/gipmf/index.htm). The Global IPM facility serves as a coordinating, consulting, advising and promoting entity for the advancement of IPM worldwide.

CGIAR System-Wide Program on IPM

The System-Wide Program on IPM (SPIPM: www.spipm.cgiar.org/) is an inter-Center initiative of the Consultative Group of International Agricultural Centers (CGIAR).

IPM Europe

The European Group for Integrated Pest Management (IPM Europe: www.ipmeurope.org) is a network for coordinating European Support to Integrated Pest Management in Research and Development. IPM Europe involves institutions of the European Commission, European Union Member States, Norway and Switzerland (the associate states) with an interest in promoting IPM in developing countries.

CABI Bioscience

CABI Bioscience (www.cabi.org/bioscience/index.htm) is an international organization with expertise in biological pest management. There are CABI Bioscience Centers in a number of countries, including Malaysia and Pakistan.

Centre for Biological Information Technology at the University of Queensland, Australia

The mission of the Centre for Biological Information Technology (www.cbit.uq.edu.au/) is to develop high-quality, innovative software products to inform, educate and train students, practitioners and others involved in agricultural and natural resource management, particularly pest management.

National IPM centers/programs

Many countries have established national IPM centers for coordinating IPM activities. For example, the US Department of Agriculture has a National IPM Network (NIPMN: www.reeusda.gov/agsys/nipmn/index.htm). India has also established a National Center for IPM (NCIPM: www.ncipm.org.in/) in New Delhi.

NSF Center for Integrated Pest Management

The National Science Foundation (NSF) sponsored Center for IPM (CIPM) located at North Carolina State University maintains internet-based information sources for a number of international, national, regional and state organizations (www.cipm.info/websites.cfm). CIPM also maintains the WWW Virtual Library for Agriculture (http://cipm.ncsu.edu/agVL/wwwvl.cfm).

ICIPE and the Africa IPM Forum

The International Center for Insect Physiology and Ecology (ICIPE: www.icipe.org) in Kenya coordinates the Africa IPM Forum. The primary mandate of ICIPE is research, capacity and institution building in integrated arthropod management. The Africa IPM Forum is a web-based forum for online IPM information sharing and discussion.

Pesticide Action Network (PAN) International

The Pesticide Action Network (PAN International: www.pan-international.org) is a globally active NGO working towards pest management. The PAN consists of a network of more than 600 participating NGOs, institutions and individuals in over 90 countries. The PAN programs focus on influencing international and national polices to reduce pesticide use and promote increased use of sustain-

Fig. 40.2 IPM model for research, development and implementation (IPM CRSP Program).

able and ecological alternatives to chemical pest control.

USAID IPM Collaborative Research Support Program

The US Agency for International Development has established a Collaborative Research Support Program (CRSP) on IPM (IPM CRSP: www.oired.vt.edu/ipmcrsp). The program includes a consortium of several public and private institutions, NGOs and national programs of selected countries in Asia, Africa, Eastern Europe and Latin America. This is a collaborative research, education/training and information exchange collaborative partnership among the USA and developing country institutions.

Much has been learned through the above mentioned initiatives, both about IPM tactics and about approaches to IPM research, diffusion and building institutional capacity. This chapter reports on lessons learned, primarily in the IPM CRSP global activities, and makes recommendations for future IPM programs.

40.2 Approach to globalizing IPM

To achieve the goals and objectives established for the IPM CRSP, the approach selected consisted of a participatory process linked to networking, institution building, private-sector involvement, technology development and technology transfer (Norton *et al.*, 2005). The model followed by the IPM CRSP is illustrated in Fig. 40.2.

40.2.1 Institutionalization of the participatory IPM process

Development and institutionalization of the participatory process has been a key factor in the success of all of the regional projects in the IPM CRSP program. The purpose of the participatory approach is to develop a research and technology transfer project that will meet the needs of all stakeholders. The success of the approach depends on how effectively the approach is able to link and enhance three key activities:

- *Problem identification* Activities involve farmer surveys, participatory appraisals, crop pest monitoring, collection of regional data and planning workshops.
- *Research* Activities focus on biological, technical and socioeconomic issues associated with a specific pest management problem.
- *Communication, extension and training activities* Included are conventional activities such as farm visits, field days and demonstration plots as well as more intensive efforts such as Farmer Field Schools.

Initial participatory IPM activities include: (1) stakeholder meetings, (2) participatory appraisals and (3) baseline surveys. The participatory appraisal is the cornerstone of the IPM CRSP (Fig. 40.3). Research, training and communication

Fig 40.3 Interviewing a tomato farmer during a participatory appraisal in Tamil Nadu, India.

activities relevant to IPM are then prioritized according to results from the participatory appraisal and a work plan is developed based on this process. Research and technology transfer activities are conducted in farmers' fields in order to accelerate the acceptance of new technology.

40.2.2 Networking

Extensive networking is a basic element in the participatory IPM approach and was intensively utilized by the IPM CRSP (Norton *et al.*, 2005). Participatory IPM involves as many stakeholders as possible, and the mechanism that provides for that participation is networking. The network approach provides a pool of expertise to meet unique problems at each site related to technology development, technology transfer, gender issues, policy instruments and quarantine problems. The networks have been a major reason for the success of the IPM programs at each regional site.

40.2.3 Institutional capacity building

The development of strong institutions, that are capable of continuing the development and transfer of IPM technology after the IPM initiative when outside support terminates, is a key component of the IPM CRSP approach to the globalization of IPM. The training of scientists (and others) builds IPM capacity within a country and a region. Capacity building involves giving an identity and visibility to IPM programs in each country so that they are appreciated and supported by the countries themselves, and thus maintained after outside-funded programs terminate.

Institution building involves a mix of training approaches including long- and short-term training. Long-term training consists of graduate degree training. Short-term training is used

Fig. 40.4 Mashed sweet gourd trap (attracts and kills fruit flies) in a cucurbit field in Bangladesh.

to empower the ability to conduct specific research tasks, thus making scientists proficient in a given technique needed for their research program.

40.2.4 Private-sector involvement

It is important to involve all relevant stakeholders at each site including the private sector. There are advantages and disadvantages in private-sector collaboration. For collaborative technology development in partnership with the private sector, intellectual property rights and ownership issues must be negotiated up-front to ensure freedom to operate and overcome technology transfer barriers once the final technology is ready for delivery to farmers (Erbisch & Maredia, 2003). The private sector plays a significant role in the trade and export of commodities such as vegetables for the niche market.

40.2.5 Technology development

IPM technology development by the IPM CRSP has stressed the necessity of a close link between the farmers and the research program, thus the participatory nature of IPM research. A systems approach is followed integrating information of various types (technical, economic, climatic, biological, etc.) and is based on an understanding of pest population dynamics, markets and policy constraints. Developing IPM packages has involved the employment of multidisciplinary and multi-institutional teams, virtually all of the critical stakeholders. Certain crops require more research for tactic development prior to technology transfer (Fig. 40.4).

40.2.6 Technology transfer

The ease of transferring technology depends on the environmental sensitivity of the technologies, and on environmental, cultural and other sources of diversity within host countries. To speed diffusion of IPM by the IPM CRSP, a multifaceted approach was employed in which all agencies were utilized: (1) traditional public extension agencies, (2) private for-profit and (3) private nonprofitable entities.

Farmers fail to adopt IPM technology for four reasons: (1) it is unavailable, (2) they are unaware of the technology or unaware that it will help them, (3) the technology is unsuitable for their farm or (4) they are skeptical of the risk/benefit of the new technologies. Numerous approaches were evaluated by the IPM CRSP for transferring knowledge and engaging farmers in IPM. These included media (radio and TV), on-farm demonstrations, field days, workshops, group meetings and Farmer Field Schools (FFSs) (Fig. 40.5). The global IPM CRSP projects have extensively utilized the FFS concept to promote the transfer of IPM technology to farmers. Through the use of this approach, and other

Fig. 40.5 Farmer Field School on potato IPM in the Ecuadorian Andes.

IPM CRSP activities, many valuable lessons were learned.

40.3 Lessons learned and recommendations to increase the impact of global IPM programs

The major constraints confronted by the IPM CRSP at the initiation of projects were: (1) few existing national vegetable and fruit IPM research programs (most programs relied primarily on pesticide applications), (2) lack of national IPM policies, (3) lack of a participatory approach that includes participation by all stakeholders, especially the growers, in the design, development and implementation of IPM packages, (4) lack of institutionalization of IPM to develop and coordinate national IPM programs, (5) lack of appropriate research programs (multidisciplinary, multi-tactic, multi-pest, etc.) to be able to be develop IPM technologies and integrate the various tactics in an IPM program, (6) need for institutional and human capacity building at most sites, (7) lack of multi-institutional collaboration and networking and (8) slow rate of adoption of IPM strategies by farmers (Norton *et al.*, 2005).

The experiences of the IPM CRSP scientists are invaluable in making recommendations for the design and implementation of future Global IPM programs. Based on the lessons learned we make the following recommendations for future global IPM programs in the developing regions of the world.

40.3.1 Participatory IPM, networking and gender issues

In the IPM CRSP projects, the participatory approach was, without a doubt, an effective method of IPM technology development and transfer. However, it is expensive on a per-farmer basis and future Global IPM programs must continue the search for more cost-effective technology transfer methods that are able to reach a much wider audience. It is also crucial that women farmers be included as full partners in the IPM training process. In Africa, women farmers are generally responsible for pest management activities. Research has shown that the technological needs and priorities of women farmers differ from those of men, but their needs are rarely addressed in the research and development of agricultural technology. Women's access to new technologies is undermined due to the failure of extension systems to establish contacts with women farmers. IPM programs must take into account the gender-related, socioeconomic factors that influence the creation, adoption and effective implementation of IPM technologies (Malena, 1995).

Strong networks are a basic element in a successful participatory IPM approach. True IPM does not exist without interdisciplinary and interagency collaboration, all working together to develop and transfer IPM technology to the grower.

In most countries, IPM programs function in isolation with very poor coordination among people working in different departments and ministries. IPM must be institutionalized to provide better planning and coordination at both the institutional and national level. The network approach provides a pool of expertise to meet the unique problems existing at each site such as technology development, technology transfer, policy instruments, export and quarantine problems and gender issues.

The international organizations have served as a very good platform for providing training and networking in IPM related areas. The IARCs have played a pivotal role in delivering improved germplasm to the national agricultural research services (NARs). Regional and global cooperation and networking (linkages/partnerships) will become increasingly important if we want to efficiently exchange and use the available IPM information. Cooperation singly between the northern and southern hemispheres and within the southern hemisphere countries will be critical to maximize worldwide IPM implementation.

40.3.2 Communication with the general public

There is a growing concern regarding the negative health effects of toxic chemical pesticides. To address the concerns of consumers, the IPM and organic methods of agricultural production are receiving increased attention (Zehnder *et al.*, 2007). In many developed and developing countries a growing segment of consumers is willing to pay higher prices for food grown through IPM and organic methods. For the greater acceptance of IPM-grown food products, educational programs and mechanisms need to be developed to educate and communicate IPM to the general public (Lagnaoui *et al.*, 2004; Ibitayo, 2006).

40.3.3 Policies

National and regional pesticide use policies are rapidly changing worldwide to reduce the reliance and availability of chemical pesticides. Treaties and conferences, such as the General Agreement on Tariffs and Trade (GATT), North American Free Trade Agreement (NAFTA) and United Nations Conference on Environment and Development (UNCED) are driving such policy changes in both developed and developing countries (Henson & Loader, 2001). For example, IPM is the preferred strategy for pest management under "Agenda 21" of the United Nations Conference on Environment and Development (UNCED, 1992). In the USA, the Food Quality Protection Act passed by Congress in 1996 has pressured many growers to implement IPM practices to cope with fewer pesticides available for use. Similar legislation to reduce dependence on chemical pesticides is under consideration in many countries that export agricultural products to the European Union, the USA and Canada; this is a major shift from the former policies which subsidized the use of agricultural chemicals.

Increased efforts are needed to promote legislation of national policies that support IPM technology development and transfer. If policies create barriers to IPM adoption, such that there is little economic incentive to adopt, there may be little return to IPM technology development and transfer activities. A national policy legislating IPM can have a significant impact on increasing the effects of IPM. Future global IPM programs should have an increased emphasis on promoting and providing expertise and direction in the development of national IPM policies and there should be greater interaction between policy makers and economists that are engaged in policy research.

40.3.4 Institutionalization of IPM

Supportive national policies are important in the institutionalization of IPM in government agencies and universities. Institutionalization of IPM takes considerable time and hence the need for long-term IPM projects if we are to achieve our objectives. Vital to the institutionalization of IPM is training, both short-term and degree training. Because of the excessive costs involved in training, future IPM programs must be innovative in developing more affordable but effective training techniques so as to meet the training needs to fully institutionalize IPM at all levels of national programs.

40.3.5 Empowerment and capacity building

The building of human and institutional capacity for the development and transfer of IPM technology is a major activity in the development of sustainable IPM programs. Institution building involves a mix of training approaches including long- and short-term training of scientific staff and farmer training. What should the mix of short- and long-term training be? What can be done to minimize the high costs of academic and farmer training?

Innovative approaches need to be established in order to cope with the high cost of degree training in developed countries. Training within the host country or a "sandwich"-type program with a portion of the training in the host country and the remainder in a developed country has been done in the IPM CRSP. Are there other types of training programs that can be utilized to produce a quality degree at minimal cost? Does distance education have a role here?

Consideration should be given to new approaches to farmer training. Farmer Field Schools (FFSs) have been shown to be effective but are too expensive to provide training to all farmers. What modifications of the FFS or other approaches can be found to meet the needs of the farmers at an affordable cost?

It must be stressed that at the initiation of an IPM program in a host country a participatory appraisal be conducted that includes an assessment of training needs at all levels from farmers, field and laboratory workers to research scientists. Based on this information a training plan can be established that meets the needs, is within the budget and can be completed in the time available prior to the end of the project contract period.

Current funding of Global IPM programs is not sufficient to meet the training needs of host countries. The lack of adequate budgets for this purpose will likely continue and thus, the only solution is the development of novel and more cost effective training methods.

40.3.6 Private-sector involvement and NGOs

Networking with the private sector and NGOs can provide vital input into technology development and transfer. The private sector plays a significant role in the trade and export of commodities such as vegetables for the niche market and in the supply of pesticides and seeds. The NGOs have been a key factor in the IPM CRSP technology transfer program with their many farmer contacts and their technology transfer activities. Collaboration with NGOs has been a way of leveraging funds to reach large numbers of farmers which without the NGOs would have not been possible. Strong links with private industry and NGOs are strongly recommended in future IPM programs.

40.3.7 Areawide IPM

More attention should be given to areawide IPM systems in future global IPM programs (Knipling, 1979, 1980). Some IPM strategies are more likely to succeed if implemented over large areas, rather than on an individual farmer field basis. It has been suggested that areawide pest management projects: (1) must be conducted on large geographic areas, (2) should be coordinated by organizations rather than by individual farmers, (3) should focus on reducing and maintaining key pest populations at acceptable low densities and (4) may involve a mandatory component to insure the necessary full participation in the program. Government extension services or private agencies should conduct areawide monitoring and provide the appropriate information on pest outbreaks to farmers on a regular basis. Weather data should be utilized in the development of predictive models for forecasting the outbreaks of pests, especially migratory pests. Training and education of farmers and extension workers in pest identification, monitoring and management approaches should be provided.

The IPM CRSP has successfully employed an areawide approach to tomato virus management in Mali via the utilization of a host-free period covering tomato fields at the village level. This project can be used as a model for other areawide vegetable IPM programs.

40.3.8 Global trade

Target crops selected for inclusion in an IPM program should be those where there can be a significant economic impact. Often those are export crops which must be free of pesticide residues and insect pests. The IPM CRSP has had a significant impact on export issues, especially green beans in Mali, snow peas in Guatemala and hot peppers in Jamaica. This has involved expertise in field surveillance, mapping and implementation of GPS/GIS technology. The establishment of a traceability system as well as web-GIS field monitoring capability should play a significant role in effective monitoring and surveillance of pests in future IPM programs. IPM scientists must be capable of providing export agencies with the training needed to handle pest problems involved in export trade including field IPM technology development and transfer and expertise in phytosanitary procedures and policy and regulatory formulation.

40.3.9 Impact assessment and program promotion

Governments, national and international donors must make IPM programs a priority and provide financial resources for continual development and implementation. Education of donors to the importance of IPM is a critical activity in being able to adequately fund IPM programs. To provide a basis on which to promote IPM programs, impact assessment studies are critical.

All IPM development programs should include an impact assessment component. Spatially referenced analytical tools are used to identify and characterize areas susceptible to specific pest problems and linked to crop loss data. This information is combined with other economic, technical and technology adoption (realized and projected) data to: (1) project where IPM programs are likely to have the greatest impacts locally, regionally and globally and (2) assess impacts of specific IPM CRSP activities that have been completed in each region. Impact assessment data should be used to promote the program to donors, government officials, growers and consumers via the media, publications and presentations and over the web.

40.4 Global IPM of the future

40.4.1 IPM technology development

IPM research in the future should seek a balance between basic and applied research. Most IPM research programs are designed as short-term programs. Landscape-level long-term ecological research projects will give a better understanding of the biological and ecological interactions within the landscapes. A thorough understanding of the biology and ecology of pests and natural enemies and their ecosystem may reveal totally new approaches for pest management.

The approach to IPM technology development must be participatory, multidisciplinary and multi-pest oriented and must integrate all available tactics, including cultural, biological control (Neuenschwander *et al.*, 2003) and genetic approaches including both biotechnology and conventional breeding. The approach must be ecologically based. Novel tactics should be explored and may include transgenic insects, biorational approaches, botanicals and management of resistance to pesticides and to resistant varieties (conventionally bred and genetically modified [GM]). Tactics that favor biodiversity should be explored.

40.4.2 Molecular markers

Research in population genetics and population ecology helps to better evaluate strategies to slow the development of resistance to insecticides and transgenic crops in insect pests of crops. Incorporation of molecular tools to address genetic parameters is highly helpful to characterize insect populations across the hosts and geographical barriers. Studies on population genomics provide a novel and necessary interface between population genetics and molecular biology.

The array of molecular marker advancements available today for entomologists has facilitated many studies that have previously been impossible. This will lead to more rational design of insecticides or transgenic host plant immunity, thereby reducing the chances of adaptation to future pest control strategies. These technologies in the field of entomology are expected to provide new direction to study insect genomes in an unprecedented way in the years to come.

40.4.3 Transgenic organisms

A transgenic organism is any living creature, such as a bacterium, plant or animal that has received a foreign gene by means of genetic engineering. *Paratransgenesis* in insects is the genetic alteration of microbes living in association with insects for various purposes (Peloquin *et al.*, 2002). The idea comes from genetic engineering where expression products of foreign engineered genes can block or eliminate the ability of the vector to transmit pathogens. Recently, Johns Hopkins University successfully transformed mosquitoes with a loss in the ability to transmit malaria (Riehle *et al.*, 2003). Most transgenic research is still in the development phase, but now is the time to address the potential role of transgenics in global IPM programs.

40.4.4 Semiochemicals

A semiochemical is a generic term used for a chemical substance or mixture that carries a message. A wide array of uses for semiochemicals has been tested over the past few decades and many practical applications have become integral components in pest management programs (Gut *et al.*, 2004). Behavior modifying chemicals such as sex pheromones were effectively employed for fruit fly monitoring and control in the IPM CRSP project in Bangladesh. The experiences gained in Bangladesh can serve as a guide in employing pheromones in future IPM projects in other countries.

40.4.5 Botanicals and biological control enterprises

Local governments and rural development programs should encourage the development of cottage industries to mass produce beneficial organisms and botanical pesticides for use in IPM programs. This may include mass production and commercialization of entomopathogenic fungi and nematodes, parasitoids, predators and botanical agents such as neem-based pesticides.

40.4.6 Biotechnology and IPM

Biotechnological developments and genetic engineering are revolutionizing agriculture. Some major benefits of this revolution will be crops that are resistant to pests and diseases, livestock that are resistant to viruses and microbes to increase resistance to frost damage.

Careful integration of transgenic crops into proven IPM systems and best cultural practices will be the most likely route to improved sustainability. It is essential to sustainability that transgenic crops be viewed as one important component of an integrated farming system.

While several crops with commercial viability have been transformed in the developed world, very little has been done to use this technology to increase food production in the harsh environments of the tropics. There is a need to use these tools for providing resistance to insects in cereals, legumes and oilseed crops that are a source of sustenance for poorer sections of the society. The advanced laboratories in developed countries, private-sector companies, international research centers and national agricultural research centers will play a major role in promoting biotechnology in the developing world (Chaturvedi & Rao, 2005; Eicher *et al.*, 2006).

The application of biotechnology has made a significant contribution in the dramatic reduction in insecticides applied per hectare to crops (Qaim & Zilberman, 2003; Brookes & Barfoot, 2006;

Fernandez-Cornejo, 2006). During the last decade, the application of innovative biotechnology has provided a foundation for the rapid adoption of IPM practices. Herbicide-tolerant crops can reduce the amount of soil cultivation and herbicide required on crops to control weeds and facilitates healthier soils through less soil disruption and reductions in residual herbicides.

Genetically modified crops contribute to the options that farmers have available and as such hectarage planted with them continues to expand worldwide (James, 2006). However, GM crops are not a silver bullet and like all technologies that help make crop protection and production more efficient, GM crops are most effective when they are utilized by integrating in IPM as key parts of sustainable pest management systems.

40.4.7 Nanotechnology

Nanotechnology will fundamentally change the concept of pest management as it provides a powerful tool to radically transform agrochemicals (Salamanca-Buentello *et al.*, 2005). Many firms are working on nano-sized capsules containing pesticides or even chemical/biological weapons designed to break open and release their contents only under certain conditions. Nanobiotechnology, the nano-scale mixing of biological and non-biological material being explored includes atomically modified seeds, incorporating non-living nanomaterials into living organisms and creating new synthetic materials incorporating biological materials.

40.4.8 Organic agriculture

Organic farming, a form of agriculture which avoids or largely excludes the use of synthetic fertilizers, pesticides and plant growth regulators, is increasing in importance and value. IPM is a key component in support of organic agriculture. A growing consumer market is naturally one of the main factors encouraging farmers to convert to organic agricultural production. Increased consumer awareness of food safety issues and environmental concerns has contributed to the growth in organic farming. Although the area currently devoted to organic farming is still limited it is evident that IPM has a major role to play in organic farming and IPM research should include the development of strategies to support organic farming (Zehnder *et al.*, 2007). Increased emphasis should be given to developing IPM programs which support this movement.

40.4.9 Sustainable agriculture

IPM is a component of sustainable agriculture (Kogan, 1998). IPM not only contributes to the sustainability of agriculture, it also serves as a model for the practical application of ecological theory and provides a paradigm for the development of other agricultural system components. The IPM paradigm provides that pests and their management exist at the interface of three multidimensional universes: ecological, socioeconomic and agricultural. IPM plays a major role in sustainable agriculture which helps farmers use resources more efficiently, protects the environment and preserves rural agriculture-based communities.

40.4.10 Global climate change

There is evidence that global climate change, if it continues, will play a critical role in the ability of insects to overwinter, on the geographic distribution or ranges of insects, on the number of generations and on their abundance in agricultural systems. Studies indicate that weeds and beneficial and harmful pest species respond to climate changes. Insects, microbes and other organisms in the environment will be responding to changes in carbon dioxide and climate. Most studies have concluded that insect pests will generally become more abundant as temperatures increase, through a number of interrelated processes, including range extensions and phenological changes, as well as increased rates of population development, growth, migration and overwintering. A gradual, continuing rise in atmospheric carbon dioxide will affect pest species directly (i.e. the carbon dioxide fertilization effect) and indirectly (via interactions with other environmental variables). Migrant pests are expected to respond more quickly to climate change than plants, and may be able to colonize newly available crops/habitats. Some leaf-feeding insects appear to do more crop damage to plants under high carbon dioxide conditions.

Most invasive and noxious weeds respond more positively to increasing carbon dioxide than do most of our cash crops. To make matters worse, research indicates that weeds are much more difficult to control with herbicides at anticipated levels of carbon dioxide. Wetter summers would tend to favor many foliar pathogens.

New pests would require the development of monitoring programs, the establishment of action thresholds and the availability of new pest control products and strategies for IPM. All of these things take time to develop and require long-term research programs.

There is a need to determine the effect of global climate change on pests and a need to develop models to predict the impact of global climate change. This evidence should be taken into account in designing long-term IPM studies.

40.4.11 Bioenergy

This is a rapidly expanding area and is causing an increase in the value of source crops. When the value of crops increase, the value of pest management practices also increase. Ethanol production from various crops is likely to increase in importance in countries on a global basis. What is the relationship between growing ethanol producing plants and pest management? First, diseases and pests that affect plant growth and yields will have an impact on the level of ethanol production per hectare. It is also possible that plants that are bred for high cellulose content may be more susceptible to pests and diseases. Another concern is that the change in agronomic practices related to production of ethanol producing crops may increase pest problems. For example, the demands of ethanol production would require extremely hardy maize (corn). William S. Niebur of DuPont says the very hardy corn needed to make ethanol could force farmers to abandon crop rotation, straining the soil: "The demand for this corn grain could be so dramatic that it would change farming practices." The economic pressure to grow maize year after year could strain the soil and allow the buildup of insects and disease. IPM specialists must be proactive and play a key role as a member of an integrated crop management team that develops pest management practices for bioenergy crop sources. Development of pest management strategies for source crops will increase the economic impact of IPM programs.

40.4.12 Communication with consumers and eco-labeling

Widespread and open communication with consumers should be a feature of all future IPM programs (Cuperus *et al.*, 1997; Govindsamy *et al.*, 1998). Consumers worldwide are increasingly demanding not only more food but high-quality and safe food, thus promoting the use of IPM tactics that do not contaminate the environment.

As businesses have come to recognize that environmental concerns may be translated into a market advantage for certain products and services, various environmental claims/labels have emerged on products and with respect to services in the marketplace (e.g. eco-friendly). These have attracted consumers looking for ways to reduce adverse environmental impacts through their purchasing choices. Communication with the consumers about the food produced through IPM practices will be important issues for promoting the importance of IPM to stakeholders and providing support to eco-labeling.

40.4.13 Biodiversity, invasive pests, bioterrorism and quarantines

Biodiversity is often a measure of the health of biological systems (Altieri, 1999; Altieri & Nicholls, 2004). Widespread introduction of invasive species by humans are threats to biodiversity. The economics of fruits and vegetables in global trade is huge and IPM specialists must play key roles in the diagnostics of pests and diseases and in the development of phytosanitary methods. Strong IPM programs are needed to battle the continued onslaught of invasive species and to be prepared to prevent and cope with the use of invasive species in bioterroristic acts.

While bioterrorist attacks in most countries may seem unlikely, it's a serious concern and preparation and vigilance can help protect a country's leading agricultural and livestock industries. Bioterrorism through the use of biological agents such as insects, weeds, plant disease agents and animal pathogens could affect both crop and livestock producers. A basic knowledge of native pests within a country is necessary to determine whether an alien species has been introduced as an act of bioterrorism. Surveys and identification of weeds, plant pathogens, insects and other pests are a basic activity in the development of an IPM program. If this information is not known, the appearance of new pests may not be recognizable, and it will not be possible to determine if deliberate bioterrorism is involved. Basic monitoring and identification studies are essential in establishing new IPM projects in the future.

If we are to reduce the potential of deliberate introduction of crop pathogens, we must be able to fingerprint pathogens and discriminate between naturally occurring disease events and those which may be deliberately introduced for harmful purposes. The effective tracking of new and emerging diseases throughout the world is critically needed. However, this is not possible because of the absence of a rapid reporting system and the great reluctance of nations to share information for fear of trade restrictions imposed by phytosanitary regulations. The web-based system currently being developed in the USA, the National Plant Diagnostic Network, is a start in addressing these deficiencies, and should be extended internationally.

40.4.14 Information technology

Information technology (IT) is a powerful tool used in networking. Broadly speaking, IT is a hybrid of computer technology, communication technology, networking technology and remote sensing technology (see Chapter 8).

The internet has a number of characteristics which make it highly suitable for disseminating IPM knowledge. The internet could be used to support decision making by IPM farmers, as well as IPM training activities. There are a number of organizations who have started to make investments in these areas. For example, the ASEAN IPM Knowledge Network (http://asean-ipm.searca.org/) is specialized, providing a on-line catalog of documents dealing with IPM in Asia. The Asia Pacific Regional Technology Center (APRTC, www.aprtc.org/), sponsored by the pesticide industry, conducts on-line courses in cotton IPM, rice IPM and vegetable IPM.

The use of IT to support pest management decision making has been summarized by Bajwa & Kogan (2001) who have provided examples of websites which are sources of information for what they call "steps in IPM," i.e. (1) identifying the pest species, (2) establishing economic injury thresholds, (3) monitoring, scouting and predicting populations and (4) selecting and applying control techniques.

Mailing lists, bulletin boards, on-line databases and libraries, web-based resource centers, organizational home pages, training and reference materials on compact disk and distance-learning courses are some forms of communicating IPM to end-users. Information technology based IPM will eventually arrive in every village in Asia and Africa, if not today, within the next two decades.

In India, the M. S. Swaminathan Research Foundation (MSSRF) and IDRC have developed a network of "Information Shops and Commercial Delivery Systems" which aim to make a profit from their services. During 1993–96, MSSRF conducted studies on the role of information in the lives of rural families in about 25 villages in the Union territory of Puducherry and Tamil Nadu. The Village Knowledge Center was started in 1998 with support from IDRC and CIDA. Realizing the importance of communication technologies through these projects, MSSRF further moved to start a National Virtual Academy (NVA) bearing the name of Jamsetji Tata (one of the India's greatest visionaries) in 2003. For the past four years, MSSRF in coordination with national and international agencies organized various participatory knowledge management workshops, training programs, and trained many rural men and women in communication technology. On February 2007, they organized the fourth convocation of Jamsetji Tata National Virtual Academy. The NVA has also developed a toolkit for helping NGOs and others interested in setting up Village Knowledge Centers. The Department of Information Technology of the Government of India has initiated a program for establishing 100 000 common service centers to cater to the needs of rural India.

The Laboratory of IPM Intelligent System Technology (IPMist), an R&D branch of the China Agricultural University, has developed many IT products that play a positive and significant role in IPM practice. Applications include multimedia technology, MIS, database, expert system, GIS, RS, GPS, modeling and simulation, image recognition, computer vision and the internet. VegePest, PestDiag and BJ-FarmKnow provide information to vegetable growers in China (www.ipmist.org). The BOLD Identification System (IDS) accepts gene sequences and returns a species-level identification for insects when one is possible (www.barcodinglife.org). In the future, molecular marker-based barcoding for pests, predators, parasites and pathogens is foreseeable.

40.4.15 Ecologically based IPM and technology adoption

The adoption level of IPM strategies by farmers has often been low and progressed at a slow pace. To increase the intended effects of IPM there is a great need for novel, economically effective technology transfer methods to reach the masses of farmers needed to be educated in the ecological principles of IPM.

Contemporary IPM programs have often been implemented with little consideration of ecosystem processes (Gliessman, 1990). Review of past literature on IPM confirms that success has come from a fundamental understanding of the ecology of complex pest interactions, rarely from a revolutionary new control tactic (Kogan, 1998; Altieri, 1999; Altieri & Nicholls, 2004). The advance of IPM to higher levels of integration will hinge on the depth of understanding of agroecosystem structure and dynamics (Landis *et al.*, 2000, 2004; Menalled *et al.*, 2004).

Ecologically based pest management (EBPM) is a profitable, safe and durable approach to controlling pests in managed ecosystems (National Research Council, 1996). The EBPM systems are built on an underlying knowledge of the managed ecosystem, including the natural processes that suppress pest populations. EBPM is a total systems approach designed to have minimal adverse effects on non-target species.

For EBPM to be successful, it must also be profitable for the grower. Growers demand safe, economical and effective tools that provide long-term management of pests, and these needs must be met before growers will implement EBPM's new tools. Because producers will implement only those pest control methods that lower economic risks and enhance profits, they will insist on assurances that biologically based tools are cost-effective and provide consistent responses.

As the agroecosystem is the overlap of ecological and socioeconomic scales, no technological innovation will be adopted unless it contributes to growers' economic goals and meets the requisites for acceptance by society (Kogan, 1986). IPM programs must be both ecologically based and participatory to meet the needs of growers. This is a cardinal principle that should be taken into account in all IPM programs.

40.5 Conclusions

A wealth of information and experience has been accumulated during the past five decades in IPM research, education and outreach worldwide. IPM has matured and the concept and principles of IPM have become widely promoted and accepted. However, losses due to insects, nematodes, weeds and diseases continue to be high in certain crops and in certain areas where IPM technology development and transfer to growers has not yet occurred, for various reasons. Also, IPM scientists and practitioners are continually facing new challenges with the advent of modern agriculture. Global trade, bioenergy, bioterrorism, biodiversity issues, areawide IPM, genetic engineering, organic agriculture and sustainable agriculture are some of the challenges calling for continued IPM research.

A constraint to the implementation of IPM is the fragmentation of Global IPM programs with most agencies "doing their own thing" and with very little collaboration and coordination among agencies. A global IPM network should be constituted to coordinate global IPM activities. Such a network would include international agencies (e.g. the World Bank, United Nations Environmental Program, FAO, International Agricultural Centers), donor countries (e.g. US Agency for International Development, Deutsche Gesellschaft für Technische Zusammenarbeit, Danish International Development Agency, Department for International Development of the UK, etc.), national IPM networks (e.g. Regional IPM Centers in the USA, National IPM Center in India and other countries), national universities with strong IPM programs in both developed and developing countries, foundations (e.g. Rockefeller, Ford, Bill and

Melinda Gates Foundations), relevant NGOs and private industry, crop-oriented USAID-supported Collaborative Research Support Programs (e.g. Sorghum, Millet and Other Grains (INTSORMIL), IPM, Bean and Cowpea, Sustainable Agriculture and Natural Resource Management (SANREM) and Peanut CRSPs) and international plant protection related scientific associations (e.g. American Phytopathological Society, Entomological Society of America, International Association for the Plant Protection Sciences, etc.). If all of the key programs and networks can collaborate, coordinate and share information and experiences with each other, tremendous progress can be made in making IPM practices a reality for the global community. Some agency needs to take the lead in the establishment of such a global network.

References

Altieri, M. A. (1999). The ecological role of biodversity in agroecosystems. *Agriculture, Ecosystems and Environment*, **74**, 19–31.

Altieri, M. A. & Nicholls, C. I. (2004). *Biodiversity and Pest Management in Agroecosystems*. Binghamton, NY: Haworth Press.

Bajwa & Kogan, M. (2001). Internet-based IPM informatics and decision support. In *Radcliffe's IPM World Textbook*, eds. E. B. Radcliffe, W. D. Hutchison & R. E. Cancelado. St. Paul. MN: University of Minnesota, available at http://ipmworld.umn.edu/chapters/Bajwa.htm.

Brookes, G. & Barfoot, P. (2006). Global impact of biotech crops: socio-economic and environmental effects in the first ten years of commercial use. *AgBioForum*, **9**, 139–151.

Chaturvedi, S. & Rao, S. R. (2005). *Biotechnology and Development: Challenges and Opportunities for Asia*. New Delhi, India: Academic Foundation.

Cuperus, G. W., Pinkston, K., Shelton, K. *et al.* (1997). *Urban Attitudes Regarding Integrated Pest Management, Environmental Issues, and the Urban Environment*. Stillwater, OK: Oklahoma State University.

Eicher, C. K., Maredia, K. M. & Sithole-Niang, I. (2006). Crop biotechnology and the African farmer. *Food Policy*, **31**, 504–527.

Erbisch, F. E. & Maredia, K. M. (2003). *Intellectual Property Rights in Agricultural Biotechnology*. Wallingford, UK: CABI Publishing.

Fernandez-Cornejo, J. (2006). *Adoption of Genetically Engineered Crops in the U.S.* Washington, DC: US Department of Agriculture, Economic Research Service.

Gliessman, S. R. (1990). *Agroecology: Researching the Ecological Basis for Sustainable Agriculture*. New York: Springer-Verlag.

Govindsamy, R., Italia, J. & Rabin, J. (1998). *Consumer Response and Behavior Towards Integrated Pest Management*, New Jersey Agricultural Experiment Station Report No. P 02137-5-98. New Brunswick, NJ: Rutgers University.

Gut, L. J., Stelinski, L. L., Thomson, D. R. & Miller, J. R. (2004). Behavior-modifying chemicals: prospects and constraints in IPM. In *Integrated Pest Management: Potential, Constraints, and Challenges*, eds. O. Koul, G. S. Dhaliwal & G. Cuperus, pp. 73–121. New York: CABI Publishing.

Henson, S. & Loader, R. (2001). Barriers to agricultural exports from developing countries: the role of sanitary and phytosanitary requirements. *World Development*, **29**, 85–102.

Ibitayo, O. O. (2006). Egyptian farmers' attitude and behavior regarding agricultural pesticides: implications for pesticide risk communication. *Risk Analysis*, **26**, 989–995.

James, C. (2006). *Global Status of Commercialized Transgenic Crops*, ISAAA Brief No. 35. Ithaca, NY: International Service for the Acquisition of Agri-Biotech Applications.

Knipling, E. F. (1979). *The Basic Principles of Insect Population Suppression and Management*, Agricultural Handbook No. 512. Washington, DC: US Department of Agriculture.

Knipling, E. F. (1980). Areawide pest suppression and other innovative concepts to cope with our more important insect pest problems. In *Minutes of the 54th Annual Meeting of the National Plant Board*, pp. 68–97. Sacramento, CA: National Plant Board.

Kogan, M. (1986). *Ecological Theory and Integrated Pest Management Practice*. New York: John Wiley.

Kogan, M. (1998). Integrated pest management: historical perspectives and contemporary developments. *Annual Review of Entomology*, **43**, 243–270.

Lagnaoui, A., Santi, E. & Santucci, F. (2004). Strategic communication for integrated pest management. Paper presented at the *Annual Conference of the International Association of Impact Assessment*, (IAIA).

Landis, D. A., Wratten, S. D. & Gurr, G. M. (2000). Habitat management to conserve natural enemies of arthropod pests in agriculture. *Annual Review of Entomolgy*, **45**, 175–201.

Landis, D. A., Menalled, F. D., Costamagna, A. C. & Wilkinson, T. K. (2004). Manipulating plant resources to enhance beneficial arthropods in agricultural landscapes. *Weed Science*, **53**, 902–908.

Malena, C. (1995). *Gender Issues in Integrated Pest Management in African Agriculture*. NRI Socio-economic Series 5.

Maredia, K. M., Dakouo, D. & Mota-Sanchez, D. (2003). *Integrated Pest Management in the Global Arena*. Wallingford, UK: CABI Publishing.

Menalled, F. D., Landis, D. A. & Dyer, L. E. (2004). Research and extension supporting ecologically based IPM systems. *Journal of Crop Improvement*, **11**, 153–174.

National Research Council (1996). *Ecologically Based Pest Management: New Solutions for a New Century*. Washington, DC: National Academy Press.

Neuenschwander, C., Borgemeister, C. & Langewald, J. (2003). *Biological Control in IPM Systems in Africa*. Wallingford, UK: CABI Publishing.

Norton, G. W., Heinrichs, E. A., Luther, G. C. & Irwin, M. E. (2005). *Globalizing Integrated Pest Management: A Participatory Research Process*. Blacksburg, VA: IPM CRSP, Virginia Tech University.

Peloquin, J. J., Lauzon, C. R., Potter, S. & Miller, T. A. (2002). Transformed bacterial symbionts reintroduced to and detected in host gut. *Current Microbiology*, **45**, 41–45.

PhilRice (2007). *Integrated Pest Management in Rice–Vegetable Cropping Sysems*. Maligaya, Science City of Muñoz, Nueva Ecija, Philippines: Philippine Rice Research Institute.

Qaim, M. & Zilberman, D. (2003). Yield effects of genetically modified crops in developing countries. *Science*, **229**, 900–902.

Rajalathi, R., Lagnaoui, A., Schillhorn-Van Veen, T. & Pehu, E. (2005). *Sustainable Pest Management: Achievements and Challenges*, Agriculture and Rural Development Report No. 32714-GLB. Washington, DC: The World Bank.

Riehle, M. A., Srinivasan, P., Moreira, C. K. & Jacobs-Lorena, M. (2003). Towards genetic manipulation of wild mosquito populations to combat malaria: advances and challenges. *Journal of Experimental Biology*, **206**, 3809.

Salamanca-Buentello, F., Persad, D. L., Court, E. B., Martin, D. K. & Daar, A. S. (2005). Nanotechnology and the developing world. *PLoS Medicine*, **2**(4), e97.

UNCED (1992). *Promoting Sustainable Agriculture and Rural Development*, Agenda 21, Chapter 14. Abdelboden, Switzerland: United Nations Conference on Environment and Development.

Zehnder, G., Gurr, G. M., Kühne, S. *et al.* (2007). Arthropod pest management in organic crops. *Annual Review of Entomology*, **52**, 57–80.

Index